STRUCTURAL DYNAMICS
IN
ENGINEERING DESIGN

STRUCTURAL DYNAMICS IN ENGINEERING DESIGN

Nuno M. M. Maia
Dario Di Maio
Alex Carrella
Francesco Marulo
Chaoping Zang
Jonathan E. Cooper
Keith Worden
Tiago A. N. Silva

WILEY

This edition first published 2024

Registered Offices
John Wiley & Sons, Inc., 111 River Street, Hoboken, NJ 07030, USA
John Wiley & Sons Ltd, The Atrium, Southern Gate, Chichester, West Sussex, PO19 8SQ, UK

For details of our global editorial offices, customer services, and more information about Wiley products visit us at www.wiley.com.

Wiley also publishes its books in a variety of electronic formats and by print-on-demand. Some content that appears in standard print versions of this book may not be available in other formats.

Library of Congress Cataloging-in-Publication Data
Hardback: 9781118770702
ePDF: 9781118770672
epub: 9781118770689
oBook: 9781118770696

Cover image: © nmcandre/Adobe Stock
Cover design by Wiley

Set in 12/14pt TimesNewRomanPSMT by Integra Software Services Pvt. Ltd, Pondicherry, India

Printed and bound by CPI Group (UK) Ltd, Croydon, CR0 4YY
C9781118770702_090724

to all who use structural dynamics to understand
and engineer the world around us

Contents

Acronyms

A

ADC	Analog to Digital Conversion
AR	Autoregressive
ARMA	Autoregressive Moving Average
ARMAX	Autoregressive Moving Average with eXhogneous inputs
ASMAC	Alternated Search Modal Assurance Criterion

C

CB	Craig-Bampton method
CDF	Cumulative Distribution Function
CMIF	Complex Mode Indicator Function
CMS	Component Mode Synthesis
CMU	Computational Model Updating
CoV	Coefficient of Variation
COMAC	Coordinate Modal Assurance Criterion
CSAC	Cross Signature Assurance Criterion
CSC	Cross Signature Correlation
CSD	Cross Power Spectral Density
CSF	Cross Signature Scale Factor

D

DAC	Digital to Analog Conversion
DIC	Digital Image Correlation
DIRS	Dynamic Improved Reduction System
DOF	Degree of Freedom

E

eCDF	empirical Cumulative Distribution Function
EFDD	Enhanced Frequency Domain Decomposition
EI	Effective Independence method
EJ	Engineering Judgement
EMA	Experimental Modal Analysis
ERA	Eigensystem Realisation Algorithm
ERA/DC	Eigensystem Realisation Algorithm using Data Correlations

F

FAAC	Frequency Amplitude Assurance Criterion

FDAC	Frequency Domain Assurance Criterion
FDD	Frequency Domain Decomposition
FE	Finite Element
FEA	Finite Element Analysis
FEM	Finite Element Method
FFT	Fast Fourier Transform
FI	Fisher Information matrix
FMFEM	Fine Mesh Finite Element Method
FRAC	Frequency Response Assurance Criterion
FRF	Frequency Response Function
FRSF	Frequency Response Scale Factor
FS-SLDV	Fast Scan Scanning Laser Doppler Velocimeter

G

GAC	Global Amplitude Criterion
GFEM	Global Finite Element Model
GSC	Global Shape Criterion
GVT	Ground Vibration Test

I

IIRS	Iterated Dynamic Improved Reduction System
IRF	Impulse Response Function
IRS	Improved Reduction System

K

KMO	Kaiser-Meyer-Olkin criterion
KS-test	Kolmogorov-Smirnov goodness-of-fit test

L

LAC	Local Amplitude Criterion
LDV	Laser Doppler Velocimeter
LHS	Latin Hypercube Sampling
LSCE	Least-Squares Complex Exponential
LSFD	Least Squares Frequency Domain

M

MAC	Modal Assurance Criterion
MBA	Modal-Based Assembly
MCMC	Markov Chain Monte-Carlo
MCS	Monte-Carlo Simulations
MDOF	Multiple Degree of Freedom
MEMS	Micro-Electro-Mechanical Systems
MIMO	Multiple Input Multiple Output
MISO	Multiple Input Single Output
MMIF	Multivariate Mode Indicator Function

ModMMIF	Modified Multivariate Mode Indicator Function
MPC	Modal Phase Collinearity index
MSF	Modal Scale Factor

N

NMPF	Normal Mode Purity Function

O

ODS	Operational Deflection Shape
OMA	Operational Modal Analysis

P

PA	Horn's Parallel Analysis
PCA	Principal Component Analysis
PDF	Probability Density Function
PID	Proportional Integral Derivative
PSD	Power Spectral Density

Q

QTH	Quasi-Steady Three-Dimensional Histogram

R

RFM	Response Function Method
RK	Runge-Kutta
RK4	Runge-Kutta of 4th order
RMS	Root Mean Square
RSS	Root Sum of Squares
RVAC	Response Vector Assurance Criterion

S

svs	Singular Values
SDE	Stochastic Differential Equation
SDOF	Single Degree of Freedom
SEREP	System Equivalent Reduction Expansion Process
SHM	Structural Health Monitoring
SLDV	Scanning Laser Doppler Velocimeter
SLE	Simultaneous Linear Equations
SIMO	Single Input Multiple Output
SISO	Single Input Single Output
SNR	Signal-to-Noise-Ratio
SPL	Sound Pressure Level
SVD	Singular Value Decomposition
SWAT	Sum of Weighted Accelerations Technique

T

TMF	Trim Mean Filtered
TR	Transmissibility

W

WEM	Whole Engine Models

Preface

The very first idea for this book came, some years ago, from Alex Carrella, who at the time was a young postdoctoral researcher, working within a University Technology Centre, integrating a group focused on applied research for a specific industry. In that case it was about vibration of helicopters. The partnership between academia and industry meant that an academic had to use the engineering pragmatism to solve some pressing issues, while practising engineers embrace the more rigorous and lengthier yet innovative practice of academia. Needless to say, the result is a fast transfer of technology to the industry and a much- needed flow of funds to academia to advance knowledge, as resources are of primary importance. For instance, in the process of preparing, carrying out and post-processing the data of a Ground Vibration Test (*GVT*) there were many questions to be answered, all within the science of structural dynamics, but related to different disciplines, each of them in a different book (or several books on the subject). A pragmatic approach would have been to have one tome with all that was needed enabling the counterpart in the industry to have a book on one's desk where he/she could dig a little deeper and have a more theoretical notion on a specific subject. Hence the idea of creating a volume to be kept on the desk of practising engineers and 'applied-researchers' for having a reference for most topics related to structural dynamics.

However, to create a book on the subject of structural dynamics particularly interesting to the industry is quite an ambitious objective to achieve, as the industry seeks the necessary knowledge to make things happen in a relatively fast way, the so-called "know-how", whereas academics explore the theoretical foundations to explain the physical phenomena, what one may call the "know-why". To find the right balance between these two perspectives is not an easy task. Although most of the co-authors of this textbook are scholars, they have the notion of the industrial environment and of the needs of those involved in the daily practice, sometimes due to some industrial experience, or because of close participation in research projects involving various types of companies.

Structural Dynamics is a vast *world* and no book can encompass the wide variety of themes. Each subject can become a book on its own. Therefore, a judicious choice had to be made and it was decided that the book would have 7 chapters, where Chapter 1 underlines the main fundamental aspects of vibration theory, from the very simple single degree of freedom system to the more general multiple degree of freedom, pointing out relevant aspects that are used in practice; Chapter 2 addresses the main practical problems that may be found in testing a structure, analysing the results and how to tackle the encountered issues in order to solve them; Chapter 3 presents the most important numerical tools that are commonly used and provides the necessary insight on how the various methods work; Chapter 4 describes in detail methods of analysing the results from dynamic tests and how to identify the dynamic properties, so to build a reliable mathematical model that represents the behaviour of a structure when in real operational conditions; Chapter 5 gives a comprehensive and solid background on the nonlinear behaviour of a system, as often the nonlinear aspects cannot be ignore by the analyst engineer; Chapter 6 describes the updating of numerical models, to improve their performance and provide better and accurate estimates of the real behaviour of the structure, either from a deterministic or from a stochastic point of view; in all these first six chapters simple examples are given, to illustrate the application of the various subjects. Finally, Chapter 7 provides some real industrial applications, with emphasis on aeronautical structures.

It is our believe that this book will be useful not only for industry, but also for students doing their master or doctorate studies. Sections identified by an asterisk mean that they may be skipped in a first reading.

Acknowledgements should be addressed to Prof. Hugo Policarpo, who helped producing most of the graphs of Chapter 1, as well as solving some text processing issues; to the work of Dr. Julian Londono-Monsalve in generating the experimental *FRFs* used in Chapter 4; to the "Aircraft Research Association Limited" and Dr. delli Carri for the test case 2 in Chapter 7; to Rolls-Royces plc. for the test articles of test cases 1 and 3; and to Dr. C. Schwingshackl for the *FE* model validations of test case 3 in Chapter 7.

The Editors,

Nuno M. M. Maia

Dario Di Maio

Alex Carrella

List of Authors

Nuno M. M. Maia
University of Lisbon, Portugal

Francesco Marulo
University of Naples Federico II, Italy

Chaoping Zang
Nanjing University of Aeronautics and Astronautics, P. R. China

Jonathan E. Cooper
University of Bristol, U.K.

Keith Worden
University of Sheffield, U.K.

Tiago A. N. Silva
Universidade Nova de Lisboa, Portugal

Dario Di Maio
University of Twente, The Netherlands

Alex Carrella
Vibration and Acoustic Consultant, Belgium

About the Companion Website

This book is accompanied by a companion website which includes a number of resources created by author for students and instructors that you will find helpful.

www.wiley.com\go\carrella\Structural Dynamics in Engineering Design

The student website includes the Figures PDF of chapter 7.

Chapter 1 Theoretical Background

1.1 Introduction

Structural dynamics is a vast area of knowledge, as it will be apparent throughout this book. In this first chapter the reader should make acquaintance with the fundamentals of the vibration phenomena, from the simplest cases to the more complex ones, in order to build a strong background with the objective of providing an easier insight to the more detailed and dedicated matters, as well as a better way to interpret the real physical problem and elaborate on a credible mathematical model. To have such a model that can reproduce with good accuracy the real dynamic behaviour of a real structure is essential when, for example, one wishes to study its response to some modification or to detect any change due to occurrence of some kind of damage.

The fundamentals of structural dynamics, from the solid mechanics viewpoint, stand on the linear theory of vibration and on the three main properties associated to a structure: mass, stiffness and damping. One has also to bear in mind the different types of motions or forces that may be measured or noticed on a structure. With those types of ingredients one can establish the dynamic equilibrium equations that allow to obtain the answers to our problem. In this chapter, one shall start by addressing some general concepts, definitions and assumptions; then, one moves to the study of the simplest case of the single degree-of-freedom (*SDOF*) system and then to the more complicated case of multiple degrees-of-freedom (*MDOF*).

1.2 Fundamental Concepts

1.2.1 Types of Signals

Thinking of something vibrating, one imagines first of all motion; this motion may be due to some induced forces or to some other induced motion. In mathematical terms or in physics, both force and motion are often referred to as signals that vary in time. The main classification distinguishes those signals that can be foreseen exactly after a certain time interval, which are named as deterministic and those that can only be estimated approximately, in statistic terms, which are called non-deterministic or random. Deterministic may be the motion of the crank-shaft mechanism or of any machine with some inner rotating shafts slightly unbalanced, while random may be the motion induced by the road onto a car wheel or by the wind on a chimney. The deterministic signals may be periodic, quasi-periodic or aperiodic. The periodic are composed by several harmonics and are quite common to find in industrial machinery in general; a gearbox, for instance, will exhibit such a response when running in steady-state conditions. A quasi-periodic signal may result from the superposition of two signals with different frequencies when those frequencies are not related through a rational quantity. As the name suggests, the resulting signal will look very much periodic, but in fact it is not exactly periodic. Finally, the aperiodic signal is a signal that does not repeat itself; it can be the force developed by a machine when it starts moving, or a bump on the road that has a certain profile

Structural Dynamics in Engineering Design, First Edition, Nuno M. M. Maia, Dario Di Maio, Alex Carrella, Francesco Marulo, Chaoping Zang, Jonathan E. Cooper, Keith Worden, and Tiago A. N. Silva.

and for a given speed of the vehicle always gives the same input, although it happens only once. Obviously, the simplest signal of all is the harmonic signal, which is a particular case of the periodic one. The non-deterministic signals may be stationary or non-stationary, the former meaning that along the considered time interval the statistical properties (mean, standard deviation) keep unchanged. Figure 1.1 illustrates the above classification.

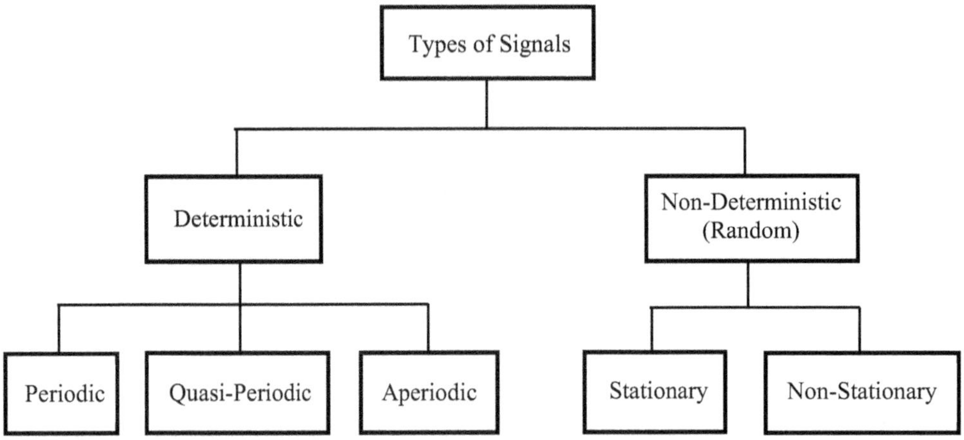

Figure 1.1 Classification of signals.

1.2.2 Degrees-of-Freedom/Discretisation

Any real structure or machine is composed by elements whose properties are distributed in a continuous manner, with some degree of deformability, as nothing is absolutely rigid; as a consequence, it has an infinite number of possibilities of motion. Each of those "possibilities of motion" is called a degree-of-freedom (*DOF*). Therefore, in reality, one has always an infinite number of *DOF*. However, in 99% of the cases it is not possible to study a complex real structure assuming an infinite number of *DOF*. One studies the machine or structure using a finite number of *DOF*, defining a model or system represented by a set of coordinates (e.g., $x_1(t)$, $x_2(t)$, etc.). In other words, the number of *DOF* of a structure is the number of independent coordinates that are necessary to describe the position of that structure at any instant of time.

The process of assigning a finite number of *DOF* to the real structure is called discretisation. Which criteria should one use to select which and how many *DOF*? The main criterion has to do with the objective of the study, unless one is limited by the available computational capacity. For instance, suppose one has a machine that moves essentially along the vertical direction. Although the machine has an infinite number of *DOF*, if one is interested in studying its global response in the vertical direction, then one only needs an *SDOF* system (Figure 1.2 (a)). The same happens if one wishes to study the dynamic response of a vehicle along the vertical direction, due to the motion imposed by the road (Figure 1.2 (b)). However, if one wishes to know the response both in the vertical direction and the angular response (pitch), a model with two *DOF* is required (Figure 1.2 (c)). And so on. In the model of Figure

1.2 (d) one has a discretisation with 7 *DOF*, the vertical translation of the body of the vehicle, 2 rotations (pitch and roll movements) and 4 movements of the wheels. If the objective is the study of the structural stresses and strains of the car body, then a very detailed model that may easily attain dozens of thousands of *DOF* may be necessary (Figure 1.2 (e)).

In the example of Figure 1.3 (a), a beam is discretised into 6 *DOF* for the study of its vertical motion, considering a set of concentrated masses, joined by linear springs (the so-called lumped masses system), whereas in Figure 1.3 (b) the same beam is divided into 6 elements, each one having 4 *DOF* (2 translations and 2 rotations). While in the case of Figure 1.3 (a) the masses are completely separated from the springs, in Figure 1.3 (b) each element has its own deformation, i.e. each element has both mass and stiffness properties. This latter approach, the finite element modeling, is normally more accurate than a corresponding lumped mass approach.

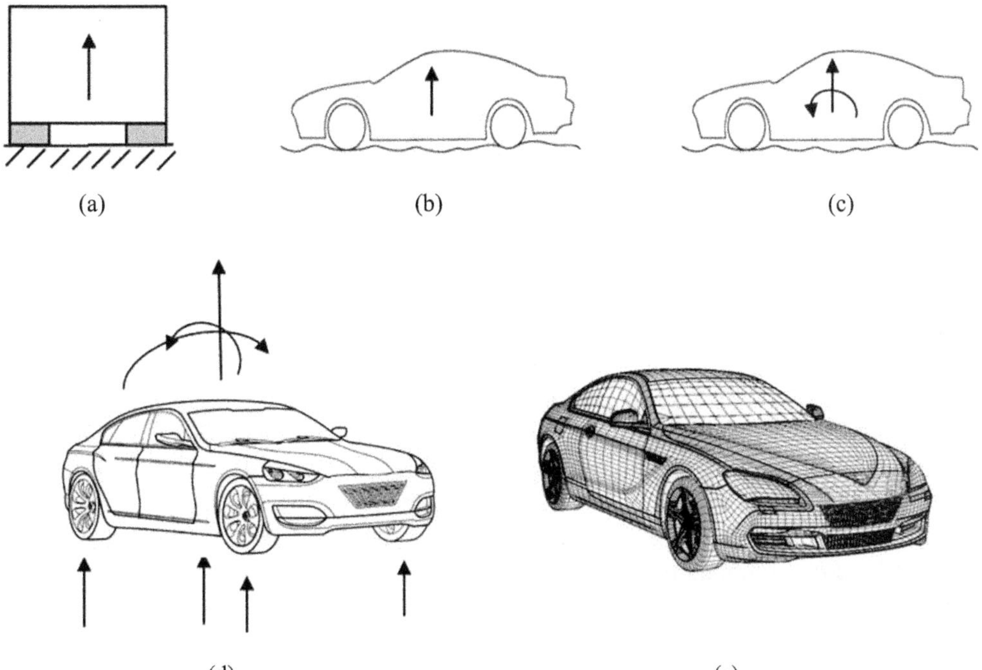

Figure 1.2 Examples of discretisation.

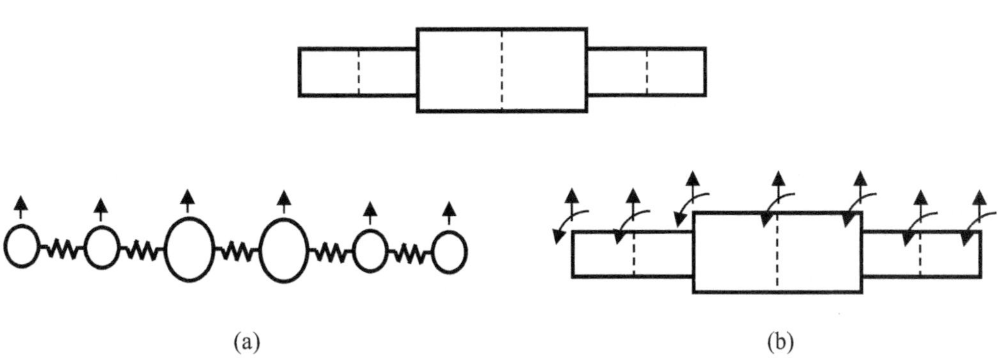

Figure 1.3 Example of lumped mass discretisation and finite element discretisation.

1.2.3 Elements of a Vibrating System

The dynamic behaviour of a structure, either vibrating freely or due to imposed loads, depends on three fundamental properties: mass, stiffness and damping. An *N DOF* system may be formed by sets of three elements, representing those properties. The fundamental set is the one of the *SDOF* system, illustrated in Figure 1.4, where one may or may not have an applied force $f(t)$.

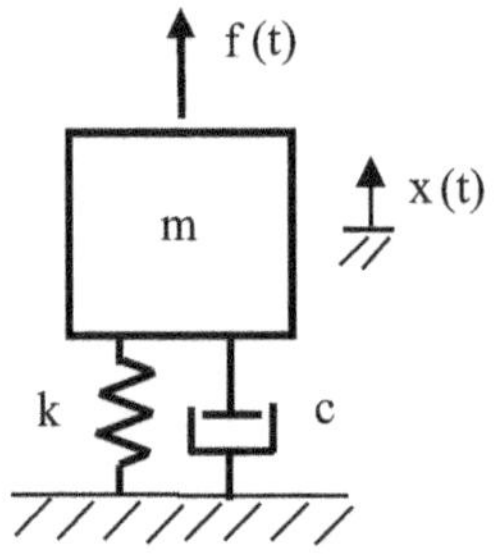

Figure 1.4 Elements of a vibrating system.

The simplest case, the *SDOF* system, is therefore constituted by a mass, a spring and a damper. This last one is often considered of the viscous type, which will also be the case in what follows.

It is very important to stress that the mass, the spring and the damper are ideal elements, i.e. they do not have material existence. They are idealised so that each one only represents its own property: the mass has just mass, it is infinitely rigid and does not dissipate any energy; all the stiffness (or flexibility) is in the spring, which is massless and does not dissipate energy; and the dissipation is represented by the damper, which is also massless and does not exhibit any stiffness. For instance, a simply supported beam in free vibration may be approximated by the mass-spring-damper model of Figure 1.4; its geometry and mass distribution lead to a certain flexibility and there is some damping due to some internal dissipation and some friction at the supports. The mass-spring-damper model is an idealisation intended to reflect its response in free vibration, and that is all one needs to know. In terms of energy, the mass only gains or supplies kinetic energy, while the spring only stores or supplies elastic energy and the damper only dissipates energy.

The Spring Element
In linear vibration theory it is assumed that the spring has a linear behaviour, so that the force, say f_k, is directly proportional to the displacement x through the stiffness coefficient k, i.e. $f_k = k\,x$. This linear relation is valid within a limited range of displacements. Beyond that range the spring tends to become stiffer (Figure 1.5 (a)). As in the large majority of the cases one is dealing with vibrations of small amplitude, the linear approach for the spring behaviour is reasonable. Otherwise, one has to use other theories or to subdivide the displacement range in various parts, assigning a different k value to each one.

The Damper Element
It is also assumed that the damper has a linear behaviour. In the viscous type of damping, the force, f_c, is directly proportional to the velocity $\dot{x}$ through the damping coefficient c, i.e.

$f_c = c\dot{x}$. As it happens with the spring, the linear assumption is only valid within a limited velocity range (Figure 1.5 (b)).

The Mass Element

When in accelerated motion, the mass reacts with an inertia force according to Newton's second law, and therefore, similarly to what happens to the spring and damper, the inertia force, f_i, is directly proportional to the acceleration $\ddot{x}$ through the mass *m*, i.e. $f_i = m\ddot{x}$. Here, the mass is always considered as a constant (Figure 1.5 (c)).

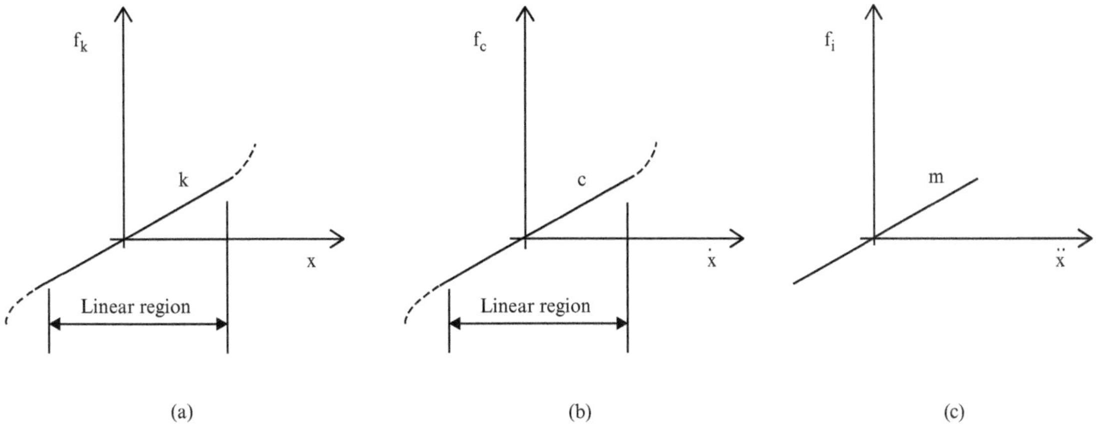

Figure 1.5 The spring, damper and mass behaviour.

The Torsional Vibration Model

So far, one has been talking about a system in translational vibration, but one may also contemplate the torsional case. The corresponding model is represented in Figure 1.6, where instead of the mass, and translational stiffness and damping, one has a disk with an inertia moment *J*, a shaft representing a spring with torsional stiffness k_t and a container with a viscous fluid representing a torsional damping coefficient c_t. Naturally, now the applied load (if any) will be a moment *M(t)* and the response coordinate will be an angle $\theta(t)$. The moments carried by each element are now $M_i = J\ddot{\theta}$ for the inertia of the disk, $M_{c_t} = c_t\dot{\theta}$ for the torsional damper and $M_{k_t} = k_t\theta$ for the torsional spring.

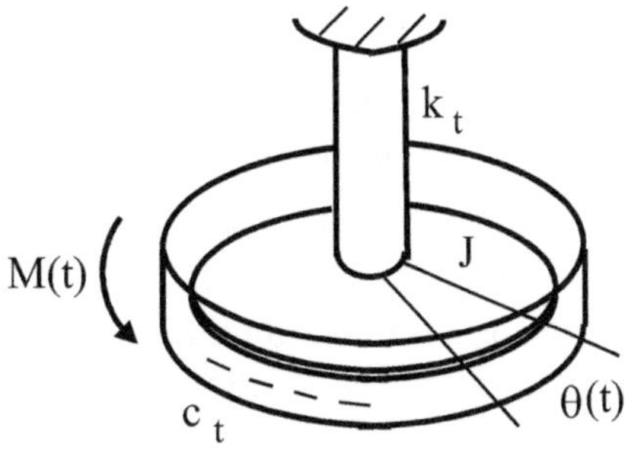

Figure 1.6 The torsional vibration model.

1.2.4 The Simple Harmonic Motion

The simple harmonic motion can be described by a *sine*, a *cosine* or a combination of both:

$$x(t) = X_1 sin\omega t + X_2 cos\omega t \tag{1.1}$$

In fact, it is not difficult to show that this expression can be written as

$$x(t) = X sin(\omega t + \phi) \tag{1.2}$$

where

$$\begin{aligned} X &= \sqrt{X_1^2 + X_2^2} \\ tan\,\theta &= \frac{X_2}{X_1} \end{aligned} \tag{1.3}$$

It is immediately concluded that the summation of two harmonic motions with the same frequency is also a harmonic motion with that frequency, even if the phase angles are different, as in Eq. (1.1). The general expression for the simple harmonic motion is therefore the one given in Eq. (1.2), where X is the amplitude, ω is the frequency in *rad/s* (also called circular frequency) and ϕ is the phase angle with respect to the initial time (t = 0). The period T is the interval of time after which the signal repeats its value and is in phase with the initial time instant taken as reference. In Hertz (cycles per second), the frequency is the inverse of the period ($f = 1/T$) and because in each cycle a 2π angle is generated, one can relate the frequencies in *rad/s* and in *Hz*: $\omega = 2\pi f$ and one also has $\omega = 2\pi/T$.

Let the displacement be given by

$$x(t) = X sin\omega t \tag{1.4}$$

The velocity and acceleration will be:

$$\dot{x}(t) = X\omega cos\omega t = X\omega sin(\omega t + \pi/2) \tag{1.5}$$

$$\ddot{x}(t) = -X\omega^2 sin\omega t = X\omega^2 sin(\omega t + \pi) \tag{1.6}$$

Thus, when one has a harmonic motion, the displacement, velocity and acceleration are all harmonic functions with the same frequency, where the amplitude is multiplied by ω each time one differentiates and the phase angle is increased by $\pi/2$, meaning – for instance – that when the displacement attains a maximum, the velocity is zero. Substituting Eq. (1.4) in Eq. (1.6) leads to the important conclusion that for harmonic motion there is a proportional relationship between the acceleration and the displacement, through the negative of the squared frequency:

$$\ddot{x}(t) = -\omega^2 x(t) \tag{1.7}$$

or

$$\ddot{x}(t) + \omega^2 x(t) = 0 \tag{1.8}$$

The summation of harmonic motions of different frequencies is not a harmonic motion, although it is periodic, i.e. one cannot define anymore a single amplitude for the signal, but there is still a time after which the signal is repeated, the period T (e.g. as in Figure 1.7).

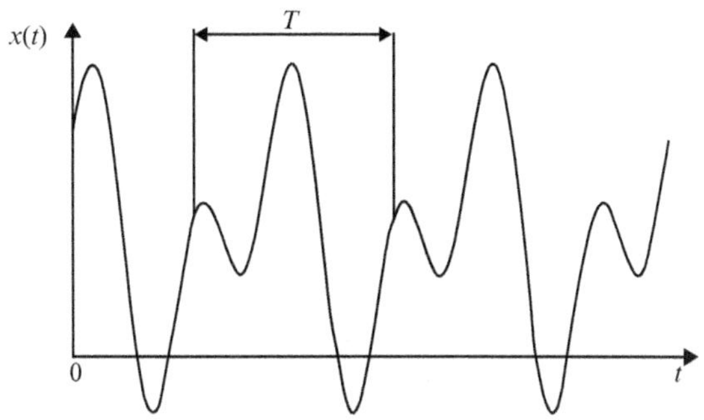

Figure 1.7 Example of a periodic signal.

1.3 Establishing the Dynamic Equilibrium Equations

To deduce the dynamic equilibrium equations one can use either the vectorial approach, i.e. using vectorial or Newtonian mechanics, based on force and/or moment balance, or the analytical approach, based on energy concepts, i.e. on Lagrangian mechanics. In what follows, one shall give a brief explanation on both approaches.

1.3.1 The Dynamic Equilibrium Equations based on Vectorial Mechanics

Using the force balance, one draws the free body diagram, where one can represent the applied as well as the reaction forces. Then, there are two possibilities, (i) either to use Newton's second law ($\sum \boldsymbol{F} = m\ddot{\boldsymbol{x}}$) or (ii) to use D'Alembert's principle ($\sum \boldsymbol{F} - m\ddot{\boldsymbol{x}} = \boldsymbol{0}$). Essentially, the latter is the same as the former; however, the physical interpretation is different, as in D'Alembert's principle one takes the inertia force as an applied force (with the appropriate signal) and therefore one states that the whole system acted by all forces including the inertia one is in equilibrium if the summation of all the forces is zero. So, in a way, D'Alembert's principle reduces the dynamic problem to a static one. From Figure 1.4, assuming that the mass moves upwards and that the displacement, velocity and acceleration are all positive along that motion, the free body diagram is the one illustrated in Figure 1.8.

The dynamic equilibrium implies:

$$f(t) - f_k - f_c - f_i = 0 \tag{1.9}$$

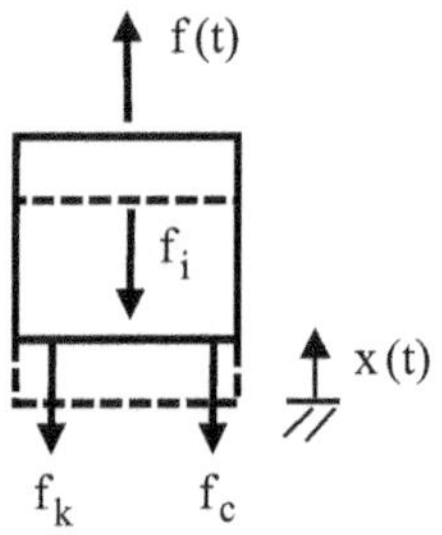

Figure 1.8 Free body diagram of the *SDOF* system.

As $f_k = k\,x$, $f_c = c\,\dot{x}$ and $f_i = m\,\ddot{x}$, after rearrangement it follows that[1]

$$m\ddot{x}+c\dot{x}+kx=f(t) \tag{1.10}$$

which is a second order, linear differential equation with constant coefficients. For many degrees of freedom, this approach may not be convenient, as one must take a lot of care concerning the directions of the various forces.

1.3.2 The Dynamic Equilibrium Equations based on Analytical Mechanics

Analytical dynamics is essentially based upon the notions of work, potential energy and kinetic energy. In this section one shall present the most important concepts and tools, in a very concise manner. Not going into too many details, one knows that there is a direct relationship between the work and the potential energy, when the forces that are acting depend only on the position, i.e, when they are conservative. In such conditions, the work is a state function, i.e. the work produced by the forces along the motion does not depend on the path that is taken, but only on the initial and final positions. For an infinitesimal displacement, one has an infinitesimal work dW and an infinitesimal potential energy dV, related by:

$$dW=-dV \tag{1.11}$$

It is also well known that there is a relation between work and kinetic energy (T). Taking again an elementary motion, one has:

$$dW=dT \tag{1.12}$$

The Principle of Conservation of Energy

From (1.11) and (1.12), it follows that $d(T+V)=0$, or

$$T+V=Constant \tag{1.13}$$

---

[1] Although the displacement, velocity and acceleration are implicit functions of time, one chooses to adopt a simplified notation, where only the applied force appears with that indication.

which is known as the principle of conservation of energy. The dynamic equilibrium equation can be deduced from

$$\frac{d}{dt}\left(T+V\right)=0 \tag{1.14}$$

For the *SDOF* system, neglecting the damping, and without any applied force (in free vibration), $T=\frac{1}{2}m\dot{x}^2$ and $V=\frac{1}{2}kx^2+mgx$. Applying Eq. (1.14) leads to:

$$m\dot{x}\ddot{x}+kx\dot{x}+mg\dot{x}=0 \tag{1.15}$$

or

$$\left(m\ddot{x}+kx+mg\right)\dot{x}=0 \tag{1.16}$$

It happens that $\dot{x}$ is zero only at the extreme positions. To guarantee that Eq. (1.16) holds for all the time instants, one has to make

$$m\ddot{x}+kx+mg=0 \tag{1.17}$$

or

$$m\ddot{x}+kx=-mg \tag{1.18}$$

The solution of Eq. (1.18) is the solution of the homogeneous equation $m\ddot{x}+kx=0$ plus the particular solution correspondent to the constant $-mg$, which is a constant initial displacement given by $-mg/k$, meaning that the system will vibrate around that new displacement due to the static weight, instead of vibrating around the zero position. As the weight is constant along the motion, it is not relevant to consider that situation and therefore one may assume that the dynamic equilibrium equation is simply:

$$m\ddot{x}+kx=0 \tag{1.19}$$

Therefore, provided that the weight remains a constant during the vibration motion, one can always forget the gravity term of the potential energy[2].

The Generalisation of the Principle of Conservation of Energy

The principle of the conservation of energy can be generalised to incorporate non-conservative forces. If there are both conservative forces ($\boldsymbol{f}_c$) and non-conservative forces ($\boldsymbol{f}_{nc}$), one may write:

$$\boldsymbol{f}=\boldsymbol{f}_c+\boldsymbol{f}_{nc} \tag{1.20}$$

Taking the dot product with $\boldsymbol{dr}$, where $\boldsymbol{r}$ is a position vector,

$$\boldsymbol{f}\cdot\boldsymbol{dr}=\boldsymbol{f}_c\cdot\boldsymbol{dr}+\boldsymbol{f}_{nc}\cdot\boldsymbol{dr} \tag{1.21}$$

[2] That is not the case, for instance, with the simple pendulum, as the component of its weight along the motion keeps changing all the time.

where each term represents infinitesimal work.

As Eq. (1.12) is valid even when non-conservative forces are included and having also into consideration Eq. (1.11), one obtains:

$$dT = -dV + \boldsymbol{f}_{nc} \cdot \boldsymbol{dr} \tag{1.22}$$

or

$$d(T+V) = \boldsymbol{f}_{nc} \cdot \boldsymbol{dr} \tag{1.23}$$

Dividing[3] by dt, leads to the generalisation of Eq. (1.14):

$$\frac{d}{dt}(T+V) = \boldsymbol{f}_{nc} \cdot \dot{\boldsymbol{r}} \tag{1.24}$$

Thus, the rate of change *w.r.t.* time of the total energy is no longer zero, but equals the dissipated power. Applying this expression to the system of Figure 1.4, this time ignoring the energy due to gravity, and recognising that the non-conservative forces are $f_{nc} = f(t) - c\dot{x}$, it follows that

$$\frac{d}{dt}\left(\frac{1}{2}m\dot{x}^2 + \frac{1}{2}kx^2\right) = \left(f(t) - c\dot{x}\right)\dot{x} \tag{1.25}$$

or

$$m\dot{x}\ddot{x} + kx\dot{x} = \left(f(t) - c\dot{x}\right)\dot{x} \tag{1.26}$$

and finally,

$$m\ddot{x} + c\dot{x} + kx = f(t) \tag{1.27}$$

The Principle of Virtual Work

Applying D'Alembert's principle, one may say that a particle of mass m_i is in dynamic equilibrium when the resultant of all the forces acting upon it, including the inertial ones, is zero:

$$\boldsymbol{F}_i - m_i\ddot{\boldsymbol{r}}_i = \boldsymbol{0} \tag{1.28}$$

Conceiving a virtual displacement[4] $\delta\boldsymbol{r}_i$ compatible with the constraints and taking the dot product, one obtains

$$\left(\boldsymbol{F}_i - m_i\ddot{\boldsymbol{r}}_i\right) \cdot \delta\boldsymbol{r}_i = 0 \tag{1.29}$$

---

[3] Note that dividing by *dt* is not the same as differentiating Eq. (1.23) with respect to time.

[4] A virtual displacement does not exist, it is imagined. While a real displacement implies the passage of time, the virtual one is thought as an alternative position at that same time instant, complying with the constraints. It is denoted by the letter δ.

As Eq. (1.28) is an equilibrium equation, so is Eq. (1.29), now in terms of virtual work of the applied forces and of the inertial ones. For the overall system (considering *N* masses), it follows that

$$\delta W = \delta W_{\substack{applied\\forces}} + \delta W_{\substack{inertial\\forces}} = \sum_{i=1}^{N} \boldsymbol{F}_i \cdot \delta \boldsymbol{r}_i + \sum_{i=1}^{N} \left(-m_i \ddot{\boldsymbol{r}}_i\right) \cdot \delta \boldsymbol{r}_i = 0 \tag{1.30}$$

Note that here one needs only to consider applied and inertial forces, no reaction forces are taken into account, as they do not produce any work. This principle can be stated as: it is a necessary and sufficient condition for a system to be in dynamic equilibrium that the work performed by all the applied and inertial forces along arbitrary virtual displacements compatible with the physical constraints be zero.

This principle can be applied to an *N DOF* system, although it is used more with *SDOF* systems, as otherwise one needs to imagine *N* virtual displacements, which may become complicated in real cases.

Hamilton's Principle

This is one of the most important integral principles in Mechanics, also known as the principle of least- action, where the "action", defined as

$$I = \int_{t_1}^{t_2} \left(T - V\right) dt \tag{1.31}$$

is a kind of measure of the balance between the kinetic and potential energies, along a given time interval. This principle states that from all the possible configurations that a system can assume when evolving from a configuration 1 at time t_1 to a configuration 2 at time t_2, the one that satisfies the dynamic equilibrium at each instant is the one that lends the action stationary (minimum, in this case) during that time interval.

Such a minimisation applies to a set, or family, of functions, i.e. one is looking for the "minimum function", it is not a classical minimisation. Therefore, what one minimises is the action, whose variation is in the same sense as the virtual displacements, i.e. at each instant one admits that the configuration of the system is one, or another that differs slightly from it, although still possible, according to the system constraints. All the different configurations should coincide at t_1 and t_2. In mathematical terms, the principle is defined as:

$$\delta I = \delta \int_{t_1}^{t_2} \left(T - V\right) dt = 0, \qquad \delta \boldsymbol{r}_i\left(t_1\right) = \delta \boldsymbol{r}_i\left(t_2\right) = \boldsymbol{0}, \qquad i = 1, \ldots N \tag{1.32}$$

This principle may be applied, more conveniently than the principle of virtual work, to *N DOF* systems. It can also be generalised to include non-conservative forces, by adding the respective virtual work term:

$$\delta I = \int_{t_1}^{t_2} \left(\delta L + \delta W_{nc}\right) dt = 0, \qquad \delta \boldsymbol{r}_i\left(t_1\right) = \delta \boldsymbol{r}_i\left(t_2\right) = \boldsymbol{0}, \qquad i = 1, \ldots N \tag{1.33}$$

where $L = T - V$ is the Lagrangian of the system.

Lagrange Equations

When dealing with systems with many degrees of freedom, the most appropriate way of deducing the dynamic equilibrium equations is by the use of Lagrange equations. These may be deduced from the principle of virtual work or from Hamilton's principle. One shall do it in a concise way, from the latter. First, one must have into account that the system may be defined by a set of physical coordinates x, although only q of them correspond to the effective number of *DOF* of the system. These are called generalised coordinates (because they can represent displacements and/or rotations).

One shall assume here that a direct geometric relation may be established between x and q (and *vice-versa*) and that some of the system constraints may be explicitly time dependent (for instance due to some known imposed displacements). In that case,

$$T=T(q,\dot{q},t) \quad \text{and} \quad V=V(q,t) \quad \Rightarrow \quad L=T-V=L(q,\dot{q},t) \tag{1.34}$$

Applying Hamilton's principle (Eq. (1.32)), it is possible to show that the Lagrange equations are given by

$$\frac{d}{dt}\left(\frac{\partial T}{\partial \dot{q}_k}\right)-\frac{\partial T}{\partial q_k}+\frac{\partial V}{\partial q_k}=0, \quad k=1, \ldots N \tag{1.35}$$

If non-conservative forces are included (Eq. (1.33)), namely forces due to viscous dampers and external applied forces, the Lagrange equations become:

$$\frac{d}{dt}\left(\frac{\partial T}{\partial \dot{q}_k}\right)-\frac{\partial T}{\partial q_k}+\frac{\partial V}{\partial q_k}+\frac{\partial \mathcal{F}}{\partial \dot{q}_k}=Q_k, \quad k=1, \ldots N \tag{1.36}$$

where $\mathcal{F}$ is Rayleigh's dissipation function and Q_k are the generalised external applied forces. This is probably the simplest way to establish the N equilibrium equations corresponding to the N *DOF* of a system.

1.4 The Single Degree-of-Freedom System

In Figure 1.4 the *SDOF* system model was represented, as the simplest of all. Although simple, there are many examples in real life that can be studied with this model, as already pointed out in Section 1.2.2. Besides, it is very important to study this model because one can learn a lot about the physical content of the vibration phenomenon, therefore facilitating the study of the multiple *DOF* case.

1.4.1 The Dynamic Equilibrium Equation

From Section 1.2, the dynamic equilibrium equation in translation was given by

$$m\ddot{x}+c\dot{x}+kx=f(t) \tag{1.37}$$

From the torsional model of Figure 1.6, one can write the corresponding equilibrium equation:

$$J\ddot{\theta}+c_t\dot{\theta}+k_t\theta=M(t) \tag{1.38}$$

Regarding Eq. (1.37), due to the existence of an applied force, it is said that the system is in *forced vibration*. If there is no applied force, the equation becomes:

$$m\ddot{x}+c\dot{x}+kx=0 \tag{1.39}$$

and the system is said to be in *free vibration*.

If the damping is weak, one may write an even simpler equation, by neglecting the damping term[5]:

$$m\ddot{x}+kx=0 \tag{1.40}$$

Eq. (1.40) is the simplest case one can have, a system vibrating freely, without damping (Figure 1.9).

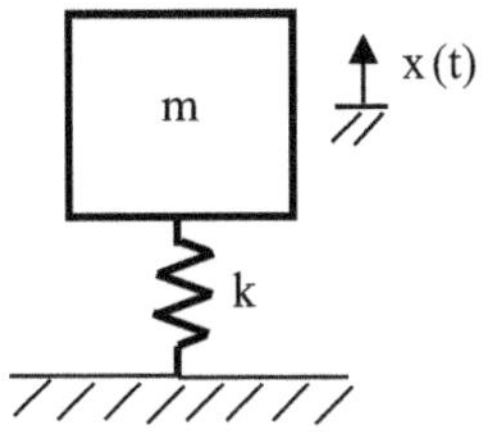

Figure 1.9 *SDOF* system model, without damping, in free vibration.

Dividing Eq. (1.40) by m, it follows that

$$\ddot{x}+\frac{k}{m}x=0 \tag{1.41}$$

From the simple harmonic motion of Section 1.2.4, one concluded that there is a relationship between the acceleration and the displacement, such that (Eq. (1.8)):

$$\ddot{x}+\omega^2 x=0 \tag{1.42}$$

where ω is the frequency of the motion. Note that Eqs. (1.41) and (1.42) are the same, which means that the solution of Eq. (1.41) is the simple harmonic motion. On other words, the *SDOF* system without damping vibrates freely in a sinusoidal way, with a frequency given by:

$$\omega_n=\sqrt{\frac{k}{m}} \tag{1.43}$$

[5] To neglect the damping can often be a good choice, because damping is difficult to evaluate and the calculations become much easier. Note that when doing so, one stays on the safe side, because the vibration amplitudes (and consequently stresses) will be higher than the ones happening in reality.

and called the *natural frequency*. For the torsional model, it is clear that the natural frequency will be given by:

$$\omega_n = \sqrt{\frac{k_t}{J}} \tag{1.44}$$

1.4.2 Equivalent Systems

To have a single *DOF* does not necessarily mean that there is only one mass, one spring and one damper. One may have many of each of those elements and still end up with a single *DOF*, depending on the constraints of the system. However, if that is the case, one may want to obtain an equivalent system, composed by equivalent mass, spring and damping elements, as well as equivalent force. For instance, in Figure 1.10 (a) one has a system composed by two masses (one in rotation, one in translation connected by a rigid massless link), two springs, one damper, subjected to an applied moment; it is clear that there is only one degree-of-freedom, as the motion of the masses are directly related. Therefore, it must be possible to establish an equivalent system in one of the coordinates, say *x*, as in Figure 1.10 (b).

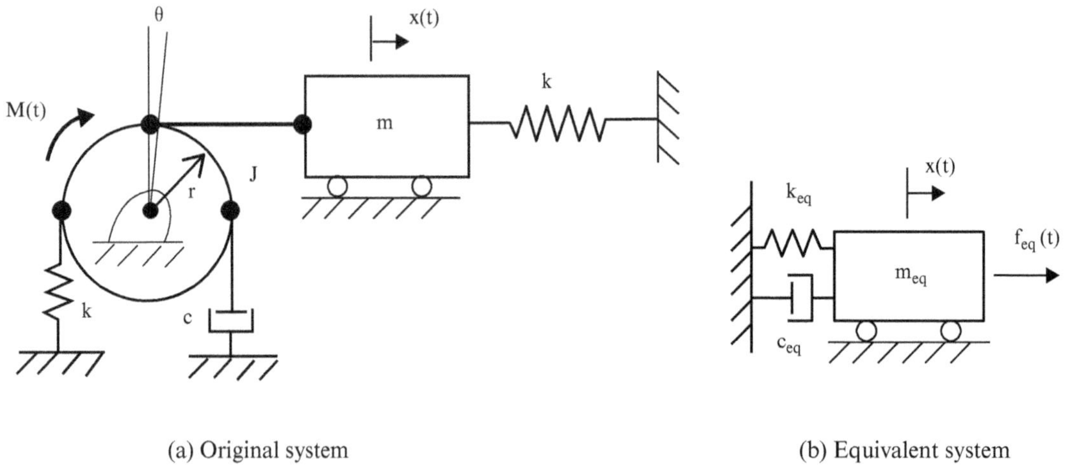

Figure 1.10 *SDOF* system with several elements and its equivalent.

Such an equivalence is accomplished by equating between the original and the equivalent systems (i) the kinetic energies to obtain the equivalent mass, (ii) the elastic energies to obtain the equivalent stiffness, (iii) the Rayleigh's dissipation function to obtain the equivalent viscous damping and (iv) the virtual work of the applied forces to obtain the equivalent force.

1.4.3 Undamped Free Vibration Response

As mentioned in Section 1.4.1, the response in free vibration without damping is harmonic, although this has been concluded in a pure physical basis. In a little more detail, one looks for the solution of the differential equation

$$m\ddot{x} + kx = 0 \tag{1.45}$$

The solution of Eq. (1.45) is sought through a transformation to the Laplace space, $x = Ce^{st}$, where s is a complex variable (Laplace's variable). Differentiating twice, one obtains $\ddot{x} = s^2 Ce^{st}$. Substituting x and $\ddot{x}$ back in Eq. (1.45), yields

$$m s^2 + k = 0 \tag{1.46}$$

which is an algebraic equation that one can easily solve, leading to:

$$s = \pm\sqrt{-k/m} = \pm i\,\omega_n \tag{1.47}$$

As there are two algebraic solutions, the response must be a linear combination of both, i.e.

$$x(t) = C_1 e^{s_1 t} + C_2 e^{s_2 t} = C_1 e^{i\omega_n t} + C_2 e^{-i\omega_n t} \tag{1.48}$$

Using Euler's formulæ, one ends up with

$$x(t) = A cos\omega_n t + iB sin\omega_n t \tag{1.49}$$

where constants A and B must be calculated using two initial conditions[6]. Let these conditions be the displacement $x(0) = x_0$ and velocity $\dot{x}(0) = \dot{x}_0$. Then, the response becomes

$$x(t) = x_0 cos\omega_n t + \frac{\dot{x}_0}{\omega_n} sin\omega_n t \tag{1.50}$$

or simply as

$$x(t) = X sin(\omega_n t + \varphi) \tag{1.51}$$

with

$$X = \sqrt{(x_0)^2 + \left(\frac{\dot{x}_0}{\omega_n}\right)^2} \quad \text{and} \quad tan\,\varphi = \frac{x_0 \omega_n}{\dot{x}_0} \tag{1.52}$$

If one just gives an initial displacement, the response (Eq. (1.50)) will be a *cosine* with amplitude x_0. If one strikes the system with a perfect impulse, which is the same as to give only an initial velocity $\dot{x}_0$ (the case of a ball colliding with another one in a perfect elastic shock, where one ball stays still and transmits to the other one all the momentum), the response will be a *sine* with amplitude $\dot{x}_0/\omega_n$. In the general case, there will be both x_0 and $\dot{x}_0$, the amplitude is X and there is a phase φ with respect to the initial zero time.

1.4.4 Viscously Damped Free Vibration Response

One is now seeking the solution of

[6] Two constants appear because one is dealing with a second order differential equation.

$$m\ddot{x} + c\dot{x} + kx = 0 \tag{1.53}$$

Once again one starts by introducing the transformation $x = Ce^{st}$, which now leads to the algebraic equation

$$ms^2 + cs + k = 0 \tag{1.54}$$

whose solutions are:

$$s_{1,2} = \frac{-c \pm \sqrt{c^2 - 4km}}{2m} \tag{1.55}$$

Three situations may happen here: either the radicand $c^2 - 4km$ is positive, or zero or negative. If it is positive, $c^2 > 4km$ and this simply means that the system "has too much damping", or in a more proper way that the dissipative forces dominate over the elastic and inertial ones and the system is said to be *over-damped.* In such a case the roots are real and so the solution $x(t) = C_1e^{s_1t} + C_2e^{s_2t}$ is the combination of two real exponents, which obviously does not correspond to a vibratory motion. For instance, if one gives to the system an initial displacement x_0 and a initial zero velocity (e.g., as in Figure 1.11 (a)), it will restore its initial position more or less quickly, depending on the amount of damping, but does not vibrate at all (Figure 1.11 (b)).

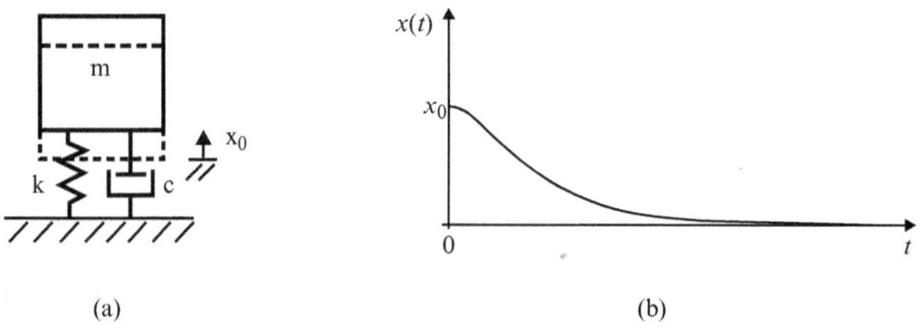

Figure 1.11 Over-damped response of an *SDOF* system.

The limit of such a situation is when the radicand is zero, i.e. $c^2 = 4km$; the roots will be real and repeated and the solution will be given by $x(t) = (C_1 + C_2t)e^{st}$, which does not represent a vibration either, but instead a decay to the original equilibrium position. The system is said to have *critical damping* and the damping coefficient value is[7]

$$c_{cr} = \sqrt{4km} = 2m\omega_n \tag{1.56}$$

[7] Note that the critical damping coefficient does not depend on the damping itself. Even for an undamped mass-spring system it is possible to calculate its critical damping value. It simply means that if one wishes that the system keeps vibrating in free conditions but with a higher value of damping, one can increase it until the critical value. Beyond that value the system will stop vibrating, simply relaxing to its original position.

In torsion,

$$c_{cr} = 2J\,\omega_n \tag{1.57}$$

Naturally, the interesting case in free vibration is when the radicand is negative, i.e. $c^2 < 4km$, meaning that the elastic and inertia forces dominate the dissipative ones. The system is called *under-damped*. The roots are complex conjugate and the solution includes complex exponential terms that can be related to *sines* and *cosines* through Euler's formulæ.

Before exploring the last case, it is important to introduce a factor that relates the actual damping coefficient to the critical damping value, called the *damping factor*[8] ξ:

$$\xi = \frac{c}{c_{cr}} = \frac{c}{2m\,\omega_n} \tag{1.58}$$

and so, for the over-damped system, $\xi > 1$; for the critical system, $\xi = 1$; and for the under-damped case, $\xi < 1$.

The advantage of working with dimensionless quantities is that any conclusion that one draws on the behaviour of the *SDOF* system will stay valid for any *SDOF* system. Let the solutions (1.55) be written in terms of the damping factor:

$$s_{1,2} = -\xi\,\omega_n \pm i\,\omega_n\sqrt{1-\xi^2} \tag{1.59}$$

As the solution is $x(t) = C_1 e^{s_1 t} + C_2 e^{s_2 t}$, it follows that

$$x(t) = e^{-\xi\omega_n t}\left(C_1 e^{\left(i\omega_n\sqrt{1-\xi^2}\right)t} + C_2 e^{\left(-i\omega_n\sqrt{1-\xi^2}\right)t} \right) \tag{1.60}$$

The damped natural frequency is defined as

$$\omega_d = \omega_n\sqrt{1-\xi^2} \tag{1.61}$$

Hence, the response is

$$x(t) = e^{-\xi\omega_n t}\left(C_1 e^{i\omega_d t} + C_2 e^{-i\omega_d t}\right) \tag{1.62}$$

One can observe that the expression within brackets is formally equal to Eq. (1.48) and therefore leads to an expression equivalent to Eq. (1.49). Thus, Eq. (1.62) can be written as

$$x(t) = e^{-\xi\omega_n t}\left(A\cos\omega_d t + iB\sin\omega_d t\right) \tag{1.63}$$

where A and B are calculated applying the initial conditions, as for the undamped case. The response becomes:

$$x(t) = e^{-\xi\omega_n t}\left(x_0\cos\omega_d t + \frac{\dot{x}_0 + \omega_n\xi x_0}{\omega_d}\sin\omega_d t \right) \tag{1.64}$$

---

[8] Also known as *damping ratio*.

Note that without damping ($\xi = 0$) Eq. (1.64) reduces to Eq. (1.50). The expression within brackets may be written as a *sine*, yielding:

$$x(t) = e^{-\xi\omega_n t} X \sin(\omega_d t + \varphi) \tag{1.65}$$

which is very similar to Eq. (1.51), though now the sinusoid (with frequency ω_d) is "squeezed" by the negative exponential that brings the response asymptotically to zero (Figure 1.12). The amplitude and phase are now given by:

$$X = \sqrt{(x_0)^2 + \left(\frac{\dot{x}_0 + \xi\omega_n x_0}{\omega_d}\right)^2} \quad \text{and} \quad \tan\varphi = \frac{x_0\omega_d}{\dot{x}_0 + \xi\omega_n x_0} \tag{1.66}$$

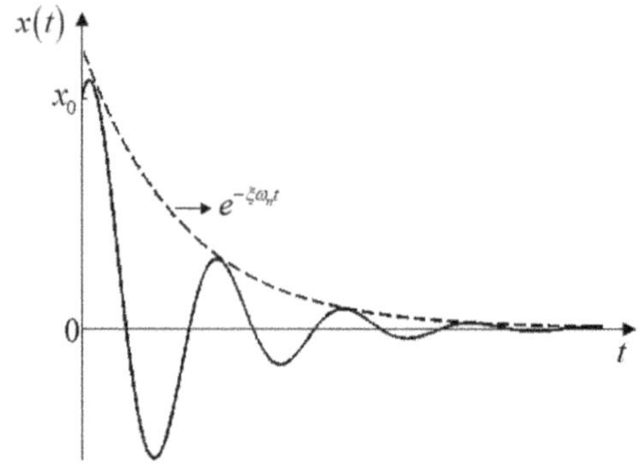

Figure 1.12 Under-damped response of an *SDOF* system.

1.4.5 Forced Vibration

When the system is actuated by some kind of input that lasts for some time, the vibration is said to be *forced*. Typically, one is talking about applied forces or applied displacements (which eventually correspond to applied forces). In this section the system will always be assumed as damped. One shall start with the simplest case of all, which is the study of the response to a simple harmonic force of constant amplitude.

Response due to a Harmonic Force of Constant Amplitude
The dynamic equilibrium equation is:

$$m\ddot{x} + c\dot{x} + kx = F \sin\omega t \tag{1.67}$$

where ω is the frequency of the applied force.

The solution is composed by two parts, the solution $x_h(t)$ of the homogeneous equation $m\ddot{x}_h + c\dot{x}_h + kx_h = 0$ and the particular solution $x_p(t)$ of $m\ddot{x}_p + c\dot{x}_p + kx_p = F \sin\omega t$, so that the complete solution is given by $x(t) = x_h(t) + x_p(t)$.

The homogeneous solution is the free vibration response (Eq. (1.65)), i.e. $x_h(t) = e^{-\xi\omega_n t} X_h sin(\omega_d t + \varphi)$; the particular solution is a combination of a *sine* and a *cosine* [9], i.e. $x_p(t) = X_{p_1} sin\omega t + X_{p_2} cos\omega t$. Differentiating once and twice and substituting in the equation, one ends up with:

$$-\omega^2 m\left(X_{p_1} sin\,\omega t + X_{p_2} cos\omega t\right) + \omega c\left(X_{p_1} cos\omega t - X_{p_2} sin\,\omega t\right) + k\left(X_{p_1} sin\,\omega t + X_{p_2} cos\omega t\right) = F\,sin\,\omega t \tag{1.68}$$

Equating the terms in *sine* and *cosine* leads to:

$$\begin{aligned}\left(k-\omega^2 m\right)X_{p_1} - \omega c X_{p_2} &= F\\ \left(k-\omega^2 m\right)X_{p_2} + \omega c X_{p_1} &= 0\end{aligned} \tag{1.69}$$

and so,

$$\begin{aligned}X_{p_1} &= \frac{F\left(k-\omega^2 m\right)}{\left(k-\omega^2 m\right)^2 + (\omega c)^2}\\ X_{p_2} &= \frac{-\omega c F}{\left(k-\omega^2 m\right)^2 + (\omega c)^2}\end{aligned} \tag{1.70}$$

The particular solution is thus given by:

$$x_p(t) = \frac{F}{\left(k-\omega^2 m\right)^2 + (\omega c)^2}\left(\left(k-\omega^2 m\right) sin\omega t - \omega c cos\omega t\right) \tag{1.71}$$

which may alternatively be simply written as

$$x_p(t) = X_p sin(\omega t - \alpha) \tag{1.72}$$

$$\text{with}\quad X_p = \frac{F}{\sqrt{\left(k-\omega^2 m\right)^2 + (\omega c)^2}} \quad\text{and}\quad \alpha = tan^{-1}\frac{\omega c}{k-\omega^2 m} \tag{1.73}$$

where the angle α represents the phase between the response and the force.

Alternative Way to obtain the Particular Solution

Instead of seeking the solution of $m\ddot{x}_p + c\dot{x}_p + kx_p = F sin\omega t$ as $x_p(t) = X_{p_1} sin\omega t + X_{p_2} cos\omega t$, one could have looked for the solution of $m\ddot{x}_p + c\dot{x}_p + kx_p = F e^{i\omega t}$, which is simply

[9] Had the damping been neglected, the particular solution would be just a *sine* (or *cosine*).

$x_p(t)=\bar{X}_p e^{i\omega t}$. This would lead to $\bar{X}_p=\dfrac{F}{k-\omega^2 m+i\,\omega c}$ and therefore to $x_p(t)=\dfrac{F}{k-\omega^2 m+i\,\omega c}e^{i\omega t}$. However, as one is looking for the case where the force is $F sin\omega t$, which is the imaginary part of $F\,e^{i\omega t}$, one should take the imaginary part of the solution, which leads to the solutions (1.72) and (1.73).

The Complete Response

In conclusion, the complete response of the system to a harmonic force of constant amplitude is:

$$x(t)=e^{-\xi\omega_n t}X_h sin(\omega_d t+\varphi)+X_p sin(\omega t-\alpha) \tag{1.74}$$

The constants X_h and φ have to be calculated from the initial conditions (displacement and velocity). Note that these are different from those in Eq. (1.66), as now the initial conditions must be applied to the complete expression (1.74). For instance, for the common case where the system is in rest before the force is applied, i.e. $x(0)=0$ and $\dot{x}(0)=0$,

$$X_h=X_p\sqrt{\left(\left(\frac{\xi\,\omega_n sin\alpha-\omega cos\alpha}{\omega_d}\right)^2+sin^2\alpha\right)} \quad \text{and} \quad tan\varphi=\frac{\omega_d\, sin\alpha}{\xi\,\omega_n sin\alpha-\omega cos\alpha} \tag{1.75}$$

An example of the complete response (1.74) when the system is initially at rest is given in Figure 1.13.

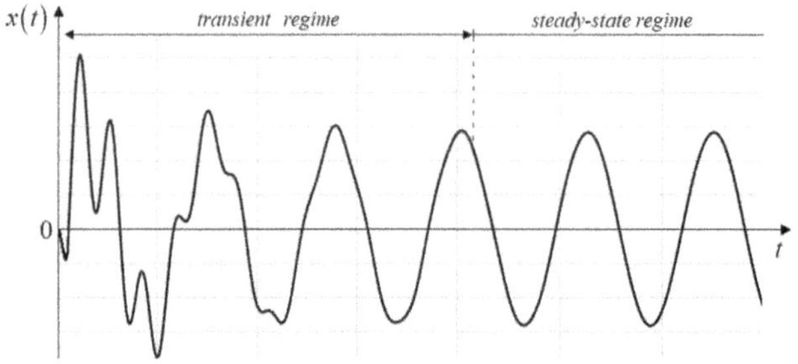

Figure 1.13 Complete response of an *SDOF* system, initially at rest.

As one may observe, there are two parts to be considered, a first one called transient regime, where both terms of the *r.h.s.* of Eq. (1.74) coexist, i.e. the term corresponding to the free vibration and the second term due to the forced vibration and a second part called steady-state regime, where the free vibration has elapsed. In most of practical applications one is only concerned with this second part, as the first one normally disappears after a few seconds. In a few cases, however, it may be important to consider the response from the very beginning, and even the first part may be the most important, as it happens in the launching of a space rocket.

The Phenomenon of Resonance

Concentrating on the steady-state response of the system, it can be written it as:

$$x(t)=\frac{F}{\sqrt{\left(k-\omega^2 m\right)^2+\left(\omega c\right)^2}}\sin\left(\omega t-\alpha\right) \tag{1.76}$$

where the phase angle α is given in the second of Eqs. (1.73). The response is harmonic and one can study the variation of the amplitude and phase, depending on the properties of the system, of the applied force and on the frequency. To do so, it is advisable to introduce dimensionless parameters, so that conclusions can be drawn in a general way, i.e. they can be valid for any *SDOF* system. One has already defined the damping factor ξ in Eq. (1.58). Defining also the frequency ratio as $\beta=\omega/\omega_n$, Eq. (1.76) may be recast as

$$x(t)=\frac{F/k}{\sqrt{\left(1-\beta^2\right)^2+\left(2\xi\beta\right)^2}}\sin\left(\omega t-\alpha\right) \tag{1.77}$$

with

$$\alpha=\tan^{-1}\frac{2\xi\beta}{1-\beta^2} \tag{1.78}$$

Defining $X_{st}=F/k$ as the amplitude that the system would exhibit if the force were applied statically, one can define the amplification or quality factor Q, as

$$Q=\frac{X}{X_{st}}=\frac{1}{\sqrt{\left(1-\beta^2\right)^2+\left(2\xi\beta\right)^2}} \tag{1.79}$$

Note that from now on it will be written X instead of X_p to simplify the notation. Q represents a measure of how the amplitude of the system varies when it is in motion, compared to the static case. Q and α are plotted in Figure 1.14 as a function of β, where one can find the undamped case as well as the damped curves, with different damping factors ξ.

As it can be observed from Figure 1.14 (a), in the static case ($\beta=0$), $Q=1$, i.e. $X=X_{st}$ and when $\beta\to\infty\Rightarrow Q\to 0$. When $\beta=1$ ($\omega=\omega_n$), one says that the system is in resonance and $Q=1/(2\xi)$. Thus, a system is in resonance when the frequency of the applied force equals its undamped natural frequency.

Note that at resonance, if the damping is neglected, $Q=\infty$. This last result must be read carefully, as one cannot forget that one is just representing the variation of amplitude against frequency, not involving the variation along the time. In other words, Figure 1.14 represents a spectrum (a continuous one, in this case), not the variation with time. If one considers the variation along the time, one concludes that the amplitude tends to infinity, but it is not (of course) instantaneously infinite. One shall discuss this matter more in detail a bit farther ahead.

(a)

(b)

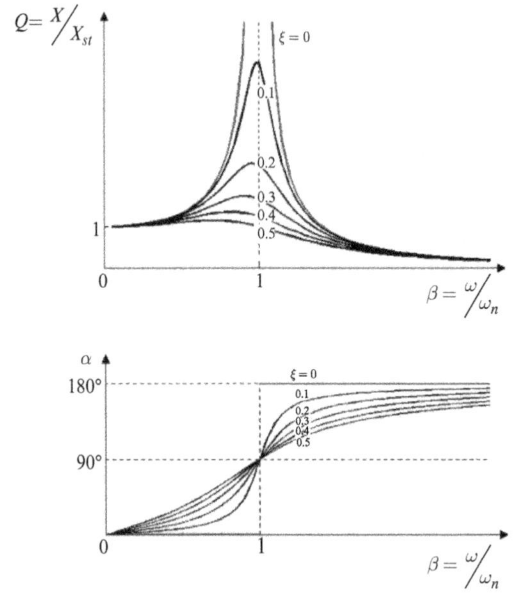

Figure 1.14 Variation of (a) the amplitude (scaled by X_{st}) and (b) phase along the frequency (scaled by ω_n).

Now, one shall discuss the variation of the phase angle (Eq. (1.78): as one can observe from Figure 1.14-b), there is a change from 0º to 180º when the system passes through the resonance. This change is slower or faster, depending on the levels of damping being higher or lower, respectively. In the limit, for the undamped case, there is a singularity at the resonance, which in practice means an abrupt change from 0º to 180º. Looking at Eq. (1.78), it is clear that:

$$\xi = 0 \Rightarrow \alpha = tan^{-1} 0 = \begin{cases} 0^\circ & for\ 1-\beta^2 > 0 \quad (\beta < 1) \\ 180^\circ & for\ 1-\beta^2 < 0 \quad (\beta > 1) \end{cases} \tag{1.80}$$

One may wonder why such a change happens. From Eqs. (1.76) and (1.77), the amplitude of the response is

$$X = \frac{F}{\sqrt{\left(k-\omega^2 m\right)^2 + \left(\omega c\right)^2}} = \frac{F/k}{\sqrt{\left(1-\beta^2\right)^2 + \left(2\xi\beta\right)^2}} \tag{1.81}$$

Neglecting the damping,

$$X = \frac{F}{\left|k-\omega^2 m\right|} = \frac{F/k}{\left|1-\beta^2\right|} \tag{1.82}$$

When $\beta < 1$, i.e. before resonance, $1-\beta^2$ is positive, which corresponds to $k-\omega^2 m$ being positive, i.e. $k > \omega^2 m$. This means that at lower frequencies, the elastic forces dominate the inertia ones. After the resonance, $k < \omega^2 m$, and the opposite situation happens, i.e. the inertia forces take over. A very simple example can illustrate this phenomenon. Observe Figure 1.15 (a), where a small ball of mass m attached to a low damping elastic string of stiffness k is put in motion at the top with a low frequency ω. The ball and elastic string have a natural frequency $\omega_n = \sqrt{k/m}$.

While the frequency ω of the imposed motion stays smaller than ω_n, the response remains in phase with a certain amplitude. If one increases the frequency ω above ω_n, the response becomes out-of-phase, i.e. when one moves our hand down the ball goes up and *vice-versa* (Figure 1.15 (b)).

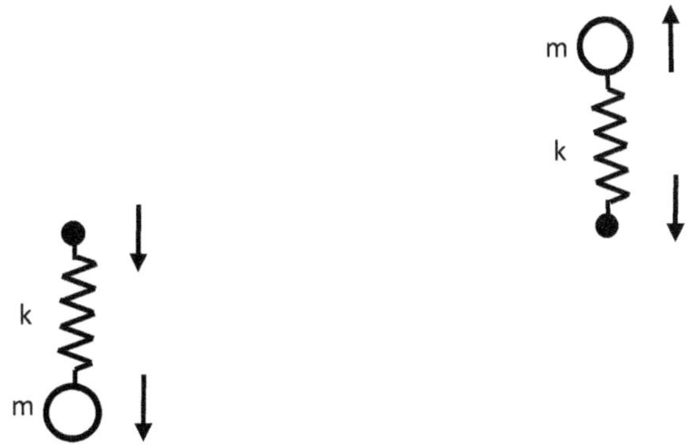

(a) $\omega < \omega_n$ (b) $\omega > \omega_n$

Figure 1.15 Response of a system before and beyond resonance.

Back to Figure 1.14 (a), it can also be observed that in general (provided that the damping is not very high), most of the curves are very close to each other, both at low and at high frequencies. At low frequencies, the amplitudes tend to be close to the static amplitude, as the system is closer to the static situation, where the response is dominated by the stiffness *k*; at higher frequencies the mass *m* has the main influence in the response; finally, the damping has its crucial role around the resonance region. It can also be observed that the maxima of the responses do not happen exactly for $\beta = 1$. If one calculates $\partial Q/\partial\beta = 0$ to find the values where the maxima occur, one will find that they happen for $\beta = \sqrt{1-2\xi^2}$.

In any case, the deviation of the maxima from 1 is due to the fact that in Figure 1.14 (a) one is displaying displacement (although in a dimensionless way) of a system with viscous damping, whose force is proportional to the velocity. Would the modulus of velocity $(\omega X/X_{st})$ be displayed and the maxima would all occur exactly at $\beta = 1$. The conclusion is that for a system with viscous damping the velocity is maximum at the resonance, not the displacement.

However, in structural dynamics, one is often dealing with small amounts of damping. For damping factors below 15%, the maxima of the response displacement still happen for β very close to 1. Note that sometimes (including in standards) the definition of resonance appears connected to the reaching of the maximum displacement, which has the consequence of having to redefine the concept according to the model of damping that is used. Here, the definition of resonance is always when the frequency of the applied force equals the undamped natural frequency of the system ($\beta = 1$).

Next, one shall return to the analysis of the response while at resonance, to understand better the evolution of the displacement along the time. One begins with Eq. (1.74) (with X instead of X_p), with the initial conditions $x(0) = 0$ and $\dot{x}(0) = 0$ that led to Eqs. (1.75):

$$x(t) = e^{-\xi\omega_n t} X \sqrt{\left(\left(\frac{\xi\,\omega_n sin\alpha - \omega cos\alpha}{\omega_d}\right)^2 + sin^2\alpha\right)} sin(\omega_d t + \varphi) + X sin(\omega t - \alpha) \quad (1.83)$$

with $tan\,\varphi = \dfrac{\omega_d\, sin\alpha}{\xi\,\omega_n sin\alpha - \omega cos\alpha}$ and $tan\,\alpha = \dfrac{2\xi\beta}{1-\beta^2}$. At the resonance, $\beta = 1$, $\alpha = \pi/2$ and $X = X_{st}/2\xi$. Therefore,

$$x(t) = \frac{X_{st}}{2\xi}\left(e^{-\xi\omega_n t}\sqrt{\left(\left(\frac{\xi\,\omega_n}{\omega_d}\right)^2 + 1\right)} sin(\omega_d t + \varphi) - cos(\omega_n t)\right) \quad \text{with} \quad tan\,\varphi = \frac{\omega_d}{\xi\,\omega_n} \quad (1.84)$$

To simplify the analysis, but without losing generality, consider a small value for ξ, like 1% or 2%. In that case, $\omega_d \approx \omega_n$, $\sqrt{\xi^2 + 1} \approx 1$ and $\varphi \approx \pi/2$. Thus, the evolution of the amplification factor along the time, at the resonance, is given by:

$$Q(t)\big|_{\beta=1} = \left.\frac{x(t)}{X_{st}}\right|_{\beta=1} \approx \frac{1}{2\xi}\left(e^{-\xi\omega_n t} - 1\right) cos(\omega_n t) \quad (1.85)$$

Plotting this function, one finds the result in Figure 1.16. As one can see, the envelop of the amplitude increases with time until it stabilizes at $1/(2\xi)$. Neglecting the damping, Eq. (1.85) leads to an undetermined result. Solving this indetermination (applying L'Hôpital's rule), yields:

$$Q(t)\big|_{\substack{\beta=1 \\ \xi=0}} = -\frac{1}{2}\omega_n t\, cos(\omega_n t) \quad (1.86)$$

The plot of Eq. (1.86) is represented in Figure 1.17 and as pointed out before, one can see that the value of the amplification factor is not infinite at the resonance, but it gradually grows towards infinity. As a consequence it can be concluded that the phenomenon of resonance takes time to reveal itself. If the forcing frequency passes rapidly by the natural frequency the system does not have time to respond and the resonance is not noticed. This can be observed sometimes when a machine is switched on and passes through a resonance without any substantial increase in the amplitude of vibration, because it went quickly to its functioning state. However, when it is switched off, it runs down slowly and it may exhibit some increase in the vibration amplitude before it comes to a total rest.

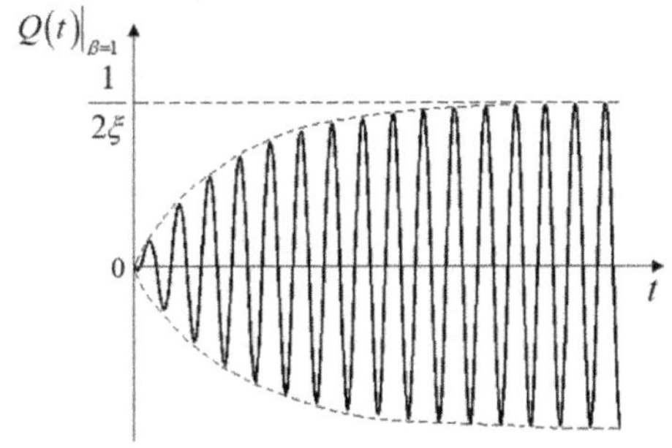

Figure 1.16 Evolution of the amplification factor along the time, at the resonance, for $\xi \neq 0$.

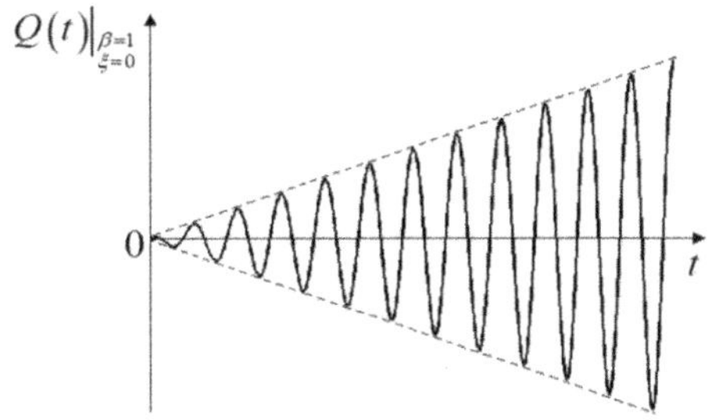

Figure 1.17 Evolution of the amplification factor along the time, at the resonance, for $\xi = 0$.

The Frequency Response Function

As seen before, the response amplitude in steady-state regime is given by

$$\bar{X} = \frac{F}{k - \omega^2 m + i\omega c} \tag{1.87}$$

The quality factor Q discussed before is useful to understand the physics of the vibration problem; however, in practice, it is preferable to use other functions, like the Frequency Response Function (*FRF*). The *FRF* is defined as the relation between the output and the input (Figure 1.18):

$$H(\omega) = \frac{\bar{X}(\omega)}{F(\omega)} = \frac{1}{k - \omega^2 m + i\omega c} \tag{1.88}$$

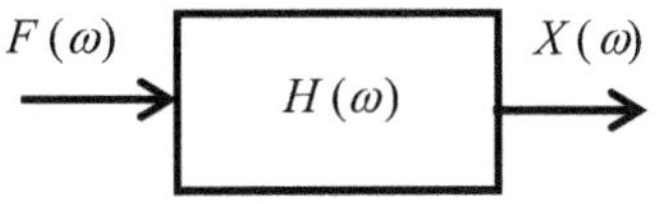

Figure 1.18 Relation between output and input (*FRF*).

The *FRF* is a complex function. Defined in this way, relating displacement and force is called receptance; alternatively, one may define *FRF* as the ratio between velocity and force (mobility) or acceleration and force (accelerance). The *FRF* represents the behaviour of the system, as it contains its dynamic properties. In an inverse problem, one measures an *FRF* and from it the dynamic properties are extracted, in order to have a model that represents the real behaviour of the system. This matter will be discussed in Chapter 4.

Response due to a Harmonic Force of Variable Amplitude

A common situation in practice is when a machine vibrates due to some internal rotating component. The vibration happens because it is not possible to have a completely balanced rotor, it always has some degree of unbalance. The model is now the one represented in Figure 1.19, where the mass m_1 rotating at the angular velocity ω with some eccentricity e represents the part of the total mass of the system ($m = m_1 + m_2$) that is unbalanced.

The rotation of m_1 produces a centrifugal force of amplitude $F = m_1 e\omega^2$. Starting from the initial horizontal position at time $t = 0$, at time t the mass m_1 has described an angle ωt; decomposing the force $\boldsymbol{F}$, one sees that the horizontal component is supported by the vertical walls, while the vertical component is the active part responsible for the motion, $F \sin\omega t = m_1 e\omega^2 \sin\omega t$.

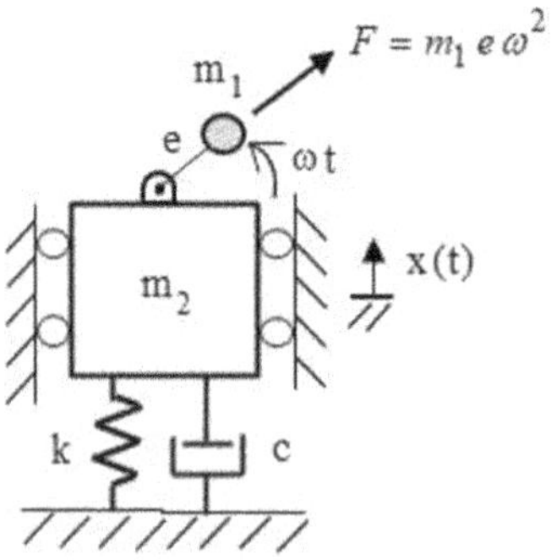

Figure 1.19 Model of an *SDOF* system with a rotating unbalanced mass.

The dynamic equilibrium equation is therefore:

$$m\ddot{x} + c\dot{x} + kx = m_1 e\omega^2 \sin\omega t \tag{1.90}$$

where $m = m_1 + m_2$.

This equation is entirely similar to Eq. (1.67), the difference is that now $F = m_1 e\omega^2$. If the frequency ω is fixed, the previous analysis for constant F holds. However, it is now of interest to analyse the evolution of the response amplitude when ω varies.

$$X = \frac{m_1 e\omega^2}{\sqrt{(k - \omega^2 m)^2 + (\omega c)^2}} = \frac{m_1 e\omega^2 / k}{\sqrt{(1 - \beta^2)^2 + (2\xi\beta)^2}} \tag{1.91}$$

Defining the ratio of the unbalanced mass to the total mass as $\mu = m_I / m$, Eq. (1.91) becomes

$$X = \frac{\mu e \beta^2}{\sqrt{\left(1-\beta^2\right)^2 + \left(2\xi\beta\right)^2}} \tag{1.92}$$

where μe represents the unbalance of the system. A new amplification factor can now be defined as

$$Q = \frac{X}{\mu e} = \frac{\beta^2}{\sqrt{\left(1-\beta^2\right)^2 + \left(2\xi\beta\right)^2}} \tag{1.93}$$

Now, $Q = 0$ when $\beta = 0$, as when the frequency is zero there is no generated force. When $\beta \to \infty$, it turns out that

$$\lim_{\beta\to\infty} Q = \lim_{\beta\to\infty} \frac{\beta^2}{\sqrt{\left(1-\beta^2\right)^2 + \left(2\xi\beta\right)^2}} = \lim_{\beta\to\infty} \frac{1}{\sqrt{\left(\frac{1}{\beta^2}-1\right)^2 + \left(\frac{2\xi}{\beta}\right)^2}} = 1 \tag{1.94}$$

This means that, as the frequency increases, $X \to \mu e$. The plot of Q against β is represented in Figure 1.20 for different values of the damping factor.

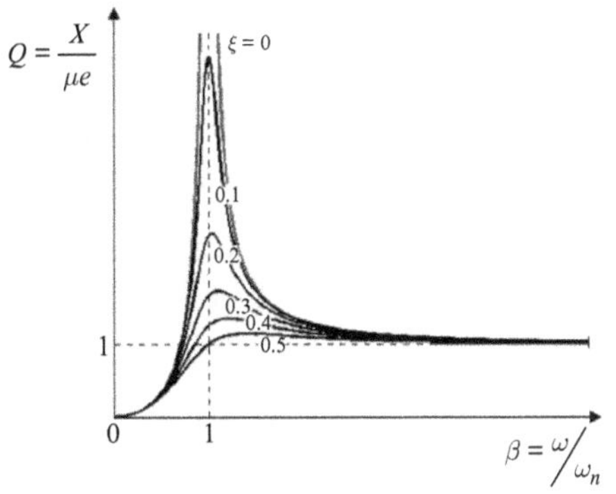

Figure 1.20 Variation of the amplification factor along the (scaled) frequency for different values of the damping factor.

Comparing to what happens in the case of constant amplitude force (Figure 1.14 (a)), one observes that there is a kind of "inversion" in the behaviour of Q. Whereas before it started in 1 and tended to zero, now it starts from zero and tends to 1, i.e. as the frequency increases the amplitude of the response tends to the value of the unbalance. This is to some extent what can be observed in a washing machine; often, at low speeds and with some clothes asymmetrically distributed, the machine vibrates considerably, but when the velocity increases the amplitude of the response tends to decrease. It can also be observed that the maxima of the curves tend to deviate towards the right, but – as before – this is a mere consequence of the type of damping that one is considering (viscous) and the type of response that one is plotting (displacement, tough in a dimensionless way). The phase angle has the same equation and therefore the same behaviour as in the case of constant amplitude force (Eq. (1.78)).

Response due to an Imposed Harmonic Displacement

Another common possibility for excitation is an applied motion, instead of an applied force. The model is given in Figure 1.21, where $y(t)$ is a known imposed harmonic displacement that, as mentioned before, can be written as $y(t)=Y\,e^{i\omega t}$.

The dynamic equilibrium of forces gives:

$$m\ddot{x}+c(\dot{x}-\dot{y})+k(x-y)=0 \qquad (1.95)$$

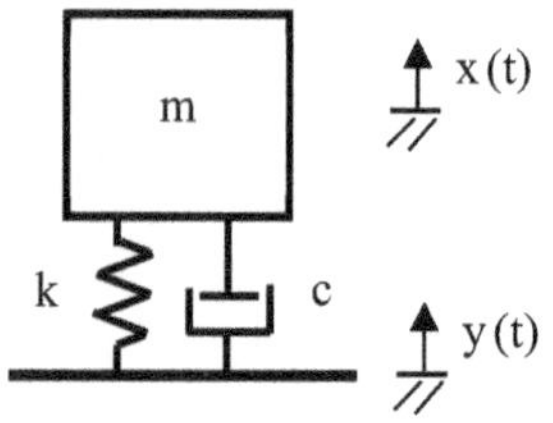

Figure 1.21 Model of an *SDOF* system with an imposed harmonic displacement at its base.

Moving the known components to the *r.h.s.* of the equation,

$$m\ddot{x}+c\dot{x}+kx=c\dot{y}+ky \qquad (1.96)$$

This is clearly equivalent to have our *SDOF* system, as before, with an applied force $f(t)=c\dot{y}+ky$. Thus, an imposed displacement ends up leading to an applied force. A typical example is a car moving along the road. The input to the car is the imposed displacement of the road, which eventually is translated into an imposed force, through the tires and suspension elements.

For $y(t)=Y\,e^{i\omega t}$, $\dot{y}(t)=i\omega Y\,e^{i\omega t}$ and the steady-state solution is of the type $x(t)=\bar{X}\,e^{i\omega t}$, where the complex amplitude $\bar{X}$ is:

$$\bar{X}=X\,e^{-i\theta}=\frac{k+i\omega c}{k-\omega^2 m+i\omega c}Y \qquad (1.97)$$

From (1.97) the amplitude and phase are:

$$X=\sqrt{\frac{k^2+(\omega c)^2}{(k-\omega^2 m)^2+(\omega c)^2}}\,Y \quad \text{and} \quad \theta=tan^{-1}\frac{\omega^3 cm}{k(k-\omega^2 m)+(\omega c)^2} \qquad (1.98)$$

or, in terms of the dimensionless quantities,

$$X = \sqrt{\frac{1+(2\xi\beta)^2}{(1-\beta^2)^2+(2\xi\beta)^2}}\,Y \quad \text{and} \quad \theta = tan^{-1}\frac{2\xi\beta^3}{1-\beta^2+(2\xi\beta)^2} \tag{1.99}$$

Transmissibility of Motion

For the case one has just seen, the ratio between the amplitude of the response and the imposed one is called transmissibility of motion. From (1.99),

$$TR = \frac{X}{Y} = \sqrt{\frac{1+(2\xi\beta)^2}{(1-\beta^2)^2+(2\xi\beta)^2}} \tag{1.100}$$

Transmissibility may be important in various applications, for instance when one wishes to evaluate the comfort of a vehicle when subjected to a rough input from the road or one needs to know whether a sensitive device is adequately isolated from its vibrating base. In most cases, when it comes to transmissibility, one does not give so much importance to the information about the phase. So, one concentrates on the evolution of the modulus of transmissibility, as in Eq. (1.100), along the (normalised) frequency. The graph *TR(β)* is presented in Figure 1.22, for different values of the damping factor.

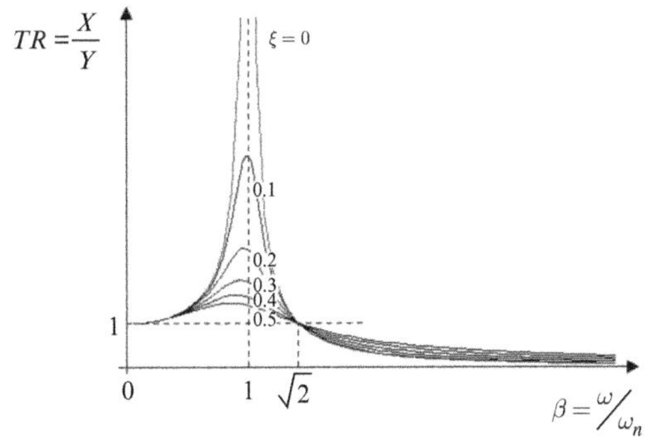

Figure 1.22 Transmissibility as a function of frequency, for different damping factors.

Several conclusions can be drawn from the observation of Figure 1.22. The first one is that for $\xi = 0$ the transmissibility coincides with the amplification factor, as defined in Eq. (1.79) and Figure 1.14 (a). Another conclusion is that all curves (for the different damping values) cross at the same point, where *TR* is unity and $\beta = \sqrt{2}$. From this point on the transmissibility is less than 1, i.e. the amplitude of the response is smaller than the imposed amplitude and one says that this is an isolated zone. In that zone the higher the damping, the higher the transmissibility, in contrast with what happens for $\beta < \sqrt{2}$. So, ideally, in the isolation zone, the best result when it comes to transmissibility is to have zero damping. However, one cannot forget that if a machine is working at a frequency in that zone it did not reached that frequency instantaneously, it had to pass through the resonance and there it should have some damping, to avoid high amplitudes (of course, as one saw before, this depends on how fast is the passage through the resonance point.

Observe now Figure 1.23, where one represents the transmissibility as a function of ω instead of β, just to make things clearer. Let also ω_f be the working frequency of a machine. It can be seen that the transmissibility is considerably high at that frequency (= 1.5), and one wishes to proceed to reduce it. Naturally, one cannot change the working frequency and so the only options are (i) to increase the damping or (ii) to decrease the natural frequency, so that ω_f will fall into the isolation region. Clearly, increasing the damping will not bring much improvement and, in any case, will definitely not make the transmissibility lower than 1.

Therefore, the best option seems to be the decrease of the natural frequency ($\omega_n = \sqrt{k/m}$), which implies either decreasing k or increasing m (or both, simultaneously). Usually, it is not practical to increase the mass. Often, the solution is to somehow modify the stiffness of the supports, choosing new ones, more flexible. This leads to the new situation illustrated in Figure 1.24, where the transmissibility value at ω_f is considerably lower (= 0.5).

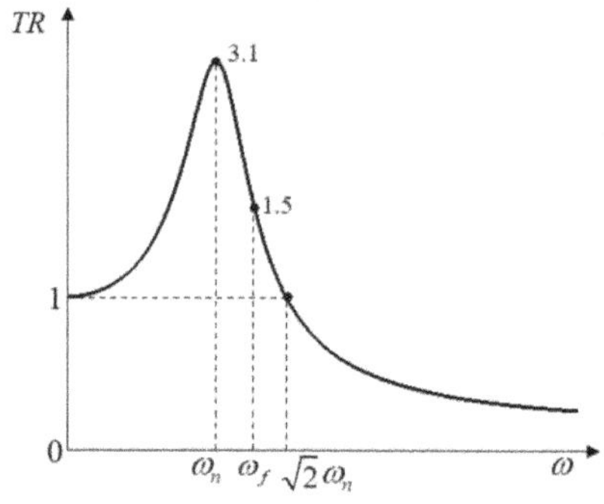

Figure 1.23 Example of a transmissibility curve as a function of frequency.

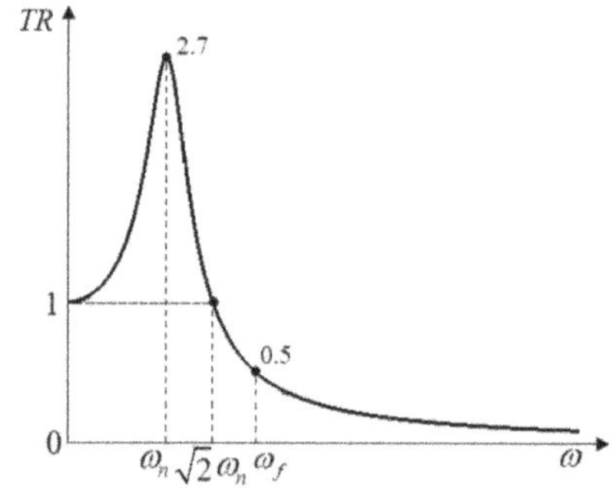

Figure 1.24 New transmissibility curve, after reducing the stiffness of the machine supports.

Transmissibility of Forces

When the *SDOF* system is subjected to a force, for instance representing some kind of machine on its elastic supports, one may wish to evaluate the amount of force that is transmitted to the ground. This may be important because if the transmitted vibration is too strong it may affect some precision machine in the neighbourhood or, at least, be the source of some unwanted noise, for instance.

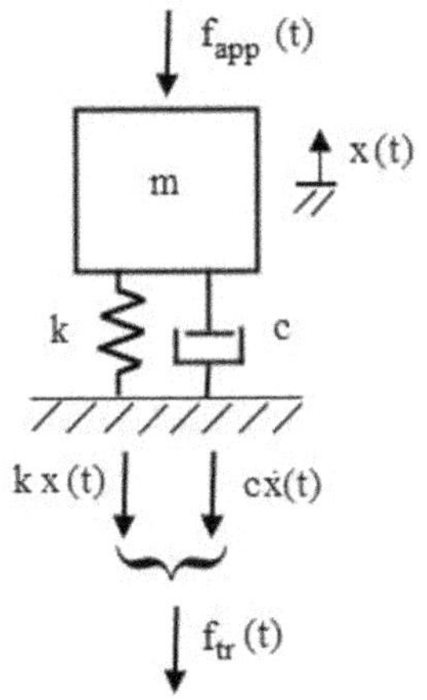

Figure 1.25 System transmitting force to the ground through the spring and damper.

For the system represented in Figure 1.25, let the applied force be harmonic, $f_{app}(t) = F_{app}\, e^{i\omega t}$. The force is transmitted to the ground through both the spring and the damper; as in steady-state regime the response is $x(t) = \bar{X}\, e^{i\omega t}$, the transmitted force is:

$$f_{tr}(t) = \bar{F}_{tr}\, e^{i\omega t} = k\,x(t) + c\,\dot{x}(t) = (k + i\omega c)\bar{X}\, e^{i\omega t} \;\Rightarrow\; \bar{F}_{tr} = (k + i\omega c)\bar{X} \qquad (1.101)$$

From Section 1.4.5, the amplitude of the steady-state response is

$$\bar{X} = \frac{F_{app}}{k - \omega^2 m + i\,\omega c} \qquad (1.102)$$

and therefore,

$$\bar{F}_{tr} = \frac{k + i\omega c}{k - \omega^2 m + i\,\omega c} F_{app} \qquad (1.103)$$

It makes sense now to define transmissibility of forces as the modulus of the ratio between the transmitted and the applied forces:

$$TR = \frac{F_{tr}}{F_{app}} = \sqrt{\frac{k^2 + (\omega c)^2}{(k - \omega^2 m)^2 + (\omega c)^2}} \qquad (1.104)$$

or, in terms of the dimensionless quantities,

$$TR = \sqrt{\frac{1 + (2\xi\beta)^2}{(1 - \beta^2)^2 + (2\xi\beta)^2}} \qquad (1.105)$$

which is exactly the expression for the transmissibility of motion (Eq. (1.100)) and so Figure 1.22 is naturally the same. Thus, one can state that everything that has been discussed and concluded about transmissibility of motion applies to the transmissibility of forces.

As for the transmissibility of motion, one can treat the transmissibility of forces as a direct problem, where the system properties and the applied force are known, to work out the transmitted force or, often more interesting, one can address the inverse problem, i.e. when one imposes that the transmitted force should not exceed a certain value and look for modifications on the system properties to reach the desired goal.

Response due to a Harmonic Force with Hysteretic Damping

The energy dissipation in a structure is modelled by damping elements. Of course, the behaviour of such elements should reflect the reality as much as possible, which can be complicated in some cases, for instance when it comes to represent the damping associated to human activities. In what structural dynamics is concerned the most common models are the viscous one, that has been studied so far and the hysteretic one. There is also the dry friction (Coulomb) model, but it will not be considered it here (and often the hysteretic model is able to substitute it quite reasonably). One can envisage more or less complicated combinations of those common models or even invent more sophisticated ones, but they are out of the scope of this section. So, besides the viscous damping model, one is going to briefly discuss the hysteretic model. The main reason to address it here is due to the fact that the viscous model, where the damping forces are directly proportional to the velocity, implies that the dissipated energy per cycle, in a harmonic forced motion, is linearly dependent on the applied frequency:

$$\Delta E_{diss./cycle} = \int_0^T f_{damp.}\, dx = \int_0^{2\pi/\omega} c\dot{x}\,\dot{x}dt \tag{1.106}$$

If $f(t) = F sin\omega t$, $x(t) = X sin(\omega t - \alpha)$ and so,

$$\Delta E_{diss./cycle} = \int_0^{2\pi/\omega} c(X\omega)^2 cos^2(\omega t - \alpha)dt = c\pi X^2\omega \tag{1.107}$$

It happens that for the most common materials used in structures it has been verified that the energy dissipated per cycle is practically constant along the frequency. Consequently, if the viscous model is used along a wide frequency range, one starts to note discrepancies between the real dynamic response of the structure and the one predicted by the model beyond a certain frequency. As the viscous model is mathematically quite “friendly”, many people like to use it anyway. However, they must keep in mind that the useful frequency range is restricted and therefore the values of the damping coefficient should be changed along the frequency.

An alternative solution is to introduce a damping force that keeps changing automatically with frequency. This is accomplished by the introduction of the hysteretic damping coefficient *d*, such that the damping force is given by:

$$f_{damp.} = \frac{d}{\omega}\dot{x} \tag{1.108}$$

In fact, the new damping coefficient is d/ω, which replaces the viscous one, *c*, with units of *Ns/m*, but usually one calls hysteretic damping coefficient to *d* itself, which has units of stiffness, *N/m*. The name "hysteretic" comes from the hysteresis phenomenon, which has to do with the loss of energy in a material, due to internal friction, when submitted to cycles of load and unload.

One concludes that when c is substituted by d/ω in Eqs. (1.106) and (1.107), the dissipated energy per cycle becomes $\Delta E_{diss/cycle} = d\,\pi X^2$, independent of the frequency. In what follows, one will see the implications of this new definition. The dynamic equilibrium equation becomes:

$$m\ddot{x} + \frac{d}{\omega}\dot{x} + kx = Fe^{i\omega t} \tag{1.109}$$

It should be noted that it only makes sense to look for the particular solution of (1.109), corresponding to the steady-state response, as the homogeneous equation includes the forcing frequency ω that does not exist in free vibration. Therefore, the hysteretic model cannot represent the free vibration condition. The steady-state response is $x(t) = \bar{X}e^{i\omega t}$ (from which one retains only the imaginary part, to get the solution to $f(t) = F\sin\omega t$). Substituting in (1.109), leads to:

$$\left(k - \omega^2 m + i\,d\right)\bar{X} = F \tag{1.110}$$

which sometimes is written as

$$\left(k^* - \omega^2 m\right)\bar{X} = F \tag{1.111}$$

where $k^* = k + i\,d$, to stress the interpretation that the system has now a complex stiffness (remember that d has units of rigidity), although this could also have been said for the viscous damping, where the complex stiffness would be $k^* = k + i\,\omega c$. A complex stiffness means, physically, that the spring element includes properties of stiffness and damping, so it is a spring that also dissipates energy. Anyway, from (1.110), the steady-state response is:

$$x(t) = Im\left(\frac{F}{k - \omega^2 m + i\,d}e^{i\omega t}\right) = \frac{F}{\sqrt{(k - \omega^2 m)^2 + d^2}}\sin(\omega t - \alpha) \tag{1.112}$$

with $\alpha = \tan^{-1}\dfrac{d}{k - \omega^2 m}$. Eq. (1.112) is the equivalent to Eq. (1.76) for the viscous damping case. One can also define a hysteretic damping factor (or loss factor) as $\eta = d/k$ and Eq. (1.112) becomes:

$$x(t) = \frac{F/k}{\sqrt{(1 - \beta^2)^2 + \eta^2}}\sin(\omega t - \alpha) \tag{1.113}$$

with $\alpha = \tan^{-1}\dfrac{\eta}{1 - \beta^2}$. The quality factor Q is now defined as

$$Q = \frac{X}{X_{st}} = \frac{1}{\sqrt{\left(1 - \beta^2\right)^2 + \eta^2}} \tag{1.114}$$

Besides the fact that the hysteretic damping cannot model the free vibration case, it has also another limitation, revealed by Eq. (1.114): $X \neq X_{st}$ when $\beta = 0$, although this only affects the static case, which is not very relevant. The plot of Q against β is shown in Figure 1.26.

The variation of Q along the frequency is very similar to the viscously damped case. The most important differences are (i) the values at $\beta = 0$ different from 1, as already mentioned and (ii) the maxima that all happen exactly at $\beta = 1$. This is because the dissipation term in Eq. (1.110) is proportional to the displacement, although in phase with the velocity (due to the multiplication by *i*). As the graph of Figure 1.26 refers to the displacement, all the maxima occur at $\beta = 1$. In summary, at resonance the velocity is maximum with the viscous model and the displacement is maximum with the hysteretic model.

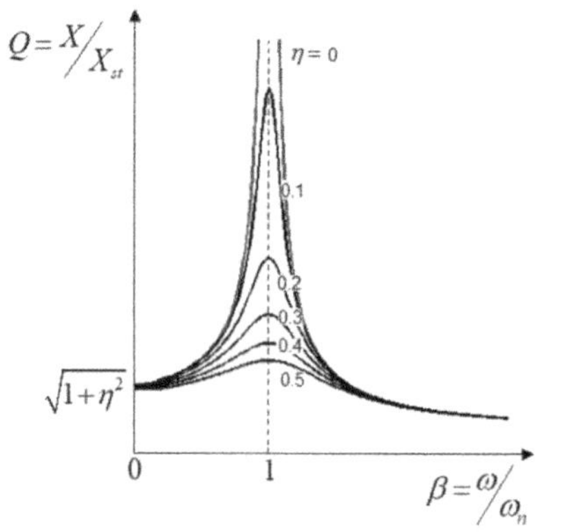

Figure 1.26 Variation of the quality factor Q against β for various values of η.

Response to a Periodic Excitation

Knowing how to calculate the response to a harmonic force (or displacement), it is very easy to generalise to the case of a periodic but non-harmonic force (or displacement). This is because any periodic function can be represented as a series of *sines* and *cosines* (Fourier series). Assuming that the system has a linear behaviour, one simply applies the superposition principle, saying that the response to a series of harmonic forces is the summation of the responses at each of the harmonic frequencies.

Consider the periodic force illustrated in Figure 1.27, with period *T*.

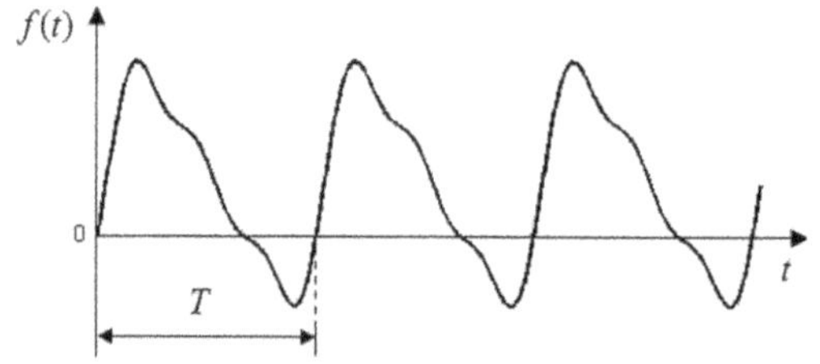

Figure 1.27 Example of a periodic force.

It is possible to write the force as a series of *sines* and *cosines*:

$$f(t)=\frac{a_0}{2}+\sum_{n=1}^{\infty}\left(a_n cos\, n\,\omega t+b_n sin\, n\,\omega t\right) \tag{1.115}$$

where $a_0=\frac{2}{T}\int_0^T f(t)dt,\quad a_n=\frac{2}{T}\int_0^T f(t)cos\, n\,\omega t\, dt,\quad b_n=\frac{2}{T}\int_0^T f(t)sin\, n\,\omega t\, dt$

Alternatively, (1.115) can be expressed in a more compact form, as

$$f(t)=F_0+\sum_{n=1}^{\infty}F_n cos\left(n\,\omega t-\varphi_n\right) \tag{1.116}$$

where $F_0=\frac{a_0}{2},\quad F_n=\sqrt{a_n^2+b_n^2},\quad \varphi_n=tan^{-1}\frac{b_n}{a_n}$

The dynamic equilibrium equation is therefore:

$$m\ddot{x}+c\dot{x}+kx=\frac{a_0}{2}+\sum_{n=1}^{\infty}\left(a_n cos\, n\,\omega t+b_n sin\, n\,\omega t\right) \tag{1.117}$$

The term $a_0/2$ represents the average value of the force, a fixed value that sometimes is not taken into account to obtain the response, as it simply implies a constant value of $x_0=a_0/2k$. It depends, however, on the applications. For instance, for the calculation of stresses in a fatigue design context, it will be important to include it, as it refers to the average stress value. The superposition of the responses for each harmonic component (see Eq. (1.77)) leads to:

$$x(t)=\frac{a_0}{2k}+\sum_{n=1}^{\infty}\frac{\left(a_n cos\, n\left(\omega t-\alpha_n\right)+b_n sin\, n\left(\omega t-\alpha_n\right)\right)/k}{\sqrt{\left(1-\beta_n^2\right)^2+\left(2\xi\,\beta_n\right)^2}} \tag{1.118}$$

where $\beta_n=\frac{n\omega}{\omega_n}$ and $\alpha_n=tan^{-1}\frac{2\xi\,\beta_n}{1-\beta_n^2}$. Each α_n represents the phase between each component of the response to the correspondent component of the force.

The procedure is in fact very simple. Once the period is identified and the force is defined over that period, all one has to do is to calculate a_0, a_n and b_n. The problem, however, may be the calculation of the integrals. If the case is like the one of Figure 1.27, one defines the force considering N small time increments Δt (see Figure 1.28), transforming the integrals into summations. At each time t_i corresponds a force F_i and the period is $T=N\Delta t$. Hence,

$$\begin{aligned} a_0&=\frac{2}{N\Delta t}\sum_{i=1}^{N}F_i\Delta t=\frac{2}{N}\sum_{i=1}^{N}F_i\\ a_n&=\frac{2}{N\Delta t}\sum_{i=1}^{N}F_i cos\, n\,\omega t_i\Delta t=\frac{2}{N}\sum_{i=1}^{N}F_i cos\frac{2\pi n}{T}t_i\\ b_n&=\frac{2}{N}\sum_{i=1}^{N}F_i sin\frac{2\pi n}{T}t_i \end{aligned} \tag{1.119}$$

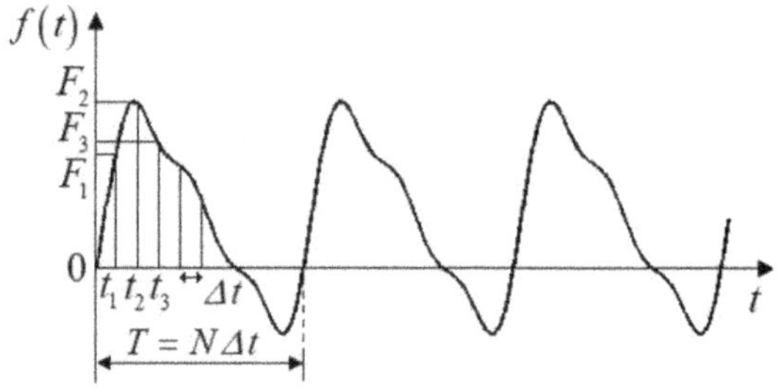

Figure 1.28 Discretisation of the applied force.

As before, one can use alternatively the complex notation, which allows even for a more compact form of Eqs.(1.115) and (1.116):

$$f(t) = Re\left[\sum_{n=0}^{\infty} F_n \,e^{i(n\omega t - \varphi_n)}\right] = Re\left[\sum_{n=0}^{\infty} \bar{F}_n \,e^{in\omega t}\right] \tag{1.120}$$

where $\bar{F}_n = F_n \,e^{-i\varphi_n}$ is the phasor of the harmonic component $n\omega$. One calls Fourier spectrum to the representation of F_n and of φ_n *versus* frequency. Applying $f(t) = \sum_{n=0}^{\infty} \bar{F}_n \,e^{in\omega t}$, one can calculate the response, given by $x(t) = \sum_{n=0}^{\infty} \bar{X}_n \,e^{in\omega t}$, from which one takes the real part (because Eq. (1.116) is written in terms of *cosine*), leading to Eq. (1.118).

Response to a Non-Periodic Excitation

A non-periodic (or aperiodic) excitation is assumed to be deterministic, as explained in Section 1.2.1. For instance, one wishes to know how the system responds to a force like the one depicted in Figure 1.29.

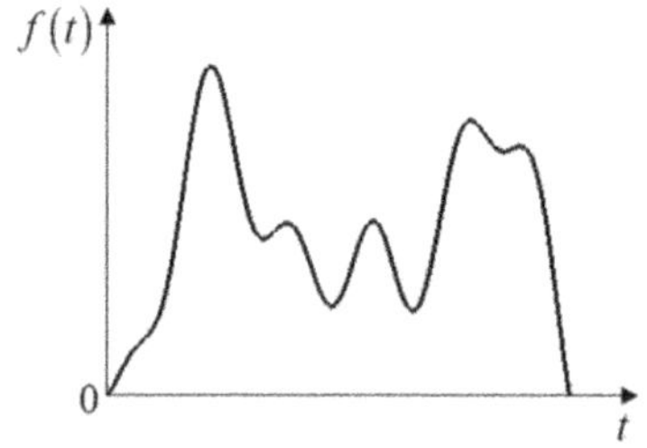

Figure 1.29 Non-periodic force.

The approach usually followed is to consider that the force is formed by a succession of impulses; if one calculates the response to each of those impulses, one can work out the response at a certain time t by superimposing all the responses to the impulses from $t = 0$ until t (see Figure 1.30).

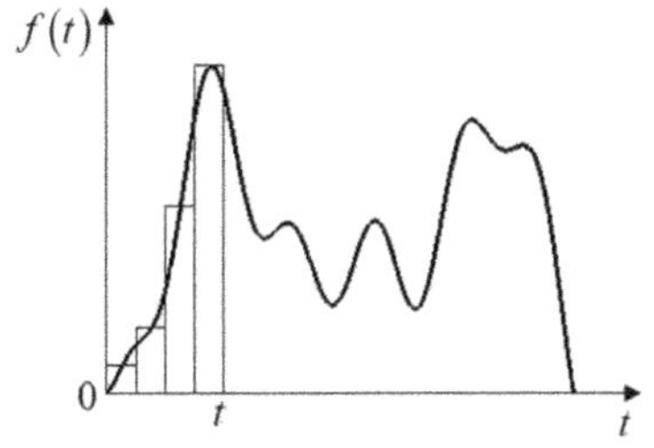

Figure 1.30 Impulses that happened until time *t*.

One starts by calculating the response to a unitary impulse, i.e. to an impulse of area equal to $1\,Ns$, as in Figure 1.31. Such a response is called *Impulse Response Function* (*IRF*) and denoted by $h(t)$.

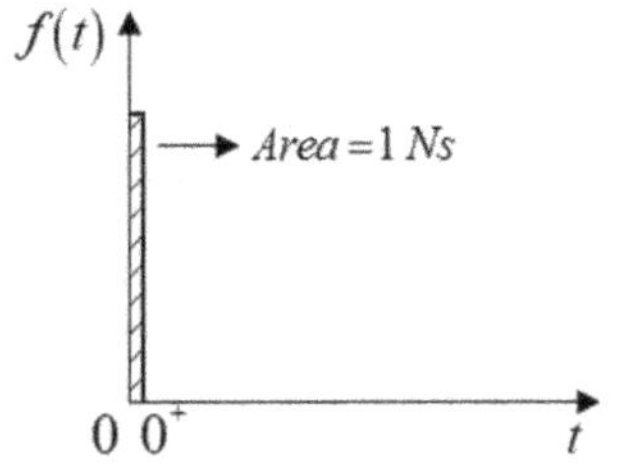

Figure 1.31 Unitary impulse at t = 0.

Immediately after the impulse, at $t = 0^+$, the system is in free vibration and the answer is given by Eq. (1.63), now in terms of $h(t)$:

$$h(t) = e^{-\xi\omega_n t}\left(A\cos\omega_d t + iB\sin\omega_d t\right) \tag{1.121}$$

where A and B must be calculated with the initial conditions at $t = 0^+$. For a pure impulse, $h\left(0^+\right) = 0$. To evaluate $\dot{h}\left(0^+\right)$ one uses the equality between the impulse and the variation of the momentum, i.e. $f\Delta t = m\Delta v = m\left(\dot{h}\left(0^+\right) - \dot{h}(0)\right)$. As the area of the impulse is unitary, $f\Delta t = 1$ and as the system is initially at rest, $\dot{h}(0) = 0$. Therefore, $\dot{h}\left(0^+\right) = 1/m$. Applying both conditions to (1.121), it follows that $A = 0$ and $B = \dfrac{1}{i\,\omega_d m}$. Thus, the *IRF* is:

$$h(t) = \frac{1}{\omega_d m} e^{-\xi\omega_n t}\sin\omega_d t \quad \text{for} \quad t > 0 \tag{1.122}$$

This is the response to a unitary impulse at $t = 0$. However, one needs to consider the summation of all the responses from 0 to a time *t*. Thus, it is necessary to know the response immediately after a unitary impulse at a generic time τ, i.e. for $t > \tau$ (see Figure 1.32).

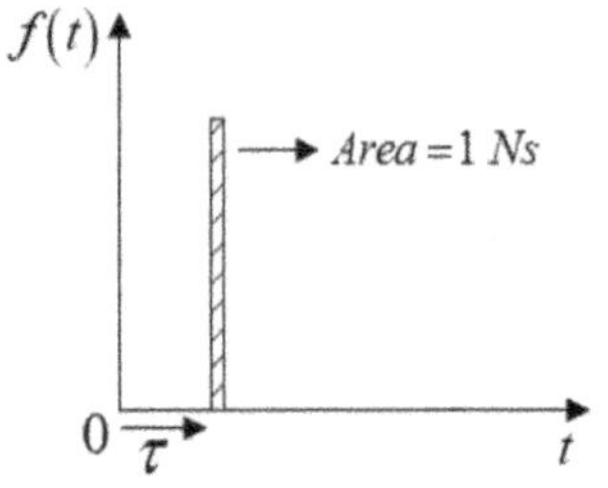

Figure 1.32 Unitary impulse at $t = \tau$.

This is easily achieved by making a change of variable, from t to $t-\tau$:

$$\begin{cases} h(t-\tau)=\dfrac{1}{\omega_d m}e^{-\xi\omega_n(t-\tau)}sin\omega_d(t-\tau) & \text{for} \quad t>\tau \\ h(t-\tau)=0 & \text{for} \quad t\leq\tau \end{cases} \tag{1.123}$$

However, it is not easy to divide a given force into unitary impulses. Normally, one simply divides it into a number of impulses, each one with an area $f(\tau)\Delta\tau$ (Figure 1.33).

The response after each impulse is therefore (now it is x instead of h):

$$\begin{cases} x(t-\tau)=\dfrac{f(\tau)\Delta\tau}{\omega_d m}e^{-\xi\omega_n(t-\tau)}sin\,\omega_d(t-\tau) & \text{for} \quad t>\tau \\ x(t-\tau)=0 & \text{for} \quad t\leq\tau \end{cases} \tag{1.124}$$

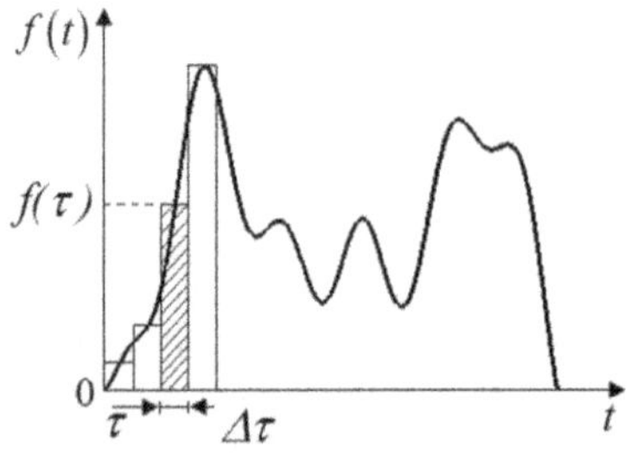

Figure 1.33 Impulse of area $f(\tau)\Delta\tau$.

Finally, the response at time t is the summation of all the responses due to each impulse happening for τ sweeping from 0 to t:

$$x(t)=\sum_{\tau=0}^{t}x(t-\tau)=\sum_{\tau=0}^{t}\frac{f(\tau)\Delta\tau}{\omega_d m}e^{-\xi\omega_n(t-\tau)}sin\,\omega_d(t-\tau) \tag{1.125}$$

When the duration of the impulse tends to an infinitesimal, i.e. $\Delta\tau\rightarrow d\tau$, the summation tends to an integral:

$$x(t)=\frac{1}{\omega_d m}\int_0^t f(\tau)e^{-\xi\omega_n(t-\tau)}sin\,\omega_d(t-\tau)d\tau \quad (1.126)$$

which is known as the convolution integral or Duhamel's integral. Note that although one has always mentioned the application of this to a non-periodic force, no restriction has been made in the mathematical derivation or on the physical interpretation concerning the nature of the applied force. In fact Eq. (1.126) is valid for any kind of deterministic force, including the simple harmonic case.

Suppose the simple case of an undamped *SDOF* system submitted to a force like the one illustrated in Figure 1.34, a suddenly applied force that stays for a while and then stops. It could be viewed as a simplified version of a block that is picked up from the floor by a crane that turns around for a while (t_0) and puts the block back again on the floor.

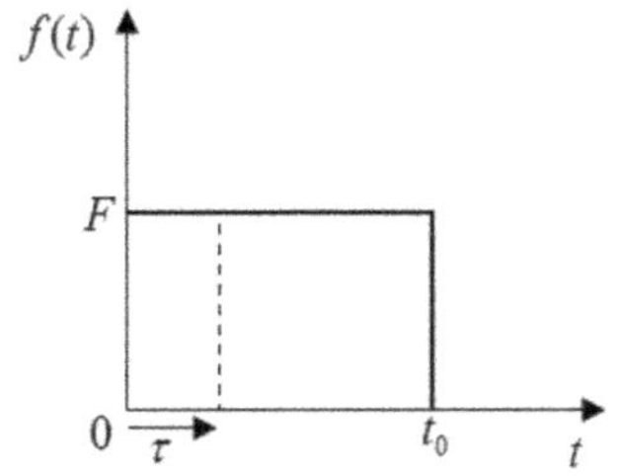

Figure 1.34 Simple non-periodic applied force.

As the damping has been neglected, $\omega_d=\omega_n$, and as $f(\tau)=F$, Eq. (1.126) yields

$$x(t)=\frac{F}{\omega_n m}\int_0^t sin\,\omega_n(t-\tau)d\tau \quad (1.127)$$

and finally, because $k=\omega_n^2 m$,

$$x(t)=\frac{F}{k}(1-cos\,\omega_n t) \quad \text{for} \quad 0\le t\le t_0 \quad (1.128)$$

The response has a mean value of F/k (the amplitude if F were applied statically) and then oscillates around that level with that same amplitude (Figure 1.35).

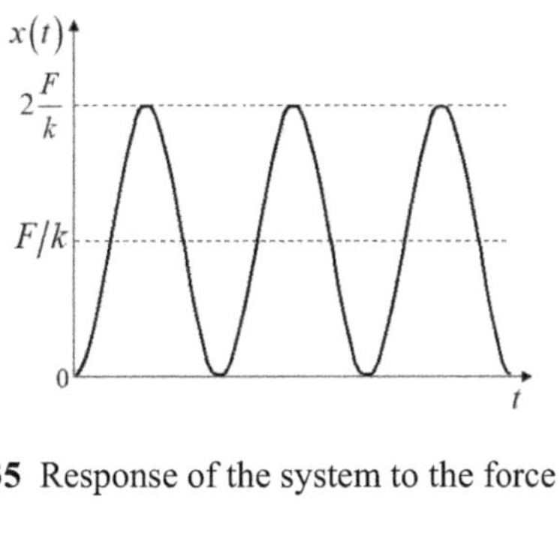

Figure 1.35 Response of the system to the force of Figure 1.34.

Immediately after t_0 the system is in free vibration, with the initial conditions (displacement and velocity) existing at t_0. Two extreme cases may happen, depending on the duration of the force. If the force ends after a natural period has elapsed, i.e. $t_0 = T_n = 2\pi/\omega_n$, the response will be zero and the system will remain still (Figure 1.36 (a)). If the force ends after half a natural period, $t_0 = T_n/2 = \pi/\omega_n$, the amplitude of the response will be $2F/k$, i.e. the double of the displacement that the system would have if the force had been applied statically (Figure 1.36 (b)). This is the worst possible case, showing that when dealing with dynamic problems, namely when designing a structural component subjected to suddenly applied forces, one should take as a safety factor the double of the value one would normally choose in a static situation.

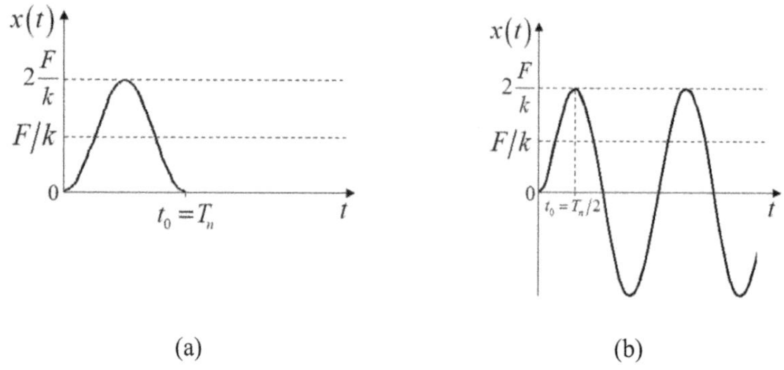

Figure 1.36 Response of the system (a) for $t_0 = T_n$ and (b) for $t_0 = T_n/2$.

Although the force that has just been considered is a very simple case, it is very useful, as if one has a more complicated situation, one can always subdivide the force into several blocks of constant force each, as in Figure 1.37 (naturally, it will be more accurate to consider blocks of trapezoids instead of rectangles). The response at $t_i \le t \le t_{i+1}$ will be:

$$x(t) = \frac{1}{\omega_n m}\left\{\int_0^{t_1} F_1 sin\,\omega_n(t-\tau)d\tau + \int_{t_1}^{t_2} F_2 sin\,\omega_n(t-\tau)d\tau + \ldots \right.$$
$$\left. \ldots + \int_{t_{i-1}}^{t_i} F_{i-1} sin\,\omega_n(t-\tau)d\tau + \int_{t_i}^{t} F_i sin\,\omega_n(t-\tau)d\tau \right\} \quad \text{for} \quad t_i \le t \le t_{i+1} \tag{1.129}$$

or

$$x(t) = \frac{1}{k}\{F_1[\cos\omega_n(t-t_1) - \cos\omega_n(t)] + F_2[\cos\omega_n(t-t_2) - \cos\omega_n(t-t_1)] + \ldots$$
$$\ldots + F_{i-1}[\cos\omega_n(t-t_i) - \cos\omega_n(t-t_{i-1})] + F_i[1-\cos\omega_n(t-t_i)]\} \quad \text{for} \quad t_i \le t \le t_{i+1} \tag{1.130}$$

If one wishes to consider damping, the integration is a bit more laborious, but one can do it in a similar way. Starting with the example of Figure 1.34, supposing a rectangular force, one has to solve:

$$x(t) = \frac{F}{\omega_d m}\int_0^t e^{-\xi\omega_n(t-\tau)} sin\,\omega_d(t-\tau)d\tau \tag{1.131}$$

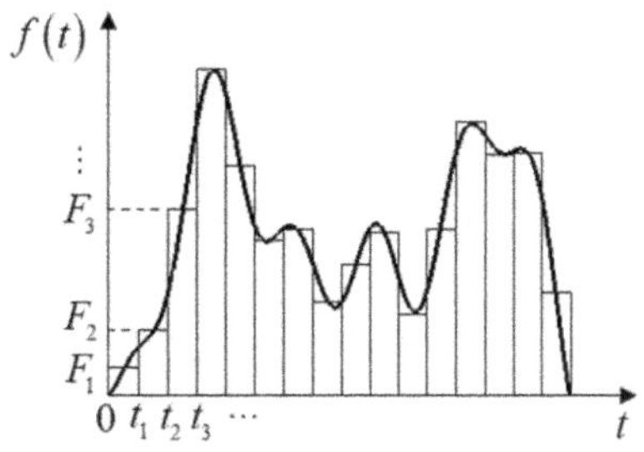

Figure 1.37 Dividing a general non-periodic force into rectangular blocks.

Solving (1.131) implies integrating by parts twice. Eventually, the following result is obtained:

$$x(t) = \frac{F}{k}\left(1 - e^{-\xi\omega_n t}\left(\cos\omega_d t + \frac{\xi}{\sqrt{1-\xi^2}}\sin\omega_d t\right)\right) \tag{1.132}$$

Then, for a general force, one would proceed as before, dividing the force into rectangular pieces.

Alternative Approach

An alternative to consider the non-periodic force as a succession of impulses is to say that it will repeat itself at infinity, i.e. that it is a periodic force with infinite period. In this sense, it would be possible to apply some kind of Fourier analysis to it. Naturally, to reproduce a non-periodic force like the one in Figure 1.29 takes a great "effort", that a series with a finite number of frequencies can hardly reproduce approximately; hence, one is led to a very rich spectrum where $\Delta\omega$ tends to zero. In fact, the spectrum will have all the frequencies, a continuous spectrum, instead of a discrete one as before. The summation turns into an integral and so, instead of a Fourier series, one ends up with a Fourier integral. For a periodic force and using complex notation, the expression was

$$f(t) = \sum_{n=0}^{\infty} \bar{F}_n \mathrm{e}^{in\omega t} \tag{1.133}$$

with

$$\bar{F}_n = \frac{2}{T}\int_0^T f(t)e^{-in\omega t}\,dt \tag{1.134}$$

Making $T \to \infty$, $\Delta\omega \to d\omega$ and the $n\omega$ discrete frequencies become a continuous function, leading to:

$$f(t) = \frac{1}{2\pi}\int_{-\infty}^{+\infty} F(\omega)\mathrm{e}^{i\omega t}\,d\omega \tag{1.135}$$

where

$$F(\omega) = \int_{-\infty}^{+\infty} f(t)\mathrm{e}^{-i\omega t}\,dt \tag{1.136}$$

Eq. (1.135) is the Fourier integral and Eqs. (1.135) and (1.136) constitute the so-called Fourier transform pair. Eq. (1.136) is also called Fourier transform and Eq. (1.135) is the inverse Fourier transform. At each frequency ω the corresponding Fourier transform of the response, $X(\omega)$, can be calculated using the *FRF* at that frequency (Eq. (1.88)):

$$X(\omega)=H(\omega)F(\omega) \tag{1.137}$$

and so, calculating the inverse Fourier transform, as in (1.135), the response is:

$$x(t)=\frac{1}{2\pi}\int_{-\infty}^{+\infty}H(\omega)F(\omega)\mathrm{e}^{i\omega t}\,d\omega \tag{1.138}$$

Relation between the FRF and the IRF

One discussed before the response to a unitary impulse, the impulse response function (*IRF*) $h(t)$. One shall see next its relation to the frequency response function (*FRF*) $H(\omega)$. The unitary impulse, also known as the Dirac δ- function, acting at a generic time τ on a function $g(t)$, is by definition,

$$\int_{-\infty}^{+\infty}\delta(t-\tau)g(t)\,dt=g(\tau) \tag{1.139}$$

Therefore, from Eq. (1.136), the Fourier transform of the unitary impulse at $\tau=0$ is:

$$F(\omega)=\int_{-\infty}^{+\infty}\delta(t-\tau)\mathrm{e}^{-i\omega t}\,dt=\mathrm{e}^{-i\omega\tau}\Big|_{\tau=0}=1 \tag{1.140}$$

Substituting in (1.138), one obtains the response to the unitary impulse

$$x(t)=\frac{1}{2\pi}\int_{-\infty}^{+\infty}H(\omega)\mathrm{e}^{i\omega t}\,d\omega \tag{1.141}$$

which is $h(t)$. Therefore,

$$h(t)=\frac{1}{2\pi}\int_{-\infty}^{+\infty}H(\omega)\mathrm{e}^{i\omega t}\,d\omega \tag{1.142}$$

Thus, it is concluded that the impulse response function, *IRF*, is the inverse Fourier transform of the frequency response function, *FRF*, which in turn is the Fourier transform of the *IRF*:

$$H(\omega)=\int_{-\infty}^{+\infty}h(t)\mathrm{e}^{-i\omega t}\,dt \tag{1.143}$$

This means that the *FRF* $H(\omega)$ and the *IRF* $h(t)$ constitute a Fourier transform pair and so it is easy to calculate one from the other.

Response to a Random Excitation

The response to a random excitation (Figure 1.38) needs a different treatment, as the signal is non-deterministic and what one has is a record with some statistical properties, like the mean value and standard deviation.

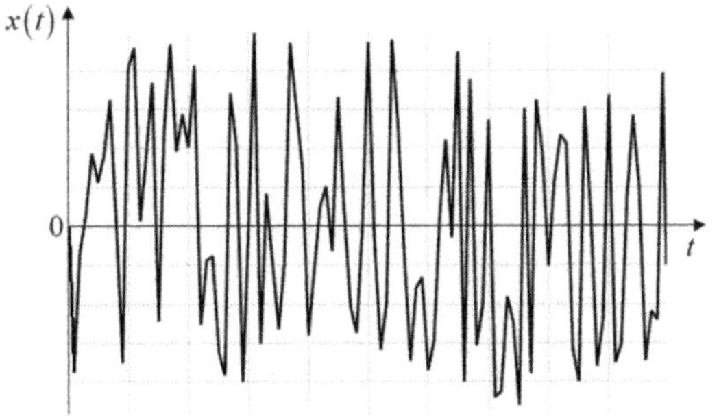

Figure 1.38 Example of a random excitation.

It is important to study this type of excitation, not only because it happens in various situations in real life, like the displacement imposed by the roughness of a road on the wheels of a vehicle, the wind blowing on a bridge or the action of an earthquake, but also because it is a type of excitation often used in the laboratory to excite a specimen, as it is very easy and fast to perform an experimental test with random excitation. It might be thought that one could make the analysis assuming an infinite periodicity, but normally this is not possible because the Dirichlet condition must be satisfied when applying Fourier transforms:

$$\int_{-\infty}^{+\infty} |f(t)|\, dt < \infty \tag{1.144}$$

It will be assumed here that the random signal is stationary and ergodic, which means that both averaging across many time history records at a given instant in time and averaging over time using just one time history give the same properties, like mean, mean square and statistical distributions. This works well in most practical cases. For instance, a long enough time record taken on a certain type of rough road surface may be considered as ergodic, as it may be assumed as typical and representative of that specific kind of roughness.

As in general one cannot use directly the recorded random input $f(t)$, one must find a function closely related to $f(t)$. Such a function is the average (or expected value) of the product $f(t)f(t+\tau)$, called the auto-correlation function $R_{ff}(\tau)$, where τ is a time shift. Thus,

$$R_{ff}(\tau) = E\left[f(t)f(t+\tau)\right] = \lim_{T\to\infty} \frac{1}{T}\int_{-T/2}^{+T/2} f(t)f(t+\tau)\,dt \tag{1.145}$$

The auto-correlation function can be interpreted as a measure of how similar a signal is to a time-shifted version of itself. In an ideal random process, with no frequency limits, the auto-correlation would be zero except for $\tau = 0$. In practice, real signals are limited in frequency and in general the auto-correlation looks like in Figure 1.39.

Therefore, instead of the original signal $f(t)$, one will study the new function $R_{ff}(\tau)$, which is real-valued, even, goes to zero as τ becomes large (in both positive and negative directions) and clearly obeys to Dirichet condition (1.144). Consequently, one can apply a Fourier transform to it:

$$S_{ff}(\omega)=\int_{-\infty}^{+\infty} R_{ff}(\tau)\mathrm{e}^{-i\omega\tau}\,d\tau \tag{1.146}$$

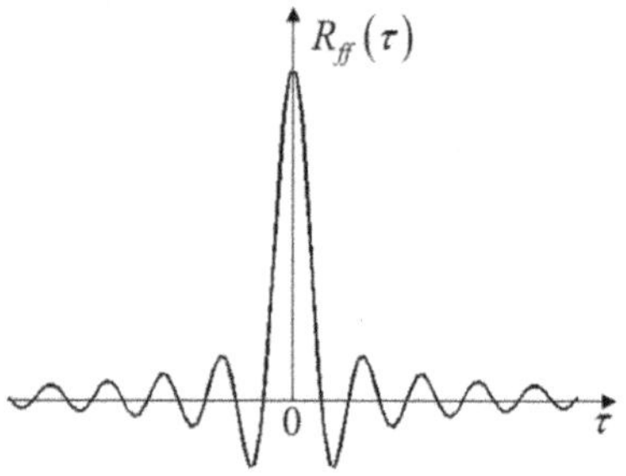

Figure 1.39 Example of an auto-correlation function.

This is known as auto-spectral density or power spectral density (*PSD*), which is also an even and real-valued function. Conversely, one can calculate its inverse transform:

$$R_{ff}(\tau)=\frac{1}{2\pi}\int_{-\infty}^{+\infty} S_{ff}(\omega)\mathrm{e}^{i\omega\tau}\,d\omega \tag{1.147}$$

Thus, the power spectral density and the auto-correlation functions constitute a Fourier transform pair, describing our stationary random process. If one sets $\tau=0$ in Eq. (1.145) and considering Eq. (1.147), it follows that:

$$R_{ff}(0)=\lim_{T\to\infty}\frac{1}{T}\int_{-T/2}^{+T/2} f^2(t)\,dt=\frac{1}{2\pi}\int_{-\infty}^{+\infty} S_{ff}(\omega)\,d\omega \tag{1.148}$$

which makes clear that the units of $S_{ff}(\omega)$ are amplitude mean square per unit frequency, justifying the designation "density".

Now, one has a frequency description of the original force $f(t)$. Similarly, one may also evaluate the *PSD* of the response $x(t)$, which is $S_{xx}(\omega)$. It may be demonstrated that the *PSD*s of the input and output are related through the *FRF* $H(\omega)$, as

$$S_{xx}(\omega)=\left|H(\omega)\right|^2 S_{ff}(\omega) \tag{1.149}$$

If $H(\omega)$ is already known, it is possible to evaluate stationary random response characteristics. However, if the objective is to extract the *FRF* $H(\omega)$ from the force and response data, Eq. (1.149) is not enough, as the information about the phase is missing. The problem is solved by generalising the concept of correlation, relating force and response, i.e. defining cross-correlation functions as

$$R_{fx}(\tau)=E\left[f(t)x(t+\tau)\right]=\lim_{T\to\infty}\frac{1}{T}\int_{-T/2}^{+T/2} f(t)x(t+\tau)\,dt \tag{1.150}$$

$$R_{xf}(\tau) = E\left[x(t)f(t+\tau)\right] = \lim_{T\to\infty}\frac{1}{T}\int_{-T/2}^{+T/2} x(t)f(t+\tau)\,dt \tag{1.151}$$

These cross-correlation functions are related through

$$R_{xf}(\tau) = R_{fx}(-\tau) \tag{1.152}$$

As a consequence, it is also possible to calculate the corresponding cross-spectral densities, which are now complex and include the information about the phase:

$$S_{fx}(\omega) = \int_{-\infty}^{+\infty} R_{fx}(\tau)\mathrm{e}^{-i\omega\tau}\,d\tau \tag{1.153}$$

$$S_{xf}(\omega) = \int_{-\infty}^{+\infty} R_{xf}(\tau)\mathrm{e}^{-i\omega\tau}\,d\tau \tag{1.154}$$

These cross-spectral densities are complex conjugates:

$$S_{xf}(\omega) = S_{fx}^{*}(\omega) \tag{1.155}$$

It can also be shown that

$$S_{fx}(\omega) = H(\omega)S_{ff}(\omega) \tag{1.156}$$

and

$$S_{xx}(\omega) = H(\omega)S_{xf}(\omega) \tag{1.157}$$

Eqs. (1.156) and (1.157) are alternative ways to determine an *FRF*, as will be seen soon.

In practice, the spectral densities are not calculated from the correlation functions; instead, they are obtained from the Fourier transforms of the response and force, because the process is much more effective. One starts with the time histories of the response and force. For instance, taking the response $x(t)$, one starts by dividing it in n_t blocks, each one with a time interval of T seconds, as in Figure 1.40.

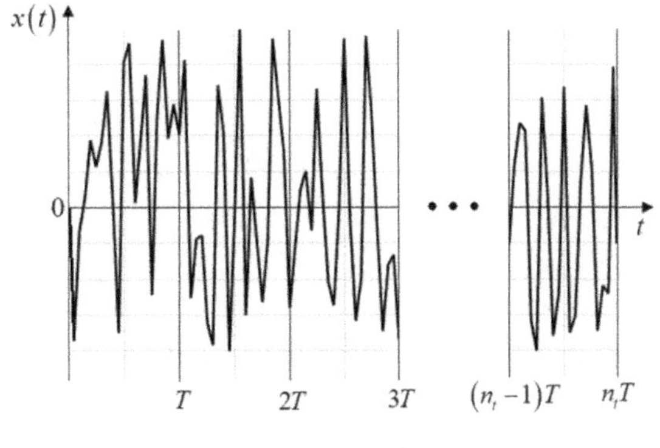

Figure 1.40 Division of the response in n_t blocks of T seconds.

Having into account that for each block the Fourier transform is

$$X_j(\omega)=\int_0^T x_j(t)e^{-i\omega t}dt \quad , \; j=1, \; ... \; n_t \tag{1.158}$$

According to Eqs. (1.145) and (1.146), one concludes that a good approximation to the power spectral density of the response is given by:

$$\hat{S}_{xx}(\omega)=\frac{1}{n_t T}\sum_{j=1}^{n_t}\left|X_j(\omega)\right|^2 \tag{1.159}$$

To calculate $X_j(\omega)$ one must deduce its discrete version. Dividing each block j into N equaly spaced intervals of Δt seconds, $T=N\Delta t$, as in Figure 1.41, one has $x(0)$, $x(\Delta t)$, $x(2\Delta t)$, ..., $x(n\Delta t)$, ..., $x((N-1)\Delta t)$.

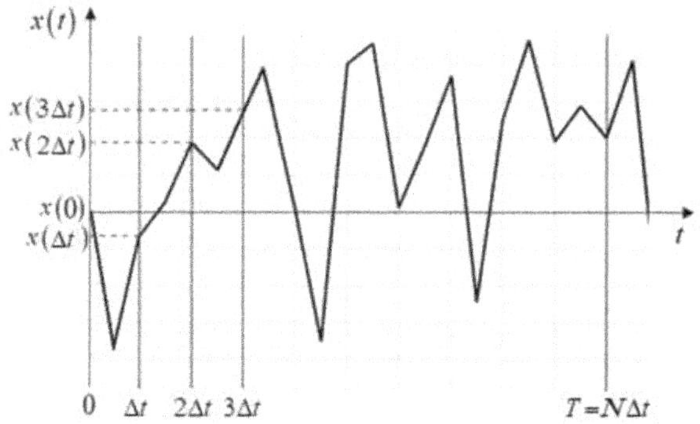

Figure 1.41 Division of a block j into N equaly spaced intervals of Δt seconds.

Let x_{jn} be the response in each interval and in each block j. The Fourier transform will generate N frequencies ω_k, and thus [10],

$$X_j(\omega_k)=\left(\sum_{n=0}^{N-1} x_{jn}e^{-i\omega_k n\Delta t}\right)\Delta t, \; j=1, \; ..., \; n_t, \text{ with } \omega_k=\frac{2\pi k}{T}=\frac{2\pi k}{N\Delta t}, \quad k=0, \; ..., \; N-1 \tag{1.160}$$

or

$$X_j(\omega_k)=\left(\sum_{n=0}^{N-1} x_{jn}\, e^{-i\frac{2\pi kn}{N}}\right)\Delta t, \; j=1, \; ..., \; n_t, \; k=0, \; ..., \; N-1 \tag{1.161}$$

Finally, the power spectral density of the response will be:

[10] This is the theoretical procedure to calculate numerically the Fourier transform of a signal. However, in practice, commercial software uses a more effective and faster technique that uses a much fewer number of numerical operations; such a technique is known as Fast Fourier Transform (*FFT*).

$$\hat{S}_{xx}(\omega_k)=\frac{1}{n_t N\Delta t}\sum_{j=1}^{n_t}\left|X_j(\omega_k)\right|^2 \ , \ k=0,\ ...,\ N-1 \tag{1.162}$$

Likewise, the power spectral density of the force will be given by:

$$\hat{S}_{ff}(\omega_k)=\frac{1}{n_t N\Delta t}\sum_{j=1}^{n_t}\left|F_j(\omega_k)\right|^2 \ , \ k=0,\ ...,\ N-1 \tag{1.163}$$

where $F_j(\omega_k)$ is the Fourier transform value corresponding to the force $f_j(n\Delta t)$. One can also infer the expression for the cross-spectral density $\hat{S}_{fx}$:

$$\hat{S}_{fx}(\omega_k)=\frac{1}{n_t N\Delta t}\sum_{j=1}^{n_t}F_j^*(\omega_k)X_j(\omega_k) \ , \ k=0,\ ...,\ N-1 \tag{1.164}$$

And from (1.155), $\hat{S}_{xf}=\hat{S}_{fx}^*$. From Eqs. (1.156) and (1.157) one has two possible estimators for calculating an *FRF*, usually known as H_1 and H_2:

$$H_1(\omega_k)=\frac{\hat{S}_{fx}(\omega_k)}{\hat{S}_{ff}(\omega_k)} \ , \ k=0,\ 1,\ ...,\ N-1 \tag{1.165}$$

$$H_2(\omega_k)=\frac{\hat{S}_{xx}(\omega_k)}{\hat{S}_{xf}(\omega_k)} \ , \ k=0,\ 1,\ ...,\ N-1 \tag{1.166}$$

Theoretically, H_1 and H_2 should provide the same result, but in practice that is not the case and therefore a kind of quality factor, known as coherence has been defined, as the ratio between both estimators:

$$\gamma^2(\omega_k)=\frac{H_1(\omega_k)}{H_2(\omega_k)}=\frac{\left|\hat{S}_{fx}(\omega_k)\right|^2}{\hat{S}_{ff}(\omega_k)\hat{S}_{xx}(\omega_k)} \ , \ k=0,\ 1,\ ...,\ N-1 \tag{1.167}$$

Ideally, the coherence should be equal to 1, but in fact it decreases near the resonances and anti-resonances (situation opposed to resonances, where the value of the *FRF* approaches zero). Section 1.4.5 presented the response to a harmonic force of constant amplitude, which allowed to understand various phenomena, including the notion of resonance. In general, the force will be neither purely harmonic nor constant, unless one does it on purpose. When performing a vibration test with a shaker one imposes a certain gain, i.e. a constant amount of energy along the frequency range. As that energy is the product of force and displacement, when the system approaches a resonance the displacement increases and the force decreases, as the system only needs to "fight" against the damping. With a very low value of force, it is natural that force transducers have more difficulty in providing a very accurate measurement, and some noise appears affecting the denominator of Eq. (1.165). At the anti-resonance frequencies something similar happens with the response, which tends to be low, making it difficult for the accelerometers to give precise measurements. This leads to errors in the numerator of Eq.

(1.166). The conclusion is that the estimator H_1 is more reliable at the anti-resonances, while H_2 is better at the resonances. Further discussion and examples on the coherence function are given in Chapter 2 and also in Chapter 5 (Section 5.3.5), where it is explained how the coherence can be used as an indicator of nonlinear behaviour.

1.5 Discrete Systems with Multiple Degrees-of-Freedom

In most cases, one will be dealing with multiple degree-of-freedom systems (*MDOF*). However, almost all the physics of the vibration phenomena can be explained under the study of the *SDOF* system. The study of the *MDOF* system is a generalisation, from the numerical point of view, and only a few new concepts will appear here. One of them is the concept of mode shape. As the *MDOF* system is described by a number of *N DOF*, i.e. by a number of *N* coordinates, it will have *N* possibilities of vibrating in a natural way and therefore it will have *N* natural frequencies. At each natural frequency the coordinates will exhibit a certain spatial configuration (shape), known as mode shape. This will be addressed further on.

1.5.1 The Dynamic Equilibrium Equation with Viscous Damping

For an *MDOF* system there will be *N* dynamic equilibrium equations that one can organise in matrix form and that will resemble very much equation (1.10) for the *SDOF* case:

$$\boldsymbol{M}\,\ddot{\boldsymbol{x}}+\boldsymbol{C}\,\dot{\boldsymbol{x}}+\boldsymbol{K}\,\boldsymbol{x}=\boldsymbol{f}(t) \tag{1.168}$$

where now one has the mass, damping and stiffness matrices $(N\times N)$ and vectors $(N\times 1)$. Due to a generalisation of the reciprocity theorem to the dynamic case, the three matrices will be symmetric (discarding the cases where there are gyroscopic forces, that will not be addressed in this book). One can obtain these equations using the Lagrange formalism presented in Section 1.3.2. For example, for the 2 *DOF* system of Figure 1.42, the equations will be:

$$\begin{aligned} m_1\ddot{x}_1+c_1\dot{x}_1+c_2(\dot{x}_1-\dot{x}_2)+k_1x_1+k_2(x_1-x_2)&=f_1(t)\\ m_2\ddot{x}_2+c_2(\dot{x}_2-\dot{x}_1)+k_2(x_2-x_1)&=f_2(t) \end{aligned} \tag{1.169}$$

or

$$\begin{bmatrix} m_1 & 0\\ 0 & m_2\end{bmatrix}\begin{Bmatrix}\ddot{x}_1\\ \ddot{x}_2\end{Bmatrix}+\begin{bmatrix} c_1+c_2 & -c_2\\ -c_2 & c_2\end{bmatrix}\begin{Bmatrix}\dot{x}_1\\ \dot{x}_2\end{Bmatrix}+\begin{bmatrix} k_1+k_2 & -k_2\\ -k_2 & k_2\end{bmatrix}\begin{Bmatrix}x_1\\ x_2\end{Bmatrix}=\begin{Bmatrix}f_1(t)\\ f_2(t)\end{Bmatrix} \tag{1.170}$$

where, as before, for simplicity, one only keeps in the force the indication that it is a function of time.

As it can be observed, the equations are coupled through the elements outside the main diagonal of the damping and stiffness matrices and therefore the solution cannot be found independently, as the response of x_1 depends on x_2 and *vice-versa*. The matrices can be diagonal or not; this depends on which coordinates are chosen to represent the system.

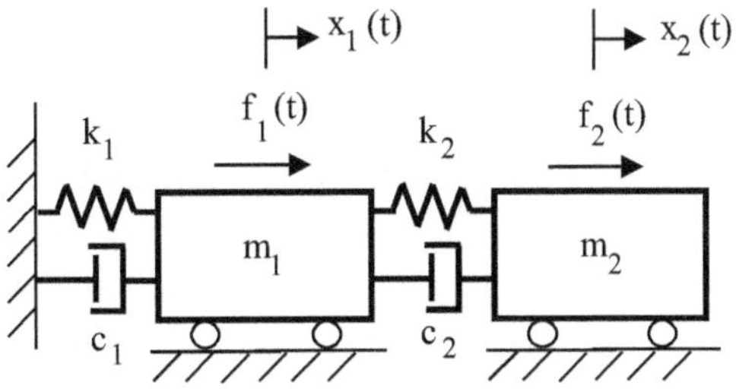

Figure 1.42 Example of a 2 *DOF* system.

Therefore, the coupling itself is not a property of the system. This point will be discussed later, when talking about transformation of coordinates.

1.5.2 Undamped Vibration Response

Natural Frequencies and Mode Shapes

The simplest case is when the system has no applied forces and the damping is neglected, i.e. one wishes to study the undamped free vibration. In this case, the equation is:

$$\boldsymbol{M}\ddot{\boldsymbol{x}} + \boldsymbol{K}\boldsymbol{x} = \boldsymbol{0} \tag{1.171}$$

As for the *SDOF* case, the solution is of the type

$$\boldsymbol{x}(t) = \psi e^{i\omega t} \tag{1.172}$$

Substituting in (1.171) leads to

$$\left(\boldsymbol{K} - \omega^2 \boldsymbol{M}\right)\psi = \boldsymbol{0} \tag{1.173}$$

which is a generalised eigenvalue and eigenvector problem, where the eigenvalues are the squared natural frequencies and the eigenvectors represent the mode shapes. Solving for ψ, it follows that

$$\psi = \left(\boldsymbol{K} - \omega^2 \boldsymbol{M}\right)^{-1} \cdot \boldsymbol{0} \tag{1.174}$$

If the inverse exists, then the solution is $\psi = \boldsymbol{0}$, the so-called trivial solution, which obviously is of no interest. Thus, for a non-trivial solution the inverse must not be possible, which implies that the determinant of $\boldsymbol{K} - \omega^2 \boldsymbol{M}$ must be zero:

$$det\left(\boldsymbol{K} - \omega^2 \boldsymbol{M}\right) = 0 \tag{1.175}$$

This leads to the characteristic equation that gives the values of the (undamped) natural frequencies ω_r, $r = 1 \ldots N$. Substituting each natural frequency in (1.173), one can evaluate the corresponding mode shape ψ_r. However, it should be noted that each solution ω_r makes the determinant equal to zero and therefore the system of equations (1.173) is rank deficient, the rank being $N-1$. This means that there is not a unique solution for each mode shape, it would be necessary an extra condition to make it unique. It is clear from Eq. (1.173) that for each ψ_r, $C\ \psi_r$ is also a solution, where C is an arbitrary non-zero constant. Thus, for each mode shape, all one can calculate is the relation between its amplitudes. The extra condition to define it uniquely is usually to impose that its norm is unitary, or other options, like the normalisation with respect to the mass matrix, as one shall see later. In any case, one knows that at each natural frequency the system exhibits a certain configuration, fixed in space (at least for the undamped case, as shall also be discussed later) and varying harmonically in time, at that frequency.

For the undamped case, the mode shapes are real. Their components may be positive or negative (or zero, of course). If they are all positive, they are all in phase, reaching their maxima or minima simultaneously. If some are positive and others negative it means that the negative ones are out of phase *w.r.t.* the positive ones, when these reach their maxima, the others reach their minima and *vice-versa*. Often the mode shapes are represented graphically and animated, to offer a clear view of how the system behaves at a particular natural frequency (Figure 1.43).

Figure 1.43 Example of the graphical representation of a mode shape.

At this moment one knows that at each natural frequency the system responds harmonically, in free vibration, with a certain configuration in space, the mode shape. However, this only happens if one gives to the system the initial conditions corresponding exactly to each mode shape. In the general case, the system will respond as a linear combination of the responses at each natural frequency and mode shape.

Here, it is important to introduce an example, simple enough to be easy to follow, but illustrative of the phenomena one is discussing. Take the undamped 2 *DOF* system of Figure 1.44 in free vibration.

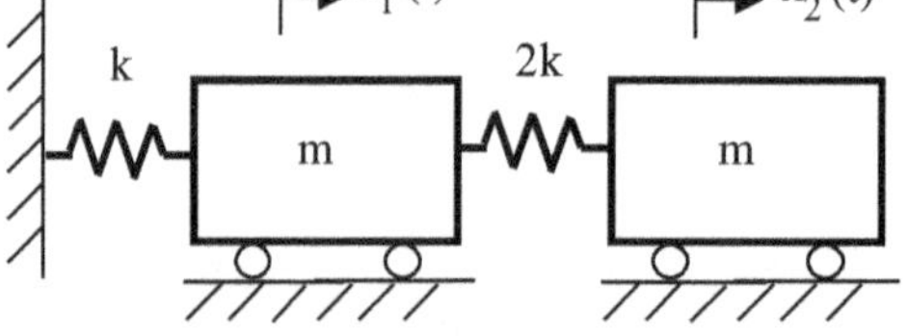

Figure 1.44 Example of an undamped 2 *DOF* system in free vibration.

In this case, the dynamic equilibrium equations are:

$$\begin{bmatrix} m & 0 \\ 0 & m \end{bmatrix} \begin{Bmatrix} \ddot{x}_1 \\ \ddot{x}_2 \end{Bmatrix} + \begin{bmatrix} 3k & -2k \\ -2k & 2k \end{bmatrix} \begin{Bmatrix} x_1 \\ x_2 \end{Bmatrix} = \begin{Bmatrix} 0 \\ 0 \end{Bmatrix} \tag{1.176}$$

The generalised eigenvalue problem of Eq. (1.173) is now

$$\begin{bmatrix} 3k - \omega^2 m & -2k \\ -2k & 2k - \omega^2 m \end{bmatrix} \begin{Bmatrix} \psi_1 \\ \psi_2 \end{Bmatrix} = \begin{Bmatrix} 0 \\ 0 \end{Bmatrix} \tag{1.177}$$

Setting the determinant to zero (Eq. (1.175)) leads to

$$m^2\omega^4 - 5km\,\omega^2 + 2k^2 = 0 \tag{1.178}$$

The natural frequencies are

$$\omega_1 = 0.662\sqrt{\frac{k}{m}} \qquad \omega_2 = 2.136\sqrt{\frac{k}{m}} \tag{1.179}$$

As each natural frequency leads to a null determinant, only one of the Eqs. (1.177) can be used to calculate the mode shapes. For mode number 1, one substitutes the first natural frequency and choosing, for instance, the first equation, it follows that:

$$(3k - 0.438k)\psi_1^{(1)} - 2k\psi_2^{(1)} = 0 \tag{1.180}$$

where the superscript stands for the mode number. The relation between the amplitudes is:

$$\frac{\psi_2^{(1)}}{\psi_1^{(1)}} = 1.281 \tag{1.181}$$

meaning that for each unit of amplitude of coordinate 1, the amplitude at coordinate 2 is 1.281 higher. One can say that the first mode is

$$\boldsymbol{\psi}_1 = C_1 \begin{Bmatrix} 1 \\ 1.281 \end{Bmatrix} \tag{1.182}$$

where C_1 is an arbitrary non-zero constant. Proceeding likewise for the second mode shape, one arrives at:

$$\boldsymbol{\psi}_2 = C_2 \begin{Bmatrix} 1 \\ -0.781 \end{Bmatrix} \tag{1.183}$$

which means that the motion of coordinates 1 and 2 are out of phase from each other, when one is at a maximum, the other is at a minimum.

Free Response

As mentioned before, the mode shapes only represent the spatial configuration, not the time response, which is harmonic at each natural frequency. Thus, the free time response of the first mode is:

$$\begin{Bmatrix} x_1^{(1)}(t) \\ x_2^{(1)}(t) \end{Bmatrix} = C_1 \begin{Bmatrix} 1 \\ 1.281 \end{Bmatrix} cos(\omega_1 t - \varphi_1) \tag{1.184}$$

and for the second mode,

$$\begin{Bmatrix} x_1^{(2)}(t) \\ x_2^{(2)}(t) \end{Bmatrix} = C_2 \begin{Bmatrix} 1 \\ -0.781 \end{Bmatrix} cos(\omega_2 t - \varphi_2) \tag{1.185}$$

where φ_1 and φ_2 are phase angles, to take into consideration possible initial velocities imposed to the system. In the general case, the system will respond in a linear combination of the responses of both mode shapes, i.e.

$$\boldsymbol{x}(t) = \begin{Bmatrix} x_1(t) \\ x_2(t) \end{Bmatrix} = C_1 \begin{Bmatrix} 1 \\ 1.281 \end{Bmatrix} cos(\omega_1 t - \varphi_1) + C_2 \begin{Bmatrix} 1 \\ -0.781 \end{Bmatrix} cos(\omega_2 t - \varphi_2) \tag{1.186}$$

Thus, there are 4 constants to evaluate, from the initial displacements and velocities imposed to the masses. If one gives only initial displacements (i.e. zero initial velocities), φ_1 and φ_2 will be both zero. If those initial displacements are $x_1(0)=1$, $x_2(0)=1.281$, i.e. coincident with the first mode shape, then $C_1 = 1$ and $C_2 = 0$, and the system responds harmonically with those amplitudes, at the first natural frequency:

$$\boldsymbol{x}(t) = \begin{Bmatrix} x_1(t) \\ x_2(t) \end{Bmatrix} = \begin{Bmatrix} 1 \\ 1.281 \end{Bmatrix} cos(\omega_1 t) \tag{1.187}$$

Likewise, if the initial displacements are $x_1(0)=1$, $x_2(0)=-0.781$, then $C_1 = 0$ and $C_2 = 1$, and the system responds harmonically with those amplitudes, at the second natural frequency:

$$\boldsymbol{x}(t) = \begin{Bmatrix} x_1(t) \\ x_2(t) \end{Bmatrix} = \begin{Bmatrix} 1 \\ -0.781 \end{Bmatrix} cos(\omega_2 t) \tag{1.188}$$

In (1.186) the constants *C* represent the contribution of each mode to the total response. As this is a sum of motions with different frequencies, the resulting motion will not be harmonic. However, it will be periodic. For a system with *N DOF*, the free vibration response will be:

$$\boldsymbol{x}(t) = C_1 \boldsymbol{\psi}_1 \, cos(\omega_1 t - \varphi_1) + \ldots + C_N \boldsymbol{\psi}_N \, cos(\omega_N t - \varphi_N) \tag{1.189}$$

or

$$\boldsymbol{x}(t) = \sum_{r=1}^{N} C_r \boldsymbol{\psi}_r \, cos(\omega_r t - \varphi_r) \tag{1.190}$$

The *2N* constants, C_r and φ_r, must be evaluated from the application of the initial conditions of displacement and velocity to (1.190). This leads to a linear system of equations to be solved for C_r and φ_r. Alternatively, one can modify Eqs. (1.190) in order to be able to introduce directly the initial conditions and obtain easily the free vibration response. One shall see this later on.

Eq. (1.190) can alternatively be written as

$$\boldsymbol{x}(t)=\begin{bmatrix}\boldsymbol{\psi}_1 & \cdots & \boldsymbol{\psi}_N\end{bmatrix}\begin{Bmatrix} C_1 \cos(\omega_1 t-\varphi_1) \\ \vdots \\ C_N \cos(\omega_N t-\varphi_N)\end{Bmatrix} \tag{1.191}$$

or in a more compact form,

$$\boldsymbol{x}(t)=\boldsymbol{\Psi}\,\boldsymbol{p}(t) \tag{1.192}$$

where $\boldsymbol{\Psi}$ is a matrix formed by the mode shapes, known as *modal matrix*. Eq. (1.192) represents a transformation of coordinates, where $\boldsymbol{x}(t)$ are the *physical coordinates* of the system and $\boldsymbol{p}(t)$ are the so-called *principal coordinates*. The reason for this last designation will be explained next.

Transformation of Coordinates

To introduce this topic it is also important to support the explanation following an example. Consider the system of Figure 1.45. Assuming that there is no motion along the horizontal direction, the system has two *DOF*, corresponding to the vertical and the angular motions.

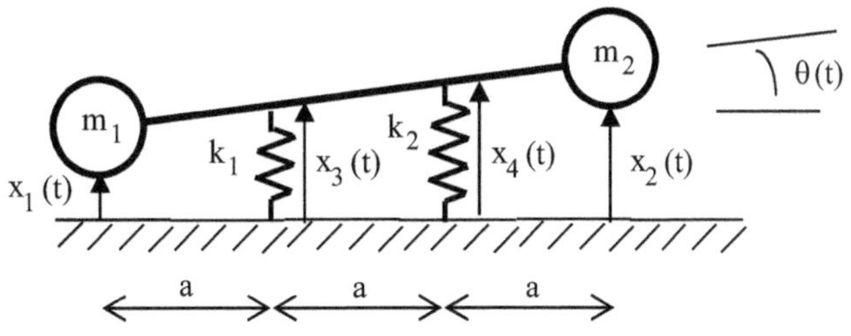

Figure 1.45 Example of a 2 *DOF* system.

To simplify, one shall also assume that the bar is rigid and massless and consider small amplitudes. To study this system, one begins by writing the kinetic and elastic energies, using the defined physical coordinates $\boldsymbol{x}$:

$$\begin{aligned} T &= \frac{1}{2}m_1\dot{x}_1^2+\frac{1}{2}m_2\dot{x}_2^2 \\ V &= \frac{1}{2}k_1 x_3^2+\frac{1}{2}k_2 x_4^2 \end{aligned} \tag{1.193}$$

As there are two *DOF*, one needs to choose two of those four coordinates. Choosing x_1 and x_2, one has to write x_3 and x_4 as a function of x_1 and x_2. From the geometry of Figure 1.45,

$$\begin{aligned} x_3 &= \frac{2}{3}x_1 + \frac{1}{3}x_2 \\ x_4 &= \frac{1}{3}x_1 + \frac{2}{3}x_2 \end{aligned} \tag{1.194}$$

The expressions of the energies become:

$$\begin{aligned} T &= \frac{1}{2}m_1\dot{x}_1^2 + \frac{1}{2}m_2\dot{x}_2^2 \\ V &= \frac{1}{2}k_1\left(\frac{2}{3}x_1 + \frac{1}{3}x_2\right)^2 + \frac{1}{2}k_2\left(\frac{1}{3}x_1 + \frac{2}{3}x_2\right)^2 \end{aligned} \tag{1.195}$$

Applying Lagrange's equations to obtain the dynamic equilibrium equations, leads to:

$$\begin{bmatrix} m_1 & 0 \\ 0 & m_2 \end{bmatrix}\begin{Bmatrix} \ddot{x}_1 \\ \ddot{x}_2 \end{Bmatrix} + \frac{1}{9}\begin{bmatrix} 4k_1 + k_2 & 2k_1 + 2k_2 \\ 2k_1 + 2k_2 & k_1 + 4k_2 \end{bmatrix}\begin{Bmatrix} x_1 \\ x_2 \end{Bmatrix} = \begin{Bmatrix} 0 \\ 0 \end{Bmatrix} \tag{1.196}$$

Here, the mass matrix is diagonal and the stiffness matrix is full, where the elements outside the main diagonal represent the coupling between both coordinates. If one had chosen coordinates x_3 and x_4 to represent the system, the result would be:

$$\begin{aligned} T &= \frac{1}{2}m_1(2\dot{x}_3 - \dot{x}_4)^2 + \frac{1}{2}m_2(-\dot{x}_3 + 2\dot{x}_4)^2 \\ V &= \frac{1}{2}k_1x_3^2 + \frac{1}{2}k_2x_4^2 \end{aligned} \tag{1.197}$$

And the equations would be:

$$\begin{bmatrix} 4m_1 + m_2 & -2m_1 - 2m_2 \\ -2m_1 - 2m_2 & m_1 + 4m_2 \end{bmatrix}\begin{Bmatrix} \ddot{x}_3 \\ \ddot{x}_4 \end{Bmatrix} + \begin{bmatrix} k_1 & 0 \\ 0 & k_2 \end{bmatrix}\begin{Bmatrix} x_3 \\ x_4 \end{Bmatrix} = \begin{Bmatrix} 0 \\ 0 \end{Bmatrix} \tag{1.198}$$

These equations are now coupled through the mass matrix. It is therefore clear that the coupling is not a property of the system, it depends on the coordinates that are chosen. If one chooses the coordinates of the masses, the mass matrix is diagonal; if one chooses the ones of the stiffnesses, the stiffness matrix is diagonal; if one had chosen any other combination, for instance, coordinates x_1 and x_3, then both matrices would be full. The question that arises is: "Is there any coordinate system that makes both matrices diagonal?" The discussion that follows responds to this question.

Going back to the kinetic and elastic energies, for an *SDOF* system they are given by $T = \frac{1}{2}m\dot{x}^2$ and $V = \frac{1}{2}kx^2$, respectively. For an *MDOF* system, those formulæ are given by:

$$T = \frac{1}{2}\dot{\boldsymbol{x}}^T \boldsymbol{M}\,\dot{\boldsymbol{x}} \qquad V = \frac{1}{2}\boldsymbol{x}^T \boldsymbol{K}\,\boldsymbol{x} \tag{1.199}$$

Changing to coordinates $\boldsymbol{y}$ through a transformation matrix $\boldsymbol{U}$, one has $\boldsymbol{x} = \boldsymbol{U}\,\boldsymbol{y}$. Substituting in (1.199), the energies referred to coordinates $\boldsymbol{y}$ are:

$$T = \frac{1}{2}\dot{\boldsymbol{y}}^T \boldsymbol{U}^T \boldsymbol{M}\,\boldsymbol{U}\,\dot{\boldsymbol{y}} = \frac{1}{2}\dot{\boldsymbol{y}}^T \boldsymbol{M}'\,\dot{\boldsymbol{y}} \qquad V = \frac{1}{2}\boldsymbol{y}^T \boldsymbol{U}^T \boldsymbol{K}\,\boldsymbol{U}\,\boldsymbol{y} = \frac{1}{2}\boldsymbol{y}^T \boldsymbol{K}'\,\boldsymbol{y} \tag{1.200}$$

Thus, when one changes coordinates, the new matrices are $\boldsymbol{M}' = \boldsymbol{U}^T \boldsymbol{M}\,\boldsymbol{U}$ and $\boldsymbol{K}' = \boldsymbol{U}^T \boldsymbol{K}\,\boldsymbol{U}$. The analysis continues in what follows.

Forced Response

Going back to the dynamic equilibrium equation (without damping), with an applied force vector,

$$\boldsymbol{M}\,\ddot{\boldsymbol{x}} + \boldsymbol{K}\,\boldsymbol{x} = \boldsymbol{f}(t) \tag{1.201}$$

Applying the transformation $\boldsymbol{x} = \boldsymbol{U}\,\boldsymbol{y}$, leads to

$$\boldsymbol{M}\,\boldsymbol{U}\,\ddot{\boldsymbol{y}} + \boldsymbol{K}\,\boldsymbol{U}\,\boldsymbol{y} = \boldsymbol{f}(t) \tag{1.202}$$

And pre-multiplying by $\boldsymbol{U}^T$,

$$\boldsymbol{U}^T \boldsymbol{M}\,\boldsymbol{U}\,\ddot{\boldsymbol{y}} + \boldsymbol{U}^T \boldsymbol{K}\,\boldsymbol{U}\,\boldsymbol{y} = \boldsymbol{U}^T \boldsymbol{f}(t) \quad \Rightarrow \quad \boldsymbol{M}'\,\ddot{\boldsymbol{y}} + \boldsymbol{K}'\,\boldsymbol{y} = \boldsymbol{f}'(t) \tag{1.203}$$

As said before, one wishes to know whether there is a particular transformation that leads to both matrices $\boldsymbol{M}'$ and $\boldsymbol{K}'$ being diagonal. This is important, because if that is possible, then a system with *N DOF* is transformed into *N SDOF* systems, much easier to handle and for which one has already studied the solution in various situations. The answer is positive, i.e. in fact there is such a transformation through the *modal matrix* $\boldsymbol{x}(t) = \boldsymbol{\Psi}\,\boldsymbol{p}(t)$ (Eq. (1.192)) and the new coordinates are called *principal coordinates*, as already mentioned. This is a consequence of the mode shapes being linearly independent. In fact, one saw before (Eq. (1.190)) that the response of the system is a linear combination of the responses of each mode shape, i.e. the mode shapes constitute a basis in an *N* dimensional space; any vector of that space is a linear combination of the mode shapes. That is why the new matrices become diagonal [11]. Therefore, one has:

$$\boldsymbol{\Psi}^T \boldsymbol{M}\boldsymbol{\Psi}\,\ddot{\boldsymbol{p}} + \boldsymbol{\Psi}^T \boldsymbol{K}\boldsymbol{\Psi}\,\boldsymbol{p} = \boldsymbol{\Psi}^T \boldsymbol{f}(t) \tag{1.204}$$

or

$$\mathcal{M}\,\ddot{\boldsymbol{p}} + \mathcal{K}\,\boldsymbol{p} = \mathcal{P}(t) \tag{1.205}$$

[11] This results from a theorem, which naturally has a proof. However, it will not be given here, to keep the explanation as clear and concise as possible.

$\mathcal{M}$, $\mathcal{K}$ and $\mathcal{P}$ are known as *modal mass matrix, modal stiffness matrix* and *modal force vector*, constituted by the modal masses m_r, modal stiffnesses k_r and modal forces $\mathcal{P}_r$, respectively. As $\mathcal{M}$ and $\mathcal{K}$ are diagonal matrices, one must solve *N* equations for coordinates ***p***, each one representing an *SDOF* system. The system becomes totally uncoupled:

$$\begin{aligned} m_1\,\ddot{p}_1 + k_1\,p_1 &= \mathcal{P}_1(t) \\ &\vdots \\ m_N\,\ddot{p}_N + k_N\,p_N &= \mathcal{P}_N(t) \end{aligned} \tag{1.206}$$

The natural frequencies do not depend on the coordinate system and thus,

$$\omega_{n_1} = \sqrt{\frac{k_1}{m_1}}, \; \ldots \; \omega_{n_N} = \sqrt{\frac{k_N}{m_N}} \tag{1.207}$$

In general, coordinates ***p*** do not have a physical meaning (except in very particular cases), as they are linear combinations of the physical coordinates ***x***. However, they are quite convenient, especially when there are aperiodic forces applied to the system. After solving (1.206) in terms of coordinates ***p***, one must simply apply the transformation $\boldsymbol{x}(\mathrm{t}) = \boldsymbol{\Psi}\boldsymbol{p}(\mathrm{t})$ to obtain the desired response in the physical coordinates ***x***.

If the applied force is harmonic there is no need for uncoupling the equations. If in Eq. (1.201) $\boldsymbol{f}(\mathrm{t}) = \bar{\boldsymbol{F}}e^{i\omega t}$, where the bar on the ***F*** means that the various components may have a distinct phase among them, the steady-state response will also be harmonic, $\boldsymbol{x}(t) = \bar{\boldsymbol{X}}e^{i\omega t}$, where the bar on the ***X*** means now the phase between the response and the force. Substituting in (1.201), it follows that

$$\left(\boldsymbol{K} - \omega^2\boldsymbol{M}\right)\bar{\boldsymbol{X}} = \bar{\boldsymbol{F}} \tag{1.208}$$

from which the complex amplitudes may be calculated by direct inversion of the dynamic stiffness matrix $\boldsymbol{Z}(\omega) = \boldsymbol{K} - \omega^2\boldsymbol{M}$. This is known as the *Direct Method*:

$$\bar{\boldsymbol{X}} = \boldsymbol{Z}(\omega)^{-1}\,\bar{\boldsymbol{F}} \tag{1.209}$$

Working out the real and imaginary parts, one can calculate the steady-state response $\boldsymbol{x}(t)$. It is important to stress here that for an *N DOF* system the responses may have phase angles other than 0° or 180°, even in an undamped situation. It depends on the phase angles between the various components of the forcing vector.

If the applied force is aperiodic, then the solution of each of Eqs. (1.206) can be sought using Duhamel's integral (see Section 1.4.5, Eq. (1.126)). Afterwards, one applies Eq. (1.192), $\boldsymbol{x}(t) = \boldsymbol{\Psi}\boldsymbol{p}(t)$ to retrieve the responses at the physical coordinates.

Orthogonality Properties and Normalisation of the Mode Shapes

One has just seen that the modal mass and modal stiffness matrices are diagonal:

$$\begin{aligned}\mathcal{M} &= \boldsymbol{\Psi}^T \boldsymbol{M} \boldsymbol{\Psi} \\ \mathcal{K} &= \boldsymbol{\Psi}^T \boldsymbol{K} \boldsymbol{\Psi}\end{aligned} \tag{1.210}$$

This means that for each mode $\boldsymbol{\psi}_r$, $m_r = \boldsymbol{\psi}_r^T \boldsymbol{M} \boldsymbol{\psi}_r$ and $k_r = \boldsymbol{\psi}_r^T \boldsymbol{K} \boldsymbol{\psi}_r$, and that any other elements outside the diagonal are zero: $m_{rs} = 0$ and $k_{rs} = 0$ for $s \neq r$. Thus, if one takes two distinct mode shapes, $\boldsymbol{\psi}_r$ and $\boldsymbol{\psi}_s$, $\boldsymbol{\psi}_r^T \boldsymbol{M} \boldsymbol{\psi}_s = 0$ and $\boldsymbol{\psi}_r^T \boldsymbol{K} \boldsymbol{\psi}_s = 0$. In summary,

$$\begin{matrix}\boldsymbol{\psi}_r^T \boldsymbol{M} \boldsymbol{\psi}_s \\ \boldsymbol{\psi}_r^T \boldsymbol{K} \boldsymbol{\psi}_s\end{matrix} \Rightarrow \begin{cases} = 0 & \text{for } s \neq r \\ \neq 0 & \text{for } s = r \end{cases} \tag{1.211}$$

These are the orthogonality properties, i.e. distinct mode shapes are orthogonal with respect to the mass and stiffness matrices. This is the generalisation of the common orthogonality between two vectors $\boldsymbol{x}$ and $\boldsymbol{y}$, which can be written as $\boldsymbol{x}^T \boldsymbol{y} = 0$ and interpreted as being orthogonal with respect to the identity matrix $\boldsymbol{I}$ ($\boldsymbol{x}^T \boldsymbol{I} \boldsymbol{y} = 0$).

As mentioned before, the mode shapes need an extra condition to be estimated uniquely. This is achieved through a normalisation, i.e. through the imposition of a certain measure for the size of the vector. One of the most used and useful normalisations is with respect to the mass matrix. For a common vector, one usually imposes a unit length, which is obtained by dividing it by its initial size. For example, if $\boldsymbol{x} \equiv \{1 \;\; ; \;\; 1\}$, its size is $\sqrt{\boldsymbol{x}^T \boldsymbol{x}} = \sqrt{2}$, so one divides each component by $\sqrt{2}$ and obtain $\{1/\sqrt{2} \;\; ; \;\; 1/\sqrt{2}\}$. This can be seen as a normalisation with respect to the unit matrix $\boldsymbol{I}$. For the mode shape, one generalises the normalisation, doing it with respect to the mass matrix $\boldsymbol{M}$, by dividing the mode shape vector $\boldsymbol{\psi}_r$ by $\sqrt{\boldsymbol{\psi}_r^T \boldsymbol{M} \boldsymbol{\psi}_r}$. Therefore, each normalised mode shape is given by:

$$\boldsymbol{\phi}_r = \frac{\boldsymbol{\psi}_r}{\sqrt{\boldsymbol{\psi}_r^T \boldsymbol{M} \boldsymbol{\psi}_r}} = \frac{\boldsymbol{\psi}_r}{\sqrt{m_r}} \tag{1.212}$$

By doing so, $\boldsymbol{\phi}_r^T \boldsymbol{M} \boldsymbol{\phi}_r = 1$. Let $\boldsymbol{\Phi}$ be the mass-normalised modal matrix; if instead of $\boldsymbol{x}(t) = \boldsymbol{\Psi} \boldsymbol{p}(t)$ one uses the transformation $\boldsymbol{x}(t) = \boldsymbol{\Phi} \boldsymbol{p}(t)$ between the physical and principal coordinates, it turns out that:

$$\boldsymbol{\Phi}^T \boldsymbol{M} \boldsymbol{\Phi} \ddot{\boldsymbol{p}} + \boldsymbol{\Phi}^T \boldsymbol{K} \boldsymbol{\Phi} \boldsymbol{p} = \boldsymbol{\Phi}^T \boldsymbol{f}(t) \tag{1.213}$$

or

$$\ddot{\boldsymbol{p}} + \boldsymbol{\Lambda} \boldsymbol{p} = \boldsymbol{\Phi}^T \boldsymbol{f}(t) \tag{1.214}$$

with $\boldsymbol{\Lambda} = \boldsymbol{\Phi}^T \boldsymbol{K} \boldsymbol{\Phi}$. As $\boldsymbol{\Phi}^T \boldsymbol{M} \boldsymbol{\Phi} = \boldsymbol{I}$, each element m_r is equal to 1 and from Eqs. (1.207) it follows that each element k_r of $\boldsymbol{\Lambda}$ becomes equal to ω_r^2. Therefore, each of the Eqs. (1.206) turns out to be:

$$\ddot{p}_r + \omega_r^2 \, p_r = \mathcal{P}_r(t) \qquad r = 1, \ldots N \tag{1.215}$$

where $\mathcal{P}_r(t)$ is now an element of $\mathcal{P}(t) = \boldsymbol{\Phi}^T \boldsymbol{f}(t)$. Attention is called to the fact that, although the same notation has been kept, the solutions $p_r(t)$ from Eq. (1.215) are different from the ones obtained with Eq. (1.206), as the coordinate transformations ($\boldsymbol{x}(t) = \boldsymbol{\Psi}\,\boldsymbol{p}(t)$ and $\boldsymbol{x}(t) = \boldsymbol{\Phi}\,\boldsymbol{p}(t)$) are different.

Undamped Free Response using the Orthogonality Properties

It has been shown that the undamped free vibration response is given by Eq. (1.190); assuming that the modes are mass-normalised, Eq. (1.190) can be recast as:

$$\boldsymbol{x}(t) = \sum_{r=1}^{N} C_r \boldsymbol{\phi}_r \, cos(\omega_r t - \varphi_r) \tag{1.216}$$

The *2N* constants, C_r and φ_r, can be evaluated from the application of the initial conditions of displacement and velocity, leading to a linear system of equations. However, one can take advantage of the orthogonality conditions to obtain the free response in a much easier and direct way. One begins by applying the initial conditions $\boldsymbol{x}(0) = \boldsymbol{x}_0$ and $\dot{\boldsymbol{x}}(0) = \dot{\boldsymbol{x}}_0$:

$$\boldsymbol{x}_0 = \sum_{r=1}^{N} C_r \boldsymbol{\phi}_r \, cos\varphi_r \tag{1.217-a}$$

$$\dot{\boldsymbol{x}}_0 = \omega_r \sum_{r=1}^{N} C_r \boldsymbol{\phi}_r \, sin\varphi_r \tag{1.217-b}$$

Taking Eq. (1.217-a) and pre-multiplying each side by $\boldsymbol{\phi}_s^T \boldsymbol{M}$, where $\boldsymbol{\phi}_s$ is one of the modes $\boldsymbol{\phi}_r$, leads to

$$\boldsymbol{\phi}_s^T \boldsymbol{M}\, \boldsymbol{x}_0 = \sum_{r=1}^{N} C_r \boldsymbol{\phi}_s^T \boldsymbol{M}\, \boldsymbol{\phi}_r \, cos\varphi_r \tag{1.218}$$

According to the orthogonality conditions and to the mass-normalisation of the mode shapes, all the terms of the summation will be zero except when *r* = *s*, where $\boldsymbol{\phi}_r^T \boldsymbol{M}\, \boldsymbol{\phi}_r = 1$. In this way, one gets rid of the summation and the result is:

$$C_r \, cos\varphi_r = \boldsymbol{\phi}_r^T \boldsymbol{M}\, \boldsymbol{x}_0 \tag{1.219}$$

In a similar way, one can work on Eq. (1.217-b), to obtain:

$$C_r \, sin\,\varphi_r = \frac{1}{\omega_r} \boldsymbol{\phi}_r^T \boldsymbol{M}\, \dot{\boldsymbol{x}}_0 \tag{1.220}$$

Returning to Eq. (1.216), one can write it as:

$$\boldsymbol{x}(t)=\sum_{r=1}^{N} C_r\,\boldsymbol{\phi}_r\left(\cos\omega_r t\cos\varphi_r+\sin\omega_r t\sin\varphi_r\right) \tag{1.221}$$

or

$$\boldsymbol{x}(t)=\sum_{r=1}^{N}\left(C_r\cos\varphi_r\,\boldsymbol{\phi}_r\cos\omega_r t+C_r\sin\varphi_r\,\boldsymbol{\phi}_r\sin\omega_r t\right) \tag{1.222}$$

Substituting (1.219) and (1.220) in (1.222), yields:

$$\boldsymbol{x}(t)=\sum_{r=1}^{N}\left(\boldsymbol{\phi}_r^T\boldsymbol{M}\,\boldsymbol{x}_0\,\boldsymbol{\phi}_r\cos\omega_r t+\frac{1}{\omega_r}\boldsymbol{\phi}_r^T\boldsymbol{M}\,\dot{\boldsymbol{x}}_0\,\boldsymbol{\phi}_r\sin\omega_r t\right) \tag{1.223}$$

Rearranging (1.223), the final result is:

$$\boldsymbol{x}(t)=\sum_{r=1}^{N}\boldsymbol{\phi}_r^T\boldsymbol{M}\left(\boldsymbol{x}_0\cos\omega_r t+\dot{\boldsymbol{x}}_0\frac{1}{\omega_r}\sin\omega_r t\right)\boldsymbol{\phi}_r \tag{1.224}$$

This expression allows to obtain the free vibration response of an *MDOF* system due to some initial conditions, in a direct way, i.e. including those conditions in the expression, without having to solve a linear system of equations to calculate the constants. Note the similarities with the expression for the free vibration of an *SDOF* system (see Eq. (1.50)).

1.5.3 Viscously Damped Vibration Response

Natural Frequencies and Mode Shapes

In free vibration the equilibrium equation is

$$\boldsymbol{M}\,\ddot{\boldsymbol{x}}+\boldsymbol{C}\,\dot{\boldsymbol{x}}+\boldsymbol{K}\,\boldsymbol{x}=\boldsymbol{0} \tag{1.225}$$

If one applies the same transformation as before, $\boldsymbol{x}(t)=\boldsymbol{\Phi}\boldsymbol{p}(t)$, it follows that

$$\ddot{\boldsymbol{p}}+\boldsymbol{\Phi}^T\boldsymbol{C}\boldsymbol{\Phi}\,\dot{\boldsymbol{p}}+\boldsymbol{\Lambda}\boldsymbol{p}=\boldsymbol{0} \tag{1.226}$$

or

$$\ddot{\boldsymbol{p}}+\mathcal{C}\,\dot{\boldsymbol{p}}+\boldsymbol{\Lambda}\boldsymbol{p}=\boldsymbol{0} \tag{1.227}$$

with $\mathcal{C}=\boldsymbol{\Phi}^T\boldsymbol{C}\boldsymbol{\Phi}$, known as *modal damping matrix*. In general, it will not be diagonal, as the mode shapes have been calculated from the free undamped solution. However, it may make sense to admit that the damping matrix is directly proportional to the stiffness and/or mass matrix, i.e. $\boldsymbol{C}=\alpha\boldsymbol{K}+\beta\boldsymbol{M}$. For instance, $\boldsymbol{C}=\alpha\boldsymbol{K}$ may be interpreted as "if a spring has a stiffness k and another has a stiffness $2k$, then if the first one has an associated damping c, the second one will have a damping $2c$". If it is reasonable to assume this kind of damping, known as *proportional* or *Rayleigh damping*, then

$$\mathcal{C}=\boldsymbol{\Phi}^T\left(\alpha\boldsymbol{K}+\beta\boldsymbol{M}\right)\boldsymbol{\Phi}=\alpha\,\boldsymbol{\Phi}^T\boldsymbol{K}\boldsymbol{\Phi}+\beta\,\boldsymbol{\Phi}^T\boldsymbol{M}\boldsymbol{\Phi}=\alpha\,\boldsymbol{\Lambda}+\beta\,\boldsymbol{I} \tag{1.228}$$

C is clearly diagonal and one has all the equations (1.227) as N *SDOF* equations. This also means that if one can assume proportional damping, than the mode shapes of the undamped case are the same for the damped case, i.e. one still has real mode shapes with phases 0° or 180° among the various coordinates.

If for some reason it is not reasonable to assume proportional damping, as it may happen if the system has clearly some localised dissipating zones not evenly distributed, than two things may happen: either the damping still has little expression and it may be possible to ignore small off-diagonal terms in the modal damping matrix or the system is "heavily" damped and one must include the damping matrix in the (now complex) eigenvalue problem. Thus, when looking for the free vibration solution of

$$\boldsymbol{M}\,\ddot{\boldsymbol{x}}+\boldsymbol{C}\,\dot{\boldsymbol{x}}+\boldsymbol{K}\,\boldsymbol{x}=\boldsymbol{0} \tag{1.229}$$

which is of the type $\boldsymbol{x}(t)=\boldsymbol{\psi}e^{st}$, the following complex eigenvalue problem is reached:

$$\left(\boldsymbol{M}\,s^2+\boldsymbol{C}\,s+\boldsymbol{K}\right)\boldsymbol{\psi}=\boldsymbol{0} \tag{1.230}$$

Usually this is solved introducing a state vector $\boldsymbol{u}(t)$:

$$\boldsymbol{u}(t)=\begin{Bmatrix}\boldsymbol{x}(t)\\ \dot{\boldsymbol{x}}(t)\end{Bmatrix}=\begin{Bmatrix}\boldsymbol{\psi}\\ s\boldsymbol{\psi}\end{Bmatrix}e^{st} \;\Rightarrow\; \dot{\boldsymbol{u}}(t)=\begin{Bmatrix}\dot{\boldsymbol{x}}(t)\\ \ddot{\boldsymbol{x}}(t)\end{Bmatrix}=\begin{Bmatrix}s\boldsymbol{\psi}\\ s^2\boldsymbol{\psi}\end{Bmatrix}e^{st}=s\,\boldsymbol{u}(t) \tag{1.231}$$

With this new variable, Eq. (1.229) is reformulated as:

$$\begin{bmatrix}\boldsymbol{C} & \boldsymbol{M}\\ \boldsymbol{M} & \boldsymbol{0}\end{bmatrix}\dot{\boldsymbol{u}}(t)+\begin{bmatrix}\boldsymbol{K} & \boldsymbol{0}\\ \boldsymbol{0} & -\boldsymbol{M}\end{bmatrix}\boldsymbol{u}(t)=\boldsymbol{0} \tag{1.232}$$

or simply

$$\boldsymbol{A}\dot{\boldsymbol{u}}(t)+\boldsymbol{B}\boldsymbol{u}(t)=\boldsymbol{0} \tag{1.233}$$

In this formulation (called state-space analysis) $\boldsymbol{A}$ and $\boldsymbol{B}$ are symmetric matrices of order $2N$, which means that now one has doubled the number of equations, but on the other hand the second order problem has been reduced to a first order one. Substituting Eqs. (1.231) in (1.233) leads to:

$$\left(s\,\boldsymbol{A}+\boldsymbol{B}\right)\begin{Bmatrix}\boldsymbol{\psi}\\ s\boldsymbol{\psi}\end{Bmatrix}=\boldsymbol{0} \tag{1.234}$$

For an underdamped system the $2N$ eigenvalues appear in complex conjugate pairs, s_r and s_r^*, similarly to what happened in the *SDOF* case (see Section 1.4.4, Eq. (1.59)). The same happens with the eigenvectors, they also appear in complex conjugate pairs, $\boldsymbol{\psi}_r$ and $\boldsymbol{\psi}_r^*$. The state-space eigenvectors are $\boldsymbol{\psi}_r'=\begin{Bmatrix}\boldsymbol{\psi}_r\\ s_r\boldsymbol{\psi}_r\end{Bmatrix}$ and $\boldsymbol{\psi}_r'^*=\begin{Bmatrix}\boldsymbol{\psi}_r^*\\ s_r^*\boldsymbol{\psi}_r^*\end{Bmatrix}$, respectively. However, the system still has

N degrees of freedom. From the real and imaginary parts of each complex eigenvalue $s_r = -\xi_r \omega_r + i\omega_r \sqrt{1-\xi_r^2}$, one can evaluate each natural frequency and damping factor, as

$$\omega_r = \sqrt{\left(Re(s_r)\right)^2 + \left(Im(s_r)\right)^2}$$
$$\xi_r = -\frac{Re(s_r)}{\omega_r} \tag{1.235}$$

Each mode shape $\boldsymbol{\psi}_r$ is complex, which means that its amplitudes have an associated phase angle that is no longer 0º or 180º, i.e. its coordinates do not reach their maxima or minima at the same time instant; as a consequence, the mode shape motion tends to wave, which can be observed in a computer screen, although naturally a static representation as in Figure 1.43 is no longer possible. Another consequence of the mode shape complexity is that the nodes or the nodal lines do not remain fixed anymore.

In terms of the state-space variables, the transformation to the principal coordinates is now $\boldsymbol{u}(t) = \boldsymbol{\Psi}'\boldsymbol{q}(t)$, where $\boldsymbol{\Psi}'$ is the state-space modal matrix, formed by $\boldsymbol{\psi}'_r$ and $\boldsymbol{\psi}'_r{}^*$. Eq. (1.233) becomes:

$$\boldsymbol{\Psi}'^T \boldsymbol{A} \boldsymbol{\Psi}' \dot{\boldsymbol{q}}(t) + \boldsymbol{\Psi}'^T \boldsymbol{B} \boldsymbol{\Psi}' \boldsymbol{q}(t) = \boldsymbol{0} \tag{1.236}$$

or

$$\boldsymbol{a}\dot{\boldsymbol{q}}(t) + \boldsymbol{b}\boldsymbol{q}(t) = \boldsymbol{0} \tag{1.237}$$

where $\boldsymbol{a}$ and $\boldsymbol{b}$ are diagonal matrices, due to the orthogonality properties. As a consequence, $\boldsymbol{\Psi}^H \boldsymbol{M} \boldsymbol{\Psi}$, $\boldsymbol{\Psi}^H \boldsymbol{C} \boldsymbol{\Psi}$ and $\boldsymbol{\Psi}^H \boldsymbol{K} \boldsymbol{\Psi}$ (^{H} stands for Hermitian transpose) are all diagonal.

Free Response

Eq. (1.237) represents a set of $2N$ uncoupled equations. Each solution is of the form $q_r(t) = \bar{Q}_r e^{s_r t}$ with $s_r = -b_r/a_r$ and so $\boldsymbol{u}(t) = \boldsymbol{\Psi}'\boldsymbol{q}(t)$ may be written as

$$\boldsymbol{u}(t) = \sum_{r=1}^{2N} \boldsymbol{\psi}'_r \bar{Q}_r e^{s_r t} \tag{1.238}$$

which is the equivalent to Eq. (1.190); although with damping and in the space-state formulation, one can still say that the free vibration response is the linear combination of the responses of each mode shape. The $2N$ coefficients $\bar{Q}_r$ depend on the initial conditions and are weighting factors representing the contribution of each mode shape to the total response; they are usually known as *modal participation factors*.

Forced Response

In forced vibration, one has to solve

$$\boldsymbol{M}\ddot{\boldsymbol{x}} + \boldsymbol{C}\dot{\boldsymbol{x}} + \boldsymbol{K}\boldsymbol{x} = \boldsymbol{f}(t) \tag{1.239}$$

As in the undamped case, when the force is harmonic, $f(t)=\bar{F}e^{i\omega t}$, the steady-state response is also harmonic, $x(t)=\bar{X}e^{i\omega t}$, and there is no need to uncouple the equations, one can use the *Direct Method*, leading to $\bar{X}=Z(\omega)^{-1}\bar{F}$, as in Eq. (1.209), but where the dynamic stiffness matrix is now given by

$$Z(\omega)=K-\omega^2 M+i\omega C \tag{1.240}$$

and the receptance matrix is its inverse:

$$H(\omega)=Z(\omega)^{-1} \tag{1.241}$$

However, it is difficult to know the damping matrix and thus it is convenient to obtain the receptance in terms of the modal properties. To do that, one returns to the state-space formulation and start with:

$$A\dot{u}(t)+Bu(t)=f'(t) \tag{1.242}$$

where $f'(t)$ is a $2N\times 1$ vector given by $\begin{Bmatrix} f(t) \\ 0 \end{Bmatrix}$. Applying again the transformation $u(t)=\Psi' q(t)$ to Eq. (1.242) and pre-multiplying by Ψ'^T, gives

$$\Psi'^T A\Psi'\dot{q}(t)+\Psi'^T B\Psi' q(t)=\Psi'^T f'(t) \tag{1.243}$$

or

$$a\dot{q}(t)+bq(t)=\Psi'^T f'(t) \tag{1.244}$$

As before, a and b are diagonal matrices, due to the orthogonality properties. Eq. (1.244) is a set of *2N* equations, each one being

$$\dot{q}_r(t)-s_r q_r(t)=\frac{1}{a_r}\psi_r'^T\begin{Bmatrix} f(t) \\ 0 \end{Bmatrix} \tag{1.245}$$

where $s_r=-b_r/a_r=-\xi_r\omega_r+i\omega_r\sqrt{1-\xi_r^2}$. Under a harmonic force $f'(t)=\begin{Bmatrix} \bar{F} \\ 0 \end{Bmatrix}e^{i\omega t}$, the steady-state response will be $q(t)=\bar{Q}e^{i\omega t}$ and thus, each equation (1.245) becomes

$$(i\omega-s_r)\bar{Q}_r=\frac{1}{a_r}\psi_r'^T\begin{Bmatrix} \bar{F} \\ 0 \end{Bmatrix} \tag{1.246}$$

and therefore,

$$\bar{Q}_r=\left(\frac{1}{i\omega-s_r}\right)\frac{1}{a_r}\psi_r'^T\begin{Bmatrix} \bar{F} \\ 0 \end{Bmatrix} \tag{1.247}$$

Recalling Eq. (1.238), now with $i\omega$ instead of s_r, one may write

$$\boldsymbol{u}(t) = \sum_{r=1}^{2N} \boldsymbol{\psi}'_r \, \bar{Q}_r \, e^{i\omega t} \tag{1.248}$$

Substituting (1.247) in (1.248), it follows that

$$\boldsymbol{u}(t) = \sum_{r=1}^{2N} \boldsymbol{\psi}'_r \left(\frac{1}{i\omega - s_r} \right) \frac{1}{a_r} \boldsymbol{\psi}'^{\,T}_r \begin{Bmatrix} \bar{\boldsymbol{F}} \\ \boldsymbol{0} \end{Bmatrix} e^{i\omega t} \tag{1.249}$$

Normalising each eigenvector $\boldsymbol{\psi}'_r$ with respect to a_r, $\boldsymbol{\phi}'_r = \boldsymbol{\psi}'_r / \sqrt{a_r}$ and thus,

$$\boldsymbol{u}(t) = \sum_{r=1}^{2N} \boldsymbol{\phi}'_r \left(\frac{1}{i\omega - s_r} \right) \boldsymbol{\phi}'^{\,T}_r \begin{Bmatrix} \bar{\boldsymbol{F}} \\ \boldsymbol{0} \end{Bmatrix} e^{i\omega t} \tag{1.250}$$

As $\boldsymbol{\phi}'_r = \begin{Bmatrix} \boldsymbol{\phi}_r \\ s_r \boldsymbol{\phi}_r \end{Bmatrix}$ and $\boldsymbol{u}(t) = \begin{Bmatrix} \boldsymbol{x}(t) \\ \dot{\boldsymbol{x}}(t) \end{Bmatrix} = \begin{Bmatrix} \bar{\boldsymbol{X}} \\ i\omega \bar{\boldsymbol{X}} \end{Bmatrix} e^{i\omega t}$, Eq. (1.250) becomes

$$\begin{Bmatrix} \bar{\boldsymbol{X}} \\ i\omega \bar{\boldsymbol{X}} \end{Bmatrix} = \sum_{r=1}^{2N} \begin{Bmatrix} \boldsymbol{\phi}_r \\ s_r \boldsymbol{\phi}_r \end{Bmatrix} \left(\frac{1}{i\omega - s_r} \right) \begin{Bmatrix} \boldsymbol{\phi}_r \\ s_r \boldsymbol{\phi}_r \end{Bmatrix}^T \begin{Bmatrix} \bar{\boldsymbol{F}} \\ \boldsymbol{0} \end{Bmatrix} \tag{1.251}$$

From Eq. (1.251), the amplitude $\bar{\boldsymbol{X}}$ is:

$$\bar{\boldsymbol{X}} = \sum_{r=1}^{2N} \frac{\boldsymbol{\phi}_r \left(\boldsymbol{\phi}_r^T \bar{\boldsymbol{F}} \right)}{i\omega - s_r} \tag{1.252}$$

where $\boldsymbol{\phi}_r^T \bar{\boldsymbol{F}}$ represents the rth *modal participation factor*. The receptance *FRF* relating the displacement response at coordinate *j* with an excitation force at coordinate *k* is therefore given by

$$H_{jk}(\omega) = \frac{\bar{X}_j}{F_k} = \sum_{r=1}^{2N} \frac{\phi_{jr} \, \phi_{kr}}{i\omega - s_r} \tag{1.253}$$

As the eigenvalues and eigenvectors come in complex conjugate pairs, (1.253) can be written as:

$$H_{jk}(\omega) = \sum_{r=1}^{N} \left(\frac{\phi_{jr} \, \phi_{kr}}{i\omega - s_r} + \frac{\phi^*_{jr} \, \phi^*_{kr}}{i\omega - s^*_r} \right) \tag{1.254}$$

The eigenvalues s_r are the poles and $\phi_{jr} \, \phi_{kr}$ are the residues. Denoting the residues as ${}_r A_{jk}$, Eq. (1.254) can still be written as:

$$H_{jk}(\omega)=\sum_{r=1}^{N}\left(\frac{{}_rA_{jk}}{\xi_r\,\omega_r+i\left(\omega-\omega_r\sqrt{1-\xi_r^2}\right)}+\frac{{}_rA_{jk}^{*}}{\xi_r\,\omega_r+i\left(\omega+\omega_r\sqrt{1-\xi_r^2}\right)}\right) \tag{1.255}$$

1.5.4 Hysteretically Damped Vibration Response

Natural Frequencies and Mode Shapes

In free vibration the equilibrium equation is now written as

$$\boldsymbol{M}\,\ddot{\boldsymbol{x}}+i\,\boldsymbol{D}\boldsymbol{x}+\boldsymbol{K}\,\boldsymbol{x}=\boldsymbol{0} \tag{1.256}$$

where $\boldsymbol{D}$ is the hysteretic damping matrix. Physically, it does not make sense to look for the free vibration response of such a model, as it would be a complex response. However, mathematically, Eq. (1.256) can be solved, the solution being of the type $\boldsymbol{x}(t)=\boldsymbol{\psi}\,e^{i\lambda t}$, leading to a complex eigenproblem:

$$\left(\boldsymbol{K}+i\,\boldsymbol{D}-\lambda^2\boldsymbol{M}\right)\boldsymbol{\psi}=\boldsymbol{0} \tag{1.257}$$

If the damping is assumed as proportional, $\boldsymbol{D}=\alpha\boldsymbol{K}+\beta\boldsymbol{M}$, and so

$$\left(\boldsymbol{K}-\lambda^2\boldsymbol{M}+i\left(\alpha\boldsymbol{K}+\beta\boldsymbol{M}\right)\right)\boldsymbol{\psi}=\boldsymbol{0} \tag{1.258}$$

The N eigenvalues λ_r^2 will be complex, but the N eigenvectors $\boldsymbol{\psi}_r$ will be real, in fact the same as for the undamped case. As the orthogonality properties hold, from Eq. (1.258) the eigenvalues can be written in the following form:

$$\lambda_r^2=\omega_r^2\left(1+i\,\eta_r\right) \tag{1.259}$$

where each natural frequency and damping factor are given, respectively, by:

$$\omega_r=\sqrt{k_r/m_r}\quad\text{and}\quad\eta_r=\alpha+\beta/\omega_r^2 \tag{1.260}$$

In the more general non-proportional case, both the eigenvalues and eigenvectors will be complex and although λ_r^2 might be written as in Eq. (1.259), the value of ω_r^2 may be slightly different, as in this case one will take ω_r^2 as the real part of λ_r^2. Due to the orthogonal properties, λ_r^2 is given by:

$$\lambda_r^2=\frac{\boldsymbol{\psi}_r^T\left(\boldsymbol{K}+i\,\boldsymbol{D}\right)\boldsymbol{\psi}_r}{\boldsymbol{\psi}_r^T\boldsymbol{M}\,\boldsymbol{\psi}_r}=\frac{k_r}{m_r} \tag{1.261}$$

where k_r and m_r are now complex quantities.

Forced Response

In forced vibration, one has to solve the following equation:

$$\boldsymbol{M}\ddot{\boldsymbol{x}}+i\boldsymbol{D}\boldsymbol{x}+\boldsymbol{K}\boldsymbol{x}=\boldsymbol{f}(t) \tag{1.262}$$

For a harmonic force $\boldsymbol{f}(t)=\bar{\boldsymbol{F}}e^{i\omega t}$, the steady-state response will be $\boldsymbol{x}(t)=\bar{\boldsymbol{X}}e^{i\omega t}$ and thus,

$$\left(\boldsymbol{K}-\omega^2\boldsymbol{M}+i\boldsymbol{D}\right)\bar{\boldsymbol{X}}=\bar{\boldsymbol{F}} \tag{1.263}$$

The receptance matrix is $\boldsymbol{H}(\omega)=\left(\boldsymbol{K}-\omega^2\boldsymbol{M}+i\boldsymbol{D}\right)^{-1}$. Now, one shall deduce the expression of each receptance element $H_{jk}(\omega)$, as it was done in Eq. (1.255) for the viscously damped case. As the eigenvectors $\boldsymbol{\psi}_r$ are independent in an N dimensional space, any vector in that space, like $\bar{\boldsymbol{X}}$, can be expressed as a linear combination of the eigenvectors, i.e.

$$\bar{\boldsymbol{X}}=\sum_{r=1}^{N}\gamma_r\boldsymbol{\psi}_r \tag{1.264}$$

Substituting in Eq. (1.263) and pre-multiplying by $\boldsymbol{\psi}_s^T$, leads to

$$\boldsymbol{\psi}_s^T\left(\boldsymbol{K}+i\boldsymbol{D}\right)\sum_{r=1}^{N}\gamma_r\boldsymbol{\psi}_r-\omega^2\boldsymbol{\psi}_s^T\boldsymbol{M}\sum_{r=1}^{N}\gamma_r\boldsymbol{\psi}_r=\boldsymbol{\psi}_s^T\bar{\boldsymbol{F}} \tag{1.265}$$

Due to the orthogonality properties (Eqs. (1.211)) and to Eq. (1.261), it follows that

$$\gamma_r\mathscr{k}_r-\omega^2\gamma_r m_r=\boldsymbol{\psi}_r^T\bar{\boldsymbol{F}} \tag{1.266}$$

Thus,

$$\gamma_r=\frac{\boldsymbol{\psi}_r^T\bar{\boldsymbol{F}}}{\mathscr{k}_r-\omega^2 m_r} \tag{1.267}$$

Substituting in Eq. (1.264),

$$\bar{\boldsymbol{X}}=\sum_{r=1}^{N}\frac{\boldsymbol{\psi}_r^T\bar{\boldsymbol{F}}\boldsymbol{\psi}_r}{\mathscr{k}_r-\omega^2 m_r} \tag{1.268}$$

Therefore, the steady-state response will be given by

$$\boldsymbol{x}(t)=\sum_{r=1}^{N}\frac{\boldsymbol{\psi}_r^T\bar{\boldsymbol{F}}\boldsymbol{\psi}_r}{\mathscr{k}_r-\omega^2 m_r}e^{i\omega t} \tag{1.269}$$

From Eqs. (1.259) and (1.261), the complex amplitude can be written as

$$\bar{\boldsymbol{X}}=\sum_{r=1}^{N}\frac{\boldsymbol{\psi}_r^T\bar{\boldsymbol{F}}\boldsymbol{\psi}_r}{m_r\left(\omega_r^2-\omega^2+i\eta_r\omega_r^2\right)} \tag{1.270}$$

Each receptance element $H_{jk}(\omega)$ relating the response at coordinate j to a single force at coordinate k can therefore be given by

$$H_{jk}(\omega) = \frac{\bar{X}_j}{\bar{F}_k} = \sum_{r=1}^{N} \frac{\psi_{jr}\,\psi_{kr}}{m_r\left(\omega_r^2 - \omega^2 + i\eta_r\omega_r^2\right)} \tag{1.271}$$

In terms of the mass-normalised mode shapes (see Eq. (1.212)), it turns out that

$$H_{jk}(\omega) = \frac{\bar{X}_j}{\bar{F}_k} = \sum_{r=1}^{N} \frac{\phi_{jr}\,\phi_{kr}}{\omega_r^2 - \omega^2 + i\eta_r\omega_r^2} \tag{1.272}$$

or simply

$$H_{jk}(\omega) = \frac{\bar{X}_j}{\bar{F}_k} = \sum_{r=1}^{N} \frac{{}_r\bar{A}_{jk}}{\omega_r^2 - \omega^2 + i\eta_r\omega_r^2} \tag{1.273}$$

where ${}_r\bar{A}_{jk}$ is the so-called modal constant, complex in general, but real if the system is undamped or proportionally damped.

1.5.5 Graphical Representation of an *FRF*

The *FRF* may be represented in various ways, often in terms of amplitude and phase, similarly to the quality factor (Figure 1.14). In order to show in a clearer way its entire dynamic range, the magnitude is usually displayed in decibel (dB), i.e. in a relative scale. In a linear scale lower levels of response cannot be observed in detail (see Figure 1.46 (a)). Therefore, in the vertical axis the values are given as

$$|H|(dB) = 20\log_{10}\left(\frac{|H|}{|H|_{ref}}\right) \tag{1.274}$$

where the reference value $|H|_{ref}$ is normally taken as 1 (m/N if it is a receptance). This kind of representation is called a Bode diagram (Figure 1.46-b)).

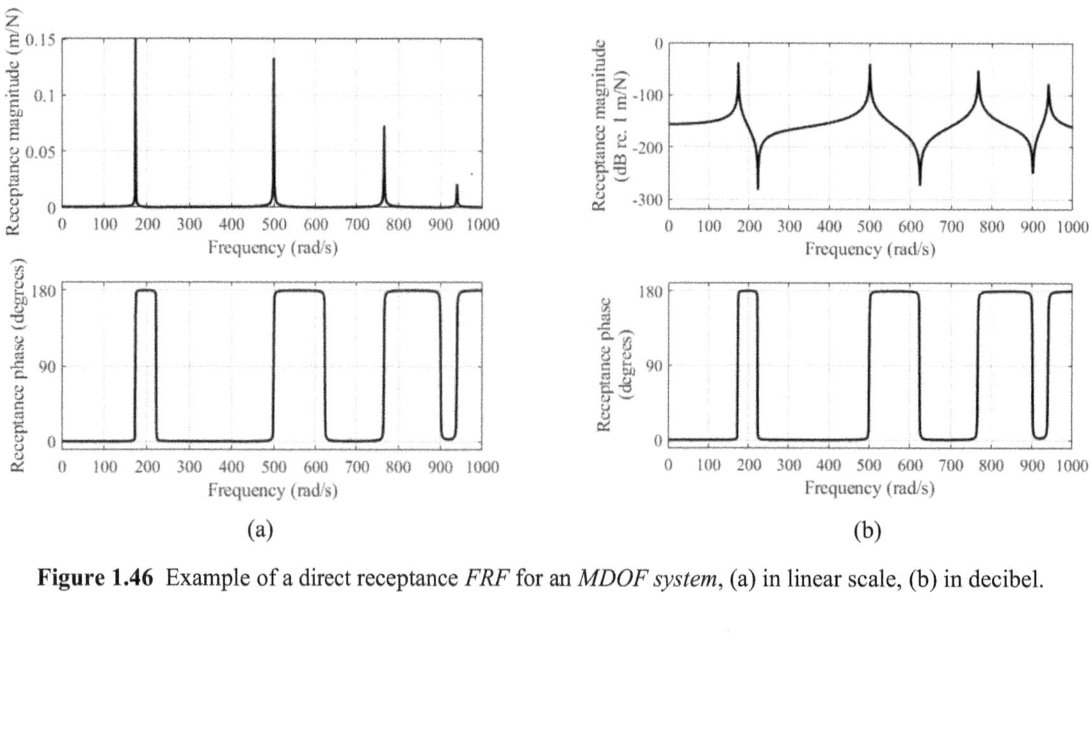

Figure 1.46 Example of a direct receptance *FRF* for an *MDOF system*, (a) in linear scale, (b) in decibel.

Figure 1.46 represents a direct *FRF*, relating the response at a *DOF* with the force at that same *DOF*. In such a type of *FRF* an anti-resonance always shows up between two peaks (resonances). In a transfer *FRF*, relating the response at one *DOF* with the force at another *DOF*, that kind of "rule" does not happen; besides anti-resonances there are also minima, as in Figure 1.47.

Alternatively, one may display the real and the imaginary parts of the *FRF* against frequency (Figure 1.48), $Re(H(\omega))$ and $Im(H(\omega))$, which may be sometimes useful. In such a case the representations must be in a linear scale, as the values may be positive or negative.

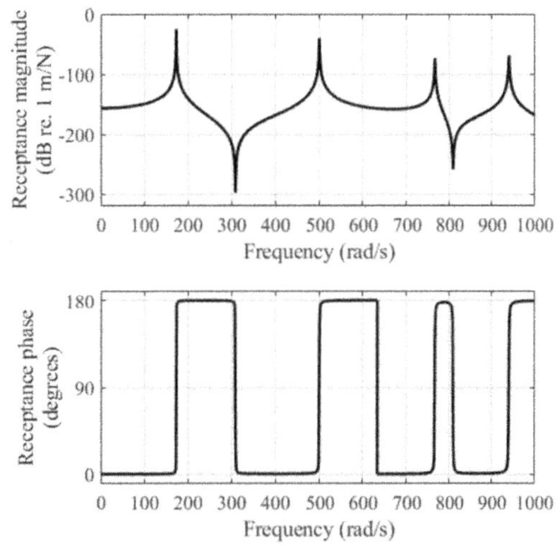

Figure 1.47 Example of a transfer *FRF* (receptance) of an *MDOF* system.

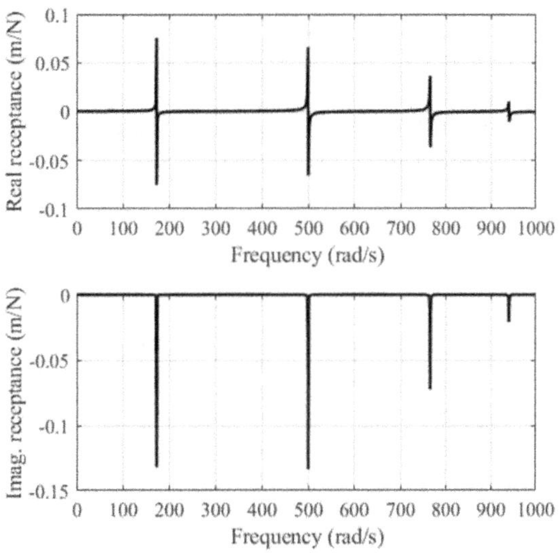

Figure 1.48 Real and Imaginary parts of an *FRF versus* frequency.

Another useful representation, namely for the identification of the dynamic properties of a system, is the so-called *Nyquist plot*, where the imaginary part is displayed against the real part (Figure 1.49). As the frequency increases, the *locus* of the amplitude is a circle around each resonance. One of the advantages is that one does not have to measure the response exactly at the resonance; a proper circle fitting and interpolation techniques can be used to identify the dynamic properties of the system. Sometimes, the display shows the various points linked by straight lines, exhibiting a polygonal aspect. In Chapter 4 this subject is discussed in more detail.

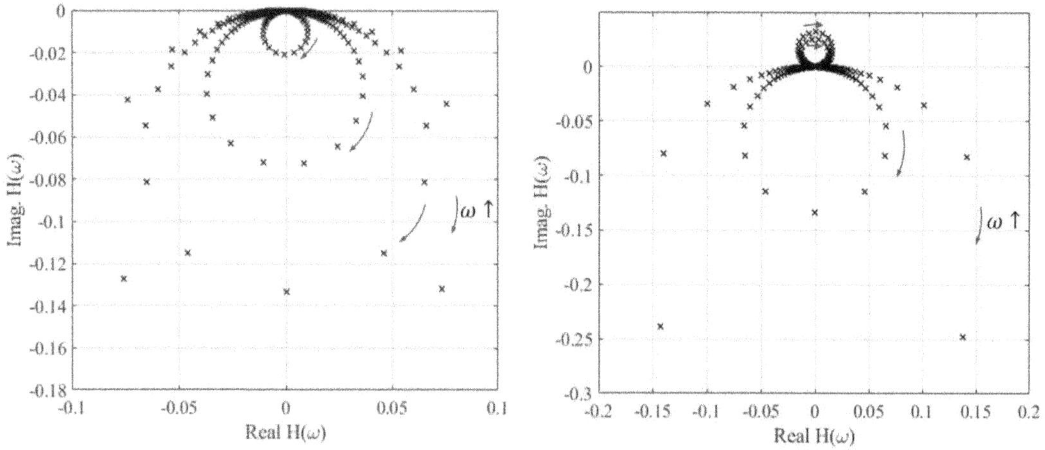

(a) (b)

Figure 1.49 Examples of Nyquist plots, (a) for a direct *FRF*, (b) for a transfer *FRF*.

1.6 Continuous Systems

Continuous systems are those which have their dynamic properties (mass, stiffness and damping) distributed continuously along the volume, in contrast with the discrete systems one has been studying so far, where those properties were concentrated at a finite number of degrees of freedom. If a structure can be studied as a continuous medium, it means that an infinite number of degrees of freedom must be taken into consideration and consequently, it will have an infinite number of natural frequencies and mode shapes. Only structures with reasonably simple geometries can be studied as such. For instance, one is talking about bars in longitudinal vibration, beams in transversal vibration, plates or shells of uniform thickness, etc.

When the structure is very complex, there is no possibility to study it in this way. So why should one care about this? If the structure is really complex, like the body-in-white of an airplane, one must undertake the study by decomposing the structure into substructures, modeling each part separately and then couple everything back together to predict the dynamic behaviour of the ensemble. Most of those substructures will be modeled numerically, others may be so complex that one has to build the models from experimental tests. However, there may be some components that are relatively simple, like bars, beams or plates, where an analytical solution can be used rather than a numerical model, which would bring unnecessary solutions of continuous systems. Here, only two simple cases will be presented, namely the free vibration of uniform [12] bars and of uniform beams. The distinction between a bar and a beam is that a bar is a structural element that resists preferentially to traction or compression, whereas a beam is a structural component that resists mostly to flexural efforts. Therefore, when studying bars one is talking about longitudinal vibrations, whereas for beams one is talking about transversal (or lateral) vibrations.

1.6.1 Free Vibration of Uniform Bars

Consider the uniform bar of Figure 1.50, with constant cross section A, Young modulus E and mass density ρ. The vibration happens along the x axis, where the displacement is $u(x,t)$. One starts by taking an infinitesimal element of length dx, with the internal force P acting on the left section, $P+dP$ on the right section, and the inertia force dF_i (Figure 1.51).

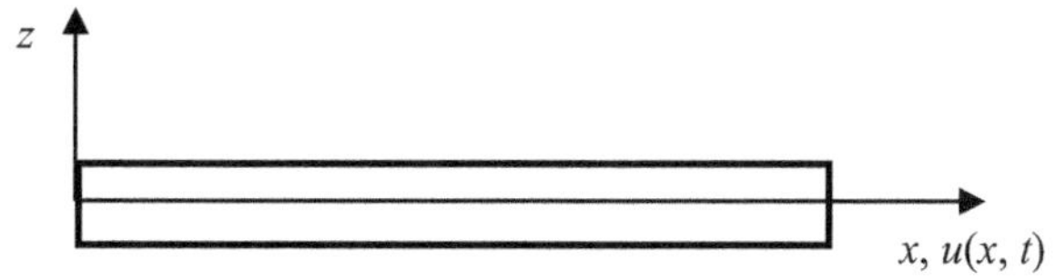

Figure 1.50 Uniform bar.

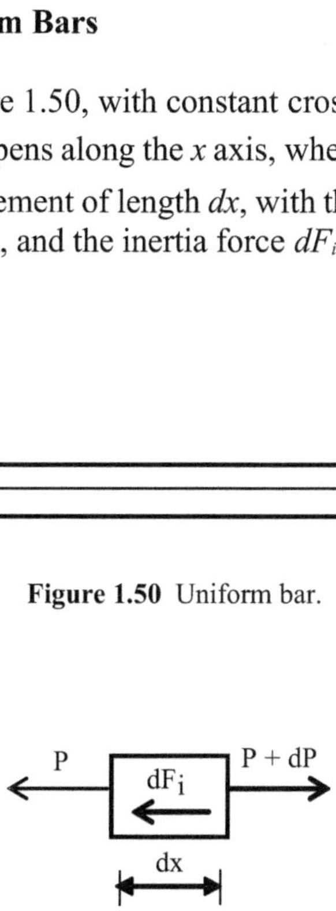

Figure 1.51 Free body diagram of an infinitesimal bar element.

From the force balance, it follows that

$$dP = dF_i \tag{1.275}$$

As $dP = \dfrac{\partial P}{\partial x}dx$ and the elementary inertia force is $dF_i = dm\dfrac{\partial^2 u}{\partial t^2} = \rho A dx\dfrac{\partial^2 u}{\partial t^2}$,

$$\frac{\partial P}{\partial x} = \rho A\frac{\partial^2 u}{\partial t^2} \tag{1.276}$$

The internal force P is the product of the stress with the area, $P = \sigma A$. From Hooke's law, $P = E\varepsilon A = EA\dfrac{\partial u}{\partial x}$. Substituting in (1.276), leads to the so-called wave equation:

---

[12] "Uniform" meaning constant cross section area.

$$c^2 \frac{\partial^2 u}{\partial x^2} = \frac{\partial^2 u}{\partial t^2} \quad \text{, with } c = \sqrt{\frac{E}{\rho}} \tag{1.277}$$

Solving the Equilibrium Equation

As (1.277) is a partial derivative equation, where u is a function of x and t, one seeks the solution using the separation of variables, meaning that u can be given by the product of a function that depends only on x, known as the shape function, by a function that depends only on t: $u(x,t) = \phi(x)W(t)$. Substituting in (1.277),

$$c^2 \frac{\partial^2 \phi(x)}{\partial x^2} W(t) = \phi(x) \frac{\partial^2 W(t)}{\partial t^2} \tag{1.278}$$

The partial derivatives may now turn into total derivatives and Eq. (1.278) may be recast as:

$$c^2 \frac{1}{\phi(x)} \frac{d^2 \phi(x)}{dx^2} = \frac{1}{W(t)} \frac{d^2 W(t)}{dt^2} \tag{1.279}$$

Now it should be noted that the left hand side does not depend on t and that the right hand side does not depend on x. However, both sides are equal and therefore they are both independent of x and t, and consequently equal to a constant. Letting that constant be a leads to two equations, one in $W(t)$, the other in $\phi(x)$:

$$\frac{1}{W(t)} \frac{d^2 W(t)}{dt^2} = a \tag{1.280}$$

$$c^2 \frac{1}{\phi(x)} \frac{d^2 \phi(x)}{dx^2} = a \tag{1.281}$$

From (1.280),

$$\frac{d^2 W(t)}{dt^2} - aW(t) = 0 \tag{1.282}$$

whose solution is of the type $W(t) = Ce^{st}$, where s is Laplace's variable. As $\frac{d^2 W(t)}{dt^2} = s^2 Ce^{st}$, upon substitution in Eq. (1.282), the roots of the algebraic equation are $s = \pm\sqrt{a}$. The solution $W(t)$ will be the linear combination of both roots:

$$W(t) = C_1 e^{\sqrt{a}\,t} + C_2 e^{-\sqrt{a}\,t} \tag{1.283}$$

If the constant a is positive, this expression can alternatively be written in terms of hyperbolic *sines* and *cosines*, which are non-periodic functions and therefore do not correspond physically to a vibratory type of motion. That is only possible if the exponents of the exponential terms are imaginary, where they can be related to *sines* and *cosines* through Euler's formulæ. Thus, a must be negative, say, $a=-d$, d being positive. The roots of the algebraic equation become imaginary: $s=\pm i\sqrt{d}$ and Eq. (1.283) turns out to be

$$W(t)=C_1e^{i\sqrt{d}t}+C_2e^{-i\sqrt{d}t} \tag{1.284}$$

which can now be written as:

$$W(t)=D_1\cos\sqrt{d}t+D_2\sin\sqrt{d}t \tag{1.285}$$

It is clear that $\sqrt{d}$ is a frequency and thus $d=\omega^2$ and $a=-\omega^2$. The time variation is therefore harmonic. One returns to Eq. (1.270), to study the variation in space, i.e. to obtain the shape function $\phi(x)$. Eq. (1.281) can now be recast as:

$$\frac{d^2\phi(x)}{dx^2}+\left(\frac{\omega}{c}\right)^2\phi(x)=0 \tag{1.286}$$

The solution is again of the type $\phi(x)=Ce^{sx}$, leading to

$$s^2+\left(\frac{\omega}{c}\right)^2=0 \tag{1.287}$$

whose roots are:

$$s=\pm i\frac{\omega}{c} \tag{1.288}$$

Similarly to $W(t)$, the shape function will be given by:

$$\phi(x)=C_1\cos\frac{\omega}{c}x+C_2\sin\frac{\omega}{c}x \tag{1.289}$$

The shape function is also harmonic, in space. As the wave equation is of the second order in *x* and in *t*, it will be necessary to introduce two boundary conditions and two initial conditions to obtain the free vibration response of the bar. Applying the boundary conditions leads to the natural frequencies, and from the shape function one obtains the mode shapes.

Natural Frequencies and Mode Shapes

To understand the procedure, one shall follow the example of a bar with both ends free and length *l*. As in this case both ends have no applied forces, the stress (σ) is zero and by Hooke's law ($\sigma=E\varepsilon=E\frac{\partial u}{\partial x}$), the strain is also zero. Thus, the boundary conditions at both ends are:

$$\left.\frac{\partial u}{\partial x}\right|_{x=0} = \left.\frac{\partial u}{\partial x}\right|_{x=l} = 0 \tag{1.290}$$

As $u(x,t) = \phi(x)W(t)$, it follows that:

$$\left.\frac{d\phi(x)}{dx}\right|_{x=0} W(t) = 0, \quad \left.\frac{d\phi(x)}{dx}\right|_{x=l} W(t) = 0 \tag{1.291}$$

As the boundary conditions do not vary along the time, it turns out that:

$$\left.\frac{d\phi(x)}{dx}\right|_{x=0} = 0, \quad \left.\frac{d\phi(x)}{dx}\right|_{x=l} = 0 \tag{1.292}$$

Differentiating (1.289),

$$\frac{d\phi(x)}{dx} = \frac{\omega}{c}\left(-C_1 sin\frac{\omega}{c}x + C_2 cos\frac{\omega}{c}x\right) \tag{1.293}$$

Applying (1.292),

$$\begin{aligned} &C_2 = 0 \\ &C_1 sin\frac{\omega}{c}l = 0 \end{aligned} \tag{1.294}$$

From (1.294) the natural frequencies are obtained from the roots of $sin\frac{\omega}{c}l = 0$. The null solution, although physically possible due to the motion of the bar as a rigid body, is not interesting in practice. The non-null solutions lead to the natural frequencies:

$$\frac{\omega}{c}l = n\pi \quad \Rightarrow \quad \omega_n = \frac{n\pi c}{l}, \qquad n = 1, 2, \ldots\infty \tag{1.295}$$

Substituting in Eq. (1.289) and because $C_2 = 0$, the mode shapes are given by:

$$\phi_n(x) = C_1^{(n)} cos\frac{n\pi}{l}x, \qquad n = 1, 2, \ldots\infty \tag{1.296}$$

Free Response

For each natural frequency the time response is harmonic:

$$u_n(x,t) = \phi_n(x) W_n(t) \tag{1.297}$$

Having into account (1.285), it turns out that

$$u_n(x,t) = cos\frac{n\pi}{l}x\left(D_1^{(n)} cos\,\omega_n t + D_2^{(n)} sin\,\omega_n t\right) \tag{1.298}$$

The response will be the superposition of the responses of each mode shape:

$$u(x,t)=\sum_{n=1}^{\infty}u_n(x,t)=\sum_{n=1}^{\infty}cos\frac{n\pi}{l}x\left(D_1^{(n)}cos\omega_n t+D_2^{(n)}sin\,\omega_n t\right) \tag{1.299}$$

where $D_1^{(n)}$ and $D_2^{(n)}$ are calculated applying the initial conditions of displacement and velocity, which are:

$$u(x,0)=u_0(x)=\sum_{n=1}^{\infty}D_1^{(n)}cos\frac{n\pi}{l}x \tag{1.300-a}$$

$$\dot{u}(x,0)=\dot{u}_0(x)=\sum_{n=1}^{\infty}\omega_n D_2^{(n)}cos\frac{n\pi}{l}x \tag{1.300-b}$$

The procedure to calculate $D_1^{(n)}$ and $D_2^{(n)}$ is similar to the one used to calculate the coefficients of Fourier series, taking advantage of the orthogonality of the *sine* and *cosine* functions. One shall derive the expression for $D_1^{(n)}$, and then present the result for $D_2^{(n)}$, as the process is entirely similar. Let *m* be one of the *n* terms in (1.300-a). Multiplying both sides by $cos\frac{m\pi}{l}x$ and integrating along the length *l*, it follows that:

$$\int_0^{\ell}u_0(x)cos\frac{m\pi}{l}x\,dx=\sum_{n=1}^{\infty}D_1^{(n)}\int_0^{l}cos\frac{n\pi}{l}x\,cos\frac{m\pi}{l}x\,dx \tag{1.301}$$

Integrating $\int_0^{l}cos\frac{n\pi}{l}x\,cos\frac{m\pi}{l}x\,dx$ twice by parts, leads to:

$$\left(1-\left(\frac{n}{m}\right)^2\right)\int_0^{l}\cos\frac{n\pi}{l}x\cos\frac{m\pi}{l}x\,dx=0 \tag{1.302}$$

When $n\neq m$ the integral is equal to zero. The only chance of being different from zero is when $n=m$, i.e. from all the terms of the summation, only one is non-zero, and thus, in that case one has to solve $\int_0^{l}cos^2\frac{n\pi}{l}x\,dx$, which is equal to $\frac{l}{2}$. Therefore, from Eq. (1.301) $D_1^{(n)}$ can be evaluated:

$$D_1^{(n)}=\frac{2}{l}\int_0^{l}u_0(x)cos\frac{n\pi}{l}x\,dx \tag{1.303}$$

Proceeding in a similar way for $D_2^{(n)}$,

$$D_2^{(n)}=\frac{2}{n\pi c}\int_0^{l}\dot{u}_0(x)cos\frac{n\pi}{l}x\,dx \tag{1.304}$$

1.6.2 Free Vibration of Uniform Beams

One shall now take the uniform beam of Figure 1.52, with constant cross section A, Young modulus E, second area moment I, length l and mass density ρ. The vibration happens along the z axis, where the displacement is $w(x,t)$.

Figure 1.52 Uniform beam in free vibration.

Taking once more an infinitesimal element of length dx in a general displaced position (see Figure 1.53), the internal efforts on the left section, where the displacement is w, are the shear force V and the bending moment M, while on the right end section, where the displacement is $w+dw$, the efforts are $V+dV$ and $M+dM$; the element is also subjected to an infinitesimal inertia force dF_i. In Figure 1.53 the represented directions of V and M are taken as positive. Needless to say, all the referred quantities are functions of both space (x) and time (t). Here the Bernoulli-Euler theory is used, i.e. one ignores the rotating inertia effect of the element, as well as the deformation due to the shear forces, assuming that the section remain plane. These effects tend to be more important as one moves to higher frequencies. A comparison between the predicted analytical dynamic response and an experimental measurement of the response indicates whether it is important to include those effects and therefore use a more accurate theory.

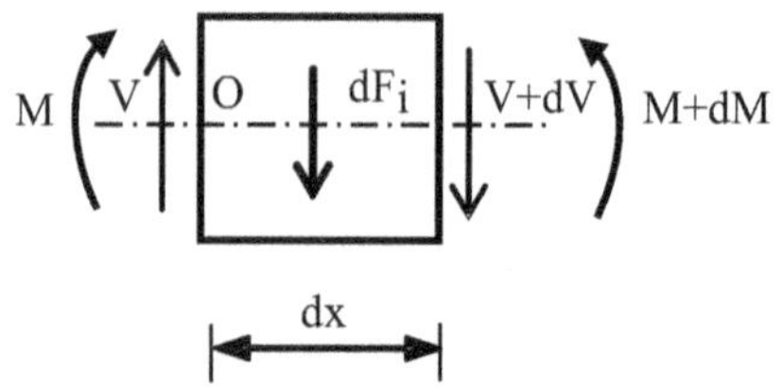

Figure 1.53 Free body diagram of an infinitesimal beam element.

For the element to be in dynamic equilibrium one must make a force and a moment balance. Summing up the forces along the z axis, it follows that:

$$V - dF_i - (V + dV) = 0 \tag{1.305}$$

As $dF_i = dm\frac{\partial^2 w}{\partial t^2} = \rho A dx \frac{\partial^2 w}{\partial t^2}$,

$$dV = \frac{\partial V}{\partial x}dx = -\rho A dx \frac{\partial^2 w}{\partial t^2} \quad \Rightarrow \quad \frac{\partial V}{\partial x} = -\rho A \frac{\partial^2 w}{\partial t^2} \tag{1.306}$$

Doing a moment balance around point O, it follows that:

$$M + dF_i \frac{dx}{2} + (V + dV)dx - (M + dM) = 0 \tag{1.307}$$

Substituting $dF_i = \rho A dx \frac{\partial^2 w}{\partial t^2}$, $dV = \frac{\partial V}{\partial x} dx$, and $dM = \frac{\partial M}{\partial x} dx$, it turns out that:

$$\rho A \frac{\partial^2 w}{\partial t^2} \frac{(dx)^2}{2} + V dx + \frac{\partial V}{\partial x}(dx)^2 - \frac{\partial M}{\partial x} dx = 0 \tag{1.308}$$

As $(dx)^2$ is much smaller than dx, one ends up with the same result known from statics, because – as mentioned before – one has neglected the rotating inertia effect of the element and the deformation due to the shear forces:

$$V = \frac{\partial M}{\partial x} \tag{1.309}$$

Substituting (1.309) in (1.306) leads to

$$\frac{\partial^2 M}{\partial x^2} = -\rho A \frac{\partial^2 w}{\partial t^2} \tag{1.310}$$

From beam theory, it is known that the bending moment is related to the curvature:

$$M = EI \frac{\partial^2 w}{\partial x^2} \tag{1.311}$$

Thus, for a beam of constant cross section in free vibration,

$$c^2 \frac{\partial^4 w}{\partial x^4} + \frac{\partial^2 w}{\partial t^2} = 0 \qquad with \qquad c = \sqrt{\frac{EI}{\rho A}} \tag{1.312}$$

which is a partial differential equation of fourth order in x and of second order in t, thus requiring four boundary conditions and two initial conditions.

Solving the Equilibrium Equation

The solution of (1.312) is sought once more by the method of separation of variables, i.e. $w(x,t) = \phi(x)W(t)$, or to simplify, $w = \phi W$, where ϕ is the shape function. Therefore,

$$c^2 \frac{1}{\phi} \frac{d^4 \phi}{dx^4} = -\frac{1}{W} \frac{d^2 W}{dt^2} \tag{1.313}$$

Taking the same rationale as for the bars, one concludes that each side of Eq. (1.313) must be equal to a constant a, leading to two equations:

$$\frac{d^2W}{dt^2} + aW = 0 \tag{1.314}$$

$$\frac{d^4\phi}{dx^4} - \frac{a}{c^2}\phi = 0 \tag{1.315}$$

The solution is of the type $W = Ae^{st}$; hence, $\frac{d^2W}{dt^2} = s^2 Ae^{st}$. Thus, upon substitution in (1.314), the algebraic equation is $s^2 + a = 0$ and the solutions are $s = \pm i\sqrt{a}$. Therefore,

$$W(t) = A_1 e^{i\sqrt{a}t} + A_2 e^{-i\sqrt{a}t} = A\cos\sqrt{a}t + B\sin\sqrt{a}t = A\cos\omega t + B\sin\omega t \tag{1.316}$$

Later, the constants A and B will have to be calculated from the initial conditions. From (1.316), $a = \omega^2$ and Eq. (1.315) becomes $\frac{d^4\phi}{dx^4} - \frac{\omega^2}{c^2}\phi = 0$. As the solution is of the type $\phi = Ce^{sx}$, it is convenient to introduce the parameter β, such that:

$$\beta^4 = \frac{\omega^2}{c^2} = \frac{\omega^2 \rho A}{EI} \tag{1.317}$$

The algebraic equation correspondent to $\frac{d^4\phi}{dx^4} - \frac{\omega^2}{c^2}\phi = 0$ is:

$$s^4 - \beta^4 = 0 \tag{1.318}$$

whose solutions are $s_1 = \beta$, $s_2 = -\beta$, $s_3 = i\beta$, and $s_4 = -i\beta$. The shape function is a linear combination of the four solutions:

$$\phi(x) = C_1 e^{\beta x} + C_2 e^{-\beta x} + C_3 e^{i\beta x} + C_4 e^{-i\beta x} \tag{1.319}$$

One can relate the complex exponentials to *sines* and *cosines* and the real exponentials to hyperbolic *sines* and *cosines*. Thus, it is often more conveniently to write instead:

$$\phi(x) = C_1 \cos\beta x + C_2 \sin\beta x + C_3 \cosh\beta x + C_4 \sinh\beta x \tag{1.320}$$

This time the shape function is not harmonic. It is necessary to apply four boundary conditions, leading to the equation of frequencies in terms of the parameter β. Once β is known, the natural frequencies can be calculated from Eq. (1.317): $\omega = \beta^2\sqrt{\frac{EI}{\rho A}}$; or, preferably, introducing the length of the beam (l):

$$\omega = (\beta l)^2 \sqrt{\frac{EI}{\rho A l^4}} \tag{1.321}$$

Natural Frequencies and Mode Shapes

As for the bars, one shall follow an example, for a better explanation. Consider a beam with simple supports at both ends. For $x=0$ and $x=l$, the vertical displacements as well as the bending moments are zero, i.e.

$$x=0 \quad \Rightarrow \quad w(0,t)=0, \quad EI\frac{\partial^2 w(0,t)}{\partial x^2}=0$$
$$x=l \quad \Rightarrow \quad w(l,t)=0, \quad EI\frac{\partial^2 w(l,t)}{\partial x^2}=0 \tag{1.322}$$

For all time, this implies that the four boundary conditions are $\phi(0)=0$, $\phi''(0)=0$, $\phi(l)=0$, and $\phi''(l)=0$. Applying these conditions to Eq. (1.310) and its second derivative, it follows that:

$$\begin{aligned}
0 &= C_1 + C_3 \\
0 &= -\beta^2 C_1 + \beta^2 C_3 \\
0 &= C_1 cos\beta l + C_2 sin\beta l + C_3 cosh\beta l + C_4 sinh\beta l \\
0 &= -C_1\beta^2 cos\beta l - C_2\beta^2 sin\beta l + C_3\beta^2 cosh\beta l + C_4\beta^2 sinh\beta l
\end{aligned} \tag{1.323}$$

The first two of Eqs. (1.323) lead to $C_1 = C_3 = 0$. Therefore, one ends up with the following matrix equation:

$$\begin{bmatrix} sin\beta l & sinh\beta l \\ -sin\beta l & sinh\beta l \end{bmatrix} \begin{Bmatrix} C_2 \\ C_4 \end{Bmatrix} = \begin{Bmatrix} 0 \\ 0 \end{Bmatrix} \tag{1.324}$$

The trivial solution $C_2 = C_4 = 0$ is of no interest, and so the non-zero solution implies that the determinant be zero:

$$2\,sin\beta l\; sinh\beta l = 0 \tag{1.325}$$

As $sinh\beta l$ is only zero when βl is zero, which would mean from (1.312) a zero natural frequency (impossible in this case, as the beam cannot have a rigid body motion), the solutions correspond to $sin\beta l = 0$, and so,

$$\beta l = n\pi, \quad n = 1,\ 2,\ \ldots\infty \tag{1.326}$$

From Eq. (1.318), the natural frequencies are:

$$\omega_n = (n\pi)^2\sqrt{\frac{EI}{\rho A l^4}}, \quad n = 1,\ 2,\ \ldots\infty \tag{1.327}$$

The mode shapes are given by the shape function, for each natural frequency. From (1.320) and because $C_1 = C_3 = 0$,

$$\phi_n(x) = C_2^{(n)} sin\beta_n x + C_4^{(n)} sinh\beta_n x \tag{1.328}$$

However, one can still relate $C_2^{(n)}$ and $C_4^{(n)}$, using one of Eqs. (1.315). For instance, from the first one,

$$sin(\beta l)_n C_2^{(n)} + sinh(\beta l)_n C_4^{(n)} = 0 \tag{1.329}$$

From (1.326), the solutions $(\beta l)_n = n\pi$ and thus $sin(\beta l)_n = 0$, which leads to $C_4^{(n)} = 0$, as here it is always true that $sinh(\beta l)_n \neq 0$. In conclusion, the mode shapes are given by:

$$\phi_n(x) = C_2^{(n)} sin\frac{n\pi}{l}x, \quad n = 1,\ 2,\ ...\infty \tag{1.330}$$

Free Response
At each natural frequency the time response is harmonic:

$$w_n(x,t) = \phi_n(x) W_n(t) \tag{1.331}$$

From (1.316) and (1.330), it follows that

$$w_n(x,t) = sin\frac{n\pi}{l}x\left(A^{(n)}cos\omega_n t + B^{(n)} sin\,\omega_n t\right), \quad n = 1,\ 2,\ ...\infty \tag{1.332}$$

The response will be the superposition of the responses of each mode shape:

$$w(x,t) = \sum_{n=1}^{\infty} w_n(x,t) = \sum_{n=1}^{\infty} sin\frac{n\pi}{l}x\left(A^{(n)}cos\,\omega_n t + B^{(n)} sin\,\omega_n t\right) \tag{1.333}$$

where the constants $A^{(n)}$ and $B^{(n)}$ will be calculated imposing initial conditions (displacement and velocity):

$$w(x,0) = w_0(x) = \sum_{n=1}^{\infty} A^{(n)} sin\frac{n\pi}{l}x \tag{1.334-a}$$

$$\dot{w}(x,0) = \dot{w}_0(x) = \sum_{n=1}^{\infty} \omega_n B^{(n)} sin\frac{n\pi}{l}x \tag{1.334-b}$$

The procedure to calculate $A^{(n)}$ and $B^{(n)}$ is entirely similar to the one explained in Section 1.6.1, leading to:

$$A^{(n)} = \frac{2}{l}\int_0^l w_0(x) sin\frac{n\pi}{l}x\,dx \tag{1.335}$$

$$B^{(n)} = \frac{2}{l\omega_n}\int_0^l \dot{w}_0(x) sin\frac{n\pi}{l}x\,dx \tag{1.336}$$

Bibliography

- Rayleigh, J. W. S., *The Theory of Sound,* Vols 1 & 2, Dover Publications, Inc., New York, 1945.
- Thomson, W. T., *Vibration Theory and Applications*, George Allen & Unwin Ltd., London, 1976.
- Warburton, G. B., *The Dynamical Behaviour of Structures*, Pergamon Press Ltd., Oxford, 1976.
- Rosenberg, R. M., *Analytical Dynamics of Discrete Systems*, Plenum Press, New York and London, 1977.
- Tse, F. S., Morse, I. E., Hinkle, R. T., *Mechanical Vibrations – Theory and Applications*, Allyn and Bacon, Inc., Boston, 1978.
- Craig Jr., R. R., *Structural Dynamics, An Introduction to Computer Methods*, Wiley, New York, 1981.
- Clough, R. W., Penzien, J., *Dynamics of Structures*, McGraw-Hill Kogakusha, Ltd., Tokyo, 1982.
- Den Hartog, J. P., *Mechanical Vibrations*, Dover Publications, Inc., New York, 1985.
- Bendat, J. S., Piersol, A. G., *Random Data – Analysis and Measurement Procedures*, Wiley, New York, 1986.
- Meirovitch, L., *Elements of Vibration Analysys*, McGraw-Hill, Inc., Singapore, 1986.
- Weaver Jr., W., Timoshenko, S. P., Young, D. H., *Vibration Problems in Engineering*, 5th Ed., Wiley, New York, 1990.
- Newland, D. E., *An Introduction to Random Vibrations, Spectral & Wavelet Analysis*, Longman Scientific & Technical, Harlow, 1993.
- Inman, D. J., *Engineering Vibration*, Prentice-Hall, New Jersey, 1994.
- Rao, S. S., *Mechanical Vibrations*, 3rd Ed., Addison-Wesley Publishing Company, Reading, 1995.
- Maia, N. M. M., Silva, J. M. M., He, J., Lieven, N. A. J., Lin, R. M., Skingle, G. W., To, W.-M., Urgueira, A. P. V., *Theoretical and Experimental Modal Analysis*, Research Studies Press Ltd., Tauton, 1997.
- Meirovitch, L., *Principles and Techniques of Vibrations*, Prentice-Hall International, Inc., New Jersey, 1997.
- Gatti, P. L., Ferrari, V., *Applied Structural and Mechanical Vibrations – Theory, Methods and Measuring Instrumentation*, E & FN Spon, London, 1999.
- Ewins, D. J., *Modal Testing, Theory, Practice and Application*, 2nd Ed., Baldock, England, 2000.
- Hart, G. C., Wong, K., *Structural Dynamics for Structural Engineers*, Wiley, New York, 2000.
- Meirovitch, L., *Fundamentals of Vibration*, McGraw-Hill, Inc., New York, 2001.
- Paz, M., Leigh, W., *Structural Dynamics, Theory and Computation*, 5th Ed., Springer, New York, 2004.
- Craig Jr., R. R., Kurdila, A. J., *Fundamentals of Structural Dynamics*, Wiley, New York, 2006.
- Radeş, M., *Mechanical Vibrations I*, Printech, Bucharest, 2006.
- Braun, S., *Discover Signal Processing, An Interactive Guide for Engineers*, Wiley, West Sussex, 2008.
- Girard, A., Roy, N., *Structural Dynamics in Industry*, ISTE Ltd., London, Wiley, New York, 2008.
- Shin, K., Hammond, J. K., *Fundamentals of Signal Processing for Sound and Vibration Engineers*, Wiley, West Sussex, 2008.

- Lalanne, C., *Mechanical Vibration and Shock Analysis, Vol.1, Sinusoidal Vibration*, ISTE Ltd., London, Wiley, New York, 2009.
- Paultre, P., *Dynamics of Structures*, ISTE Ltd., London, Wiley, New York, 2010.
- Radeş, M., *Mechanical Vibrations II*, Printech, Bucharest, 2010.
- Avitabile, P., *Modal Testing: A Practitioner's Guide*, The Society for Experimental Mechanics and Wiley, New York, 2018.

Chapter 2 Vibration Testing and Analysis

2.1 Introduction

A fruitful attitude when starting an experimental activity is to consider the vibration testing as an investigation process, where one should put together all the clues, just like a detective does, trying to have a reasonable meaning for all of them, with specific priorities. Vibration testing, or any type of testing, should not rely only or essentially on sophisticated software or complex data acquisition systems. That kind of equipment is extremely powerful and can be very helpful, but the behaviour of the structure under test should be broadly understood assuming that such instrumentations are unavailable. Such an engineering approach leads to increase the number of potential clues towards the best solution, for instance checking points, define the range of some specific parameters, and opens the mind to imagine simple models to be used for a clear interpretation of the results, or hand calculations for an easy verification of of how a specific parameter could affect the results of the measurements.

Furthermore, except the rare case of a very well-known structural behaviour, a literature survey of results of test conditions used by other experimenters for similar conditions should be conducted, just to be ready to cope with problems evidenced by others, or, at least, to define the boundaries of our own testing.

Based on the personal experience, one should be able to weigh all that information, to define a framework for a specific test, to include the acquired data in a meaningful context and to end up with robust, reliable and fully justified results.

Vibration testing is not just measurements, data acquisition and software manipulation (*tap* and *click* actions). Some people may believe that a vibration testing is not related to understanding how the structure behaves, but that it is sufficient to install a consistent number of sensors, sending everything to a computer and that's it! This is not an engineering attitude. Do not believe that it is enough to acquire as many data as possible. Additional warning should be addressed to the industries. Generally, they have a technical department asking to a laboratory department to perform the test. And usually there is no interaction between those two offices, which is a big mistake.

The analysis and the interpretation of the measurements are the most tricky and enjoyable activities, as well as having them correlating with a plausible behaviour of the structure under test. This is a typical engineering activity, where one should be critical with both simulation and testing, and repeat some specific, more defined tests, when doubts emerge.

In addition to those quite general advices, some others, a bit more detailed, are relative to the specific organisation of the test. However, field testing or even ground testing are not – in general – the ideal situations to concentrate on problems that have not an easy and immediate solution. To avoid falling down in such situations, one should try to organise and to prevent all the possible difficulties and circumstances which tend to create stressing environments. If

Structural Dynamics in Engineering Design, First Edition, Nuno M. M. Maia, Dario Di Maio, Alex Carrella, Francesco Marulo, Chaoping Zang, Jonathan E. Cooper, Keith Worden, and Tiago A. N. Silva.

something unfortunate happens, the best solution would be to stop testing, to solve the inconvenient or the unexpected problem in a dedicated proper location and later, when an acceptable solution has been found, resume testing. Most of the times this is not possible, because of time (and money) constraints. Again, an engineering approach is helpful. Depending on the severity of the unexpected situation, one should choose (i) to stop the test (the most severe case), (ii) to take note of the problem for later investigation (the most probable solution), or (iii) to continue the test without caring about the problem (the easiest solution, not always the best).

In preparing for a vibration test, at least one person of the experimental team should have a clear understanding of the features of the instrumentation to be used (sensors, data acquisition systems, hardware and software), of the objective for which the vibration test is required and also which use will be given to the results. This is extremely important, otherwise it is like having a dialogue between two persons who speak different languages and neither of them understands the language of the other.

Based on such and other similar considerations, the purpose of this chapter is to provide the reader with some practical information backed by a few mathematical tools. There are detailed books for signal processing that the interested reader should consult for an insight on methods and procedures. There are also books and manuals which perfectly describe how to stick a sensor on a structure, what is a calibration mass and so on. The background idea of this chapter is to provide some understanding of the main features that a young engineer/researcher should have to set or choose when undertaking a measurement (like sensors, sampling frequency, type of excitation, filtering and windowing, data analysis and so on). They can be crucial when it comes to experimental dynamics and one choice or another can produce different and even wrong results.

The following section deals with the test set-up and some principles that intuitively may come from daily life. Some basic understanding of the signal processing is the main part of section 3, on how to acquire valid and meaningful data. The process of data analysis of the measured data is summarised in section 4, while section 5 is focused on post-processing the data to present to someone who has not been involved in the measurements, but has the need to collect several information for taking specific decisions.

2.2 Test Set-up

The very basic decisions to take when setting up a measurement are (Figure 2.1):

1) which quantity to be measured;
2) which sensor is able to measure that quantity;
3) which system will read what the sensor has measured.

Based on the experience coming from both simulation and testing, it is always very important and usually time saving to include a verification step to make sure that the measurement (or the result, in case of numerical simulation) is correct. This approach could be called "Navigator's Principle". Nowadays, when faced with a new *GPS*-based navigator system, it is a good and intuitive practice to start driving through very well-known routes and conditions, to become fully confident on how the navigation information works.

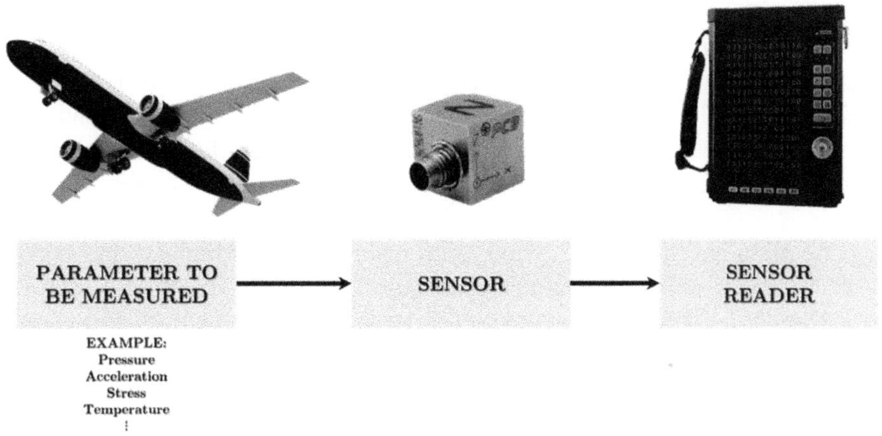

Figure 2.1 Basic definitions for setting up a measurement.

This preliminary practice allows the user to correctly understand and interpret the different information available and where such an information is located on the screen, much better and quicker than just reading a flat user's manual. Such exercises generate a subjective, though strong confidence between the user and the measurement system, which is the basis for relaxed driving when the route is unknown.

A similar approach should be followed when starting with lab or field testing. Firstly, one should measure something that one knows well, so that one can predict its response. The result of such measurement will confirm the reliability of the measurement chain, the ability of the user in performing the test and, possibly, the tolerance to be adopted.

This procedure seems to be so straightforward, that – very often – is neglected. During a test campaign started rapidly for various reasons, it may happen that some doubts and uncertainties come out, and a check appears as the only way to put things in order. If the check fails, all the previous measurements are useless and one understands why the "Navigator Principle" is time-saving.

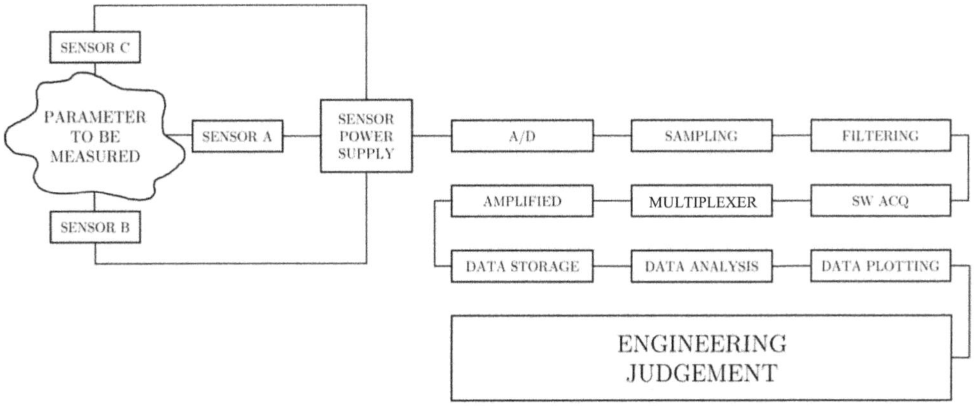

Figure 2.2 More complete set-up and instrumentation for reliable measurements.

Back to the basics of measurements shown in Figure 2.1, novice experimenters should be aware that usually life is not that easy and for correct measurements, other instrumentation and software (sometimes complex) are needed, see Figure 2.2. The list of equipment presented in Figure 2.2 is not always required or it could be insufficient. It depends on the type of measurement to be carried out. Modern equipment includes all of them in tiny "magic" boxes that are claimed to help the user as much as possible, to make life easier.

The list of equipment in Figure 2.2 ends with a simple box, which (so far) cannot be included in any hardware or software package. It is the user ability in making investigation, listing priorities, weighting measurements to end up with decisions for improving performances that is the main, direct or indirect, expected outcome of any test session. Maybe the future will substitute such last box by some kind of artificial intelligence, but it is believed that for many years to come, Engineering Judgement (*EJ*) will make the difference between long, boring, sometime useless test sessions and reliable and efficient measurements. *EJ* comes from the experience of previous tests, but also from a *trial and error* approach on how the measured component or element should behave and if a simple model can match the measurements, if not in exact values, at least as a trend. *EJ* relies strongly on:

- mathematical tools (domain definition, filtering, transforming) which may help in interpreting acquired data;
- the physics of the phenomenon (structure, fluid, electricity, chemistry and so on);
- the behaviour of the parameters to be measured (steady, unsteady, linear, non-linear, highly non-linear);
- the equipment used for the measurements, ability to exercise the intuition (sometimes the fantasy and new ideas) for understanding what the measurements are telling us, sometimes changing the location of the sensors, altering some parameter in the hardware and/or software or *inventing* some other way to collect clues that may help confirming initial hypotheses.

Experience and *EJ* tells us that when one modifies test or numerical conditions for understanding trends, only one parameter should be changed at a time and by a small amount, otherwise it will be really difficult to assess its influence.

Open literature, specialised journals, free technical publications are available for insights at any single element and aspect of sensor principles, data acquisition and data analysis. In what follows, some highlight on these topics will be given with the attempt to evidence tips and tricks that might be useful.

2.3 Fundamentals of Data Acquisition

The role of transducers is to convert a physical parameter (temperature, pressure, displacement, etc.) into an electrical signal. An incredible variety of sensors are available, with different precision, sensitivity, dynamic range and so on. With the miniaturisation of the electronic hardware, transducers are now capable of storing directly on a memory card the digitised signal useful for future post-processing analysis. All the sensors are based on a specific principle which directly, or indirectly, *senses* the physical parameter and converts it into a proportional electrical quantity.

The most common sensor is certainly the accelerometer. They are now so popular, inexpensive and miniaturised that they are even used in many very common devices, such as smartphones, tablets, and video game controllers. The accelerometers can be categorised in three different types, each designed to efficiently function in a specific environment. These three types are: piezoelectric, piezoresistive and capacitive.

The piezoelectric accelerometer is based on the piezoelectric property of certain materials (generation of an electrical charge when deformed) to sense changes in acceleration. Piezoelectric accelerometers are most commonly used in vibration and shock measurement.

The piezoresistive accelerometers are much less sensitive than piezoelectric accelerometers, and they are generally used for vehicle crash testing. A piezoresistive accelerometer increases its resistance proportionally to the amount of pressure applied to it.

Capacitive accelerometers are the most commonly used, namely those referred to as Micro-Electro-Mechanical Systems (*MEMS*) accelerometers, nowadays very much used in smartphones and vehicles. They tend to be very small and very low weight. Due to their size and affordability, they are also used by zoologists (to track the movement of animals in the wild), engineers (especially in collision experiments) and factories (remote monitoring of machinery vibration).

There are many other static and dynamic sensors like load cells, microphones, pressure gauges, strain gauges and so on. Some of them are quite inexpensive and broadly used in many applications. Generally, they require specific power supply and amplification for improving the signal-to-noise ratio. Sometimes the equipment associated to the sensor is called signal conditioning system and it may be more complex and costly than the sensor itself.

One of the most expensive sensors is the Laser Doppler Vibrometer (*LDV*) and its complementary Scanning Laser Doppler Vibrometer (*SLDV*). The great advantage of *LDV* is its non-contact measurement. It can be employed in very critical, sometimes dangerous, environment and is based on a phase relation between two very precise light signals. A reference signal is compared to another signal, which after impacting the vibrating object changes its phase and frequency and allows to identify the dynamic parameter (vibration velocity amplitude and frequency, in this case) by extracting it from the Doppler shift of the frequency and phase due to the motion of the surface.

Recently, the use of Digital Image Correlation (*DIC*) is receiving great attention, [2.1 – 2.3], based on pixels analysis and its correlation among different conditions, which is able to identify object motion, stress distribution and so on. The main advantages are referred to be the ability to measure simultaneously a portion of a structure (not single points) with a non-contact approach and, in some cases, even with inexpensive tools. Complex digital signal analysis is required, but some good software and nice applications are freely downloadable from the web.

Before starting any dynamic measurement, it may be very useful to have a look at the data acquisition system, to have a clear idea of the setting of the instruments and to verify whether or not they fit specific needs requested by the ongoing measurements.

One important parameter of the data acquisition system is its dynamic range. It describes the range of the input signal levels that can be correctly measured simultaneously, or, in other words, the ability to measure low amplitude signals in the presence of high amplitude signals. Human hearing, for example, has a very high dynamic range. Human ears are capable of hearing anything from a quiet murmur to the sound of the loudest rock concert. Such a difference can exceed 80 dB or more. The real problem is to perform these feats of perception, both extremes, simultaneously. The use of decibel (*dB*) as unit for the dynamic range is again requested by the need to fit big variations inside "small" numbers. For some specific advantages, some authors suggest the use of the ratio of full-scale signal and noise power per square root of Hertz (dBFS/$\sqrt{Hz}$) claiming that it describes better the real achievable dynamic range of the system and makes easier comparison among different systems.

Another important parameter characterising a data acquisition system is its *ADC* (Analog to Digital Conversion) sampling frequency (in some cases even its counterpart, *DAC*, Digital to Analog Conversion). *ADC* (or *DAC*) are processes that allow computers to interact with analog signals. Digital data are different from analog data at least for two main reasons, sampling and quanti*sation*, which are the digital effect on the *x*-axis and *y*-axis, respectively. These two operations restrict the information of the original analog signal. The digitised signal has a finite number of data, while the analog signal has infinite values, even when confined in a specific time-frame, Figure 2.3.

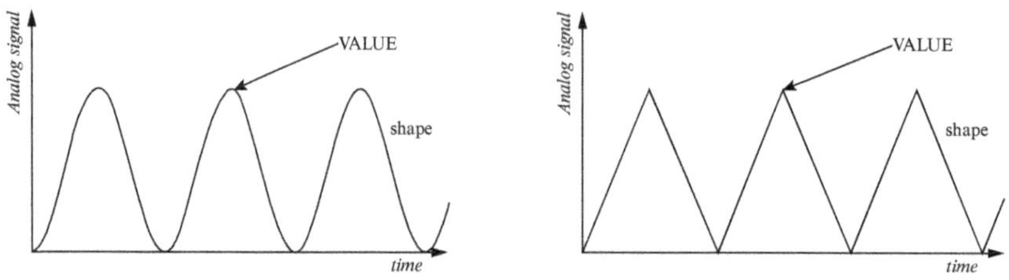

Figure 2.3 Examples of analog signals with different shapes and values.

The point is to understand what information must be retained and what information can be lost. The easiest answer to this question is to retain all the information that will allow to reconstruct the analog signal within an acceptable tolerance. To better understand this point, it is important to have a closer look at the sampling frequency, number of bits and type of filter to convert between the analog and digital data.

Figure 2.4 shows examples of sampling frequency. It is important to highlight that reconstruction of the signal does not mean *visual* similarity between the original and the reconstructed signal, but similarity of the *physical* content, which could be frequency content, power content, statistical content and so forth.

The sampling frequency f_s is the number of samples per second. Figure 2.5 gives some examples of sampling. Figure 2.5 (a) is the sampling of a constant (DC value) analog signal. This is the simplest case of easy and correct signal reconstruction from digital samples. Being the frequency of the original signal equal to zero, any value of the sampling frequency greater than zero is acceptable for a correct reconstruction. Figure 2.5 (b) is a 10 Hz sine wave (representing the analog signal) sampled at 150 Hz. This means that there are 150/10 = 15 samples taken over a complete period. These digital (discrete) samples represent accurately the original signal because these are no other sinusoids that can produce the same samples.

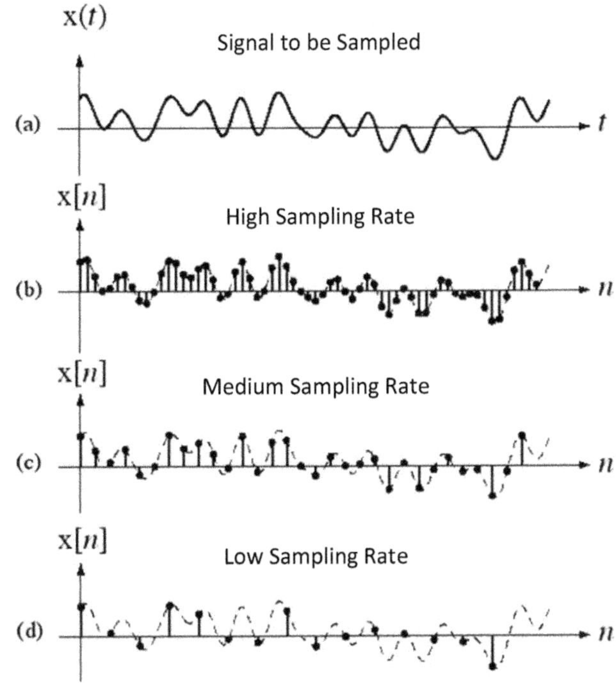

Figure 2.4 Examples of sampling rates.

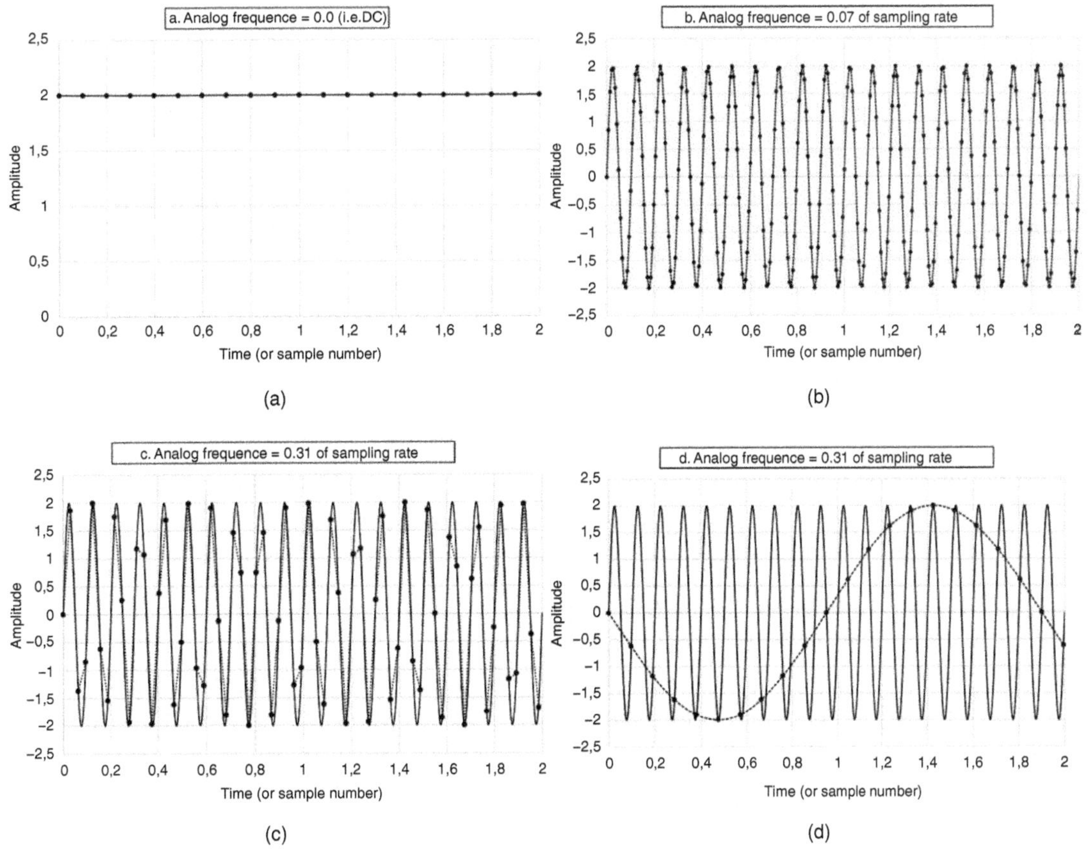

Figure 2.5 Examples of proper and improper sampling.

In Figure 2.5 (c) the sinusoid is sampled with a sampling frequency equal to 32 Hz. The digital samples now are so sparse they do not seem to *visually* reproduce the original (analog) wave. Though it may look strange, again it can be proved that no sine wave other than the 10 Hz can produce the same samples. This allows to conclude that the frequency content is preserved, even if the *visual* comparison does not correlate perfectly.

Figure 2.5 (d) is an example of improper sampling, since the sine wave is sampled with a sampling frequency equal to 12 Hz. It is also visually clear that this is an improper sampling of the original signal because a different sine wave can reproduce the same samples. The original sine wave represents itself as another (alias) sine. This phenomenon is therefore called aliasing (the original sine has another identity, or has lost its true identity).

From these results, the sampling Nyquist-Shannon theorem comes out: Suppose the highest frequency of a signal be $f_{\max}$ then a proper sampling requires a sampling frequency f_s equal or higher than $2\times f_{\max}$ (the Nyquist frequency).

With regard to the quantisation operation (digital conversion of the *y*-axis), it may be useful to take a look at the two stages process: sample and hold, and quantisation. This double action will encode the original signal into bits. Figure 2.6 gives a graphical representation of this process. The sample hold takes the value of the signal at a sampled time and maintains (holds) this value up to the next sampled time. As a result the signal is still analogic, but it looks like *stairs*, preparing it to be quantified. Next, each flat region in the sampled signal is translated (rounded) to the nearest member of a set of discrete values. This quantised signal may be encoded into bits; a sequence of 12 bits, for example, allows a set of $2^{12}=4096$ numbers (levels or quanta). If the signal has an analog variation from 0 to 5 volt, such continuous (infinite) interval may be quantified, with a 12 bit *ADC*, in 4096 numbers with a maximum error for each level of $\frac{5.0}{4096}=1.22\times10^{-3}$ Volt.

The main effect of quantisation is the introduction of a random error, ranging from $-1/2$ to $+1/2$ of the quantum level, with zero mean and standard deviation $1/\sqrt{12}$ of the quantum level. This random error will add to whatever noise is present in the analog signal and will decrease with an increasing number of bits.

For newcomers in the field of dynamic testing it may be *relaxing* to know that modern equipment are able to take automatic care of all, and even more, of these aspects. But this pleasant consideration does not allow any test engineer from neglecting proper ckecking of his/her own equipment, from the very beginning of the measurement chain (sensors, transducers) to its very end (software driving the acquisition and the analysis) especially if multiple options and configurations are allowed.

Before becoming an ordered sequence of numbers, an analog signal may undergo several operations like amplifying, filtering, multiplexing, isolation, excitation, linearisation, just to name a few. These operations are grouped into the *signal conditioning* expression which is important to know in order to get the best from the transducer and prepare the analog signal to become a correct and reliable ordered sequence of numbers.

Data acquisition hardware acts as an interface between the computer and the outside world. Specific care must be taken in order to maintain the same amount of information of the original signal, and to avoid, on the one hand to lose physical meaning and on the other hand to have the digitisation process adding non-physical information.

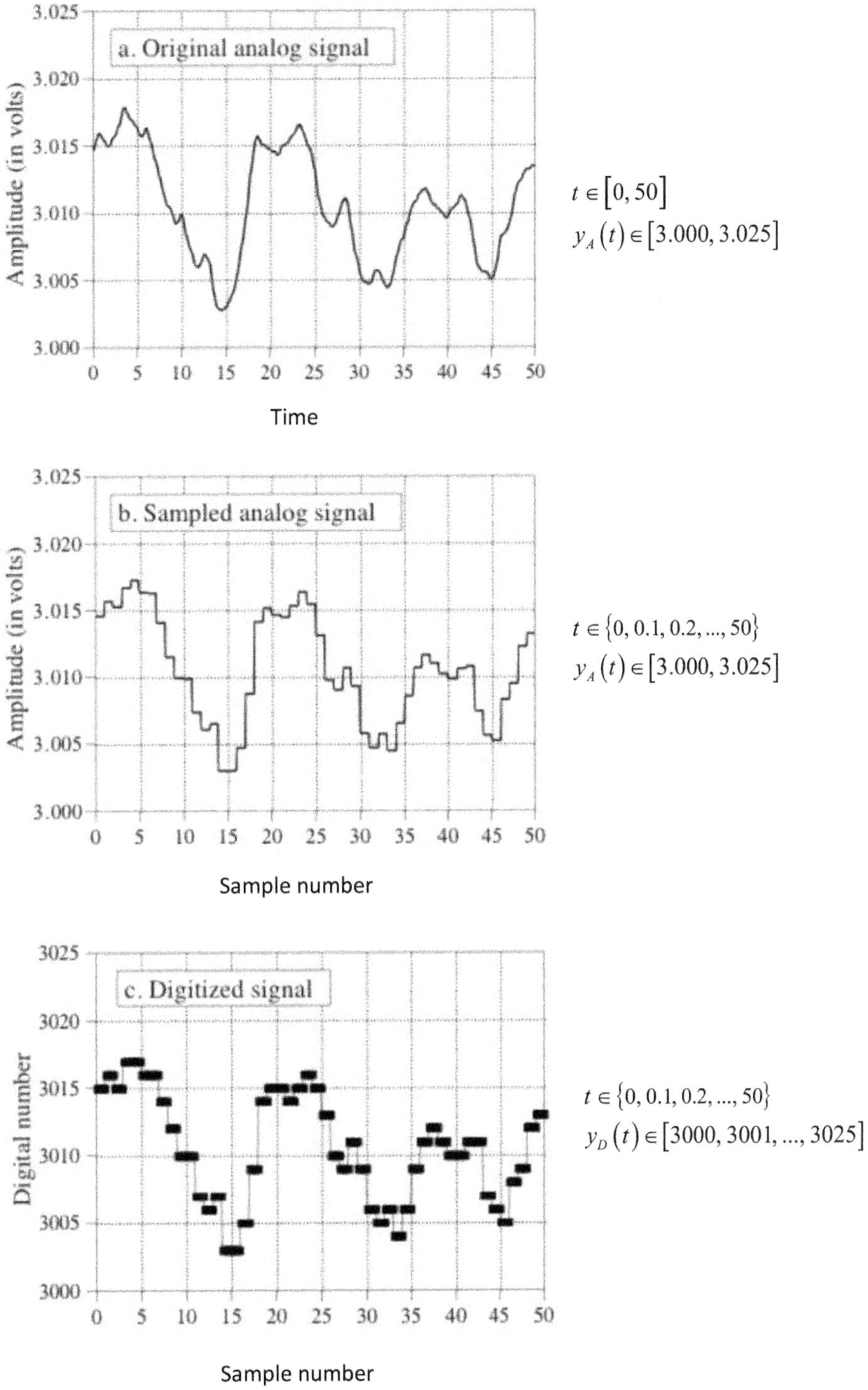

Figure 2.6 Digitising process (sampling and holding, and quantisation).

Suppose an example with a simple sine wave that could be a component of the analog signal, as in Figure 2.7. The digitisation process treats the acquired signal in a certain time-frame; suppose also that signal has been correctly sampled and quantified, as in Figure 2.8. The analog signal of Eq. (2.1), becomes the digital signal of Eq. (2.2):

$$y = \sin 2\pi f t \tag{2.1}$$

$$y_i = \sin 2\pi f t_i \tag{2.2}$$

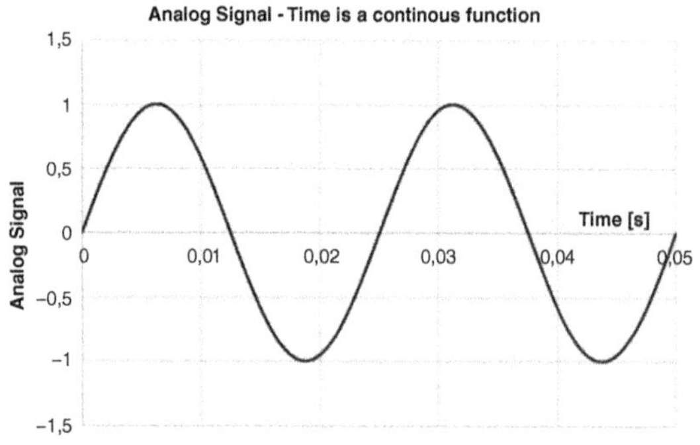

Figure 2.7 Analog signal – t is a continuous function.

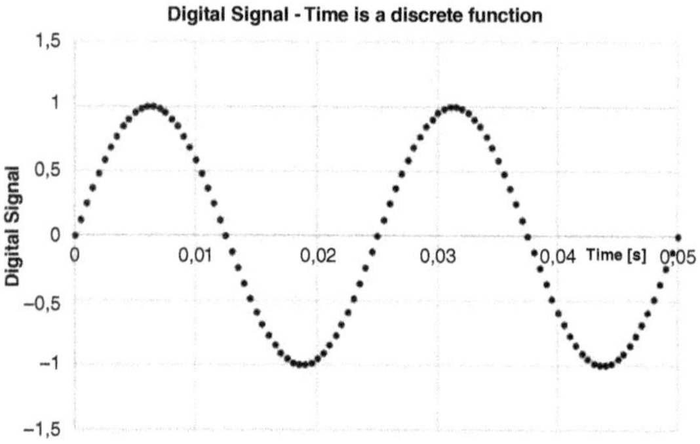

Figure 2.8 Digital signal – t_i is a discrete function.

The period T of the sine wave (seconds per cycle) is the reciprocal of the frequency

$$T = \frac{1}{f} \tag{2.3}$$

Defining

N = number of time intervals in the data segment
f_s = sampling rate in samples/second

the value of $\Delta t = t_{i+1} - t_i$ is equal to $1/f_s$ and the total signal duration T_t is given by $T_t = N\Delta t = N/f_s$. The number of periods, N_T that are *captured* in the recorded signal is

$$N_T = \frac{T_t}{T} = \frac{Nf}{f_s} \tag{2.4}$$

Now, as long as N_T is an integer member, the digitised sine wave is called bin-centred, which means that an exact number of periods are included in the digitised signal, i.e. the beginnings and the ends match perfectly. It must be pointed out that this is never the case, or, at least, if it happens for one component of the signal, it cannot happen for another one and usually real signals have many different components.

For example, in the case of a sampling rate $f_s = 1024$ samples/second and acquiring $N = 1024$ samples, from Eq. (2.4), the number of periods is equal to the frequency, but changing the value of N, or equivalently, changing the duration of the acquisition time, N_T is not an exact number anymore and the original signal is changed by something which is not physical, but only relative to digitisation process.

Figure 2.9 shows the effect of the discontinuity at the end of the sine and at the beginning of the next one, both when the duration of the signal is less (95 or 98 %) than the exact duration or higher (101, 102, 105%) than it. The effect of the non-matching ends created by the definition of periodicity may be attenuated (not really cured) by the use of "windows", starting from zero and ending to zero and therefore removing jumps at the extremes of a digitised signal.

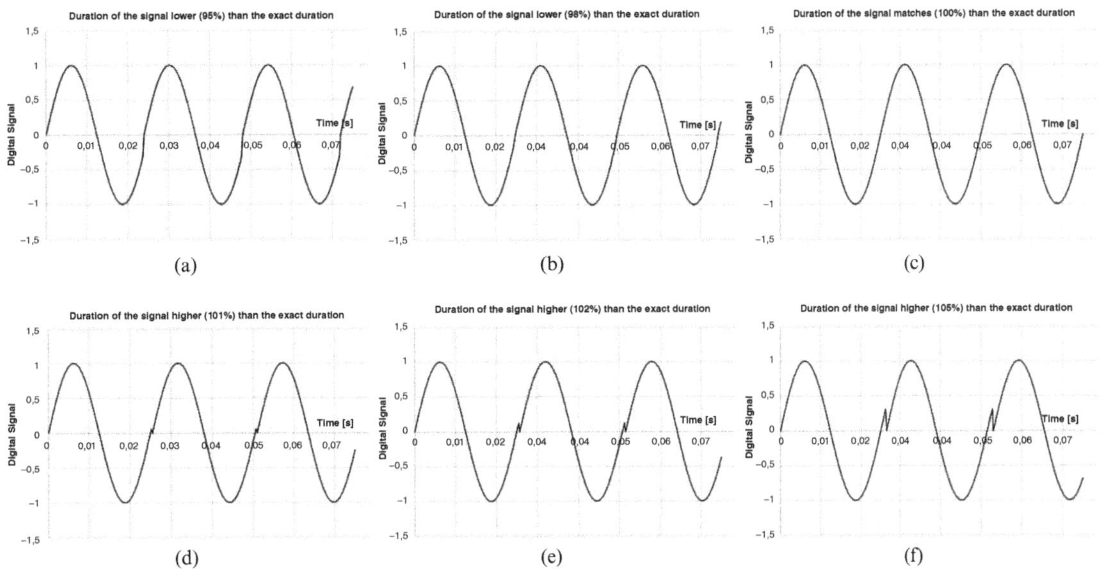

Figure 2.9 Effect of reconstruction of the signal when the acquisition time does not fit exactly an integer number of periods of the signal.

Windows are shapes like a hill, bumping at the centre and smoothly going to zero at the ends. Multiplying these window shapes, term by term, with a segment of data, the "tail" problem is eliminated and, at least, the ends match the beginnings, Figure 2.10. Many windows are described in the literature. Figure 2.11 shows six of them, among the most popular ones.

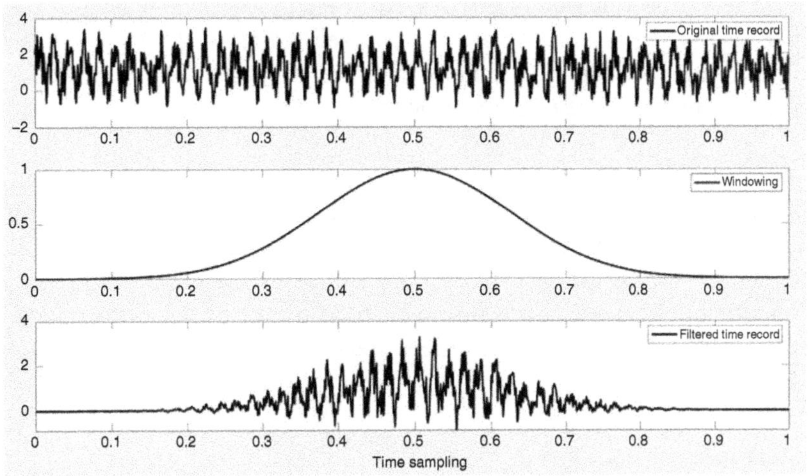

Figure 2.10 Windowing effect on a typical time-history acquisition.

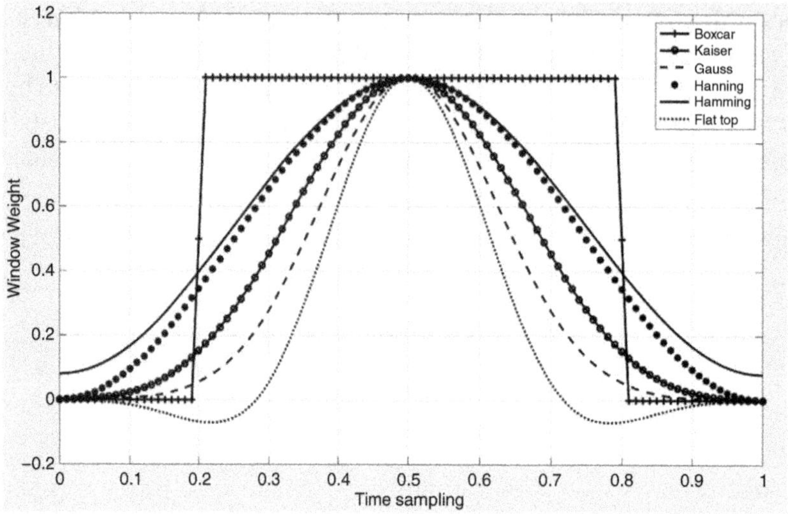

Figure 2.11 Example of six different windows used for cancelling the "tail" problem.

It is now interesting to discuss the effect of using different windowing when analysing data in the frequency domain, applying a Fast Fourier Transform[1] (*FFT*) to the time-histories of Figure 2.9. It should be pointed out that the simulated data is a simple sinusoidal function with 10 Hz frequency, no added noise, zero average and multiplying amplitude equal to 1. It is like having a sinusoid running from $-\infty$ to $+\infty$ (periodic function as requested by the theory of the Fourier Transform), capturing a time-frame and analysing it. In capturing such a time-frame, one cannot be sure to include an integer number of periods, but it can happen that the time-frame includes more or less than an exact number of periods.

In Figure 2.12, like the notation used in Figure 2.9, 100% stands for an exact number of periods inside the acquisition time, while 98% or 102% means that the last period is not completed or the very beginning of a new one has started.

The *theoretical* case is the plot of Figure 2.12 (c) for the 100% condition. No windowing has been applied, the acquisition time is exact. The *FFT* applied to this time-history, which replicates the perfect sine function, from $-\infty$ to $+\infty$, gives back what is expected: a peak amplitude equal to 1 at 10 Hz, which are the characteristics of the signal generated for this analysis. The other plots of Figure 2.12 show what happens to the same process in the case of shorter or longer acquisition time than the exact one.

As expected, if the difference between the exact acquisition time (which, obviously, cannot be known *a priori*) and the used acquisition time increases, the effect on the resulting *FFT* is more evident. Such effect is more pronounced on the amplitude than for the value of the frequency. Figure 2.12 (f) shows that for the 105 % case, the value of the amplitude is reduced more than 30% and the resulting peak frequency is not so sharp as for the previous cases and it may create some doubt in the precise identification of the value of the frequency.

---

[1] The Fourier Transform leads to the calculation and representation of a time signal in terms of its frequency content. The Fast Fourier Transform (*FFT*) is an ingenious numerical technique that makes the calculations much faster.

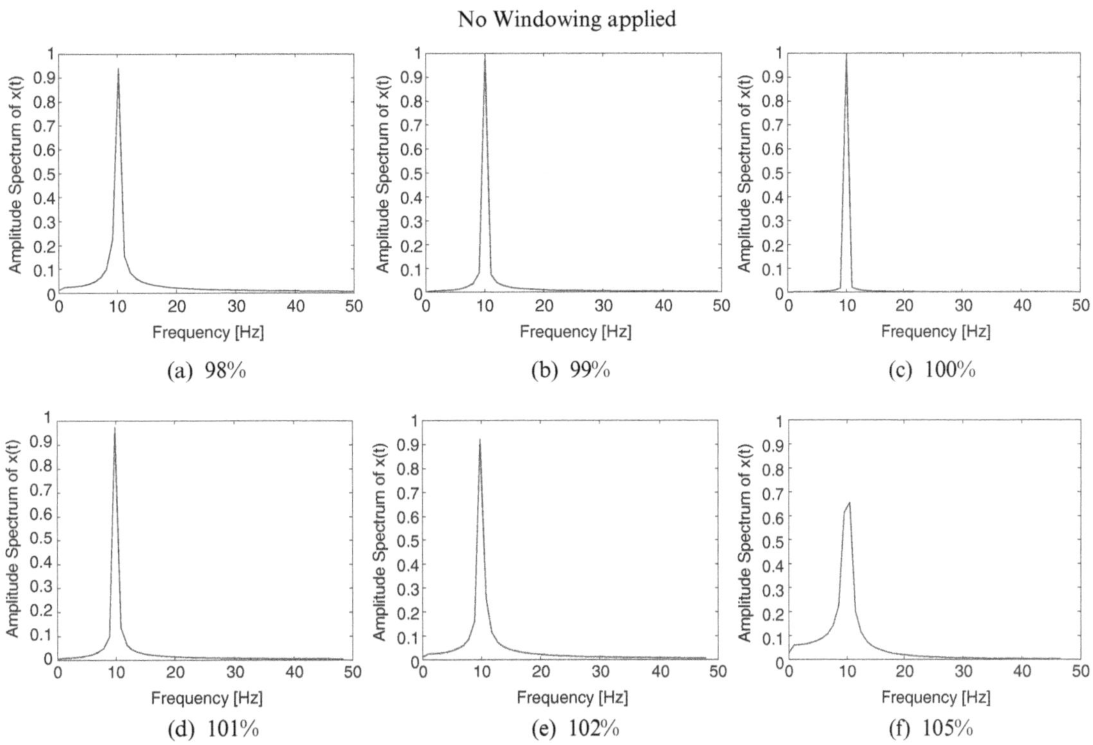

Figure 2.12 *FFT* Analysis of the corresponding signals of Figure 2.9 – No windowing applied.

Figures 2.13 and 2.14 show the result of using a window for minimising the "non-matching end" problem. These examples are presented, for the sake of brevity, only for two windows: the Hanning window, which is one of the most used and the Flat-Top window, which, in this particular case, leads to a different shape for the expected peak at the vibrating frequency.

The first evident effect of applying a window for reducing the "tail effect" turns out when looking at the amplitude of the computed *FFT*, which is usually reduced, while the value of the frequency is better kept. Clearly, different windows result in a different way of keeping these results.

Each group of these figures, from 2.12 to 2.14, should be clearly compared with its 100% case. A nice consideration comes out from these curves and it is related to the estimation of the natural frequency. Such estimation is almost independent of the applied window. The resulting shape of the frequency response function is affected by the selected window, but the value of the natural frequency remains independent from it, or at least inside the tolerances which can be accepted from an experimental test. Such a consideration is not true for the damping estimation. This parameter is one of the most critical and difficult factors to be estimated. Moreover, the use of windows affects its evaluation much more than the natural frequency and its small (acceptable) variability is also depending on damping. This is generally true, being the dissipation a function of micro and macro parameters of a specific structure. From the material composition to structural assembly, different dissipation mechanisms arise, depending on the type, entity and location of the excitation, local temperature, type of constraints and so on.

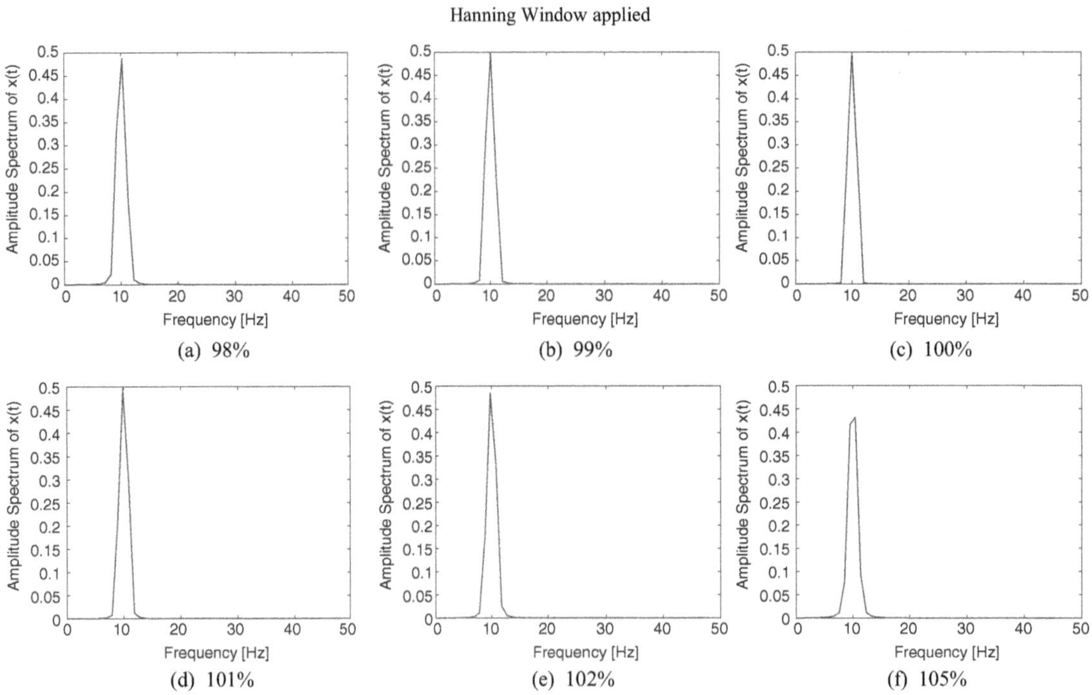

Figure 2.13 *FFT* Analysis of the corresponding signals of Figure 2.9 – Hanning window applied.

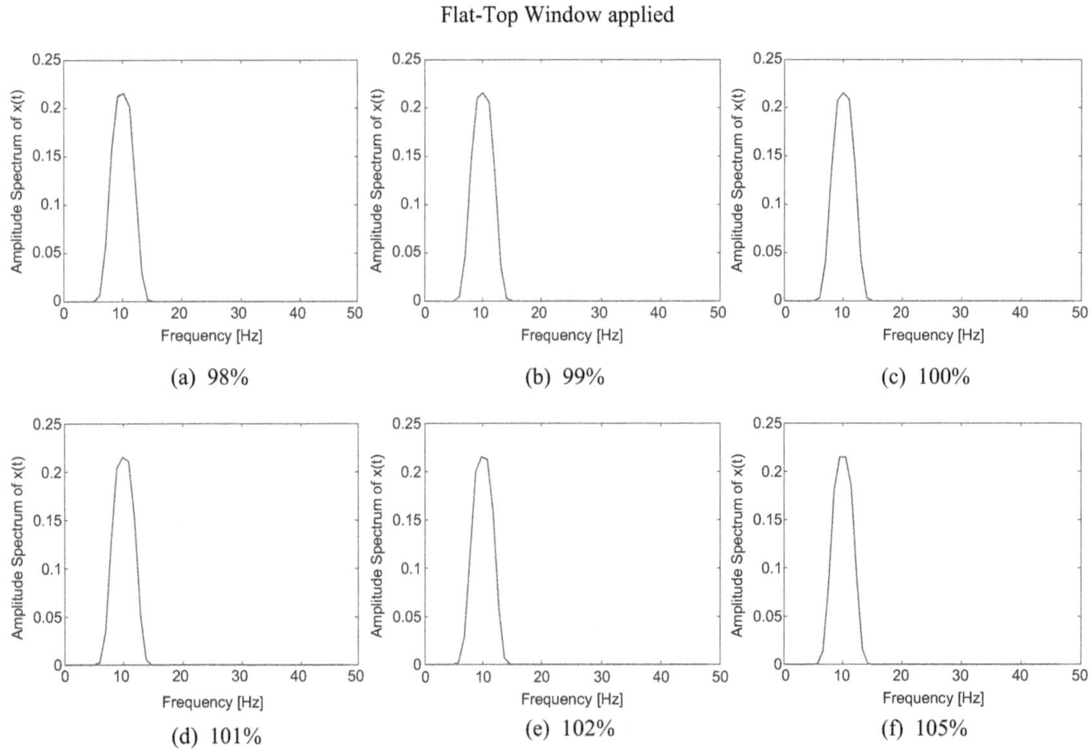

Figure 2.14 *FFT* Analysis of the corresponding signals of Figure 2.9 – Flat-Top window applied.

During a vibration test these aspects are influencing the damping behavior together with the signal processing algorithm. With reference to the windowing effect on damping estimation, one should remember that windowing is used for reducing the effects of the leakage, forcing to zero a non-periodic signal at the beginning and at the end of the time record. It should be sufficient only this consideration to understand that this process is adding some artificial damping to the acquired signal. Depending on the type of window, such (necessary) mathematical damping varies from quite a small amount when using, for example, a boxcar window and quite a long time acquisition frame (this will reduce the effect of windowing), compared to an exponential window. In the simple case of a theoretical *SDOF* using an impulse force for its excitation, one may find the following outcomes (Table 2.1):

Table 2.1 Effect of time record and windowing on damping ratio estimation.

Item #	Acquisition time [s]	Window	f_n [Hz]	ξ
Theory	---	---	10.00	0.01
1	1.60	Boxcar	9.95	0.0133
2	1.60	Exponential	10.0	0.0495
3	3.20	Boxcar	10.0	0.0104
4	3.20	Exponential	10.0	0.0299

Depending on the type of excitation (*sine*, *sine sweep*, *random* and so on), the time duration of the acquisition and the window, one may have different results on modal parameters estimation, less pronounced on the natural frequencies, but much more on damping values. A good selection of such parameters depends also on the precision one is expecting or is acceptable from a specific test.

It may be worth noting that no particular window is always better than another and sometimes windowing might be not necessary. If one chooses a type of excitation signal that is periodic within the time length (like periodic random, for instance) there may be no need for a window, because the result could have the opposite effect. It may depend on the type of signal, test and its purpose, which could be also different for the input and the output signal. When using hammer testing, for example, one should use a flat-top window around the force sign to be sure that nothing else but the input force enters the system; and use an exponential window on the response signal, to be sure that the response dies out at the end of the time length. It is also important to remember that this kind of window in particular may have a significant effect on increasing the damping, perhaps implying a correction. Again, it may be worth checking with known results, for getting full confidence on the identified parameters. This is even more true if one tries to benefit from the available option of modern analysers, which allow for user-defined windows.

The less sensitive window to the acquisition length of time appears to be the flat-top window which gives back almost the same plot for both the 100% and the 105% cases. On the other hand, flat-top window is affecting, in any situation, the value of the amplitude more than the other windows.

All the windows reduce the effect of the duration of the acquisition time when compared with the "no window case", both in amplitude and frequency. From such a kind of analysis it is possible to have a more precise idea on how the windowing process may affect the acquired data in terms of frequency, amplitude and signal-to-noise ratio, and to gain more confidence and have a more rational basis to make a decision on the selection of a specific window.

The developed example on the windowing effect may be also considered as an application of the "Navigator Principle" presented before. At the starting point everything is well-known (type of function, amplitude, frequency) and therefore it is easy to *simulate* what could be the effect of a digital acquisition (sampling frequency, different duration of the acquisition time, quantisation effects and so on) and finding a possible or a best solution for mitigating the unwanted effects.

The main objective of the manipulation of the original data is to extract the principal characteristics of the signal, usually hidden in the time domain, though revealed by transforming the data to a different domain. The transfer to another domain, as discussed here from time to frequency domain, requires some attention to avoid that the digital process might introduce or change values of the parameters. In this regard, a numerical simulation of the procedure which is going to be used may result very helpful in gaining confidence on the resulting data.

In conclusion, it does not exist one window which is valid "for any season" and this is the reason why the numerical simulation is invoked several times. This is generally true for all those cases where more than one option is still valid. Usually, the choice of one window compared to another is based on familiarity, type of testing, or any other acceptable reason, but it should be remembered that they could affect the results of the testing. For example, in the previous discussions it has been noted that one may have a slight modification on the value of the natural frequency, or the damping value may be quite strongly affected by one window compared to another. Depending on the required precision on a specific parameter, it could be useful to compare results obtained from different identification procedures. It is also a matter of available time from testing to analysis, often too short.

2.4 Understanding and Analysing Measured Data

The focus of this paragraph is addressed to give information and suggestions on how to deal with time domain acquired data. A typical situation is the availability of measurements, but less obvious is the analysis of these experimental data and what they can tell us. It should be remembered that the digital world may give us a lot of information, but extracting what is really important and needed is still a matter for an expert dynamicist.

Therefore, this section deals with the analysis of measured data, running through several classical methodologies, which are typically used when performing dynamic testing. It offers an excellent opportunity for using the methodology of *simulating* the test condition to be able to set up all the necessary requirements for obtaining a good measurement and reliable results. The methodology can be extended to many other variations than those presented here, and this may be the trick for any new test.

In the following, reference will be made to both single degree of freedom (*SDOF*) and multiple degree of freedom (*MDOF*) systems (Figure 2.15), which are taken as examples of systems to be measured.

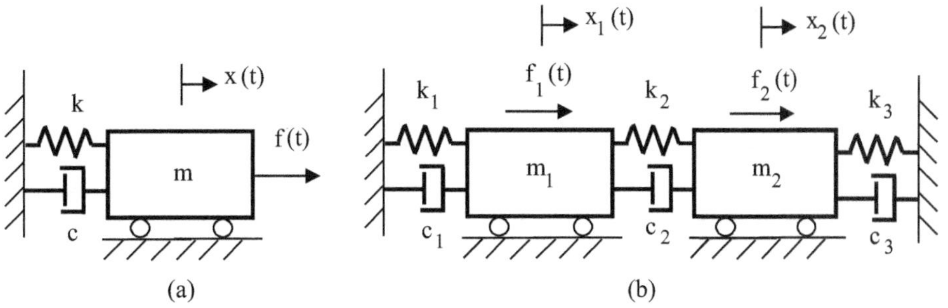

Figure 2.15 (a) Single Degree of Freedom system and (b) Multiple Degree of Fredom system.

These models should be considered with extreme care by any experimenters because they represent the *key-point* for understanding the dynamical phenomena. Symbolically the equation of motion of these *simple* systems is exactly the same as for any other (self-adjoint) system:

$$\boldsymbol{M}\ddot{\boldsymbol{x}} + \boldsymbol{C}\dot{\boldsymbol{x}} + \boldsymbol{K}\boldsymbol{x} = \boldsymbol{F} \tag{2.5}$$

with the dimensions of the arrays varying from 1 (*SDOF*) to N (*MDOF*), no matter how big N is (2, 3 or millions).

Correct modelling or trend analysis, most of the times, may result very useful to be done with reduced models (small value of N) easily manageable, and then transferred to much bigger models (high value of N). This allows increased confidence for very complex models and ability to understand the behaviour of part or of the complete systems. Simple models (both *SDOF* or *MDOF* with low value of N) may be considered like a toolbox, while the experimental engineer is the expert craftsman able to choose the right tool for any specific operation.

Potential realistic situations requiring measurements and analysis may be broadly subdivided in two main categories: possibility to measure input and output (typically experimental modal analysis – *EMA*) or possibility to measure the output only (generally named operational modal analysis – *OMA*). Sometimes these measurements are used for setting-up a "validated" numerical model which can be used for further calculations or for design optimisation. Other times these analyses are requested for showing compliance with requirements or for identifying reasons for unexpected behaviours and proposing retrofit for reliable solutions.

Motivations may come out for improving performances, extending service life, ensuring absence of critical conditions, increasing comfort, reducing noise emitted towards crowded communities, driving to greener and more sustainable systems.

Consider a first simple example of simulating a dynamic testing, using an *SDOF* with the following parameters:

$$\begin{aligned} m &= 10\,kg \\ k &= 4\times10^5\times\pi^2\,N/m \\ c &= 40\times\pi\,Ns/m \end{aligned} \quad \Rightarrow \quad \begin{aligned} f_n &= \frac{1}{2\pi}\sqrt{\frac{k}{m}} = 100\,Hz \\ \xi &= \frac{c}{2\sqrt{k\,m}} = \frac{40\times\pi}{4\times\pi\times1000} = 0.01 \end{aligned} \tag{2.6}$$

Obviously in the real case one does not know the modal parameters (natural frequency and damping factor) and often is even difficult to properly define the physical values of m, c and k. It is now necessary to excite the system, to measure its response and to identify its modal parameters. In the case of an *N DOF* system only the excited mode shapes can be identified.

With an *SDOF* system, any dynamic excitation is suitable, but with an *MDOF* system one should be careful on the type and location of the excitation. Depending on the possibility to excite the structure at a single point or at several points (and the testing time constraints), it is advisable to repeat the measurements changing the position of the excitation (shaker or hammer). Such procedure will increase the probability to excite (and consequently to identify) all the mode-shapes in a specific frequency range. Very well excited mode shapes will be more easily identified. From a theoretical standpoint, a simple impulse is able to evidence many mode-shapes, but from a practical point of view, without transferring sufficient energy to the structure, only noisy measurements are obtained, often meaningless or, even worst, driving towards wrong results.

Back to our simple *SDOF* system, a first excitation system is impact testing, usually obtained using an instrumented hammer, equipped with a force transducer to measure the impact transferred to the system. The system response is measured by a transducer (often an accelerometer). Both signals are sent to a data acquisition system and next to a software for the identification of the dynamic properties of the system. Figure 2.16 summarizes the main steps of this procedure, assuming no uncertainties through the whole process.

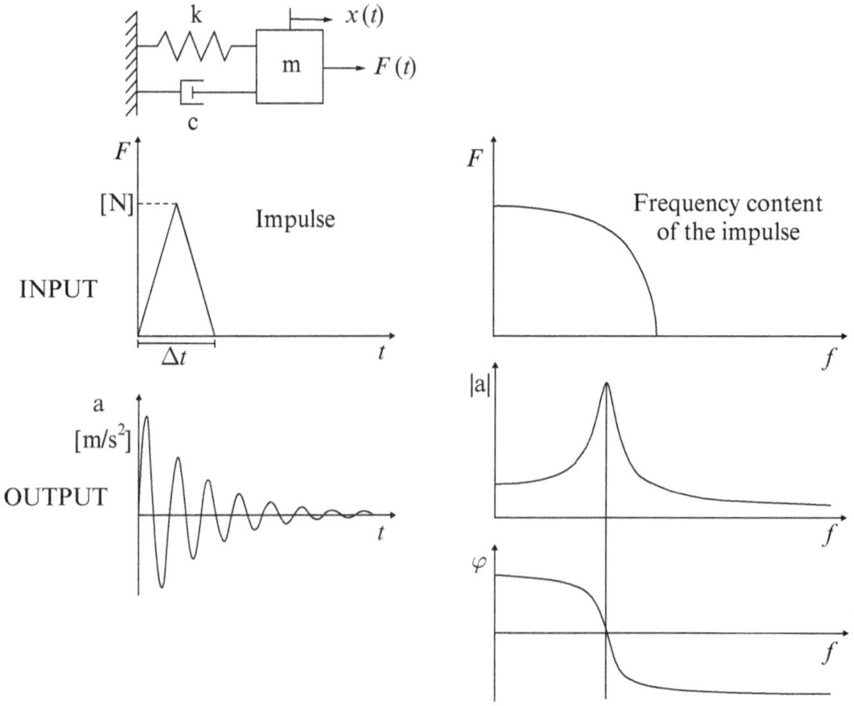

Figure 2.16 Impulse excitation and subsequent system identification for an *SDOF* system.

This *ideal* simulation allows for the identification of the system parameters both in frequency domain (half-power points)[2] and time domain (logarithmic decrement analysis)[3], with the following results:

[2] See Section 4.2.2 for details.
[3] See Section 4.2.3 for details.

	$f_n[Hz]$	ξ
Theory	100.00	0.0100
Time domain	100.00	0.0100
Frequency domain	99.61	0.0085

It is very interesting to observe, even in this very simple case, how the results of time domain are closer to the theoretical results than those obtained from the frequency domain. The time domain is the *natural*, *intuitive* domain and it does not require any manipulation of the results (or the acquired data). The real problem is that the time domain results are almost never as *clean* or those presented in Figure 2.16. The passage to the frequency domain is often necessary, as it will be clear later on, for *cleaning* a time-history not easily readable. However, it should be remembered that the Fourier transform process is a linear process and therefore if the system has strong non-linearities, the quality of the results may be affected by the transformation process. With the help of simple systems (not necessarily the *SDOF* system) and their corresponding numerical simulations, those kind of issues together with their pros and cons may be checked allowing more confidence and reliability on the obtained results.

The simple *SDOF* simulation, reported in its *ideal* representation in Figure 2.16, may also result very useful for exploring more realistic situations. Again the comparison with the theory is made through the natural frequency and the damping estimation (no more parameters for an *SDOF* system).

Consider the classical case of a double impulse, which should not be ignored when using hammer testing. It is known that one should avoid double hits because they might lead to wrong identification, especially if it is performed in time domain. Only part of the measured response signal should be analysed, but in a real case it might become almost impossible to extract meaningful results from part of the signal.

Moving to the frequency domain, considering also that one is generally doing averages, if one of the impulses is a double hit, it may spoil the final results. Although in practice double hits should be avoided, it is interesting to see what happens if we deliberately impose a double hit, as in Figure 2.17 (a). This is a clear example of how the numerical simulation can help in understanding real situations and practical examples.

Impulse excitation is frequently used the dynamic identification of bridges, measuring the response obtained from a heavy lorry running over a prescribed bump. Assuming a theoretical impulse function may result in a strong simplification, especially thinking that at least two wheel axes are passing over the bump and therefore it becomes almost impossible to avoid double hits.

As in Figure 2.17 (a), it is clear that the input spectrum is limited in frequency (as truly happens for the case of the lorry exciting the bridge), but even so, the frequency response function is good enough, despite the input has some potential uncertainty.

Figure 2.17 (b) shows the effect of adding some noise to the double impulse force. In this case the time domain analysis is almost impossible, while the results from the frequency domain remain very clear and are comparable to the previous cases. This example shows the real *power* of the Fourier transform. The same cannot be said when the noise is added only to the output, Figure 2.17 (c). In this case both time domain and frequency domain results obtained from simple algorithms are very confusing and more complex algorithms should be used to get more stable results.

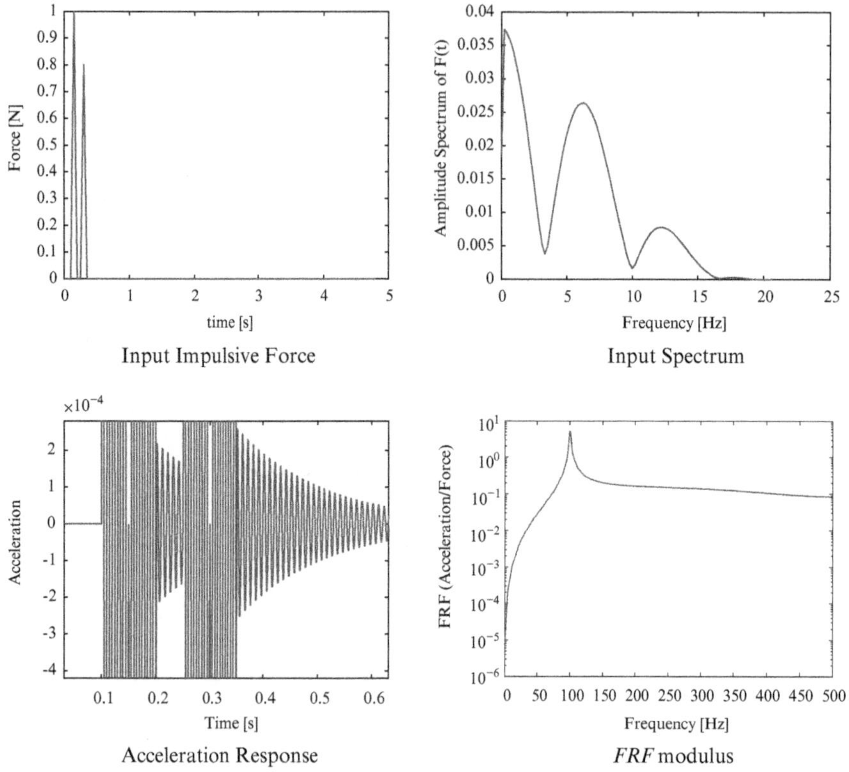

Figure 2.17 (a) Double impulse.

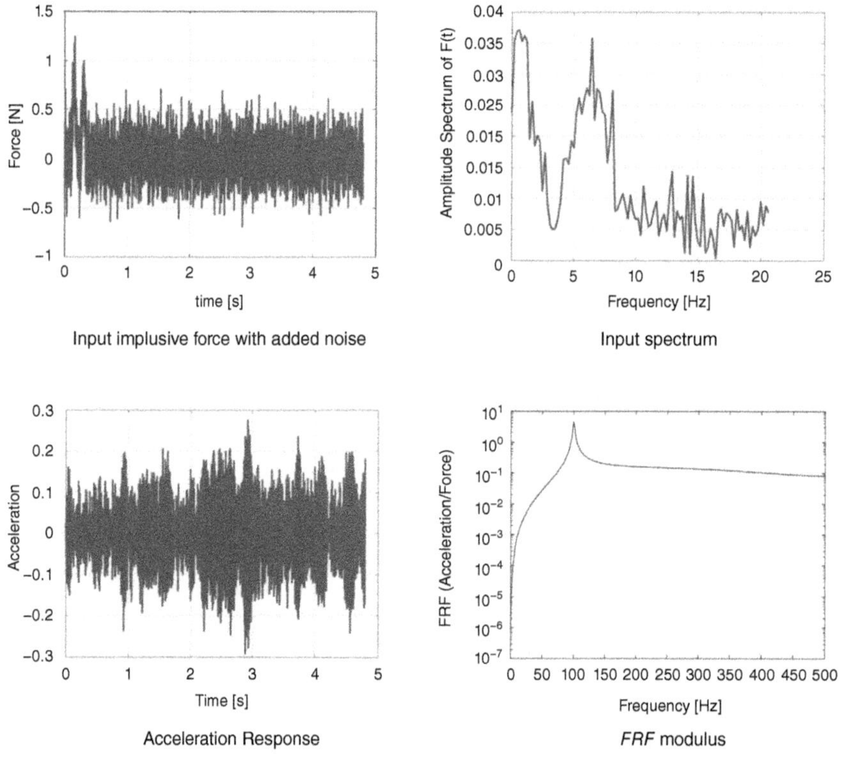

Figure 2.17 (b) Double impulse and noise added only to the input.

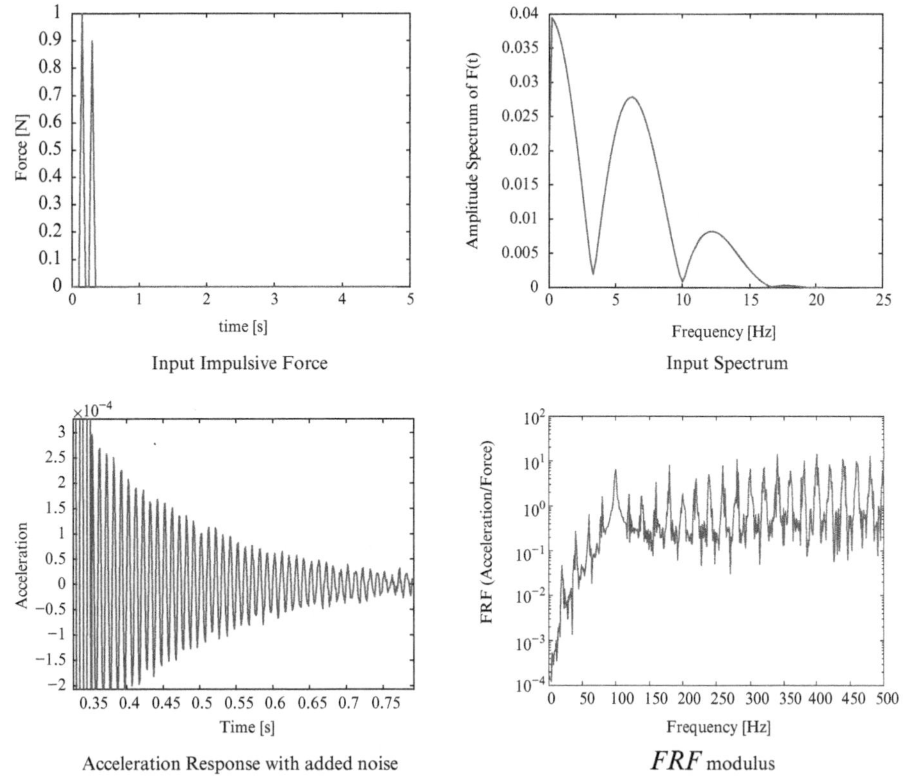

Figure 2.17 (c) Double impulse and noise added only to the output.

Another interesting case is changing the type of input, passing from impulse to sine sweep. In this case the forcing function is varying sinusoidally with the frequency sweeping between two values selected by the user. Additionally, the sweep rate (linear or logarithmic) may also be selected by the user.

Figure 2.18 (a), still referring to the initial *SDOF* system, shows the efficiency of this excitation function resulting in very close values of the system parameters ($f_n = 99.61\,Hz$, $\xi = 0.0078$) to the theoretical ones. But again, it is important to select the correct options of this type of excitation, otherwise the system identification might result very difficult or even wrong (see Figure 2.18 (b).

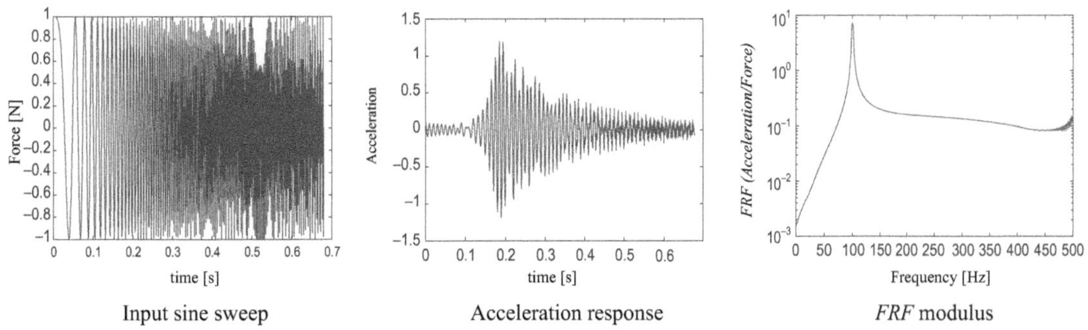

Figure 2.18 (a) Slow rate sine sweep input excitation.

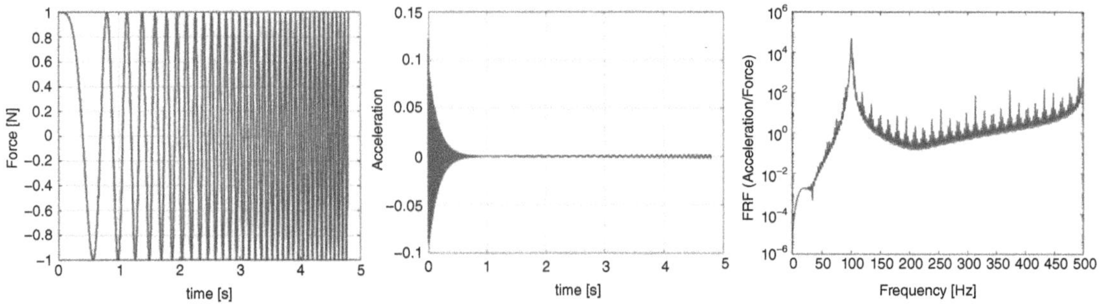

Input sine sweep   Acceleration response   *FRF* modulus

Figure 2.18 (b) Fast rate sine sweep input excitation.

Usually, the value of the natural frequency is quite stable, but for the damping identification the variation could result very wide. The results presented so far are just simple examples used to show on the one hand how it can be hard to find reliable parameters or, on the other hand, how useful can a simulation tool be to help tuning the correct option and selecting the right tools from the workbox.

Moving now towards a multiple degree of freedom (*MDOF*) system, it offers the possibility to create another *simulation tool* based on popular softwares that are used worldwide in design offices. Most of these very complex softwares, such as *ABAQUS*©, *ANSYS*©, *NASTRAN*©, and so on, are based on the Finite Element Method (*FEM*) and, typically, *FE* models have been used for designing, optimising and correlating numerical and experimental results. One should remember that even a very complex and detailed Finite Element Model remains a "model" of the *true* behaviour of the structure, and even the experimental measurements may be affected by a series of errors.

A common joke reports that the structure does not know numerical simulation methods or experimental measurements, but "lives" in its world which one is trying to understand, to simulate and to measure for the benefit of the final product.

Another good practice is represented by putting together data coming from different models and those coming from the laboratory and to verify how they correlate, and which are the trends for improving correlation and reliability of the models.

In the following the measurement data are simulated via *FE* models and their analysis is performed through classical methods. The focus again will be given on the correct application of the methodologies. The confidence on the reliability of the results comes from the comparison with parameters that are known, with the hope that the acquired knowledge may positively help in the correct identification of structures, or, in general, systems whose parameters need to be identified.

Example: Consider the *MDOF* system of Figure 2.19, with 4 *DOFs*. The equations of motion are given by Eq. (2.5):

$$\boldsymbol{M}\ddot{\boldsymbol{x}}+\boldsymbol{C}\dot{\boldsymbol{x}}+\boldsymbol{K}\boldsymbol{x}=\boldsymbol{F}$$

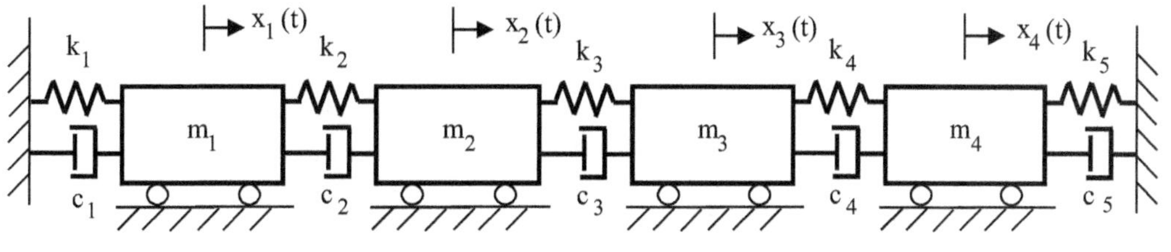

Figure 2.19 - Four degrees of freedom system.

The system matrices are:

$$\boldsymbol{M}=\begin{bmatrix} m_1 & 0 & 0 & 0 \\ 0 & m_2 & 0 & 0 \\ 0 & 0 & m_3 & 0 \\ 0 & 0 & 0 & m_4 \end{bmatrix} \quad \boldsymbol{C}=\begin{bmatrix} c_1+c_2 & -c_2 & 0 & 0 \\ -c_2 & c_2+c_3 & -c_3 & 0 \\ 0 & -c_3 & c_3+c_4 & -c_4 \\ 0 & 0 & -c_4 & c_4+c_5 \end{bmatrix}$$

$$\boldsymbol{K}=\begin{bmatrix} k_1+k_2 & -k_2 & 0 & 0 \\ -k_2 & k_2+k_3 & -k_3 & 0 \\ 0 & -k_3 & k_3+k_4 & -k_4 \\ 0 & 0 & -k_4 & k_4+k_5 \end{bmatrix}$$

Assuming $m_1 = m_2 = m_3 = m_4 = 10\,kg$, $k_1 = k_2 = k_3 = k_4 = k_5 = 10^4\ N/m$ and $c_1 = c_2 = c_3 = c_4 = c_5 = 2\,Ns/m$,

the classical complex eigenvalue problem gives the following results:

$$\{\omega\}=\begin{Bmatrix} 19.54 \\ 37.17 \\ 51.17 \\ 60.15 \end{Bmatrix} \text{[rad/s]} \quad \{\xi\}=\begin{Bmatrix} 0.0039 \\ 0.0074 \\ 0.0102 \\ 0.0120 \end{Bmatrix} \quad \boldsymbol{\Phi}=\begin{bmatrix} 0.1176 & 0.1902 & 0.1902 & 0.1176 \\ 0.1902 & 0.1176 & -0.1176 & -0.1902 \\ 0.1902 & -0.1176 & -0.1176 & 0.1902 \\ 0.1176 & -0.1902 & 0.1902 & -0.1176 \end{bmatrix}$$

where the mode shapes have been normalised with respect to the mass matrix, i.e. $\boldsymbol{\Phi}^T \boldsymbol{M} \boldsymbol{\Phi} = \boldsymbol{I}$.

Note that in this case the mode shapes are real because the damping is proportional (to the stiffness matrix in this example); however, in general, the damping will be non-proportionally distributed, inducing phase angles different from either 0° or 180° between the degrees of freedom of the system and thus making the mode shapes complex.

Assume now that measurements have been undertaken on such an *MDOF* system and an identification is required. Several options may be considered, and, depending on the choice, different identification algorithms can be used, either in the time domain or in the frequency domain, as explained in Chapter 4. Some examples are given in the following with the aim to evidence potential differences, user selectable options and how they influence the measurements.

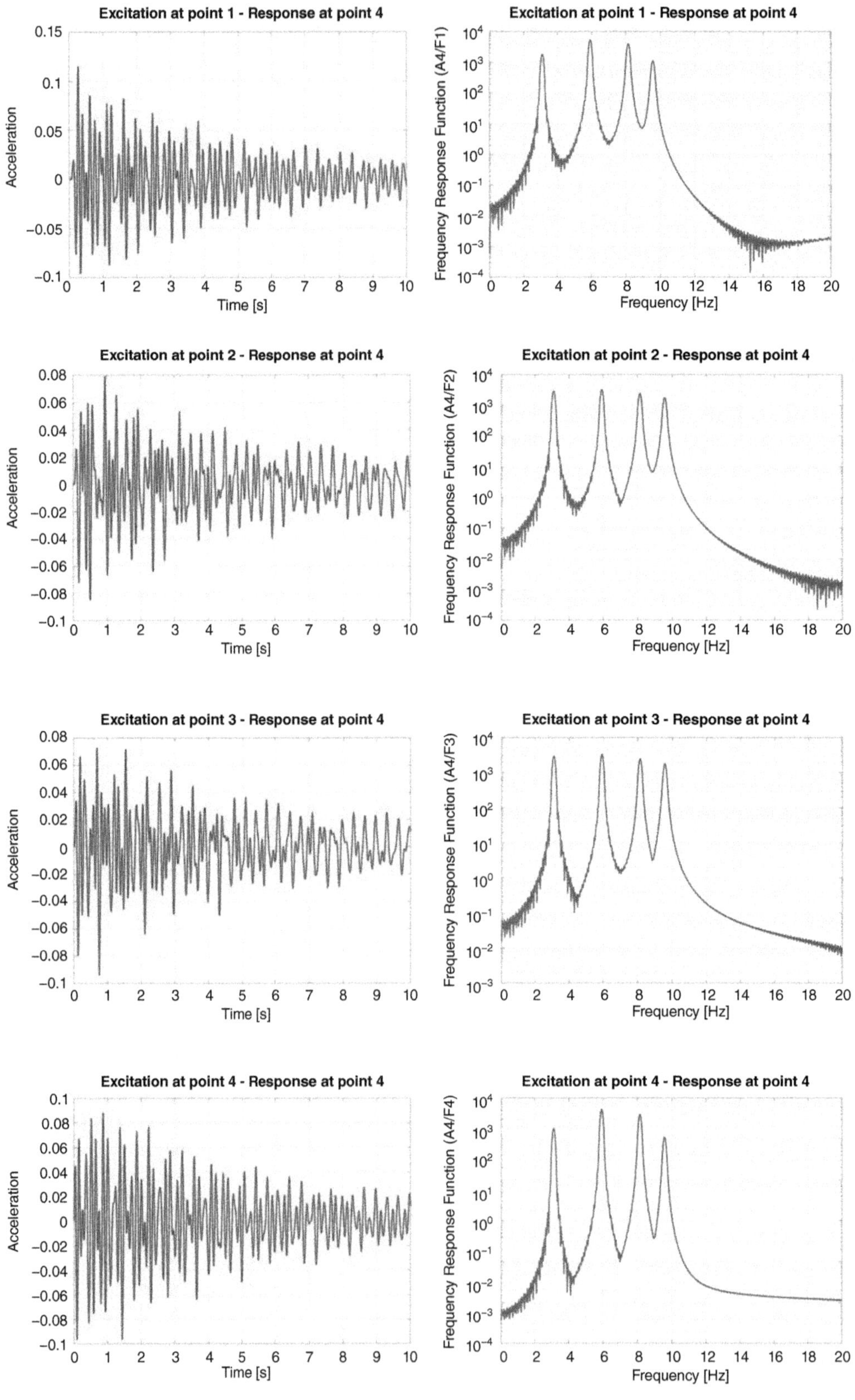

Figure 2.20 Time-histories (left) and corresponding frequency response functions (right) for impact testing (Fixed Sensor – Moving Excitation – Light damping).

The first example refers to impact testing, neglecting for the sake of brevity, the cases of double hits, or input/output noise which may add uncertainties to the results. Impact testing on an *MDOF* system may require the decision of measuring always at the same location and impacting at different points, or *vice versa*, impacting at the same point and measuring at different locations.

Figure 2.20, referring to the system of Figure 2.19, shows the option of moving the excitation, maintaining the "measurement" point fixed (number 4). The simulation is relative to a lightly damped system using a Hanning window for the estimation of the *FRFs*.

As previously noted, the "most popular" window, the Hanning window, has been used for obtaining the estimation of the frequency response functions. Just to make an example on what can be the effect of windowing for such a very simple case, one can refer to the last time-history (excitation at point 4 and response at point 4) using a different window. Figure 2.21 shows the results, which in terms of identifying the natural frequencies may still be acceptable, but it provides wrong results for the damping estimation.

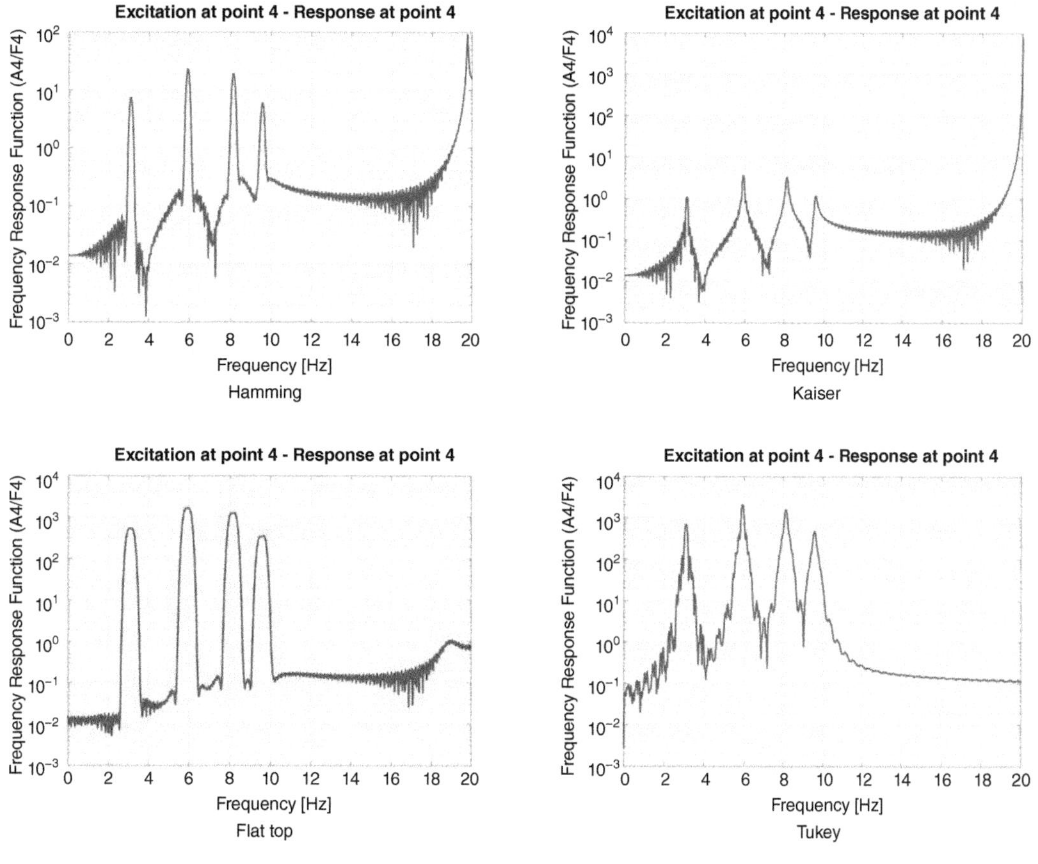

Figure 2.21 Estimation of the frequency response functions using different windows.

Coming back to the Hanning window, it can be verified that changing the procedure, fixing the excitation and moving the sensor, does not change the results. This is a significative outcome, showing the validity of the reciprocity principle and a linear behaviour of the system.

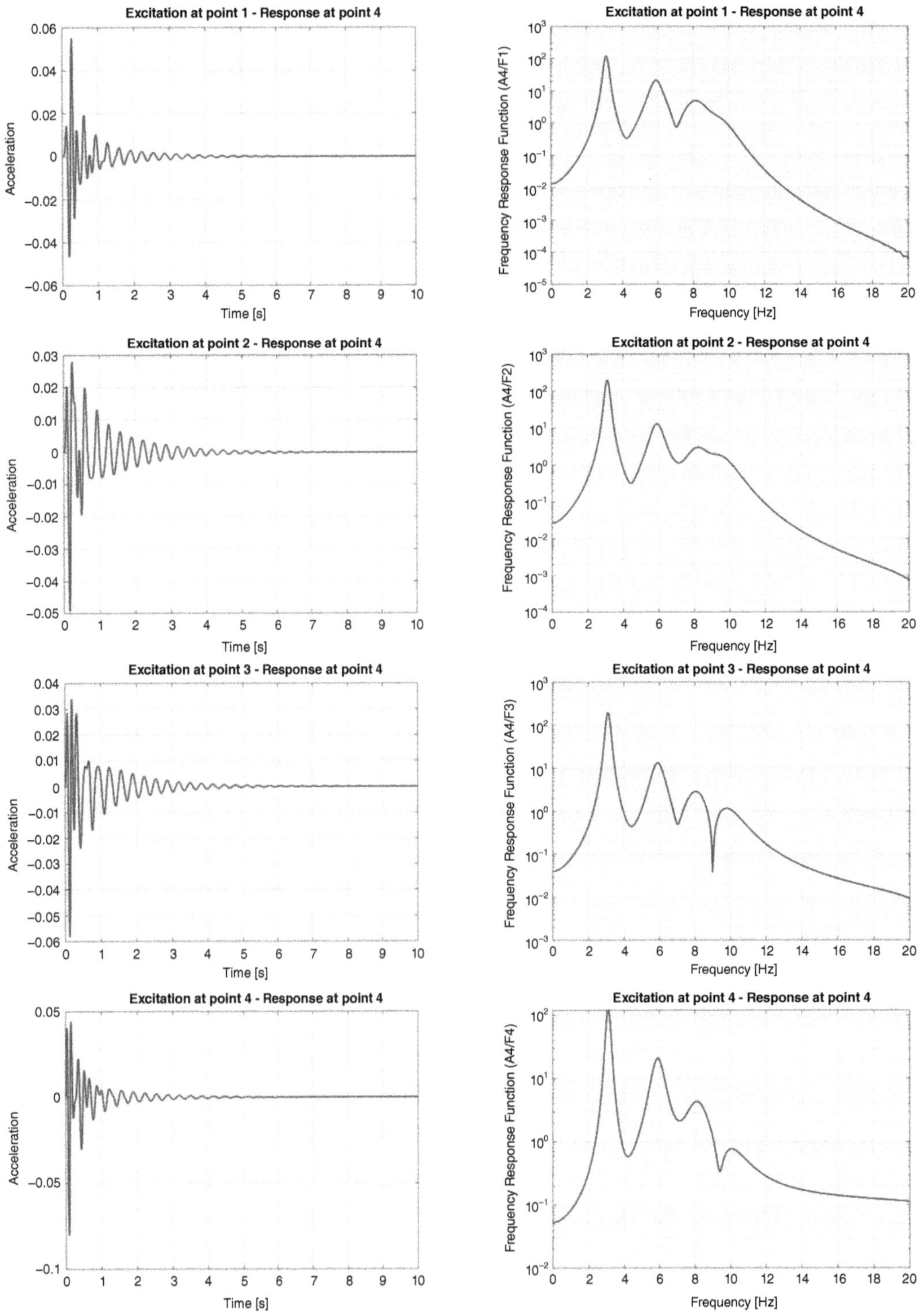

Figure 2.22 Time-histories (left) and corresponding frequency response functions (right) for impulse testing (Fixed Sensor – Moving Excitation - Heavy damping).

As expected, increasing the damping makes the shapes of the curves change, as in Figure 2.22, where the same conditions of Figure 2.20 are presented, but considering a much heavier damped system, by changing the values of c_i from 2.0 to 40.0 [*Ns/m*]: there is a slight modification on the values of the natural frequencies, but a substantial increase on the values of the damping factors (about 20 times higher):

$$\{\omega\} = \begin{Bmatrix} 19.54 \\ 37.17 \\ 51.17 \\ 60.15 \end{Bmatrix} \text{ [rad/s]} \qquad \{\xi\} = \begin{Bmatrix} 0.078 \\ 0.149 \\ 0.206 \\ 0.242 \end{Bmatrix}$$

The new completely different time and frequency behaviour is presented in Figure 2.22.

In this case, the four natural frequencies (information that one is not supposed to have) are not always evident and in some cases they tend to be disguised. This result may affect the selection of the identification method and the linear behaviour of the structure. Adding some uncertainty (noise from measurements, not rigid constraints, imperfect excitation, and so on) coming from field testing, may suggest the use of a different method, less convenient and practical than impact testing, but more effective.

2.4.1 Validating Experimental Measurements: the Coherence Function

A very helpful function, already introduced in Chapter 1, is the coherence function. It relates how well the output of the structure under test is correlated to (or "is coherent with") the input to the structure itself. This function helps during the measurements because it gives, frequency by frequency, an indication of the quality of the frequency response function that is under acquisition. As already defined in Chapter 1, the coherence (or ordinary coherence, sometimes normalised coherence) is expressed by the following relationships:

$$\gamma^2(\omega) = \frac{H_1(\omega)}{H_2(\omega)} = \frac{\left|\hat{S}_{fx}(\omega)\right|^2}{\hat{S}_{ff}(\omega)\hat{S}_{xx}(\omega)} \tag{2.7}$$

where $H_1(\omega)$ and $H_2(\omega)$ are two alternative estimators for the *FRF*, as defined in Chapter 1 (Eqs. (1.165) and (1.166)); $\hat{S}_{ff}(\omega)$ and $\hat{S}_{xx}(\omega)$ are, respectively, the numerically computed power spectral densities of the force and the response and $\hat{S}_{fx}(\omega)$ is the cross-spectral density, also defined in Chapter 1 (Eqs. (1.162), (1.163) and (1.164)).

The coherence is a real function valued between zero and one. When it is zero, the measured output is not correlated to the input, at that frequency. If the value of the coherence is close to 1, the output is almost exclusively depending on the input forcing function and not polluted by other sources.

Figure 2.23 shows four different conditions depending on the presence of unwanted noise which is polluting the measurement on input and/or output. The results are based on the previous *MDOF* system lightly damped, excited at *DOF* x_4 and "measured" at the same point. In the case of absence of noise, the coherence function is almost 1 everywhere except at the resonant frequencies, where the ratio between big and small values tends to reduce it. The presence of noise is more critical when it is applied to the output signal. Such a condition is the one that mostly degrades the coherence.

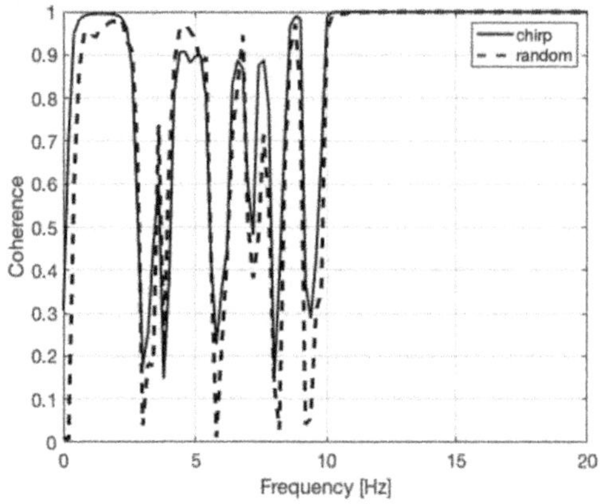

No noise both input and output

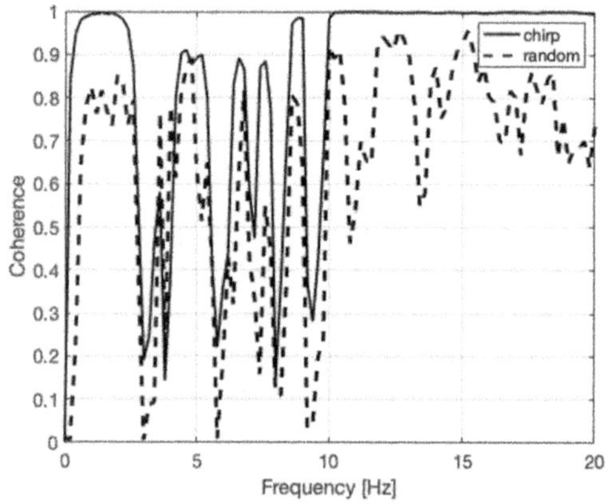

Noise on input only

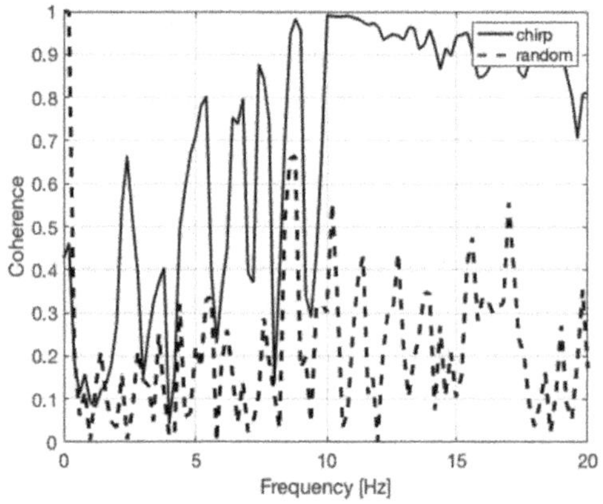

Noise on output only

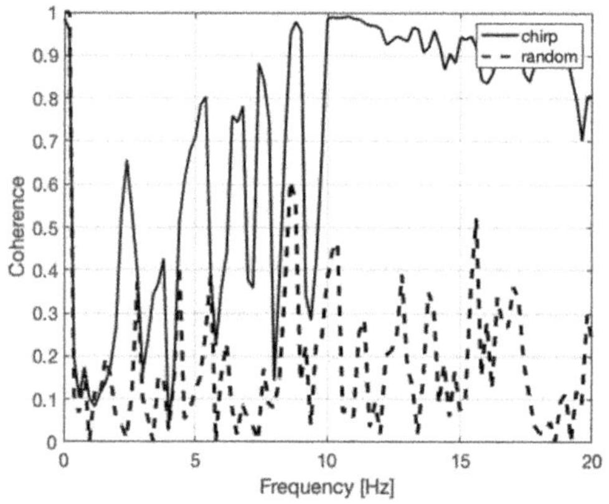

Noise on both input and output

Figure 2.23 Coherence function calculation for an *MDOF* system with chirp and random excitation and noise in case of polluted input and/or output signals.

It should be noted that, in practical measurements, the coherence function is based on statistical averages of the spectral densities. If one computes the coherence based on a single measurement, it will result equal to unity, even in the presence of noise, because no information is available to correlate input and output. After several measurements (at least two of them, but usually much more) are processed for an averaging process, the coherence can detect the lack of correlation, if any.

Figure 2.23 top right shows the case in which the noise is simulated on the input excitation only. The time history of the output suffers from the input noise, but the response is correlated to the input (no added noise to the output) and therefore the coherence function is still acceptable (close to the unitary value), quite similar to the case of absence of noise.

This is not true anymore when the input and the output time histories are not fully correlated, which may happen for both cases of noise on the output signal only, Figure 2.23 bottom left, or, similarly when the signals are polluted both on input and output, Figure 2.23 bottom right. The coherence function shows the effect of the lack of correlation between input and output.

During real testing, the coherence function could be used as a monitor function for an immediate feedback on the quality of the measurements, and in some cases, take proper actions for its improvement. For example, increasing the gain in the force power amplifier may also generally improve the coherence and thus the *FRF*, resulting in a better system identification. Or one may accept low coherence because the measurement involves points close to constraints, or complex geometries, locally damped points and so on.

In order to show how the coherence function can help for assessing the validity of the measured data, a simple cantilever beam has been prepared and tested. Figure 2.24 shows the test specimen including the positions of the input (load cell) and the output (accelerometer) sensors.

This simple structure has been excited by an electrodynamic shaker driven by a signal generator. In the examples presented herein, chirp (sine sweep) and random (white noise) signals have been used in a frequency range encompassing the first natural frequencies of the cantilever beam.

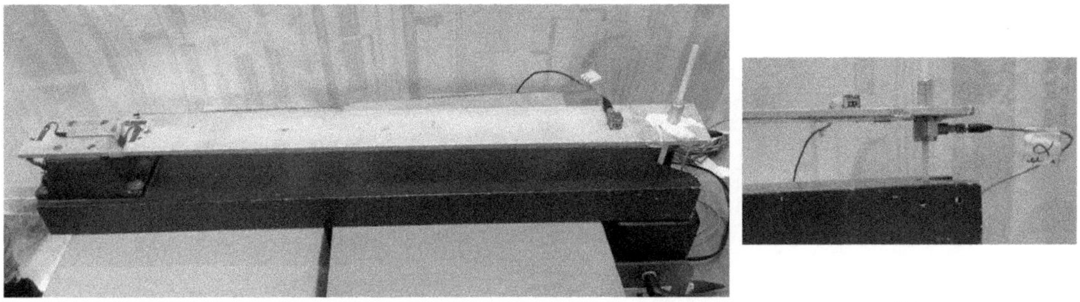

(a) (b)

Figure 2.24 (a) Cantilever beam used for *FRF* and coherence measurements and (b) position of the sensors.

Figures 2.25 and 2.26 show the results of the measurements, in the case of high frequency resolution (1024 points) and low frequency resolution (512 points), respectively. Both figures present the measured accelerance *FRF* (top) and the corresponding coherence (bottom), for chirp (left) and random (right) excitation.

These measurements have been obtained using the same number of averages (16) and they show that, in this simple case, chirp excitation offers a *cleaner* view of the *FRF* for both resolutions, compared to the random excitation, while the coherence function runs close to unity for almost all the measured frequency range, except close to the natural frequencies where it drops down as expected. In the case of random excitation, this behaviour is much more evident.

The plots obtained with chirp excitation and low frequency resolution (top and bottom left side of Figure 2.26) do not lose information of the *FRF*, but they do for the coherence function. The drops in correspondence with the natural frequencies are not evident anymore. The same conclusions do not apply for the random excitation (right side of Figures 2.25 and 2.26). The corresponding diagrams are less affected by the different frequency resolution even though in all cases they show a *crispier* behavior. The low frequency resolution is usually less time consuming, but – depending on the excitation function, it might not be the best choice. On the other hand, random excitation appears to give more stable behavior, but it could result more difficult to analyse for more complex structures or in the case of close natural frequencies and mode shapes.

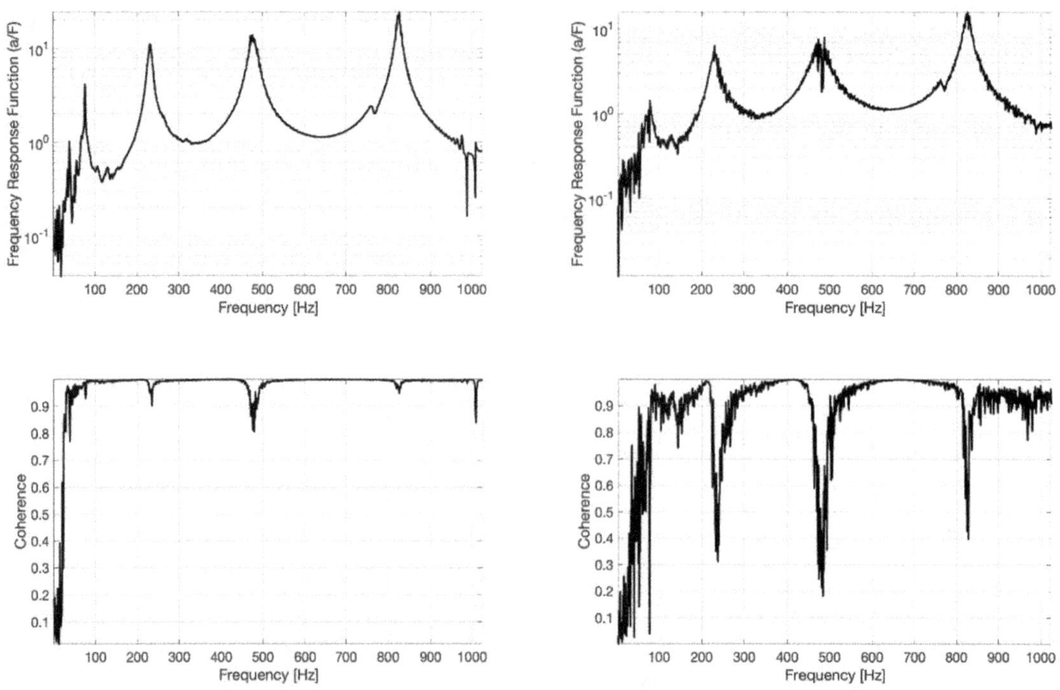

Figure 2.25 High frequency resolution. Chirp excitation (left) and random excitation (right).

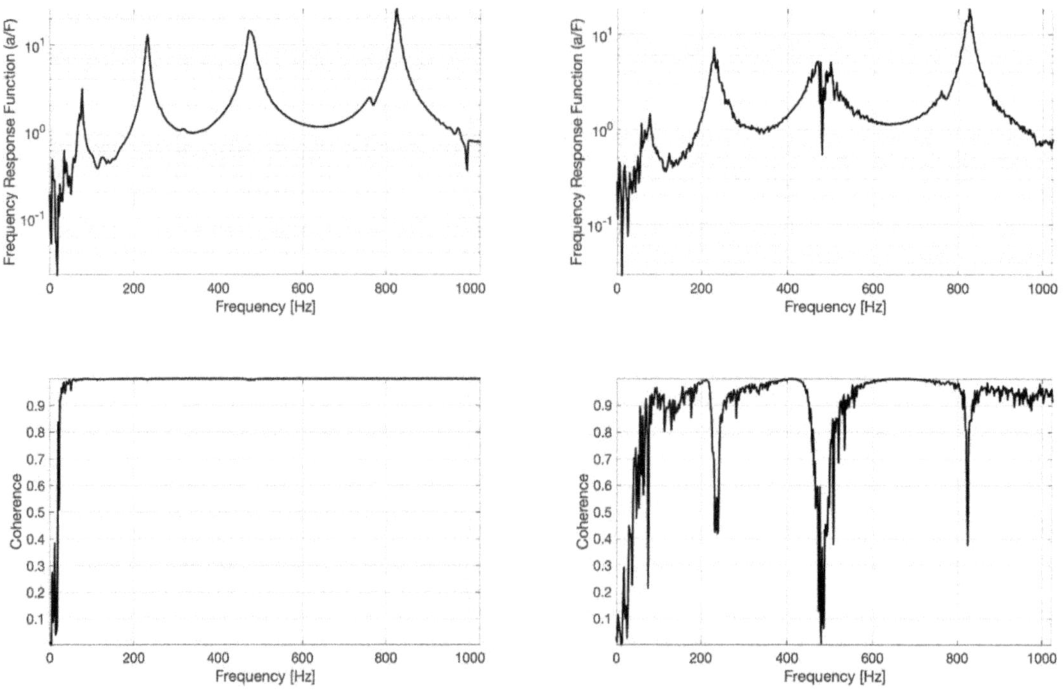

Figure 2.26 Low frequency resolution. Chirp excitation (left) and random excitation (right).

Additional parameters may be considered like the number of averages; increasing them tends to improve the measurements and consequently improve the coherence; beyond a certain value, though, the improvement is not visible. It has been fixed in this example, but it could be an additional factor to be left to the experience and time availability of the test engineer. In general, the overlapping of the time records (for example 50%) also improves the results.

An important conclusion can be drawn at this point and it refers to the fact that it does not exist the *best* method for measuring the dynamic behavior of a structure. One should never forget that the structure does not *know* the mathematical models. It just behaves as a result of the given excitation. One has created tools (numerical and experimental models) for interpreting and identifying its behaviour and depending on the specific situation, one should be able to select the best method for that specific case.

In general, the modal testing process is not an easy task and it should be performed putting together and harmonising information coming from different perspectives. This process may justify the need for including the step "Engineering Judgement" in the block diagram of Figure 2.2.

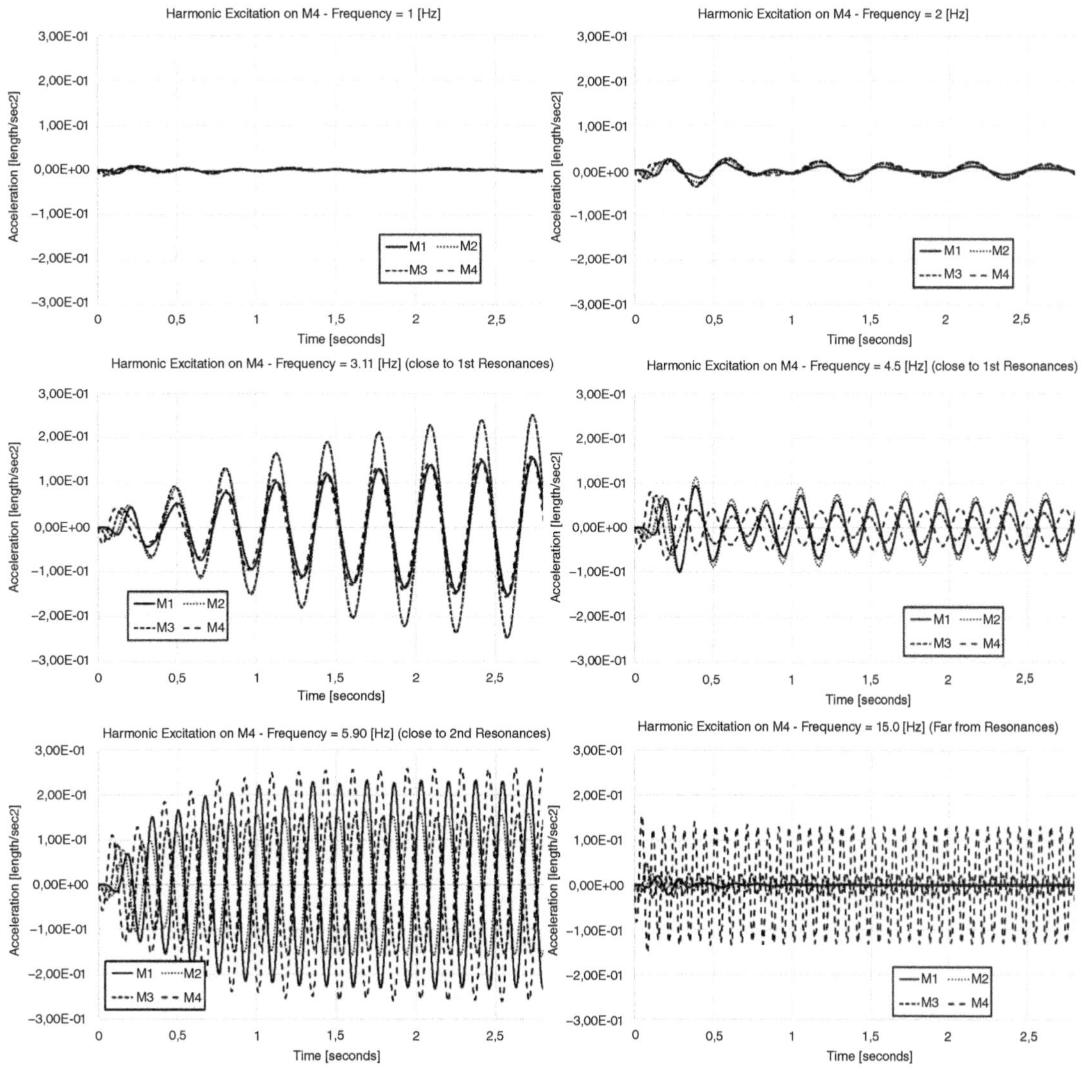

Figure 2.27 Time-histories measured on a damped 4 *DOF* system sinusoidally excited at different frequencies.

2.4.2 Time-Response to Harmonic, Stepped-Sine, Excitation

When dealing with a complex structure it may be worth to come back to the simple theory of the harmonic excitation and remembering that the structure will respond with a filtered harmonic answer. This old, but still valid, approach conflicts with the necessity for long testing time, generally not allowed. Exciting the previous 4 *DOFs* damped system of Figure 2.19 with a sinusoidal force at several frequencies, applied – for example – to mass no. 4, one obtains the results presented in Figure 2.27.

Naturally, Figure 2.27 confirms that far from the transient time duration, the system responds with the same frequency of the excitation, with the amplitude and phase among the *DOF*s depending on the vicinity of the excitation frequency to a system resonance. Plotting the amplitude and the phase shift of each *DOF* versus the frequency, the transformation into the frequency domain is obtained. The system can be easily identified with a better signal-to-noise ratio, which results in a more precise identification.

This procedure, known as stepped-sine, allows the assessment and verification of potential non-linearities, but it clearly requires a much higher testing duration than more sophisticated algorithms. It can be suggested as a companion method only for very peculiar conditions and short frequency range.

2.4.3 Other Types of Excitation

The use of the frequency sweep excitation, in the case of an *MDOF* system, is a technique that attempts to take advantage of both good excitation and a wide frequency range of analysis (instead of a single frequency per time). An example of a sine sweep excitation is given in Figure 2.28 and it was used on mass #4 of the *MDOF* system of Figure 2.19, for simulating the results presented in the next figures.

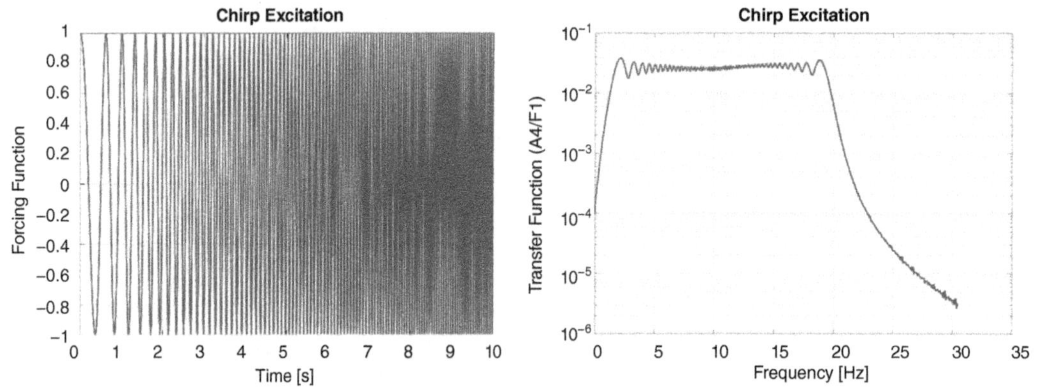

Figure 2.28 Linear sine sweep excitation in time (left) and frequency (right) domains.

The equation of sine sweep excitation assuming a linear sweep of the frequency between two values is given by:

$$F(t) = F_0 \, sin(2\pi f_i(t)) \tag{2.8}$$

where the instantaneous frequency sweep $f_i(t)$ is given by $f_i(t) = f_0 + \beta t$, with $\beta = \frac{f_1 - f_0}{t_1 - t_0}$ and f_0 is the frequency $f_i(t)$ at time t_0 and f_1 is the frequency $f_i(t)$ at time t_1. Other expressions for $f_i(t)$, with quadratic sweep, logarithmic, etc., may be used for the frequency varying between f_0 and f_1.

For pointing out some definition and examples, the following recalls the meaning of sine sweep, multi-sine and chirp excitation. They are not exactly the same thing, but very similar and sometimes they can be confused. Some clarification on random signals is also given.

Sine Sweep

Structures and components may be mounted on shaker tables and subjected to sine sweep base excitation. The essence of a sine sweep test is that the base excitation input consists of a single frequency at any given time. The frequency itself, however, varies with time. The sine sweep test may begin at a low frequency and then sweep to a high frequency, or vice-versa. Some specifications require several cycles, where one cycle is defined as from low to high frequency and then from high back to low frequency.

The specification might require either a linear or a logarithmic sweep rate. The sweep will spend greater time at the lower frequency end if the sweep is logarithmic. Usually the amplitude is kept constant, but specifications might require that the amplitude vary with frequency.
Some potential test objectives are:

- Identify natural frequencies and amplification factors or damping ratios;
- Perform sine sweep before and after random vibration test to determine if any parts loosened, etc.;
- Check for linearity of stiffness and damping by varying the input amplitude;
- Workmanship screen for defective parts and solder joints;
- Represent an actual environment such as a rocket motor oscillation;
- *NASA/GSFC* typically uses sine sweep vibration for spacecraft testing.

Using sine sweep excitation, the responses at each *DOF* of the system in Figure 2.19 are presented in Figures 2.29 and 2.30 for the undamped and damped conditions, respectively. Such results should be compared with those presented respectively in Figures 2.20 and 2.22.

It should be mentioned that the same algorithm has been used for both these analyses, especially referring to sampling frequency, windowing (Hanning window in these cases) and frequency response function estimation. Both excitation methods show the obvious effect of damping. For the low damping case the peaks at the natural frequencies are clearly defined, with the mode shapes well excited by both forcing functions. Increasing the damping, the sine sweep excitation tends to smear out the peaks more than the impulse force. The companion time-histories justify this behaviour, being the energy of the forcing function captured by a reduced number of modes. Indeed the time history of Figure 2.22 (excitation at point 4 – response at point 4, for example) shows the presence, at its beginning, of several frequencies, while its companion of Figure 2.30 evidences the dominance of a damped frequency for most of the time, obscuring the effect of the other frequencies.

Pure sine excitation, when dealing with highly damped structures, may represent the solution for avoiding misinterpretation of the dynamic behaviour of a system.

The damping effect is obvious from Figures 2.29 and 2.30. The damped system tends to spread out the natural frequencies, flattening the peaks, making much more difficult to identify them. In these cases, the use of time domain identification methods may result more effective, even if the algorithms can result more complex.

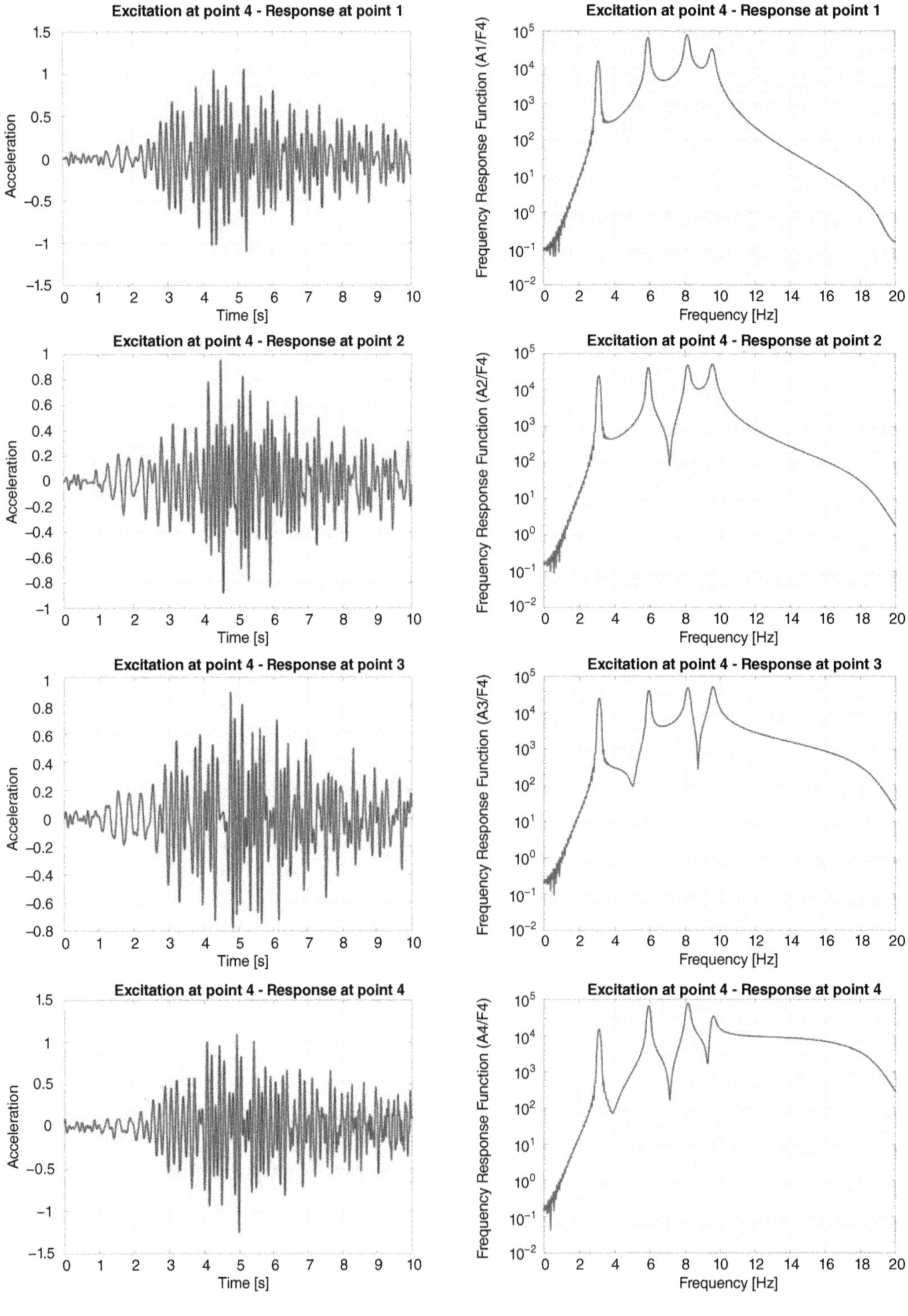

Figure 2.29 Time-histories (left) and corresponding *FRFs* (right) for the undamped *MDOF* system.

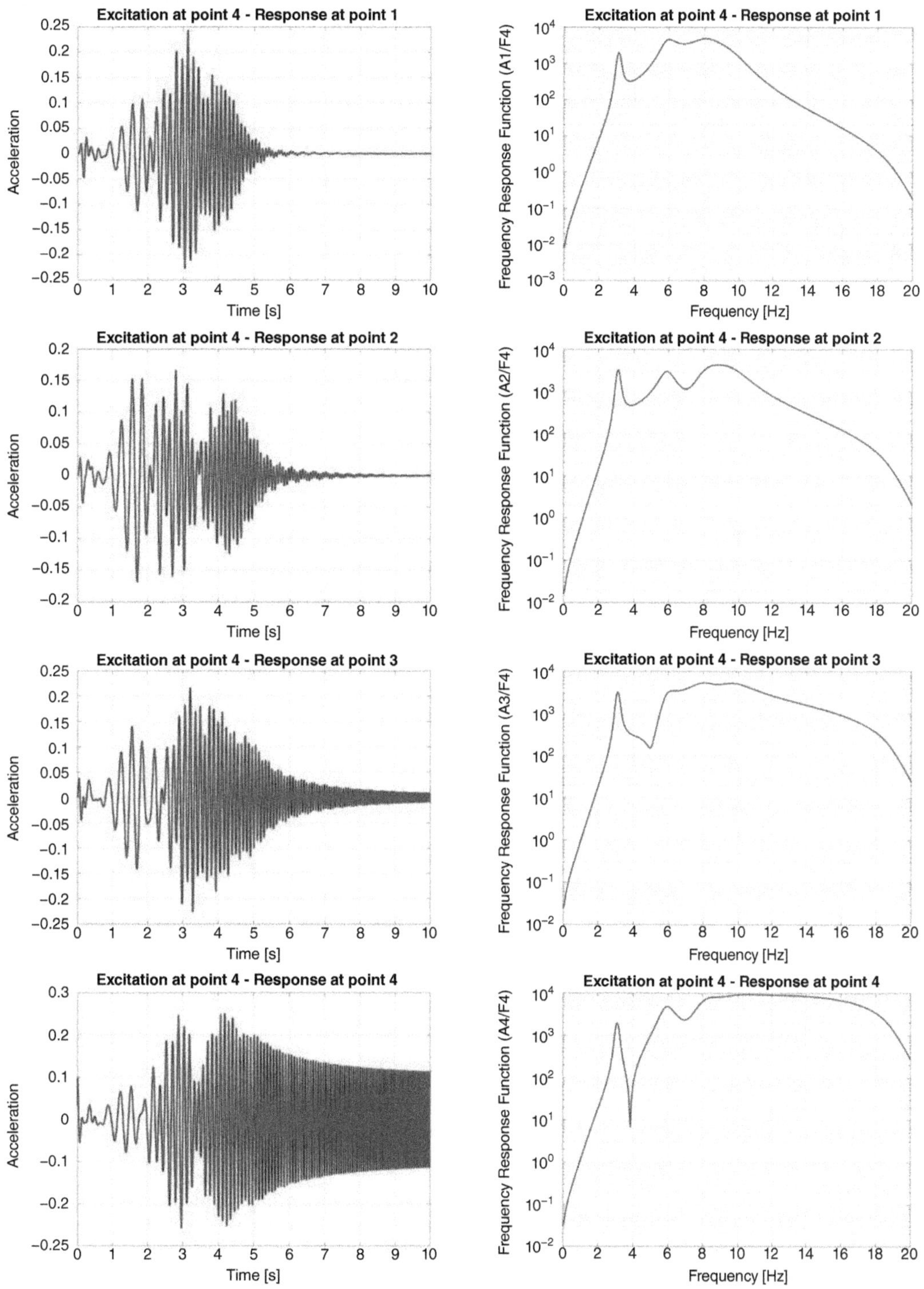

Figure 2.30 Time-histories (left) and corresponding *FRFs* (right) for the damped *MDOF* system.

Multi-Sine Sweep

The Multi Frequency Sine method involves use of multiple true sine tones swept simultaneously across the frequency range to excite all resonances. The primary advantage of this technique is that it significantly reduces the required test duration in such a way that the control is not compromised. Each sine tone utilizes a high quality tracking filter to ensure that testing is done at the correct level.

The Multi Frequency Sine control software controllers enables multi frequency sine testing without sacrificing control accuracy and performance. The time reduction comes from dividing the sweep frequency range into multiple intervals. For example, four sine tones swept simultaneously across the same frequency range at the same sweep rate will reduce the time required to produce the same fatigue by 75 percent compared to a single frequency swept sine test. This can reduce a typical test that takes 84 hours per axis to just 21 hours per axis.

Chirp

A chirp is a signal in which the frequency increases (up-chirp) or decreases (down-chirp) with time. In some sources, the term chirp is used interchangeably with sine sweep. Chirp appears as a generalisation of the sine sweep excitation. In a linear-frequency chirp, or simply linear chirp, the instantaneous frequency varies exactly linearly with time:

$$f(t) = ct + f_0 \tag{2.9}$$

where f_0 is the starting frequency (at time $t = 0$), and c is the chirp rate assumed constant: $c = \dfrac{f_1 - f_0}{T}$, where f_1 is the final frequency; T is the time it takes to sweep from f_0 to f_1.

In a geometric chirp, also called an exponential chirp, the frequency of the signal varies exponentially with time. If two points in the waveform are chosen, t_1, and t_2, and the time interval between them $(t_2 - t_1)$ is kept constant, the frequency ratio $f(t_2)/f(t_1)$ will also be constant. The frequency of the signal is given by:

$$f(t) = f_0 k^t \tag{2.10}$$

where f_0 is the starting frequency (at time $t = 0$), and k is the rate of exponential change in frequency. Unlike the linear chirp, which has a constant chirpiness, an exponential chirp has an exponentially increasing frequency rate:

$$k = \left(\frac{f_1}{f_0}\right)^{\frac{1}{T}} \tag{2.11}$$

According to the classification, chirp excitation is the most general and complete definition.

Random, Pseudo-Random and Periodic Random

Other excitation signals are commonly used in addition to those so far discussed. In general, one can classify the excitation functions into four categories: steady-state, random, periodic and transient. It should be pointed out that each type of excitation has its own advantages and drawbacks, and it should be selected depending on the test requirements and post-processing opportunities (software, available time, acceptable precision and so on).

Referring to random signals, sometimes there is some confusion between random, pseudo-random and periodic random excitation. The random, or pure random, signal is a stationary, ergodic signal whose random time history is generated with the constraint of a Gaussian probability distribution. In general, its frequency content includes all the frequencies in the band of interest. A pseudo-random signal is again a stationary ergodic signal, but containing in its frequency spectrum only integer multiples of the *FFT* frequency increment, with constant amplitude and random phase. It can have several advantages when compared to the random signal. It is more cost-effective in the sense that it allows a more accurate estimate of the structural response over a broader range of frequencies using a smaller amount of time. Indeed, it requires a smaller number of averages for reducing uncertainties in the frequency response and has a better signal-to-noise ratio. The main disadvantage compared to the random signal is that the loading sequence is repeatable and there can be peak resonances that are not excited because they do not correspond to one of the pseudo-random frequencies. On the other hand, periodic random signal has a frequency spectrum which is random in amplitude and phase, always from a stationary ergodic signal. The excitation signal is periodically exciting the structure whose response is periodic with respect to the sample period.

2.4.4 A Different Representation: the Lissajous Curves

An example of different plots of the same time-histories of Figure 2.25 is given in Figure 2.31. The previous measurements are now presented as the acceleration of one *DOF* versus the acceleration of another *DOF*. Such polar plot (each point is also a function of another hidden parameter; time in this case) is called Lissajous curve (or figure), from the name of the French physicist Jules Antoine Lissajous. They give another representation of the resonance (harmonic excitation of 3.11 and 5.90 Hz, in this case) and the relative phase among the degrees of freedom, when approximating a straight line with a specific slope.

In Figure 2.31 the scales of the axes have been kept constant, therefore it is possible to identify the vicinity of a resonant frequency by the amplitude of the curves and their average slopes give also the phase relationships of the *DOFs*. Lissajous figures are often used in ground vibration testing (*GVT*), where the information about the phase can be particularly useful for force appropriation[4], to detect which responses are in quadrature (90°), to identify normal mode shapes. Table 2.2, synthesizes the main characteristics of the most used excitation functions in modal testing:

Table 2.2 Characteristics of the most used excitation functions.

	Sine	**Impact**	**Sine sweep**	**Multi-sine sweep/Chirp**	**Pseudo Random**	**Random**
Minimise leakage	No	Yes	Yes	Yes	Yes	Yes
Signal-to-noise	Very high	Low	High	High	Fair	Fair
RMS to Peak Ratio	High	Low	High	High	Fair	Fair
Test measurement time	Very long	Very good	Long	Long	Very good	Fair
Controlled frequency content	Yes	No	Yes	Yes	Yes	Yes
Controlled amplitude content	Yes	No	Yes	Yes	Yes	No
Characterise nonlinearlity	Yes	No	Yes	Yes	Yes	No

[4] See Chapter 4, Section 4.6.

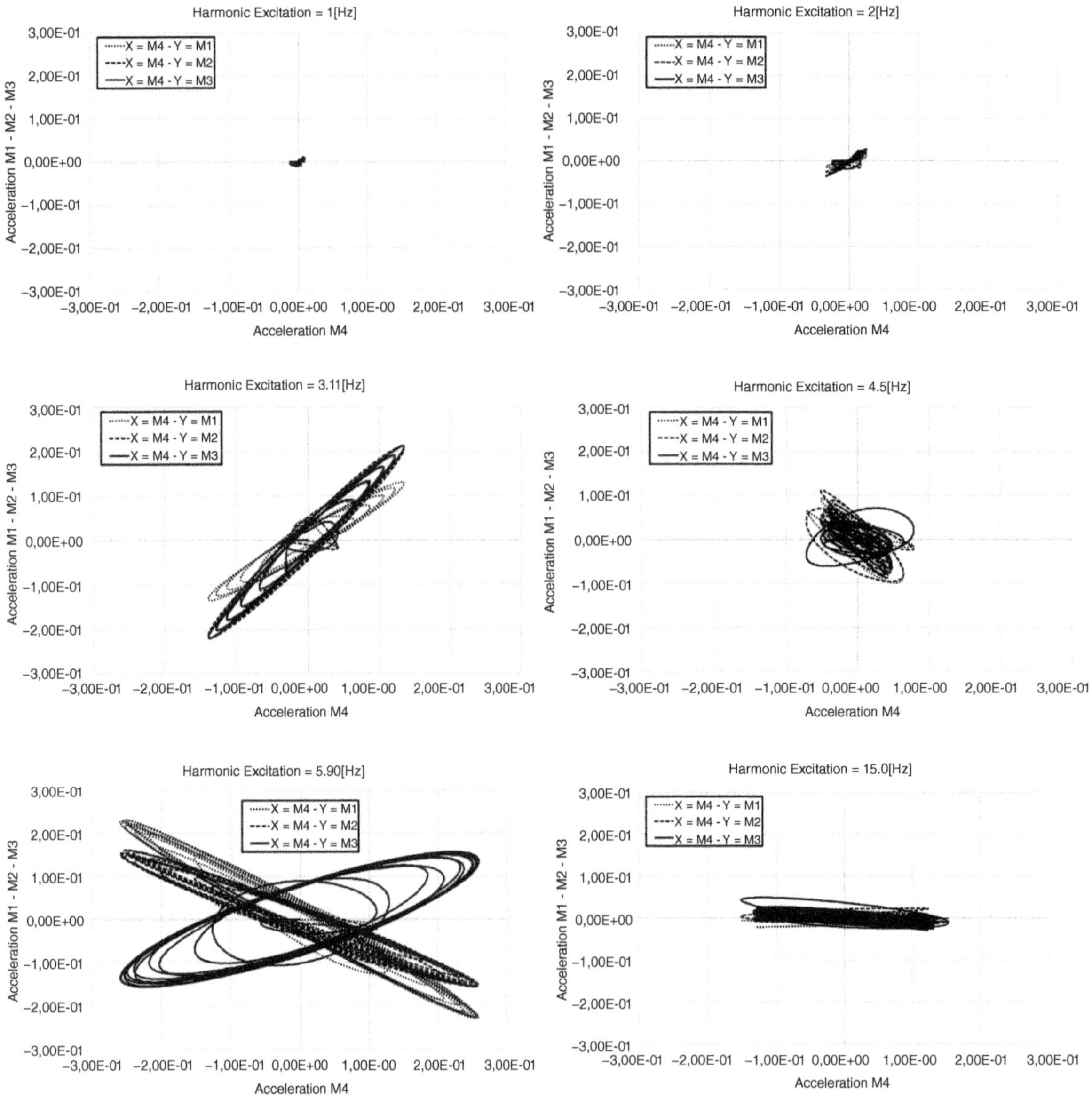

Figure 2.31 Lissajous plots of the time-histories of Figure 2.27 assuming in this example the acceleration of mass #4 as the reference.

2.4.5 Hammer Testing *versus* Shaker Testing

To obtain the vibration characteristics of a structure implies the use of an excitation system able to efficiently allow the vibration of what one wishes to measure.

Hammer testing uses an impact hammer to excite the structure under test. The impulse force applied to the structure is categorised as a broadband range excitation because it contains energy up to a certain maximum frequency value. The hammer impact test is quick to set up and to carry out, and thus is widely used. However, it does have its limitations and there are some concerns to be considered. The hammer tip or its material plays a big role in determining the upper frequency required by the test. On the one hand, a soft tip leads to a relatively lower level impact with a wider time length and short frequency range, because it does not provide too much energy. It may be appropriate for testing a light structure. On the other hand, a harder tip provides a shorter time impulse and a wider frequency range, which can be checked looking at the auto power spectrum. It could be useful for a heavier structure. The following graph, Figure 2.32, illustrates this in the time and frequency domains. It is advisable to choose a hammer tip material with an auto spectrum decay less than 6 dB from the upper frequency of the set-up.

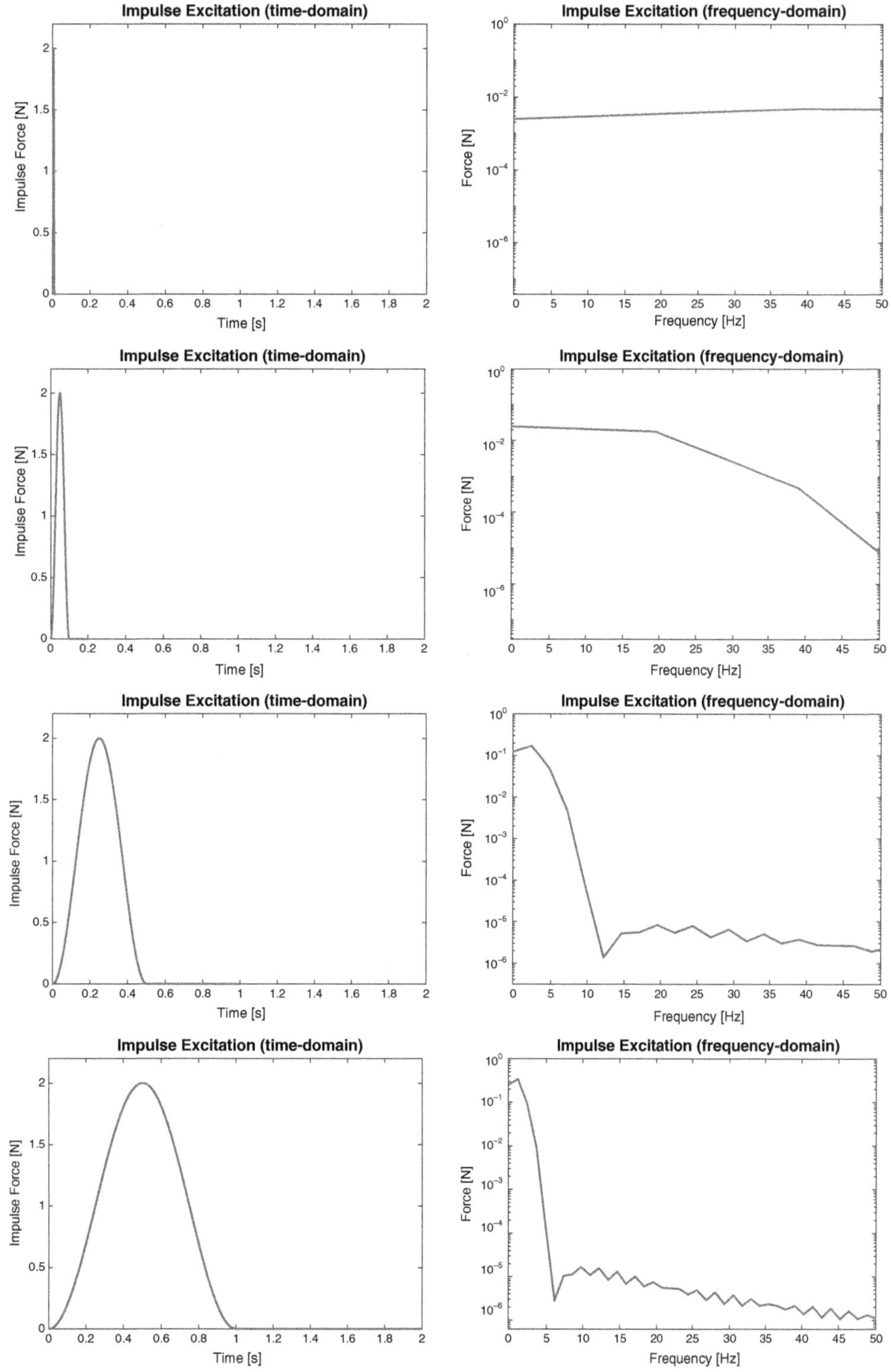

Figure 2.32 Time-histories (left) of several impulses with different duration and corresponding auto power spectrum. Increasing the time duration, reduces the frequency content of the excitation. The resulting magnitude of the force, in the frequency domain, is a function of such frequency range.

Shaker testing requires a modal shaker to excite the structure under test. The shaker is driven by an output channel of the dynamic signal analyser, which – though a power amplifier – allows the shaker to properly excite the structure. Users can select from a list of waveform types to excite the structure under test. This testing method is usually applied to complex or large scale structural testing. Compared to the hammer impact test, it is easier to replicate. On the other hand, the modal shaker testing method requires more hardware (i.e., modal shaker(s), stinger, more input and output channels from the dynamic signal analyser) and an experienced user. During a shaker testing, the excitation point remains fixed, while measurements are taken by one or more sensors roving along the structure. Being time independent measurements, the number of employed sensors can range from one to hundreds, depending on the capability of the acquisition system. Based on the number of shakers, the vibration test may be categorised as *SIMO* (Single Input Multiple Output) or *MIMO* (Multiple Input Multiple Output).

Hammer testing is generally easier to set up, while shaker testing allows a better definition of the forcing function and the possibility to measure nonlinear behaviour of the structures.

A final consideration should be devoted to dynamical data measured from environmentally excited structures, or measurements in operational conditions. The identification of the dynamic properties of a structure is important for a numerical-experimental correlation, but – in general – it cannot tell whether the result of the identification has critical or unstable conditions. Putting the structure in its operating environment can tell us which identified natural frequency (or frequencies) and corresponding mode shape (or mode shapes) are more important. Furthermore, unexpected behaviour might come out, possibly not measured during ground or lab testing, just because it was considered as not critical.

Measurements in real operational conditions are usually very important for several reasons, which may broadly summarised as:

- Confirming lab or ground testing under controlled conditions;
- Completing the correlation with numerical models;
- Assessing the linear behaviour of the structures;
- Ensuring the absence of unexpected and undesired behaviours.

The main difficulty of measurements in operational conditions comes from the impossibility to measure the input. Usually it comes from road, wind, turbulence, random noise and so on. The system identification may take advantage on the knowledge coming from previous measurements and numerical models. The real problem comes out when, from the operational conditions, unwanted vibrations come out, usually affecting the performance of the structure or of a specific component. In these cases, retrofit solutions are implemented, offering a partial, not optimised, solution. Magic sticks do not work in structural dynamics. The next paragraph will give some preliminary insight for time series data, evidencing the importance to understand and to drive each step of the data analysis.

2.5 Tips and Tricks for Dynamical Data Analysis

Time series analysis is not a new science and its numerous approaches for the interpretation of the underlying phenomena may be overwhelming to the experiment investigator who is more interested in analysing his experimental results than in knowing how to calculate and use a Fourier transform. This paragraph gives an overview of the simplest data analysis techniques that may be initially employed when dealing with dynamical measurements. Such summary is provided for easier understanding of the results obtained from acceleration data together with their dynamic definition and giving examples of data interpretation when necessary.

In the analysis of time series, certain restrictions are imposed by the length of the data window being analyzed and by the sampling rate, f_s, used when digitising continuous data. For a time series segment of length T seconds (N total points), the fundamental period of the segment is assumed to be T, even though the series is not necessarily periodic. This periodicity assumption is intrinsic to the calculation of the Fourier transform, which is the basis for all spectral analysis herein discussed. The finest frequency resolution obtainable is $dF = f_s/N$. A lower value of dF is considered a better resolution than a higher value of dF.

A constant sampling frequency f_s equal to 100 samples per second is assumed in the following examples, while the time series segment T may be variable, depending on the possibility of reaching and maintaining a specific measurement condition. These conditions usually happen in real measurements, fixing the sampling frequency and measuring for a time duration which is not a priori known.

2.5.1 Time Domain Analysis

The time domain data analysis techniques are acceleration *versus* time, interval average acceleration *versus* time, interval root-mean-square (*RMS*) acceleration *versus* time, quasi-steady three-dimensional histogram (*QTH*) of acceleration data and trim mean filtered (*TMF*) acceleration *versus* time. All these options, except the histogram, can be presented on a per axis basis or as a vector magnitude. Table 2.3 provides an overview of the time domain analysis techniques discussed here.

Table 2.3 Time domain analysis techniques.

Analysis Technique	Units	Use
Acceleration *vs.* time	g or fraction of g	Precise accounting of the variation of acceleration magnitude as a function of time
Interval Avg. Acc. *vs.* time	g or fraction of g	Indication of net accelerations lasting for a time period ≥ parameter interval
Interval *RMS* Acc. *vs.* time	g or fraction of g	Measure of oscillatory content in data
Trim Mean Acc. *vs.* time	g or fraction of g	To smooth data and reject transient, higher magnitude contribution
Quasi-steady 3-D Histogram	Percentage of time and g or fraction of g	Summarises the acceleration vector magnitude and direction

Acceleration *versus* Time

Acceleration magnitude *versus* time plots show acceleration in units of g *versus* time. One begins with a specific measurement which yields the most precise accounting for the variation of acceleration magnitude as a function of time. The required data are all hidden in this time history and they need to be discovered by an accurate investigation. Understanding and highlighting the hidden parameters of the system does not give a solution for undesired behaviours, but it is a good starting point. The length of time represented in a plot is usually determined by the focus of the investigation.

Average Acceleration in a Time Interval *versus* Time

A plot of the average acceleration in a time interval (also called moving average) in units of g *versus* time gives an indication of the net accelerations that last for a number of seconds equal to or greater than the interval parameter. Shorter duration, high amplitude accelerations can also be detected with this type of plot. However, the exact timing and magnitude of specific acceleration events cannot be extracted. The average acceleration in a time interval for the x-axis is defined as

$$x_{avg_k} = \frac{1}{M}\sum_{i=1}^{M} x_{(k-1)M+i} \quad \text{with} \quad k = 1,2,\dots N/M \tag{2.12}$$

where N is the number of acceleration data points and M the number of points to consider in each average.

Averaging tends to smooth the appearance of the data. The higher the value of M, the higher the artificial damping introduced numerically into the time series. The choice of such a value is strictly depending on what one likes to enhance from the *original* measurement.

Root-Mean-Square Acceleration in a Time Interval *versus* Time

A plot of the root-mean-square acceleration in a time interval in units of g *versus* time gives a measure of the oscillatory content in the acceleration data. For the period of time considered, this quantity gives an indication of the time-averaged power in the signal due to purely oscillatory acceleration sources. The *RMS* acceleration in a time interval for the x-axis, provided the subtraction of the static average, is defined as

$$x_{RMS_k} = \sqrt{\frac{1}{M}\sum_{i=1}^{M}\left(x_{(k-1)M+i}\right)^2} \quad \text{with} \quad k = 1,2,\dots N/M \tag{2.13}$$

Trim Mean Acceleration *versus* Time

A trim mean filter is applied to raw measurement data to reject transient and higher magnitude accelerations. For example, one may wish to smooth the data to achieve an estimate of the quasi-steady accelerations experienced in-flight. It is an indication of the static value of the acceleration and should be a number close to 1 for the z-axis accelerations and 0 for the x and y-axis accelerations. Confirmation of these values is a check of the expected right measurement.

Quasi-steady Three-dimensional Histogram

The quasi-steady three-dimensional histogram analysis can be used for displaying a summary of the acceleration vector magnitude and alignment projected on three orthogonal planes. It requires that the physical parameter (acceleration) be measured along the three orthogonal axes.

Examples of Analysis in Time Domain

Next, one presents some examples of time domain analysis of the definitions and relative results discussed previously. The starting point in this case is a known and simulated measurement, in order to correlate the information of the measurement and what results from the application of the previous defined algorithms. Consider the following function:

$$x_i = A + B_1 sin(2\pi f_1 t_i + \varphi_1) + B_2 sin(2\pi f_2 t_i + \varphi_2) + B_3 sin(2\pi f_3 t_i + \varphi_3) + B_4 RND(t_i) \quad (2.14)$$

where *RND* is a random numbers function generator, usually producing simulated noise data ranging from 0 to 1, with the following values:

$$\begin{array}{lll} A = 1.0 & f_1 = 5.0\ Hz & \varphi_1 = 0.0 \\ B_1 = 0.8 & f_2 = 7.0\ Hz & \varphi_2 = \pi/4 \\ B_2 = 1.2 & f_3 = 15.0\ Hz & \varphi_3 = \pi/6 \\ B_3 = 1.5 & t_i = i\Delta t,\ i = 1, 2, \ldots, N & \\ B_4 = 0.4 & \Delta t = 0.01\ s & N \geq 6000 \end{array}$$

The simulated measurement is represented in Figure 2.33.

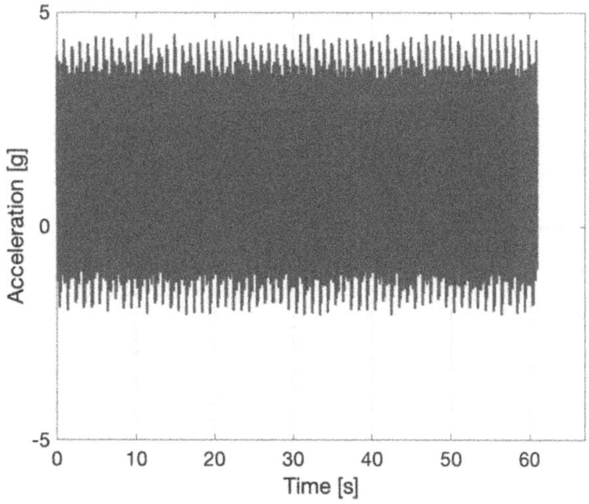

Figure 2.33 Simulated acceleration measurement.

Imagining this time-history as basic (raw) data, the plots of Figure 2.34 show how the moving average algorithm modifies the original time history. Such an algorithm is usually used for smoothing the original data, reducing peaks and valleys and for identifying trends. In fact, from the plots of Figure 2.34 it is clear the effect of the moving average algorithm, keeping the original value of the average, but smoothing the variations and reducing the range. As it happens for all data analysis algorithms, it is important to know how the algorithm affects the original data and draw considerations and conclusions accordingly. Such effects are also reflected in the values presented in the relative Table 2.4, where a comparison of the computed statistical scalar parameters is given, using a different step number (10 and 50) for interval average.

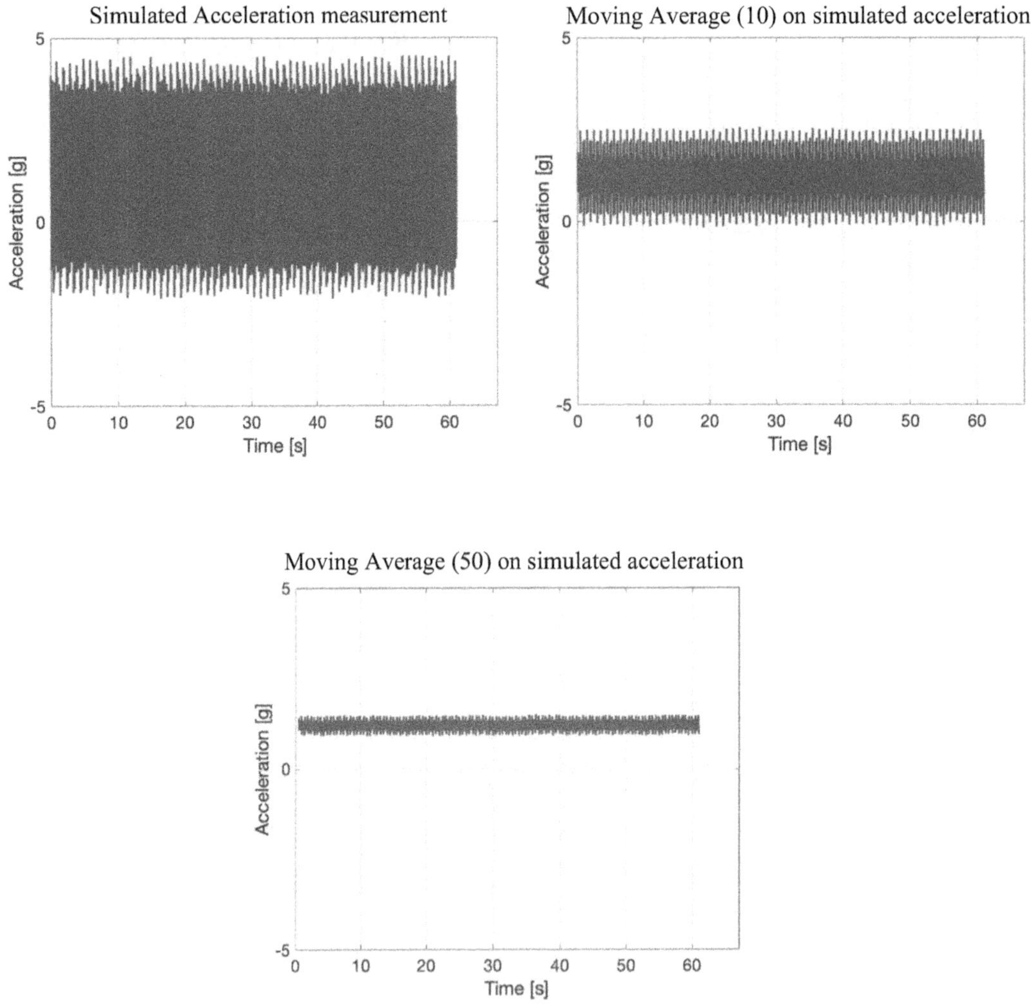

Figure 2.34 Moving average effect on raw data.

Table 2.4 Statistical data of time history.

	Basic	**Mov. Avg. 10**	**Mov. Avg. 50**
Average	1.2003	1.1999	1.2001
St. Dev.	1.4754	0.5344	0.1170
Min	-2.0835	-0.1707	0.9093
Max	4.5075	2.5713	1.4945
Range	6.5910	2.7420	0.5852

2.5.2 Frequency Domain Analysis

Transformation of data to the frequency domain is done to gain more insight about the measured environment and to help identifying acceleration sources. Analyses that transform data into the frequency domain result in displays of acceleration power spectral density *versus* frequency, cumulative *RMS* acceleration *versus* frequency, one third octave band *RMS* acceleration *versus* frequency, and power spectral density *versus* frequency and time (spectrogram). Fourier transformation is a common tool of studying time series data in the frequency domain. The fundamental information obtained from Fourier transformation is the relative magnitudes of the various sinusoidal signals that compose the time series. A knowledge of the predominant disturbing frequencies is often of interest to the user. The Fourier transformation is the basis for the power spectral density and other frequency domain quantities presented in the following subsections. Table 2.5 provides an overview of the frequency domain analysis techniques discussed here.

Table 2.5 Frequency domain analysis techniques.

Analysis Technique	Units	Use
Power Spectral Density *vs.* frequency	g^2/Hz	Estimate of the distribution of energy with respect to frequency
Cumulative *RMS* Acceleration *vs.* frequency	g_{RMS}	Quantifies contributions of spectral components to the overall *RMS* acceleration level for the time period studied in cumulative fashion
RMS Acc. *vs.* frequency	g_{RMS}	Quantifies contributions of spectral components per frequency bin
Spectrogram	g^2/Hz	Road map of how acceleration signals vary with respect to both time and frequency

Power Spectral Density *versus* Frequency

As explained in Chapter 1, the definition of the power spectral density can be generalised to discrete time variables x_n. As in the previous time-histories plots, for a finite window of $1 \leq n \leq N$ with the signal sampled at discrete times $x_n = x(n\Delta t)$ for a total measurement period $T = N\Delta t$, a single estimate of the *PSD* can be obtained through a summation rather than an integration:

$$\hat{S}_{xx}(\omega) = \frac{(\Delta t)^2}{T}\left|\sum_{n=1}^{N} x_n e^{-i\omega n\Delta t}\right|^2 \tag{2.15}$$

As before, the actual *PSD* is achieved when *N* (and thus *T*) approaches infinity and the expected value is formally applied. In a real-world application, one would typically average this single-measurement *PSD* over many trials to obtain a more accurate estimate of the theoretical *PSD* of the physical process underlying each individual measurement.

This method for computation of the *PSD* is consistent with Parseval's theorem, which states that the *RMS* value of a time signal is equal to the square root of the integral of the *PSD* across the frequency band represented by the original signal:

$$x_{RMS} = \sqrt{\frac{1}{T}\int_0^T x^2(t)\,dt} = \sqrt{\int_0^\infty PSD(f)\,df} \tag{2.16}$$

Cumulative *RMS* Acceleration *versus* Frequency

A plot of the cumulative *RMS* acceleration in units of g_{RMS} *versus* frequency quantifies in a cumulative fashion the contributions of spectral components to the overall *RMS* acceleration level for the time frame spanned. Therefore, vertical steps in plots of this type indicate discrete frequencies (or a narrow band of frequencies) that contribute significantly to the acceleration environment, while plateaus are indicative of relatively quiet portions of the spectrum. The *x*-axis cumulative *RMS* acceleration is computed as follows:

$$x_{cum\ RMS_k} = \sqrt{\sum_{i=1}^{k} PSD_x(i)}\ , \quad k = 1, 2, \ldots L \tag{2.17}$$

where L is equal to $(N-1)/2$ if N is odd and is equal to $(N/2)+1$ if N is even.

Root-Mean-Square Acceleration *versus* Frequency

By dividing the spectrum into k equal-width frequency bands and computing the g_{RMS} level for each of these bands, it is possible to plot these values to display the *RMS* acceleration as a function of frequency. The width of these bands is defined by the user and the results of this type of analysis can be presented in tabular or graphical form.

One Third Octave Band *RMS* Acceleration *versus* Frequency

This analysis is a specialised version of the graphical representation of g_{RMS} values computed for specified proportional bandwidth frequency bands. The analysis quantifies the *RMS* acceleration level contributed by each of the one third octave frequency bands. The one third octave band *RMS* acceleration is computed as

$$accel_{RMS_k} = \sqrt{\sum_{i=f_{low(k)}}^{f_{high(k)}} RSS\left(PSD(i)\right)}\ , \quad k = 1, 2, \ldots L \tag{2.18}$$

where *RSS* stands for Root Sum of Squares.

Power Spectral Density *versus* Frequency and Time (Spectrogram)

Spectrograms provide a road map of how acceleration signals vary with respect to both time and frequency. To produce a spectrogram, *PSDs* are computed for successive intervals of time. The *PSDs* are oriented vertically on a page such that frequency increases from bottom to top. *PSDs* from successive time slices are aligned horizontally across the page, such that time increases from left to right. Each time-frequency bin is imaged as a colour or grey scale corresponding to the logarithm of the *PSD* magnitude at that time and frequency. Spectrograms are particularly useful in identifying when certain activities begin and end.

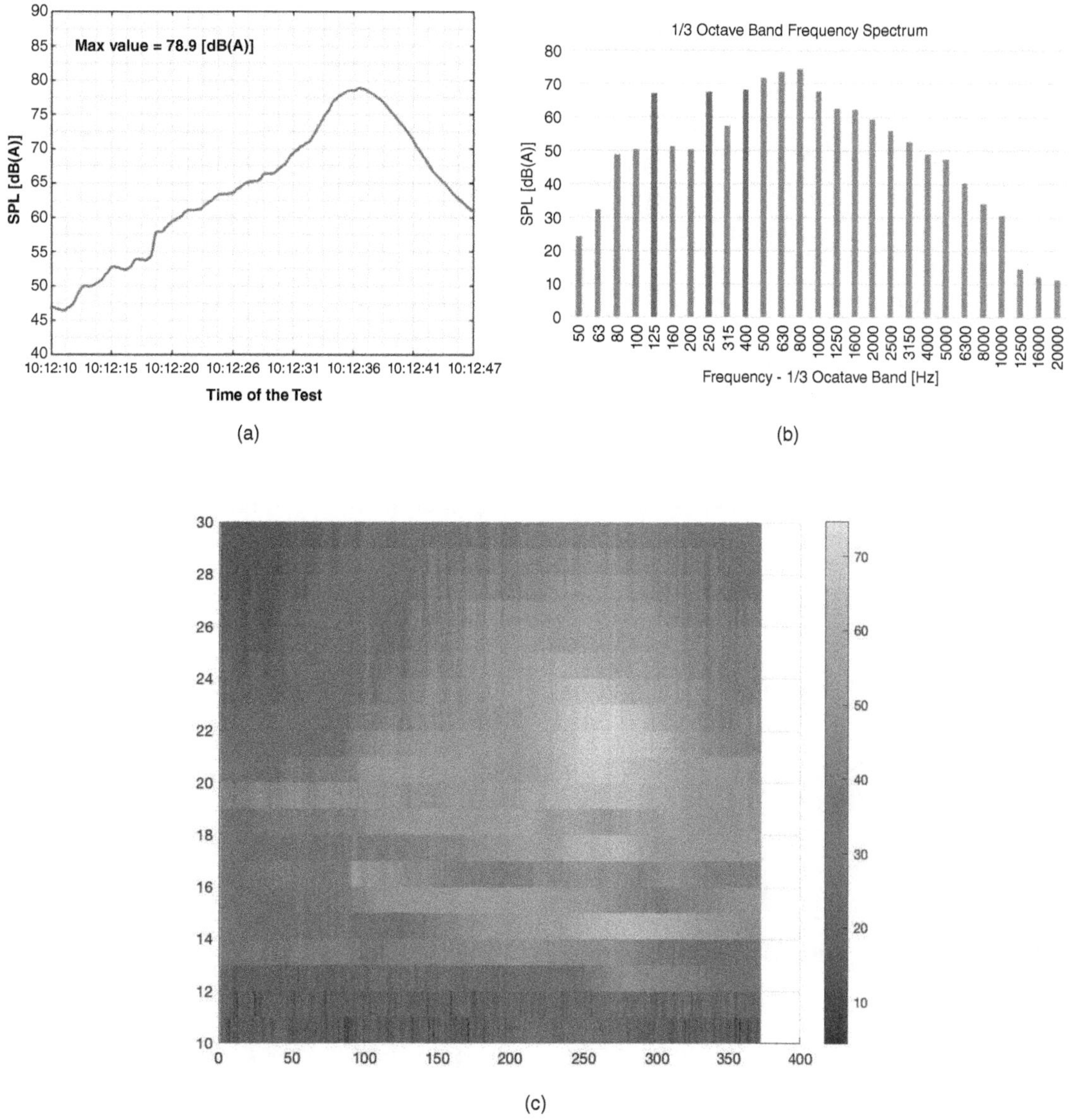

Figure 2.35 (a) Measured Sound Pressure Level; (b) corresponding one-third Octave Band Spectrum; (c) Spectrogram[5].

[5] The one-third octave frequencies on the y-axis of the spectrogram should be read as

To the value	10	11	12	13	14	15	16
corresponds the frequency	**50**	**63**	**80**	**100**	**125**	**160**	**200**
To the value	17	18	19	20	21	22	23
corresponds the frequency	**250**	**315**	**400**	**500**	**630**	**800**	**1000**
To the value	24	25	26	27	28	29	---
corresponds the frequency	**1250**	**1600**	**2000**	**2500**	**3150**	**4000**	**---**

Naturally, spectrograms can represent other quantities, not only *PSDs*. The plots of Figure 2.35 show an example of the use of a spectrogram starting from a measured Sound Pressure Level (*SPL*) emitted by a propeller airplane over a ground installed microphone. The plot of Figure 2.35 (a) shows the measured *SPL* with the *classical bell curve* which represents the airplane flying towards the microphone and then moving away from it. The corresponding average one-third octave band is reported in Figure 2.35 (b) where the first three blade passage frequencies are evidenced. Finally, Figure 2.35 (c) gives the spectrogram of this measurement. On the x-axis there is the time, on the y-axis there are the one-third octave frequencies and the grey scale corresponds to the *SPL*. From it, it is easy to identify the time of happening of a specific parameter (maximum *SPL* in this example), how a specific frequency (or a frequency band, in this example) varies with the time. For example, the band number 14 (125 Hz) becomes evident when the airplane is close to the microphone. Different behaviour is for band no. 19 (400 Hz) which is already evident at the beginning of the recording.

Examples of Analysis in the Frequency Domain

With reference to the same simulated data presented for time domain analysis, here the opportunities offered by the basic frequency domain analysis are presented. Here, one gives some examples of frequency domain analysis for the definitions and relative results discussed previously.

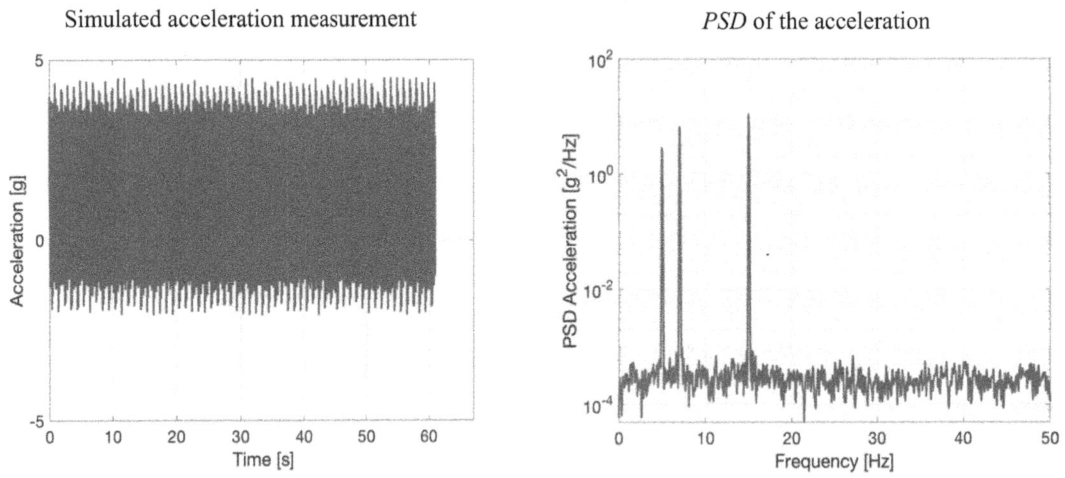

Figure 2.36 Acceleration time history and relative Power Spectral Density.

It is clear from Figure 2.36 that the frequency content of the signal cannot be extracted from the time domain plot, whereas it is obvious from the frequency analysis. The frequencies 5, 7 and 15 Hz, which were the values for the three sinusoidal functions, are clear in the *PSD* plot, together with their amplitudes (0.8, 1.2 and 1.5, respectively), though with some error, as the accuracy depends on the sampling frequency.

The cumulative distribution is easily computed from the power spectral density and in the case presented here this is shown in Figure 2.37.

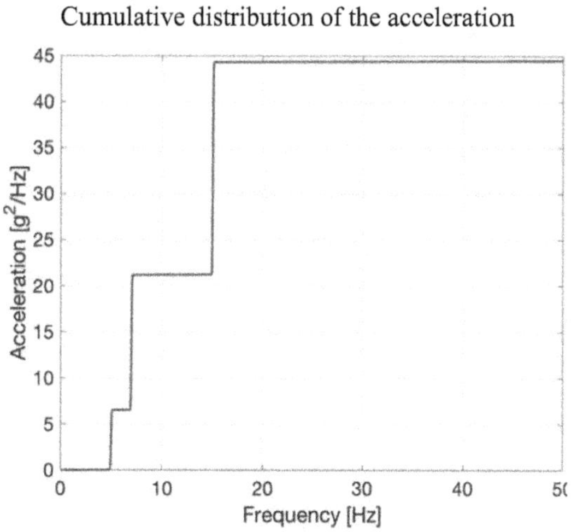

Figure 2.37 Cumulative distribution of the *PSD* of Figure 2.36.

The plot of this function tells us up to which frequency there is an increasing value of the parameter. In the case of Figure 2.37, nothing changes after around 15 Hz. This conclusion, in this simple case could be evident also from the *PSD*, but this is not always the case. It is also interesting to note that the maximum value of the cumulative distribution is related to the standard deviation of the original signal by the following expression:

$$x_{RMS} = \sqrt{\frac{1}{T}\int_0^T x^2(t)\,dt} = \sqrt{\int_0^\infty PSD(f)\,df} = \sqrt{44.47 \times 0.049} = 1.476 \tag{2.19}$$

where the value of 44.47 is the maximum of Figure 2.37 and 0.049 is the corresponding frequency interval resulting from the calculation. The final result is about the same of the basic standard deviation reported in Table 2.4.

References

[2.1] McAndrew, A., *An Introduction to Digital Image Processing with Matlab*, School of Computer Science and Mathematics Victoria University of Technology, 2004.

[2.2] Burger, W., Burge M. J., *Digital Image Processing*, Second Edition, Springer-Verlag, London, 2016.

[2.3] Sutton, M. A., Orteu J. J., Schreier H., *Image Correlation for Shape, Motion and Deformation Measurements*, Springer Verlag-US, London, 2009, ISBN 978-0-387-78746-6.

Further Reading

- Rogers, M. J. B., Hrovat, K., MsPherson, K., Moskowitz M. E., Reckart T., *Accelerometer Data Analysis and Presentation Techniques*, NASA TM-113173, September 1997.

- Gaberson, H. A., *A Comprehensive Windows Tutorial*, Sound and Vibration, March 2006.
- Newland, D. E., *An Introduction to Random Vibrations, Spectral & Wavelet Analysis*, Dover Publications, 2005

Chapter 3 Numerical Methods

3.1 Introduction

Numerical methods take an important role in structural dynamics, namely when it comes to solve problems related to complex structures. Therefore – and without intending to be exhaustive – it is important to give a brief explanation of some of the most important ones. They are introduced in the following four parts:

1. **Approximation methods in time domain**. The traditional ways for the numerical approximation of the time domain response are briefly described. Those include the following methods: Central Difference, Runge-Kutta, Houbolt, Wilson and Newmark-β. A damped single degree-of-freedom (*SDOF*) system in forced vibration is used for demonstration of these applications in Section 3.2.

2. **Approximation methods for natural frequencies**. Some methods for estimating natural frequencies of the vibration system are introduced such as Dunkerley's, Rayleigh's, Ritz' and Holzer's.

3. **Matrix methods**. In order to obtain both the natural frequencies and mode shapes in structural dynamics, matrix decomposition methods are also discussed. These are the Bisection method, Sturm Sequences, *MATLAB* Roots method, Cholesky decomposition, Matrix iteration, Jacobi's method, Singular Value Decomposition (*SVD*) and Principle Component Analysis (*PCA*).

4. **Finite element method**. Although the classical mechanical methods mentioned above can help understanding well the nature of dynamic problems, the finite element method nowadays has become very popular in solving dynamic engineering problems, especially for complicated structures. Some classical numerical methods also lay foundations for the solution of the finite element equations. Therefore, the most practical numerical calculation method, the finite element method, is also briefly introduced at the end of this chapter.

For a better understanding of these methods, *MATLAB* source codes are provided in Appendix A.

3.2 Approximation Methods in Time Domain

The time domain responses of a vibration system, which can be computed analytically or numerically, are essential information for the analysis of the structural dynamic behaviour. Generally, the analytical process of solving time responses of a vibration system tends to be

Structural Dynamics in Engineering Design, First Edition, Nuno M. M. Maia, Dario Di Maio, Alex Carrella, Francesco Marulo, Chaoping Zang, Jonathan E. Cooper, Keith Worden, and Tiago A. N. Silva.

complicated. The numerical approximation methods are widely used since they are easy to be implemented. To give a brief introduction on the traditional ways of numerical approximation of the time domain response, five different methods are described in this section. A damped *SDOF* system in forced vibration is taken as an example to demonstrate the use of these methods.

3.2.1 Central Difference Method

The Central Difference method uses the truncated Taylor expansion of the displacement response to acquire the approximations in an iterative approach. An *SDOF* forced vibration system with mass m, stiffness k and damping c is shown in Figure 3.1.

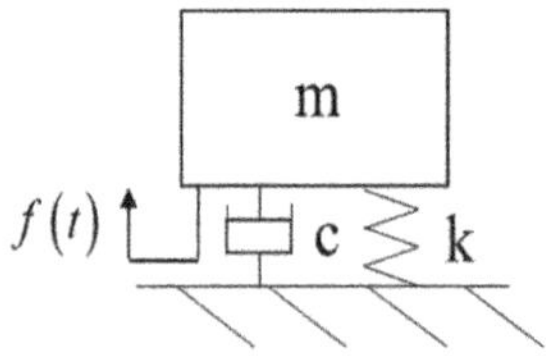

Figure 3.1 *SDOF* system.

It has the following dynamic equilibrium equation under the excitation force $f(t)$:

$$m\ddot{x}+c\dot{x}+kx=f(t) \tag{3.1}$$

Consider a harmonic excitation. As explained in Chapter 1, Section 1.4.5, the response is a combination of both the free vibration response and the steady state response. Thus, the response will be initially unsteady and gradually converges to the steady harmonic. The response may look like the one shown in Figure 3.2.

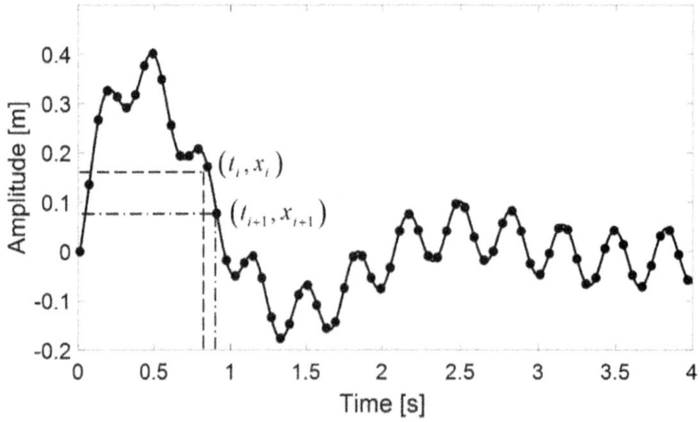

Figure 3.2 Time response.

The response line in Figure 3.2 is the continuous response of the system. However, in a traditional *PC* the response is stored in a discrete way, shown as black points in Figure 3.2. One of the discrete time responses at time instant t_i, the focus of the dashed lines, is recorded as $x(t_i)$ and simplified as x_i. The focus of the dotted-dashed lines means response x_{i+1} at time instant t_{i+1}. The time response approximation aims at using the information of x_i to acquire the response x_{i+1}, and repeat the above step to acquire the whole set of points of the concerned responses.

The Central Difference approximation method for the time response starts by expanding x_i through its *Forward-* and *Backward Taylor expansions*, as in Eq. (3.2):

$$\begin{aligned} x_{i+1} &= x_i + \Delta t\,\dot{x}_i + \frac{\Delta t^2}{2}\ddot{x}_i + \frac{\Delta t^3}{6}\dddot{x}_i + \cdots \\ x_{i-1} &= x_i - \Delta t\,\dot{x}_i + \frac{\Delta t^2}{2}\ddot{x}_i - \frac{\Delta t^3}{6}\dddot{x}_i + \cdots \end{aligned} \tag{3.2}$$

where Δt is the time increment, $\dot{x}_i$ is the velocity at time instant t_i, $\ddot{x}_i$ is the acceleration at that same instant, etc. Then, subtracting the backward expansion from the forward expansion, leads to the velocity of the system at time t_i:

$$\dot{x}_i = \frac{1}{2\Delta t}(x_{i+1} - x_{i-1}) + O(3) \tag{3.3}$$

where $O(3)$ means the differential residues with the highest order of 3. Adding up now the forward and backward expansions, yields the value for the acceleration:

$$\ddot{x}_i = \frac{1}{\Delta t^2}(x_{i+1} + x_{i-1} - 2x_i) - O(4) \tag{3.4}$$

Truncating Eqs. (3.3) and (3.4) at order 2, the central difference approximation of the velocity and acceleration at time t_i are obtained as a function of the displacements:

$$\begin{aligned} \dot{x}_i &\simeq \frac{1}{2\Delta t}(x_{i+1} - x_{i-1}) \\ \ddot{x}_i &\simeq \frac{1}{\Delta t^2}(x_{i+1} + x_{i-1} - 2x_i) \end{aligned} \tag{3.5}$$

Substituting Eq. (3.5) into the original equilibrium equation (3.1), the result is the following recurrence formula:

$$x_{i+1} = \frac{2\Delta t^2 f(t_i) - (2k\Delta t^2 - 4m)x_i - (2m - c\Delta t)x_{i-1}}{2m + c\Delta t} \tag{3.6}$$

Therefore, the displacement response at time t_{i+1} can be easily approximated from the displacements at instants t_i and t_{i-1} and the force at t_i. Then, by iteratively solving Eq. (3.6), the response in the concerned time range can be obtained, although one must be aware that this is not a self-starting method. In fact, let the initial displacement and velocity be x_0 and $\dot{x}_0$, respectively; the response at time t_2, according to Eq. (3.6), is:

$$x_2 = \frac{2\Delta t^2 f(t_1) - (2k\Delta t^2 - 4m)x_1 - (2m - c\Delta t)x_0}{2m + c\Delta t} \tag{3.7}$$

However, x_2 cannot be directly obtained, as x_1 is not known yet. Thus, to start the central difference program, x_1 can be approximated by the first order Taylor expansion of the initial parameters:

$$x_1 = x_0 + \Delta t\,\dot{x}_0 \tag{3.8}$$

Alternatively, one can use the second order approximation: from the first of Eq. (3.2),

$$x_1 = x_0 + \Delta t\,\dot{x}_0 + \frac{\Delta t^2}{2}\ddot{x}_0 \tag{3.9}$$

where $\ddot{x}_0$ can be obtained from the equilibrium equation (3.1):

$$m\ddot{x}_0 + c\dot{x}_0 + kx_0 = f(0) \quad \Rightarrow \quad \ddot{x}_0 = \frac{1}{m}\left(f(0) - c\dot{x}_0 - kx_0\right) \tag{3.10}$$

After that, the program can start automatically and find all of the response points at the intended discrete time instants.

Example 3.2.1 Time response approximation by the Central Difference method

A numerical case is used here to help the reader understanding the central difference method. An *SDOF* system with mass 1kg, stiffness 10 N/m, damping 2N/(m/s) is used for simulation. The initial displacement and velocity are set as 0 m and 2 m/s, respectively. The excitation is set as $20\cos\left((2\pi\times 3)t\right)$. The equilibrium equation is:

$$\begin{aligned} &\ddot{x} + 2\dot{x} + 10x = 20\cos(6\pi t) \\ &x_0 = 0 \\ &\dot{x}_0 = 2 \end{aligned} \tag{3.11}$$

According to Eq. (3.6), it follows that:

$$x_{i+1} = \frac{40\Delta t^2\cos(6\pi t_i) - (20\Delta t^2 - 4)x_i - (2 - 2\Delta t)x_{i-1}}{2 + 2\Delta t} \tag{3.12}$$

and from Eq. (3.8), $x_1 = 2\Delta t$. The *MATLAB* code is programmed according to Eq. (3.10) and are given in Appendix A (code 1).

Three cases have been studied, with the intervals between two adjacent time instants set as 0.05s, 0.01s and 0.001s. The program is then started to simulate the responses. The results, together with the analytical one, are overlaid in Figure 3.3. It can be seen that the responses during the initial five seconds are unsteady; beyond that, the responses tend gradually to be almost steady. After the first five seconds, the responses calculated from all the three steps of approximation match the analytical response perfectly. One can observe that during the first 5 seconds the responses are unsteady and during the last 3 seconds the responses are steady. Then, the mean differences between the analytical response and numerical responses are calculated, and the results are:

$$Dif = \frac{1}{N}\sum_{i=1}^{N} \frac{abs\left(x_{pre}(t_i) - x_{real}(t_i)\right)}{x_{real}(t_i)}$$

$$Dif_{unsteady} = \{46.63\% \quad 6.977\% \quad 0.2814\%\}$$

$$Dif_{steady} = \{11.0912\% \quad 0.4861\% \quad 0.2451\%\}$$

One can see that the differences between the unsteady parts are larger than the differences of the steady parts. Besides, when the time interval decreases, the response discrepancies between the central difference approximation and the analytical response gradually reduce. When the time interval is reduced to 0.001s, the mean differences of the steady and unsteady parts are reduced to nearly 1%. Thus, the central difference method converges to the related analytical result when the time interval is small.

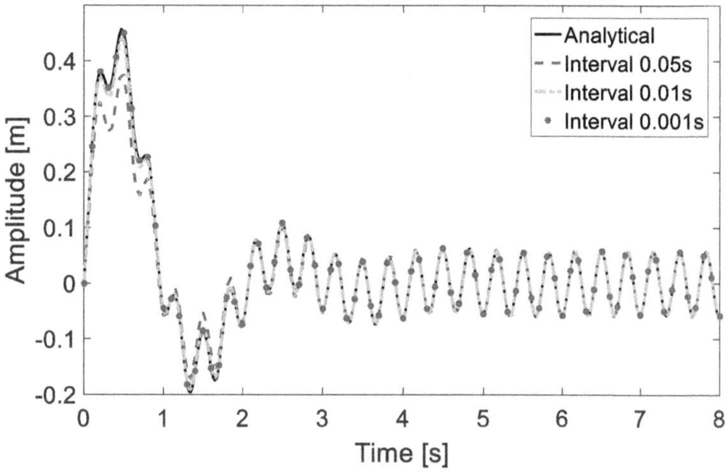

Figure 3.3 Central difference method: comparison between the analytical response and the numerical ones, for different time intervals.

The Central Difference method can be easily extended to *MDOF* systems. The expressions for the velocity and acceleration are entirely similar to those seen before (Eq. 3.5), now in terms of vectors:

$$\dot{\boldsymbol{x}}_i \simeq \frac{1}{2\Delta t}\left(\boldsymbol{x}_{i+1} - \boldsymbol{x}_{i-1}\right)$$
$$\ddot{\boldsymbol{x}}_i \simeq \frac{1}{\Delta t^2}\left(\boldsymbol{x}_{i+1} + \boldsymbol{x}_{i-1} - 2\boldsymbol{x}_i\right) \tag{3.13}$$

The equation of motion at time t_i is therefore given by:

$$\frac{1}{\Delta t^2}\boldsymbol{M}\left(\boldsymbol{x}_{i+1} + \boldsymbol{x}_{i-1} - 2\boldsymbol{x}_i\right) + \frac{1}{2\Delta t}\boldsymbol{C}\left(\boldsymbol{x}_{i+1} - \boldsymbol{x}_{i-1}\right) + \boldsymbol{K}\boldsymbol{x}_i = \boldsymbol{f}\left(t_i\right) \tag{3.14}$$

from which the response can be obtained recursively:

$$\boldsymbol{x}_{i+1} = \left(\frac{1}{\Delta t^2}\boldsymbol{M} + \frac{1}{2\Delta t}\boldsymbol{C}\right)^{-1}\left(\boldsymbol{f}\left(t_i\right) - \left(\boldsymbol{K} - \frac{2}{\Delta t^2}\boldsymbol{M}\right)\boldsymbol{x}_i - \left(\frac{1}{\Delta t^2}\boldsymbol{M} - \frac{1}{2\Delta t}\boldsymbol{C}\right)\boldsymbol{x}_{i-1}\right) \tag{3.15}$$

As before, due to the non self-start nature of the method, one must evaluate the response $\boldsymbol{x}_1$, which can be found from the second order Taylor expansion

$$\boldsymbol{x}_1 = \boldsymbol{x}_0 + \Delta t\,\dot{\boldsymbol{x}}_0 + \frac{\Delta t^2}{2}\ddot{\boldsymbol{x}}_0 \tag{3.16}$$

where $\ddot{\boldsymbol{x}}_0$ can be obtained from the equilibrium equation:

$$\ddot{\boldsymbol{x}}_0 = \boldsymbol{M}^{-1}\left(\boldsymbol{f}\left(0\right) - \boldsymbol{C}\dot{\boldsymbol{x}}_0 - \boldsymbol{K}\boldsymbol{x}_0\right)$$

The recursive application of Eq. (3.15) can now start from $\boldsymbol{x}_2$ on.

3.2.2 Runge-Kutta's Method

Runge-Kutta's theory, which is based on the truncated Taylor expansions, can also be used for the approximation of the forced vibration time response. This is a self-starting method that only uses the first order Taylor expansion for the computation.

Differently from the Central Difference method, the Runge-Kutta method uses both the velocity and displacement for the time response approximation, as shown in Figure 3.4, for the same previous example.

This method works with the state space control equation, so the *SDOF* forced vibration system in Eq. (3.1) is transformed to its state-space form:

$$\begin{Bmatrix} \dot{x} \\ \dot{v} \end{Bmatrix} = \begin{Bmatrix} v \\ \dfrac{f\left(t\right) - cv - kx}{m} \end{Bmatrix} \tag{3.17}$$

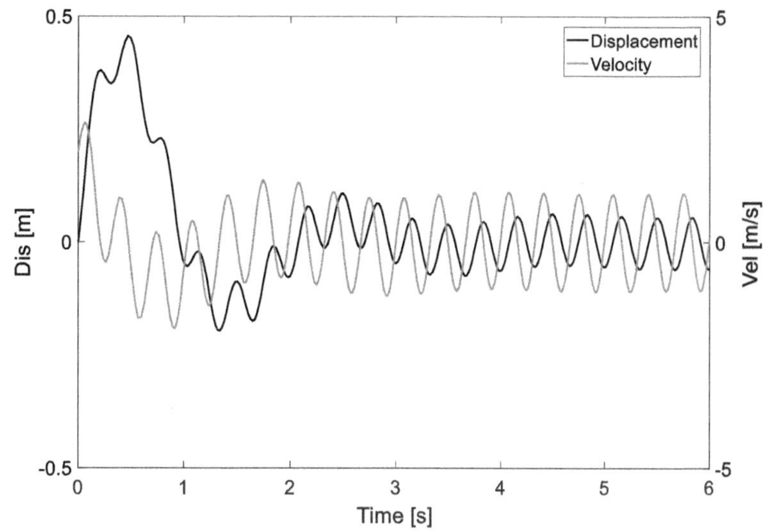

Figure 3.4 Displacement and velocity.

where v is the velocity of the system, and $\dot{v}$ expresses the acceleration. Then, the velocity and displacement are expanded by the forward Taylor expansion at time instant t_i and truncated at the first order:

$$\begin{aligned} v_{i+1} &\simeq v_i + \Delta t\,\dot{v}_i \\ x_{i+1} &\simeq x_i + \Delta t\,\dot{x}_i \end{aligned} \tag{3.18}$$

Substituting Eq. (3.18) in the state-space equation, leads to the response at time instant t_{i+1}:

$$\begin{aligned} x_{i+1} &= x_i + \Delta t\, v_i \\ v_{i+1} &= v_i + \Delta t\,\frac{f(t_i) - cv_i - kx_i}{m} \end{aligned} \tag{3.19}$$

Eq. (3.19) is the simplified control equation by the Runge-Kutta method and can self-start. For instance, by inputting initial values x_0 and v_0 to Eq. (3.19), the responses at time instant t_i are directly obtained:

$$\begin{aligned} x_1 &= x_0 + \Delta t\, v_0 \\ v_1 &= v_0 + \Delta t\,\frac{f(t_0) - c\,v_0 - k\,x_0}{m} \end{aligned} \tag{3.20}$$

Therefore, the concerned time responses can be further iteratively acquired.

The method just described only uses the displacement and velocity at time instant t_i for the derivation of the simplified equilibrium equation. By using more terms in the control equation, a higher order Runge-Kutta method with better precision can be obtained. The most commonly used Runge-Kutta method is the 4th order Runge-Kutta (*RK4*), usually for the time response approximation of a nonlinear system and the derivation is similar to the description given above. Therefore, it will not be presented here; besides, in Chapter 5 it will be further addressed, in the context of nonlinear systems and in Appendix C a dedicated *MATLAB* program is presented.

Example 3.2.2 Time response approximation by the Runge-Kutta method

In order to better understand the method, the numerical case of Eq. (3.11) is utilised here for analysis. As the *MATLAB* tool already provides the solver of the traditional *RK4* method, called '*ODE45*', here only the state space control equation of Eq. (3.17) does need to be derived. The result is:

$$\begin{aligned} v &= \dot{x} \\ \dot{v} &= 20\cos(6\pi t) - 2\dot{x} - 10x \end{aligned} \tag{3.21}$$

After that, the *MATLAB* code of the Runge-Kutta approximation combined with *ODE45* solver is programmed. See Appendix A for details (code 2).

The program is then executed, and the result is shown in Figure 3.5. One can see that the Runge-Kutta approximation matches the analytical result well and proves its effectiveness.

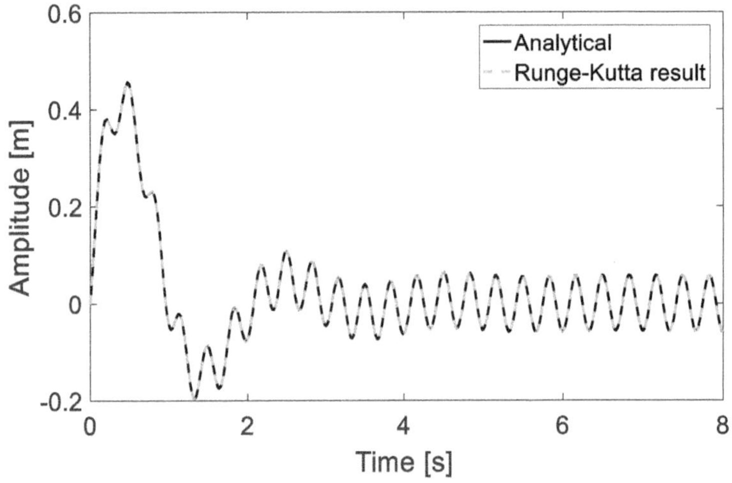

Figure 3.5 Comparison between the analytical result and the Runge-Kutta result.

As one can see, *RK4* works very well. Besides, it is easy to generalise it to *MDOF* systems. In such a case, it is only necessary to change the equilibrium equation from *SDOF* to *MDOF* and to give initial parameters; the *ODE45* solver can directly calculate the result.

3.2.3 Houbolt's Method

The Houbolt method also uses the truncated Taylor series for response approximation. However, it is different from the Central Difference and Runge-Kutta's methods. It uses the first three truncated Taylor expansions and works on displacements related to four adjacent equally spaced time instants, $t_{i-2}=t_i-2\Delta t$, $t_{i-1}=t_i-\Delta t$, t_i and $t_{i+1}=t_i+\Delta t$, given by x_{i-2}, x_{i-1}, x_i and x_{i+1}, respectively, as shown in Figure 3.6.

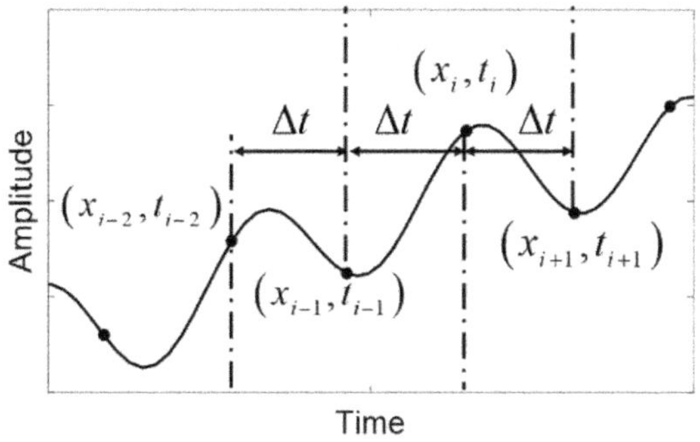

Figure 3.6 Equally spaced grid points.

Firstly, Houbolt's method starts by expanding the response at time t_{i+1} to order 3 to represent the response x_i at time instant t_i:

$$x(t)\approx x(t+\Delta t)+\dot{x}(t+\Delta t)(-\Delta t)+\frac{\ddot{x}(t+\Delta t)}{2}(-\Delta t)^2+\frac{\dddot{x}(t+\Delta t)}{6}(-\Delta t)^3 \tag{3.22}$$

Simplifying Eq. (3.22), one obtains:

$$x_i\approx x_{i+1}-\dot{x}_{i+1}\Delta t+\frac{\ddot{x}_{i+1}}{2}(\Delta t)^2-\frac{\dddot{x}_{i+1}}{6}(\Delta t)^3 \tag{3.23}$$

Secondly, the response x_{i-1} is approximated by x_{i+1} according to the 3rd order Taylor expansion:

$$x_{i-1}\approx x_{i+1}-2\dot{x}_{i+1}\Delta t+2\ddot{x}_{i+1}(\Delta t)^2-\frac{4\dddot{x}_{i+1}}{3}(\Delta t)^3 \tag{3.24}$$

Thirdly, the response x_{i-2} is also approximated by x_{i+1}:

$$x_{i-2}\approx x_{i+1}+\dot{x}_{i+1}(-3\Delta t)+\frac{\ddot{x}_{i+1}}{2}(-3\Delta t)^2+\frac{\dddot{x}_{i+1}}{6}(-3\Delta t)^3 \tag{3.25}$$

Now, one can eliminate $\dddot{x}_{i+1}$ between Eqs. (3.23) and (3.24), leading to

$$2(\Delta t)^2\ddot{x}_{i+1}-6\Delta t\,\dot{x}_{i+1}=-7x_{i+1}+8x_i-x_{i-1} \tag{3.26}$$

and eliminating again $\dddot{x}_{i+1}$ between Eqs. (3.24) and (3.25), one obtains

$$-18(\Delta t)^2 \ddot{x}_{i+1} + 30\Delta t \dot{x}_{i+1} = 19x_{i+1} - 27x_{i-1} + 8x_{i-2} \tag{3.27}$$

One ends up with a system of equations, from which $\dot{x}_{i+1}$ and $\ddot{x}_{i+1}$ can be expressed in terms of x_{i-2}, x_{i-1}, x_i and x_{i+1}:

$$\dot{x}_{i+1} = \frac{1}{6\Delta t}\left(11x_{i+1} - 18x_i + 9x_{i-1} - 2x_{i-2}\right) \tag{3.28}$$

$$\ddot{x}_{i+1} = \frac{1}{(\Delta t)^2}\left(2x_{i+1} - 5x_i + 4x_{i-1} - x_{i-2}\right) \tag{3.29}$$

Then, considering the equilibrium equation for the response at time instant t_{i+1},

$$m\ddot{x}_{i+1} + c\dot{x}_{i+1} + kx_{i+1} = f_{i+1} \tag{3.30}$$

the displacement x_{i+1} is obtained by substituting Eqs. (3.28) and (3.29) in (3.30):

$$x_{i+1} = \frac{f_{i+1} + \left(\frac{5}{(\Delta t)^2}m + \frac{3}{\Delta t}c\right)x_i - \left(\frac{4}{(\Delta t)^2}m + \frac{3}{2\Delta t}c\right)x_{i-1} + \left(\frac{1}{(\Delta t)^2}m + \frac{1}{3\Delta t}c\right)x_{i-2}}{\frac{2}{(\Delta t)^2}m + \frac{11}{6\Delta t}c + k} \tag{3.31}$$

where x_{i+1} is approximated by its related responses. However, it is apparent that this method is also non self-starting. For example, suppose one wishes to evaluate x_3:

$$x_3 = \frac{f_3 + \left(\frac{5}{(\Delta t)^2}m + \frac{3}{\Delta t}c\right)x_2 - \left(\frac{4}{(\Delta t)^2}m + \frac{3}{2\Delta t}c\right)x_1 + \left(\frac{1}{(\Delta t)^2}m + \frac{1}{3\Delta t}c\right)x_0}{\frac{2}{(\Delta t)^2}m + \frac{11}{6\Delta t}c + k} \tag{3.32}$$

One knows x_0, but x_1 and x_2 are not known and the algorithm cannot start. Thus, a low-order Taylor approximation method like the central difference one should be conducted first to acquire x_1 and x_2. After that, the Houbolt method can proceed iteratively.

Example 3.2.3 Time response approximation using Houbolt's method

The *SDOF* system in Section 3.2.1 is used for verification of Houbolt's method and the *MATLAB* code is given in Appendix A (code 3). The result is overlaid with the analytical solution (Figure 3.7) and it is clear that they almost coincide with each other, showing the effectiveness of the method.

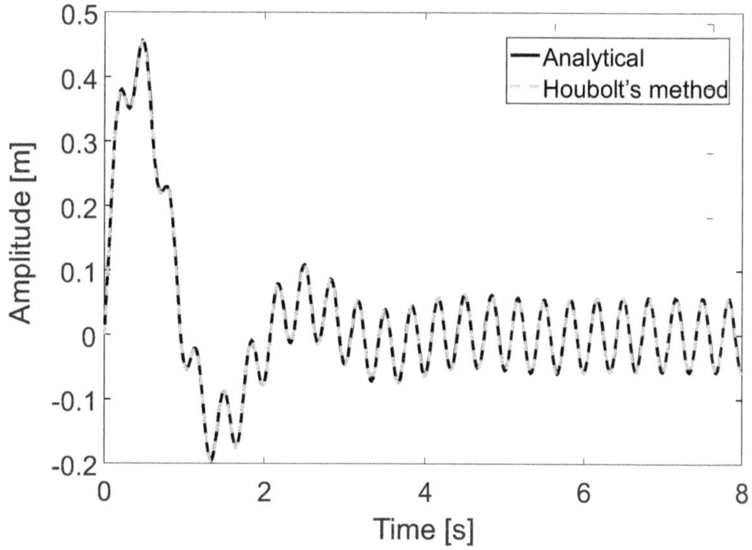

Figure 3.7 Response of the *SDOF* system using the Houbolt method.

The Houbolt method can also be easily implemented for *MDOF* systems. Firstly, one considers the equation of motion of an *MDOF* system at t_{i+1}:

$$\boldsymbol{M}\ddot{\boldsymbol{x}}_{i+1} + \boldsymbol{C}\dot{\boldsymbol{x}}_{i+1} + \boldsymbol{K}\boldsymbol{x}_{i+1} = \boldsymbol{F}_{i+1} \tag{3.33}$$

By substituting Eqs. (3.28) and (3.29) in Eq. (3.33), it follows that:

$$\boldsymbol{x}_{i+1} = \left[\frac{2}{(\Delta t)^2}\boldsymbol{M} + \frac{11}{6\Delta t}\boldsymbol{C} + \boldsymbol{K}\right]^{-1} \left\{\boldsymbol{F}_{i+1} + \left(\frac{5}{(\Delta t)^2}\boldsymbol{M} + \frac{3}{\Delta t}\boldsymbol{C}\right)\boldsymbol{x}_i - \left(\frac{4}{(\Delta t)^2}\boldsymbol{M} + \frac{3}{2\Delta t}\boldsymbol{C}\right)\boldsymbol{x}_{i-1} \right.$$
$$\left. + \left(\frac{1}{(\Delta t)^2}\boldsymbol{M} + \frac{1}{3\Delta t}\boldsymbol{C}\right)\boldsymbol{x}_{i-2}\right\} \tag{3.34}$$

Eq. (3.34) gives the *MDOF* Houbolt formulation. The previous notes on the non self-starting character of the method concerning the *SDOF* system apply here as well, i.e one can use the *MDOF* central difference method to start the Houbolt iteration process.

3.2.4 Wilson's Method

The Wilson method is also based on the Taylor expansion. However, it works on the acceleration of the system, instead of displacement and velocity. It assumes the acceleration to have a linear increment in the time interval from t_i to $t_{i+\theta} = t_i + \theta\Delta t$, and uses time integrals to represent the displacement and velocity for response calculation, where $\theta \geq 1$; the schematic diagram is shown in Figure 3.8. Such method is also called $Wilson-\theta$ and coincides with Newmark linear acceleration method when $\theta = 1$. In practice, θ should always be greater than 1 and for good accuracy and stability one should make $\theta = 1.42$.

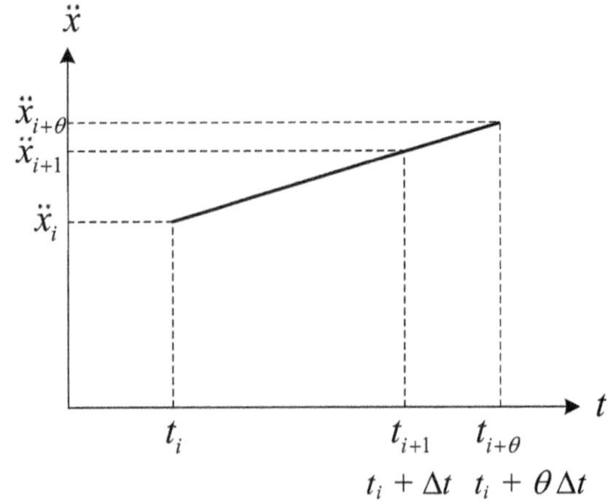

Figure 3.8 Linear acceleration assumption of the Wilson method.

To calculate the response $x(t_i+\tau)$ at any time τ between t_i and $t_{i+\theta}$, one begins by calculating the acceleration, which can be approximated by:

$$\ddot{x}(t_i+\tau)=\ddot{x}_i+\frac{\tau}{\theta\Delta t}(\ddot{x}_{i+\theta}-\ddot{x}_i) \quad 0\le\tau\le\theta\Delta t \tag{3.35}$$

The velocity is obtained by integrating the acceleration from 0 to τ:

$$\dot{x}(t_i+\tau)=\int_0^\tau \ddot{x}_i+\frac{\tau}{\theta\Delta t}(\ddot{x}_{i+\theta}-\ddot{x}_i)d\tau=\dot{x}_i+\ddot{x}_i\tau+\frac{\tau^2}{2\theta\Delta t}(\ddot{x}_{i+\theta}-\ddot{x}_i) \tag{3.36}$$

And the displacement is obtained by integrating Eq. (3.36):

$$x(t_i+\tau)=\int_0^\tau \dot{x}_i+\ddot{x}_i\tau+\frac{\tau^2}{2\theta\Delta t}(\ddot{x}_{i+\theta}-\ddot{x}_i)d\tau=x_i+\dot{x}_i\tau+\frac{1}{2}\ddot{x}_i\tau^2+\frac{\tau^3}{6\theta\Delta t}(\ddot{x}_{i+\theta}-\ddot{x}_i) \tag{3.37}$$

Next, Wilson's method needs to evaluate the displacement and velocity at time instant $t_{i+\theta}$, and uses those results to obtain the responses at t_{i+1}. So, let $\tau=\theta\Delta t$. The displacement and velocity at $t_{i+\theta}$ are approximated as:

$$\dot{x}_{i+\theta}=\dot{x}(t_i+\theta\Delta t)=\dot{x}_i+\frac{\theta\Delta t}{2}(\ddot{x}_{i+\theta}+\ddot{x}_i) \tag{3.38}$$

$$x_{i+\theta}=x(t_i+\theta\Delta t)=x_i+\dot{x}_i\theta\Delta t+\frac{(\theta\Delta t)^2}{6}(\ddot{x}_{i+\theta}+2\ddot{x}_i) \tag{3.39}$$

Rearranging Eq. (3.39), the result is:

$$\ddot{x}_{i+\theta}=\frac{6}{(\theta\Delta t)^2}(x_{i+\theta}-x_i)-\frac{6}{\theta\Delta t}\dot{x}_i-2\ddot{x}_i \tag{3.40}$$

Next, by substituting Eq. (3.40) into Eq. (3.38), the velocity $\dot{x}_{i+\theta}$ is given by:

$$\dot{x}_{i+\theta} = \frac{3}{\theta\Delta t}\left(x_{i+\theta} - x_i\right) - 2\dot{x}_i - \frac{\theta\Delta t}{2}\ddot{x}_i \tag{3.41}$$

However, $x_{i+\theta}$ is unknown. To solve this problem, one uses the equilibrium equation at $t_{i+\theta}$:

$$m\ddot{x}_{i+\theta} + c\dot{x}_{i+\theta} + kx_{i+\theta} = \tilde{f}_{i+\theta} \tag{3.42}$$

where $\tilde{f}_{i+\theta}$ is a "projected" load, equal to $f_i + \theta\left(f_{i+1} - f_i\right)$. Next, $x_{i+\theta}$ is solved according to Eqs. (3.40) and (3.41):

$$x_{i+\theta} = \frac{f_i + \theta\left(f_{i+1} - f_i\right) + \left(\dfrac{6}{\left(\theta\Delta t\right)^2}m + \dfrac{3}{\theta\Delta t}c\right)x_i + \left(\dfrac{6m}{\theta\Delta t} + 2c\right)\dot{x}_i + \left(2m + \dfrac{\theta\Delta t}{2}c\right)\ddot{x}_i}{\dfrac{6}{\left(\theta\Delta t\right)^2}m + \dfrac{3}{\theta\Delta t}c + k} \tag{3.43}$$

Substituting Eq. (3.43) into Eqs. (3.40) and (3.41), the acceleration $\ddot{x}_{i+\theta}$ and the velocity $\dot{x}_{i+\theta}$ can be obtained. Afterwards, they are substituted back into Eq. (3.35) to get the acceleration at t_{i+1}:

$$\ddot{x}_{i+1} = \ddot{x}_i + \frac{1}{\theta}\left(\ddot{x}_{i+\theta} - \ddot{x}_i\right) \tag{3.44}$$

The result is:

$$\ddot{x}_{i+1} = \frac{6}{\theta^3\Delta t^2}\left(x_{i+\theta} - x_i\right) - \frac{6}{\theta^2\Delta t}\dot{x}_i + \left(1 - \frac{3}{\theta}\right)\ddot{x}_i \tag{3.45}$$

The velocity $\dot{x}_{i+1}$ and the displacement x_{i+1} are obtained having into account Eqs. (3.36) and (3.37), respectively:

$$\dot{x}_{i+1} = \dot{x}_i + \frac{\Delta t}{2}\left(\ddot{x}_{i+1} + \ddot{x}_i\right) \tag{3.46}$$

$$x_{i+1} = x_i + \dot{x}_i\Delta t + \frac{\left(\Delta t\right)^2}{6}\left(\ddot{x}_{i+1} + 2\ddot{x}_i\right) \tag{3.47}$$

Afterwards, the other concerned responses are approximated by repeating the above steps. Note that, since Wilson's method only uses the response at t_i, it is self-starting and does not need to appeal to another method.

Example 3.2.4 Time response approximation with Wilson's method

Finally, the effectiveness of Wilson's method is verified using the case presented in Section 3.2.1. This is programmed in Appendix A (code 4). The result is illustrated in Figure 3.9 and shows that the approximated response is nearly the same as the analytical one.

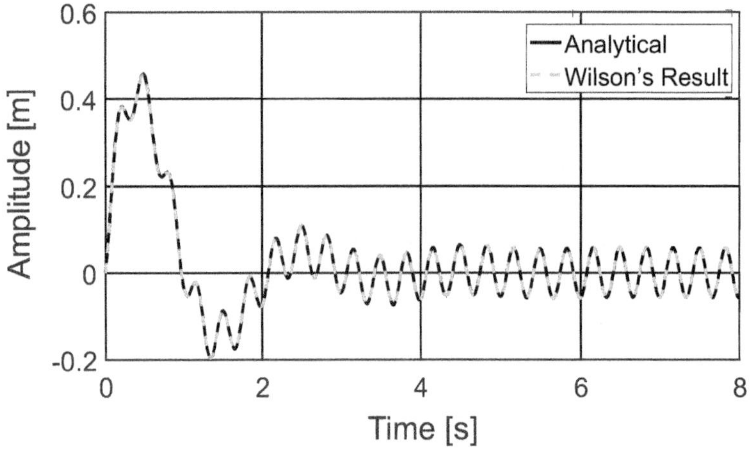

Figure 3.9 Response of the single *DOF* system determined by Wilson's method.

The implementation of Wilson's method to *MDOF* cases is also simple. Firstly, following Eq. (3.42), one considers the *MDOF* response at $t_{i+\theta}$:

$$\boldsymbol{M}\ddot{\boldsymbol{x}}_{i+\theta} + \boldsymbol{C}\dot{\boldsymbol{x}}_{i+\theta} + \boldsymbol{K}\boldsymbol{x}_{i+\theta} = \tilde{\boldsymbol{F}}_{i+\theta} \tag{3.48}$$

where $\tilde{\boldsymbol{F}}_{i+\theta}$ is the excitation vector and $\tilde{\boldsymbol{F}}_{i+\theta} = \boldsymbol{F}_i + \theta\left(\boldsymbol{F}_{i+1} - \boldsymbol{F}\right)_i$. It is then expanded according to Eqs. (3.42) and (3.43):

$$\boldsymbol{x}_{i+\theta} = \left[\frac{6}{(\theta\Delta t)^2}\boldsymbol{M} + \frac{3}{\theta\Delta t}\boldsymbol{C} + \boldsymbol{K}\right]^{-1}\left\{\boldsymbol{F}_i + \theta\left(\boldsymbol{F}_{i+1} - \boldsymbol{F}_i\right) + \left(\frac{6}{(\theta\Delta t)^2}\boldsymbol{M} + \frac{3}{\theta\Delta t}\boldsymbol{C}\right)\boldsymbol{x}_i + \left(\frac{6\boldsymbol{M}}{\theta\Delta t} + 2\boldsymbol{C}\right)\dot{\boldsymbol{x}}_i + \left(2\boldsymbol{M} + \frac{\theta\Delta t}{2}\boldsymbol{C}\right)\ddot{\boldsymbol{x}}_i\right\} \tag{3.49}$$

Then, $\ddot{\boldsymbol{x}}_{i+1}$ at time t_{i+1} can be obtained following the same steps as for the *SDOF* system:

$$\ddot{\boldsymbol{x}}_{i+1} = \ddot{\boldsymbol{x}}_i + \frac{1}{\theta}\left(\ddot{\boldsymbol{x}}_{i+\theta} - \ddot{\boldsymbol{x}}_i\right) \tag{3.50}$$

$$\ddot{\boldsymbol{x}}_{i+1} = \frac{6}{\theta^3\Delta t^2}\left(\boldsymbol{x}_{i+\theta} - \boldsymbol{x}_i\right) - \frac{6}{\theta^2\Delta t}\dot{\boldsymbol{x}}_i + \left(1 - \frac{3}{\theta}\right)\ddot{\boldsymbol{x}}_i \tag{3.51}$$

The velocity $\ddot{\boldsymbol{x}}_{i+1}$ and displacement $\boldsymbol{x}_{i+1}$ at time t_{i+1} are derived accordingly and result is:

$$\dot{\boldsymbol{x}}_{i+1} = \dot{\boldsymbol{x}}_i + \frac{\Delta t}{2}\left(\ddot{\boldsymbol{x}}_{i+1} + \ddot{\boldsymbol{x}}_i\right) \tag{3.52}$$

$$\boldsymbol{x}_{i+1} = \boldsymbol{x}_i + \dot{\boldsymbol{x}}_i\Delta t + \frac{(\Delta t)^2}{6}\left(\ddot{\boldsymbol{x}}_{i+1} + 2\ddot{\boldsymbol{x}}_i\right) \tag{3.53}$$

The Wilson method for *MDOF* system can be easily programmed to obtain the desired responses.

3.2.5 The Newmark-β Method

This method assumes an incremental relationship between acceleration and time and solves the simplified equilibrium equation in terms of acceleration to obtain the time response approximation. Since the relationship between acceleration and time is assumed, several Newmark-β methods can be derived under different types of relationships. The formulation here is the linear acceleration-based Newmark-β method, but other formulations can be derived following the same steps.

Firstly, the Newmark-β method works with the accelerations, as shown in Figure 3.10, and aims to approximate the acceleration in the concerned time range.

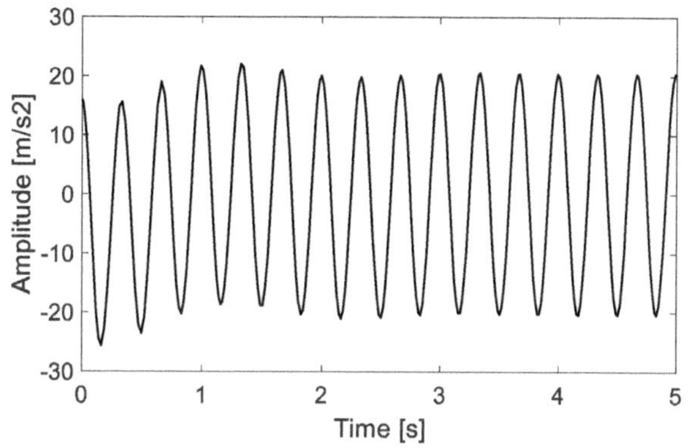

Figure 3.10 Acceleration response.

Then, assuming a linear relationship between acceleration and time, the response at time t can be predicted by the accelerations at time instants t_i and t_{i+1}:

$$\ddot{x}(t) = \ddot{x}_i + \frac{\ddot{x}_{i+1} - \ddot{x}_i}{\Delta t}(t - t_i) \tag{3.54}$$

Then, integrating the acceleration, the displacement and velocity and time t are obtained:

$$\int_{t_i}^{t} \ddot{x}(t)\,dt = \int_{t_i}^{t} \ddot{x}_i\,dt + \int_{t_i}^{t} \frac{\ddot{x}_{i+1} - \ddot{x}_i}{\Delta t}(t - t_i)\,dt \tag{3.55}$$

Developing (3.55),

$$\left[\dot{x}(t)\right]_{t_i}^{t} = \dot{x}(t) - \dot{x}(t_i) = \ddot{x}_i \int_{t_i}^{t} dt + \frac{\ddot{x}_{i+1} - \ddot{x}_i}{\Delta t} \int_{t_i}^{t} (t - t_i)\,dt \tag{3.56}$$

And finally,

$$\dot{x}(t)=\dot{x}_i+\ddot{x}_i(t-t_i)+\frac{\ddot{x}_{i+1}-\ddot{x}_i}{2\Delta t}(t-t_i)^2 \tag{3.57}$$

Similarly, integrating (3.57), it turns out that

$$x(t)=x_i+\dot{x}_i(t-t_i)+\frac{1}{2}\ddot{x}_i(t-t_i)^2+\frac{\ddot{x}_{i+1}-\ddot{x}_i}{6\Delta t}(t-t_i)^3 \tag{3.58}$$

The velocity and displacement at t_{i+1} are then given by:

$$\begin{aligned}\dot{x}_{i+1}&=\dot{x}_i+\frac{\ddot{x}_{i+1}+\ddot{x}_i}{2}\Delta t\\ x_{i+1}&=x_i+\dot{x}_i\Delta t+\frac{1}{3}\ddot{x}_i\Delta t^2+\frac{\ddot{x}_{i+1}}{6}\Delta t^2\end{aligned} \tag{3.59}$$

Substituting Eq. (3.59) in the equilibrium equation (3.1) at time instant t_i, the acceleration at time t_{i+1} is obtained:

$$\ddot{x}_{i+1}=\frac{f(t_{i+1})-\frac{c}{m}\left(\dot{x}_i+\frac{1}{2}\ddot{x}_i\Delta t\right)-\frac{k}{m}\left(x_i+\dot{x}_i\Delta t+\frac{1}{2}\ddot{x}_i\Delta t^2\right)}{1+\frac{1}{2}\frac{c}{m}\Delta t+\frac{1}{6}\frac{k}{m}\Delta t^2} \tag{3.60}$$

The recursive application of the above equation leads to the concerned time responses.

Example 3.2.5 Time response approximation of the Newmark-β method

Finally, the Newmark-β method is also simulated and compared with the analytical result; the *MATLAB* code is given in Appendix A (code 5) and the result is shown in Figure 3.11, where one can see that the numerical response fits the analytical one very well, showing its effectiveness.

3.2.6 Numerical Case Comparison: *SDOF* System

Now that the various methods have been introduced, the results of the Central Difference, Runge-Kutta's, Houbolt's and Wilson's methods are compared and for all the cases the time interval used for the calculations has been set as 0.001s. The overlay is shown in Figure 3.12. It can be seen that when the time interval of the methods is small enough, the responses of the different numerical methods will be the same, showing their efficiency.

3.3 Approximation Methods for Natural Frequencies

In vibration analysis, natural frequencies are important intrinsic properties of an *MDOF* system, which need to be extracted from the equilibrium equation in an accurate way. However, the computational cost of solving the complex *MDOF* vibration equations directly is typically high.

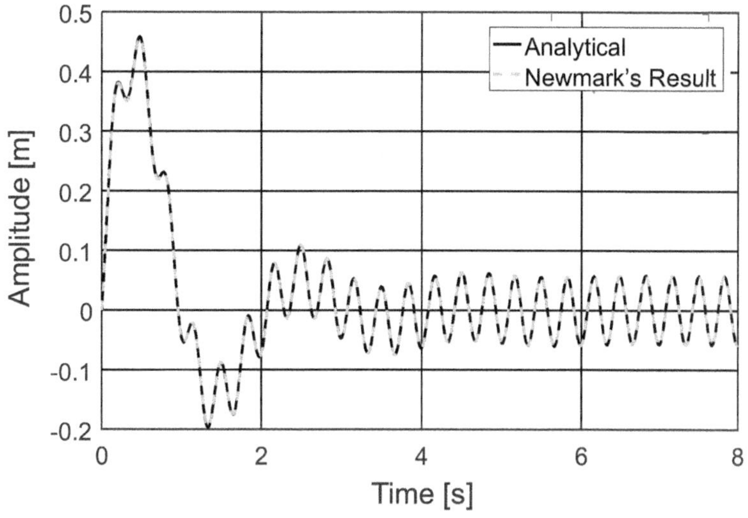

Figure 3.11 Response of the *SDOF* system using the Newmark-β method.

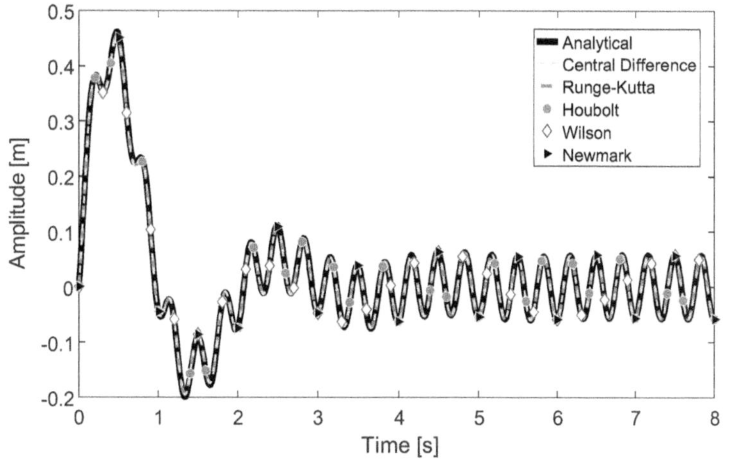

Figure 3.12 Time response approximation: comparison of results.

Sometimes, the preliminary use of an approximation method with less calculation time shows its superiority; there are several methods widely used in the field of vibrations. To give a brief introduction, four methods, namely Dunkerley's, Rayleigh's, Ritz' and Holzer's, are described in this section.

3.3.1 Dunkerley's Method

Dunkerley's method can be applied to a lightly damped system, i.e. whenever it is possible to neglect the damping. It is useful for estimating the fundamental natural frequency of an *MDOF* system, whenever its value is considerably lower than the remaining natural frequencies. To derive the method, a general undamped N degree of freedom system is shown in Figure 3.13. Its equilibrium equation is given as:

$$\boldsymbol{M}\ddot{\boldsymbol{x}} + \boldsymbol{K}\boldsymbol{x} = \boldsymbol{0} \tag{3.61}$$

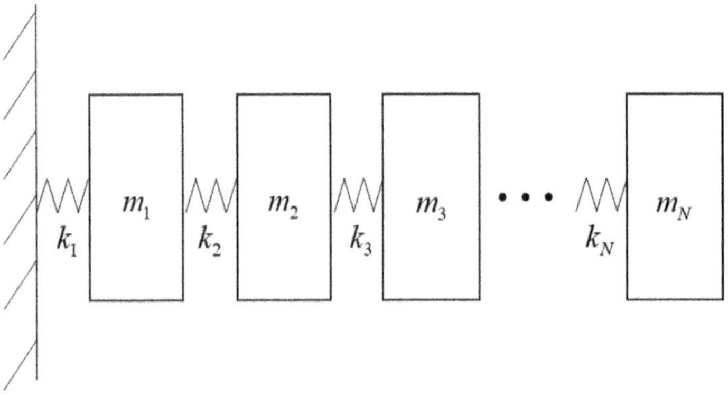

Figure 3.13 Example of an N degree-of-freedom system.

The system shown in Figure 3.13 has N natural frequencies and N related mode shapes, which have potential information for the vibration analysis. As explained in Chapter 1, Section 1.5.2, the free vibration solution of Eq. (3.61) implies the solving of a generalised eigenvalue and eigenvector problem. By setting the determinant of the resulting dynamic stiffness matrix of (3.61) to zero, the natural frequencies can be found:

$$\left|\boldsymbol{K} - \omega^2 \boldsymbol{M}\right| = 0 \tag{3.62}$$

Then, pre-multiplying by the inverse of the stiffness matrix and dividing by ω^2, leads to

$$\left|\frac{1}{\omega^2}\boldsymbol{I} - \boldsymbol{A}\boldsymbol{M}\right| = 0 \tag{3.63}$$

where $\boldsymbol{A}$ is the flexibility matrix. For a lumped parameter system, Eq. (3.63) has the following form:

$$\left|\frac{1}{\omega^2}\begin{bmatrix} 1 & & & \\ & 1 & & \\ & & \ddots & \\ & & & 1 \end{bmatrix} - \begin{bmatrix} a_{11} & a_{12} & \cdots & a_{1N} \\ a_{21} & a_{22} & \cdots & a_{2N} \\ \vdots & \vdots & \ddots & \vdots \\ a_{N1} & a_{N2} & \cdots & a_{NN} \end{bmatrix}\begin{bmatrix} m_1 & & & \\ & m_2 & & \\ & & \ddots & \\ & & & m_N \end{bmatrix}\right| = 0 \tag{3.64}$$

This can be recast as:

$$\begin{vmatrix} \dfrac{1}{\omega^2} - a_{11}m_1 & a_{12}m_2 & \cdots & a_{1N}m_N \\ a_{21}m_1 & \dfrac{1}{\omega^2} - a_{22}m_2 & \cdots & a_{2N}m_N \\ \vdots & \vdots & \ddots & \vdots \\ a_{N1}m_1 & a_{N2}m_2 & \cdots & \dfrac{1}{\omega^2} - a_{NN}m_N \end{vmatrix} = 0 \tag{3.65}$$

Expanding, it follows that:

$$\begin{aligned} &\left(\frac{1}{\omega^2}\right)^N - \left(a_{11}m_1 + a_{22}m_2 + \ldots + a_{NN}m_N\right)\left(\frac{1}{\omega^2}\right)^{N-1} \\ &+ \left(a_{11}a_{22}m_1m_2 + a_{11}a_{33}m_1m_3 + \ldots + a_{N-1,N-1}a_{NN}m_{N-1}m_N\right. \\ &\left. - a_{12}a_{21}m_1m_2 - \ldots - a_{N-1,N}a_{N,N-1}m_{N-1}m_N\right)\left(\frac{1}{\omega^2}\right)^{N-2} - \ldots = 0 \end{aligned} \tag{3.66}$$

This is a polynomial equation of Nth degree in $1/\omega^2$ and can be further represented in terms of its N roots:

$$\begin{aligned} &\left(\frac{1}{\omega^2} - \frac{1}{\omega_1^2}\right)\left(\frac{1}{\omega^2} - \frac{1}{\omega_2^2}\right)\cdots\left(\frac{1}{\omega^2} - \frac{1}{\omega_N^2}\right) \\ &= \left(\frac{1}{\omega^2}\right)^N - \left(\frac{1}{\omega_1^2} + \frac{1}{\omega_2^2} + \ldots + \frac{1}{\omega_N^2}\right)\left(\frac{1}{\omega^2}\right)^{N-1} - \ldots = 0 \end{aligned} \tag{3.67}$$

Equating the terms in $\left(1/\omega^2\right)^{N-1}$ between Eqs. (3.66) and (3.67), it turns out that

$$\frac{1}{\omega_1^2} + \frac{1}{\omega_2^2} + \ldots + \frac{1}{\omega_N^2} = a_{11}m_1 + a_{22}m_2 + \ldots + a_{NN}m_N \tag{3.68}$$

With the assumption that ω_2 to ω_N are much higher than ω_1, Eq. (3.68) can be simplified to:

$$\frac{1}{\omega_1^2} \approx a_{11}m_1 + a_{22}m_2 + \ldots + a_{NN}m_N \tag{3.69}$$

The fundamental frequency can then be obtained accordingly. Besides, as this approximation eliminates high order natural frequencies, it is always lower than the real one.

Example 3.3.1 Fundamental frequency of a 3 *DOF* system

To have a better understanding of Dunkerley's method, the 3 *DOF* system shown in Figure 3.14 is used for verification.

The flexibility matrix required for the application of Dunkerley's formula is given by

$$A = \begin{bmatrix} a_{11} & a_{12} & \cdots & a_{1N} \\ a_{21} & a_{22} & \cdots & a_{2N} \\ \vdots & \vdots & \ddots & \vdots \\ a_{N1} & a_{N2} & \cdots & a_{NN} \end{bmatrix} = K^{-1} \tag{3.70}$$

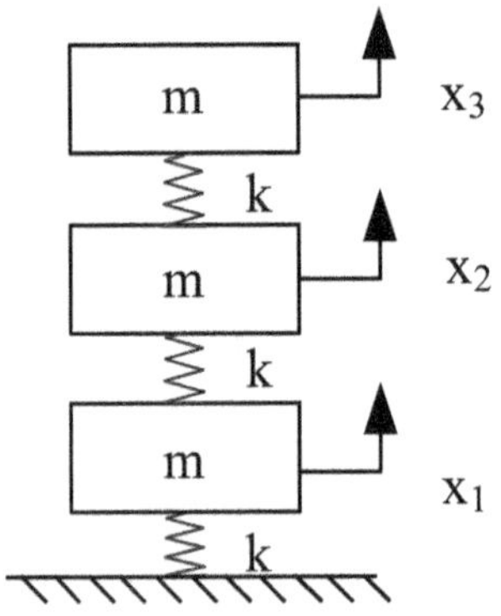

Figure 3.14 3 *DOF* system.

The stiffness matrix of the 3 *DOF* system is:

$$K = k\begin{bmatrix} 2 & -1 & 0 \\ -1 & 2 & -1 \\ 0 & -1 & 1 \end{bmatrix} \tag{3.71}$$

and the mass matrix is:

$$M = m\begin{bmatrix} 1 & 0 & 0 \\ 0 & 1 & 0 \\ 0 & 0 & 1 \end{bmatrix} \tag{3.72}$$

Accordingly, the flexibility matrix is given by:

$$A = K^{-1} = \frac{1}{k}\begin{bmatrix} 1 & 1 & 1 \\ 1 & 2 & 2 \\ 1 & 2 & 3 \end{bmatrix} \tag{3.73}$$

and thus,

$$AM = \frac{m}{k}\begin{bmatrix} 1 & 1 & 1 \\ 1 & 2 & 2 \\ 1 & 2 & 3 \end{bmatrix} \tag{3.74}$$

The approximate solution of the fundamental frequency by Dunkerley's formula is:

$$\omega_1 \approx \sqrt{\frac{1}{a_{11}m_1 + a_{22}m_2 + \ldots + a_{NN}m_N}} = \sqrt{\frac{k}{6m}} \approx 0.408\sqrt{\frac{k}{m}} \tag{3.75}$$

The analytical solution of the fundamental frequency for this 3 *DOF* system is $0.445\sqrt{k/m}$, which is slightly larger than the approximate solution.

3.3.2 Rayleigh's Method

Rayleigh's method is used to estimate the value of a natural frequency, once a reasonable approximation for the correspondent mode shape is given; in principle, it can be applied to obtain the approximate value of any natural frequency, although being typically used to the evaluation of the first natural frequency, as most of the times it is not easy to find a good approximation for mode shapes higher than the first one. Rayleigh's method is based on the principle of conservation of energy and therefore can be applied to lightly damped systems, precisely those where one can neglect the damping.

It is known that the free vibration response of the system at each natural frequency ω_r is harmonic, i.e. $\boldsymbol{x}_r(t) = \psi_r sin\omega_r t$ and thus the maximum displacement is given by the mode shape ψ_r and the maximum velocity by $\omega_r \psi_r$. Therefore, the maximum kinetic and elastic energies at each natural frequency are given, respectively, by

$$\begin{aligned} T_{\max_r} &= \frac{1}{2}\omega^2 \psi_r^T \boldsymbol{M} \psi_r \\ V_{\max_r} &= \frac{1}{2}\psi_r^T K \psi \end{aligned} \tag{3.76}$$

Since, from the principle of conservation of energy, $T_{\max_r} = V_{\max_r}$, the natural frequency ω_r is given by:

$$\omega_r = \sqrt{\frac{\psi_r^T \boldsymbol{K} \psi_r}{\psi_r^T \boldsymbol{M} \psi_r}} \tag{3.77}$$

Obviously, when the mode shape is the exact one, the result will be equal to the exact natural frequency ω_r. However, in an actual situation, the mode shape will be approximated by engineering experience, so the approximated natural frequency will be different to some extent from the real one. In fact, it will always be higher. Actual engineering experience shows that using the static response under uniformly distributed excitation as a mode shape will usually lead to a good approximation for the first natural frequency.

Example 3.3.2 Fundamental frequency approximation by Rayleigh's method

An example is used here to help the reader understanding the method. The model is the 3 *DOF* system, shown in Figure 3.. In the model, *m* is set as 1 kg and *k* is set as 10 N/m. From the analytical result, the first natural frequency is 1.4073 rad/s. Then, since the stiffnesses and masses are the same, one can try for the static response the vector $\{1 \quad 1.7 \quad 2\}$. The Rayleigh formula is then established according to Eq. (3.77):

$$\omega_1 = \sqrt{\frac{\{1 \quad 1.7 \quad 2\}\begin{bmatrix} 20 & -10 & 0 \\ -10 & 20 & -10 \\ 0 & -10 & 10 \end{bmatrix}\begin{Bmatrix} 1 \\ 1.7 \\ 2 \end{Bmatrix}}{\{1 \quad 1.7 \quad 2\}\begin{bmatrix} 1 & 0 & 0 \\ 0 & 1 & 0 \\ 0 & 0 & 1 \end{bmatrix}\begin{Bmatrix} 1 \\ 1.7 \\ 2 \end{Bmatrix}}} \tag{3.78}$$

The *MATLAB* code for this simulation is then programmed and shown in Appendix A (code 6).

Running the program, the result is 1.4151 rad/s.. The relative error between the approximated natural frequency and the real one is 0.55% showing its effectiveness. After that, another simulation aims at investigating the quality of the Rayleigh method for approximating the 2nd natural frequency. When the 2nd mode is excited, there is typically a point that does not move (a node). Thus, three responses are used, $[1 \quad 0 \quad -1]^T, [1 \quad -1 \quad 0]^T, [0 \quad -1 \quad 1]^T$; the *MATLAB* code is given in Appendix A (code 7).

The results are shown in Figure 3.15. One can see that for the three assumed response vectors, the approximated 2nd natural frequency varies rapidly. The real 2nd natural frequency is 3.9433 rad/s and the relative errors between the three approximated frequencies and the real one are 1.69%, 39.03% and 26.91%, respectively. The differences vary largely. Thus, using Rayleigh's method for the estimation of higher natural frequencies requires a careful selection of the assumed mode shape.

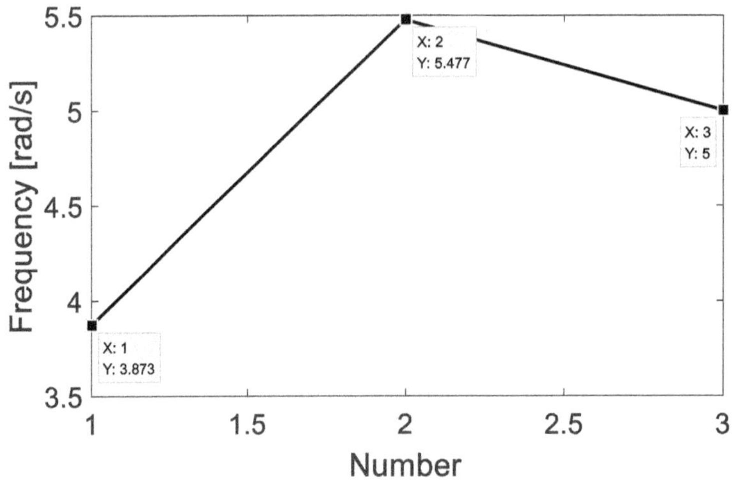

Figure 3.15 Approximate results for the 2nd natural frequency.

3.3.3 Ritz' Method

The Ritz method can be seen as a model reduction method. Compared with the Rayleigh method, it does not need a very strict approximation of the mode shape vectors and can be used for approximating the mode shapes and modal frequencies simultaneously. It uses several low order assumed mode shape vectors to lower the *DOFs* of the system. The reduced order model can then be solved easily to acquire the low order natural frequencies and mode shapes.

Firstly, for a general N degree of freedom system (Eq. (3.61)), the first r mode shape vectors are assumed as the Ritz basis:

$$\boldsymbol{\Psi} = \begin{bmatrix} \psi_1 & \cdots & \psi_r \end{bmatrix} \tag{3.79}$$

where $\boldsymbol{\Psi}$ is the assumed modal matrix and ψ_i is its ith column vector, representing the ith assumed modal vector. Then, the response $\boldsymbol{x}$ is the summation of the assumed modal vectors:

$$\boldsymbol{x} = \sum_{i=1}^{r} \psi_i q_i = \boldsymbol{\Psi q} \tag{3.80}$$

where $\boldsymbol{q}$ is the modal response vector, and q_i is the ith modal response.

Afterwards, the *DOFs* of Eq. (3.61) can be reduced from N to r by using the assumed mode shape matrix:

$$\bar{\boldsymbol{M}}_{r\times r}\ddot{\boldsymbol{q}}_{r\times 1} + \bar{\boldsymbol{K}}_{r\times r}\boldsymbol{q}_{r\times 1} = \left(\boldsymbol{\Psi}^T_{r\times N}\boldsymbol{M}_{N\times N}\boldsymbol{\Psi}_{N\times r}\right)\ddot{\boldsymbol{q}}_{r\times 1} + \left(\boldsymbol{\Psi}^T_{r\times N}\boldsymbol{K}_{N\times N}\boldsymbol{\Psi}_{N\times r}\right)\boldsymbol{q}_{r\times 1} = \boldsymbol{0} \tag{3.81}$$

where $\bar{\boldsymbol{M}}$ and $\bar{\boldsymbol{K}}$ are the new mass and stiffness matrices with dimensions $r\times r$.

However, since the assumed modal vectors are typically different from the real ones, $\bar{\boldsymbol{M}}$ and $\bar{\boldsymbol{K}}$ in general will not be diagonal matrices. Then, using the generalised eigenvalue theory, Eq. (3.81) is recast as:

$$\left(\bar{\boldsymbol{K}} - \omega^2\bar{\boldsymbol{M}}\right)\boldsymbol{q} = \boldsymbol{0} \tag{3.82}$$

Since the number of *DOFs* of Eq. (3.82) is smaller than the original number of *DOFs*, the problem is easily solved, leading to new natural frequencies $\bar{\omega}$ and mode shapes $\boldsymbol{q}$ that can be gathered in a new modal matrix $\bar{\boldsymbol{q}}$ that can be used to obtain the new approximated mode shape vectors, which are closer to the real ones than the assumed Ritz basis:

$$\bar{\varphi} = \boldsymbol{\Psi}\bar{\boldsymbol{q}} = \begin{bmatrix} \bar{\varphi}_1 & \cdots & \bar{\varphi}_r \end{bmatrix} \tag{3.83}$$

where $\bar{\varphi}$ is the approximated modal matrix of the first r modes.

Example 3.3.3 Fundamental frequency approximation by Ritz' theory

The model shown in Figure 3.14 is used here to help the reader understanding the Ritz method.

Firstly, the first two mode shape vectors are assumed to constitute the Ritz basis. The first mode shape is taken as the static response under a uniformly distributed force, and the second mode shape is assumed to contain the phase reverse characteristic:

$$\boldsymbol{\Psi} = \begin{bmatrix} 1 & 1 \\ 1.7 & 0 \\ 2 & -1 \end{bmatrix} \tag{3.84}$$

From Eq. (3.81), the model is then reduced by the Ritz basis:

$$\begin{bmatrix} 7.89 & -1 \\ -1 & 2 \end{bmatrix} \ddot{\boldsymbol{q}} + \begin{bmatrix} 15.8 & 0 \\ 0 & 30 \end{bmatrix} \boldsymbol{q} = \boldsymbol{0} \tag{3.85}$$

To solve this function, a *MATLAB* code is programmed in Appendix A (code 8). Applying that code, the assumed natural frequencies and mode shape vectors are obtained:

$$\bar{\omega}_1 = 1.4083\ rad/s,\ \bar{\omega}_2 = 4.0211\ rad/s,\ \bar{\boldsymbol{q}} = \begin{bmatrix} 1 & 1 \\ -0.0762 & 6.9130 \end{bmatrix} \tag{3.86}$$

The relative errors between the approximated natural frequencies and the real ones are 0.045% and 2.07%, respectively, which shows the effectiveness of the method. From Eqs. (3.83),

$$\bar{\boldsymbol{\varphi}} = \begin{bmatrix} 1 & 1 \\ 1.7 & 0 \\ 2 & -1 \end{bmatrix} \begin{bmatrix} 1 & 1 \\ -0.0762 & 6.9130 \end{bmatrix} = \begin{bmatrix} 0.9238 & 7.9130 \\ 1.7000 & 1.7000 \\ 2.0762 & -4.9130 \end{bmatrix} \tag{3.87}$$

Normalising the mode shapes so that they have the first element as 1,

$$\bar{\boldsymbol{\varphi}} = \begin{bmatrix} 1 & 1 \\ 1.8402 & 0.2148 \\ 2.2475 & -0.6209 \end{bmatrix} \tag{3.88}$$

As mentioned before, one finds that, after the Ritz approximation, the new computed modal vectors are closer to the real ones than the original assumed modal vectors. This can be checked using the Modal Assurance Criterion[1] (*MAC*), the most used tool to quantify the correlation between mode shapes. The closer *MAC* is to 1, the closer are the vectors. Its expression is:

$$MAC = \frac{\left| \left(\psi_{exact}^T \psi_{assumed} \right) \right|^2}{\left(\psi_{exact}^T \psi_{exact} \right) \left(\psi_{assumed}^T \psi_{assumed} \right)} \tag{3.89}$$

Knowing that the first two exact mode shapes are

[1] See also Chapter 6, Section 6.1.3.

$$\boldsymbol{\Psi}_{exact} = \begin{bmatrix} 1 & 1 \\ 1.8010 & 0.4450 \\ 2.2460 & -0.8010 \end{bmatrix} \tag{3.90}$$

One calculates the *MAC* between the vectors in expressions (3.84) and (3.90) and between vectors in (3.88) and (3.90). The results are:

$$MAC_{exact}^{initial\_assumed} = 0.9984,\ 0.8816 \quad MAC_{exact}^{Ritz\_result} = 0.9999,\ 0.9634 \tag{3.91}$$

These results confirm the effectiveness of the Ritz method.

3.3.4 Holzer's Method

Holzer's method can be used for the evaluation of natural frequencies and mode shapes of many different kinds of systems, in free or forced vibration, free or fixed ends, with or without damping, branched systems, etc. It follows a trial and error scheme and sets a threshold to find the intended result.

To make things easy, the explanation will be based on the *n DOF* semi-definite lumped parameter system shown in Figure 3.16.

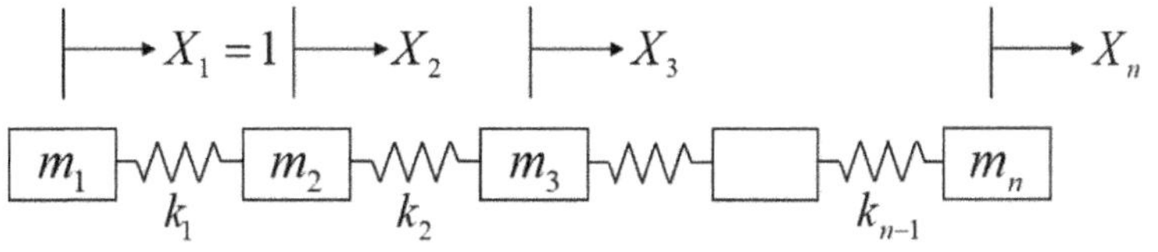

Figure 3.16 Semi-definite spring-mass system.

Firstly, the equations of motion are established:

$$\begin{cases} m_1\ddot{x}_1 + k_1(x_1 - x_2) = 0 \\ m_2\ddot{x}_2 + k_1(x_2 - x_1) + k_2(x_2 - x_3) = 0 \\ \quad\vdots \\ m_i\ddot{x}_i + k_{i-1}(x_i - x_{i-1}) + k_i(x_i - x_{i+1}) = 0 \\ \quad\vdots \\ m_n\ddot{x}_n + k_{n-1}(x_n - x_{n-1}) = 0 \end{cases} \tag{3.92}$$

As at each natural frequency the motion of mass m_i is harmonic with amplitude X_i, the equilibrium equations become:

$$\begin{cases} \omega^2 m_1 X_1 = k_1(X_1 - X_2) \\ \omega^2 m_2 X_2 = k_1(X_2 - X_1) + k_2(X_2 - X_3) \\ \vdots \\ \omega^2 m_i X_i = k_{i-1}(X_i - X_{i-1}) + k_i(X_i - X_{i+1}) \\ \vdots \\ \omega^2 m_n X_n = k_{n-1}(X_n - X_{n-1}) \end{cases} \tag{3.93}$$

The process starts by taking a trial frequency ω decided by the user and setting X_1 as 1. The remaining variables can be solved accordingly:

$$\begin{cases} X_2 = X_1 - \dfrac{\omega^2}{k_1} m_1 X_1 \\ X_3 = X_2 - \dfrac{\omega^2}{k_2} \sum_{j=1}^{2} m_j X_j \\ \vdots \\ X_i = X_{i-1} - \dfrac{\omega^2}{k_{i-1}} \sum_{j=1}^{i-1} m_j X_j \\ \vdots \\ X_n = X_{n-1} - \dfrac{\omega^2}{k_{n-1}} \sum_{j=1}^{n-1} m_j X_j \end{cases} \tag{3.94}$$

Then, according to Eq. (3.93), the residual term related to the equation of *DOF* n is calculated as the resultant force:

$$F = \omega^2 m_n X_n - k_{n-1}(X_n - X_{n-1}) = \omega^2 \sum_{j=1}^{n} m_j X_j \tag{3.95}$$

In theory, if the trial frequency is correct, the residual term should be zero, as the system is in free vibration. In practice, a small number ε is set as the threshold and new trial frequencies are given until $abs(F) < \varepsilon$ is satisfied. When that happens, the related amplitude vector $\{X_1 \ \cdots \ X_n\}$ is the approximated mode shape. After that the process continues, to find higher natural frequencies and mode shapes.

For other types of systems, the derivation of Holzer's formulation can be conducted in a similar way. For instance, in a system with a fixed end the displacement amplitude at that end should be zero, so the resultant amplitude at the fixed end can be used to judge the correctness of the trial frequency.

Example 3.3.4 Natural frequencies and mode shapes of a 3-*DOF* system

To give a better understanding of Holzer's method, the 3 *DOF* system of Figure 3.17 is taken as an example. The stiffness coefficients k between the masses are set as 100 *N/m* and each mass m is 1 *kg*. The *MATLAB* code is given in Appendix A (code 9).

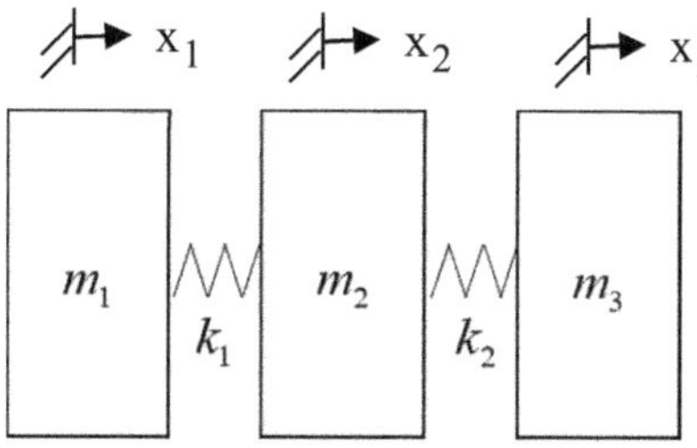

Figure 3.17 3 *DOF* free-free dynamic system.

The equilibrium equations are:

$$\begin{cases} m_1\ddot{x}_1 + k_1(x_1 - x_2) = 0 \\ m_2\ddot{x}_2 + k_1(x_2 - x_1) + k_2(x_2 - x_3) = 0, \\ m_3\ddot{x}_3 + k_2(x_3 - x_2) = 0 \end{cases} \quad \begin{matrix} m_1 = m_2 = m_3 = 1 \\ k_1 = k_2 = 100 \end{matrix} \tag{3.96}$$

Then, according to Holzer's formulation, the response vector under the trial frequency ω and the related resultant force are:

$$\begin{cases} X_2 = X_1 - \dfrac{\omega^2}{k_1} m_1 X_1 \\ X_3 = X_2 - \dfrac{\omega^2}{k_2}\left(m_1 X_1 + m_2 X_2\right) \\ F = \omega^2\left(m_1 X_1 + m_2 X_2 + m_3 X_3\right) \end{cases} \tag{3.97}$$

Then, the trial frequency is set to vary from 0 to $2\pi \times 3$ rad/s and the resultant force is computed. The approximated natural frequencies and mode shapes are separately given in Figure 3.18 and Figure 3.19, respectively. One can find that the calculated frequencies and mode shapes fit the exact ones, showing the validity of Holzer's method.

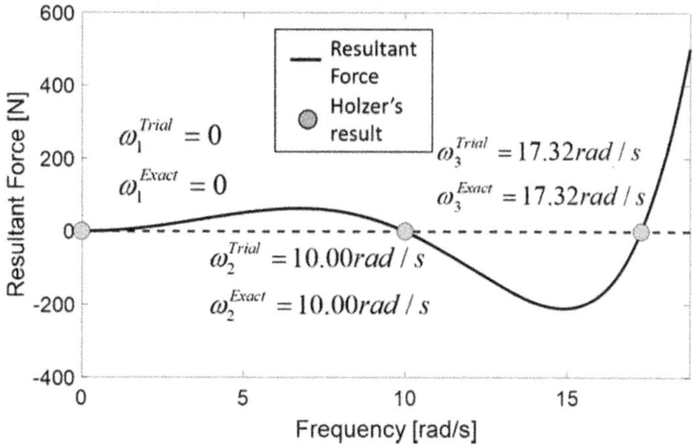

Figure 3.18 Resultant force versus frequency and natural frequencies.

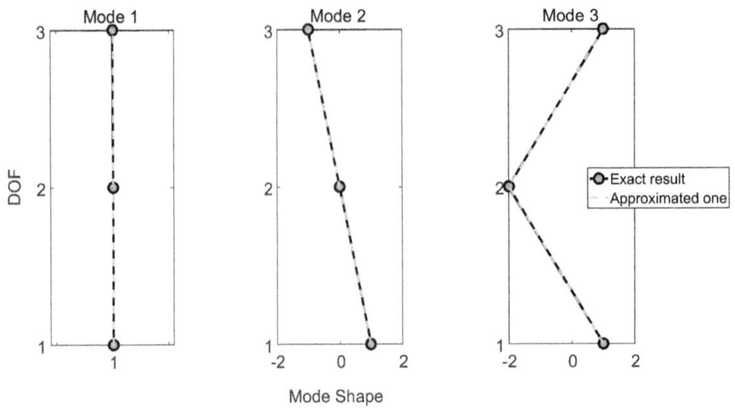

Figure 3.19 Mode shapes.

3.4 Matrix Methods

For an *MDOF* system, the eigenproblem is solved to obtain the natural frequencies and mode shapes, which can be further used to obtain the response by the superposition principle. Thus, the matrix methods for solving the eigenvalues and eigenvectors are important and widely used in *MDOF* systems. In this section, eight methods, called Bisection, Sturm Sequences, *MATLAB ROOTS*, Matrix Iteration, Jacobi's, Cholesky decomposition, Singular Value Decomposition (*SVD*), and Principal Component Analysis (*PCA*), are described in detail.

3.4.1 Bisection Method

The bisection method is an iterative approach to solve the eigenvalues of an *MDOF* system. For the *MDOF* equilibrium equation shown in Eq. (3.61), its natural frequencies and related modal shapes are the eigenvalues and eigenvectors of the system, respectively. As explained in Chapter 1 (Section 1.5.2), the generalised eigenproblem is given by:

$$(\boldsymbol{K}-\lambda\boldsymbol{M})\boldsymbol{\psi}=\boldsymbol{0} \tag{3.98}$$

where λ represents the eigenvalues of the system, the ith eigenvalue λ_i being equal to the square of the natural frequency, ω_i^2, and $\boldsymbol{\psi}$ are the eigenvectors, each one being the correspondent mode shape. Then, since solving the above equation requires the determinant of $(\boldsymbol{K}-\lambda\boldsymbol{M})$ being set to zero, the characteristic equation of the *MDOF* system is then derived as:

$$R(\lambda)=\lambda^N+a_1\lambda^{N-1}+\cdots+a_{N-1}\lambda+a_N=0 \tag{3.99}$$

where $a_1,\cdots,a_N$ are the polynomial coefficients derived from the determinant.

Then, the range of the eigenvalues is selected to start the bisection iteration. This can be easily done by calculating $R(\lambda)$ with λ increasing in steps and drawing a picture by using *MATLAB* to ascertain such areas with the following property:

$$\lambda_i \in \left[a_0^i, b_0^i\right] \quad \text{and} \quad R\left(a_0^i\right) \times R\left(b_0^i\right) < 0 \tag{3.100}$$

where a_0^i and b_0^i are the left and right boundaries of the selected range for eigenvalue λ_i. Eq. (3.100) means that there is one eigenvalue in the selected eigenvalue range, and the product of the values of the boundary points is negative. After that, the first bisection iteration is conducted:

$$\begin{aligned} & c_0^i = \frac{a_0^i + b_0^i}{2} \\ & if\left(R\left(c_0^i\right) \times R\left(a_0^i\right) < 0\right) \quad \begin{matrix} a_1^i = a_0^i \\ b_1^i = c_0^i \end{matrix} \\ & else\left(R\left(c_0^i\right) \times R\left(b_0^i\right) < 0\right) \quad \begin{matrix} a_1^i = c_0^i \\ b_1^i = b_0^i \end{matrix} \end{aligned} \tag{3.101}$$

where c_0^i is the initial middle point for eigenvalue λ_i. Eq. (3.101) means ascertaining the interval where the solution exists from the two intervals divided equally. Then, iteratively conduct the above step, until the residue of the middle point reaches convergence at iteration k:

$$R\left(c_k^i\right) < T \;\Rightarrow\; \lambda_i = c_k^i \tag{3.102}$$

where T is the convergence threshold set by the researcher. After the convergence is reached, the middle point of the kth iteration is assigned as the ith eigenvalue λ_i. Next, conduct the above steps for the rest of the eigenvalues, and the natural frequencies will be given as $\omega_i = \sqrt{\lambda_i}$, $i = 1, 2, 3, \cdots, N$. Afterwards, substitute the eigenvalues into Eq. (3.98) to obtain the mode shapes:

$$\left(\boldsymbol{K} - \lambda_i \boldsymbol{M}\right)\psi_i = \boldsymbol{0}, \quad i = 1, 2, 3, \cdots, N \tag{3.103}$$

3.4.2 Sturm Sequences Method

The Sturm Sequences method is also an iterative approach to obtain the natural frequencies from the dynamic matrix. It works like the bi-section method with a trial and error scheme, but can obtain a more complete result, because it can judge the number of eigenvalues when evaluating them.

Firstly, for an *MDOF* dynamic system of order *N*, its characteristic equation p is obtained from the determinant of the matrix $\left(\boldsymbol{K} - \lambda \boldsymbol{M}\right)$:

$$p: \; \left|\mathbf{K} - \lambda \mathbf{M}\right| = 0 \tag{3.104}$$

The leading principal minors of such determinant, $p_1 \sim p_N$, can form a Sturm sequence with the basis of p_N and p_{N-1}:

$$\begin{aligned} f_0 &= Simplify(p_N) \\ f_1 &= Simplify(p_{N-1}) \\ &\vdots \\ f_N &= 1 \end{aligned} \tag{3.105}$$

where '*Simplify*' means simplifying the polynomial equation by setting the coefficient of the highest order term as 1. Then, a left open right closed interval $(A, B]$ is set and the value of Eq. (3.105) is computed based on the interval border values A and B:

$$\begin{aligned} \boldsymbol{f}(A) &= \left[f_0(A) \quad f_1(A) \quad \cdots \quad f_N(A)\right] \\ \boldsymbol{f}(B) &= \left[f_0(B) \quad f_1(B) \quad \cdots \quad f_N(B)\right] \end{aligned} \tag{3.106}$$

The computed sequence is checked and, if the ith element in the sequence equals zero, it is excluded:

$$\boldsymbol{f}^i_{exclude} = \left[f_0 \quad \cdots \quad f_{i-1} \quad f_{i+1} \quad \cdots \quad f_N\right] \quad if: f_i = 0 \tag{3.107}$$

Afterwards, the number of changes in the signatures of the consecutive elements of the computed sequences are evaluated:

$$\begin{aligned} V(A) &= V\left(\left[sign(f_0(A)) \quad sign(f_1(A)) \quad \cdots \quad sign(f_N)\right]\right) \\ V(B) &= V\left(\left[sign(f_0(B)) \quad sign(f_1(B)) \quad \cdots \quad sign(f_N)\right]\right) \end{aligned} \tag{3.108}$$

where V means iteratively conducting the following judgment from f_0 to f_{N-1}:

$$sign(f_i) = \begin{cases} 1 & f_i > 0 \\ -1 & f_i < 0 \end{cases} \qquad V = \begin{cases} V+1 & \dfrac{sign(f_i)}{sign(f_{i+1})} = -1 \\ V & \dfrac{sign(f_i)}{sign(f_{i+1})} = 1 \end{cases} \tag{3.109}$$

The number of the eigenvalues in the interval $(A, B]$ is then obtained:

$$N_{eigenvalue} = V(A) - V(B) \tag{3.110}$$

where $N_{eigenvalue}$ means the number of the eigenvalues. If $N_{eigenvalue}$ equals N, all the eigenvalues are in the set interval and $(A, B]$ is then separated into many small intervals through the *MATLAB* function '*linspace*':

$$\boldsymbol{C} = \{c_1 \quad c_2 \quad \cdots \quad c_k\} = linspace(A, B, k) \tag{3.111}$$

The small intervals where each eigenvalue exists are then ascertained through the property of the Sturm sequence:

$$if\left(V(c_l)-V(c_{l+1})=1\right) \quad \lambda_i \in (c_l, c_{l+1}] \tag{3.112}$$

Afterwards, the above separation and judgment step is iteratively conducted for each $(c_l, c_{l+1}]$ until the required precision is reached. The eigenvalue together with the related natural frequency can be then obtained:

$$\lambda_i = c_{l+1} \qquad \omega_i = \sqrt{\lambda_i} \tag{3.113}$$

Example 3.4.2 Sturm sequences method

For a better understanding of this method, the 3 *DOF* model of Figure 3.14 is set as an example and the related code is given in Appendix A (code 10), where the stiffness coefficient k is set as 10 N/m and the mass m is set as 1 kg. The leading principal minors of the eigenvalue determinant are:

$$\begin{aligned} p_1 &= |20-\lambda| = -\lambda+20 \\ p_2 &= \begin{vmatrix} 20-\lambda & -10 \\ -10 & 20-\lambda \end{vmatrix} = \lambda^2 - 40\lambda + 300 \\ p_3 &= \begin{vmatrix} 20-\lambda & -10 & 0 \\ -10 & 20-\lambda & -10 \\ 0 & -10 & 10-\lambda \end{vmatrix} = -\lambda^3 + 50\lambda^2 - 600\lambda + 1000 \end{aligned} \tag{3.114}$$

The Sturm sequence is then established as:

$$\begin{aligned} f_0 &= \lambda^3 - 50\lambda^2 + 600\lambda - 1000 \\ f_1 &= \lambda^2 - 40\lambda + 300 \\ f_2 &= \lambda - 20 \\ f_3 &= 1 \end{aligned} \tag{3.115}$$

The interval is initially set as $(0, 40]$ and the element signature changes are evaluated according to the border values, which are given in Table 3.1.

Table 3.1 The computed result of the Sturm sequence by the initially set interval.

	f_0	f_1	f_2	f_3	V
A=0	-1000	+300	-20	+1	3
B=40	+7000	+300	+20	+1	0

One can find that $V(A)-V(B)$ equals 3, so the eigenvalues are all in the set interval. Then, $(0, 40]$ is further separated as 10001 points by the function '*linspace*' and 10000 small intervals are formed by the adjacent points. Their element signature adjustments are then evaluated and given in Figure 3.20.

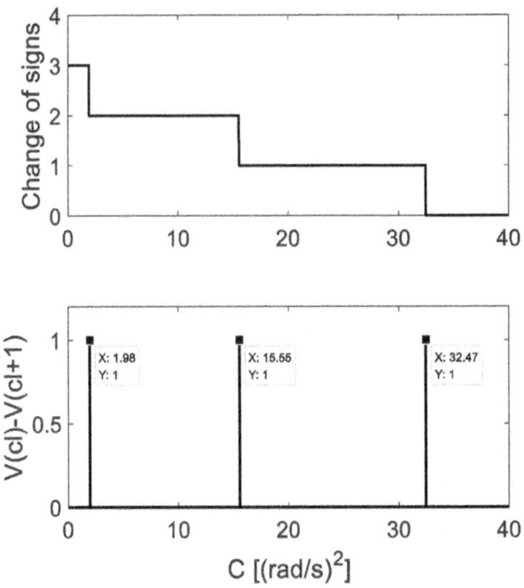

Figure 3.20 The computed changes of the element signatures of the small intervals.

One can find that the change of the signs reduces by one at 1.98 rad/s^2, 15.55 rad/s^2 and 32.47 rad/s^2. So, the eigenvalues exist within the three small intervals around the three values:

$$\begin{aligned} \lambda_1 &\in (1.9800, 1.9840] \\ \lambda_2 &\in (15.5480, 15.5520] \\ \lambda_3 &\in (32.4680, 32.4720] \end{aligned} \tag{3.116}$$

Each interval of Eq. (3.116) is separated again into 10000 parts to repeat the above step and obtain the related natural frequencies. The results are compared with the exact natural frequencies and given in Table 3.2. The computed results completely match the exact ones, showing the excellent performance of the Sturm sequence procedure.

Table 3.2 Comparison of the Sturm sequence results with the exact ones.

Natural frequency [rad/s]	Sturm sequence	Exact
1	1.4073	1.4073
2	3.9433	3.9433
3	5.6982	5.6982

3.4.3 *MATLAB ROOTS* Method

ROOTS is a function provided by *MATLAB* that can be directly used for the calculation of the eigenvalues.

For the characteristic equation (3.99), its polynomial coefficients are written as a vector:

$$\{1 \quad a_1 \quad \cdots \quad a_N\} \tag{3.117}$$

It is then input to the function '*ROOTS*' in *MATLAB*:

$$\{\lambda_1 \quad \lambda_2 \quad \cdots \quad \lambda_N\} = ROOTS(\{1 \quad a_1 \quad \cdots \quad a_N\}) \tag{3.118}$$

The function directly outputs the roots of the characteristic equation, i.e. the eigenvalues, from which one calculates the natural frequencies; substitution in Eq. (3.103) provides the mode shapes.

Example 3.4.3 *MATLAB ROOTS* method

The previous 3 *DOF* model of Figure 3. is used again here to help understanding this method. For that system, the characteristic equation is:

$$\lambda^3 - 50\lambda^2 + 600\lambda - 1000 = 0 \tag{3.119}$$

This is a 3rd order polynomial equation, and its coefficient vector is $\{1 \quad -50 \quad 600 \quad -1000\}^T$. Such vector can be directly input into the *ROOTS* function:

$$\{\lambda_1 \quad \lambda_2 \quad \lambda_3\} = ROOTS(\{1 \quad -50 \quad 600 \quad -1000\}) \tag{3.120}$$

The *MATLAB* code is programmed in Appendix A and the simulation is conducted (code 11). The result is shown in Table 3.3. One can see that the natural frequencies are directly solved and acquired, which shows its effectiveness.

Table 3.3 *ROOTS* result.

Natural frequency [rad/s]	*ROOTS*
1	1.4073
2	3.9432
3	5.6982

3.4.4 Cholesky's Decomposition

Cholesky's method works directly with the mass and stiffness matrices of the system and is used to obtain its eigenvalues and eigenvectors.

Firstly, *MATLAB* has a tool called 'eig', which can be used to solve the eigenvalues and eigenvectors of a symmetric matrix. In order to use such a tool, Eq. (3.98) is changed to the following form:

$$\left(\boldsymbol{M}^{-1}\boldsymbol{K} - \lambda\boldsymbol{I}\right)\boldsymbol{\psi} = \boldsymbol{0} \tag{3.121}$$

where the matrix $\boldsymbol{M}^{-1}\boldsymbol{K}$ can be solved to acquire the natural frequencies and mode shapes. However, although $\boldsymbol{M}$ and $\boldsymbol{K}$ are symmetric, $\boldsymbol{M}^{-1}\boldsymbol{K}$ is usually asymmetric and cannot be applied to the 'eig' function directly.

To solve this problem, the Cholesky decomposition is used to build a matrix $\boldsymbol{U}$ ($\boldsymbol{U} = chol(\boldsymbol{M})$) and thus decompose the mass matrix such that:

$$\boldsymbol{M} = \boldsymbol{U}^T\boldsymbol{U} \tag{3.122}$$

Next, changing the variable from $\boldsymbol{\psi}$ to $\boldsymbol{\varphi}$ through the transformation $\boldsymbol{\psi} = \boldsymbol{U}^{-1}\boldsymbol{\varphi}$ and substituting in (3.98),

$$(\boldsymbol{K} - \lambda\boldsymbol{M})\boldsymbol{U}^{-1}\boldsymbol{\varphi} = \boldsymbol{0} \quad \Leftrightarrow \quad (\boldsymbol{K}\boldsymbol{U}^{-1} - \lambda\boldsymbol{M}\boldsymbol{U}^{-1})\boldsymbol{\varphi} = \boldsymbol{0} \tag{3.123}$$

Pre-multiplying by $\boldsymbol{U}^{-T}$ it follows that:

$$(\boldsymbol{U}^{-T}\boldsymbol{K}\boldsymbol{U}^{-1} - \lambda\boldsymbol{U}^{-T}\boldsymbol{M}\boldsymbol{U}^{-1})\boldsymbol{\varphi} = \boldsymbol{0} \tag{3.124}$$

Substituting (3.122) in (3.124), one obtains

$$(\boldsymbol{U}^{-T}\boldsymbol{K}\boldsymbol{U}^{-1} - \lambda\boldsymbol{I})\boldsymbol{\varphi} = \boldsymbol{0} \tag{3.125}$$

As the stiffness matrix is symmetric, $\boldsymbol{U}^{-T}\boldsymbol{K}\boldsymbol{U}^{-1}$ is also symmetric, and so the *MATLAB* function 'eig' can be applied ($eig(\boldsymbol{U}^{-T}\boldsymbol{K}\boldsymbol{U}^{-1})$); the eigenvalues λ and the eigenvectors $\boldsymbol{\varphi}$ are calculated, and the eigenvectors $\boldsymbol{\psi}$ are recovered from $\boldsymbol{\psi} = \boldsymbol{U}^{-1}\boldsymbol{\varphi}$.

Example 3.4.4 Cholesky's decomposition

Once again, the 3 *DOF* system of Figure 3. is used to illustrate the application of the method. The three steps are:

$$\boldsymbol{U} = Cholesky\left(\begin{bmatrix} 1 & 0 & 0 \\ 0 & 1 & 0 \\ 0 & 0 & 1 \end{bmatrix}\right)$$

$$[\lambda, \boldsymbol{\varphi}] = eig\left(\boldsymbol{U}^{-T}\begin{bmatrix} 20 & -10 & 0 \\ -10 & 20 & -10 \\ 0 & -10 & 10 \end{bmatrix}\boldsymbol{U}^{-1}\right) \tag{3.126}$$

$$\boldsymbol{\psi} = \boldsymbol{U}^{-1}\boldsymbol{\varphi}$$

The *MATLAB* code is programmed in Appendix A for this example (code 12). The natural frequencies are given in Table 3.4 and compared with the *ROOTS* method. Both results match perfectly, showing the effectiveness of these numerical methods. An advantage of Cholesky's method over the *ROOTS* or Bisection methods is that it directly provides the mode shapes, as shown in Figure 3.21.

Table 3.4 Natural frequencies: comparison between *ROOTS* and Cholesky's method.

Natural frequency [rad/s]	*ROOTS*	Cholesky
1	1.4073	1.4073
2	3.9432	3.9433
3	5.6982	5.6982

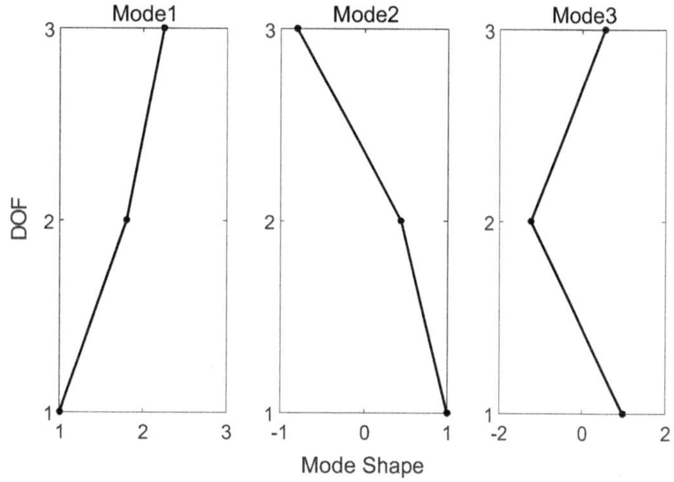

Figure 3.21 Mode shapes.

3.4.5 Matrix Iteration Method

The Matrix Iteration method is used to compute eigenvalues and eigenvectors of systems well separated modes. Such method follows an iteration scheme and is implemented by trial and error.

Firstly, for an *n DOF* system, a trial mode shape $\tilde{\psi}_1$ is set and can be represented as a linear combination of *n* real mode shape vectors since they form a basis in an *n*-dimensional space:

$$\tilde{\psi}_1 = c_1\psi_1 + c_2\psi_2 + ... + c_n\psi_n \tag{3.127}$$

where c_1 to c_n are the coefficients of the linear combination. Let the matrix $\boldsymbol{E} = \boldsymbol{K}^{-1}\boldsymbol{M}$ be pre-multiplied to Eq. (3.127):

$$\boldsymbol{E}\tilde{\psi}_1 = c_1\boldsymbol{E}\psi_1 + c_2\boldsymbol{E}\psi_2 + ... + c_n\boldsymbol{E}\psi_n \tag{3.128}$$

$\boldsymbol{E}\psi_i$ can be further simplified; as $\left(\boldsymbol{K} - \omega_i^2\boldsymbol{M}\right)\psi_i = \boldsymbol{0}$,

$$\left(K^{-1}M-\frac{1}{\omega_i^2}I\right)\psi_i=\mathbf{0}\;\Rightarrow\;E\psi_i=\frac{1}{\omega_i^2}\psi_i \tag{3.129}$$

Substituting Eq. (3.129) into Eq. (3.128), the result is:

$$\tilde{\psi}_2=E\tilde{\psi}_1=\frac{c_1}{\omega_1^2}\psi_1+\frac{c_2}{\omega_2^2}\psi_2+\ldots+\frac{c_n}{\omega_n^2}\psi_n \tag{3.130}$$

$E\tilde{\psi}_1$ is taken as the second trial shape $\tilde{\psi}_2$ and, according to this scheme, the third trial shape $\tilde{\psi}_3$ is obtained by repeating the above pre-multiplication:

$$\tilde{\psi}_3=E\tilde{\psi}_2=\frac{c_1}{\omega_1^4}\psi_1+\frac{c_2}{\omega_2^4}\psi_2+\ldots+\frac{c_n}{\omega_n^4}\psi_n \tag{3.131}$$

Then, iteratively conducting this trial shape updating to a relatively large number r, the result is:

$$\tilde{\psi}_{r+1}=E\tilde{\psi}_r=\frac{c_1}{\omega_1^{2r}}\psi_1+\frac{c_2}{\omega_2^{2r}}\psi_2+\ldots+\frac{c_n}{\omega_n^{2r}}\psi_n \tag{3.132}$$

Due to the assumption that the system does not exhibit any close modes and the fact that $\omega_1<\omega_2\cdots<\omega_n$, the natural frequencies will vary a lot after they are powered at order r; as a consequence,

$$\frac{1}{\omega_1^{2r}}\gg\frac{1}{\omega_2^{2r}}\gg\ldots\gg\frac{1}{\omega_n^{2r}} \tag{3.133}$$

Thus, $\tilde{\psi}_{r+1}$ can be further simplified as:

$$\tilde{\psi}_{r+1}\approx\frac{c_1}{\omega_1^{2r}}\psi_1 \tag{3.134}$$

and therefore, $\tilde{\psi}_r$ will have a similar form:

$$\tilde{\psi}_r\approx\frac{c_1}{\omega_1^{2(r-1)}}\psi_1 \tag{3.135}$$

Comparing Eqs. (3.134) and (3.135), one can find the fundamental natural frequency by computing the square root of the division of the corresponding elements of $\tilde{\psi}_r$ and $\tilde{\psi}_{r+1}$:

$$\omega_1\approx\sqrt{\frac{\tilde{\psi}_r^i}{\tilde{\psi}_{r+1}^i}},\quad\text{for any } i=1,2,\ldots,n \tag{3.136}$$

where i represents the ith component of the mode shapes.

After several iterations, the mode shape ψ_1 can be obtained either by $\tilde{\psi}_{r+1}$ or $\tilde{\psi}_r$, as these are proportional to each other (through ω_1^2, as in Eq. 3.136). After the fundamental natural frequency and mode shape are found, the 2nd order natural frequency and mode shape are explored. However, the influence of the fundamental mode should be eliminated from the matrix $\boldsymbol{E}$. Otherwise, the final result will always be ω_1 and $\boldsymbol{\psi}_1$.

The elimination procedure starts with the mass-normalisation of $\boldsymbol{\psi}_1$. Let the normalised $\boldsymbol{\psi}_1$ be $\boldsymbol{\phi}_1$. Then,

$$\boldsymbol{\phi}_1 = \frac{\boldsymbol{\psi}_1}{\sqrt{\boldsymbol{\psi}_1^T \boldsymbol{M} \boldsymbol{\psi}_1}} \tag{3.137}$$

The influence of mode 1 can then be removed by the mass-normalised mode shape and the result is called a deflated matrix $\boldsymbol{E}_2$, such that:

$$\boldsymbol{E}_2 = \boldsymbol{E}_1 - \frac{1}{\omega_1^2} \boldsymbol{\phi}_1 \boldsymbol{\phi}_1^T \boldsymbol{M} \tag{3.138}$$

where $\boldsymbol{E}_1 = \boldsymbol{E}$, and $\boldsymbol{E}_2$ has the eigenvalues 0, and ω_2^2 to ω_n^2. In fact, as any response vector $\tilde{\boldsymbol{\phi}}$ is a linear combination of the mode shapes,

$$\tilde{\boldsymbol{\phi}} = c_1 \boldsymbol{\phi}_1 + c_2 \boldsymbol{\phi}_2 + \ldots + c_n \boldsymbol{\phi}_n = \sum_{r=1}^{n} c_r \boldsymbol{\phi}_r \tag{3.139}$$

from (3.138),

$$\boldsymbol{E}_2 \tilde{\boldsymbol{\phi}} = \boldsymbol{E}_1 \sum_{r=1}^{n} c_r \boldsymbol{\phi}_r - \frac{1}{\omega_1^2} \boldsymbol{\phi}_1 \sum_{r=1}^{n} c_r \boldsymbol{\phi}_1^T \boldsymbol{M} \boldsymbol{\phi}_r \tag{3.140}$$

Due to the orthogonality properties[2] and the mass-normalisation (Eq. 3.137)),

$$\boldsymbol{E}_2 \tilde{\boldsymbol{\phi}} = \boldsymbol{E}_1 \sum_{r=1}^{n} c_r \boldsymbol{\phi}_r - \frac{1}{\omega_1^2} c_1 \boldsymbol{\phi}_1 \tag{3.141}$$

And because $\boldsymbol{E}_1 \boldsymbol{\phi}_1 = \frac{1}{\omega_1^2} \boldsymbol{\phi}_1$ (Eq. 3.129), it follows from (3.141) that

$$\boldsymbol{E}_2 \tilde{\boldsymbol{\phi}} = \boldsymbol{E}_1 \sum_{r=2}^{n} c_r \boldsymbol{\phi}_r \tag{3.142}$$

meaning that $\boldsymbol{E}_2$ no longer contains information about the first natural frequency. Conducting the matrix iteration process on $\boldsymbol{E}_2$, ω_2 and $\boldsymbol{\phi}_2$ can be obtained. The procedure can be repeated to obtain the remaining natural frequencies and mode shapes. For ω_3 and $\boldsymbol{\phi}_3$ one would use

$$\boldsymbol{E}_3 = \boldsymbol{E}_2 - \frac{1}{\omega_2^2} \boldsymbol{\phi}_2 \boldsymbol{\phi}_2^T \boldsymbol{M} \tag{3.143}$$

[2] See Chapter 1, Section 1.5.2, under 'Orthogonality Properties and Normalisation of the Mode Shapes', for details.

and in general, the deflated matrices will be given by:

$$\boldsymbol{E}_s = \boldsymbol{E}_{s-1} - \frac{1}{\omega_{s-1}^2}\boldsymbol{\phi}_{s-1}\boldsymbol{\phi}_{s-1}^T\boldsymbol{M} \quad \text{for} \quad s = 2, \ldots n \tag{3.144}$$

Example 3.4.5 Natural frequencies of a 3 *DOF* system

To give a better understanding of the matrix iteration method, the 3 *DOF* system shown in Figure 3. is again used for verification, comparing the natural frequencies and mode shapes calculated by the iteration method with the analytical ones. The *MATLAB* code is programmed in Appendix A for this example (code 13).

Solution: The analytical natural frequencies of the 3 *DOF* system are:

$$\omega_1 = 0.445\sqrt{\frac{k}{m}}, \quad \omega_2 = 1.247\sqrt{\frac{k}{m}}, \quad \omega_3 = 1.802\sqrt{\frac{k}{m}} \tag{3.145}$$

1st Step: estimating the fundamental frequency and mode shape.

The matrix $\boldsymbol{E}$ of this 3 *DOF* system is:

$$\boldsymbol{E} = \boldsymbol{K}^{-1}\boldsymbol{M} = \frac{m}{k}\begin{bmatrix} 1 & 1 & 1 \\ 1 & 2 & 2 \\ 1 & 2 & 3 \end{bmatrix} \tag{3.146}$$

Assuming the trial vector $\tilde{\boldsymbol{\psi}}_1$ as

$$\tilde{\boldsymbol{\psi}}_1 = \begin{Bmatrix} 1 \\ 1 \\ 1 \end{Bmatrix} \tag{3.147}$$

the second trial vector can be determined by Eq. (3.130):

$$\tilde{\boldsymbol{\psi}}_2 = \boldsymbol{E}\tilde{\boldsymbol{\psi}}_1 = \frac{m}{k}\begin{Bmatrix} 3 \\ 5 \\ 6 \end{Bmatrix} \tag{3.148}$$

After this iteration, the fundamental frequency can be approximated from the components of $\tilde{\boldsymbol{\psi}}_1$ and $\tilde{\boldsymbol{\psi}}_2$:

$$\omega_1 \approx \sqrt{\frac{\tilde{\psi}_1^1}{\tilde{\psi}_2^1}} = \sqrt{\frac{k}{3m}} \tag{3.149}$$

ω_1 is approximated with more iterations and the variations of the results are shown in Table 3.5.

Table 3.5 Example of the Matrix Iteration method for the 1st natural frequency and mode shape.

Iterations(i)	Trial vector $\tilde{\psi}_i$	Trial vector $\tilde{\psi}_{i+1}$	ω_1
1	$\begin{Bmatrix} 1 \\ 1 \\ 1 \end{Bmatrix}$	$\frac{m}{k}\begin{Bmatrix} 3 \\ 5 \\ 6 \end{Bmatrix}$	$0.5773\sqrt{\frac{k}{m}}$
2	$\frac{m}{k}\begin{Bmatrix} 3 \\ 5 \\ 6 \end{Bmatrix}$	$\frac{m^2}{k^2}\begin{Bmatrix} 14 \\ 25 \\ 31 \end{Bmatrix}$	$0.4629\sqrt{\frac{k}{m}}$
3	$\frac{m^2}{k^2}\begin{Bmatrix} 14 \\ 25 \\ 31 \end{Bmatrix}$	$\frac{m^3}{k^3}\begin{Bmatrix} 70 \\ 126 \\ 157 \end{Bmatrix}$	$0.4472\sqrt{\frac{k}{m}}$
⋮	⋮	⋮	⋮
7	$\frac{m^6}{k^6}\begin{Bmatrix} 8997 \\ 16212 \\ 20216 \end{Bmatrix}$	$\frac{m^7}{k^7}\begin{Bmatrix} 45425 \\ 81853 \\ 102069 \end{Bmatrix}$	$0.44504\sqrt{\frac{k}{m}}$
8	$\frac{m^7}{k^7}\begin{Bmatrix} 45425 \\ 81853 \\ 102069 \end{Bmatrix}$	$\frac{m^8}{k^8}\begin{Bmatrix} 229347 \\ 413269 \\ 515338 \end{Bmatrix}$	$0.44504\sqrt{\frac{k}{m}}$

It can be seen that the mode shape and the natural frequency converged after 8 iterations. Thus, the 1st natural frequency and related mode shape are given as:

$$\omega_1 = 0.44504\sqrt{\frac{k}{m}} \tag{3.150}$$

$$\psi_1 = 229347\begin{Bmatrix} 1 \\ 1.8019 \\ 2.2470 \end{Bmatrix} \text{ or, apart from the constant 229347, simply } \begin{Bmatrix} 1 \\ 1.8019 \\ 2.2470 \end{Bmatrix} \tag{3.151}$$

<u>2nd Step</u>: estimating the 2nd natural frequency and mode shape.

According to Eq. (3.137), ψ_1 must be mass-normalised in the first place:

$$\phi_1 = \frac{\begin{Bmatrix} 1 \\ 1.8019 \\ 2.2470 \end{Bmatrix}}{\sqrt{m\{1 \quad 1.8019 \quad 2.2470\}\begin{bmatrix} 1 & 0 & 0 \\ 0 & 1 & 0 \\ 0 & 0 & 1 \end{bmatrix}\begin{Bmatrix} 1 \\ 1.8019 \\ 2.2470 \end{Bmatrix}}} = \begin{Bmatrix} 0.32799 \\ 0.59101 \\ 0.73698 \end{Bmatrix} m^{-1/2} \tag{3.152}$$

The deflated matrix $\boldsymbol{E}_2$ is then constructed as

$$\boldsymbol{E}_2 = \boldsymbol{E}_1 - \frac{1}{\omega_1^2}\boldsymbol{\phi}_1\boldsymbol{\phi}_1^T\boldsymbol{M} = \frac{m}{k}\left(\begin{bmatrix}1 & 1 & 1\\ 1 & 2 & 2\\ 1 & 2 & 3\end{bmatrix} - \frac{1}{0.44504^2}\begin{Bmatrix}0.32799\\ 0.59101\\ 0.73698\end{Bmatrix}\{0.32799 \quad 0.59101 \quad 0.73698\}\right)$$

$$= \frac{m}{k}\begin{bmatrix}0.45687 & 0.02131 & -0.22041\\ 0.02131 & 0.23645 & -0.19910\\ -0.22041 & -0.19910 & 0.25776\end{bmatrix} \tag{3.153}$$

One can choose any trial vector to start with. Taking again $\tilde{\boldsymbol{\psi}}_1$ as

$$\tilde{\boldsymbol{\psi}}_1 = \begin{Bmatrix}1\\ 1\\ 1\end{Bmatrix} \tag{3.154}$$

the iteration process is conducted and the results are given in Table 3.6.

Table 3.6 Example of the Matrix Iteration method for the 2st natural frequency and mode shape.

Iterations(i)	Trial vector $\tilde{\boldsymbol{\psi}}_i$	Trial vector $\tilde{\boldsymbol{\psi}}_{i+1}$	ω_1
1	$\begin{Bmatrix}1\\ 1\\ 1\end{Bmatrix}$	$\frac{m}{k}\begin{Bmatrix}0.25776\\ 0.05866\\ -0.16175\end{Bmatrix}$	$1.9697\sqrt{\frac{k}{m}}$
2	$\frac{m}{k}\begin{Bmatrix}0.25776\\ 0.05866\\ -0.16175\end{Bmatrix}$	$\frac{m^2}{k^2}\begin{Bmatrix}0.15466\\ 0.05157\\ -0.11019\end{Bmatrix}$	$1.2910\sqrt{\frac{k}{m}}$
3	$\frac{m^2}{k^2}\begin{Bmatrix}0.15466\\ 0.05157\\ -0.11019\end{Bmatrix}$	$\frac{m^3}{k^3}\begin{Bmatrix}0.09604\\ 0.03743\\ -0.07276\end{Bmatrix}$	$1.2690\sqrt{\frac{k}{m}}$
$\vdots$	$\vdots$	$\vdots$	$\vdots$
11	$\frac{m^{10}}{k^{10}}\begin{Bmatrix}0.00423\\ 0.00188\\ -0.00339\end{Bmatrix}$	$\frac{m^{11}}{k^{11}}\begin{Bmatrix}0.00272\\ 0.00121\\ -0.00218\end{Bmatrix}$	$1.24704\sqrt{\frac{k}{m}}$
12	$\frac{m^{11}}{k^{11}}\begin{Bmatrix}0.00272\\ 0.00121\\ -0.00218\end{Bmatrix}$	$\frac{m^{12}}{k^{12}}\begin{Bmatrix}0.00175\\ 0.00078\\ -0.00140\end{Bmatrix}$	$1.24701\sqrt{\frac{k}{m}}$

The iteration process converges after 12 iterations and the second modal frequency and mode shape are given by:

$$\omega_2 = 1.24701\sqrt{\frac{k}{m}} \tag{3.155}$$

$$\psi_2 = \begin{Bmatrix} 0.00175 \\ 0.00078 \\ -0.00140 \end{Bmatrix}, \text{ or taking the first component as 1, } \psi_2 = \begin{Bmatrix} 1 \\ 0.4457 \\ -0.8000 \end{Bmatrix} \tag{3.156}$$

3rd Step: estimating the 3rd natural frequency and mode shape.

Repeating the iteration process to for mode 3, the iteration converges after 4 iterations and the related natural frequency and mode shape are given by:

$$\omega_3 = 1.80194\sqrt{\frac{k}{m}} \tag{3.157}$$

$$\psi_3 = \begin{Bmatrix} 0.00097 \\ -0.00121 \\ 0.00054 \end{Bmatrix}, \text{ or } \psi_3 = \begin{Bmatrix} 1 \\ -1.2474 \\ 0.5567 \end{Bmatrix} \tag{3.158}$$

Table 3.7 and Figure 3.22 compare the results of the natural frequencies and mode shapes with the analytical ones, where it is clear that the errors are very small.

Table 3.7 Comparison between the analytical natural frequencies and those from the Matrix Iteration method.

Natural frequency	Analytical	Matrix Iteration method
ω_1	$0.445\sqrt{\frac{k}{m}}$	$0.44504\sqrt{\frac{k}{m}}$
ω_2	$1.247\sqrt{\frac{k}{m}}$	$1.24701\sqrt{\frac{k}{m}}$
ω_3	$1.802\sqrt{\frac{k}{m}}$	$1.80194\sqrt{\frac{k}{m}}$

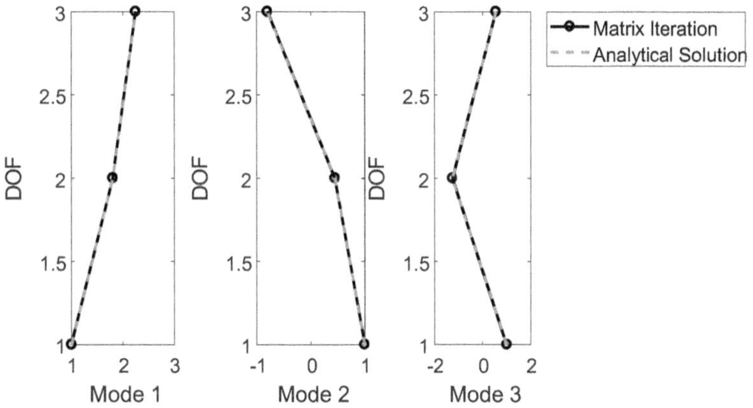

Figure 3.22 Mode shapes.

3.4.6 Jacobi's Method

Jacobi's method is also an iterative method for the calculation of natural frequencies and mode shapes. It uses orthogonal rotation matrices $\boldsymbol{R}$ with the objective of modifying the matrix $\boldsymbol{E}$ (defined in the previous section as $\boldsymbol{K}^{-1}\boldsymbol{M}$) step by step, leading eventually to a matrix $\boldsymbol{R}^T\boldsymbol{E}\boldsymbol{R}$ totally diagonal, from which the natural frequencies and mode shapes can be easily extracted. Attention should be called to the fact that Jacobi's method only works with symmetric matrices, which – in general – is not the case of $\boldsymbol{K}^{-1}\boldsymbol{M}$.

In the first place, let the rotation matrix $\boldsymbol{R}$ be defined as:

$$\boldsymbol{R}=\begin{bmatrix} 1 & 0 & & & & & \\ 0 & 1 & & & & & \\ & & \ddots & & & & \\ & & & cos\theta & & -sin\theta & \\ & & & & \ddots & & \\ & & & sin\theta & & cos\theta & \\ & & & & & & \ddots \\ & & & & & & & 1 \end{bmatrix} \begin{matrix} \\ \\ \\ i^{th}\ \text{row} \\ \\ j^{th}\ \text{row} \\ \\ \\ \end{matrix} \tag{3.159}$$

$$\qquad i^{th}\ \text{column} \quad j^{th}\ \text{column}$$

Matrix $\boldsymbol{R}$ is like the identity matrix, except at rows and columns i and j. Note that in a matrix transformation like $\boldsymbol{R}^T\boldsymbol{E}\boldsymbol{R}$, $\boldsymbol{R}$ induces a pure rotation, since its determinant is equal to 1. The resulting elements of $\boldsymbol{R}^T\boldsymbol{E}\boldsymbol{R}$, at rows and columns i and j are:

$$\tilde{e}_{ii}=e_{ii}cos^2\theta+2e_{ij}sin\theta cos\theta+e_{jj}sin^2\theta \tag{3.160}$$

$$\tilde{e}_{ij}=\tilde{e}_{ji}=\left(e_{jj}-e_{ii}\right)sin\theta cos\theta+e_{ij}\left(cos^2\theta-sin^2\theta\right) \tag{3.161}$$

$$\tilde{e}_{jj}=e_{ii}sin^2\theta-2e_{ij}sin\theta cos\theta+e_{jj}cos^2\theta \tag{3.162}$$

In Eqs. (3.160) to (3.162), e_{ii}, e_{ij}, e_{ji}, e_{jj} are the elements of $\boldsymbol{E}$ in positions (i, i), (i, j), (j, i), and (j, j), respectively and $\tilde{e}_{ii}$, $\tilde{e}_{ij}$, $\tilde{e}_{ji}$, $\tilde{e}_{jj}$ are the resulting elements of matrix $\boldsymbol{R}^T\boldsymbol{E}\boldsymbol{R}$. The final objective is that $\boldsymbol{R}^T\boldsymbol{E}\boldsymbol{R}$ becomes diagonal. The procedure can start by choosing an off-diagonal element of $\boldsymbol{E}$ that one wishes to put to zero. For instance, if the chosen element is e_{35}, then matrix $\boldsymbol{R}$ should be the identity matrix except for the elements $R_{33}=cos\theta$, $R_{35}=-sin\theta$, $R_{53}=sin\theta$ and $R_{55}=cos\theta$. And one must choose the angle θ so that $\tilde{e}_{35}$ is zero. In general, from Eq. (3.161),

$$\tilde{e}_{ij}=\left(e_{jj}-e_{ii}\right)sin\theta cos\theta+e_{ij}\left(cos^2\theta-sin^2\theta\right)=0 \tag{3.163}$$

As $cos2\theta=cos^2\theta-sin^2\theta$ and $sin2\theta=2sin\theta cos\theta$, one should choose θ such that

$$tan2\theta=\frac{2e_{ij}}{e_{ii}-e_{jj}} \tag{3.164}$$

After one of the off-diagonal elements related to i and j is reduced to zero, the new matrix is $\boldsymbol{R}_1^T \boldsymbol{E}\boldsymbol{R}_1$. Then the process is repeated to make other off-diagonal elements to be zero. A second step leads to a new matrix $\boldsymbol{R}_2^T \boldsymbol{R}_1^T \boldsymbol{E}\boldsymbol{R}_1\boldsymbol{R}_2$. Such an iteration induces a new zero off-diagonal element, but also has a side effect: the previous zero off-diagonal element changes to non-zero. However, repeating the above steps to a considerable number of iterations will reduce such side effects. Therefore, the Jacobi's iteration process should be conducted to a large order r:

$$\begin{aligned} \boldsymbol{E}_1 &= \boldsymbol{R}_1^T \boldsymbol{E}\boldsymbol{R}_1 \\ \boldsymbol{E}_2 &= \boldsymbol{R}_2^T \boldsymbol{R}_1^T \boldsymbol{E}\boldsymbol{R}_1\boldsymbol{R}_2 \\ \boldsymbol{E}_3 &= \boldsymbol{R}_3^T \boldsymbol{R}_2^T \boldsymbol{R}_1^T \boldsymbol{E}\boldsymbol{R}_1\boldsymbol{R}_2\boldsymbol{R}_3 \\ &\vdots \\ \boldsymbol{E}_r &= \boldsymbol{R}_r^T \cdots \boldsymbol{R}_2^T \boldsymbol{R}_1^T \boldsymbol{E}\boldsymbol{R}_1\boldsymbol{R}_2 \cdots \boldsymbol{R}_r = \boldsymbol{R}^T \boldsymbol{E}\boldsymbol{R} \end{aligned} \tag{3.165}$$

with $\boldsymbol{R} = \boldsymbol{R}_1\boldsymbol{R}_2 \cdots \boldsymbol{R}_r$.

After the Jacobi iteration is considered as having converged, the natural frequencies and mode shapes can be obtained directly. As $(\boldsymbol{K} - \omega_i^2 \boldsymbol{M})\boldsymbol{\psi}_i = \boldsymbol{0}$, $\left(\boldsymbol{K}^{-1}\boldsymbol{M} - \frac{1}{\omega_i^2}\boldsymbol{I}\right)\boldsymbol{\psi}_i = \boldsymbol{0}$ and so, $\left(\boldsymbol{E} - \frac{1}{\omega_i^2}\boldsymbol{I}\right)\boldsymbol{\psi}_i = \boldsymbol{0}$, meaning that the natural frequencies are given by the inverse of the square root of the diagonal values of $\boldsymbol{R}^T \boldsymbol{E}\boldsymbol{R}$ and the mode shapes are the columns of $\boldsymbol{R}$:

$$\boldsymbol{R} = \begin{bmatrix} \boldsymbol{\psi}_1 & \cdots & \boldsymbol{\psi}_n \end{bmatrix} \tag{3.166}$$

Example 3.4.6 Natural frequencies and mode shapes of a 3 *DOF* system using Jacobi's Method

Once more the 3 *DOF* system of Figure 3.14 is used for verification. The *MATLAB* code is programmed in Appendix A for this example (code 14).

Solution: The analytical natural frequencies are:

$$\omega_1 = 0.445\sqrt{\frac{k}{m}}, \quad \omega_2 = 1.247\sqrt{\frac{k}{m}}, \quad \omega_3 = 1.802\sqrt{\frac{k}{m}} \tag{3.167}$$

The matrix $\boldsymbol{E}$ is:

$$\boldsymbol{E} = \boldsymbol{K}^{-1}\boldsymbol{M} = \frac{m}{k}\begin{bmatrix} 1 & 1 & 1 \\ 1 & 2 & 2 \\ 1 & 2 & 3 \end{bmatrix} \tag{3.168}$$

<u>1st iteration</u>:

Firstly, e_{12} (for example) is chosen to be reduced to zero; the angle is:

$$\theta_1 = \frac{1}{2} tan^{-1} \frac{2e_{12}}{e_{11} - e_{22}} = \frac{1}{2} tan^{-1} \frac{2}{1-2} = -0.55357\, rad \tag{3.169}$$

and the iteration is conducted accordingly, calculating R_{11}, R_{12}, R_{21} and R_{22}:

$$\boldsymbol{R}_1 = \begin{bmatrix} 0.85065 & 0.52573 & 0 \\ -0.52573 & 0.85065 & 0 \\ 0 & 0 & 1 \end{bmatrix} \tag{3.170}$$

$$\boldsymbol{E}_1 = \boldsymbol{R}_1^T \boldsymbol{E} \boldsymbol{R}_1 = \frac{m}{k} \begin{bmatrix} 0.381967 & 0 & -0.20081 \\ 0 & 2.61803 & 2.22703 \\ -0.20081 & 2.22703 & 3 \end{bmatrix} \tag{3.171}$$

2nd iteration:

Now, let e_{13} of $\boldsymbol{E}_1$ be the chosen one:

$$\theta_2 = \frac{1}{2} tan^{-1} \frac{2e_{13}}{e_{11} - e_{33}} = \frac{1}{2} tan^{-1} \frac{2\times(-0.20081)}{0.381967-3} = 0.07611\, rad \tag{3.172}$$

$$\boldsymbol{R}_2 = \begin{bmatrix} 0.99711 & 0 & -0.07604 \\ 0 & 1 & 0 \\ 0.07604 & 0 & 0.99711 \end{bmatrix} \tag{3.173}$$

$$\boldsymbol{E}_2 = \boldsymbol{R}_2^T \boldsymbol{E}_1 \boldsymbol{R}_2 = \frac{m}{k} \begin{bmatrix} 0.36665 & 0.16934 & 0 \\ 0.16934 & 2.61803 & 2.22059 \\ 0 & 2.22059 & 3.01531 \end{bmatrix} \tag{3.174}$$

It is confirmed here that putting e_{13} to zero had the consequence of increasing e_{12} from zero to 0.16934. But this will be attenuated, as the iterations progress.

3rd iteration:

Next, e_{23} of $\boldsymbol{E}_2$ is chosen to be reduced to zero:

$$\theta_3 = \frac{1}{2} tan^{-1} \frac{2e_{23}}{e_{22} - e_{33}} = \frac{1}{2} tan^{-1} \frac{2\times 2.22059}{2.61803-3.01531} = -0.74079\, rad \tag{3.175}$$

$$\boldsymbol{R}_3 = \begin{bmatrix} 1 & 0 & 0 \\ 0 & 0.73794 & 0.67487 \\ 0 & -0.67487 & 0.73794 \end{bmatrix} \tag{3.176}$$

$$\boldsymbol{E}_3 = \boldsymbol{R}_3^T \boldsymbol{E}_2 \boldsymbol{R}_3 = \frac{m}{k}\begin{bmatrix} 0.36665 & 0.12496 & 0.11428 \\ 0.12496 & 0.58722 & 0 \\ 0.11428 & 0 & 5.04613 \end{bmatrix} \tag{3.177}$$

As one can see, now it is element e_{13} that changed to non-zero, but it can also be observed that the element e_{12}, although still different from zero, has decreased. So, Jacobi's iteration must be continued and, when the iterative process is taken up to the 9th order, the result is:

$$\boldsymbol{E}_9 = \boldsymbol{R}_9^T \boldsymbol{E}_8 \boldsymbol{R}_9 = \frac{m}{k}\begin{bmatrix} 0.30798 & 7\times10^{-14} & -2\times10^{-22} \\ 7\times10^{-14} & 0.64310 & 3\times10^{-8} \\ -2\times10^{-22} & 3\times10^{-8} & 5.04892 \end{bmatrix} \tag{3.178}$$

The off-diagonal elements are all much smaller than the diagonal ones and the convergence has been reached. The natural frequencies and the mode shapes can now be extracted and compared with the analytical ones (see Table 3.8 and Figure 3.23). The agreement between the computed results and the exact ones is very good.

Table 3.8 Comparison between the analytical natural frequencies and those from Jacobi's method.

Natural frequency	Analytical	Jacobi's method
ω_1	$0.445\sqrt{\frac{k}{m}}$	$0.44504\sqrt{\frac{k}{m}}$
ω_2	$1.247\sqrt{\frac{k}{m}}$	$1.24698\sqrt{\frac{k}{m}}$
ω_3	$1.802\sqrt{\frac{k}{m}}$	$1.80194\sqrt{\frac{k}{m}}$

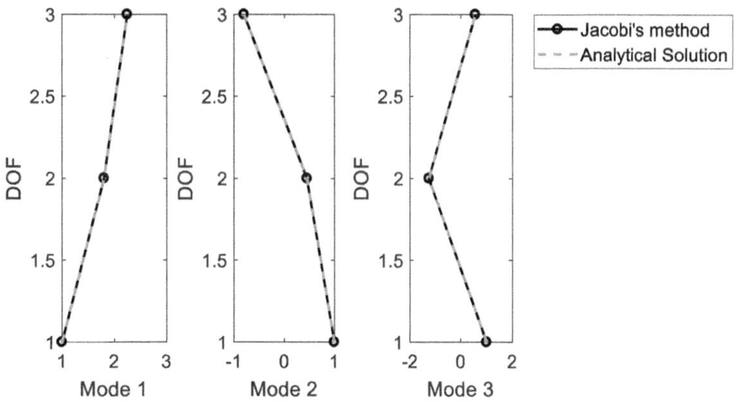

Figure 3.23 Mode shapes.

3.4.7 Singular Value Decomposition

The Singular Value Decomposition (*SVD*) was invented independently by Beltrami and Jordan (in 1873-74) for square matrices and further developed by Eckart and Young for general rectangular matrices.

First, the standard problem of eigenvalues and eigenvectors is given as:

$$(\boldsymbol{A}-\lambda \boldsymbol{I})\boldsymbol{x}=\boldsymbol{0} \tag{3.179}$$

Once the n eigenvalues and eigenvectors are known, matrix $\boldsymbol{A}$ can be decomposed as

$$\boldsymbol{A}=\boldsymbol{W}\boldsymbol{\Lambda}\boldsymbol{W}^T \tag{3.180}$$

where $\boldsymbol{W}$ is an $n\times n$ matrix containing the n eigenvectors, and $\boldsymbol{\Lambda}$ is an $n\times n$ diagonal matrix containing the n eigenvalues. This is called the Spectral Decomposition.

If $\boldsymbol{A}$ is not a square matrix it can still be decomposed in a similar way. It is the so-called Singular Value Decomposition. So, unlike the Spectral Decomposition mentioned above, the *SVD* does not require the matrix $\boldsymbol{A}$ to be a square matrix. One uses the idea of orthogonal transformation to derive the *SVD*.

Assume that $\boldsymbol{A}$ is an $m\times n$ matrix with rank *r*, $rank(\boldsymbol{A})=r$ and let $\boldsymbol{V}$ form an orthogonal basis, such that:

$$\boldsymbol{V}=(\boldsymbol{v}_1\ \boldsymbol{v}_2\ \cdots\ \boldsymbol{v}_r),\quad \boldsymbol{v}_i^T\boldsymbol{v}_j=\delta_{ij} \tag{3.181}$$

Applying a transformation through the matrix $\boldsymbol{A}$, the resulting matrix will be

$$\boldsymbol{A}\boldsymbol{V}=(\boldsymbol{A}\boldsymbol{v}_1\ \boldsymbol{A}\boldsymbol{v}_2\ \cdots\ \boldsymbol{A}\boldsymbol{v}_r) \tag{3.182}$$

For the vectors of $\boldsymbol{AV}$ to be orthogonal, then

$$(\boldsymbol{A}\boldsymbol{v}_i)^T(\boldsymbol{A}\boldsymbol{v}_j)=\begin{cases} 0 & for\quad i\neq j \\ \neq 0 & for\quad i=j \end{cases} \tag{3.183}$$

Therefore, for $i\neq j$, one has

$$\boldsymbol{v}_i^T\left(\boldsymbol{A}^T\boldsymbol{A}\right)\boldsymbol{v}_j=0 \quad \text{and} \quad \boldsymbol{v}_i^T\boldsymbol{v}_j=0 \tag{3.184}$$

This is true if the following standard eigenproblem is verified:

$$\left(\boldsymbol{A}^T\boldsymbol{A}-\lambda_i\boldsymbol{I}\right)\boldsymbol{v}_i=0 \tag{3.185}$$

or

$$\left(\boldsymbol{A}^T\boldsymbol{A}\right)\boldsymbol{v}_i=\lambda_i\boldsymbol{v}_i \tag{3.186}$$

Pre-multiplying by $\boldsymbol{v}_i^T$ and considering (3.181),

$$\boldsymbol{v}_i^T\left(\boldsymbol{A}^T\boldsymbol{A}\right)\boldsymbol{v}_i = \lambda_i \tag{3.187}$$

Normalising the vectors $\boldsymbol{A}\boldsymbol{v}_i$, one obtains

$$\boldsymbol{u}_i = \frac{\boldsymbol{A}\boldsymbol{v}_i}{\|\boldsymbol{A}\boldsymbol{v}_i\|} = \frac{\boldsymbol{A}\boldsymbol{v}_i}{\sqrt{(\boldsymbol{A}\boldsymbol{v}_i)^T(\boldsymbol{A}\boldsymbol{v}_i)}} = \frac{\boldsymbol{A}\boldsymbol{v}_i}{\sqrt{\lambda_i}} \tag{3.188}$$

Thus,

$$\boldsymbol{A}\boldsymbol{v}_i = \sqrt{\lambda_i}\,\boldsymbol{u}_i \tag{3.189}$$

In matrix form,

$$\boldsymbol{A}\boldsymbol{V} = \boldsymbol{U}\boldsymbol{\Sigma} \tag{3.190}$$

where $\boldsymbol{V} = \left(\boldsymbol{v}_1\ \boldsymbol{v}_2\ \cdots\ \boldsymbol{v}_r\right)$, $\boldsymbol{\Sigma} = \begin{bmatrix} \sigma_1 & & & \\ & \sigma_2 & & \\ & & \ddots & \\ & & & \sigma_r \end{bmatrix}$, $\boldsymbol{U} = \left(\boldsymbol{u}_1\ \boldsymbol{u}_2\ \cdots\ \boldsymbol{u}_r\right)$, $\sigma_i = \sqrt{\lambda_i}$.

As matrices $\boldsymbol{U}$ and $\boldsymbol{V}$ are orthogonal, they satisfy

$$\boldsymbol{U}^T\boldsymbol{U} = \boldsymbol{I},\quad \boldsymbol{V}^T\boldsymbol{V} = \boldsymbol{I} \tag{3.191}$$

and from (3.190), the *SVD* of an $m \times n$ matrix $\boldsymbol{A}$ is defined as[3]:

$$\boldsymbol{A} = \boldsymbol{U}\boldsymbol{\Sigma}\boldsymbol{V}^T \tag{3.192}$$

where $\boldsymbol{U}$ is an $m \times m$ matrix of the left singular vectors, $\boldsymbol{\Sigma}$ is an $m \times n$ diagonal matrix of singular values and $\boldsymbol{V}$ is an $n \times n$ matrix of the right singular vectors. Naturally, The Spectral Decomposition and the *SVD* are closely related. If one calculates $\boldsymbol{A}\boldsymbol{A}^T$ and $\boldsymbol{A}^T\boldsymbol{A}$, the result is

$$\boldsymbol{A}\boldsymbol{A}^T = \boldsymbol{U}\boldsymbol{\Sigma}\boldsymbol{V}^T\boldsymbol{V}\boldsymbol{\Sigma}\boldsymbol{U}^T \tag{3.193}$$

$$\boldsymbol{A}^T\boldsymbol{A} = \boldsymbol{V}\boldsymbol{\Sigma}\boldsymbol{U}^T\boldsymbol{U}\boldsymbol{\Sigma}\boldsymbol{V}^T \tag{3.194}$$

Due to the relations (3.191),

---

[3] If matrix $\boldsymbol{A}$ is complex, then $\boldsymbol{U}$ and $\boldsymbol{V}$ are also complex and $\boldsymbol{A} = \boldsymbol{U}\boldsymbol{\Sigma}\boldsymbol{V}^H$, where the superscript ^{H} stands for *Hermitian* (complex conjugate transpose). In that case, $\boldsymbol{U}^H\boldsymbol{U} = \boldsymbol{I},\ \boldsymbol{V}^H\boldsymbol{V} = \boldsymbol{I}$.

$$\boldsymbol{A}\boldsymbol{A}^T = \boldsymbol{U}\boldsymbol{\Sigma}^2\boldsymbol{U}^T \tag{3.195}$$

$$\boldsymbol{A}^T\boldsymbol{A} = \boldsymbol{V}\boldsymbol{\Sigma}^2\boldsymbol{V}^T \tag{3.196}$$

Eqs. (3.195) and (3.196) are the Spectral Decompositions of $\boldsymbol{A}\boldsymbol{A}^T$ and $\boldsymbol{A}^T\boldsymbol{A}$, respectively, where $\boldsymbol{U}$ are the left eigenvectors, $\boldsymbol{V}$ the right eigenvectors and $\boldsymbol{\Sigma}^2$ contains the squared singular values. Comparing to Eq. (3.180), $\boldsymbol{\Lambda} = \boldsymbol{\Sigma}^2$, i.e. the eigenvalues correspond to the squared singular values.

The elements of $\boldsymbol{\Sigma}$, the singular values, denoted as σ_i, are usually sorted in decreasing order. If the rank of the matrix $\boldsymbol{A}$ is r, there will be $m-r$ singular values that are zero (or very close to zero) and can be ignored. The resulting version of the *SVD* is known as *compact SVD*[4]. For the *compact SVD*, one has

$$\boldsymbol{U}^{\mathrm{T}}\boldsymbol{U} = \boldsymbol{V}^{\mathrm{T}}\boldsymbol{V} = \boldsymbol{I}_{r\times r} \tag{3.197}$$

Due to the rounding error, instead of the exact rank, the approximate rank is defined as the maximum value of r such that

$$\frac{\sigma_r}{\sigma_1} \geq 10^{-15} \tag{3.198}$$

Using such a condition, the compact *SVD* is implemented in the *MATLAB* program in Appendix A (code 15).

Example 3.4.7-1

Perform the *SVD* of the following matrix:

$$\boldsymbol{A} = \begin{bmatrix} 2 & 0 & 2 \\ 2 & 1 & 3 \\ 3 & 0 & 3 \\ 4 & 0 & 4 \\ 6 & 0 & 6 \end{bmatrix}$$

One can execute $[\boldsymbol{U}, \boldsymbol{\Sigma}, \boldsymbol{V}]$= svd ($\boldsymbol{A}$) in *MATLAB* to obtain the matrices $\boldsymbol{U}$ and $\boldsymbol{V}$ quickly:

$$\boldsymbol{U} = \begin{bmatrix} -0.2367 & -0.0742 & 0.9303 & 0.2338 & -0.1355 \\ -0.2990 & 0.9542 & 0 & 0 & 0 \\ -0.3551 & 0.1113 & -0.0276 & -0.6617 & -0.6503 \\ -0.4734 & -0.1484 & -0.3616 & 0.6830 & -0.3957 \\ -0.7102 & -0.2225 & -0.0552 & -0.2024 & 0.6341 \end{bmatrix}$$

[4] Also known sometimes as "Skinny *SVD*"

$$\boldsymbol{\Sigma} = \begin{bmatrix} 11.9429 & 0 & 0 \\ 0 & 1.1693 & 0 \\ 0 & 0 & 0 \\ 0 & 0 & 0 \\ 0 & 0 & 0 \end{bmatrix}$$

$$\boldsymbol{V} = \begin{bmatrix} -0.6943 & -0.4297 & -0.5774 \\ -0.0250 & 0.8161 & -0.5774 \\ -0.7193 & 0.3864 & 0.5774 \end{bmatrix} ; \quad \boldsymbol{V}^T = \begin{bmatrix} -0.6943 & -0.0250 & -0.7193 \\ -0.4297 & 0.8161 & 0.3864 \\ -0.5774 & -0.5774 & 0.5774 \end{bmatrix}$$

The results for the singular values are: $\sigma_1 = 11.9429$, $\sigma_2 = 1.1693$ and $\sigma_3 = 0$. In the compact version of the *SVD*, σ_3 is ignored by truncation.

Besides solving linear equations, the *SVD* has other applications such as the condition number of a matrix. The condition number is an important basis to judge ill-conditioned matrices. After the *SVD* calculation, the condition number can be expressed as $\sigma_{\max} / \sigma_{\min}$. The *SVD* can also solve the Moore-Penrose pseudo-inverse matrix, which can be found in many books on matrix theory.

Example 3.4.7-2

Let the example of the 3 *DOF* system of Section 3.3.1, Figure 3.14, be used once more, to show how the *SVD* method works. Now, it is assumed that the system vibrates under a harmonically external excitation, with a frequency of 1 rad/s, and k and m are equal to 1. The dynamic equilibrium equation is:

$$\boldsymbol{M}\ddot{\boldsymbol{x}} + \boldsymbol{K}\boldsymbol{x} = \boldsymbol{F} sin\omega t$$

$$\boldsymbol{K} = \begin{bmatrix} 2 & -1 & 0 \\ -1 & 2 & -1 \\ 0 & -1 & 1 \end{bmatrix}, \quad \boldsymbol{M} = \begin{bmatrix} 1 & 0 & 0 \\ 0 & 1 & 0 \\ 0 & 0 & 1 \end{bmatrix}, \quad \boldsymbol{F} = \begin{Bmatrix} 0 \\ 0.5 \\ 0 \end{Bmatrix}$$

It is easy to use the Runge-Kutta method (see Section 3.2.2) to solve the time domain responses on m_1 , m_2 and m_3. Let these three signals be arranged into a matrix:

$$\boldsymbol{X} = \begin{bmatrix} \boldsymbol{x}_1 \\ \boldsymbol{x}_2 \\ \boldsymbol{x}_3 \end{bmatrix}_{3\times n}$$

where $\boldsymbol{x}_i$, $i = 1, 2, 3$ denotes the time domain response of the ith mass, and n is the number of data points. In this example, the time is taken from 0 to 100 seconds, with 2161 data points ($n = 2161$). The time signals $\boldsymbol{x}_1, \boldsymbol{x}_2, \boldsymbol{x}_3$ are shown in Figure 3.24.

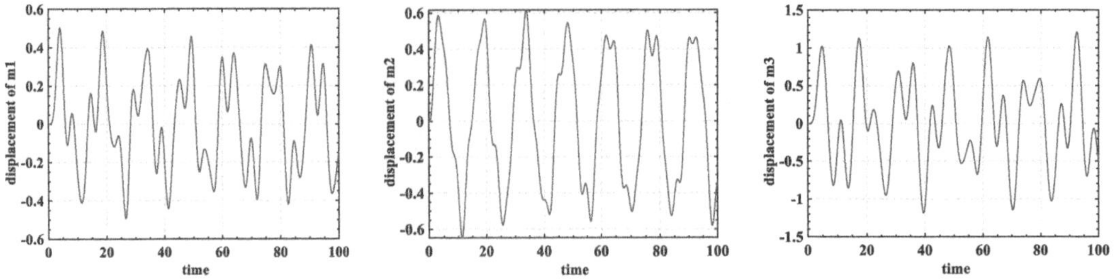

Figure 3.24 Time signals x_1, x_2, x_3.

The *SVD* is performed on matrix $\boldsymbol{X}$, $\boldsymbol{X} = \boldsymbol{U\Sigma V}^T$. In this case, $\boldsymbol{U}$, $\boldsymbol{\Sigma}$ and $\boldsymbol{V}$ are as follows:

$$\boldsymbol{U} = \begin{bmatrix} \boldsymbol{u}_1 & \boldsymbol{u}_2 & \boldsymbol{u}_3 \end{bmatrix} = \begin{bmatrix} -0.2585 & -0.5465 & 0.7966 \\ -0.4647 & -0.6526 & -0.5985 \\ -0.8469 & 0.5249 & 0.0853 \end{bmatrix}_{3\times 3}$$

$$\boldsymbol{\Sigma} = \begin{bmatrix} \sigma_1 & 0 & 0 & 0 & \cdots & 0 \\ 0 & \sigma_2 & 0 & 0 & \cdots & 0 \\ 0 & 0 & \sigma_3 & 0 & \cdots & 0 \end{bmatrix} = \begin{bmatrix} 30.3064 & 0 & 0 & 0 & \cdots & 0 \\ 0 & 12.5360 & 0 & 0 & \cdots & 0 \\ 0 & 0 & 4.2348 & 0 & \cdots & 0 \end{bmatrix}_{3\times n}$$

$$\boldsymbol{V} = \begin{bmatrix} \bullet \end{bmatrix}_{n\times n}$$

where vectors $\boldsymbol{u}_i$ have been normalised to unit length.

One can also solve for the eigenvalues and eigenvectors of the system:

$$\left(\boldsymbol{K} - \omega_i^2 \boldsymbol{M}\right)\boldsymbol{\phi}_i = \boldsymbol{0} \quad i = 1,2,3$$

where $\boldsymbol{\phi}_i$ is the eigenvector corresponding to ω_i. The mass-normalised mode shape matrix $\boldsymbol{\Phi} = \begin{bmatrix} \boldsymbol{\phi}_1 & \boldsymbol{\phi}_2 & \boldsymbol{\phi}_3 \end{bmatrix}$ and corresponding eigenvalues of the 3 *DOF* system are

$$\boldsymbol{\Phi} = \begin{bmatrix} \boldsymbol{\phi}_1 & \boldsymbol{\phi}_2 & \boldsymbol{\phi}_3 \end{bmatrix} = \begin{bmatrix} -0.328 & 0.737 & -0.591 \\ -0.591 & 0.328 & 0.737 \\ -0.737 & -0.591 & -0.328 \end{bmatrix}, \quad \omega_1^2 = 0.1981, \quad \omega_2^2 = 1.555, \quad \omega_3^2 = 3.247$$

Comparing $\boldsymbol{U}$ and $\boldsymbol{\Phi}$, one obtains the following:

$$\begin{aligned} \boldsymbol{u}_1 &= 0.9836\boldsymbol{\phi}_1 + 0.1576\boldsymbol{\phi}_2 + 0.0881\boldsymbol{\phi}_3 \\ \boldsymbol{u}_2 &= 0.1781\boldsymbol{\phi}_1 - 0.9270\boldsymbol{\phi}_2 - 0.3301\boldsymbol{\phi}_3 \\ \boldsymbol{u}_3 &= 0.0296\boldsymbol{\phi}_1 + 0.3404\boldsymbol{\phi}_2 - 0.9398\boldsymbol{\phi}_3 \end{aligned}$$

u_i , $i = 1, 2, 3$ is the ith column of U . Each column of U is a linear combination of the 3 mode shapes of the system. The U matrix contains complete modal information. One can also see that the projection of u_1 is dominated by ϕ_1 , and the projections of u_2 and u_3 by ϕ_2 and ϕ_3 , respectively. The relations of u_1 , u_2 and u_3 with the mode shapes ϕ_1 , ϕ_2 and ϕ_3 are compared in Figure 3.25.

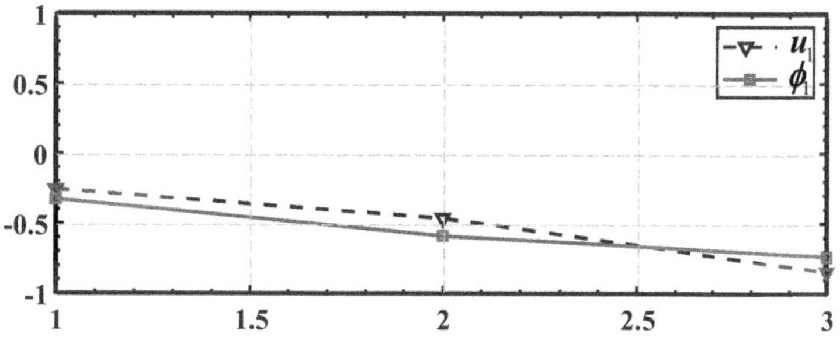

First mode shape

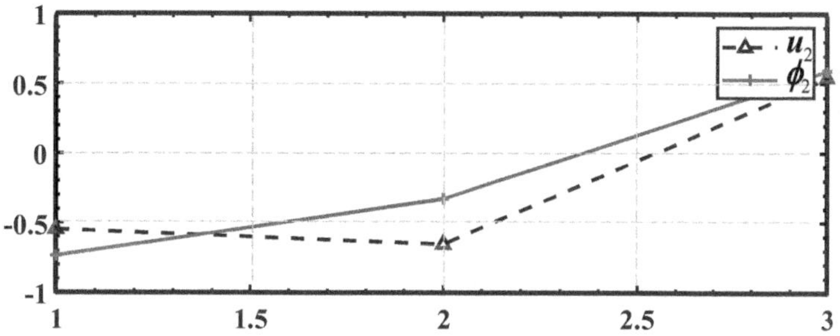

Second mode shape

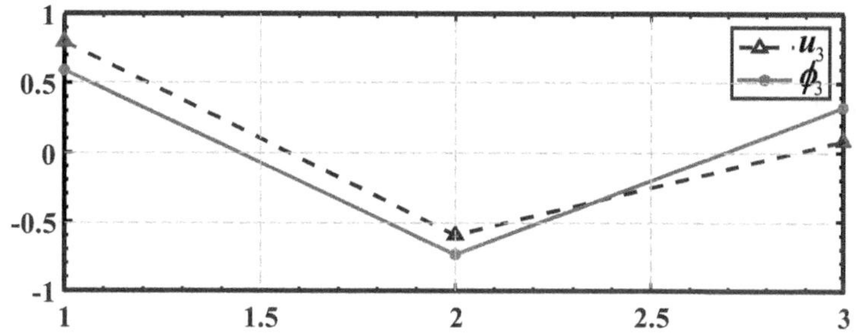

Third mode shape

Figure 3.25 Mode shape comparison.

The first 3 columns of matrix V are shown in Figure 3.26. Each column of V is also a linear combination of signals x_i $(i = 1, 2, 3)$ in the time domain. For this example, the first 3 columns of $V = \begin{pmatrix} v_1 & v_2 & v_3 & \cdots & v_n \end{pmatrix}$ are:

$$\boldsymbol{v}_1 = -0.0089\boldsymbol{x}_1 - 0.0157\boldsymbol{x}_2 - 0.0272\boldsymbol{x}_3$$
$$\boldsymbol{v}_2 = -0.0416\boldsymbol{x}_1 - 0.0487\boldsymbol{x}_2 + 0.0417\boldsymbol{x}_3$$
$$\boldsymbol{v}_3 = 0.1926\boldsymbol{x}_1 - 0.1462\boldsymbol{x}_2 + 0.0212\boldsymbol{x}_3$$

These 3 columns $\boldsymbol{v}_1$, $\boldsymbol{v}_2$ and $\boldsymbol{v}_3$ correspond to σ_1, σ_2 and σ_3. One can also write $\boldsymbol{x}_1$, $\boldsymbol{x}_2$ and $\boldsymbol{x}_3$ as a combination of signals $\boldsymbol{v}_i$ $(i = 1, 2, 3)$:

$$\boldsymbol{x}_1 = -8.3626\boldsymbol{v}_1 - 7.1165\boldsymbol{v}_2 + 3.2686\boldsymbol{v}_3$$
$$\boldsymbol{x}_2 = -14.7191\boldsymbol{v}_1 - 8.3394\boldsymbol{v}_2 - 2.4814\boldsymbol{v}_3$$
$$\boldsymbol{x}_3 = -25.5325\boldsymbol{v}_1 + 7.1421\boldsymbol{v}_2 + 0.3628\boldsymbol{v}_3$$

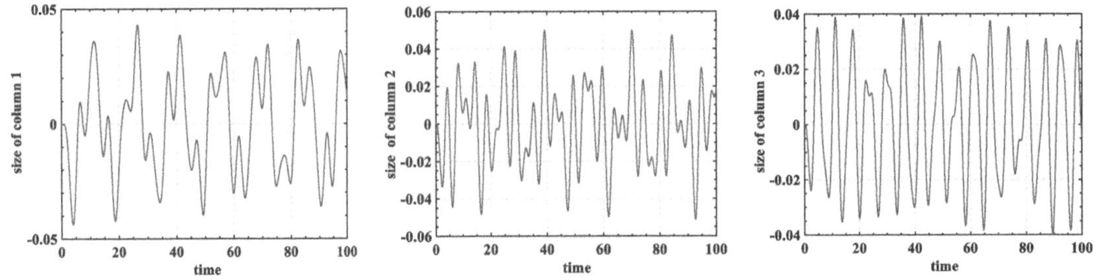

Figure 3.26 Size of the first three columns of $\boldsymbol{V}$ ($\boldsymbol{v}_1$, $\boldsymbol{v}_2$ and $\boldsymbol{v}_3$).

The reconstructed signals in the time domain at the positions of $\boldsymbol{x}_1$, $\boldsymbol{x}_2$ and $\boldsymbol{x}_3$ are plotted in Figure 3.27. One can see that the reconstructed signals are the same as the original signals.

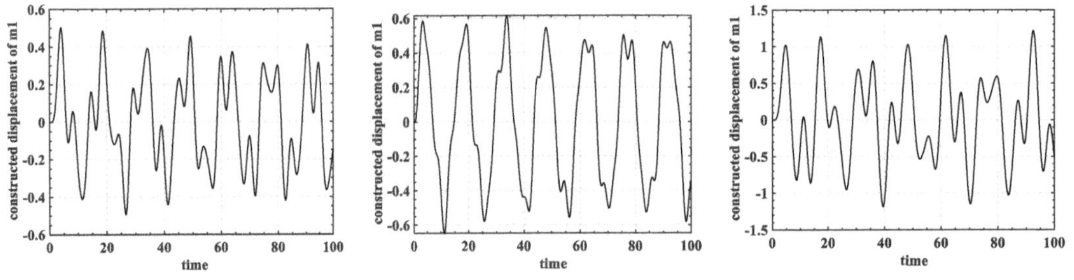

Figure 3.27 Reconstructed signals based on $\boldsymbol{v}_1$, $\boldsymbol{v}_2$ and $\boldsymbol{v}_3$.

Example 3.4.7-3 Using the *SVD* to solve a set of overdetermined equations.

$$\boldsymbol{A}_{m\times n}\boldsymbol{x}_{n\times 1} = \boldsymbol{b}_{m\times 1}, \quad m > n$$

Decomposing $\boldsymbol{A}$,

$$\boldsymbol{U\Sigma V}^T\boldsymbol{x} = \boldsymbol{b} \quad \Rightarrow \quad \boldsymbol{\Sigma V}^T\boldsymbol{x} = \boldsymbol{U}^T\boldsymbol{b}$$

or

$$\Sigma q = f \quad \text{with} \quad q = V^T x, \;\; f = U^T b$$

This is a decoupled system of equations, and the solution q can be obtained in the least-squares sense, from which,

$$x = Vq$$

3.4.8 Principal Component Analysis

Principal Component Analysis (*PCA*) is closely related to both the Spectral Decomposition and the *SVD*. In many fields of research and application, it is usually necessary to observe data containing multiple variables, collect a large amount of data, and then analyse them to find rules. Multivariable large data sets will undoubtedly provide rich information for research, but also increase the workload of data collection to a certain extent. So, there is a need to reduce the amount of data, while reducing to a minimum the loss of information contained in the original data. *PCA* is one of the most widely used data dimensionality reduction algorithms. As the name suggests, it is meant to find the most important characteristics in the data and replace the original data with those most important characteristics.

In this context, it is important to begin by defining the concept of covariance. Based on probability theory, the covariance between two variables is given by:

$$cov(\boldsymbol{X},\boldsymbol{Y}) = \frac{1}{n-1}\sum_{i=1}^{n}(x_i - \tilde{x})(y_i - \tilde{y}) \tag{3.199}$$

where $\tilde{x}$ and $\tilde{y}$ are the mean values of $\boldsymbol{X}$ and $\boldsymbol{Y}$, respectively.

However, things can be simplified, by making the mean of the variables $\boldsymbol{X}$ and $\boldsymbol{Y}$ equal to zero. So, one subtracts the mean $\tilde{x}$ from each element in $\boldsymbol{X}$, and the mean $\tilde{y}$ from each of the elements in $\boldsymbol{Y}$. The covariance matrix of $\boldsymbol{X}$ and $\boldsymbol{Y}$ can be written as

$$\begin{bmatrix} cov(\boldsymbol{X},\boldsymbol{X}) & cov(\boldsymbol{X},\boldsymbol{Y}) \\ cov(\boldsymbol{Y},\boldsymbol{X}) & cov(\boldsymbol{Y},\boldsymbol{Y}) \end{bmatrix} = \frac{1}{n-1}\begin{bmatrix} \sum_{i=1}^{n}(x_i')^2 & \sum_{i=1}^{n} x_i' y_i' \\ \sum_{i=1}^{n} y_i' x_i' & \sum_{i=1}^{n}(y_i')^2 \end{bmatrix} \tag{3.200}$$

where $x_i' = x_i - \tilde{x}$ and $y_i' = y_i - \tilde{y}$.

The matrix on the right-end-side of (3.200) is the sample matrix AA^T with zero mean, where $\boldsymbol{A} = \begin{bmatrix} x_1' & x_2' & \cdots & x_n' \\ y_1' & y_2' & \cdots & y_n' \end{bmatrix}$.

The covariance between the original features $\boldsymbol{X}$ and $\boldsymbol{Y}$ is not zero, and its covariance matrix is an ordinary symmetric matrix. Therefore, one wishes to represent features $\boldsymbol{X}$ and $\boldsymbol{Y}$ in terms of a different basis, so that the new features have no relation with each other. Inevitably, these new features must be orthogonal to each other. This is when the *SVD* comes into play, or rather the Spectral Decomposition, applied to the covariance matrix. As the covariance matrix is

$$\boldsymbol{C} = \frac{1}{n-1}\boldsymbol{A}\boldsymbol{A}^T \tag{3.201}$$

its Spectral Decomposition can be written as

$$\boldsymbol{C} = \boldsymbol{U}\boldsymbol{\Lambda}\boldsymbol{U}^T \tag{3.202}$$

The most significant r eigenvalues are arranged from large to small, and the corresponding eigenvectors are taken as row vectors to form the matrix $\boldsymbol{P}$. The data is then converted to the new space constituted by the r eigenvectors, $\begin{bmatrix} \breve{\boldsymbol{X}} \\ \breve{\boldsymbol{Y}} \end{bmatrix} = \boldsymbol{P}\begin{bmatrix} \boldsymbol{X}' \\ \boldsymbol{Y}' \end{bmatrix}$.

This is the *PCA* process of 2 groups of data $\boldsymbol{X}$ and $\boldsymbol{Y}$. For multiple groups of data (other than 2 groups of data), the covariance matrix and eigen-decomposition matrix can be obtained by simply repeating Eqs. (3.201) and (3.202). Next, some examples are given, to help understanding the procedure.

Example 3.4.8-1

Assuming that the object of the study has two characteristic attributes ***X*** and ***Y***, the data sampling results are as follows:

X	2	2	4	8	4
Y	2	6	6	8	8

First, one represents the data in a two-dimensional plane.

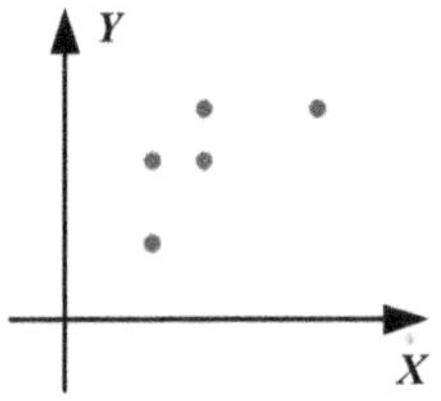

Combining the distribution of feature variables ***X*** and ***Y***, one can find that features ***X*** and ***Y*** of the 5 samples show positive correlation, and the data have influence on each other. Data dimension reduction needs to consider the following two aspects:

- removing the correlation between features, as the idea is to create another set of features to describe the sample, so that the new features be unrelated to each other;

- in the new mutually independent feature set, discard unimportant features and retain fewer ones, so as to achieve dimensionality reduction of data and avoid as much as possible loss of information.

First of all, the original data are treated with zero mean; as $\tilde{x} = 4$ and $\tilde{y} = 6$, the data become

X'	-2	-2	0	4	0
Y'	-4	0	0	2	2

The covariance matrix is $\boldsymbol{C} = \frac{1}{4}\begin{bmatrix} 24 & 16 \\ 16 & 24 \end{bmatrix}$. By solving the eigenvectors of the covariance matrix ***C***, one can obtain two newly linearly independent eigenprojection bases. The eigenvalue matrix of the covariance matrix ***C*** is $\boldsymbol{\Lambda} = \begin{bmatrix} 10 & 0 \\ 0 & 2 \end{bmatrix}$, and the normalised eigenvector matrix is $\boldsymbol{U} = \frac{1}{\sqrt{2}}\begin{bmatrix} 1 & -1 \\ 1 & 1 \end{bmatrix}$.

The directions of the original features are the positive direction of the *X*-axis and the positive direction of the *Y*-axis respectively, and the original features are related to each other. The two newly constructed features, in directions $\begin{Bmatrix} 1/\sqrt{2} \\ 1/\sqrt{2} \end{Bmatrix}$ and $\begin{Bmatrix} -1/\sqrt{2} \\ 1/\sqrt{2} \end{Bmatrix}$, are independent with zero covariance. Next, one transforms the sample values of the original features to the two newly projection bases to obtain a new set of data, where $\boldsymbol{P} = \boldsymbol{U}^T$. The new data matrix is

$$\begin{bmatrix} \breve{\boldsymbol{X}} \\ \breve{\boldsymbol{Y}} \end{bmatrix} = \boldsymbol{P}\begin{bmatrix} \boldsymbol{X'} \\ \boldsymbol{Y'} \end{bmatrix} = \frac{1}{\sqrt{2}}\begin{bmatrix} 1 & 1 \\ -1 & 1 \end{bmatrix}\begin{bmatrix} -2 & -2 & 0 & 4 & 0 \\ -4 & 0 & 0 & 2 & 2 \end{bmatrix} = \frac{1}{\sqrt{2}}\begin{bmatrix} -6 & -2 & 0 & 6 & 2 \\ -2 & 2 & 0 & -2 & 2 \end{bmatrix}$$

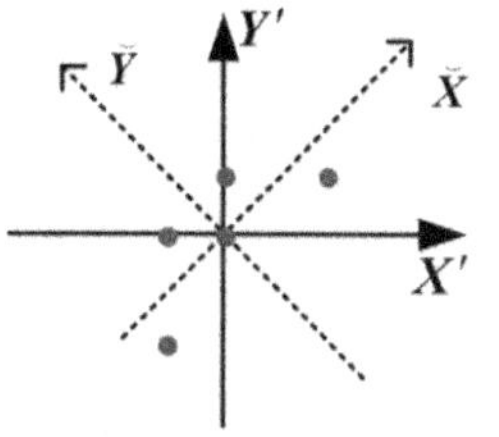

For the data one chooses to retain, the variance is taken as the standard judgment. The larger the variance is, the greater the dispersion degree of data distribution in the feature and the more the contained information. On the contrary, if the variance of the data is small and the distribution is concentrated, it indicates that the information contained is small. Naturally, one chooses to keep the most informative features.

The variance of the data in the direction $\begin{Bmatrix} 1/\sqrt{2} \\ 1/\sqrt{2} \end{Bmatrix}$ is 10, while the variance of the data in the direction $\begin{Bmatrix} -1/\sqrt{2} \\ 1/\sqrt{2} \end{Bmatrix}$ is 2. If one wishes to reduce the dimension of the data, one should keep the eigenvalue in the direction $\begin{Bmatrix} 1/\sqrt{2} \\ 1/\sqrt{2} \end{Bmatrix}$. At this point, one has completed the extraction of the principal components; if one wishes to make the reduction to one-dimensional data, the final value of the 5 samples is $\frac{1}{\sqrt{2}}\{-6 \quad -2 \quad 0 \quad 6 \quad 2\}$. The basis vector that remain is $\begin{Bmatrix} 1/\sqrt{2} \\ 1/\sqrt{2} \end{Bmatrix}$. In this example, there were originally two features, and one chooses to keep the first feature as the principal component. The principal component contribution rate is $\frac{\lambda_1}{\lambda_1 + \lambda_2} = \frac{10}{12} = 83.3\%$. Thus, in the process of dimensionality reduction, onr compresses the data by 50%, but retains 83.3% of the information.

Example 3.4.8-2

Take again the example of the 3 *DOF* system in Section 3.3.1 (Figure 3.14). For the sake of discussion, one writes again the dynamic equilibrium equation:

$$\boldsymbol{M}\ddot{\boldsymbol{x}} + \boldsymbol{K}\boldsymbol{x} = \boldsymbol{F} sin\omega t$$

$$\text{with} \quad \boldsymbol{K} = \begin{bmatrix} 2 & -1 & 0 \\ -1 & 2 & -1 \\ 0 & -1 & 1 \end{bmatrix}, \quad \boldsymbol{M} = \begin{bmatrix} 1 & 0 & 0 \\ 0 & 1 & 0 \\ 0 & 0 & 1 \end{bmatrix}, \quad \boldsymbol{F} = \begin{Bmatrix} 0 \\ 0.5 \\ 0 \end{Bmatrix}, \quad \omega = 1$$

Arranging the three time domain signals into a matrix,

$$\boldsymbol{X} = \begin{bmatrix} \boldsymbol{x}_1 \\ \boldsymbol{x}_2 \\ \boldsymbol{x}_3 \end{bmatrix}_{3\times n} = \begin{bmatrix} x_{ij} \end{bmatrix}$$

where, $\boldsymbol{x}_i$, $i = 1, 2, 3$ denotes the time domain response of the ith mass, and n is the number of data points. In this example, one takes the time from 0 to 100 seconds, and one can get 2161 data points ($n = 2161$).

According to the definition of *PCA*, the mean covariance of matrix $\boldsymbol{X}$ is calculated:

$$\boldsymbol{C} = \frac{1}{n-1}\boldsymbol{X}'\boldsymbol{X}'^T, \text{ where } \boldsymbol{X}' = \begin{bmatrix} x'_{ij} \end{bmatrix} = x_{ij} - \frac{1}{n}\sum_{j=1}^{n} x_{ij} \quad , \quad i = 1, 2, 3$$

Next, one should compute the eigenvalues and eigenvectors of ***C***,

$$C = U\Lambda U^T$$

where U is the eigenvector matrix and Λ is the eigenvalue matrix. In this example, one has:

$$C = \begin{bmatrix} 0.0607 & 0.0806 & 0.0759 \\ 0.0806 & 0.1352 & 0.1460 \\ 0.0759 & 0.1460 & 0.3258 \end{bmatrix}, \quad U = \begin{bmatrix} 0.2727 & 0.5442 & 0.7934 \\ 0.4800 & 0.6377 & -0.6024 \\ 0.8338 & -0.5451 & 0.0873 \end{bmatrix}$$

$$\Lambda = \begin{bmatrix} 0.4346 & 0 & 0 \\ 0 & 0.0792 & 0 \\ 0 & 0 & 0.0079 \end{bmatrix}$$

The eigenvectors and corresponding eigenvalues for this 3 *DOF* system are:

$$\Phi = \begin{bmatrix} \phi_1 & \phi_2 & \phi_3 \end{bmatrix} = \begin{bmatrix} -0.328 & 0.737 & -0.591 \\ -0.591 & 0.328 & 0.737 \\ -0.737 & -0.591 & -0.328 \end{bmatrix}, \quad \omega_1^2 = 0.1981, \ \omega_2^2 = 1.555, \ \omega_3^2 = 3.247$$

It turns out that

$$\begin{aligned} u_1 &= 0.0315\phi_1 + 0.3356\phi_2 - 0.9415\phi_3 \\ u_2 &= -0.1536\phi_1 + 0.9324\phi_2 + 0.3271\phi_3 \\ u_3 &= -0.9876\phi_1 - 0.1344\phi_2 - 0.0809\phi_3 \end{aligned}$$

where u_i (i = 1, 2, 3) is the ith column of U. Like the *SVD* method, each column of the eigenmatrix U is a linear combination of 3 mode shapes of the system. The U matrix contains complete modal information. The projection of u_1 is also dominated by the first mode shape, ϕ_1, of the system, and the projections of u_2 and u_3 by the second and third mode shapes, ϕ_2 and ϕ_3 respectively. The comparison is shown in Figure 3.28.

Matrix $C = \frac{1}{n-1} X' X'^T$ is a 3×3 matrix, reflecting the information of the spatial positions of points; instead, one may compute the following covariance matrix:

$$\hat{C} = \frac{1}{n-1} X'^T X'$$

In this case, the covariance matrix is a $n \times n$ matrix, containing the complete time domain information. Next, one can compute the eigenvalues and eigenvectors of $\hat{C}$,

$$\hat{C} = V\hat{\Lambda}V^T$$

and one obtains

$$V = \begin{bmatrix} v_1 & v_2 & v_3 & \cdots & v_n \end{bmatrix}_{n \times n}$$

$$\hat{\boldsymbol{\Lambda}} = \begin{bmatrix} 0.4346 & 0 & 0 & \cdots & 0 \\ 0 & 0.0792 & 0 & \cdots & 0 \\ 0 & 0 & 0.0079 & \cdots & 0 \\ \vdots & \vdots & \vdots & \ddots & \vdots \\ 0 & 0 & 0 & \cdots & 0 \end{bmatrix}_{n \times n}$$

It is clear that the most significant eigenvalues of $\hat{\boldsymbol{\Lambda}}$ are the same as for $\boldsymbol{\Lambda}$. The first 3 columns of $\boldsymbol{V}$, $\boldsymbol{v}_1$, $\boldsymbol{v}_2$ and $\boldsymbol{v}_3$ are the principal components, shown in Figure 3.29.

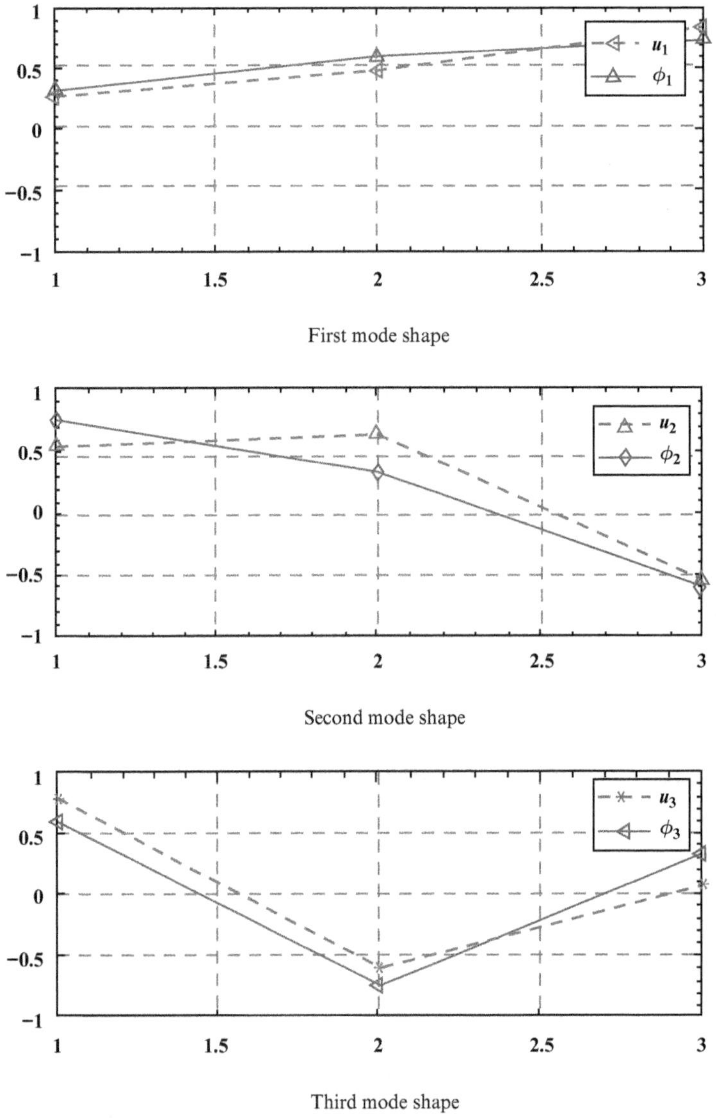

Figure 3.28 Mode shape comparison.

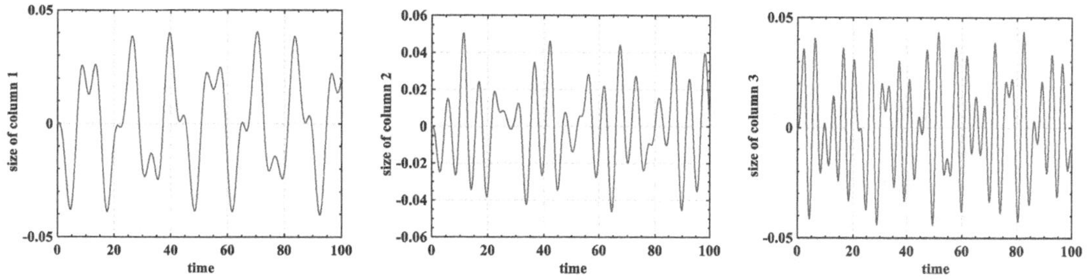

Figure 3.29 The size of the first three columns of V (v_1, v_2 and v_3).

Similarly, the first 3 columns of V can be written as a linear combination of x_i ($i = 1, 2, 3$):

$$\begin{aligned} v_1 &= -0.0089x_1 - 0.0156x_2 - 0.0272x_3 \\ v_2 &= -0.0416x_1 - 0.0487x_2 + 0.0417x_3 \\ v_3 &= -0.1926x_1 + 0.1462x_2 - 0.0212x_3 \end{aligned}$$

and conversely,

$$\begin{aligned} x_1 &= -8.3602v_1 - 7.1200v_2 - 3.2672v_3 \\ x_2 &= -14.7151v_1 - 8.3436v_2 + 2.4852v_3 \\ x_3 &= -25.5631v_1 + 7.1300v_2 - 0.3549v_3 \end{aligned}$$

The original signals x_1, x_2 and x_3, reconstructed using v_1, v_2 and v_3 are shown in Figure 3.30. As one can see, similarly to the *SVD*, the reconstructed signal is consistent with the original one.

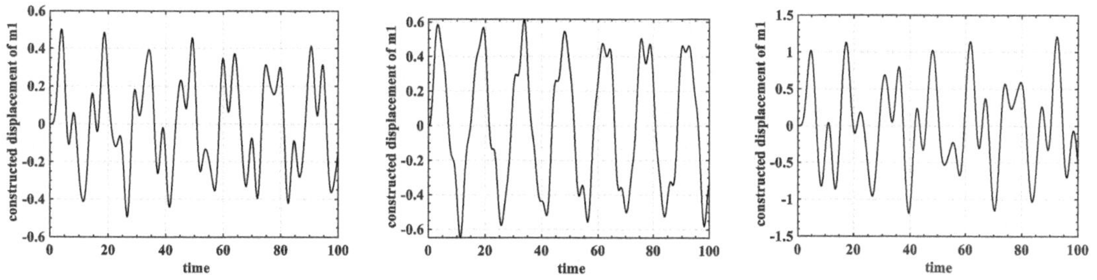

Figure 3.30 Reconstructed signals based on v_1, v_2 and v_3.

3.5 The Finite Element Method

As a numerical analysis tool, the Finite Element Method (*FEM*) has played an important role in promoting the development of modern science and technology. Its idea can be traced back to a long time ago, although the name of "finite element" was put forward by Clough in 1960. Euler, a Swiss mathematician and engineer, used a method similar to the modern finite element method to solve the equilibrium problem of a straight bar under the action of an axial force, by the end of the 18th century. But even in the 1960's people could not overcome the difficulties of large amount of computing, due to the lack of powerful computing tools of those times.

The popularisation of electronic computers had provided a solid foundation for the development of finite elements until the 1980's. Since then, the theoretical research, engineering application and software have developed vigorously, and related academic monographs appeared constantly. Nowadays, the finite element method has been widely used in almost all fields, such as solid mechanics, fluid mechanics, heat conduction, electromagnetism, acoustics etc., to solve all kinds of field distribution problems, as well as the multi-physical fields problem.

The mathematical idea of *FEM* is discrete approximation, that is, the solution domain is divided into a series of small portions named finite elements. This approach can be very convenient to deal with complex problems, because the nature of the problem can be very much simplified in each unit after subdivision.

In this section, the basic concept and calculation flow of the finite element method are introduced briefly. The interested reader is invited to study more advanced textbooks in this area if more complex elements and situations arise. Here, one concentrates on the explanation of the simple spring element as a starting point, and then the rod and the beam elements are also introduced.

Here, to enhance practicality, the introduction of the theories and formulas of *FEM* is kept to a minimum, and the content is intended only to give the reader an understanding of the philosophy of the method.

3.5.1 Basic Idea of the *FEM*

For a complex engineering structure, such as a car body or an aircraft wing subjected to an aerodynamic load distribution, it is difficult to derive an analytical expression through a theoretical calculation. Then one must rely on finite analysis to obtain a numerical solution.

The basic idea of *FEM* is to break a complex structure into some simple elements, or to assemble some elements into a complex structure in a certain way. Compared to the whole structure, the analysis of the element can be considered as simple. This idea is also very common in our lives, for example, when a child assembles some blocks to form an object; the object can be regarded as an engineering complex structure, and the individual blocks can be regarded as elements.

Figure 3.31 illustrates a finite element model of the rotor structure of an aero-engine, where the complex structure is divided into specific elements, forming a mesh.

3.5.2 General Procedure for Finite Element Analysis

The analysis and calculation of *FEM*, known as Finite Element Analysis (*FEA*), is generally divided into 6 steps:

1 – Discrete solution region. The first step of *FEM* is to divide the solving region (or structure) into several small pieces by a grid or mesh, that is, into finite elements. Subdivision and approximation are the basic concepts of *FEM*.

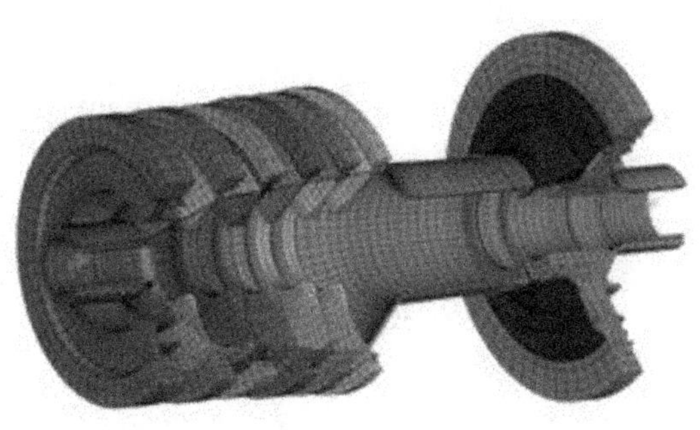

Figure 3.31 Example of a finite element model.

2 – Select the interpolation function (or displacement mode). The displacement distribution in the element is assumed as some simple function of the coordinates. Polynomials are often chosen as interpolation functions for constant variables because polynomials are easy to integrate or differentiate.

3 – Analyse the (mechanical) characteristics of the element. The matrix equation of the element characteristics (element stiffness matrix and mass matrix) is determined by using a physical direct method, variational principle, or weighted remainder method. The derivation of the stiffness and mass matrices of the element is the core of the element characteristic analysis.

4 – Assemble all the elements to establish the equilibrium equations for the entire system. This assembly process includes two aspects: one is to assemble the stiffness and mass matrices of each element into the stiffness and mass matrices of the whole system; second, the equivalent nodal force vectors acting on each element are assembled into the overall load array.

5 – Solve the assembled equations of the system.

6 – Perform additional calculations as needed. For example, the node displacement has been obtained, and then the strain and stress are calculated according to their relationship. Another example, given the pressure in fluid mechanics, the velocity and flow rate can be solved.

To explain the *FEM*, one starts with a simple spring element (Figure 3.32).

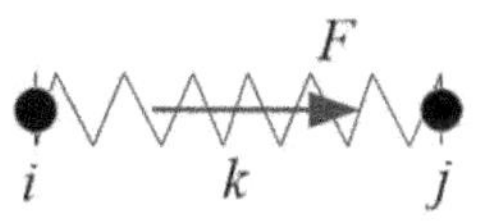

Figure 3.32 Sping element.

The spring element has two nodes i and j. The nodal displacements and nodal forces are u_i, u_j, f_i, f_j respectively, and the spring stiffness is k. Note that all vector directions are specified with a positive direction to the right. The relationship between the spring force (F) and the elongation (δ) is linear in the case of small deformations, so one has

$$F = k\delta, \quad \delta = u_j - u_i \tag{3.203}$$

Consider the balance of forces at the nodes of the spring element. At node i,

$$f_i = -k\left(u_j - u_i\right) = ku_i - ku_j \tag{3.204}$$

At the node j,

$$f_j = k\left(u_j - u_i\right) = ku_j - ku_i \tag{3.205}$$

In matrix form,

$$\begin{bmatrix} k & -k \\ -k & k \end{bmatrix} \begin{Bmatrix} u_i \\ u_j \end{Bmatrix} = \begin{Bmatrix} f_i \\ f_j \end{Bmatrix} = \begin{Bmatrix} F_i \\ F_j \end{Bmatrix}$$

or

$$\boldsymbol{ku} = \boldsymbol{F} \tag{3.206}$$

where, $\boldsymbol{k}$ is the element stiffness matrix, $\boldsymbol{u}$ is the nodal displacement vector and $\boldsymbol{F}$ is the external force vector.

So, Eq. (3.206) is the nodal force balance equation of the spring element. In practice, to characterise the actual structure, there will be multiple elements that will need to be assembled together, for example, take the two springs system as shown in Figure 3.33.

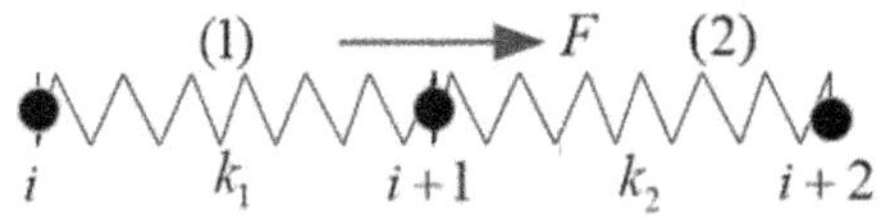

Figure 3.33 Assembly of two spring elements.

For element 1, one has

$$\begin{bmatrix} k_1 & -k_1 \\ -k_1 & k_1 \end{bmatrix} \begin{Bmatrix} u_i \\ u_{i+1} \end{Bmatrix} = \begin{Bmatrix} f_i^{(1)} \\ f_{i+1}^{(1)} \end{Bmatrix} \tag{3.207}$$

Note here that the superscript parenthesis $^{(\,)}$ represents the element number. Likewise, for element 2,

$$\begin{bmatrix} k_2 & -k_2 \\ -k_2 & k_2 \end{bmatrix} \begin{Bmatrix} u_{i+1} \\ u_{i+2} \end{Bmatrix} = \begin{Bmatrix} f_{i+1}^{(2)} \\ f_{i+2}^{(2)} \end{Bmatrix} \tag{3.208}$$

The stiffness matrices of elements 1 and 2 can be expanded as

$$\begin{bmatrix} k_1 & -k_1 & 0 \\ -k_1 & k_1 & 0 \\ 0 & 0 & 0 \end{bmatrix} \begin{Bmatrix} u_i \\ u_{i+1} \\ u_{i+2} \end{Bmatrix} = \begin{Bmatrix} f_i^{(1)} \\ f_{i+1}^{(1)} \\ 0 \end{Bmatrix} \tag{3.209}$$

and

$$\begin{bmatrix} 0 & 0 & 0 \\ 0 & k_2 & -k_2 \\ 0 & -k_2 & k_2 \end{bmatrix} \begin{Bmatrix} u_i \\ u_{i+1} \\ u_{i+2} \end{Bmatrix} = \begin{Bmatrix} 0 \\ f_{i+1}^{(2)} \\ f_{i+2}^{(2)} \end{Bmatrix} \tag{3.210}$$

Superposition of Eqs. (3.209) and (3.210) leads to:

$$\begin{bmatrix} k_1 & -k_1 & 0 \\ -k_1 & k_1+k_2 & -k_2 \\ 0 & -k_2 & k_2 \end{bmatrix} \begin{Bmatrix} u_i \\ u_{i+1} \\ u_{i+2} \end{Bmatrix} = \begin{Bmatrix} f_i^{(1)} \\ f_{i+1}^{(1)} + f_{i+1}^{(2)} \\ f_{i+2}^{(2)} \end{Bmatrix} = \begin{Bmatrix} F_i \\ F_{i+1} \\ F_{i+2} \end{Bmatrix} \tag{3.211}$$

It can be proved that Eq. (3.211) is the same as that obtained by force equilibrium. One can alternatively use the principle of minimum potential energy based on the variational method to obtain the above result. The variational principle is the basis of *FEM*. Here, only a brief analysis will be made, considering the balance between mathematical foundation and practical application.

Consider the strain energy of the spring system shown in Figure 3.33:

$$U = \frac{1}{2}k_1\left(u_{i+1}-u_i\right)^2 + \frac{1}{2}k_2\left(u_{i+2}-u_{i+1}\right)^2 \tag{3.212}$$

Eq. (3.212) can be written as

$$U = \frac{1}{2}\left(\{u_i \quad u_{i+1}\}\begin{bmatrix} k_1 & -k_1 \\ -k_1 & k_1 \end{bmatrix}\begin{Bmatrix} u_i \\ u_{i+1} \end{Bmatrix} + \{u_{i+1} \quad u_{i+2}\}\begin{bmatrix} k_2 & -k_2 \\ -k_2 & k_2 \end{bmatrix}\begin{Bmatrix} u_{i+1} \\ u_{i+2} \end{Bmatrix}\right) \tag{3.213}$$

Connecting both terms of (3.213) as it was done before to obtain expression (3.211),

$$U = \frac{1}{2}\{u_i \quad u_{i+1} \quad u_{i+2}\}\begin{bmatrix} k_1 & -k_1 & 0 \\ -k_1 & k_1+k_2 & -k_2 \\ 0 & -k_2 & k_2 \end{bmatrix}\begin{Bmatrix} u_i \\ u_{i+1} \\ u_{i+2} \end{Bmatrix} \tag{3.214}$$

The potential energy of the external force is

$$\Omega = -F_i u_i - F_{i+1}u_{i+1} - F_{i+2}u_{i+2} = -\{u_i \quad u_{i+1} \quad u_{i+2}\}\begin{Bmatrix} F_i \\ F_{i+1} \\ F_{i+2} \end{Bmatrix} \tag{3.215}$$

Thus, the total potential energy of the system is

$$\Pi = U + \Omega = \frac{1}{2}\{u_i \quad u_{i+1} \quad u_{i+2}\}\begin{bmatrix} k_i & -k_i & 0 \\ -k_i & k_i + k_{i+1} & -k_{i+1} \\ 0 & -k_{i+1} & k_{i+1} \end{bmatrix}\begin{Bmatrix} u_i \\ u_{i+1} \\ u_{i+2} \end{Bmatrix} - \{u_i \quad u_{i+1} \quad u_{i+2}\}\begin{Bmatrix} F_i \\ F_{i+1} \\ F_{i+2} \end{Bmatrix} \tag{3.216}$$

When the potential energy of is a minimum, the system will be in stable equilibrium (principle of minimum potential energy). So, one takes partial derivatives of the energy with respect to the nodal displacements in Eq. (3.216):

$$\frac{\partial \Pi}{\partial u_i} = 0, \quad \frac{\partial \Pi}{\partial u_{i+1}} = 0, \quad \frac{\partial \Pi}{\partial u_{i+2}} = 0 \tag{3.217}$$

and the result is the same as Eq. (3.211).

If node i is fixed and a force Q in the positive direction is applied at nodes $i+1$ and $i+2$,

$$u_i = 0, \quad F_{i+1} = F_{i+2} = Q \tag{3.218}$$

Substituting Eq. (3.218) into Eq. (3.211),

$$\begin{bmatrix} k_1 & -k_1 & 0 \\ -k_1 & k_1 + k_2 & -k_2 \\ 0 & -k_2 & k_2 \end{bmatrix}\begin{Bmatrix} 0 \\ u_{i+1} \\ u_{i+2} \end{Bmatrix} = \begin{Bmatrix} F_i \\ Q \\ Q \end{Bmatrix} \tag{3.219}$$

This is equivalent to solve

$$\begin{bmatrix} k_1 + k_2 & -k_2 \\ -k_2 & k_2 \end{bmatrix}\begin{Bmatrix} u_{i+1} \\ u_{i+2} \end{Bmatrix} = \begin{Bmatrix} Q \\ Q \end{Bmatrix} \quad \text{and} \quad F_i = -k_1 u_{i+1} \tag{3.220}$$

From which,

$$\begin{Bmatrix} u_{i+1} \\ u_{i+2} \end{Bmatrix} = \begin{Bmatrix} 2Q / k_1 \\ 2Q / k_1 + Q / k_2 \end{Bmatrix} \quad \text{and} \quad F_i = -2Q \tag{3.221}$$

This exemplifies the basic solution process of the *FEM* above.

The spring element is the simplest element in the *FEM*. In the following sections one will look at more complex and commonly used elements.

3.5.3 Bars and Trusses

An element that only bears the axial force but cannot bear the bending moment is called a bar element. The bar element is a one-dimensional element. Considering general truss structures, no matter how a global coordinate system is selected, there is always a bar that is inclined in that global coordinate system. However, unnecessary complexity will be added if the stiffness matrix of the element is developed under the condition that the element axis is not parallel to the coordinate axis. Therefore, one starts by supposing the local coordinate system coincident with the global coordinate system, with the element aligned with the axis x, as shown in Figure 3.34; this is used to establish the standard stiffness matrix of the bar element. Afterwards, the element will be considered in a general position and the stiffness matrix with respect to the global coordinate system is obtained by a linear transformation.

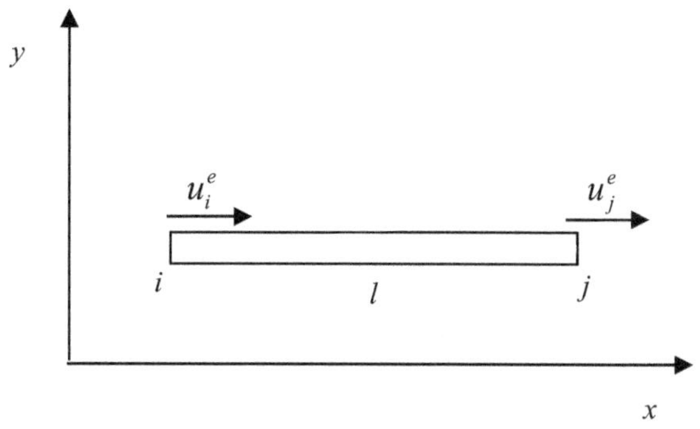

Figure 3.34 Uniform bar element of length l.

Both experimental and theoretical analysis prove that the displacement within the bar subjected to axial forces is linear. Therefore, the displacement function within the bar can be obtained directly. The relationship between the displacement u^e in a bar of length l and the coordinate x can be expressed in the local coordinate system as follows:

$$u^e = \left(1 - \frac{x}{l}\right) u_i^e + \frac{x}{l} u_j^e \tag{3.222}$$

In matrix form,

$$u^e = \left\{ 1 - \frac{x}{l} \quad \frac{x}{l} \right\} \begin{Bmatrix} u_i^e \\ u_j^e \end{Bmatrix} \tag{3.223}$$

or

$$u^e = \boldsymbol{N}\boldsymbol{u}^e \tag{3.224}$$

For a 1-dimensional element, the strain – as a function of displacement – is:

$$\varepsilon = \frac{du^e}{dx} = \frac{d\boldsymbol{N}}{dx}\boldsymbol{u}^e = \frac{1}{l}\{-1 \quad 1\} \begin{Bmatrix} u_i^e \\ u_j^e \end{Bmatrix} \tag{3.225}$$

Let $\boldsymbol{B} = \frac{d\boldsymbol{N}}{dx} = \frac{1}{l}\{-1 \quad 1\}$. Then,

$$\varepsilon = \boldsymbol{B}\boldsymbol{u}^e \tag{3.226}$$

Consider the strain energy stored in the bar,

$$U = \frac{1}{2}\int_V \sigma^T \varepsilon dV \tag{3.227}$$

As $\sigma = E\varepsilon = E\boldsymbol{B}\boldsymbol{u}^e$, where E is the Young's modulus,

$$U = \frac{1}{2}\int_V \sigma^T \varepsilon dV = \frac{1}{2}(\boldsymbol{u}^{eT}\boldsymbol{B}^T E\boldsymbol{B}\boldsymbol{u}^e)dV = \frac{1}{2}\boldsymbol{u}^{eT}\left[\int_V (\boldsymbol{B}^T E\boldsymbol{B})dV\right]\boldsymbol{u}^e = \frac{1}{2}\boldsymbol{u}^{eT}\boldsymbol{k}^e\boldsymbol{u}^e \tag{3.228}$$

where $\boldsymbol{k}^e$ is the stiffness matrix of the element.

The potential energy of applied external forces is

$$\Omega = -F_i^e u_i^e - F_j^e u_j^e = -\begin{Bmatrix} F_i^e & F_j^e \end{Bmatrix}\begin{Bmatrix} u_i^e \\ u_j^e \end{Bmatrix} = -\boldsymbol{F}^e\boldsymbol{u}^e \tag{3.229}$$

The total potential energy is given by

$$\Pi = U + \Omega \tag{3.230}$$

and therefore, according to the principle of minimum potential energy,

$$\frac{\partial \Pi}{\partial \boldsymbol{u}} = \boldsymbol{k}^e\boldsymbol{u}^e - \boldsymbol{F}^e = \boldsymbol{0} \tag{3.231}$$

The stiffness matrix of the 1-dimensional bar member in the coordinate system of Figure 3.34 is:

$$\boldsymbol{k}^e = \int_V \boldsymbol{B}^T E\boldsymbol{B}dV = A\int_0^l \boldsymbol{B}^T E\boldsymbol{B}dx = AE\int_0^l \frac{1}{l^2}\begin{Bmatrix} -1 \\ 1 \end{Bmatrix}\begin{Bmatrix} -1 & 1 \end{Bmatrix}dx = \frac{AE}{l}\begin{bmatrix} 1 & -1 \\ -1 & 1 \end{bmatrix} \tag{3.232}$$

where A is the cross-section area of the bar.

Substituting in (3.231), the equilibrium equation of the element can be written in the following form:

$$\frac{EA}{l}\begin{bmatrix} 1 & -1 \\ -1 & 1 \end{bmatrix}\begin{Bmatrix} u_i^e \\ u_j^e \end{Bmatrix} = \begin{Bmatrix} F_i^e \\ F_j^e \end{Bmatrix} \tag{3.233}$$

Now, consider the local coordinate system x^e - y^e (where in general the element stands) rotated with respect to the global system by an angle θ^e, as shown in Figure 3.35. The transformation from the local coordinate system to the global coordinate system can be completed through a coordinate rotation, such that

$$\begin{aligned} x^e &= x\cos\theta^e + y\sin\theta^e \\ y^e &= -x\sin\theta^e + y\cos\theta^e \end{aligned} \tag{3.234}$$

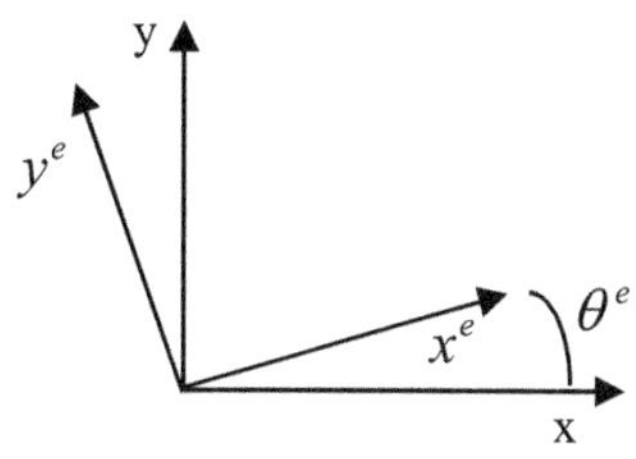

Figure 3.35 Coordinate transformation.

where, the superscript ^{e} indicates that it belongs to the local coordinate system of the element. Writing in matrix form,

$$\begin{Bmatrix} x^e \\ y^e \end{Bmatrix} = \begin{bmatrix} cos\theta^e & sin\theta^e \\ -sin\theta^e & cos\theta^e \end{bmatrix} \begin{Bmatrix} x \\ y \end{Bmatrix} = \boldsymbol{\Gamma}^e \begin{Bmatrix} x \\ y \end{Bmatrix} \tag{3.235}$$

where $\boldsymbol{\Gamma}^e = \begin{bmatrix} cos\theta^e & sin\theta^e \\ -sin\theta^e & cos\theta^e \end{bmatrix}$ is the rotation matrix.

However, as the axis of the bar is not parallel anymore to the x axis, its displacement will have projection components in the x-direction and y-direction as well. In this case, v_i^e and v_j^e should be added to the displacement vector of the element. The displacement column vector of the nodes becomes $\{u_i^e \quad v_i^e \quad u_j^e \quad v_j^e\}^T$. In the local coordinate system, the stiffness matrix of the element can be extended as

$$\boldsymbol{k}^e = \frac{EA}{l} \begin{bmatrix} 1 & 0 & -1 & 0 \\ 0 & 0 & 0 & 0 \\ -1 & 0 & 1 & 0 \\ 0 & 0 & 0 & 0 \end{bmatrix} \tag{3.236}$$

The displacement vector for each element can be expressed as

$$\boldsymbol{d}^e = \begin{Bmatrix} u_i^e \\ v_i^e \\ u_j^e \\ v_j^e \end{Bmatrix} = \begin{bmatrix} \boldsymbol{\Gamma}^e & \boldsymbol{0} \\ \boldsymbol{0} & \boldsymbol{\Gamma}^e \end{bmatrix} \begin{Bmatrix} u_i \\ v_i \\ u_j \\ v_j \end{Bmatrix} = \begin{bmatrix} cos\theta^e & sin\theta^e & 0 & 0 \\ -sin\theta^e & cos\theta^e & 0 & 0 \\ 0 & 0 & cos\theta^e & sin\theta^e \\ 0 & 0 & -sin\theta^e & cos\theta^e \end{bmatrix} \begin{Bmatrix} u_i \\ v_i \\ u_j \\ v_j \end{Bmatrix} = \boldsymbol{R}^e \boldsymbol{d} \tag{3.237}$$

where $\boldsymbol{R}^e$ is the transformation matrix of the element.

Next, an explanation is given on how to use the transformation matrix to transform the expression of the stiffness matrix from the local coordinate system to the global coordinate system.

The equilibrium equation of the element established in the local coordinate system is

$$\boldsymbol{k}^e \boldsymbol{d}^e = \boldsymbol{F}^e \tag{3.238}$$

The displacement vector of the element, from Eq. (3.219) is $\boldsymbol{d}^e = \boldsymbol{R}^e \boldsymbol{d}$ and likewise, for the load vector,

$$\boldsymbol{F}^e = \boldsymbol{R}^e \boldsymbol{F} \tag{3.239}$$

Substituting Eq. (3.237) and (3.239) in Eq. (3.238), one has

$$\boldsymbol{k}^e \boldsymbol{R}^e \boldsymbol{d} = \boldsymbol{R}^e \boldsymbol{F} \tag{3.240}$$

from which,

$$\left(\boldsymbol{R}^e\right)^{-1} \boldsymbol{k}^e \boldsymbol{R}^e \boldsymbol{d} = \boldsymbol{F} \tag{3.241}$$

Defining $\boldsymbol{K}^e = \left(\boldsymbol{R}^e\right)^{-1} \boldsymbol{k}^e \boldsymbol{R}^e$, leads to:

$$\boldsymbol{K}^e \boldsymbol{d} = \boldsymbol{F} \tag{3.242}$$

Since the coordinate transformation matrix is orthogonal, $\left(\boldsymbol{R}^e\right)^T = \left(\boldsymbol{R}^e\right)^{-1}$ and so, in practical applications, one uses:

$$\boldsymbol{K}^e = \left(\boldsymbol{R}^e\right)^T \boldsymbol{k}^e \boldsymbol{R}^e \tag{3.243}$$

and finally, one ends up with the stiffness of the element with respect to the global coordinate system:

$$\boldsymbol{K}^e = \frac{EA}{l} \begin{bmatrix} cos^2\theta^e & cos\theta^e sin\theta^e & -cos^2\theta^e & -cos\theta^e sin\theta^e \\ cos\theta^e sin\theta^e & sin^2\theta^e & -cos\theta^e sin\theta^e & -sin^2\theta^e \\ -cos^2\theta^e & -cos\theta^e sin\theta^e & cos^2\theta^e & cos\theta^e sin\theta^e \\ -cos\theta^e sin\theta^e & -sin^2\theta^e & cos\theta^e sin\theta^e & sin^2\theta^e \end{bmatrix} \tag{3.244}$$

In addition to the stiffness matrix of the element, the mass matrix is also needed to solve the dynamics problem. There are two types of mass matrices: lumped mass matrix and consistent mass matrix. The bar element is used to explain the meaning of both types. As shown in Figure 3.36, a lumped mass element concentrates the masses into the extremities of the bar.

The lumped mass matrix of a one-dimensional bar element can be written as:

$$\boldsymbol{m}_l^e = \begin{bmatrix} \dfrac{\rho Al}{2} & 0 \\ 0 & \dfrac{\rho Al}{2} \end{bmatrix} \tag{3.245}$$

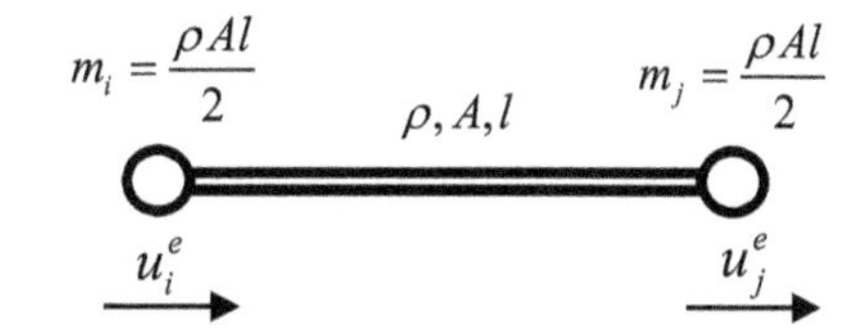

Figure 3.36 Lumped mass element.

Considering also the vertical degrees of freedom, the lumped mass matrix of the element is

$$\boldsymbol{m}_l^e = \frac{\rho Al}{2} \boldsymbol{I}_{4\times4} \tag{3.246}$$

The consistent mass matrix can be calculated from the kinetic energy:

$$T = \frac{1}{2}\int_V \left(\dot{u}^e\right)^2 dm = \frac{1}{2}\int_0^l \rho A\left(\dot{u}^e\right)^2 dx \tag{3.247}$$

As $u^e = \boldsymbol{N}\boldsymbol{u}^e$ (Eq. (3.224), $\dot{u}^e = \boldsymbol{N}\dot{\boldsymbol{u}}^e$ and Eq. (3.247) can be written as:

$$T = \frac{1}{2}\int_0^l \rho A\left(\dot{\boldsymbol{u}}^e\right)^T \boldsymbol{N}^T\boldsymbol{N}\dot{\boldsymbol{u}}^e dx = \frac{1}{2}\left(\dot{\boldsymbol{u}}^e\right)^T\left(\int_0^l \rho A\boldsymbol{N}^T\boldsymbol{N}dx\right)\dot{\boldsymbol{u}}^e \tag{3.248}$$

Thus, the consistent mass matrix is:

$$\boldsymbol{m}_c^e = \int_0^l \rho A\boldsymbol{N}^T\boldsymbol{N}dx \tag{3.249}$$

Considering just the 2 *DOFs* along the x direction, $\boldsymbol{N} = \left\{1-\frac{x}{l} \quad \frac{x}{l}\right\}$ and so,

$$\boldsymbol{m}_c^e = \int_0^l \rho A\begin{Bmatrix}1-\frac{x}{l}\\ \frac{x}{l}\end{Bmatrix}\left\{1-\frac{x}{l} \quad \frac{x}{l}\right\}dx = \frac{\rho Al}{6}\begin{bmatrix}2 & 1\\ 1 & 2\end{bmatrix} \tag{3.250}$$

If the axis of the rod element is not parallel to the x axis, although it is one-dimensional, it also has projection components in the x and y directions. The displacement column vector of the nodes becomes $\left\{u_i^e \quad v_i^e \quad u_j^e \quad v_j^e\right\}^T$, and the consistent mass matrix of the element is

$$\boldsymbol{m}_c^e = \frac{\rho Al}{4}\begin{bmatrix}2 & 0 & 1 & 0\\ 0 & 2 & 0 & 1\\ 1 & 0 & 2 & 0\\ 0 & 1 & 0 & 2\end{bmatrix} \tag{3.251}$$

Applying a transformation similar to the one used for the stiffness matrix (Eq.(3.243)), the lumped and consistent mass matrices under the global coordinate system are, respectively,

$$M_l^e = \left(R^e\right)^T m_l^e R^e = \frac{\rho A l}{2}\begin{bmatrix} cos^2\theta^e & cos\theta^e sin\theta^e & 0 & 0 \\ cos\theta^e sin\theta^e & sin^2\theta^e & 0 & 0 \\ 0 & 0 & cos^2\theta^e & cos\theta^e sin\theta^e \\ 0 & 0 & cos\theta^e sin\theta^e & sin^2\theta^e \end{bmatrix} \quad (3.252)$$

$$M_c^e = \left(R^e\right)^T m_c^e R^e = \frac{\rho A l}{6}\begin{bmatrix} 2cos^2\theta^e & 2cos\theta^e sin\theta^e & cos^2\theta^e & cos\theta^e sin\theta^e \\ 2cos\theta^e sin\theta^e & 2sin^2\theta^e & cos\theta^e sin\theta^e & sin^2\theta^e \\ cos^2\theta^e & cos\theta^e sin\theta^e & 2cos^2\theta^e & 2cos\theta^e sin\theta^e \\ cos\theta^e sin\theta^e & sin^2\theta^e & 2cos\theta^e sin\theta^e & 2sin^2\theta^e \end{bmatrix}$$

(3.253)

Here, an example is given to illustrate the solving process of the bar element, which is given in the form of a *MATLAB* code. If the reader does not understand the program statement, he can enter the "help command" in the *MATLAB* software command window, to view the relevant help files.

Example 3.5.3 Natural frequencies of a bar

Determine the natural frequency of a free bar, like in Figure 3.37. The bar is divided into 8 elements. The elastic modulus is 198 *GPa*, the cross-sectional area is 0.0016 m^2 and the mass density is 7850 kg/m^3.

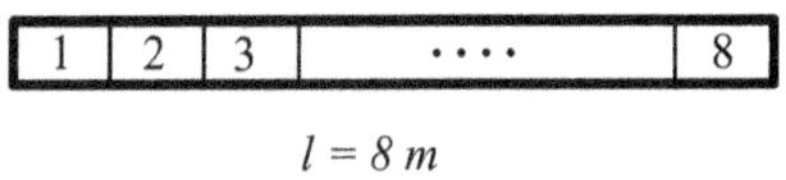

Figure 3.37 Finite element discretisation of a free-free bar.

One can execute the *MATLAB* code16 in Appendix A. The results of the program are given in Table 3.9.

Table 3.9 Natural frequencies of the bar of Figure 3.37.

Mode	Natural frequency (rad/s)	Exact (rad/s)
1	0.000	0.00
2	1984	1975
3	4046	3970
4	6261	6003
5	8698	8093

The global mass matrix is taken as an example to illustrate the assembly process of the elements (Figure 3.38), according to Eq. (3.250); the assembly process for the stiffness matrix is similar.

$$\begin{matrix} & \begin{matrix} u_1 & u_2 & \cdots & u_9 \end{matrix} \\ \begin{matrix} u_1 \\ u_2 \\ \vdots \\ u_9 \end{matrix} & \begin{bmatrix} \textit{global mass} \\ \textit{matrix} \end{bmatrix} \end{matrix} = \frac{\rho Al}{6} \underbrace{\left(\begin{bmatrix} 2 & 1 & 0 & 0 & \cdots & 0 \\ 1 & 2 & 0 & 0 & \cdots & 0 \\ 0 & 0 & 0 & 0 & \cdots & 0 \\ \vdots & \vdots & \vdots & \vdots & \ddots & \vdots \\ 0 & 0 & 0 & 0 & \cdots & 0 \end{bmatrix} + \begin{bmatrix} 0 & 0 & 0 & 0 & \cdots & 0 \\ 0 & 2 & 1 & 0 & \cdots & 0 \\ 0 & 1 & 2 & 0 & \cdots & 0 \\ \vdots & \vdots & \vdots & \vdots & \ddots & \vdots \\ 0 & 0 & 0 & 0 & \cdots & 0 \end{bmatrix} + \cdots \right)}_{\textit{total of 8 element expansion matrices}}$$

Figure 3.38 Element assembly process.

3.5.4 Beams

Beams and frames are frequently used in constructions and engineering equipment, like buildings, lifting equipment and vehicles. Beams are slender structural members subjected primarily to transverse loads. A beam is geometrically similar to a bar in that its longitudinal dimension is significantly larger than the two transverse dimensions. Unlike bars, the deformation in a beam is predominantly due to bending in transverse directions. Such a bending-dominated deformation is the primary mechanism for a beam to resist to transverse loads.

The beam structure can be analysed as a one-dimensional element with two end nodes. The element deformation is the transverse displacement and the rotation angle, as shown in Figure 3.39.

The Bernoulli–Euler hypothesis can be adopted assuming that the shear deformation of the beam is not considered and that the inertia due to the rotational deformation is neglected. More elaborated models can be envisaged, where those effects are taken into account, but for the purpose of this section, one shall restrict to this simpler case.

According to the Bernoulli–Euler hypothesis, the section perpendicular to the centre line of the beam before deformation remains perpendicular to the centre line of the beam after deformation, as shown in Figure 3.40.

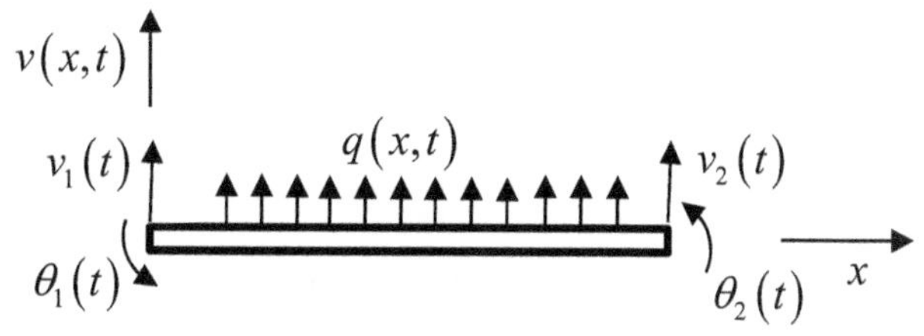

Figure 3.39 Beam element.

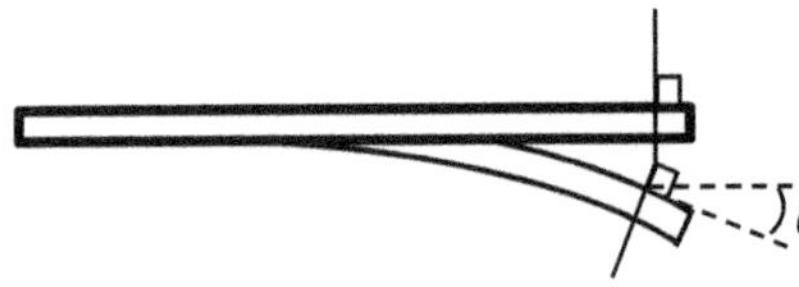

Figure 3.40 Bernoulli-Euler beam.

The equation of motion of the beam in bending, according to the above assumptions, is as follows:

$$\rho A \frac{\partial^2 v}{\partial t^2} + \frac{\partial^2}{\partial x^2}\left(EI \frac{\partial^2 v}{\partial x^2} \right) = q(x,t) \tag{3.254}$$

where E, I, ρ and A are, respectively, the Young's modulus, the second area moment of the cross section, the mass density and the cross section area; $v(x,t)$ is the deflection (transversal displacement) of the neutral axis of the beam and $q(x,t)$ is an applied transversal distributed load.

One needs to deduce the mass matrix, the stiffness matrix and the force vector with respect to the degrees of freedom of the element. Consequently, the equilibrium equation will be:

$$\begin{bmatrix} m_{11} & m_{12} & m_{13} & m_{14} \\ m_{21} & m_{22} & m_{23} & m_{24} \\ m_{31} & m_{32} & m_{33} & m_{34} \\ m_{41} & m_{42} & m_{43} & m_{44} \end{bmatrix} \begin{Bmatrix} \ddot{v}_1(t) \\ \ddot{\theta}_1(t) \\ \ddot{v}_2(t) \\ \ddot{\theta}_2(t) \end{Bmatrix} + \begin{bmatrix} k_{11} & k_{12} & k_{13} & k_{14} \\ k_{21} & k_{22} & k_{23} & k_{24} \\ k_{31} & k_{32} & k_{33} & k_{34} \\ k_{41} & k_{42} & k_{43} & k_{44} \end{bmatrix} \begin{Bmatrix} v_1(t) \\ \theta_1(t) \\ v_2(t) \\ \theta_2(t) \end{Bmatrix} = \begin{Bmatrix} F_1(t) \\ M_1(t) \\ F_2(t) \\ M_2(t) \end{Bmatrix} \tag{3.255}$$

where $F_1(t)$, $M_1(t)$, $F_2(t)$ and $M_2(t)$ are the forces and moments applied at $v_1(t)$, $\theta_1(t)$, $v_2(t)$ and $\theta_2(t)$, respectively, due to the applied distributed load $q(x,t)$.

Eq. (3.255) can be written in a more compact form:

$$\boldsymbol{m}^e \ddot{\boldsymbol{d}}^e(t) + \boldsymbol{k}^e \boldsymbol{d}^e(t) = \boldsymbol{f}^e(t) \tag{3.256}$$

with

$$\boldsymbol{d}^e(t) = \begin{Bmatrix} v_1(t) \\ \theta_1(t) \\ v_2(t) \\ \theta_2(t) \end{Bmatrix} \quad \text{and} \quad \boldsymbol{f}^e(t) = \begin{Bmatrix} F_1(t) \\ M_1(t) \\ F_2(t) \\ M_2(t) \end{Bmatrix} \tag{3.257}$$

The deduction of the stiffness matrix will be developed from the strain energy of the element:

$$U^e = \frac{1}{2} \int_0^l EI \left(\frac{\partial^2 v}{\partial x^2} \right)^2 dx \tag{3.258}$$

where the displacement of the element is taken as a polynomial function:

$$v(x,t)=a_0(t)+a_1(t)x+a_2(t)x^2+a_3(t)x^3 \tag{3.259}$$

The rotation angle is its partial derivative with respect to *x*:

$$\theta(x,t)=\frac{\partial v(x,t)}{\partial x}=a_1(t)+2a_2(t)x+3a_3(t)x^2 \tag{3.260}$$

At $x=0$ and $x=l$, one has

$$\begin{aligned}
v(0,t)&=a_0(t)=v_1(t)\\
\theta(0,t)&=a_1(t)=\theta_1(t)\\
v(l,t)&=a_0(t)+a_1(t)l+a_2(t)l^2+a_3(t)l^3=v_2(t)\\
\theta(l,t)&=a_1(t)+2a_2(t)l+3a_3(t)l^2=\theta_2(t)
\end{aligned} \tag{3.261}$$

Writing (3.261) in matrix form,

$$\boldsymbol{d}^e(t)=\begin{Bmatrix} v_1(t)\\ \theta_1(t)\\ v_2(t)\\ \theta_2(t)\end{Bmatrix}=\begin{bmatrix}1&0&0&0\\0&1&0&0\\1&l&l^2&l^3\\0&1&2l&3l^2\end{bmatrix}\begin{Bmatrix}a_0(t)\\a_1(t)\\a_2(t)\\a_3(t)\end{Bmatrix} \tag{3.262}$$

Solving Eq. (3.262), it follows that

$$v(x,t)=N_1(x)v_1(t)+N_2(x)\theta_1(t)+N_3(x)v_2(t)+N_4(x)\theta_2(t)=\boldsymbol{N}(x)\boldsymbol{d}^e(t) \tag{3.263}$$

where $\boldsymbol{N}(x)=\{N_1(x)\quad N_2(x)\quad N_3(x)\quad N_4(x)\}$, with

$$\begin{aligned}
&N_1(x)=1-3\left(\frac{x}{l}\right)^2+2\left(\frac{x}{l}\right)^3, \quad N_2(x)=\left(\left(\frac{x}{l}\right)-2\left(\frac{x}{l}\right)^2+\left(\frac{x}{l}\right)^3\right)l\\
&N_3(x)=3\left(\frac{x}{l}\right)^2-2\left(\frac{x}{l}\right)^3, \quad N_4(x)=\left(-\left(\frac{x}{l}\right)^2+\left(\frac{x}{l}\right)^3\right)l
\end{aligned} \tag{3.264}$$

$N_i(x)$ are the *Hermite shape functions*. Changing variable such that $\xi=\frac{x}{l}$,

$$v(\xi,t)=N_1(\xi)v_1(t)+N_2(\xi)\theta_1(t)+N_3(\xi)v_2(t)+N_4(\xi)\theta_2(t)=\boldsymbol{N}(\xi)\boldsymbol{d}^e(t) \tag{3.265}$$

with $\boldsymbol{N}(\xi)=\{N_1(\xi)\quad N_2(\xi)\quad N_3(\xi)\quad N_4(\xi)\}$, and

$$\begin{aligned}
&N_1(\xi)=1-3\xi^2+2\xi^3, \quad N_2(\xi)=(\xi-2\xi^2+\xi^3)l\\
&N_3(\xi)=3\xi^2-2\xi^3, \quad N_4(\xi)=(-\xi^2+\xi^3)l
\end{aligned} \tag{3.266}$$

In terms of the new variable, the strain energy becomes:

$$U^e = \frac{1}{2}\int_0^1 EI \frac{1}{l^3}\left(\frac{\partial^2 v(\xi,t)}{\partial \xi^2}\right)^2 d\xi \tag{3.267}$$

From Eq. (3.265),

$$\frac{\partial^2 v(\xi,t)}{\partial \xi^2} = N_1''(\xi)v_1(t) + N_2''(\xi)\theta_1(t) + N_3''(\xi)v_2(t) + N_4''(\xi)\theta_2(t) = \boldsymbol{N}''(\xi)\boldsymbol{d}^e(t) \tag{3.268}$$

with

$$\begin{aligned} N_1''(\xi) &= -6+12\xi\,, \quad N_2''(\xi) = (-4+6\xi)l \\ N_3''(\xi) &= 6-12\xi\,, \quad N_4''(\xi) = (-2+6\xi)l \end{aligned} \tag{3.269}$$

Substituting Eq. (3.268) in Eq. (3.267) and simplifying the notation, it follows that:

$$U^e = \frac{1}{2}\int_0^1 \frac{EI}{l^3}(\boldsymbol{d}^e)^T (\boldsymbol{N}'')^T \boldsymbol{N}''\boldsymbol{d}^e \, d\xi \tag{3.270}$$

or

$$U^e = \frac{1}{2}(\boldsymbol{d}^e)^T \left(\frac{EI}{l^3}\int_0^1 (\boldsymbol{N}'')^T \boldsymbol{N}'' \, d\xi\right)\boldsymbol{d}^e \tag{3.271}$$

As $U^e = \frac{1}{2}(\boldsymbol{d}^e)^T \boldsymbol{k}^e \boldsymbol{d}^e$, the stiffness matrix of the element is given by:

$$\boldsymbol{k}^e = EI\frac{1}{l^3}\int_0^1 (\boldsymbol{N}'')^T \boldsymbol{N}'' d\xi \tag{3.272}$$

Substituting expressions (3.269) in Eq. (3.272) and doing the calculations, one arrives finally at:

$$\boldsymbol{k}^e = \frac{EI}{l^3}\begin{bmatrix} 12 & 6l & -12 & 6l \\ 6l & 4l^2 & -6l & 2l^2 \\ -12 & -6l & 12 & -6l \\ 6l & 2l^2 & -6l & 4l^2 \end{bmatrix} \tag{3.273}$$

For the mass matrix, the deduction is similar, but starting from the kinetic energy of the element:

$$T^e = \frac{1}{2}\int_0^l \rho A\left(\frac{\partial v(x,t)}{\partial t}\right)^2 dx = \frac{1}{2}\int_0^1 \rho A\left(\frac{\partial v(\xi,t)}{\partial t}\right)^2 l \, d\xi \tag{3.274}$$

As $\frac{\partial v(\xi,t)}{\partial t} = \boldsymbol{N}(\xi)\frac{\partial(\boldsymbol{d}^e(t))}{\partial t} = \boldsymbol{N}(\xi)\dot{\boldsymbol{d}}^e(t)$, one obtains, after simplifying the notation,

$$T^e = \frac{1}{2}\int_0^1 \rho A \left(\dot{\boldsymbol{d}}^e\right)^T \boldsymbol{N}^T \boldsymbol{N} \dot{\boldsymbol{d}}^e l \, d\xi = \frac{1}{2}\left(\dot{\boldsymbol{d}}^e\right)^T \left(\rho A l \int_0^1 \boldsymbol{N}^T \boldsymbol{N} \, d\xi\right) \dot{\boldsymbol{d}}^e \tag{3.275}$$

As $T^e = \frac{1}{2}\left(\dot{\boldsymbol{d}}^e\right)^T \boldsymbol{m}^e \dot{\boldsymbol{d}}^e$, the mass matrix of the element is given by:

$$\boldsymbol{m}^e = \rho A l \int_0^1 \boldsymbol{N}^T \boldsymbol{N} \, d\xi \tag{3.276}$$

Substituting expressions (3.266) in Eq. (3.276) and doing the calculations, one arrives finally at:

$$\boldsymbol{m}^e = \frac{\rho A l}{420} \begin{bmatrix} 156 & 22l & 54 & -13l \\ 22l & 4l^2 & 13l & -3l^2 \\ 54 & 13l & 156 & -22l \\ -13l & -3l^2 & -22l & 4l^2 \end{bmatrix} \tag{3.277}$$

To evaluate the generalised forces at the nodes of the element, one equals the virtual work of the applied force to the virtual work of the generalised forces:

$$\begin{aligned} \delta W^e = \int_0^l q(x,t)\delta v(x,t)\,dx &= F_1(t)\delta v_1(t) + M_1(t)\delta\theta_1(t) + F_2(t)\delta v_2(t) + M_2(t)\delta\theta_2(t) \\ &= \left(\boldsymbol{f}^e(t)\right)^T \delta \boldsymbol{d}^e(t) \end{aligned} \tag{3.278}$$

From Eq. (3.263),

$$\delta v(x,t) = \boldsymbol{N}(x)\delta\boldsymbol{d}^e(t) \tag{3.279}$$

Substituting Eq. (3.279) in (3.278) leads to

$$\int_0^l q(x,t)\boldsymbol{N}(x)\,\delta\boldsymbol{d}^e(t)\,dx = \left(\int_0^l q(x,t)\boldsymbol{N}(x)\,dx\right)\delta\boldsymbol{d}^e(t) = \left(\boldsymbol{f}^e(t)\right)^T \delta\boldsymbol{d}^e(t) \tag{3.280}$$

and because the virtual displacements are arbitrary and independent, it turns out that

$$\boldsymbol{f}^e(t) = \int_0^l q(x,t)\left(\boldsymbol{N}(x)\right)^T dx \tag{3.281}$$

or, in terms of the variable ξ,

$$\boldsymbol{f}^e(t) = \int_0^1 q(\xi,t)\left(\boldsymbol{N}(\xi)\right)^T l\,d\xi \tag{3.282}$$

Explicitly, each of the generalised forces of the element will be:

$$\begin{aligned} F_1(t) &= \int_0^1 q(\xi,t) N_1(\xi)\, l\, d\xi \\ M_1(t) &= \int_0^1 q(\xi,t) N_2(\xi)\, l\, d\xi \\ F_2(t) &= \int_0^1 q(\xi,t) N_3(\xi)\, l\, d\xi \\ M_2(t) &= \int_0^1 q(\xi,t) N_4(\xi)\, l\, d\xi \end{aligned} \tag{3.283}$$

For instance, when the force $q(x,t)$ on the beam is a uniformly distributed sinusoidal load $q_0 \sin\omega t$, the generalised forces are:

$$\begin{aligned} F_1(t) &= q_0 \sin\omega t \int_0^1 \left(1 - 3\xi^2 + 2\xi^3\right) l\, d\xi = \frac{1}{2} q_0 l \sin\omega t \\ M_1(t) &= q_0 \sin\omega t \int_0^1 \left(\xi - 2\xi^2 + \xi^3\right) l^2\, d\xi = \frac{1}{12} q_0 l^2 \sin\omega t \\ F_2(t) &= q_0 \sin\omega t \int_0^1 \left(3\xi^2 - 2\xi^3\right) l\, d\xi = \frac{1}{2} q_0 l \sin\omega t \\ M_2(t) &= q_0 \sin\omega t \int_0^1 \left(-\xi^2 + \xi^3\right) l^2\, d\xi = -\frac{1}{12} q_0 l^2 \sin\omega t \end{aligned} \tag{3.284}$$

Thus, the generalised applied force vector is:

$$\boldsymbol{f}^e(t) = \frac{q_0}{12} \begin{Bmatrix} 6l \\ l^2 \\ 6l \\ -l^2 \end{Bmatrix} \sin\omega t \tag{3.285}$$

When there is a concentrated load on the beam at $\xi = \xi_0$, for instance $P(\xi,t) = P_0 \sin\omega t$, the force vector is

$$\boldsymbol{f}^e(t) = \int_0^1 P(\xi,t)\delta(\xi - \xi_0) \begin{Bmatrix} N_1 \\ N_2 \\ N_3 \\ N_4 \end{Bmatrix} l\, d\xi = P_0\, l \begin{Bmatrix} N_1(\xi_0) \\ N_2(\xi_0) \\ N_3(\xi_0) \\ N_4(\xi_0) \end{Bmatrix} \sin\omega t \tag{3.286}$$

where $\delta(\xi - \xi_0)$ is the Dirac's distribution acting on $P(\xi,t)$. After defining the mass and stiffness matrices, as well as the force vector for each element, these are assembled together to obtain the global matrices and force vector. This assembling procedure has into account the connectivity between the degrees of freedom of the elements. The next example illustrates the procedure for the stiffness matrix.

Example 3.5.5 Stiffness matrix of a clamped beam at both ends.

The beam shown in Figure 3.41 is clamped at both ends and acted upon by the force P and moment M at the midspan. It is discretised in 2 finite elements. In static conditions, one needs

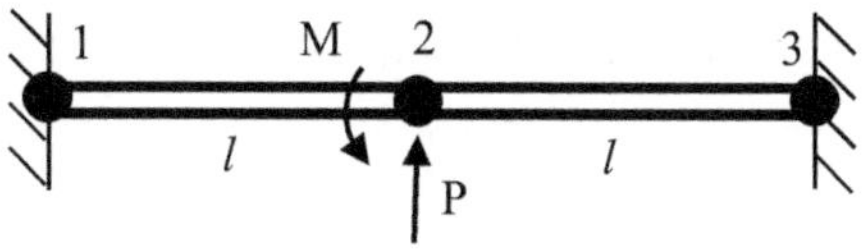

Figure 3.41 Clamped beam at both ends.

the global stiffness matrix to evaluate the deflection and rotation at the central node and the reaction forces and moments at both ends.

The stiffness matrices of the two elements are:

$$\boldsymbol{k}^{(1)} = \frac{EI}{l^3}\begin{array}{c} \begin{matrix} v_1 & \theta_1 & v_2 & \theta_2 \end{matrix} \\ \begin{bmatrix} 12 & 6l & -12 & 6l \\ 6l & 4l^2 & -6l & 2l^2 \\ -12 & -6l & 12 & -6l \\ 6l & 2l^2 & -6l & 4l^2 \end{bmatrix} \end{array}$$

$$\boldsymbol{k}^{(2)} = \frac{EI}{l^3}\begin{array}{c} \begin{matrix} v_2 & \theta_2 & v_3 & \theta_3 \end{matrix} \\ \begin{bmatrix} 12 & 6l & -12 & 6l \\ 6l & 4l^2 & -6l & 2l^2 \\ -12 & -6l & 12 & -6l \\ 6l & 2l^2 & -6l & 4l^2 \end{bmatrix} \end{array}$$

w and ϕ are the deflection and angle at each node, respectively.

As each node has 2 *DOFs*, each element has 4 *DOFs* (Figure 3.42).

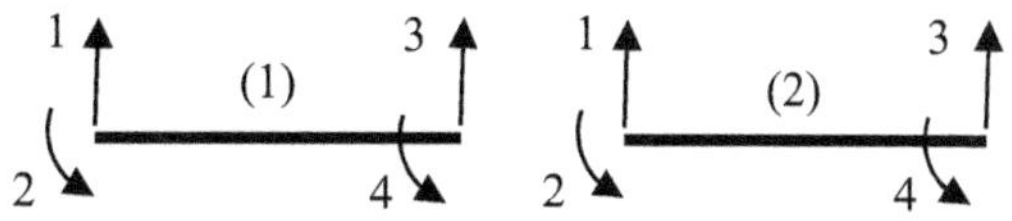

Figure 3.42 2 finite elements, each one with 4 *DOFs*.

The total number of *DOFs* of the beam (without any constraints) is 6 (Figure 3.43).

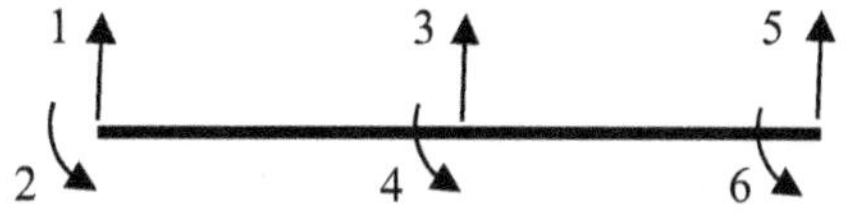

Figure 3.43 Complete beam, with 6 *DOFs*.

There must be compatibility between both elements, that share the *DOFs* corresponding to the global *DOFs* 3 and 4. Let the global stiffness matrix be $\boldsymbol{K}$ (a 6x6 matrix). Comparing Figures 3.42 and 3.43, one must state that:

$$K_{11}=k_{11}^{(1)}\;;\;K_{12}=k_{12}^{(1)}\;;\;K_{13}=k_{13}^{(1)}\;;\;K_{14}=k_{14}^{(1)}\;;\;K_{15}=0\;;\;K_{16}=0$$
$$K_{22}=k_{12}^{(1)}\;;\;K_{23}=k_{23}^{(1)}\;;\;K_{24}=k_{24}^{(1)}\;;\;K_{25}=0\;;\;K_{26}=0$$
$$K_{33}=k_{33}^{(1)}+k_{11}^{(2)}\;;\;K_{34}=k_{34}^{(1)}+k_{12}^{(2)}\;;\;K_{35}=k_{13}^{(2)}\;;\;K_{36}=k_{14}^{(2)}$$
$$K_{44}=k_{44}^{(1)}+k_{22}^{(2)}\;;\;K_{45}=k_{23}^{(2)}\;;\;K_{46}=k_{24}^{(2)}$$
$$K_{55}=0\;;\;K_{56}=0$$
$$K_{66}=0$$

The assembly of the elements is graphically illustrated in Figure 3.44. The *DOFs* 3 and 4 are v_2 and θ_2 of the global matrix:

$$\boldsymbol{Kd}=\frac{EI}{l^3}\begin{bmatrix} 12 & 6l & -12 & 6l & 0 & 0 \\ 6l & 4l^2 & -6l & 2l^2 & 0 & 0 \\ -12 & -6l & 24 & 0 & -12 & 6l \\ 6l & 2l^2 & 0 & 8l^2 & -6l & 2l^2 \\ 0 & 0 & -12 & -6l & 12 & -6l \\ 0 & 0 & 6l & 2l^2 & -6l & 4l^2 \end{bmatrix}\begin{Bmatrix} v_1 \\ \theta_1 \\ v_2 \\ \theta_2 \\ v_3 \\ \theta_3 \end{Bmatrix}=\begin{Bmatrix} F_1 \\ M_1 \\ F_2 \\ M_2 \\ F_3 \\ M_3 \end{Bmatrix} \tag{3.287}$$

(columns labelled $v_1\;\theta_1\;v_2\;\theta_2\;v_3\;\theta_3$)

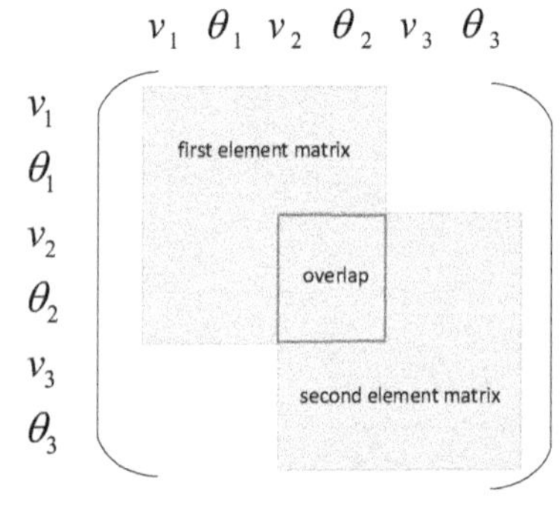

Figure 3.44 Element assembly.

The load and boundary conditions are:

$$F_2=-P, M_2=M, v_1=v_3=\theta_1=\theta_3=0$$

By solving equation (3.287), the results are:

$$\begin{Bmatrix} v_2 \\ \theta_2 \end{Bmatrix}=\frac{l}{24EI}\begin{Bmatrix} -Pl^2 \\ 3M \end{Bmatrix}$$

$$\begin{Bmatrix} F_1 \\ M_1 \\ F_3 \\ M_3 \end{Bmatrix} = \frac{EI}{l^3}\begin{bmatrix} -12 & 6l \\ -6l & 2l^2 \\ -12 & -6l \\ 6l & 2l^2 \end{bmatrix}\begin{Bmatrix} w_2 \\ \phi_2 \end{Bmatrix} = \frac{EI}{l^3}\begin{Bmatrix} 0.5P+0.75M/l \\ 0.25PL+0.25M \\ 0.5P-0.75M/l \\ -0.25PL+0.25M \end{Bmatrix}$$

In general, and as previously seen for the bars, beams may be inclined and thus, after defining them in their local coordinate system, it is necessary to transform them to a global coordinate system.

Assume that the local coordinate system is represented by x and y, and the global coordinate system is represented by $\bar{x}$ and $\bar{y}$, as shown in Figure 3.45.

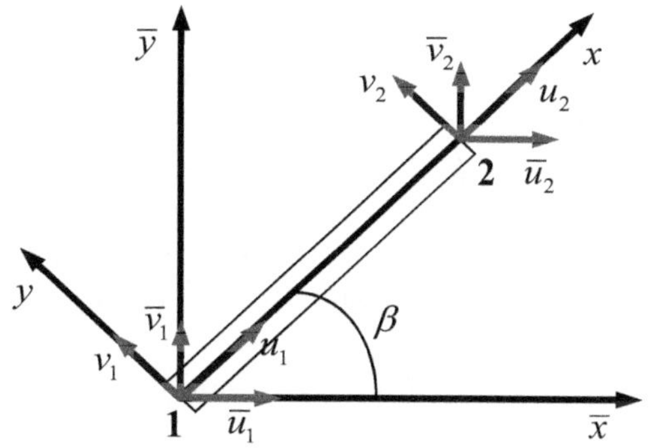

Figure 3.45 Global *versus* local coordinate systems.

The angular displacements (rotations) are equal in both coordinate systems, i.e. $\theta = \bar{\theta}$. Thus, the transformation between the local and the global coordinate systems is

$$\begin{Bmatrix} x \\ y \\ \theta \end{Bmatrix} = \begin{bmatrix} cos\beta & sin\beta & 0 \\ -sin\beta & cos\beta & 0 \\ 0 & 0 & 1 \end{bmatrix}\begin{Bmatrix} \bar{x} \\ \bar{y} \\ \bar{\theta} \end{Bmatrix} \tag{3.288}$$

Therefore, the transformation of the element node displacements between the local and the global coordinate systems is

$$\begin{Bmatrix} u_1 \\ v_1 \\ \theta_1 \\ u_2 \\ v_2 \\ \theta_2 \end{Bmatrix} = \begin{bmatrix} cos\beta & sin\beta & 0 & 0 & 0 & 0 \\ -sin\beta & cos\beta & 0 & 0 & 0 & 0 \\ 0 & 0 & 1 & 0 & 0 & 0 \\ 0 & 0 & 0 & cos\beta & sin\beta & 0 \\ 0 & 0 & 0 & -sin\beta & cos\beta & 0 \\ 0 & 0 & 0 & 0 & 0 & 1 \end{bmatrix}\begin{Bmatrix} \bar{u}_1 \\ \bar{v}_1 \\ \bar{\theta}_1 \\ \bar{u}_2 \\ \bar{v}_2 \\ \bar{\theta}_2 \end{Bmatrix} \tag{3.289}$$

or in a more compact form,

$$d^e = R^e \bar{d}^e \tag{3.290}$$

The new mass matrix, stiffness matrix and force vector of the element with respect to the global coordinate system are

$$\bar{M}^e = \left(R^e\right)^T m^e R^e \quad \bar{K}^e = \left(R^e\right)^T k^e R^e \quad \bar{F}^e = \left(R^e\right)^T f^e \tag{3.291}$$

After assembling all the beam elements, the equilibrium equation of the overall beam structure can be obtained as

$$\bar{M}\ddot{\bar{d}} + \bar{K}\bar{d} = \bar{F}(t) \tag{3.292}$$

Bibliography

- Houbolt, J. C., *A Recurrence Matrix Solution for the Dynamic Response of Elastic Aircraft*, Journal of the Aeronautical Sciences, 17(9), pp. 540-550, 1950.
- Clough, R. W., *The Finite Element Method in Plane Stress Analysis*, Proceedings of the 2nd ASCE Conference on Electronic Computation, Pittsburgh Pa., USA, Sept. 8-9, 1960.
- Wei, M., *Perturbation Theory for the Eckart-Young-Mirsky Theorem and the Constrained Total Least Squares Problem, in* Linear Algebra and its Applications, 280.2-3, pp. 267-287, 1998.
- Chen, X., Liu, Y., *Finite Element Modeling and Simulation with ANSYS Workbench*, CRC Press, 2018.
- Vega, J. M., Le Clainche, S., *Higher Order Dynamic Mode Decomposition and its Applications*, Chapter 2 - Higher Order Dynamic Mode Decomposition, Academic Press, Elsevier Inc., London, UK, pp. 29-83, 2021.

Chapter 4 Linear System Identification

4.1 Introduction

Modal tests are performed to determine the modal parameters - natural frequencies, damping ratios, mode shapes and modal masses/amplitudes - of a structure from the measurement and analysis of experimental vibration data. Modal testing [4.1, 4.2] is undertaken for a number of different reasons; often there is a need to validate computational or mathematical models of a structure before entry into service, or vibration problems that need trouble shooting, or there might be a requirement to optimise the dynamic behaviour of a system.

The process is usually performed in four stages, as shown in Figures 4.1 and 4.2, (i) determining the Frequency Response Functions (*FRFs*) (itself an identification process) from measured time signals, identifying (sometimes called estimating) the modal parameters – often undertaken in two stages, (ii) frequencies and damping ratios and then, (iii) the mode shapes and modal amplitudes or masses. Finally, (iv) the estimated model is validated by comparing with the experimental measurements. Care should be taken to confirm the quality of the entire process, particularly confirming that the data are of a standard suitable to use system identification methods (poor data can only lead to inferior estimates).

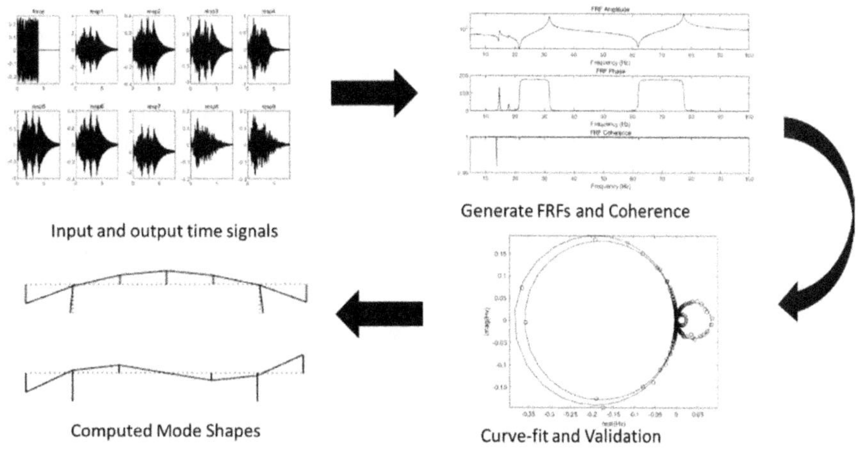

Figure 4.1 Typical Modal Testing and Analysis Process.

4.1.1 Outline of the Chapter

This chapter will consider approaches that are used in the second and third stages of the above process, usually referred to as System Identification or Modal Parameter Estimation methods, and will consider possible options, depending upon what data are available and the required outcomes from the analysis. These outcomes could range from simply determining the natural frequencies of a fan, through to performing a full modal test of an aircraft to identify many of

Structural Dynamics in Engineering Design, First Edition, Nuno M. M. Maia, Dario Di Maio, Alex Carrella, Francesco Marulo, Chaoping Zang, Jonathan E. Cooper, Keith Worden, and Tiago A. N. Silva.

the natural frequencies, damping ratios, mode shapes and modal contributions so that the Finite Element model can be validated.

The underlying mathematical models and analysis for a range of different system identification methods are discussed and their use illustrated, highlighting the different approaches that are available, their pros and cons, and also how to get the best results from them. Application of the techniques will be demonstrated using a set of experimental *FRFs*.

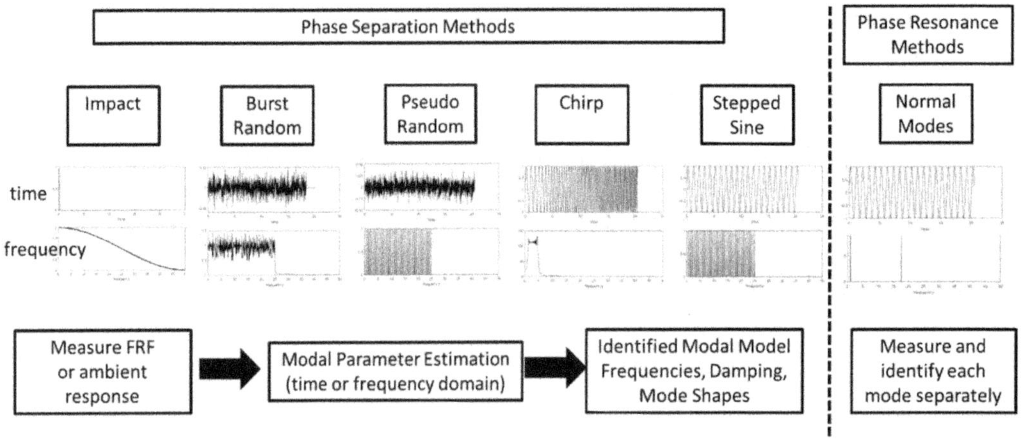

Figure 4.2 Types of excitation and analysis method.

4.1.2 Assumptions about the Measured Data

The system identification methods described in the following sections are formulated with the assumptions that the set of *FRFs*, or transient / ambient time responses, have been acquired with the following conditions satisfied:

- Linearity – there are no nonlinearities in the structure being tested – often requires that a low excitation level is used so that the deflections are small. Confirmation of these requirements can be found by ensuring that the homogeneity (*FRFs* from different excitation amplitudes should overlay) and reciprocity hold;

- Stationarity – the modal parameters of the structure do not change throughout the duration of the test – e.g. an aircraft rapidly changing speed or altitude, or a rocket rapidly burning fuel, would break this assumption;

- Controllability – all the modes of interest have been adequately excited – this can depend on where on the structure the excitation is applied and whether it has sufficient frequency content and amplitude;

- Observability – the positioning of the transducers (accelerometers, laser vibrometers, displacement probes, strain gauges) is such that the motion of all modes of interest are adequately measured;

- Noise – corruption (noise) of the data can arise from a number of sources such as poor instrumentation, inadequate excitation signals or test procedures, or a noisy environment, etc. It does not matter how sophisticated your identification analysis is, very noisy data will result in poor results, so it is inherently assumed that the signal-to-noise ratio is high. Of course, if you get things wrong with your analysis, having good data does not necessarily mean that good results will be found.

- Ambient data – a further assumption employed when the input is not measured is that the spectrum of the input is "*white*", i.e. the frequency content of the excitation is "*flat*" over the frequency range of interest, and that there is sufficient energy to excite the modes of interest;

If any of the above assumptions are not satisfied, then it is likely that inferior parameter estimates will be obtained.

The data may be characterised into various categories depending upon the test set-up, which excitation method has been used and how many points on the structure are measured:

- Single-Input Single-Output (*SISO*) - the response at a single point on the structure due to the excitation at a different point. If these two points are concurrent, then this is known as a "driving point";

- Single-Input Multiple-Output (*SIMO*) – simultaneous measurement at many points on a structure subjected to a particular excitation, e.g. hammer, shaker or ambient excitation (this final case is where the input is unknown, e.g. wind excitation of a bridge is considered as a single input);

- Multiple-Input Single-Output (*MISO*) – often occurs for "roving hammer" excitation where different input points are used to excite responses at many measurement stations. Through application of the principle of reciprocity, *SIMO* and *MISO* data sets can be interchanged;

- Multiple Input Multiple-Output (*MIMO*) – Simultaneous excitation at several points on the structure and measurement at many others – typically in an aircraft Ground Vibration Test (*GVT*) 4 – 6 shakers will be used simultaneously to excite the structure (e.g. using uncorrelated random excitation) and many 100s of accelerometers used to measure the response to the these inputs.

All system identification methods can be used on all of the above data sets, but the more sophisticated approaches are better suited to deal with large *MIMO* data sets or when greater accuracy is required.

4.1.3 Categories of System Identification Methods

It is convenient to categorise the various system identification methods in terms of the underlying philosophy and theoretical basis, the assumptions that are made, and the complexity of the estimation process.

Phase Separation *versus* Phase Resonance Methods

There are two main approaches to performing modal tests, as shown in Figure 4.2. The original techniques developed in the 1960s, known as *phase resonance, force appropriation* or *normal mode testing methods,* employ multiple shakers operated at single frequencies to isolate and excite individual modes via optimised combinations of amplitude and phase for each shaker. These methods arguably give the most accurate modal estimates, but are much more difficult to implement. They will be discussed in more detail in Section 4.6.

The 1970s saw the development of the more common *phase separation* identification methods. applied directly to data generated via broad frequency range excitation methods, e.g. random, swept sine, chirp or ambient excitation techniques as shown in Figure 4.2. The analyses employed for these techniques can be formulated in either the time or frequency domains.

The different excitation strategies for phase separation and phase resonance methods are illustrated in Figure 4.3.

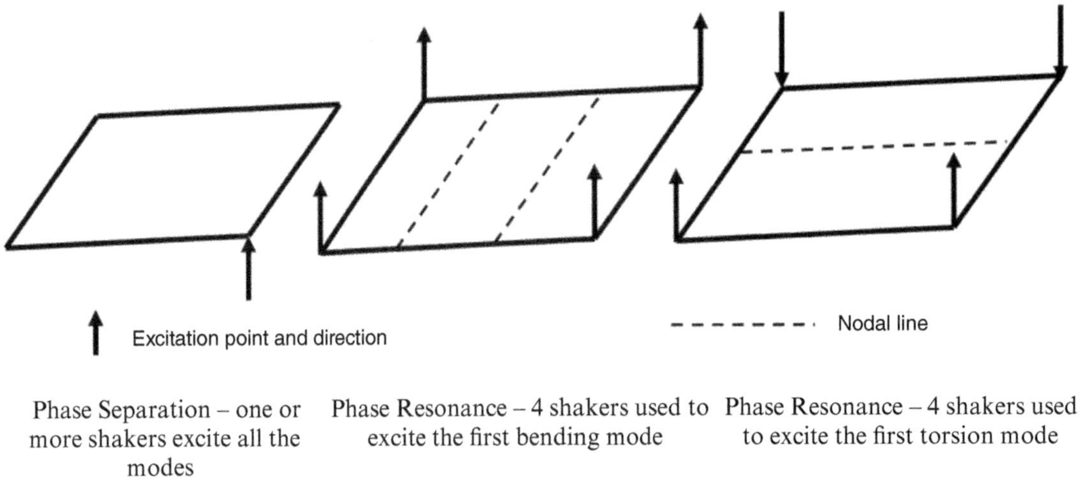

Figure 4.3 Phase Separation and Phase Resonance excitation approaches.

Phase Separation Methods

It is convenient to have a range of analysis tools available depending upon the desired goal of the investigation and whether a quick or more detailed / accurate analysis is required.

Single versus Multiple Degree of Freedom Methods

- Single Degree of Freedom (*SDOF*) methods are used with the assumption that each mode can be analysed separately and that there is no interference from other modes. They are only used for *SISO* data sets, which could be a subset of a *MIMO* data set, known as a *local* estimate. They are fast and simple to use, but usually less accurate than *MDOF* approaches; however, if the modes are well-separated (i.e. they do not overlap each other to any great degree – half-power points being visible for all peaks is a reasonable rule of thumb to follow) then they can be a good way to quickly determine natural frequencies and give a reasonable estimate of the damping levels and mode shapes for *SIMO* data sets. These methods are unable to identify repeated modes (those with the same natural frequency which might occur in symmetrical structures) and give poor estimates for close modes.

- Multiple Degree of Freedom (*MDOF*) methods are implemented by curve-fitting all modes at once, so there is no requirement for the modes not to interfere with each other. They have much more complicated formulations than *SDOF* methods and require additional interpretation of the results. An advantage of such an approach is that *SIMO* and *MIMO* data sets can be dealt with concurrently, including repeated modes for *MIMO* data sets, producing what are known as *global* estimates.

Figure 4.4 illustrates the different analysis approaches for using *SDOF* and *MDOF* methods.

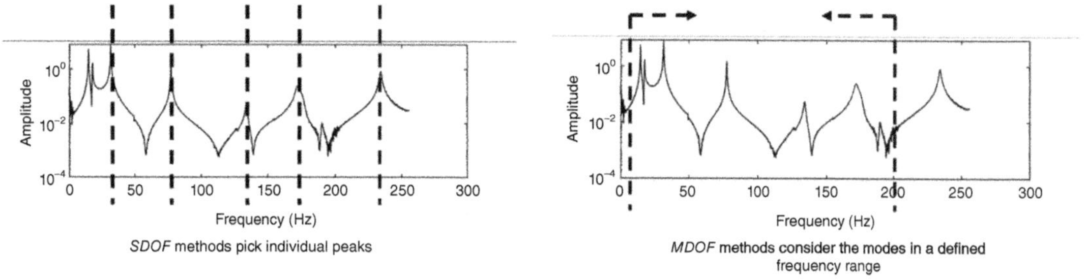

Figure 4.4 Differences between *SDOF* and *MDOF* methods.

4.1.4 Mathematical Models

All system identification methods are implemented with the assumption that the input / output relationship of the test data is formulated as a mathematical model obeying the assumptions described above. All curve-fitting approaches are used to identify the parameters that make up the assumed model. Typically, the models are based in either the frequency or time domains.

Frequency Domain Models

The basis of the frequency models are Frequency Response Functions (*FRFs*) which relate the input and output at a given harmonic frequency. A single *FRF* between force f_k at point *k* and response y_l at point *l* on a structure usually takes the form (for displacement data) for *N* modes as

$$H_{kl}(\omega)=\frac{y(\omega)}{f(\omega)}=\sum_{r=1}^{N}\left(\frac{R_{kl_r}}{i\omega-\lambda_r}+\frac{R^*_{kl_r}}{i\omega-\lambda^*_r}\right) = \sum_{r=1}^{N}\frac{1}{m_r\left(\omega_r^2-\omega^2+i2\xi_r\omega_r\omega\right)} \tag{4.1}$$

where $\lambda_r=-\xi_r\omega_r\pm i\omega_r\sqrt{1-\xi_r^2}$ is the *pole* and $R_r=\dfrac{1}{i2m_r\omega_{d_r}}$ is the *residue*, with $\omega_{dr}=\omega_r\sqrt{1-\xi_r^2}$ the damped natural frequency.

For *MIMO* data (*NI* input stations and *NO* measurement stations), the model in (4.1) takes a similar structure, but is now formulated in a matrix form such that

$$\boldsymbol{H}(\omega)=\sum_{r=1}^{N}\left(\frac{\boldsymbol{R}_r}{i\omega-\lambda_r}+\frac{\boldsymbol{R}^*_r}{i\omega-\lambda^*_r}\right) = \sum_{r=1}^{N}\frac{\boldsymbol{A}_r+i\omega\boldsymbol{B}_r}{\left(\omega_r^2-\omega^2+i2\xi_r\omega_r\omega\right)} \tag{4.2}$$

where for each frequency ω, $\boldsymbol{H}$ and $\boldsymbol{R}_r$ are complex $(NO\times NI)$ matrices.

Another well-known form of the *FRF* is written as

$$\boldsymbol{H}(\omega)=\sum_{r=1}^{N}\left(\frac{Q_r\psi_r\psi_r^T}{i\omega-\lambda_r}+\frac{Q_r^*\psi_r^*\psi_r^H}{i\omega-\lambda_r^*}\right) = \sum_{r=1}^{N}\left(\frac{\psi_r\boldsymbol{L}_r^T}{i\omega-\lambda_r}+\frac{\psi_r^*\boldsymbol{L}_r^H}{i\omega-\lambda_r^*}\right) \tag{4.3}$$

where ψ are the mode shapes and $\boldsymbol{L}_r = \mathrm{Q}_r\psi_r$ is the modal participation factor for the rth mode, a measure of how well a force at a particular excitation point is able to excite the structure, and Q_r is a scaling factor. The superscript ^{H} stands for Hermitian transpose.

It is often useful to perform an analysis within a particular frequency range and, consequently, the effects of all of the modes above and below this range, known as *residuals*, need to be accounted for; otherwise, the parameter estimates closest to the boundaries will be very likely to contain errors due to the "tails" from the out-of-range terms, as seen in Figure 4.5 for a 3 *DOF* system. Consider Eq. 4.1 for displacement data (*receptance*), then it can be found that the *FRF* for each individual mode tends to a *constant value* for $\omega << \omega_r$ and *drops away towards zero* for $\omega >> \omega_r$ as seen in Fig. 4.5 (a); this behaviour reverses for an *accelerance FRF*.

Consequently, the lower (stiffness) and upper (mass) residuals can be added to the *FRF* model (displacement data), shown in Fig. 4.5 (b), such that

$$H_{kl}(\omega)=\sum_{r=1}^{N}\left(\frac{R_{kl_r}}{i\omega-\lambda_r}+\frac{R_{kl_r}^*}{i\omega-\lambda_r^*}\right) + UR_{kl} - \frac{LR_{kl}}{\omega^2} \tag{4.4}$$

The form of the residuals changes depending upon whether displacement, velocity or acceleration measurements are being made. To perform an accurate identification these out-of-range residuals must also be curve-fitted.

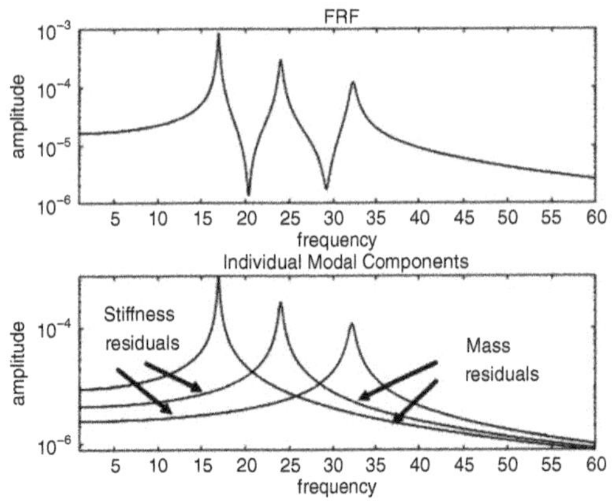

Figure 4.5 (a) Stiffness and Mass residual tails from individual modes.

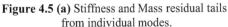

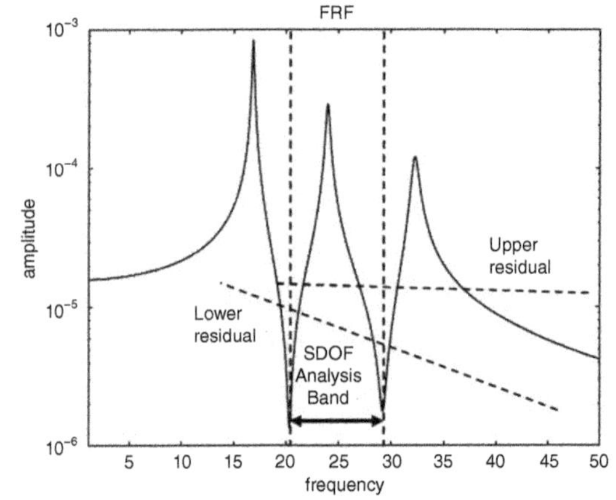

Figure 4.5 (b) Effect of the residuals on a Single Mode.

Knowing that for the *MIMO* case the input $\boldsymbol{F}$ and output $\boldsymbol{Y}$ are related by the expression $\boldsymbol{HF}=\boldsymbol{Y}$, one can define the power spectra [4.3] of the input $\boldsymbol{S}_{FF}$ ($NI\times NI$) and response $\boldsymbol{S}_{YY}$ ($NO\times NO$) as[1]

[1] See also Chapter 1, under "Response to a random excitation".

$$S_{FF}(\omega)=F(\omega)F^{H}(\omega) \quad \text{and} \quad S_{YY}(\omega)=Y(\omega)Y^{H}(\omega) \tag{4.5}$$

where the superscript ^{H} stands for Hermitian transpose.

It can then be shown that

$$S_{YY}=HS_{FF}H^{H} \tag{4.6}$$

For ambient excitation cases, where there is no measured input and the input is assumed to be "white", so S_{FF} is constant, combining (4.3) and (4.6) leads to

$$S_{YY}(\omega)=\sum_{r=1}^{N}\left(\frac{d_r\psi_r L_r^T}{i\omega-\lambda_r}+\frac{d_r^*\psi_r^* L_r^H}{i\omega-\lambda_r^*}+\frac{d_r\psi_r L_r^T}{-i\omega-\lambda_r}+\frac{d_r^*\psi_r^* L_r^H}{-i\omega-\lambda_r^*}\right) \tag{4.7}$$

so apart from the scaling factors d_r, the same information about the poles and the mode shapes is available as for the *FRFs*.

Multivariate Mode Indicator Function

Mode Indicator Functions can be used to determine how many modes there are in a system, to complement what can be deduced from inspection of the *FRFs*. They can be used to determine repeated roots, provided that there are enough excitation points, and are particular useful when there is a high modal density. They also have important consequences for multi shaker force appropriation, i.e. phase resonance methods, which will be considered later in this chapter.

The Multivariate Mode Indicator Function (*MMIF*) [4.4] is perhaps the best-known approach and can be computed from the *FRFs* as follows. Consider the standard relationship between the input $\boldsymbol{F}$ and output $\boldsymbol{Y}$ at each frequency via the *FRF* $\boldsymbol{H}$; then, writing in terms of real and imaginary parts, gives

$$H(\omega)F(\omega)=Y(\omega) \quad \rightarrow \quad HF=Y \quad \rightarrow \quad H_R F+iH_I F=Y_R+iY_I \tag{4.8}$$

where in a matrix format $\boldsymbol{H}$, for each frequency ω, has dimensions $NO\times NI$, $\boldsymbol{F}$ is $NI\times 1$, and $\boldsymbol{Y}$ is $NO\times 1$, with the subscripts $_R$ and $_I$ standing for real and imaginary, respectively.

For a normal mode to be excited at a particular frequency, the force vector must be such that the real part of the response is as small as possible compared to the total response (see Section 4.6 on *Phase Resonance* methods later on for more details).

Defining the 2 norm of the real and total responses as

$$\|Y_R\|^2=Y_R^{\mathrm{T}}MY_R \quad \text{and} \quad \|Y_R+\mathrm{i}Y_I\|^2=Y_R^{\mathrm{T}}MY_R+Y_I^{\mathrm{T}}MY_I \tag{4.9}$$

where $\boldsymbol{M}$ is the mass matrix (or some estimate of the local mass distribution – if this is not available then simply using the identity matrix is a reasonable alternative) - then the *MMIF* can be found, after some algebra, by minimising the function

$$\alpha=\frac{\|Y_R\|^2}{\|Y_R+iY_I\|^2}=\frac{F^T AF}{F^T(A+B)F} \tag{4.10}$$

w.r.t. $\boldsymbol{F}$ where $\boldsymbol{A} = \boldsymbol{H}_R^T \boldsymbol{M} \boldsymbol{H}_R$, $\boldsymbol{B} = \boldsymbol{H}_I^T \boldsymbol{M} \boldsymbol{H}_I$ and $\boldsymbol{A}$ and $\boldsymbol{B}$ are both $NI \times NI$ matrices.

This optimisation problem, which is equivalent to minimising the size of the real part divided by the total function, can be shown to be the same as finding, for measured $\boldsymbol{A}$ and $\boldsymbol{B}$ terms, the smallest eigenvalue λ of the generalised eigenproblem

$$\boldsymbol{A}\boldsymbol{F} = \left(\boldsymbol{A} + \boldsymbol{B}\right)\boldsymbol{F}\lambda \tag{4.11}$$

where the *MMIFs* are the *NI* eigenvalues computed at every frequency value. By plotting the *MMIF* eigenvalues *versus* frequency, as shown in Figure 4.6, drops in the smallest values corresponds to a mode. Should more than one of the *MMIF* eigenvalues tend to zero at a certain frequency, then this indicates that there are coincident, or very close, modes. Here, only one of the *MMIF* eigenvalues drops to zero at each resonance, indicating a single mode.

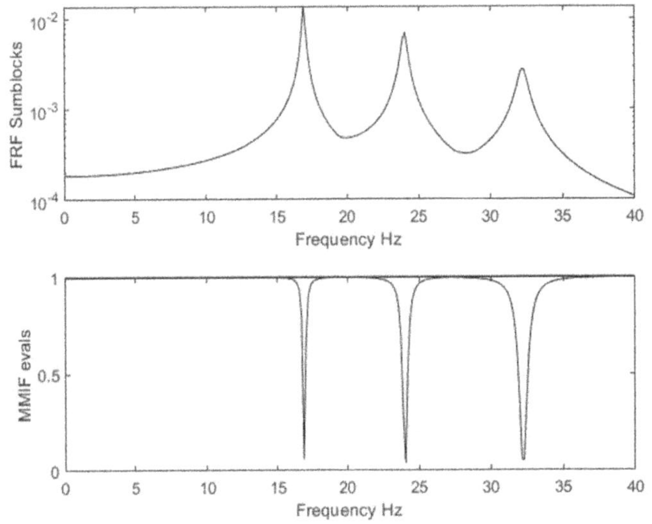

Figure 4.6 Summed *FRFs* and corresponding *MMIF* eigenvalues.

Complex Mode Indicator Function

An alternative indicator function, the Complex Mode Indicator Function (*CMIF*) [4.5] follows from Eq. (4.3), reproduced here:

$$\boldsymbol{H}(\omega) = \sum_{r=1}^{N}\left(\frac{Q_r \psi_r \psi_r^T}{i\omega - \lambda_r} + \frac{Q_r^* \psi_r^* \psi_r^H}{i\omega - \lambda_r^*}\right) = \sum_{r=1}^{N}\left(\frac{\psi_r \boldsymbol{L}_r^T}{i\omega - \lambda_r} + \frac{\psi_r^* \boldsymbol{L}_r^H}{i\omega - \lambda_r^*}\right) \tag{4.12}$$

which in matrix form becomes, for each frequency,

$$\boldsymbol{H}(\omega) = \boldsymbol{\Phi}\, diag\left(\frac{A_r}{i\omega - \lambda_r}\right)\boldsymbol{L} \tag{4.13}$$

where $\boldsymbol{\Phi}$ is a $NO \times 2N$ matrix of mass normalised mode shapes, $\boldsymbol{L}$ is a $2N \times NO$ matrix of modal participation factors, with A_r and λ_r being a scaling factor and pole of each mode, respectively.

Now, if the compact (often called "economy" or "skinny") *SVD* [4.6], as described in Chapter 3, is taken for the *FRF* matrix at every frequency, one obtains

$$\boldsymbol{H}(\omega) = \boldsymbol{U\Sigma V}^H \tag{4.14}$$

where the dimensions of $\boldsymbol{U}, \boldsymbol{\Sigma}, \boldsymbol{V}$ are $NO \times NI$, $NI \times NI$ and $NI \times NI$, respectively (assuming that $NO > NI$, which is usually the case) and these terms can be related to Eq. (4.13). The singular values of $\boldsymbol{H}$ are called the *CMIFs*[2]. Figure 4.7 shows how a peak in the *CMIFs* corresponds to each of the modes on the *FRFs*. Care should be taken with the interpretation of the *CMIFs* as they are not scaled, unlike the *MMIFs*. Considering the middle plot, the apparent peaks in the second *CMIF* (at 20Hz and 28Hz) and third *CMIF* (at 24Hz) (dashed and dotted lines respectively) are due to the cross over between the different mode traces and consequently do not correspond to modes. It is usual to define a "tracked" *CMIF* which follows the response from each mode (tails for each trace in the "tracked" plot follow the respective peak), as seen in the bottom plot. It is only when there is more than one peak at a particular frequency that coincident modes are indicated.

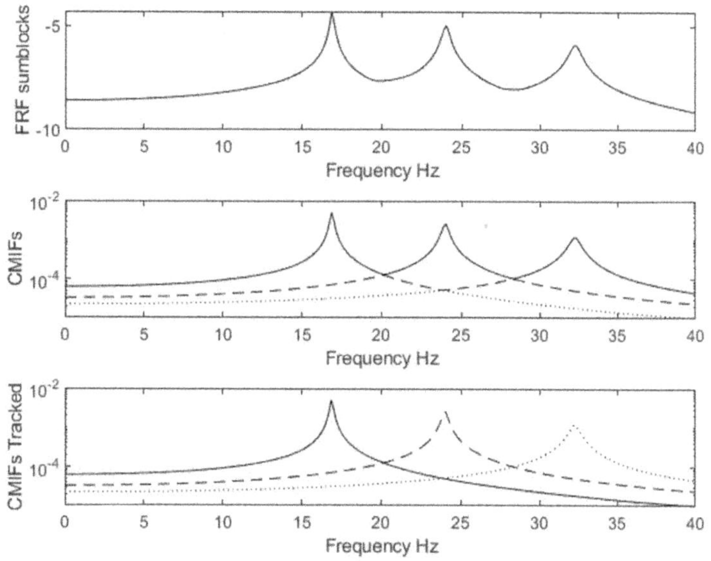

Figure 4.7 Sum of *FRFs* and corresponding *CMIF* singular values (untracked and tracked).

Relationship between the Time and Frequency Domains

The Fast Fourier Transform (*FFT*) was developed in the late 1960s [4.7, 4.8] and provided the ability to quickly transform data between the time and frequency domains, and in the opposite direction using its inverse. Its implementation in modal testing is illustrated in Figure 4.8. As well as being the fundamental tool for formulating the *FRFs*, the duality between the *FRF* and the Impulse Response Function (*IRF*) is extremely useful and enables analysis in both the time and frequency domains[3].

---

[2] In some applications the *CMIF* is defined as the eigenvalues of $\boldsymbol{H}^H\boldsymbol{H}$.

[3] See Chapter 1, under "*Relation between the FRF and the IRF*"

Care must be taken in using the *FFT* as its formulation assumes that the time signals considered are a repeating sample of an infinite periodic signal. The most practical way to ensure that this condition is met is (i) to either window the data [4.8], e.g. Hanning, (same window applied to both the input and the output, except for hammer testing, where one uses a transient window for the force and an exponential window for the response), or (ii) to use excitation signals where the input is stopped some time before the end of the data acquisition time window (e.g. a burst random input signal) so that the start and end of the input and output signals are zero. Such an approach ensures that the periodicity assumption is met and that errors (known as "leakage") do not occur [4.8, 4.9]. In Chapter 2, Sections 2.3 and 2.4, the subject of "windowing" is discussed.

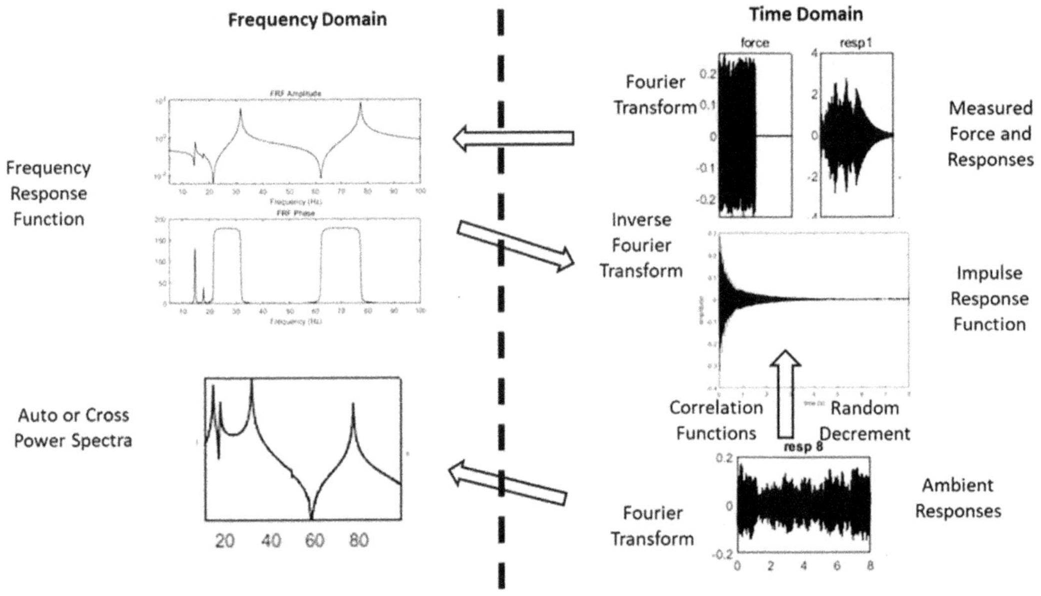

Figure 4.8 Relationship between the Time and Frequency Domain data.

Time Domain Models

Considering the generic system shown in Figure 4.9, the typical formation for the relationship between the force (input) $\boldsymbol{f}$ (t) and the response $\boldsymbol{y}$ (t) in the time domain is the well-known 2nd order equation [4.10]

$$\boldsymbol{M}\ddot{\boldsymbol{y}} + \boldsymbol{C}\dot{\boldsymbol{y}} + \boldsymbol{K}\boldsymbol{y} = \boldsymbol{f} \tag{4.15}$$

where for an N degree of freedom system, $\boldsymbol{M}$, $\boldsymbol{C}$ and $\boldsymbol{K}$ are $N \times N$ mass , damping and stiffness matrices, respectively, and $\boldsymbol{f}$ and $\boldsymbol{y}$ are force and displacement vectors.

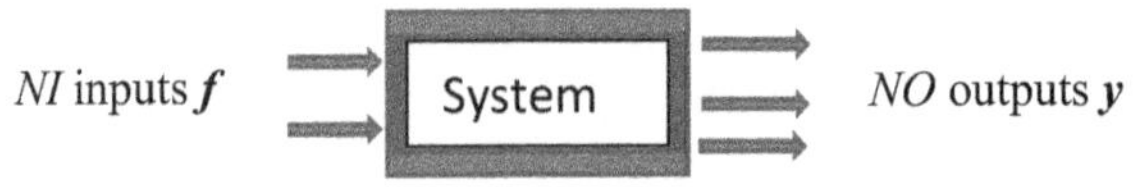

Figure 4.9 Schematic of relationship between inputs (shakers) and outputs (accelererometers).

Moving into a discrete time formulation, the relationship between a single input $f_j = f(j\Delta t)$, at time instant j, and single output y_j positions for an *MDOF* system can be represented as an Autoregressive Moving Average with eXhogneous inputs (*ARMAX*) model [4.11] such that

$$y_j + a_1 y_{j-1} + \dots + a_{2N} y_{j-2N} = b_1 f_{j-1} + \dots + b_{2N} f_{j-2N} + \varepsilon_j, \quad j = 2N+1, \dots npts \qquad (4.16)$$

where the a_i and b_i coefficients represent the system and ε_j is the residual error. Making the determinant of the homogeneous equation equal to zero, one can calculate the roots of the resulting characteristic equation, based upon the a_i terms,

$$v_r^{2N} + a_1 v_r^{2N-1} + \dots + a_{2N-1} v_r + a_{2N} = 0 \qquad (4.17)$$

where

$$v_r = e^{\lambda_r \Delta t}, \; r = 1, \dots 2N \text{ with } \lambda_r = -\xi_r \omega_r \pm i\omega_r \sqrt{1-\xi_r^2} \qquad (4.18)$$

leading to the system natural frequencies and damping ratios.

If the input signal is "white", then the *ARMAX* model can be rewritten as the Autoregressive Moving Average (*ARMA*) model [4.11]

$$y_j + a_1 y_{j-1} + \dots + a_{2N} y_{j-2N} = b_1 \varepsilon_{j-1} + \dots + b_{2N} \varepsilon_{j-2N} + \varepsilon_j, \quad j = 2N+1, \dots npts \qquad (4.19)$$

The *ARMA* and *ARMAX* model can be extended into a vector or matrix format so that *SIMO* and *MIMO* data can be accommodated.

By far the most common way to use time domain methods is to consider the response to an impulse where the time history is a summation of exponentially damped sine waves in the form

$$\begin{aligned} y(t) &= \sum_{r=1}^{N} A_r e^{-\xi_r \omega_r t} sin(\omega_{dr} t + \phi_r) \\ &= \sum_{r=1}^{N} A_r e^{(-\xi_r \omega_r + i\omega_{dr})t} + \sum_{r=N+1}^{2N} A_r^* e^{(-\xi_r \omega_r - i\omega_{dr})t} = \sum_{r=1}^{N} A_r e^{\lambda_r t} + \sum_{r=N+1}^{2N} A_r^* e^{\lambda_r^* t} \end{aligned} \qquad (4.20)$$

This response can be obtained in a number of ways:

1. apply an impulse or step-release to the system;
2. the inverse Fourier Transform of the *FRF* is the Impulse Response Function;
3. use the response to an unmeasured random input to generate an Impulse Response Function [4.12, 4.13].

The *ARMA* for a *SISO* model can be written for an impulse response by removing the input terms so that the so-called Autoregressive model (*AR*) is obtained as

$$y_j + a_1 y_{j-1} + \dots + a_{2N} y_{j-2N} = 0, \quad j = 2N+1, \dots npts \qquad (4.21)$$

where the a_i terms are unknown coefficients.

The solution of each of Eqs. (4.21) can be shown to take the form

$$y_j = c_1 e^{j\lambda_1} + c_2 e^{j\lambda_2} + \; \dots \; + c_{2N} e^{j\lambda_{2N}}$$
$$= \{c_1 \;\; c_2 \;\; \dots \;\; c_{2N}\} \begin{bmatrix} e^{\lambda_1} & 0 & & 0 \\ 0 & e^{\lambda_2} & & \\ & & \ddots & \\ 0 & & & e^{\lambda_{2N}} \end{bmatrix}^j \begin{Bmatrix} 1 \\ 1 \\ \vdots \\ 1 \end{Bmatrix} = \mathbf{c}^T e^{\Lambda^j} \mathbf{I}_{2N\times1} \tag{4.22}$$

where the c_i terms are constants that need to be computed to determine the amplitude of each mode; for an oscillating mechanical system the poles and constant terms occur in complex conjugate pairs.

For the *SIMO* case the responses become part of a (*NO* x *1*) vector multiplied by the same scalar autoregressive coefficients such that

$$\mathbf{y}_j + \mathbf{y}_{j-1} a_1 + \; \dots \; + \mathbf{y}_{j-2N} a_{2N} = \mathbf{0} \;\; , \; j = 2N+1, \; \dots npts \tag{4.23}$$

and for *MIMO* experiments (multiple input cases) the *AR* equation becomes

$$\mathbf{Y}_j + \mathbf{Y}_{j-1} \mathbf{A}_1 + \; \dots \; + \mathbf{Y}_{j-2N} \mathbf{A}_{2N} = \mathbf{0} \;\; , \; j = 2N+1, \; \dots npts \tag{4.24}$$

where $\mathbf{Y}_j$ is now a ($NO \times NI$) matrix and the autoregressive constants $\mathbf{A}_i$ are ($NI \times NI$) block elements.

For the *MIMO* case, Eq. (4.22) then takes the form for the jth time instant as

$$\mathbf{Y}_{j+1} = \mathbf{C} e^{\Lambda^j} \mathbf{B} \tag{4.25}$$

where $\Lambda = diag[\lambda_1, \lambda_2, \dots, \lambda_s]$ and $\mathbf{C}$ and $\mathbf{B}$ are ($NO \times s$) and ($s \times NI$) matrices respectively, where s is an integer such that $s \times NI = 2N$. Eq. (4.25) is equivalent to the well-known Markov formulation [4.14]

$$\mathbf{Y}_{j+1} = \mathbf{C}\mathbf{A}^j \mathbf{B} = \mathbf{C}\boldsymbol{\Psi}^T e^{\Lambda^j} \boldsymbol{\Psi}\mathbf{B} \tag{4.26}$$

where $\mathbf{A}$ is the system state matrix, $\mathbf{B}$ and $\mathbf{C}$ are the standard input and output state-space matrices, Λ is the diagonal matrix of the system poles and $\boldsymbol{\Psi}$ is a matrix containing the mode shapes.

4.1.5 Example: the Wing-Pylon Model

The system identification methods will be illustrated using *FRFs* generated using the structure shown in Figure 4.10, which is a very simplified and idealised aircraft Wing-Pylon model [4.15]. Although the structure has been designed to explore the effect of non-linearities, the test data that is considered used a very low level of excitation and consequently can be considered as linear. Two shakers were applied at stations 3 and 5 (see Figure 4.11) and accelerometers were positioned at stations 1 – 9 so in total 18 *FRFs* were measured. 20 averages using 50% burst random (random input for 50% of the time window and then no input for the rest of the window) excitation were used for an 8 seconds time window with a bandwidth of 256 Hz, as shown in Figure 4.12, leading to *FRFs* with a frequency resolution of 0.125 Hz, as seen in Figures 4.13.

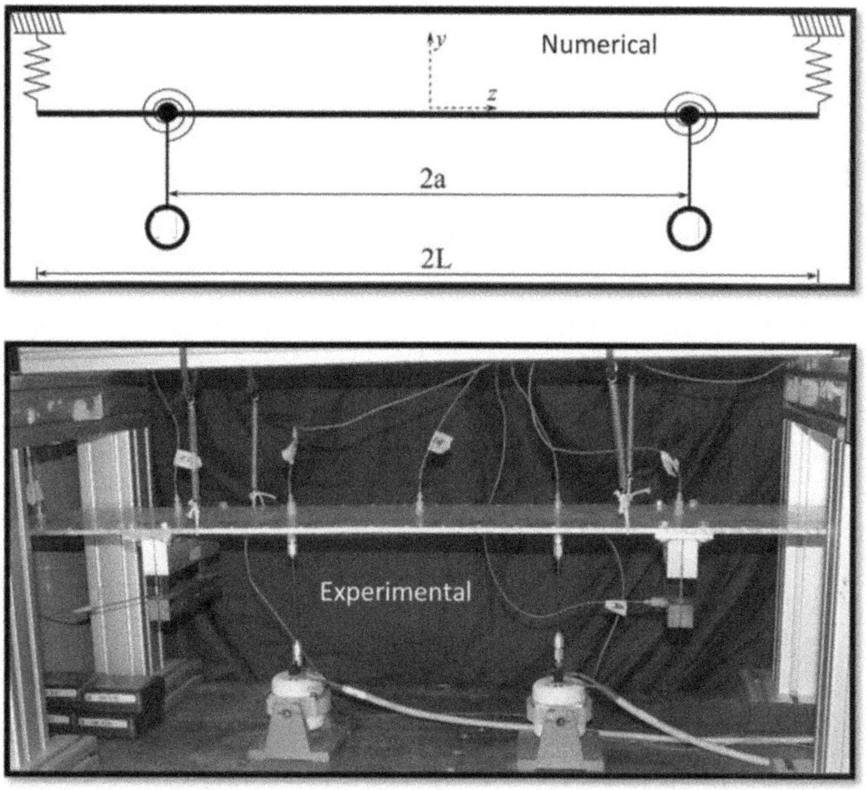

Figure 4.10 Wing-Pylon test structure.

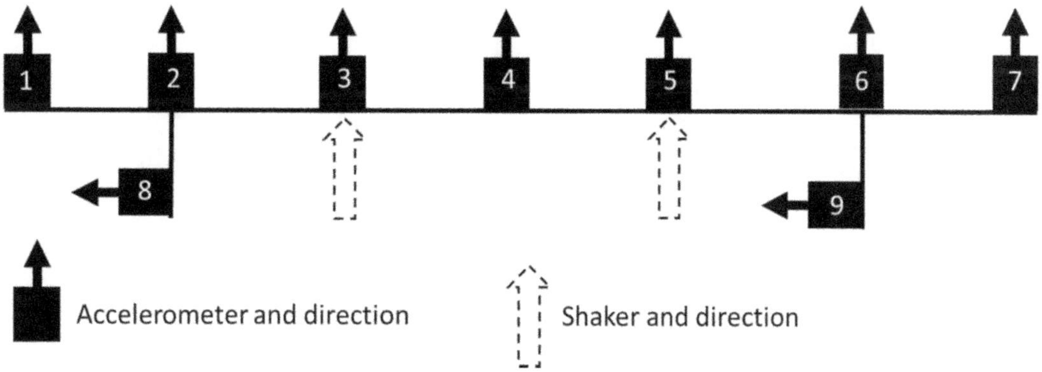

Figure 4.11 Test Stations - Excitations at positions 3 and 5, Accelerometers at positions 1 – 9.

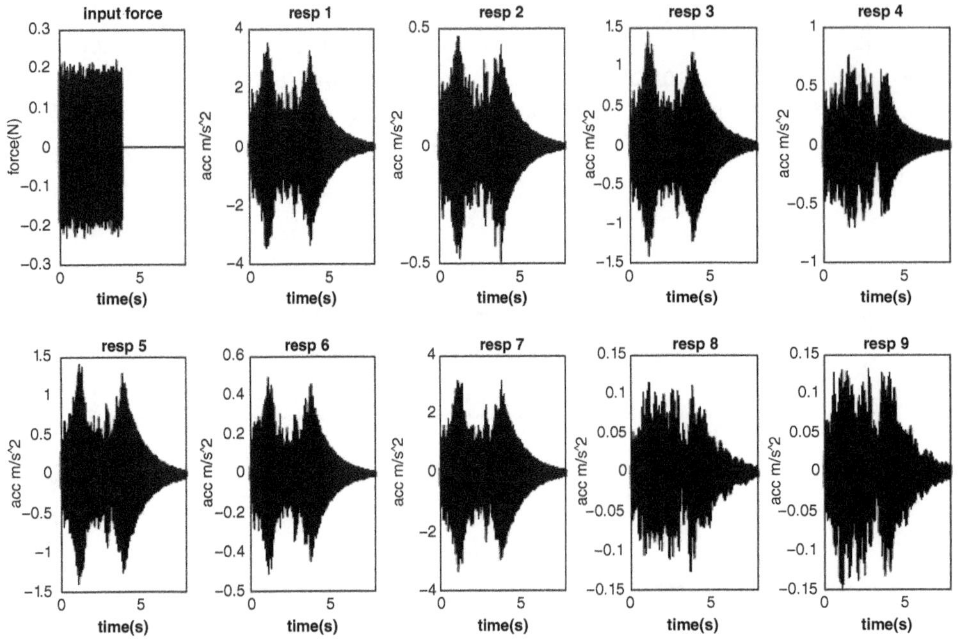

Figure 4.12 Sample force input and acceleration responses.

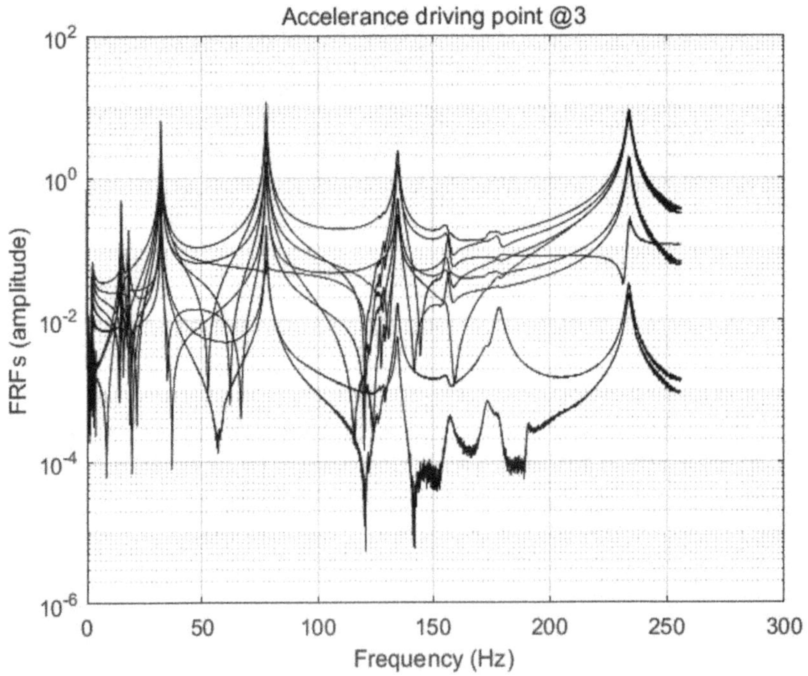

Figure 4.13 (a) Measured *FRFs* for reference 3.

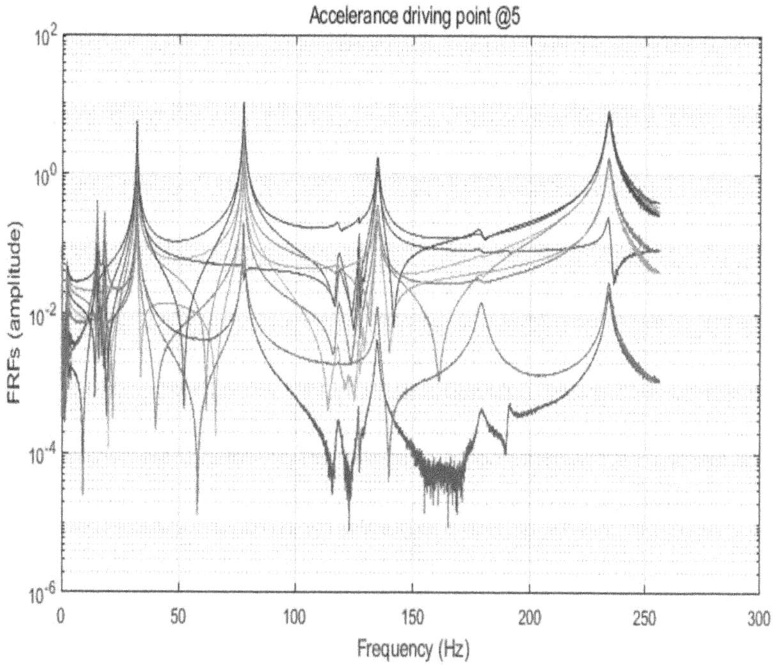

Figure 4.13 (b) Measured *FRFs* for reference 5.

Initial Interpretation of the *FRFs*

Even before starting to perform system identification on the *FRFs* one can get an idea of what results should be obtained, and how good are the data sets that are being used.

Consider the *driving point* (input #3, output #3) *FRF* in Figure 4.14 and the corresponding coherence plot. It can be seen from the *FRF* peaks that there are 3 dominant modes in the frequency range of 0 → 100Hz at approximately 15, 32 and 78 Hz which can be considered to behave as individual single degree of freedom systems. These peaks correspond to rapid changes of phase with a value at resonance of ~ +90 degrees. The coherence at these modes is almost one, indicating a good quality measurement. The dips in the coherence correspond to the anti-resonances, which is fine as they are not of interest for the modal estimates.

There is potentially another mode around 18 Hz, but it can be seen that it is small compared to the mode just below it and there are no clear half-power points and the phase does not go through +/- 90 degrees, all of which indicate that an *MDOF* curve-fit would be required to get a good identification. It is possible that other transducers might pick up this mode more clearly from different points on the structure.

A further indication of the quality of the data is to note that as this is a *driving point* FRF, the input and the output are from the same point on the structure. All of the phase values are between 0 and 180 degrees which is to be expected for a *driving point FRF*.

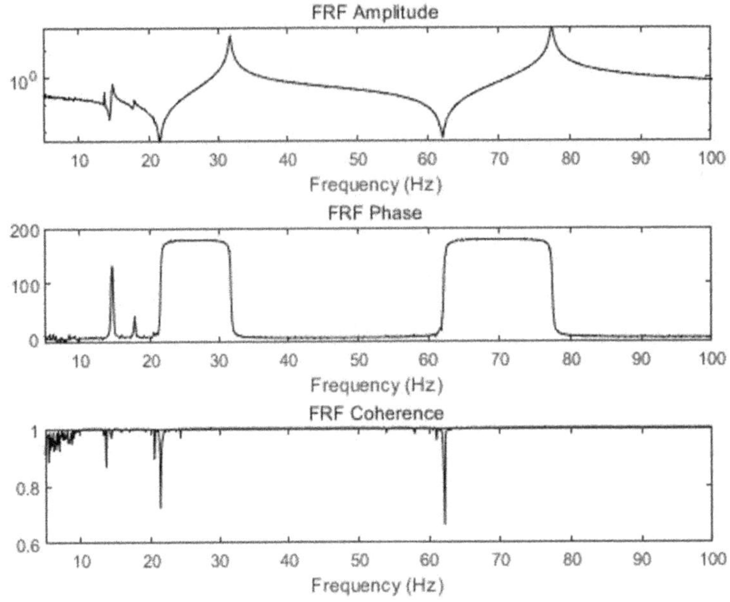

Figure 4.14 Sample *FRF* and Coherence plot.

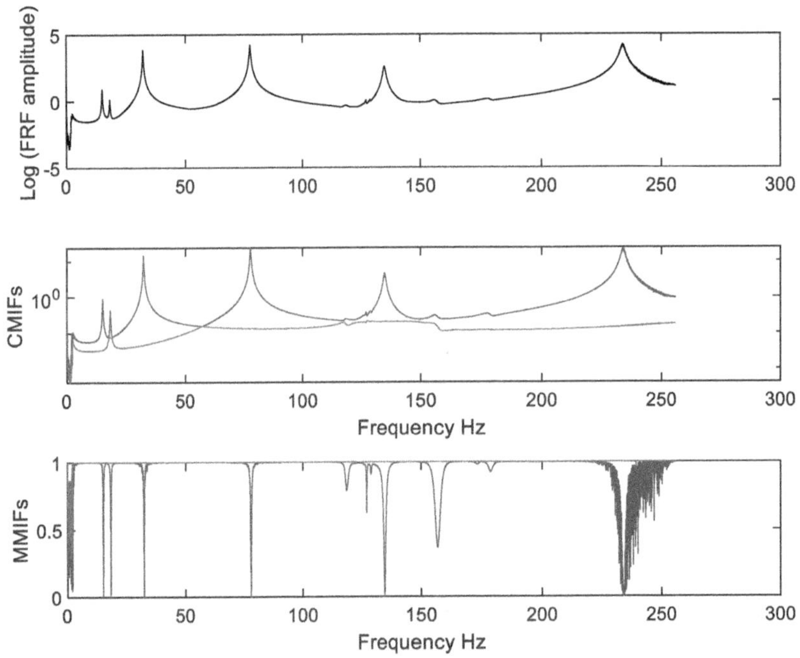

Figure 4.15 Summed *FRFs* and corresponding *CMIFs* and *MMIFs*.

Consider the sum of all the *FRFs* and the corresponding *MMIF* eigenvalues and *CMIF* singular values shown in Figure 4.15 up to 250Hz. The correspondence of the peaks in the *FRFs* and the first *MMIF* eigenvalue is clearly seen, with the drop not being so dramatic for the modes that are not so clearly excited. Even the two engine modes, which only appear on a few of the *FRFs*, are clearly indicated. The second *MMIF* eigenvalue does not drop down, indicating that only a single shaker is required to excite all of the modes. Similar findings are shown for the untracked *CMIFs* with the main modes being clearly identified. In all cases each identified mode is clearly a single mode.

A further indication of the quality of the data is to overlay reciprocal *FRFs*. It can be seen in Figure 4.16 that there is a good correspondence between both of them, indicating not only the quality of the acquired data but helping to show that there is little nonlinearity in the responses. Only the peaks are of interest, so any differences at the anti-resonances can be ignored. It is apparent on the Nyquist plot that the highest frequency (234 Hz) mode has a noisy appearance, potentially reducing the quality of any curve-fit. Note that as one is using a viscous damping model, the Nyquist plots are not exactly circles for the receptance, although the differences are very small, when the damping is light (as is the case here).

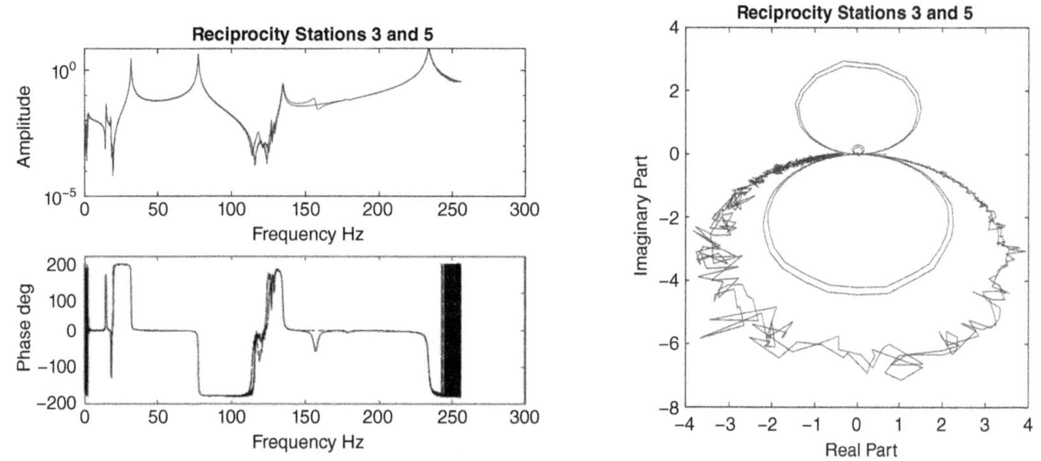

Figure 4.16 Reciprocity *FRFs* for Stations 3 and 5.

Further *FRFs* for an input at station 3 and measurements at positions 1, 3 and 9 are shown in Figure 4.17. Considering the Bode plots it can be deduced that in the frequency range of interest (0 → 250 Hz) there are six main modes (14.9 Hz, 18.1 Hz, 32.0 Hz, 77.7 Hz, 140 Hz and 235 Hz) but two of them are only visible from position 9, as they are modes that relate to the engine motions. It can be deduced that the modes are well-separated as none of the peaks in the Bode plots are close to each other and the resonances occur with approximately either 90° or -90° phase (this is the case for displacement or acceleration data and implies that the real part of the response is zero at resonance). The Bode plots using a log scale for the amplitude axis indicate some modes that have very small amplitudes. Note the driving point *FRF* (same excitation and measurement points) at position 3 in Figure 4.17 as the phase is between 0 and 180 degrees on the Bode plots and all the circles are on the same side of the real axis and that also there is an anti-resonance between each resonance peak. These characteristics are typical behaviour for driving point measurements.

The Nyquist plots also provide useful information, with each circle corresponding to the frequency points close to the resonance of each mode; they should be used in conjunction with the Bode plots. For each mode, the circles are constructed in a clockwise direction with increasing frequency, but the shape of the plots does not directly give frequency information.

The different phasing of the modes is clearly represented by the modes being either above or below the real axis (as one is dealing with the receptance). The somewhat jagged response on the largest circle is due to measurement noise on the highest frequency considered, but is not apparent on the Bode plots. If the frequency resolution is not big enough, particularly for the higher frequency modes, then instead of a circle, a very jagged or even triangular shape is found, which is inadequate for frequency and damping estimation.

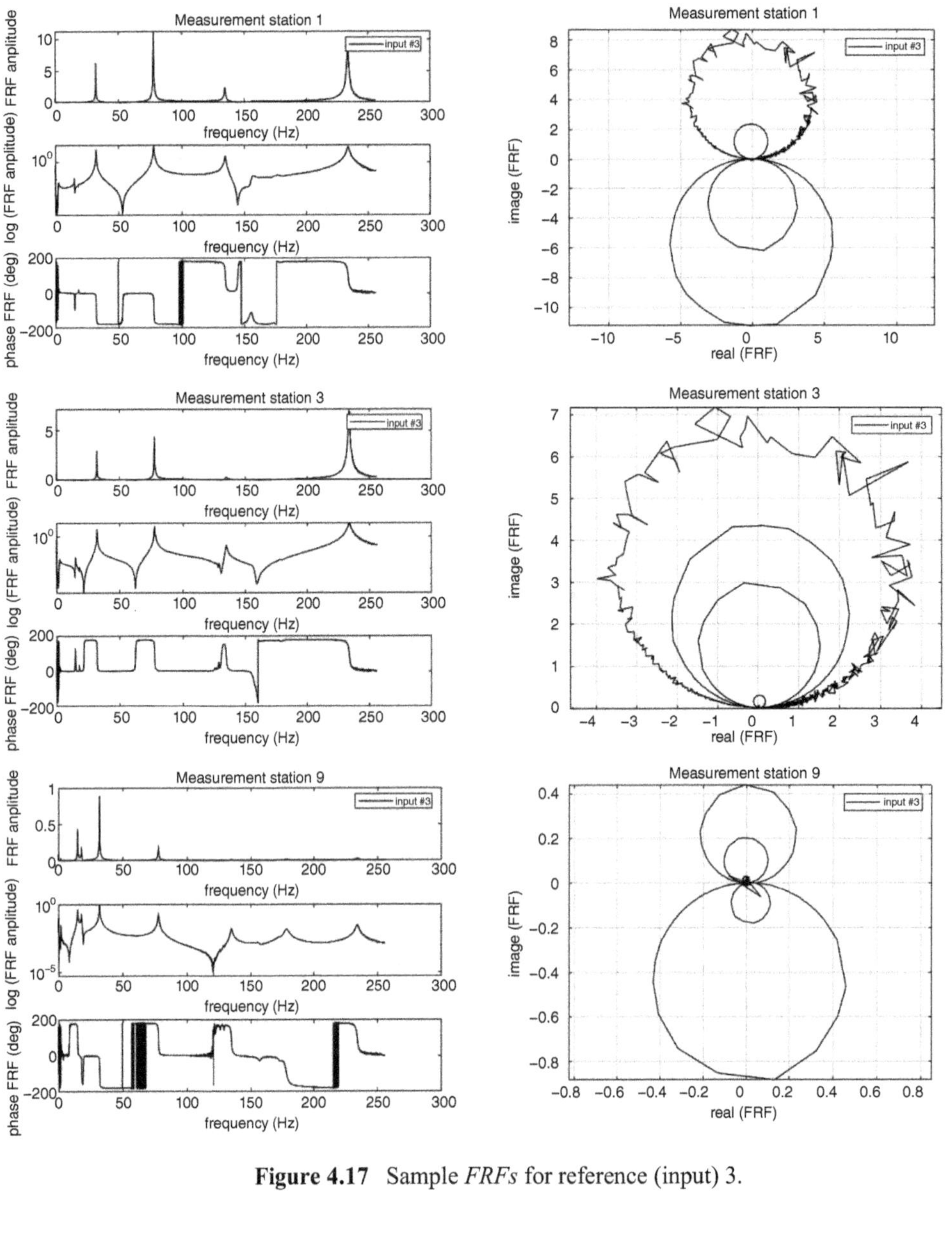

Figure 4.17 Sample *FRFs* for reference (input) 3.

Considerations on the Impulse Response Functions

Application of the inverse Fourier Transform to the *FRFs* gives the *IRFs*, as shown in Figure 4.18 for all 18 available *FRFs*. It can be seen that for the total time history most of the responses decay to zero very quickly in the total time window and therefore it is good practice to truncate the time window so that most of the zero response has been eliminated, as shown in Figure 4.19. Although it is difficult to deduce detailed information from the *IRFs*, it is clear that the responses from measurement stations 8 and 9 contain different modes, an effect of the pylon modes visible in the *FRFs*.

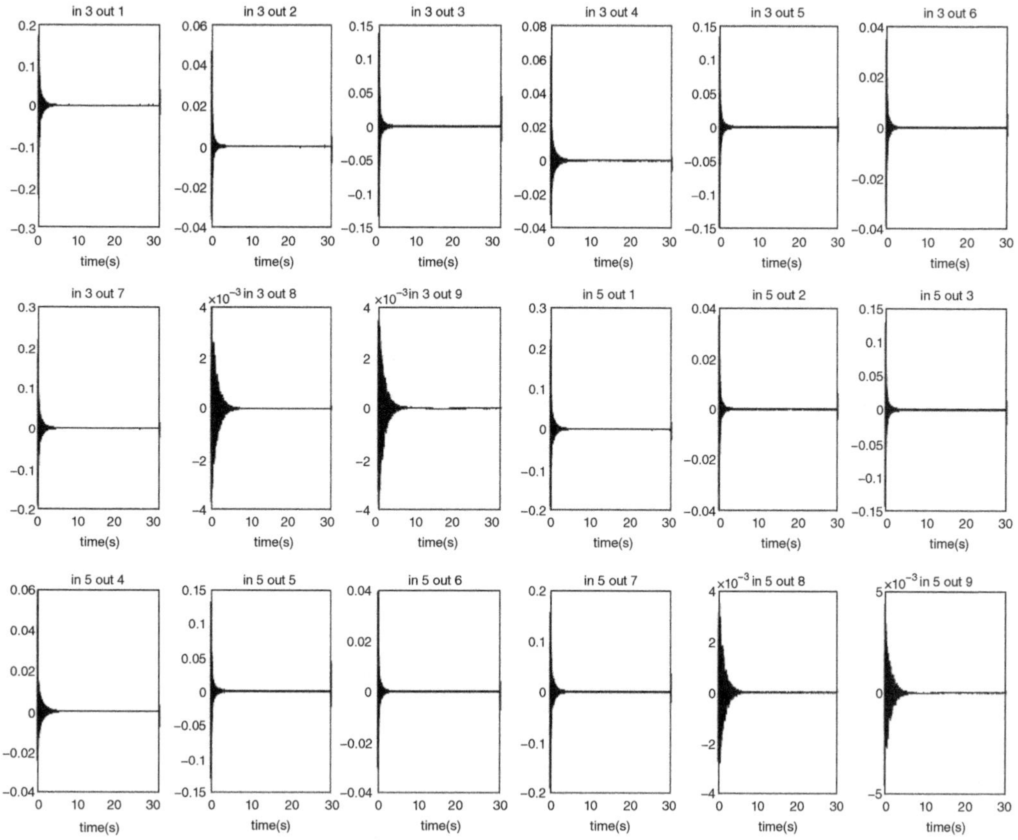

Figure 4.18 Raw Impulse Response Functions.

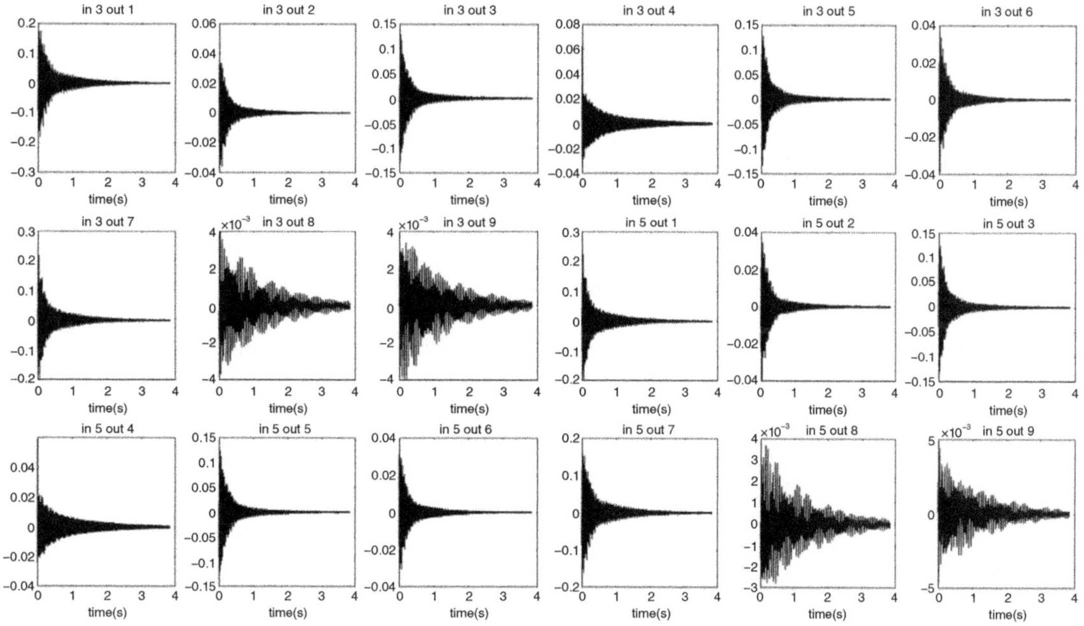

Figure 4.19 Truncated Impulse Response Functions.

4.2 System Identification Methods

4.2.1 Time Domain or Frequency Domain?

Considering that a lightly damped mode has a long-time domain *IRF*, but a very narrow peak in the frequency domain, and conversely a heavily damped mode will have a wide frequency domain peak but a short time domain response, leads to a good rule of thumb: to use time domain methods for lightly damped systems, and the frequency domain for heavily damped systems; see some suggested damping ratio ranges in Table 4.1. Note that there is an overlap in the ranges covered by the two analysis approaches and also that the developers of the *PolyMAX* method [4.16] state that it can be used for all frequency ranges. The user is encouraged to examine time and frequency domain representations of the data before starting any analysis.

A further point of consideration in the frequency domain is whether the frequency resolution is small enough[4]. In the frequency domain the resolution is $\Delta f = 1/\left(npts \times \Delta t\right)$ where *npts* are the number of time points and Δt is the sampling interval. Looking at the Nyquist plot, if the resonance peak is very "spikey" then this is a strong indication that the frequency resolution is poor, and a time domain method is preferable.

Conversely, if there is a high sampling rate then the *Nyquist frequency* may be much greater than the highest frequency of the modes of interest. In such a scenario the time domain analysis is likely to be much more sensitive to noise [4.18] and consequently *decimation* is recommended to artificially reduce the sampling rate by missing out data points from the time sequence (i.e. take the 1st, 3rd, 5th time points etc) to around 4 times the highest frequency of interest. Such a procedure must be undertaken using a *low pass filter* to avoid any *aliasing* issues.

The system identification methods considered in this section are detailed in Table 4.2 and are classified as being formulated in the frequency or time domains, and whether they are applied to *SDOF*, *MDOF* or *Ambient* data sets.

Table 4.1 Preferred analysis methods depending upon damping ratio levels.

Time domain	$\xi < 8\%$
Frequency domain	$\xi > 1\%$

Table 4.2 Phase Separation System Identification Methods.

	Frequency Domain	Time Domain
Single *DOF*	Half-power points Circle Fitting Inverse *FRF* fitting	Logarithmic Decrement
Multiple *DOF*	Nonlinear fitting Rational Fraction Polynomial *PolyMAX*	*LSCE* Polyreference *ERA*
Ambient	Auto/Cross Spectra *PolyMAX* Frequency Domain Decomposition Enhanced Frequency Domain Decomposition	Correlations / Random Decrement *ERA* *ERA/DC*

[4] See Chapter 2, Section 2.3 for more details.

4.2.2 Single Degree of Freedom – Frequency Domain

Peak Picking / Half-Power Points

The simplest method of determining the natural frequencies, damping ratios and mode shapes is to use the Peak Picking and Half-Power Points approach. It must be remembered that the results, particularly the damping values, will be subjected to a degree of uncertainty, especially if the modes start having significant interactions. However, the approach has much to be recommended as a means to get a good idea of the characteristics of the system before using more sophisticated approaches.

Consider the *FRF*, and the resulting *FRF* magnitude, shown in Eq. (4.27), for a single degree of freedom system with natural frequency ω_n and viscous damping ratio ξ in terms of whether the displacement (receptance), velocity (mobility) or acceleration (accelerance) is measured. It is simple to transform from velocity data to displacement (or acceleration) through division (or multiplication) of each frequency line by $i\omega$, taking care close to the zero frequency.

$$
\begin{aligned}
H_d(\omega) &= \frac{1}{m\left(\omega_n^2 - \omega^2 + i2\xi\omega_n\omega\right)} \rightarrow \left|H_d(\omega)\right| = \frac{1}{m\sqrt{\left(\omega_n^2 - \omega^2\right)^2 + 4\xi^2\omega_n^2\omega^2}} \\
H_v(\omega) &= \frac{i\omega}{m\left(\omega_n^2 - \omega^2 + i2\xi\omega_n\omega\right)} \rightarrow \left|H_v(\omega)\right| = \frac{\omega}{m\sqrt{\left(\omega_n^2 - \omega^2\right)^2 + 4\xi^2\omega_n^2\omega^2}} \\
H_a(\omega) &= \frac{-\omega^2}{m\left(\omega_n^2 - \omega^2 + i2\xi\omega_n\omega\right)} \rightarrow \left|H_a(\omega)\right| = \frac{\omega^2}{m\sqrt{\left(\omega_n^2 - \omega^2\right)^2 + 4\xi^2\omega_n^2\omega^2}}
\end{aligned}
\tag{4.27}
$$

Assuming that the system under test is lightly damped it is usual, and easiest, to assume that the peak of the *FRF* corresponds to the natural frequency of that mode.

For the mobility plot (velocity is the measured quantity) an excitation at the natural frequency gives a magnitude, see Figure 4.20, of

$$
\left|H_v(\omega)\right| = \frac{\omega_n}{2m\xi\omega_n^2} = \frac{1}{2m\xi\omega_n} \tag{4.28}
$$

Now, defining the half-power points as the points on either side of the resonant frequency where the *FRF* amplitude is $\left|H_v(\omega)\right|/\sqrt{2}$ (equivalent to a drop of 3 *dB* on a log scale plot), then the amplitude at the half-power points is found as

$$
\begin{aligned}
&\left|H_v(\omega)\right|_{HP} = \frac{1}{2\sqrt{2}m\xi\omega_n} = \frac{\omega_{HP}}{m\sqrt{\left(\omega_n^2 - \omega_{HP}^2\right)^2 + 4\xi^2\omega_n^2\omega_{HP}^2}} \\
&\rightarrow \frac{1}{8\xi^2\omega_n^2} = \frac{\omega_{HP}^2}{\left(\omega_n^2 - \omega_{HP}^2\right)^2 + 4\xi^2\omega_n^2\omega_{HP}^2} \\
&\rightarrow \left(\omega_n^2 - \omega_{HP}^2\right)^2 = 4\xi^2\omega_n^2\omega_{HP}^2 \rightarrow 2\xi\omega_n\omega_{HP} = \pm\left(\omega_n^2 - \omega_{HP}^2\right)
\end{aligned}
\tag{4.29}
$$

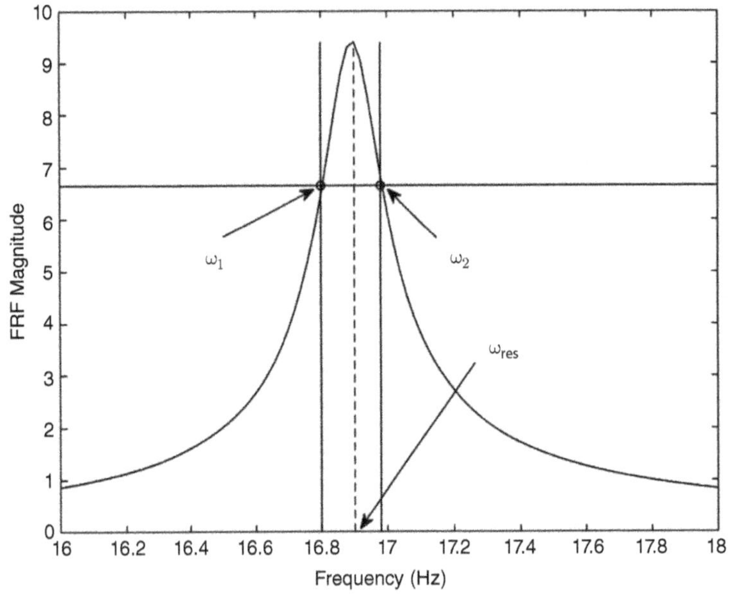

Figure 4.20 Schematic of Single *DOF* Peak Picking / Half-Power Points.

There are two solutions for the half-power point frequencies ω_1 and ω_2 that occur on either side of the resonance. Writing the above expression for both of these points, gives

$$2\xi\omega_n\omega_1 = \omega_n^2 - \omega_1^2 \quad \text{and} \quad 2\xi\omega_n\omega_2 = \omega_2^2 - \omega_n^2 \tag{4.30}$$

and adding both equations together leads to the expression

$$2\xi\omega_n(\omega_1 + \omega_2) = \omega_2^2 - \omega_1^2 = (\omega_2 + \omega_1)(\omega_2 - \omega_1) \tag{4.31}$$

and therefore, the damping ratio can be found as

$$\xi = \frac{\omega_2 - \omega_1}{2\omega_n} \tag{4.32}$$

This analysis is exact for all types of measurement data as long as the system damping is small.

So, by determining the value of the resonance frequency and those of the half-power points, it is possible to calculate the damping value. Note that it has been assumed that none of the modes are "close", that the half-power points are clearly visible and also that although (4.32) is valid only if one uses the mobility, for lightly damped systems the result gives very reasonable estimates for all data types. In practice the frequency resolution will mean that exact half-power points are not found on the *FRFs* and therefore *extrapolation* must be used to estimate them (shown as circles in Figure 4.21).

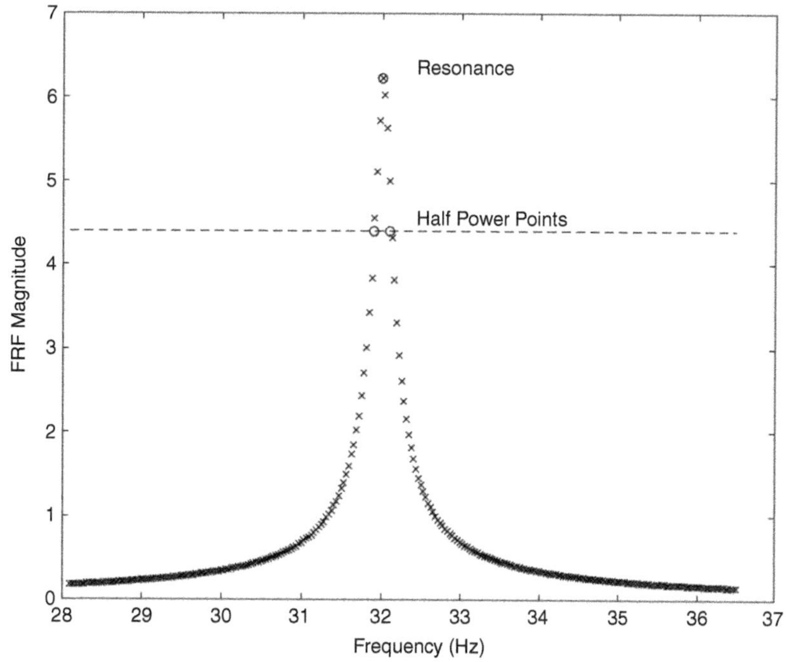

Figure 4.21 Sample measured *FRF* amplitude.

Example: Half-Power Points application

Consider the Bode plot shown in Figure 4.21. Taking the maximum value at the resonant frequency, then

$$f_n = 32 \text{ Hz} \qquad Amp. = 6.22 \text{ g/N}$$
$$\text{Amp.}/\sqrt{2} = 4.399 \text{ g/N}$$

Frequency range of the extrapolated half-power points $= 32.10 - 31.90 = 0.2$ Hz

$$\rightarrow \xi = \frac{0.2}{2\text{x}32} = 0.0031 = 0.31\%$$

This approach can then be extended to consider a range of *FRFs* from which not only the frequencies and damping ratios can be identified, and averaged over all the *FRFs*, but it is also possible to deduce the mode shapes. Considering the Wing-Pylon test structure and considering the 9 *FRFs* from the input at position 3, the frequencies and damping ratios for the first 4 modes are shown in Table 4.3. It can be seen that whereas there are estimates for modes 3 and 4 for all of the transducers (except number 4), this is not the case for the first two modes, which are made up primarily of engine motions and consequently only transducers 8 and 9 pick them up (a good example of observability). By taking the *FRF* amplitude at resonance and taking into account whether the phase is (or in practice is close to) either + 90° or – 90° (the shaded elements) the size of the mode shape at a particular position can be plotted. In the – 90° phase cases simply multiply the magnitude by – 1. The mode shapes can then be sketched out, as seen in Figure 4.22.

Table 4.3 Peak picking analysis results. Resonant Freq (Hz), Damping (%), Amplitude(g/N), Phase (degs)

FRF	Mode 1				Mode 2				Mode 3				Mode 4			
	Freq.	Damp.	Ampl.	Phase	Freq.	Damp.	Ampl.	Phase	Freq.	Damp.	Ampl.	Phase	Freq.	Damp.	Ampl.	Phase
1									32.0	0.31	6.22	-82.7	77.7	0.26	11.3	-81.4
2									32.0	0.33	0.47	-80.0	77.7	0.25	1.49	97.8
3									32.0	0.30	3.00	96.6	77.7	0.26	4.36	98.4
4									32.0	0.30	4.42	96.7	77.7			
5									32.0	0.30	2.96	96.6	77.7	0.26	4.46	-80.5
6									32.0	0.32	0.30	-82.0	77.7	0.26	1.66	-81.0
7									32.0	0.30	5.65	-83.3	77.7	0.26	10.63	99.14
8	14.9	0.72	0.49	-87.5	18.09	0.72	0.15	-111.5	32.0	0.31	0.81	97.99	77.7	0.26	0.18	99.8
9	14.91	0.69	0.44	90.75	18.06	0.70	0.18	-78.5	32.0	0.31	0.89	-81.5	77.7	0.27	0.20	99.9
Av	**14.9**	**0.71**			**18.07**	**0.71**			**32.0**	**0.31**			**77.7**	**0.26**		

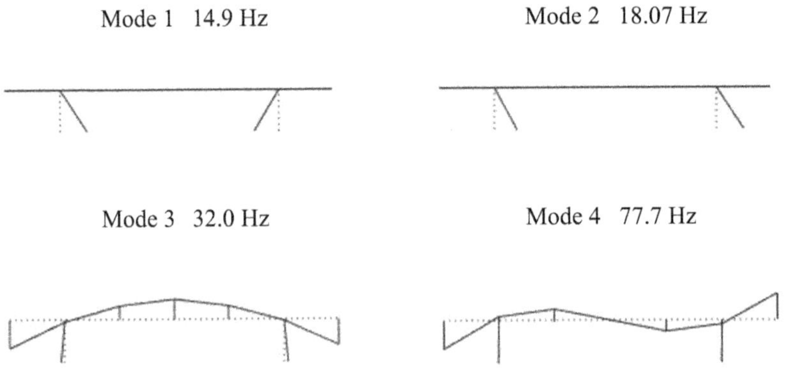

Figure 4.22 Estimated mode shapes from Peak Picking.

Circle Fitting

Peak Picking combined with the Half-Power points approach is an extremely quick and useful way to obtain estimates of frequency, damping and mode shapes, but it also suffers from being very sensitive to measurement noise. The damping predictions are particularly sensitive to errors, especially for low levels of damping where determining the exact position of the half-power points can be problematic. The next step that will be considered in developing more accurate identification approaches is to note that the Nyquist plots for each well separated mode are represented by a circle (make sure that the real and imaginary axes are plotted using the same scale). For compliance or accelerance data these circles appear above (or below) the real axis if their phase at resonance is – 90° (+ 90°). For mobility, the circles are positioned either to the right or left of the imaginary axis.

Starting with the mobility *FRF* for an *SDOF* system with viscous damping

$$H_v(\omega) = \frac{i\omega}{m\left(\omega_n^2 - \omega^2 + i2\xi\omega_n\omega\right)} = \frac{i\omega}{\left(k - m\omega^2 + i\omega c\right)} \tag{4.33}$$

then the real and imaginary parts can be written as

$$\mathrm{Re}\left(H_v(\omega)\right) = \frac{\omega^2 c}{\left(k - m\omega^2\right)^2 + \omega^2 c^2} \quad \text{and} \quad \mathrm{Im}\left(H_v(\omega)\right) = \frac{\omega\left(k - m\omega^2\right)}{\left(k - m\omega^2\right)^2 + \omega^2 c^2} \tag{4.34}$$

which, after some mathematical manipulation, can be shown to be in the form

$$\left(\mathrm{Re}\left(H_v(\omega)\right)-\frac{1}{2c}\right)^2+\left(\mathrm{Im}\left(H_v(\omega)\right)\right)^2=\left(\frac{1}{2c}\right)^2 \tag{4.35}$$

This expression is in the well-known form of a circle on the Nyquist plot with centre $\left(\frac{1}{2c},0\right)$ and radius $\frac{1}{2c}$ as illustrated in Figure 4.23 (a).

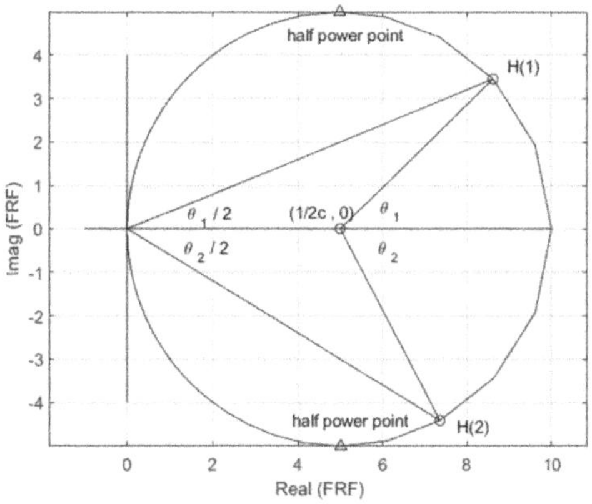

Figure 4.23 (a) Mobility circle for a single *DOF*.

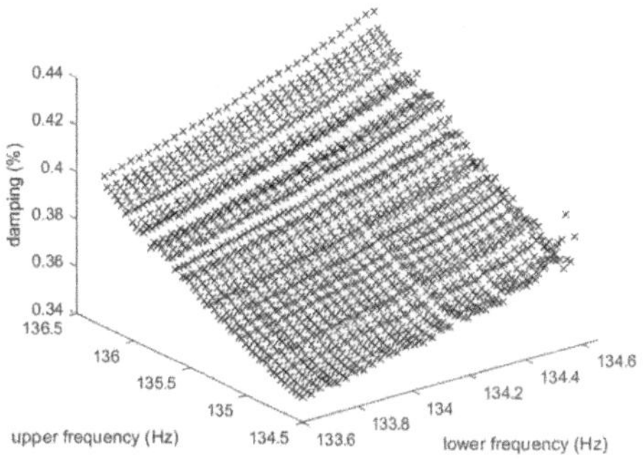

Figure 4.23 (b) Damping values from different pairs of circle fit points.

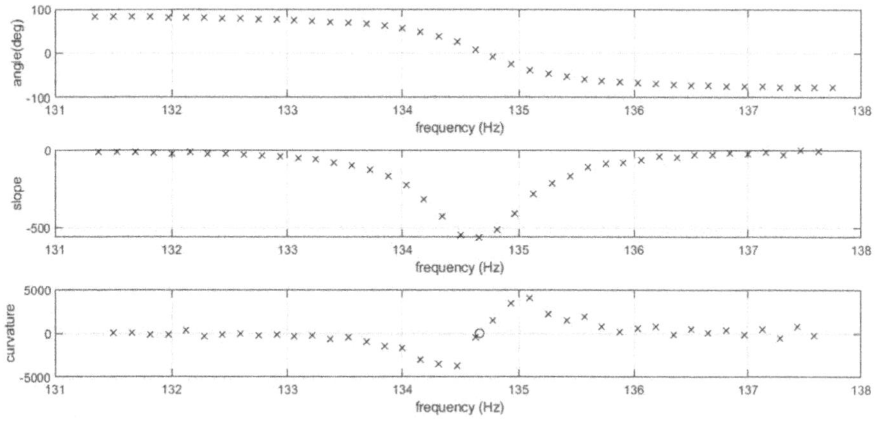

Figure 4.23 (c) Angle, slope and curvature for circle plot going through resonance O.

Considering Figure 4.23 (a), then it can be seen that

$$tan\left(\frac{\theta}{2}\right)=\frac{\mathrm{Im}(H)}{\mathrm{Re}(H)}=\frac{\omega\left(k-\omega^2 m\right)}{\omega^2 c}=\frac{\omega_n^2-\omega^2}{2\xi\omega_n\omega}\ \text{, as }\ \omega_n^2=\frac{k}{m}\ \text{ and }\ \xi=\frac{c}{2m\omega_n} \tag{4.36}$$

Hence, knowing the resonant frequency it is possible to compute the damping ratio from a single measurement point at frequency ω such that

$$\xi = \frac{\left(\omega_n^2 - \omega^2\right)\mathrm{Re}(H)}{2\omega_n \omega \,\mathrm{Im}(H)} \tag{4.37}$$

It can also be shown that the points on the Nyquist circle for $\theta / 2 = \pm 45^\circ$ correspond to the Bode plot half-power points. A similar, but not exact, analysis can be performed for acceleration or displacement data; however, it is straightforward to transform into velocity data by dividing or multiplying, respectively, by $i\omega$.

Considering individual points on the circle is very likely to produce wildly varying estimates, so strategies that contain as many points as possible produce an averaged estimate that minimises the error. A common approach is to consider any two measured points on the circle (not necessarily the half-power points) ω_1 and ω_2 where $\omega_1 < \omega_n < \omega_2$. Then,

$$tan\left(\frac{\theta_1}{2}\right) = \frac{\omega_n^2 - \omega_1^{\,2}}{2\xi\omega_n\omega_1} \quad \text{and} \quad tan\left(\frac{\theta_2}{2}\right) = \frac{\omega_2^2 - \omega_n^{\,2}}{2\xi\omega_n\omega_2} \tag{4.38}$$

and adding these two expressions together gives

$$tan\left(\frac{\theta_1}{2}\right) + tan\left(\frac{\theta_2}{2}\right) = \frac{\omega_2\left(\omega_n^2 - \omega_1^{\,2}\right) + \omega_1\left(\omega_2^2 - \omega_n^{\,2}\right)}{2\xi\omega_n\omega_1\omega_2} \tag{4.39}$$

from which the damping ratio can be calculated as

$$\xi = \frac{\omega_2\left(\omega_n^2 - \omega_1^{\,2}\right) + \omega_1\left(\omega_2^2 - \omega_n^{\,2}\right)}{2\omega_n\omega_1\omega_2\left(tan\left(\frac{\theta_1}{2}\right) + tan\left(\frac{\theta_2}{2}\right)\right)} \tag{4.40}$$

Figure 4.23 (b) shows the range of different damping ratios that are estimated from different pairings of frequency points below and above the resonant frequency; the user is recommended to take the average value of the identified damping ratios.

An even better approach is to fit a circle through all of the points around the circle. Writing out Eq. (4.37) into a matrix equation for each frequency point, assuming that the resonance frequency ω_n is known, it follows that

$$\begin{Bmatrix} 2\omega_n\omega_1 \,\mathrm{Im}(H_1) \\ 2\omega_n\omega_2 \,\mathrm{Im}(H_2) \\ \vdots \\ 2\omega_n\omega_L \,\mathrm{Im}(H_L) \end{Bmatrix} \xi = \begin{Bmatrix} \left(\omega_n^2 - \omega_1^2\right)\mathrm{Re}(H_1) \\ \left(\omega_n^2 - \omega_2^2\right)\mathrm{Re}(H_2) \\ \vdots \\ \left(\omega_n^2 - \omega_L^2\right)\mathrm{Re}(H_L) \end{Bmatrix} \rightarrow \boldsymbol{p}\xi = \boldsymbol{q} \tag{4.41}$$

and then the least-squares solution for the damping value is found as

$$\xi = \left(\boldsymbol{p}^{\mathrm{T}}\boldsymbol{p}\right)^{-1}\boldsymbol{p}^{\mathrm{T}}\boldsymbol{q} \tag{4.42}$$

The absolute value of the modal amplitude is taken as the radius of the circle, but the amplitude can either be positive or negative depending on the phase of the mode. Here it is assumed that the exact resonance frequency is known, whereas there are likely to be some errors, but these are no worse than from the peak-picking approach and the damping estimates that are obtained are much more accurate.

To compute the radius and centre of the circle, which defines the amplitude of the mode, letting $r = 1/(2c)$, $H_R = \mathrm{Re}(H_V)$ and $H_I = \mathrm{Im}(H_V)$, then from (4.35),

$$\left(H_R(\omega) - r\right)^2 + H_I^2(\omega) = r^2 \;\Rightarrow\; H_R^2(\omega) - 2H_R(\omega)r + H_I^2(\omega) = 0 \tag{4.43}$$

So, for L frequency points this equation can be written as

$$\begin{Bmatrix} H_R^2(\omega_1) + H_I^2(\omega_1) \\ H_R^2(\omega_2) + H_I^2(\omega_2) \\ \vdots \\ H_R^2(\omega_L) + H_I^2(\omega_L) \end{Bmatrix} = \begin{Bmatrix} 2H_R(\omega_1) \\ 2H_R(\omega_2) \\ \vdots \\ 2H_R(\omega_L) \end{Bmatrix} r \;\rightarrow\; \boldsymbol{p}r = \boldsymbol{q} \tag{4.44}$$

and so the amplitude term is found using the least-squares approach as before such that

$$r = \left(\boldsymbol{p}^{T}\boldsymbol{p}\right)^{-1}\boldsymbol{p}^{T}\boldsymbol{q} \tag{4.45}$$

Instead of simply taking the resonance frequency as that corresponding to the maximum *FRF* amplitude, a more accurate value can be determined by noting that angle θ_1, and by implication the angle subtended from the circle origin $\theta_1/2$, have their greatest change going through the resonance as seen in Figure 4.23 (c). By considering the curvature of the angle $\theta_1/2$ *versus* frequency, interpolation can be used to determine the frequency at the zero crossing.

The above analysis has been made assuming that there are no effects from out of range modes, that the *FRF* circles have an origin on the Nyquist plot at (0,0) and there is no rotation of the local axes of the circle. A range of methods exist [4.2] to extend the circle fitting technique to deal with such effects but they will not be considered here. Instead, the reader is recommended to make use of one of the many *MDOF* methods considered later in this chapter.

Example: Circle Fitting Application

The above methodology was applied to the *FRF* between input 3 and output 1, shown in Figure 4.17. Table 4.4 and Figure 4.24 show the resulting identified modal parameters and regenerated curve fit, and it can be seen that there is a good correlation with the half-power points results and a very good comparison between the estimated model and the experimental data.

Table 4.4 Modal Estimates from Half-Power points, Circle Fitting and Inverse Fit methods.

	Mode number	1	2	3	4
Half-Power points	Natural Frequency (Hz)	32.0	77.77	134.1	234.1
	Damping Ratio (%)	0.31	0.26	0.41	0.285
Circle Fit	Natural Frequency (Hz)	32.0	77.66	134.7	233.9
	Damping Ratio (%)	0.31	0.26	0.38	0.32
Inverse Fit	Natural Frequency (Hz)	32.0	77.69	134.8	234.0
	Damping Ratio (%)	0.36	0.23	0.42	0.37

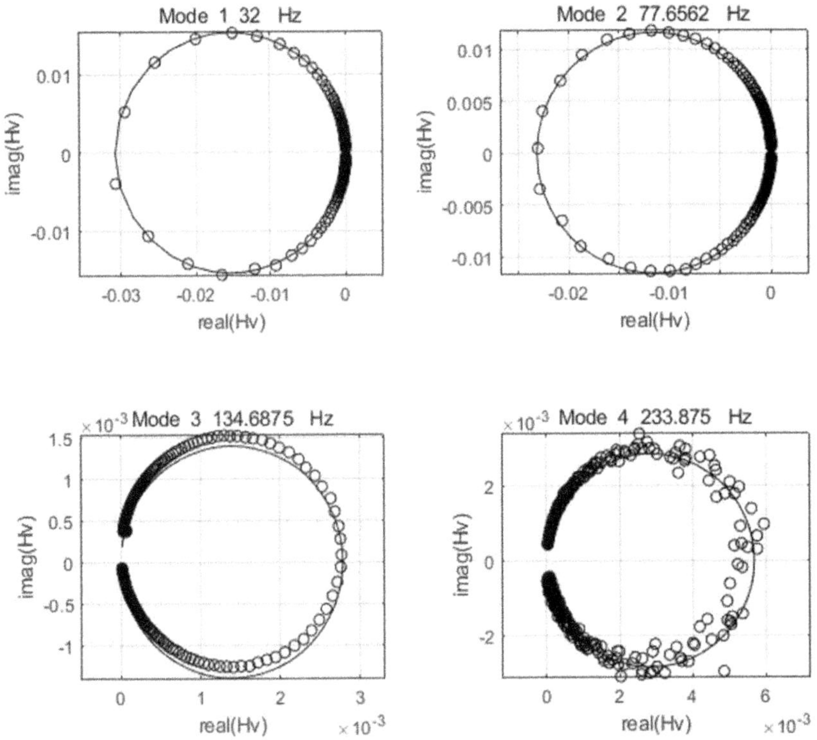

Figure 4.24 Circle Fitting: comparison of regenerated model and experimental (mobility) data.

Inverse Fit Method

A further simple *SDOF* frequency domain approach is based on the observation [4.19] that the inverse of the compliance[5] *FRF* is written as

$$H_{inv}(\omega) = m\left(\omega_n^2 - \omega^2 + i2\xi\omega_n\omega\right) \tag{4.46}$$

and thus the real and imaginary parts if the inverse *FRF* are found as

$$\mathrm{Re}\left(H_{inv}(\omega)\right) = m\left(\omega_n^2 - \omega^2\right) \quad \text{and} \quad \mathrm{Im}\left(H_{inv}(\omega)\right) = 2m\xi\omega_n\omega \tag{4.47}$$

[5] Compliance (or dynamic compliance in this case) is the same as Receptance.

Considering Figure 4.25 and noting that the real part can be plotted as a straight line versus ω^2 and that the imaginary part versus ω is also a straight line, then an identification strategy follows as:

i. determine where the real part straight line plot fit crosses zero, giving the resonant frequency ω_n;

ii. the slope of the real part straight line fit gives the modal mass m;

iii. determine the slope of the imaginary straight line fit which is equal to $2m\xi\omega_n$ and hence the damping value can be found.

Modal estimates for the Wing-Pylon model can be seen in Table 4.4. A regenerated curve-fit for the 32 Hz mode is shown in Figure 4.26, where the single mode that has been identified is clearly visible.

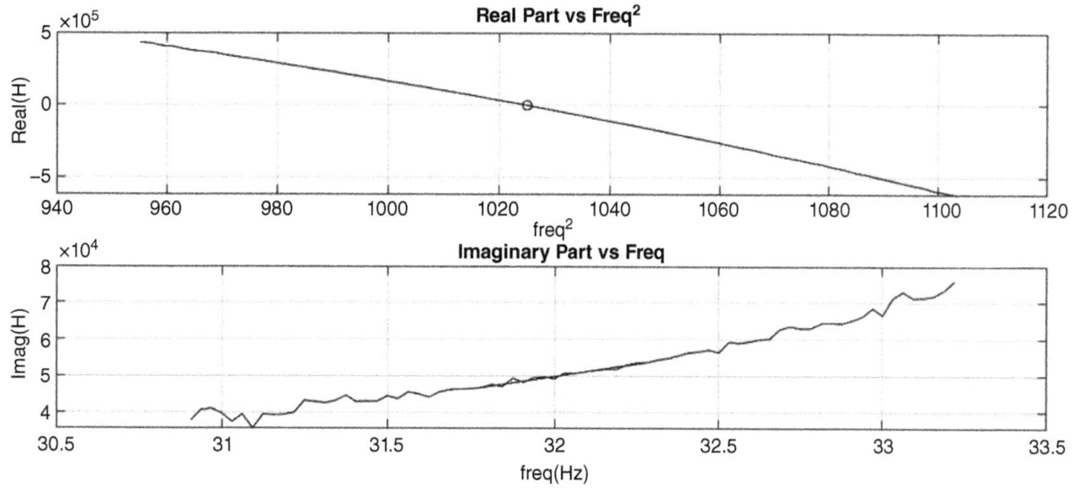

Figure 4.25 Real and Imaginary Plots for the Inverse Fit method.

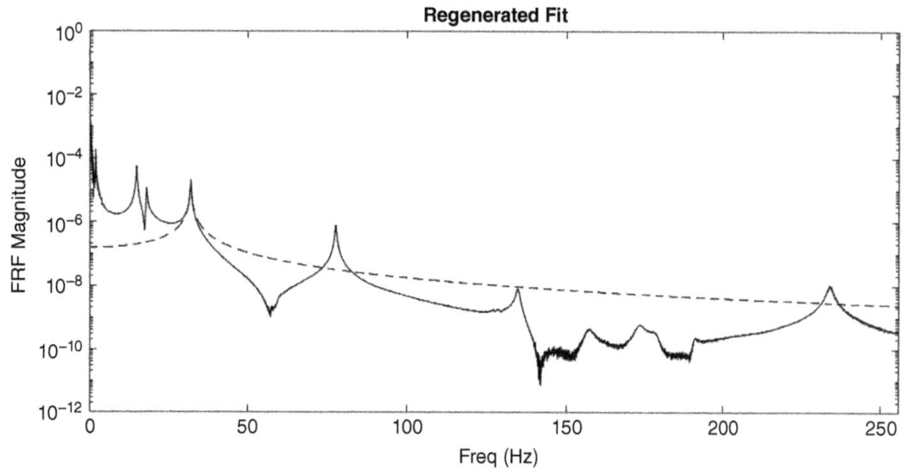

Figure 4.26 Regenerated Inverse Fit plot for the 32 Hz mode.

Short Note on the Solution of Overdetermined Simultaneous Linear Equations

A common tool in a lot of curve-fitting algorithms is the solution of overdetermined (more rows than columns) simultaneous linear equations (*SLEs*) in the form

$$\boldsymbol{A}\boldsymbol{x}=\boldsymbol{b} \tag{4.48}$$

where matrix $\boldsymbol{A}$ and the vectors $\boldsymbol{x}$ (unknown) and $\boldsymbol{b}$ have dimensions $p \times m$, $m \times 1$ and $p \times 1$, respectively and $p > m$.

As $\boldsymbol{A}$ is not square one cannot simply pre-multiply each side by its inverse, so a very common approach is to form the Moore-Penrose inverse [4.20] by pre-multiplying the equation by $\boldsymbol{A}^{\mathrm{T}}$ and, as $\boldsymbol{A}^{\mathrm{T}}\boldsymbol{A}$ can be inverted (dimensions $m \times m$), the solution follows as

$$\boldsymbol{A}^{\mathrm{T}}\boldsymbol{A}\boldsymbol{x}=\boldsymbol{A}^{\mathrm{T}}\boldsymbol{b} \;\rightarrow\; \boldsymbol{x}=\left(\boldsymbol{A}^{\mathrm{T}}\boldsymbol{A}\right)^{-1}\boldsymbol{A}^{\mathrm{T}}\boldsymbol{b} \tag{4.49}$$

The above approach is referred to as a least-squares solution as it minimises the squared error δ^2 between $\boldsymbol{A}\boldsymbol{x}$ and $\boldsymbol{b}$ where

$$\boldsymbol{\delta}=\boldsymbol{A}\boldsymbol{x}-\boldsymbol{b} \;\rightarrow\; \delta^2=\left(\boldsymbol{A}\boldsymbol{x}-\boldsymbol{b}\right)^{\mathrm{T}}\left(\boldsymbol{A}\boldsymbol{x}-\boldsymbol{b}\right)=\boldsymbol{x}^{\mathrm{T}}\boldsymbol{A}^{\mathrm{T}}\boldsymbol{A}\boldsymbol{x}-\boldsymbol{b}^{\mathrm{T}}\boldsymbol{A}\boldsymbol{x}-\boldsymbol{x}^{\mathrm{T}}\boldsymbol{A}^{\mathrm{T}}\boldsymbol{b}+\boldsymbol{b}^{\mathrm{T}}\boldsymbol{b} \tag{4.50}$$

Minimising δ^2 with respect to $\boldsymbol{x}$, i.e. making $\partial\left(\delta^2\right)/\partial\boldsymbol{x}=\boldsymbol{0}$,

$$2\boldsymbol{A}^{\mathrm{T}}\boldsymbol{A}\boldsymbol{x}-2\boldsymbol{A}^{\mathrm{T}}\boldsymbol{b}=\boldsymbol{0} \tag{4.51}$$

from which one obtains $\boldsymbol{x}=\left(\boldsymbol{A}^{\mathrm{T}}\boldsymbol{A}\right)^{-1}\boldsymbol{A}^{\mathrm{T}}\boldsymbol{b}$, as in (4.49).

The reader is cautioned that this approach squares the condition number [4.6] of the $\boldsymbol{A}$ matrix, which makes the computation of the inverse much more susceptible to errors, and can result in erroneous estimates for large order lightly damped systems. A better approach, from a speed and accuracy viewpoint, is to solve the equations without explicitly computing the inverse using either of the following two methods.

QR Decomposition [4.6]

Defining the ($p\times m$) $\boldsymbol{A}$ matrix as the decomposition $\boldsymbol{A}=\boldsymbol{Q}\boldsymbol{R}$, where $\boldsymbol{Q}$ is a ($p\times m$) orthogonal matrix ($\boldsymbol{Q}^T\boldsymbol{Q}=\boldsymbol{I}$) and $\boldsymbol{R}$ is a ($m\times m$) upper triangular matrix, then the solution of $\boldsymbol{A}\boldsymbol{x}=\boldsymbol{b}$ is found as

$$\boldsymbol{A}\boldsymbol{x}=\boldsymbol{b} \;\rightarrow\; \boldsymbol{Q}\boldsymbol{R}\boldsymbol{x}=\boldsymbol{b} \;\rightarrow\; \boldsymbol{Q}^T\boldsymbol{Q}\boldsymbol{R}\boldsymbol{x}=\boldsymbol{Q}^T\boldsymbol{b}\rightarrow\; \boldsymbol{R}\boldsymbol{x}=\boldsymbol{Q}^T\boldsymbol{b} \tag{4.52}$$

and as $\boldsymbol{R}$ is upper triangular, it is a simple task to compute the unknown $\boldsymbol{x}$ without needing to invert $\boldsymbol{R}$.

Singular Value Decomposition (*SVD*)

As described in Chapter 3, a very common approach is to decompose $\boldsymbol{A}$ (p x m with $p > m$) using the compact or "economy" *SVD* [4.6] such that

$$\boldsymbol{A}=\boldsymbol{U}\boldsymbol{\Sigma}\boldsymbol{V}^T \tag{4.53}$$

with $\boldsymbol{U}\,(p\times m)$ and $\boldsymbol{V}\,(m\times m)$ being orthogonal matrices with the properties $\boldsymbol{U}^T\boldsymbol{U}=\boldsymbol{I}_{m\times m}$ and $\boldsymbol{V}^T\boldsymbol{V}=\boldsymbol{I}_{m\times m}$, and $\boldsymbol{\Sigma}$ $(m\times m)$ is a diagonal matrix containing the singular values in descending order of magnitude (they cannot be negative, but can be zero).

The solution to $\boldsymbol{Ax}=\boldsymbol{b}$ then follows as

$$\boldsymbol{Ax}=\boldsymbol{b}\;\rightarrow\;\boldsymbol{U\Sigma V}^T\boldsymbol{x}=\boldsymbol{b}\;\rightarrow\;\left(\boldsymbol{V\Sigma}^{-1}\boldsymbol{U}^T\right)\boldsymbol{U\Sigma V}^T\boldsymbol{x}=\left(\boldsymbol{V\Sigma}^{-1}\boldsymbol{U}^T\right)\boldsymbol{b}\;\rightarrow\;\boldsymbol{x}=\boldsymbol{V\Sigma}^{-1}\boldsymbol{U}^T\boldsymbol{b} \tag{4.54}$$

Solution for Rank Deficient Systems

A further advantage of using matrix decomposition methods is that the equations can be solved even if the number of columns in the $\boldsymbol{A}$ matrix is greater than the number of linearly independent columns (in modal analysis terms this situation occurs when the number of columns is greater than twice the number of modes in the system under test, sometimes called an overspecified solution). In this case, matrix $\boldsymbol{A}$ is known as rank deficient and it is impossible to find the inverse of $\boldsymbol{A}^T\boldsymbol{A}$ (although for a real experimental test there is likely to be enough noise and model uncertainty for the inverse to be computed).

Here, the *SVD* is used as a means to solve such rank deficient systems; however, note that a similar approach can be taken with the *QR* decomposition. The singular values (*svs*) can be used to determine the rank of a matrix, which is equal to the number of non-zero *svs*. Remember that the *svs* are always non-negative and listed in descending order.

Assuming that there are $s<m$ non-zero *svs*, then matrix $\boldsymbol{A}$ can be decomposed such that

$$\begin{aligned}\boldsymbol{A}_{p\times m}&=\boldsymbol{U}_{p\times m}\boldsymbol{\Sigma}_{m\times m}\boldsymbol{V}^T_{m\times m}=\begin{bmatrix}\boldsymbol{U}_{1_{p\times s}} & \boldsymbol{U}_{0_{p\times(m-s)}}\end{bmatrix}\begin{bmatrix}\boldsymbol{\Sigma}_{1_{s\times s}} & 0\\ 0 & \boldsymbol{\Sigma}_{0_{(m-s)\times(m-s)}}\end{bmatrix}\begin{bmatrix}\boldsymbol{V}^T_{1_{s\times m}}\\ \boldsymbol{V}^T_{0_{(m-s)\times m}}\end{bmatrix}\\ &=\boldsymbol{U}_1\boldsymbol{\Sigma}_1\boldsymbol{V}_1^T+\boldsymbol{U}_0\boldsymbol{\Sigma}_0\boldsymbol{V}_0^T\approx\boldsymbol{U}_{1_{p\times s}}\boldsymbol{\Sigma}_{1_{s\times s}}\boldsymbol{V}^T_{1_{s\times m}}\end{aligned} \tag{4.55}$$

where $\boldsymbol{\Sigma}_1$ is a diagonal $(s\times s)$ matrix of non-zero *svs* and $\boldsymbol{\Sigma}_0$ is a diagonal $(m-s)\times(m-s)$ matrix of zero *svs* (in practice these are often not exactly zero, but very small).

The solution of the *SLEs* simply follows using those terms in the decomposition that relate to the non-zero *svs* such that

$$\boldsymbol{Ax}=\boldsymbol{b}\quad\rightarrow\quad\boldsymbol{U}_1\boldsymbol{\Sigma}_1\boldsymbol{V}_1^T\boldsymbol{x}=\boldsymbol{b}\quad\rightarrow\quad\boldsymbol{x}=\boldsymbol{V}_1\boldsymbol{\Sigma}_1^{-1}\boldsymbol{U}_1^T\boldsymbol{b} \tag{4.56}$$

4.2.3 Single Degree of Freedom – Time Domain

The alternative to the frequency domain approaches is to perform the identification in the time domain. A fair estimate of the frequency and damping of a dominant mode can be achieved by using a simple impulse or step relaxation excitation and then analysing the resultant decaying response. Free decays also follow from the termination of a "chirp" (rapid frequency sweep), burst random or resonant dwell excitation (but remember that any attached shaker will influence the response of the system.) For *MDOF* systems a further approach is to generate the *FRFs*, then window around each mode of interest and use the inverse *FFT* to generate the *SDOF* impulse response function. Such a methodology works well as long as the modes are well separated.

Logarithmic Decrement

For a linear *SDOF* system with viscous damping, the free decay [4.10] will oscillate at the damped natural frequency $\omega_d = \omega_n\sqrt{1-\xi^2}$ bounded by an exponential envelope $e^{-\xi\omega_n t}$.

Consider the free decay of a single *DOF* system shown in Figure 4.27; from Chapter 1, this is given by

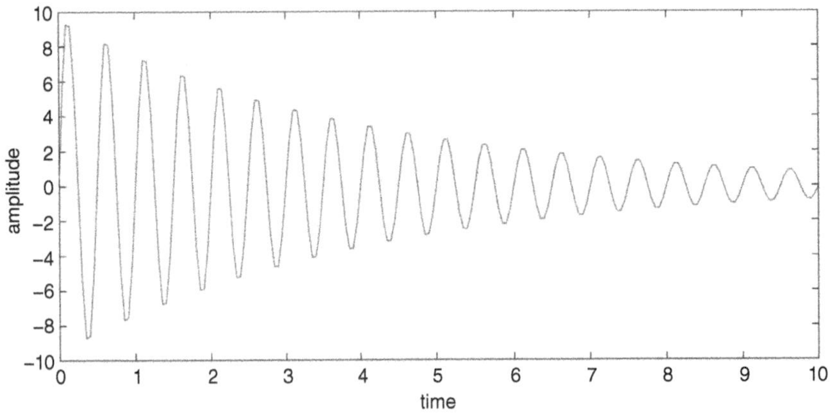

Figure 4.27 *SDOF* free decay.

$$y(t) = Ae^{-\xi\omega_n t}sin(\omega_d t + \varphi) \tag{4.57}$$

ω_d can be found from the average spacing between zero crossings with the same slope or the peaks, or count zero crossings, and for low damping ($\xi < 20\%$) this can be assumed to be the same as ω_n.

To compute the damping, if the amplitudes of the jth and $(j+1)^{th}$ peaks are defined by Y_j and Y_{j+1}, which are separated by time T_d, where $T_d = \dfrac{2\pi}{\omega_n\sqrt{1-\xi^2}}$, then the logarithmic decrement is defined as

$$\delta = ln\left(\frac{Y_j}{Y_{j+1}}\right) = ln\left(\frac{Ae^{-\xi\omega_n t_j}}{Ae^{-\xi\omega_n(t_j+T_d)}}\right) = ln\left(e^{\xi\omega_n T_d}\right) = \xi\omega_n T_d \tag{4.58}$$

Thus,

$$\delta = \frac{2\pi\xi}{\sqrt{1-\xi^2}} \approx 2\pi\xi \text{ for low damping} \tag{4.59}$$

and hence the damping can be computed.

Instead of choosing consecutive peaks, errors due to digitalisation and measurement noise can be minimised by selecting the reduction over *N* peaks such that

$$\delta_N = ln\left(\frac{Y_j}{Y_{j+N}}\right) = ln\left(\frac{Ae^{-\xi\omega_n t_j}}{Ae^{-\xi\omega_n(t_j+NT_d)}}\right) = ln\left(e^{\xi\omega_n NT_d}\right) = N\xi\omega_n T_d \tag{4.60}$$

Thus,

$$\delta_N = \frac{N2\pi\xi}{\sqrt{1-\xi^2}} \approx 2\pi\xi N \text{ for low damping} \tag{4.61}$$

As a rule of thumb to get an idea of the damping level, consider the number of cycles $N_{1/2}$ it takes to get to half amplitude, thus

$$\begin{aligned} \delta_{N_{1/2}} &= ln\left(\frac{1}{1/2}\right) = ln(2) = 0.693 = 2\pi\xi N_{1/2} \\ \rightarrow \xi &= \frac{0.11}{N_{1/2}} \end{aligned} \tag{4.62}$$

So for instance, if it takes 11 cycles to achieve half amplitude, the damping ratio is 0.11 / 11 = 0.01 = 1%.

Finally, an even more accurate approach is to take the logarithm of the expression for the peaks of the decay response. As $Y(t) = Ae^{-\xi\omega_n t}$, then

$$ln(Y(t)) = lnA - \xi\omega_n t \tag{4.63}$$

and, as can be seen in Figure 4.28, the slope of the straight line joining the logarithm of the peaks is $-\xi\omega_n$ and the amplitude can be found from the response value at $t = 0$.

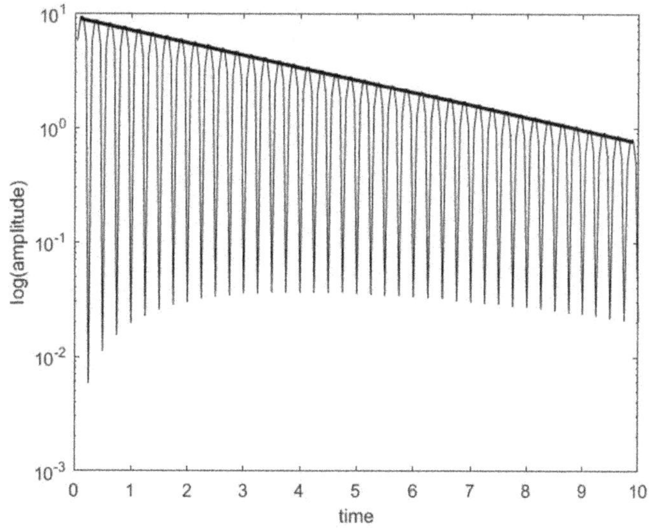

Figure 4.28 Straight line fit of $ln(|y(t)|)$.

The logarithmic decrement approach is particularly good for applications involving very lightly damped, well separated, modes.

4.2.4 Effect of Multiple Modes – When can *SDOF* Methods not be used?

All the methods that have been considered so far have been used with the assumption that each mode does not interact with any others; however, when the modes are "close", i.e. they have a significant interaction, the shapes of the *FRFs* differ compared to those of "well separated" modes. Consider Figure 4.29, where for a well separated 2 mode system (5 Hz, 1% damping, amplitude = 1 and 10 Hz, 1% damping, amplitude = -1) both the Bode and Nyquist plots clearly indicate how each *SDOF* can be identified. However, when the natural frequencies are set much closer, (9.5 Hz, 1% damping, amplitude = 1 and 10 Hz, 2% damping, amplitude = -1) the *FRF* shape changes. Considering the Bode plot, it might be thought that there is only one mode and an *SDOF* analysis will break down; however, the Nyquist plot shows how the two circles have merged into non-circular shapes. It can be seen that the *MMIF* eigenvalue plots still do a very good job of indicating both modes and the resonant frequencies.

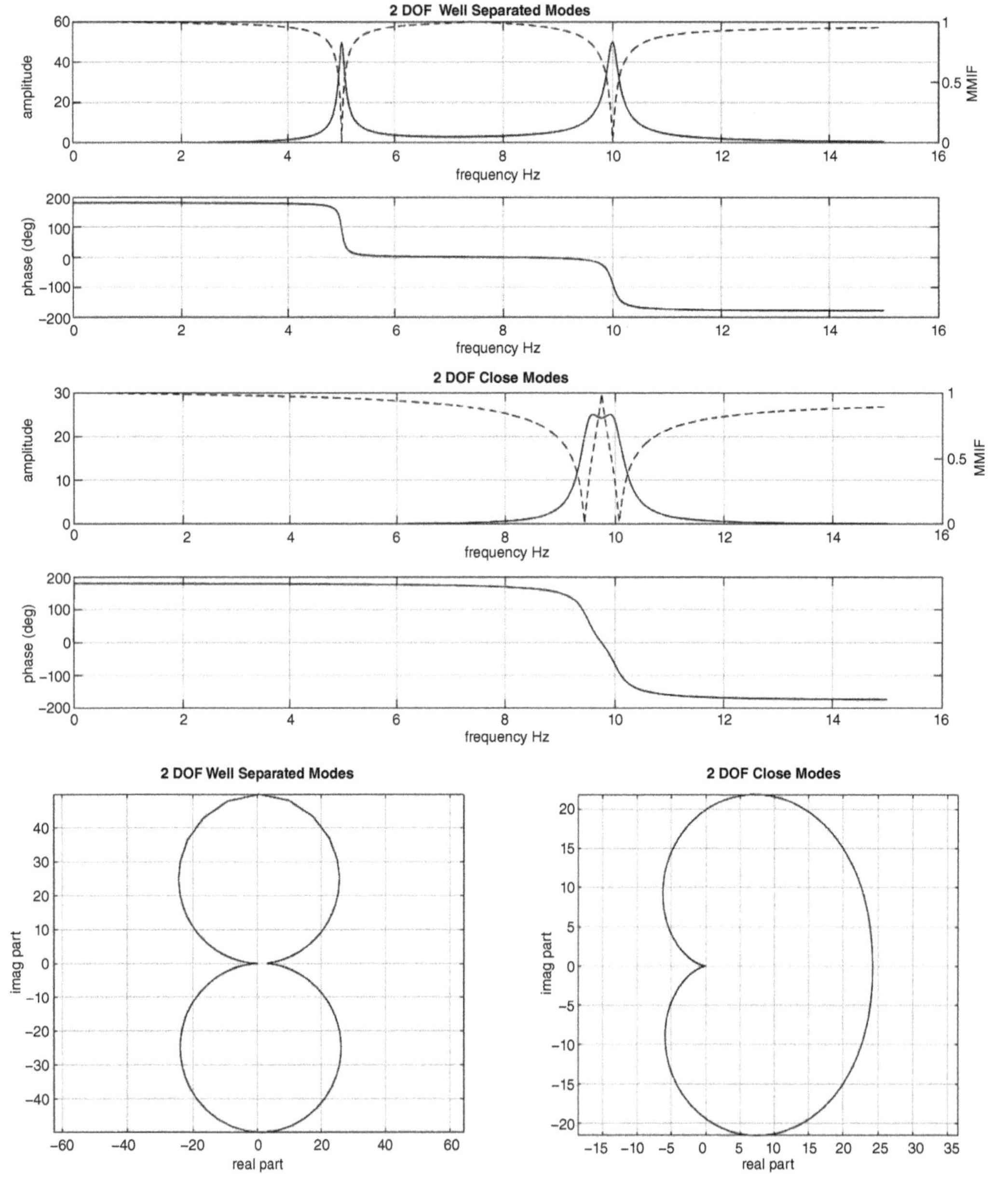

Figure 4.29 *FRFs* (Bode and Nyquist plots) for 2 mode systems – well separated and close modes.

Observe the time histories of Figures 4.30 (a) and 4.30 (b), which have been generated by selecting a frequency window around each of the well-separated peaks and then Inverse Fourier transforming to obtain the Impulse Response Functions. In these figures, both of the windowed modes show that the *SDOF* has been clearly isolated and the slope of the log (amplitude) plot is a straight line.

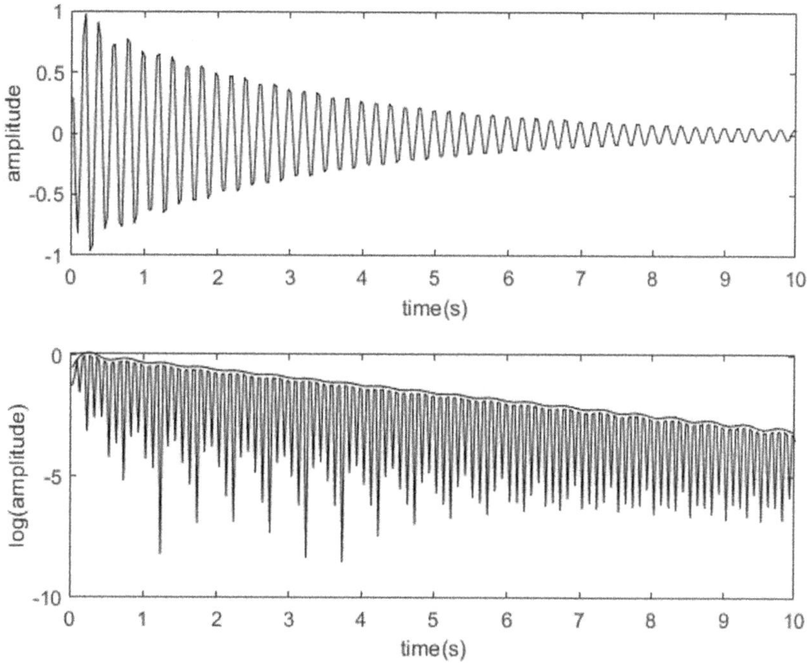

Figure 4.30 (a) Time histories of the lower windowed mode (well separated modes).

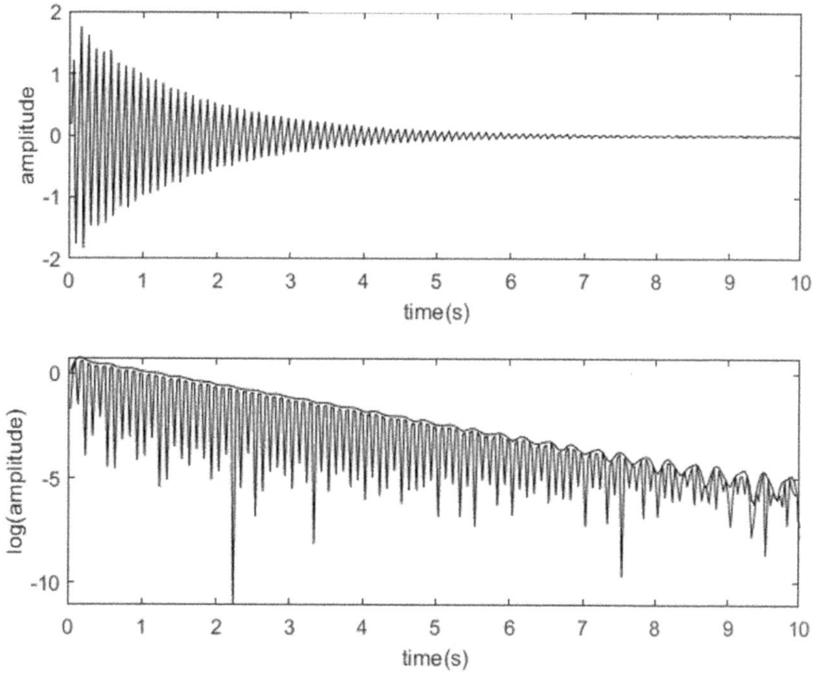

Figure 4.30 (b) Time histories of the upper windowed mode (well separated modes).

When two or more modes cannot be isolated it is clear, see Figure 4.30 (c), that the decay response contains more than one mode and the Logarithmic Decrement method cannot be used.

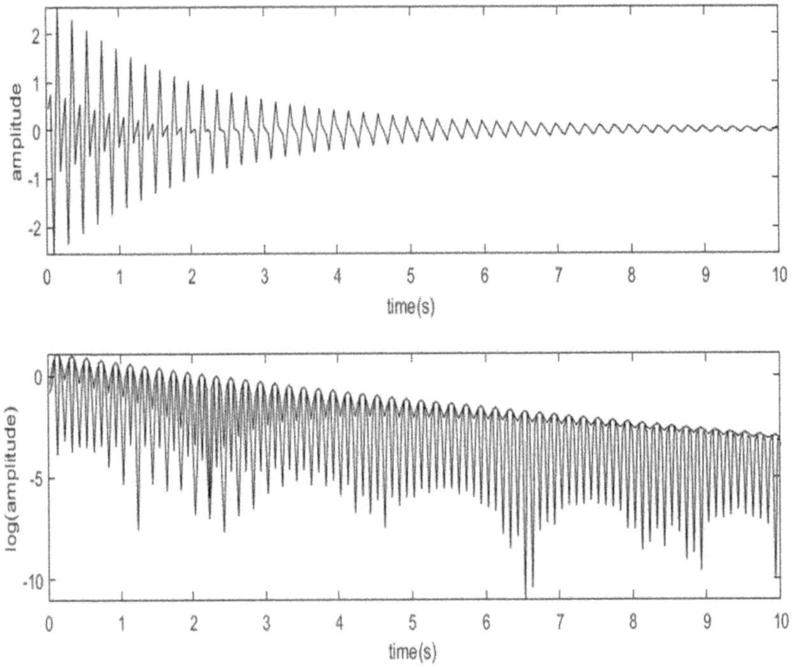

Figure 4.30 (c) Close modes – window across both modes.

4.3 *MDOF* Frequency Domain Methods

The limitations of the *SDOF* methods has been discussed previously, showing how they can be used to give initial, but not accurate, parameter estimates. More sophisticated *MDOF* curve-fitting approaches should be used if a more rigorous identification is needed.

4.3.1 Nonlinear *FRF* Curve-Fit

A brute force approach to determining the modal parameters is to perform a nonlinear (iterative) curve-fit to the *FRF* data. Such an iterative approach [4.21] is dependent upon using high-quality starting values to obtain good parameter estimates, otherwise convergence problems can occur.

Consider the *SISO FRF* model for N modes including, for generality, complex upper and lower residuals, such that

$$H(\omega)=H_R(\omega)+iH_I(\omega)=-\frac{LR_R}{\omega^2}-i\frac{LR_I}{\omega}+\sum_{r=1}^{N}\frac{A_r+i\omega B_r}{\omega_r^2-\omega^2+i2\omega\omega_r\xi_r}+UR_R+i\omega UR_I$$

$$=-\frac{LR_R}{\omega^2}+UR_R+\sum_{r=1}^{N}\frac{A_r\left(\omega_r^2-\omega^2\right)+2B_r\omega^2\omega_r\xi_r}{\left(\omega_r^2-\omega^2\right)^2+4\omega^2\omega_r^2\xi_r^2}+i\sum_{r=1}^{N}\frac{\omega B_r\left(\omega_r^2-\omega^2\right)-2A_r\omega\omega_r\xi_r}{\left(\omega_r^2-\omega^2\right)^2+4\omega^2\omega_r^2\xi_r^2}-i\frac{LR_I}{\omega}+i\omega UR_I$$

(4.64-a)

and there are now complex terms to represent the lower ($\omega_r << \omega$) and upper ($\omega << \omega_r$) residuals and also the residues of each mode. Closer inspection shows that the upper residual must be real and so $UR_I = 0$. Different formulations of this *FRF* could be used depending upon the measurement types that have been made.

Separating $H(\omega)=\sum_{r=1}^{N}\frac{A_r + i\omega B_r}{\omega_r^2-\omega^2+i2\omega\omega_r\xi_r}$ into its real and imaginary parts,

$$H(\omega)=H_R(\omega)+iH_I(\omega)=\sum_{r=1}^{N}\frac{A_r\left(\omega_r^2-\omega^2\right)+2B_r\omega^2\omega_r\xi_r}{\left(\omega_r^2-\omega^2\right)^2+4\omega^2\omega_r^2\xi_r^2}+i\sum_{r=1}^{N}\frac{\omega B_r\left(\omega_r^2-\omega^2\right)-2A_r\omega\omega_r\xi_r}{\left(\omega_r^2-\omega^2\right)^2+4\omega^2\omega_r^2\xi_r^2}$$

or

$$H(\omega)=\sum_{r=1}^{N}\frac{A_r\left(\omega_r^2-\omega^2\right)+2B_r\omega^2\omega_r\xi_r+i\omega\left(B_r\left(\omega_r^2-\omega^2\right)-2A_r\omega_r\xi_r\right)}{\omega_r^4-2\omega^2\omega_r^2+\omega^4+4\omega^2\omega_r^2\xi_r^2} \tag{4.64-b}$$

Making use of the Newton-Raphson method, one minimises the sum of the squares of the difference between the measured *FRFs* and the fitted model $\mathrm{e}(\omega)$, where

$$\begin{aligned} e(\omega) &= H_R(\omega)+iH_I(\omega)+\frac{LR_R}{\omega^2}-UR_R \\ &-\sum_{r=1}^{N}\frac{A_r\left(\omega_r^2-\omega^2\right)+2B_r\omega^2\omega_r\xi_r}{\left(\omega_r^2-\omega^2\right)^2+4\omega^2\omega_r^2\xi_r^2}-i\sum_{r=1}^{N}\frac{\omega B_r\left(\omega_r^2-\omega^2\right)-2A_r\omega\omega_r\xi_r}{\left(\omega_r^2-\omega^2\right)^2+4\omega^2\omega_r^2\xi_r^2}+i\frac{LR_I}{\omega} \end{aligned} \tag{4.65}$$

Summing for all the L frequencies, the squared error is $E=\sum_{i=1}^{L}e(\omega_i)^* e(\omega_i)$. E is a function of the unknowns $\left(LR_R, UR_R, LR_I, A_r, B_r, \omega_r, \xi_r\right)$. Let $\boldsymbol{x}$ represent the vector of the unknowns. The minimisation of the error leads to a set of nonlinear equations:

$$\frac{\partial E(\boldsymbol{x})}{\partial x_j}=f_j(\boldsymbol{x})=0 \quad , \quad j=1,\ldots 4N{+}4 \tag{4.66}$$

where x_j represents each of the unknowns. For each of these equations, initials values $\boldsymbol{x}_0$ are needed, so that in a first order approximation,

$$f_j(\boldsymbol{x})=f_j(\boldsymbol{x}_0)+\left.\frac{\partial \mathrm{f}_j}{\partial \boldsymbol{x}}\right|_{\boldsymbol{x}_0}\delta\boldsymbol{x} \quad , \quad j=1,\ldots 4N{+}4 \tag{4.67}$$

Substituting (4.66) in (4.67), leads to

$$\left.\frac{\partial E(\boldsymbol{x})}{\partial x_j}\right|_{\boldsymbol{x}_0}+\left.\frac{\partial^2 E(\boldsymbol{x})}{\partial x_j\partial\boldsymbol{x}}\right|_{\boldsymbol{x}_0}\delta\boldsymbol{x}=0 \quad , \quad j=1,\ldots 4N{+}4 \tag{4.68}$$

From (4.68), the increment $\delta\boldsymbol{x}$ is calculated and the variables are updated to the next iteration, until convergence is reached.

Considering the Wing-Pylon structure, Figure 4.31 shows a typical fit that was achieved using the nonlinear frequency domain approach. To ensure convergence good estimates of the initial guesses of the parameters are required, and it is also better to divide up the data into smaller frequency ranges consisting of a few modes, rather than trying to cover all the modes at once. The initial values for the residuals can be taken as zero. It is also possible to re-write this fit to include all the *FRFs* in one go rather than the individual fits considered here. Very good fits were achieved with this nonlinear *FRF* approach with the estimates shown in Table 4.5.

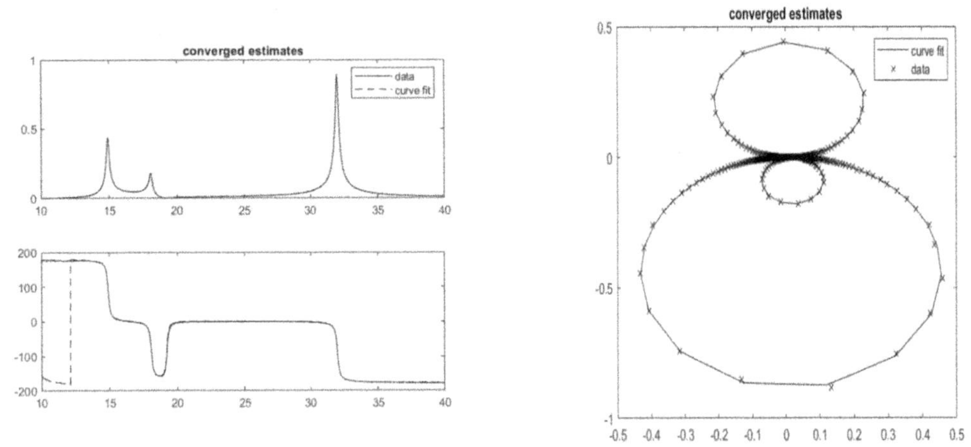

Figure 4.31 Curve-fitting for the Wing-Pylon model using a nonlinear *FRF* fit.

Table 4.5 Parameter estimates using a nonlinear *FRF* curve fitting approach.

	Mode 1		Mode 2		Mode 3		Mode 4	
	ω (Hz)	ξ (%)	ω (Hz)	ξ (%)	ω (Hz)	ξ (%)	ω (Hz)	ξ (%)
H_{11}	14.91	0.68	18.07	0.65	32.01	0.35	77.69	0.25
H_{21}	14.91	0.69	18.07	0.65	32.01	0.35	77.69	0.25
H_{31}	14.91	0.68	18.08	0.64	32.01	0.35	77.69	0.25
H_{41}	14.91	0.68	18.05	0.71	32.01	0.35	77.69	0.25
H_{51}	14.91	0.68	18.07	0.64	32.01	0.35	77.69	0.25
H_{61}	14.91	0.67	18.07	0.63	32.01	0.35	77.69	0.25
H_{71}	14.91	0.68	18.07	0.62	32.01	0.35	77.69	0.25
H_{81}	14.91	0.68	18.08	0.63	32.01	0.35	77.69	0.25
H_{91}	14.91	0.68	18.07	0.63	32.01	0.35	77.69	0.25
Av	**14.91**	**0.68**	**18.07**	**0.64**	**32.01**	**0.35**	**77.69**	**0.25**

4.3.2 Rational Fraction Polynomial Method

So far the methods that have been considered require a good idea of how many modes there are, either via *SDOF* or nonlinear iterative *MDOF* approaches, and what are the corresponding natural frequencies. The more common *Phase Separation* approaches consider a frequency band of interest and performs an analysis where the exact number of modes is initially unknown (although the analyst should have a good idea of what to expect). The generic identification process described here is used for most frequency and time domain *Phase Separation* methods.

Consider the usual expression seen before for the *FRF* of a *SISO* system with residuals such that

$$H(\omega)=\sum_{r=1}^{N}\left(\frac{R_r}{i\omega-\lambda_r}+\frac{R_r^*}{i\omega-\lambda_r^*}\right)+UR_R-\frac{LR_R}{\omega^2}-i\frac{LR_I}{\omega} \tag{4.69}$$

then this expression can be linearised into a rational fraction polynomial form, a frequency domain version of (4.16), for either *SISO*, *SIMO* or *MIMO* datasets in the frequency domain. For the *SISO* case such that for the general case of parameter $p = 2N$ (twice the number of assumed modes),

$$\begin{aligned}H(\omega)&=B(\omega)A^{-1}(\omega)\\&=\left(\beta_p(i\omega)^p+\beta_{p-1}(i\omega)^{p-1}+\ldots+\beta_0(i\omega)^0\right)\left(\alpha_p(i\omega)^p+\alpha_{p-1}(i\omega)^{p-1}+\ldots+\alpha_0(i\omega)^0\right)^{-1}\end{aligned} \tag{4.70}$$

where α_r provide the poles and β_r give the mode shape information. Note that there is no specific residual term included in this expression so extra terms need to be included in (4.70) to allow for this effect.

Rewriting Eq. (4.70) as

$$\begin{aligned}\boldsymbol{\alpha}_0H(\omega)=&-\boldsymbol{\alpha}_pH(\omega)(i\omega)^p-\boldsymbol{\alpha}_{p-1}H(\omega)(i\omega)^{p-1}\ldots-\boldsymbol{\alpha}_1H(\omega)(i\omega)\\&+\boldsymbol{\beta}_p(i\omega)^p+\boldsymbol{\beta}_{p-1}(i\omega)^{p-1}\ldots+\boldsymbol{\beta}_1(i\omega)+\boldsymbol{\beta}_0\end{aligned} \tag{4.71}$$

and by normalising to the α_0 term, this formulation enables the unknown coefficients to be expressed in a single vector for L frequency lines such that

$$\begin{Bmatrix}H(\omega_1)\\H(\omega_2)\\\vdots\\H(\omega_L)\end{Bmatrix}=\begin{bmatrix}H(\omega_1)(i\omega_1)^p & H(\omega_1)(i\omega_1)^{p-1} & \ldots & H(\omega_1)(i\omega_1) & (i\omega_1)^p & (i\omega_1)^{p-1} & \ldots & 1\\H(\omega_2)(i\omega_2)^p & H(\omega_2)(i\omega_2)^{p-1} & \ldots & H(\omega_2)(i\omega_2) & (i\omega_2)^p & (i\omega_2)^{p-1} & \ldots & 1\\\vdots & \vdots & & \vdots & \vdots & \vdots & & \vdots\\H(\omega_L)(i\omega_L)^p & H(\omega_L)(i\omega_L)^{p-1} & \ldots & H(\omega_L)(i\omega_L) & (i\omega_L)^p & (i\omega_L)^{p-1} & \ldots & 1\end{bmatrix}\begin{Bmatrix}-\alpha_p\\-\alpha_{p-1}\\\vdots\\-\alpha_1\\\beta_p\\\beta_{p-1}\\\vdots\\\beta_0\end{Bmatrix} \tag{4.72}$$

which can be solved in a least-squares sense, minimising the sum of the squares of the error $B(\omega)-A(\omega)H(\omega)$, to determine the unknown values α,β. For a *MIMO* system the α,β terms are $NI\times NI$ matrices, otherwise they are scalars. One issue with this formulation is that the data matrix involves terms with very high and also low powers of ω which can lead to conditioning problems in the identification process. The *Global Rational Fraction Polynomial* method [4.22] was developed to try and circumvent these problems through the use of orthogonal polynomials in the formulation.

The procedure for performing the identification takes the following three steps:

1. Determine the poles (and hence the frequencies and damping ratios) for a range of mathematical model orders;
2. Use a stabilisation diagram to select the frequency and damping values for each mode;
3. Estimate the mode shapes and upper / lower residuals.

It should be noted that the above process is followed by nearly all the Phase Separation methods.

Pole Identification

For a *SISO* polynomial model containing *2N* terms, which would correspond to *N* modes, the roots of the corresponding characteristic polynomial from the denominator of Eq. (4.70) becomes

$$\alpha_0\lambda^{2N} + \alpha_1\lambda^{2N-1} + \quad \dots \quad + \alpha_{2N-1}\lambda + \alpha_{2N} = 0 \tag{4.73}$$

where (from equations 4.2) the roots are formed as $\lambda_r = -\xi_r\omega_r \pm i\omega_r\sqrt{1-\xi_r^2}$. These roots lead to the natural frequencies and damping ratios:

$$\begin{aligned} &\lambda_r = -\xi_r\omega_r \pm i\omega_r\sqrt{1-\xi_r^2} = p_r \pm iq_r \\ &\rightarrow p_r = -\xi_r\omega_r \qquad q_r = \omega_r\sqrt{1-\xi_r^2} \\ &\rightarrow \omega_r = \sqrt{p_r^2 + q_r^2} \quad \text{and} \quad \xi_r = -\frac{p_r}{\sqrt{p_r^2 + q_r^2}} \end{aligned} \tag{4.74}$$

For *MIMO* systems the α terms are $NI \times NI$ matrices, and so there would be $s = 2N \times NI$ roots of the characteristic equation

$$\boldsymbol{Y}_{\mathrm{j}} + \boldsymbol{\alpha}_1\boldsymbol{Y}_{\mathrm{j-1}} + \quad \dots \quad + \boldsymbol{\alpha}_{\mathrm{S-1}}\boldsymbol{Y}_{\mathrm{j-S+1}} + \boldsymbol{\alpha}_{\mathrm{S}}\boldsymbol{Y}_{\mathrm{j-S}} = \mathbf{0} \tag{4.75}$$

A more efficient, and accurate, way to do this is to substitute Eq. (4.25) into (4.73) giving the expression

$$e^{\Lambda^S}\boldsymbol{B} + e^{\Lambda^{S-1}}\boldsymbol{B}\boldsymbol{\alpha}_1 + \quad \dots \quad + e^{\Lambda}\boldsymbol{B}\boldsymbol{\alpha}_{S-1} + \boldsymbol{B}\boldsymbol{\alpha}_S = \mathbf{0} \tag{4.76}$$

and this equation can be reformulated into the eigenvalue form

$$\begin{bmatrix} -\boldsymbol{\alpha}_1^T & -\boldsymbol{\alpha}_2^T & \dots & \dots & -\boldsymbol{\alpha}_{s-1}^T & -\boldsymbol{\alpha}_s^T \\ \boldsymbol{I} & 0 & & \dots & \dots & 0 \\ 0 & \boldsymbol{I} & & & & \vdots \\ \vdots & \vdots & \ddots & & 0 & 0 \\ 0 & 0 & \dots & \boldsymbol{I} & 0 & 0 \\ 0 & 0 & \dots & \dots & \boldsymbol{I} & 0 \end{bmatrix} \begin{bmatrix} \boldsymbol{B}^T e^{\Lambda^{S-1}} \\ \boldsymbol{B}^T e^{\Lambda^{S-2}} \\ \vdots \\ \boldsymbol{B}^T e^{\Lambda} \\ \boldsymbol{B}^T \end{bmatrix} = \begin{bmatrix} \boldsymbol{B}^T e^{\Lambda^{S-1}} \\ \boldsymbol{B}^T e^{\Lambda^{S-2}} \\ \vdots \\ \boldsymbol{B}^T e^{\Lambda} \\ \boldsymbol{B}^T \end{bmatrix} e^{\Lambda} \tag{4.77}$$

The eigenvalues of the above sparse Hessenberg matrix give the poles, where $\Lambda = diag\left[\lambda_1, \lambda_1, \ldots, \lambda_{2N}\right]$ and the eigenvectors give the modal participation factors $\boldsymbol{B}$.

4.3.3 Stability Plots – How Many Modes are there?

The mathematical models that have been used enable the estimation of the system poles and hence the frequencies and dampings. However, the problem arises when dealing with real structures in that it is unknown exactly how many modes are present in the data sets and how much "noise" might be present. A further difficulty is that if the curve-fit is performed using the exact number of modes that are present in the system, then "bias" (a systematic error) will occur [4.18] due to the use of the least-squares method as a fundamental element of most methods, and also residual errors from the out of analysis band modes. Slightly different results will be obtained when considering different input sequences, data lengths, frequency ranges, mathematical model orders, analysis methods and whether *SISO*, *SIMO* or *MIMO* data sets are chosen. Typically, the frequency estimates will vary by much less than those of the mode shapes and the damping factors, with the latter being particularly sensitive to variations in the data sets or analysis approach chosen.

To overcome these problems it has become common-place to compare the modal parameter solutions obtained using a number of different mathematical model orders on the same data set and to assume that the true system parameters will not vary by very much for each solution, whereas there will be much more variation in the spurious (i.e. have no physical meaning) modes. This approach has been shown to produce good estimates, without any rigourous mathematical proof, and is now standard. A methodology is required to be able to distinguish between the system (true) modes and the mathematical modes (spurious) that are also estimated.

A number of tools are available for the engineer to help determining the true parameter estimates, and the use of engineering common-sense (e.g. do the results agree with what was expected?) is also very useful. It should also be remembered that there will always be some variation, particularly for noisy data sets, and determining the amount of this deviation can help to give some confidence on the accuracy of the identification.

The key tool is the so called "Stability Plot" [4.23] shown in Figure 4.32, where the estimated modal parameters are plotted for different mathematical model orders (one mode, two modes, etc.) and then the results for one order are compared with the previous. If the parameters change within some pre-defined limits, as shown in Table 4.6, then the point is represented by a number of different symbols. In some commercial applications coloured letters are used. For very noisy data sets (e.g. wind tunnel or flight flutter testing) it might be required to relax the stabilisation criteria. If the identified estimates are not permissible, e.g. out of frequency range or negative damping, then no symbol is shown.

Considering the stability plot in Figure 4.32, using the stabilization criteria in Table 4.6, the summed *FRFs* and the *MMIF* eigenvalues indicate that there are modes around 134 Hz, 309 Hz, 390 Hz, 484 Hz and 562 Hz, and the trends of the stabilised poles support this hypothesis. The stability plot was computed using fits between 2 and 32 modes (the model order); these are typical ranges to choose and the user needs to ensure that large enough model orders are employed to ensure stabilised frequency and damping estimates although stability might be achieved with a smaller model order. A number of columns of 'O's can be seen which indicate a set of estimates that vary very little between the different model orders, and these indicate a

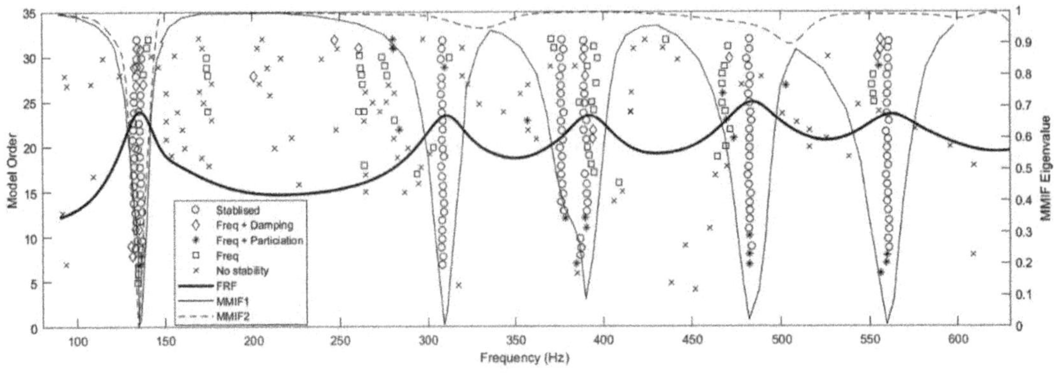

Figure 4.32 Stability Plot with summed *FRFs* and *MMIF* eigenvalues.

Table 4.6 Typical Stability Plot Stabilisation Parameters.

Parameter	Maximum Variation	Symbol
Unpermissable value		None
Unstabilised		X
Frequency	1%	□
Frequency and participation factors	1% and 3%	*
Frequency and damping	1% and 5%	◊
All parameters stabilised	All of the above	O

mode. There are a number of other symbols on the plot, some showing partial stabilisation on the way towards full stabilistion, but the others relate to spurious estimates. Estimates that are obviously wrong, e.g. negative damping factors are simply ignored and not plotted. Typically, the frequency values will stabilise first, followed by the participation factors and then the damping values.

The modes indicated by the stabilised columns are reinforced by comparison with the peaks of the summed *FRF* curve, and also the dips in the *MMIF* eigenvalues. For the lowest mode it can be seen that both (this data set was obtained using simultaneous excitation from two shakers, hence there are two *MMIFs*) the *MMIF* eigenvalues drop down, indicating two close modes, and careful inspection shows that there are actually two very close columns of stabilised values.

Consider the frequency range 380Hz – 390 Hz where, although the block *FRF* indicates a single mode, there are two columns of stabilised poles, one of which corresponds to a kink in the first *MMIF* eigenvalue. It is likely that there is a mode around 380 Hz, but it does not appear on all the *FRFs*. A further analysis should be performed on a subset of *FRFs* that include this mode.

The user selects one of the stabilised poles from each of the columns of 'O's representing each mode. There is no need to select modes from the same model order as some modes will stabilise before others, and it is considered best practice to pick one of the lower order stabilised values. All the modes in Figure (4.32) stabilise between a model order of 10 and 15. For much larger mathematical models, it is possible that poles will "split" and there will be two poles fitting a single peak or the stabilised column stops, as can be seen beyond a model order of 25 on the two close modes close to 134 Hz. Engineering judgement should be employed to consider whether the identified mode shapes and damping values seem sensible.

If the identified results still seem poor, consider fitting a smaller frequency bandwidth, ensuring that the boundaries are chosen at distinct points between the *FRF* peaks and check that the time decays (for time domain methods) that are being used are sensible. Interpreting stability plots can sometimes require a lot of engineering judgement.

4.3.4 Mode Shape Estimation – Least-Squares Frequency Domain

Having identified the natural frequencies and damping ratios of the system modes then, considering Eq. (4.64) for the *FRF* model with complex residuals, it can be seen that the unknowns in the denominator have been computed and so the unknown residue and residual terms are now linear in their formulation [4.24]. A linear approach is then feasible to determine the remaining terms in the *FRF* formulation. Considering (4.64) for a single mode, and then separating real and imaginary parts one obtains the expressions

$$\begin{aligned} H_R(\omega) &= \frac{A_r\left(\omega_r^2-\omega^2\right)+2B_r\xi_r\omega_r\omega^2}{\left(\omega_r^2-\omega^2\right)^2+\left(2\xi_r\omega_r\omega\right)^2}+UR_R-\frac{LR_R}{\omega^2} \\ H_I(\omega) &= \frac{\left(\omega B_r\left(\omega_r^2-\omega^2\right)-2A_r\xi_r\omega_r\omega\right)}{\left(\omega_r^2-\omega^2\right)^2+\left(2\xi_r\omega_r\omega\right)^2}-\frac{LR_I}{\omega} \end{aligned} \tag{4.78}$$

and, defining the denominator as $D(\omega)=\left(\omega_r^2-\omega^2\right)^2+\left(2\xi_r\omega_r\omega\right)^2$, a set of simultaneous linear equations can be formulated for L frequencies such that

$$\begin{Bmatrix} H_R(\omega_1) \\ H_I(\omega_1) \\ \vdots \\ H_R(\omega_L) \\ H_I(\omega_L) \end{Bmatrix} = \begin{bmatrix} 1 & -\frac{1}{\omega_1^2} & \frac{\left(\omega_r^2-\omega_1^2\right)}{D(\omega_1)} & \frac{2\xi_r\omega_r\omega_1^2}{D(\omega_1)} & 0 \\ 0 & 0 & -\frac{2\xi_r\omega_r\omega_1}{D(\omega_1)} & \frac{\omega_1\left(\omega_r^2-\omega_1^2\right)}{D(\omega_1)} & -\frac{1}{\omega_1} \\ \vdots & \vdots & \vdots & \vdots & \vdots \\ 1 & -\frac{1}{\omega_L^2} & \frac{\left(\omega_r^2-\omega_L^2\right)}{D(\omega_L)} & \frac{2\xi_r\omega_r\omega_L^2}{D(\omega_L)} & 0 \\ 0 & 0 & -\frac{2\xi_r\omega_r\omega_L}{D(\omega_L)} & \frac{\omega_L\left(\omega_r^2-\omega_L^2\right)}{D(\omega_L)} & -\frac{1}{\omega_L} \end{bmatrix} \begin{Bmatrix} UR_R \\ LR_R \\ A_r \\ B_r \\ LR_I \end{Bmatrix} \tag{4.79}$$

which, having supplied the previously identified natural frequencies and damping ratios, can be solved to find the unknowns A_r, B_r and other residual terms. This is the Least-Squares Frequency Domain (*LSFD*) process.

The user can decide whether to use the upper and lower residuals, or to set them as real values, but it is recommended to use the full equations. The number of modes included in such a fit can be varied so that subsets of the modes are considered rather than employing the full range in one go. It should be stressed that that good initial estimates of the natural frequencies and damping ratios must be provided to get accurate mode shapes.

Having computed these parameters, the user can then consider the comparison of the measured and fitted *FRFs* to see how good the fit is and whether any modes have been missed or are badly fitted. Figure 4.33 shows an example of these plots, and in this case the correlation between data and curve-fit is excellent. To improve the analysis, using a small frequency range with fewer modes is recommended. It is good practice to explore the identified mode shapes to see whether they make sense from a physical point of view.

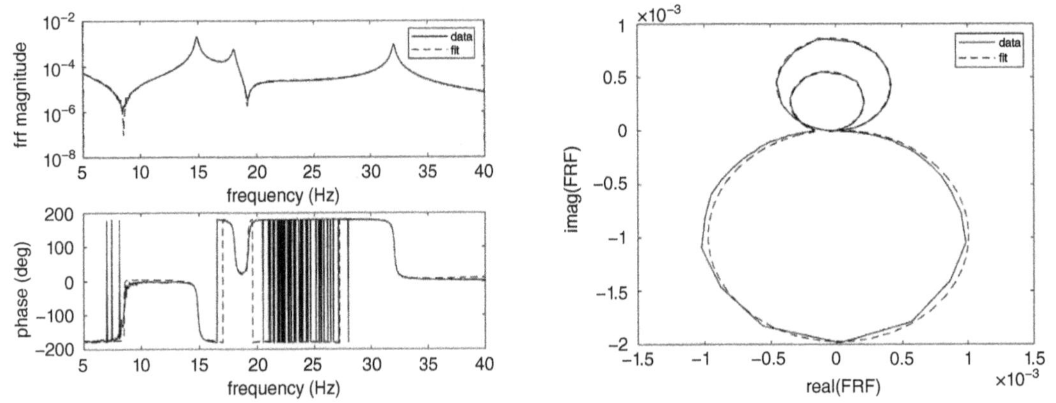

Figure 4.33 Comparison of measured *FRFs* and *LSFD* fitted models.

4.3.5 *PolyMAX*

The *PolyMAX* method [4.16] was introduced to improve upon previous frequency domain methods, as with the Least-Squares Complex Frequency Domain [4.17], a variant of the Least-Squares Complex Exponential. One of the main advantages over those other methods is to provide much clearer and easier to interpret stability plots, commonly used for many Phase Separation methods. This benefit is due to a key difference in the formulation, which is based on the z domain (a frequency domain method derived from a discrete time model) as opposed to the usual Laplace domain expression (a frequency domain formulation derived from a continuous time model).

Making use of the z transformation $z = e^{i\omega\Delta t}$, the Rational Fraction expansion for a *SISO FRF* becomes

$$H(\omega) = \left(\beta_p z^p + \beta_{p-1} z^{p-1} + \ldots + \beta_0\right)\left(\alpha_p z^p + \alpha_{p-1} z^{p-1} + \ldots + \alpha_0\right)^{-1} \tag{4.80}$$

or

$$\alpha_0 H(\omega) = -\alpha_p H(\omega) z^p - \alpha_{p-1} H(\omega) z^{p-1} \ldots - \alpha_1 H(\omega) z + \beta_p z^p + \beta_{p-1} z^{p-1} \ldots + \beta_1 H(\omega) z + \beta_0 \tag{4.81}$$

By normalising to the α_0 term and considering L frequencies, this formulation enables the unknowns to be expressed in a single block vector such that

$$\begin{Bmatrix} H(\omega_1) \\ H(\omega_2) \\ \vdots \\ H(\omega_L) \end{Bmatrix} = \begin{bmatrix} H(\omega_1)z_1^p & H(\omega_1)z_1^{p-1} & \cdots & H(\omega_1)z_1 & z_1^p & z_1^{p-1} & \cdots & 1 \\ H(\omega_2)z_2^p & H(\omega_2)z_2^{p-1} & \cdots & H(\omega_2)z_2 & z_2^p & z_2^{p-1} & \cdots & 1 \\ \vdots & \vdots & & \vdots & \vdots & \vdots & & \vdots \\ H(\omega_L)z_L^p & H(\omega_L)z_L^{p-1} & \cdots & H(\omega_L)z_L & z_L^p & z_L^{p-1} & \cdots & 1 \end{bmatrix} \begin{Bmatrix} -\alpha_p \\ -\alpha_{p-1} \\ \vdots \\ -\alpha_1 \\ \beta_p \\ \beta_{p-1} \\ \vdots \\ \beta_0 \end{Bmatrix} \tag{4.82}$$

which can be solved in a least-squares sense to determine the unknown values α, β.

The same three step approach as described above for the Rational Fraction Polynomial method can then be used to determine the natural frequencies, damping ratios, modal participation factors and mode shapes.

However, as the α_k parameters are not required for the mode shape computations, it makes sense to reformulate the identification formulation so that only the α_k parameters are found using the following approach. Writing the solution equations (4.82) in a block format such that

$$\boldsymbol{H} = \begin{bmatrix} \boldsymbol{H}_\alpha & \boldsymbol{H}_\beta \end{bmatrix} \begin{bmatrix} \boldsymbol{\alpha} \\ \boldsymbol{\beta} \end{bmatrix} \tag{4.83}$$

then the least-squares solution is found from

$$\begin{bmatrix} \boldsymbol{H}_\alpha^T \\ \boldsymbol{H}_\beta^T \end{bmatrix} \boldsymbol{H} = \begin{bmatrix} \boldsymbol{H}_\alpha^T \\ \boldsymbol{H}_\beta^T \end{bmatrix} \begin{bmatrix} \boldsymbol{H}_\alpha & \boldsymbol{H}_\beta \end{bmatrix} \begin{bmatrix} \boldsymbol{\alpha} \\ \boldsymbol{\beta} \end{bmatrix}$$
$$\rightarrow \boldsymbol{H}_\alpha^T \boldsymbol{H} = \boldsymbol{H}_\alpha^T \boldsymbol{H}_\alpha \boldsymbol{\alpha} + \boldsymbol{H}_\alpha^T \boldsymbol{H}_\beta \boldsymbol{\beta} \quad \text{and} \quad \boldsymbol{H}_\beta^T \boldsymbol{H} = \boldsymbol{H}_\beta^T \boldsymbol{H}_\alpha \boldsymbol{\alpha} + \boldsymbol{H}_\beta^T \boldsymbol{H}_\beta \boldsymbol{\beta} \tag{4.84}$$

The second equation can be rewritten in terms of $\boldsymbol{\beta}$, giving

$$\boldsymbol{\beta} = \left(\boldsymbol{H}_\beta^T \boldsymbol{H}_\beta\right)^{-1} \left(\boldsymbol{H}_\beta^T \boldsymbol{H} - \boldsymbol{H}_\beta^T \boldsymbol{H}_\alpha \boldsymbol{\alpha}\right) = \left(\boldsymbol{H}_\beta^T \boldsymbol{H}_\beta\right)^{-1} \boldsymbol{H}_\beta^T \left(\boldsymbol{H} - \boldsymbol{H}_\alpha \boldsymbol{\alpha}\right) \tag{4.85}$$

and so $\boldsymbol{\alpha}$ can be found by substituting $\boldsymbol{\beta}$ into the first equation such that

$$\boldsymbol{H}_\alpha^T \boldsymbol{H} = \boldsymbol{H}_\alpha^T \boldsymbol{H}_\alpha \boldsymbol{\alpha} + \boldsymbol{H}_\alpha^T \boldsymbol{H}_\beta \left(\boldsymbol{H}_\beta^T \boldsymbol{H}_\beta\right)^{-1} \boldsymbol{H}_\beta^T \left(\boldsymbol{H} - \boldsymbol{H}_\alpha \boldsymbol{\alpha}\right)$$
$$\rightarrow \boldsymbol{H}_\alpha^T \boldsymbol{H} = \left(\boldsymbol{H}_\alpha^T \boldsymbol{H}_\alpha - \boldsymbol{H}_\alpha^T \boldsymbol{H}_\beta \left(\boldsymbol{H}_\beta^T \boldsymbol{H}_\beta\right)^{-1} \boldsymbol{H}_\beta^T \boldsymbol{H}_\alpha\right) \boldsymbol{\alpha} + \boldsymbol{H}_\alpha^T \boldsymbol{H}_\beta \left(\boldsymbol{H}_\beta^T \boldsymbol{H}_\beta\right)^{-1} \boldsymbol{H}_\beta^T \boldsymbol{H} \tag{4.86}$$
$$\rightarrow \boldsymbol{\alpha} = \left(\boldsymbol{H}_\alpha^T \boldsymbol{H}_\alpha - \boldsymbol{H}_\alpha^T \boldsymbol{H}_\beta \left(\boldsymbol{H}_\beta^T \boldsymbol{H}_\beta\right)^{-1} \boldsymbol{H}_\beta^T \boldsymbol{H}_\alpha\right)^{-1} \left(\boldsymbol{H}_\alpha^T \boldsymbol{H} - \boldsymbol{H}_\alpha^T \boldsymbol{H}_\beta \left(\boldsymbol{H}_\beta^T \boldsymbol{H}_\beta\right)^{-1} \boldsymbol{H}_\beta^T \boldsymbol{H}\right)$$

The implementation of the *PolyMAX* method follows the same three stages as the other Phase Separation methods: determine the poles for several different mathematical model orders, use a stabilisation diagram to select the frequency and damping values for each mode, and then estimate the mode shapes and upper / lower residuals. Figure 4.34 and Table 4.7 show the stability plot and estimated modal parameters for the wing/engine model using the *PolyMAX* method; the four modes in the frequency range of interest are clearly identified.

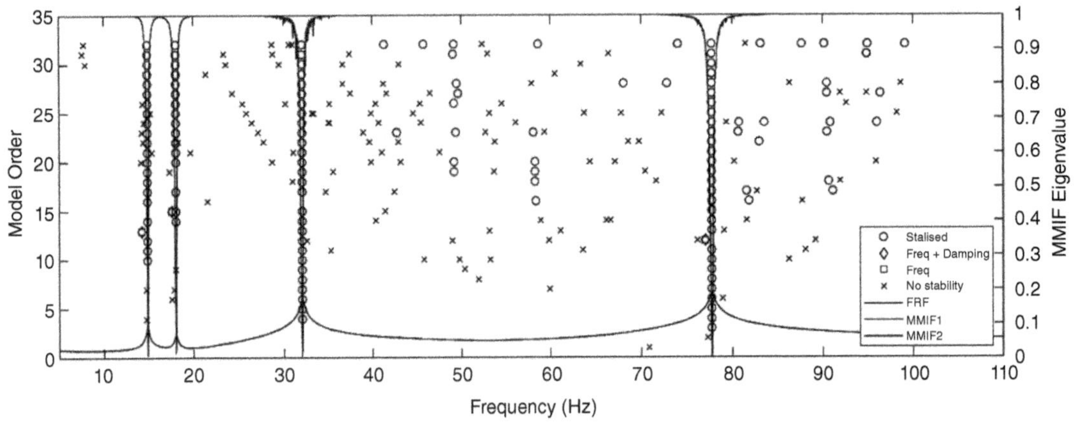

Figure 4.34 Stability plot for Wing-Pylon data using *PolyMAX*.

Table 4.7 Estimated natural frequencies and damping ratios using *PolyMAX*.

Mode	1	2	3	4
Natural Frequency (Hz)	14.75	17.82	31.73	77.38
Damping ratio (%)	0.54	0.59	0.30	0.17

4.4 *MDOF* Time Domain Methods

A common approach to determine the modal parameters is to perform the analysis in the time domain using the Impulse Response Functions generated from the inverse Fourier Transforms of the *FRFs*; however, other approaches (e.g. step release, ambient responses) can be used – see Figure 4.8. Such methods can be formulated in *SISO*, *SIMO* or *MIMO* versions, where the multiple inputs related to different excitation cases and the curve-fits are performed, assuming the responses are the summation of exponentially damped sinusoids.

4.4.1 Extended Logarithmic Decrement Method

An extension to the classical logarithmic decrement method [4.25, 4.26] is to modify it to enable the identification of modal parameters from the decay response of a lightly damped system ($\omega_n \approx \omega_d$) containing two modes.

Consider the following response where each mode is lightly damped and the modes are either totally in-phase or out of phase with each other, such that

$$y(t) = Ae^{-\xi_1\omega_1 t}cos\omega_1 t + Be^{-\xi_2\omega_2 t}cos\omega_2 t \quad (4.87)$$

where $\omega_2 \geq \omega_1$. It can be shown that this expression can be reformulated as

$$y(t) = \left(Ae^{-\xi_1\omega_1 t} + Be^{-\xi_2\omega_2 t}\right)cos\omega_p t\, cos\omega_m t + \left(Ae^{-\xi_1\omega_1 t} - Be^{-\xi_2\omega_2 t}\right)sin\omega_p t\, sin\omega_m t \quad (4.88)$$

with $\omega_p = \dfrac{\omega_1 + \omega_2}{2}$ and $\omega_m = \dfrac{\omega_2 - \omega_1}{2}$.

Examining the typical two *DOF* transient response shown in Figure 4.35, the presence of "beating" results in a series of peaks and troughs that do not uniformly decrease in magnitude. The term *minor* will be used to refer to the peaks and troughs of the response time history, and *major* will be used to refer to the peaks and troughs of the response envelope. Closer inspection of the response also reveals that the spacing of the peaks is not uniform.

Eq. (4.88) implies that the major peaks have amplitudes bounded by

$$y_p = Ae^{-\xi_1\omega_1 t} + Be^{-\xi_2\omega_2 t} \quad (4.89)$$

and the major troughs have amplitudes bounded by

$$y_t = Ae^{-\xi_1\omega_1 t} - Be^{-\xi_2\omega_2 t} \quad (4.90)$$

The dotted lines in Figure 4.35 indicate the major peaks and major troughs.

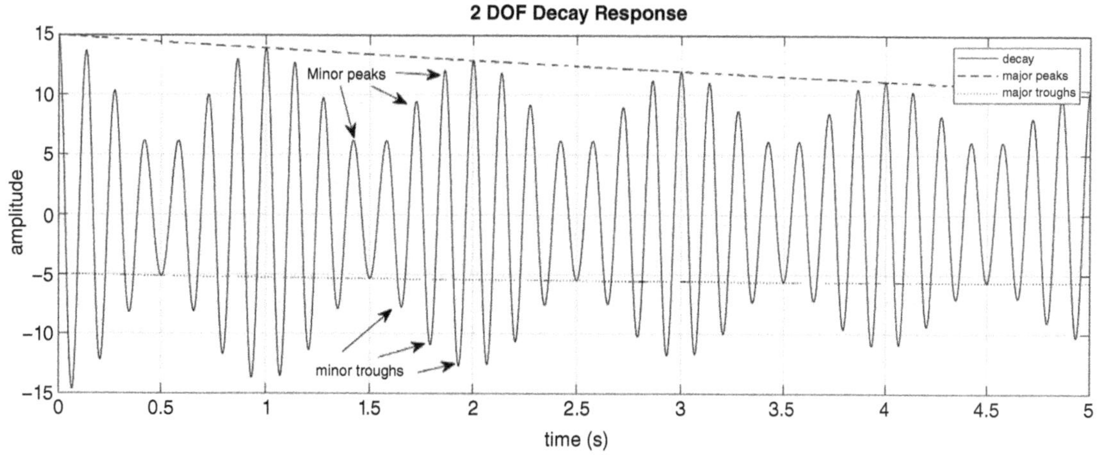

Figure 4.35 Transient response of a 2 *DOF* system, showing major peaks and troughs.

The absolute peak and trough values are used to construct the curves joining the major peaks and troughs, but they may be positive or negative, depending upon the amplitudes and frequencies of the system.

Natural Frequency Estimation

The natural frequencies ω_1 and ω_2 can be determined from the following observations:

1. The beating occurs at a frequency equal to the difference between the two frequencies $\omega_2 - \omega_1$;

2. The number of peaks between the major peaks gives the frequency of the largest mode;

3. Starting at one of the major peaks, the time interval between the minor peaks will *increase* moving towards the next minor trough if the second frequency is *greater* than the one found in point 2 above or *decrease* if the second frequency is *smaller*.

Damping and Amplitude Estimates

Considering the curves connecting the major peaks y_p and major troughs y_t shown in Figure 4.36, interpolation can be used to determine the values of both at the same time instant. By adding and subtracting the values of the major troughs and peaks from Eqs. (4.89) and (4.90), one obtains

$$y_p + y_t = 2Ae^{-\xi_1\omega_1 t} \text{ and } y_p - y_t = 2Be^{-\xi_2\omega_2 t} \tag{4.91}$$

Thus, in an approach analogous to the logarithmic decrement method, the logarithm of these equations can be written as

$$ln\left(y_p + y_t\right) = ln\left(2A\right) - \xi_1\omega_1 t \text{ and } ln\left(y_p - y_t\right) = ln\left(2B\right) - \xi_2\omega_2 t \tag{4.92}$$

So, by taking a number of points, a straight line fit to the slope can be used to find the damping ratios and the intersection with the y axis the amplitudes. Care has to be taken so that the largest absolute values of the major peaks and troughs are used, and also that any negative amplitude terms are estimated correctly.

Example

The transient response of a 2 *DOF* system with modal parameters listed in Table 4.8 is shown in Figure 4.35. By considering the absolute value of the amplitude plot, shown in Figure 4.36, it is relatively straightforward to determine the major peaks and troughs and these are tabulated in Table 4.9., and Table 4.10. displays the minor peaks. These values can then be interpolated as in Figure 4.37 so that the major peak and trough values at the same time instance can be found. Finally, by adding or subtracting these extrapolated values, the plots shown in Figure 4.38 are achieved, with the logarithmic plots being used to find the damping values.

Table 4.8 Parameters of a 2 *DOF* system.

	Natural Frequency (Hz)	Damping Ratio (%)	Amplitude
Mode 1	7	0.1	10
Mode 2	8	0.3	5

Table 4.9 Amplitudes and time of the major peaks and troughs.

Positive Peaks		Positive Troughs		Negative Peaks		Negative Troughs	
Amplitude	Time(s)	Amplitude	Time(s)	Amplitude	Time(s)	Amplitude	Time(s)
15.00	0.00	6.162	0.582	15.00	0.00	6.162	0.582
13.870	1.00	6.161	1.580	13.870	1.00	6.161	1.580
12.857	2.00	6.147	2.579	12.857	2.00	6.147	2.579
11.944	3.00	6.114	3.723	11.944	3.00	6.114	3.723
11.122	4.00			11.122	4.00		

Table 4.10 First seven minor peaks of the response.

Amplitude	15.000	13.66	10.309	6.162	6.162	10.025	12.927
Time(s)	0.0000	0.136	0.273	0.418	0.582	0.726	0.864
Interval		0.136	0.137	0.145	0.164	0.144	0.138

1. The beating interval (time between major peaks) is 1 second, therefore the difference between the two frequencies is 1 Hz.
2. There are seven peaks between the major peaks, consequently the natural frequency of the largest amplitude mode is 7 Hz.
3. Considering Table 4.10. the time interval between the minor peaks increases, so the second mode frequency is greater than the largest amplitude mode by the inverse of the beating interval, thus 7 Hz + 1 Hz = 8 Hz.
4. Considering Figure 4.38, and taking into account the closeness of the frequencies, it can be seen that the largest amplitude mode has the lowest slope and hence less damping. Fitting the straight lines to the log plots gives the damping ratio and amplitude values shown in Table 4.11.

Table 4.11 Identified parameters of the 2 *DOF* system using the Extended Log Dec method.

	Natural frequency (Hz)	Damping Ratio (%)	Amplitude
Mode 1	7	0.094	10.03
Mode 2	8	0.34	5.04

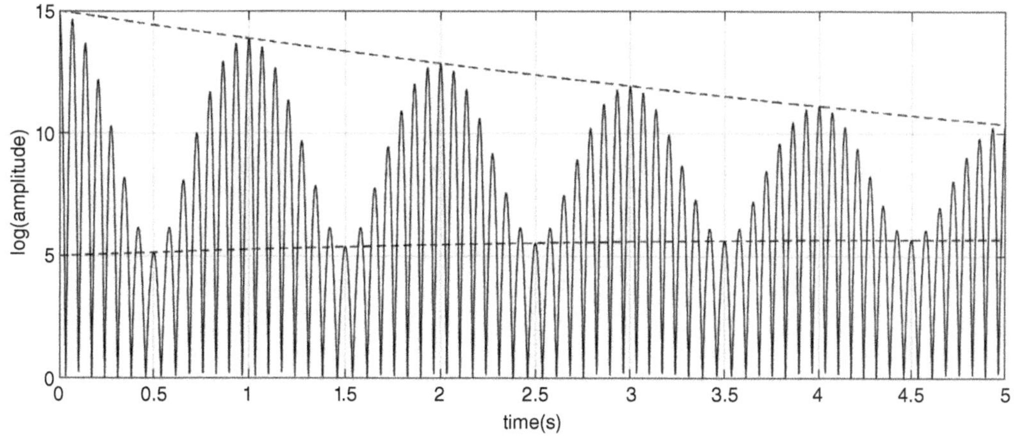

Figure 4.36 Selection of major peaks and troughs using the absolute value of the response.

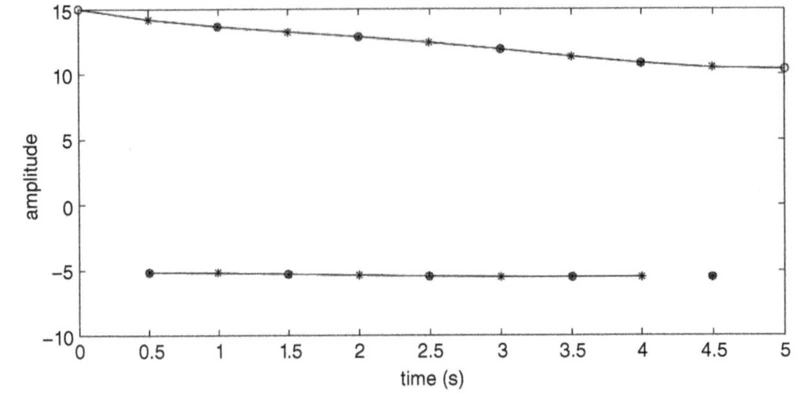

Figure 4.37 Interpolated major peak and trough values.

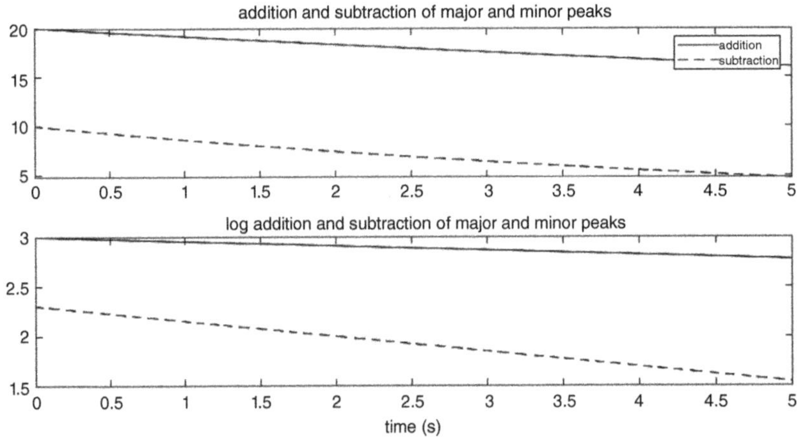

Figure 4.38 Addition and subtraction of major and minor peaks and troughs

4.4.2 Least-Squares Complex Exponential

The Least-Squares Complex Exponential (*LSCE*) method [4.24], based upon the Prony method [4.27] (originally developed in the 18th century) was the forerunner of the many time domain system identification methods. It follows the three stage process that is typical of such phase separation methods, as discussed previously for the Rational Fraction Polynomial and *PolyMAX* methods.

Consider the *SISO* case, the *MDOF* impulse response can be written as a summation of exponentially damped sine waves [4.18, 4.28], as seen previously (4.20), but now in the form

$$y_j = \sum_{k=1}^{2N} C_k e^{(-\xi_k \omega_k + i\omega_{dk}) j\Delta t} \tag{4.93}$$

which can be written as the $2N^{\text{th}}$ order difference equation (or autoregression)

$$y_j + a_1 y_{j-1} + a_2 y_{j-2} \quad \cdots \quad + a_{2N} y_{j-2N} = 0 \tag{4.94}$$

where

$$y_j = \sum_{k=1}^{2N} C_k e^{\lambda_k j \Delta t} = \sum_{k=1}^{2N} C_k v_k^j \tag{4.95}$$

and C_k occurs in complex conjugate pairs and λ_k are the corresponding eigenvalues.

There is a natural extension to the *SIMO* case, where the y_j terms in Eq. (4.94) become vectors. In both cases, it can be shown that this difference equation has the related characteristic polynomial

$$v_r^{2N} + a_1 v_r^{2N-1} + \quad \ldots \quad + a_{2N-1} v_r + a_{2N} = 0 \tag{4.96-a}$$

with $v_r = e^{\lambda_r \Delta t}$ and $\lambda_r = -\xi_r \omega_r \pm i\omega_r \sqrt{1-\xi_r^2}$ as before, leading to estimates of the natural frequencies and damping ratios in a slightly different way, as one is now dealing with the time domain, to that described above in Eq. (4.74) such that

$$\begin{aligned}
& v_r = v_R \pm i v_I = \mathrm{e}^{\left(-\xi_r \omega_r \pm i\omega_r \sqrt{1-\xi_r^2}\right)\Delta t} \\
& \rightarrow ln\left(v_R \pm i v_I\right) = ln\left(\sqrt{v_R^2 + v_I^2}\right) \pm i\, tan^{-1}\left(\frac{v_I}{v_R}\right) = p_r \pm i q_r = \left(\lambda_R \pm i\lambda_I\right)\Delta t \\
& \rightarrow p_r = -\xi_r \omega_r \Delta t \qquad q_r = \omega_r \sqrt{1-\xi_r^2}\,\Delta t \\
& \rightarrow \omega_r = \frac{\sqrt{p_r^2 + q_r^2}}{\Delta t} \quad \text{and} \quad \xi_r = -\frac{p_r}{\sqrt{p_r^2 + q_r^2}}
\end{aligned} \tag{4.96-b}$$

Step 1 - Estimation of *AR* Coefficients, Natural Frequencies and Damping Ratios
Expanding the above difference Eq. (4.94) for L data points and combining into a matrix form, gives

$$\begin{Bmatrix} y_{2N+1} \\ y_{2N+2} \\ \vdots \\ y_L \end{Bmatrix} = \begin{bmatrix} y_{2N} & y_{2N-1} & \cdots & y_1 \\ y_{2N+1} & y_{2N} & \cdots & y_2 \\ \vdots & \vdots & & \vdots \\ y_{L-1} & y_{L-2} & \cdots & y_{L-2N} \end{bmatrix} \begin{Bmatrix} -a_1 \\ -a_2 \\ \vdots \\ -a_{2N} \end{Bmatrix} + \begin{Bmatrix} \varepsilon_{2N+1} \\ \varepsilon_{2N+2} \\ \vdots \\ \varepsilon_L \end{Bmatrix}$$

or

$$\boldsymbol{y} = \boldsymbol{\Xi}\,\boldsymbol{\theta} + \boldsymbol{\varepsilon} \tag{4.97}$$

where $\boldsymbol{\varepsilon}$ is a vector containing error terms to allow for noise, nonlinearities or other deficiencies in the model. The structure of Ξ, where the diagonal terms sloping downwards to the right are the same, is known as a Toeplitz matrix.

There must be at least as many rows as unknown parameters and ideally many more points are chosen. The time window can be chosen to focus on the modes of interest, higher frequency modes will decay much quicker than the low frequency modes. Also make sure that the very low amplitude responses are not included as they are very likely to include lots of noise and little contribution from the modes – see Figures 4.18 - resulting in inaccurate estimates. Figure 4.39 shows a typical time domain response and its corresponding frequency content for one the beam pylon structure responses that includes the pylon modes.

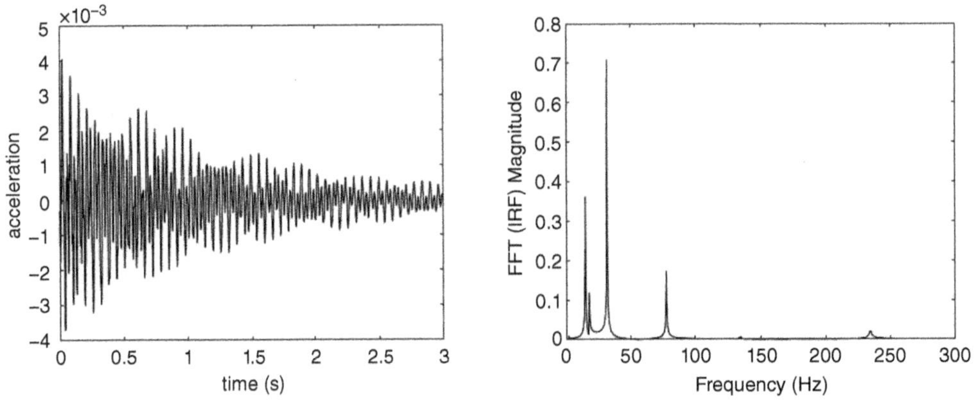

Figure 4.39 Typical impulse response - time and frequency content.

Eq. (4.97) can be solved in a least-squares sense to find the unknown vector of *AR* coefficients, such that

$$\boldsymbol{\theta} = \left(\boldsymbol{\Xi}^{\mathrm{T}}\boldsymbol{\Xi}\right)^{-1}\boldsymbol{\Xi}^{\mathrm{T}}\boldsymbol{y} \tag{4.98}$$

although, as discussed previously, it is better not to explicitly solve the pseudo-inverse and to use approaches such as the *QR* or *SVD* decompositions to solve the simultaneous linear equations. (e.g. use the 'backslash' command in *MATLAB*)

The roots of Eq. (4.96) can be solved using any polynomial root solving method, but it is better practice to find the eigenvalues of the matrix (for *N* modes) as discussed previously in Section 4.3.2. For example, rewriting Eq. (4.94) into the matrix formulation

$$\begin{bmatrix} y_{2N+1} & y_{2N} & \cdots & y_2 \\ y_{2N+2} & y_{2N} & \cdots & y_3 \\ \vdots & \vdots & & \vdots \\ y_L & y_{L-1} & \cdots & y_{L-2N+1} \end{bmatrix} = \begin{bmatrix} y_{2N} & y_{2N-1} & \cdots & y_1 \\ y_{2N+1} & y_{2N} & \cdots & y_2 \\ \vdots & \vdots & & \vdots \\ y_{L-1} & y_{L-2} & \cdots & y_{L-2N} \end{bmatrix} \begin{bmatrix} -a_1 & 1 & 0 & \cdots & 0 \\ -a_2 & 0 & 1 & \cdots & 0 \\ \vdots & & & \ddots & \vdots \\ \vdots & \vdots & \vdots & & 1 \\ -a_{2N} & 0 & 0 & \cdots & 0 \end{bmatrix}$$

or

$$\boldsymbol{\Xi}_1 = \boldsymbol{\Xi}_0 \boldsymbol{Q} \tag{4.99}$$

enables the eigenvalues of the sparse Hessenberg matrix $\boldsymbol{Q}$ to be used to obtain the system poles.

Step 2 – Determination of the System Poles using Stability Plots

As with the frequency domain phase separation methods described previously, the most suitable mathematical models size must be determined to eliminate the bias that is inherent in an exact order solution. Figure 4.40 (a) shows a different form of stability plot whereby the frequencies that are computed for each mathematical model order are displayed and it can clearly be seen where the columns of stabilised frequencies correspond to the spectral frequencies at 14.9 Hz, 17.6 Hz, 77.7 Hz, 134.7 Hz and 234.0 Hz. Those modes that have a smaller spectral content typically take longer to stabilise, with the 17.6 Hz taking longest as a larger mathematical model is required to separate it from the much larger surrounding modes.

A further plot that can be used to determine the amount of scatter on the modal estimates is shown in Figure 4.40 (b) where all of the frequencies identified on the stability chart are plotted against their corresponding damping values. Clusters of identified poles can be seen showing the identified modes. Note that there is always much more scatter on the damping values. Choosing the centre of each frequency/damping cluster in Figure 4.40 (b) eliminates the stability plot issue of determining which pole to select from each stabilised column in Figure 4.40 (a) and also gives an indication of the robustness of the frequency and damping estimates.

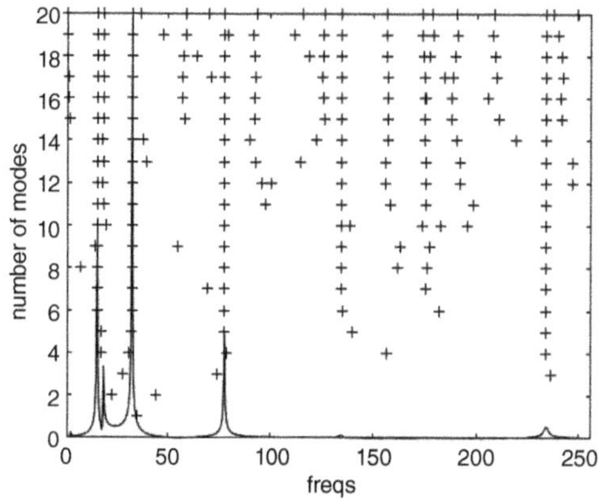

Figure 4.40 (a) Estimated frequencies *versus* mathematical model order.

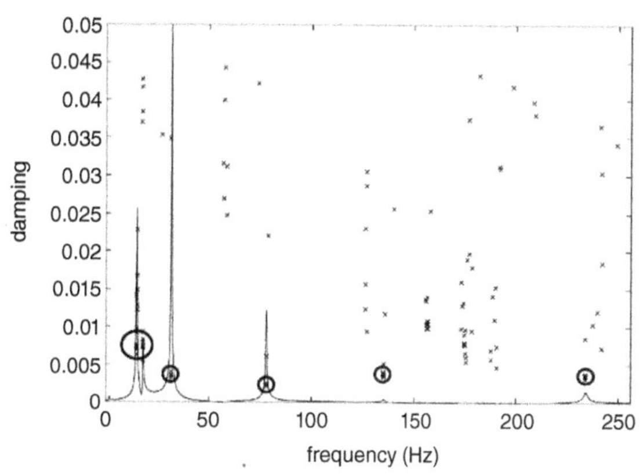

Figure 4.40 (b) Estimated frequencies *versus* damping ratios.

A further tool to aid the analyst is the *Least-Squares Error Plot* which indicates the error for the fit to the decay response for each model order, as shown in Figure 4.41 (a), which can be used to determine how many dominant modes there are in the data. Note that the greater the model order that is used the better the fit will be until, in the extreme case, the fit goes through every point. The trends show that there is an "elbow" (or "dog-leg") which indicates the number of significant modes, in this case is 8. Note that if this value was assumed to be the best model order, then the mode at 17.6 Hz would be missed.

A similar approach to determine the main contributing modes is to take the *SVD* of the largest (i.e. biggest model order) Ξ matrix in Eq. (4.97). Theoretically, the rank of the system is defined by the number of non-zero singular values (*svs*), however in practice on real data sets there will be a number of close to non-zero *svs*. Figure 4.41 (b) shows plots of the *svs* and their cummulative sum, normalised so that their sum is 100%. Beyond a value of 10 the *svs* all become very close to zero and this corresponds to 98.5% of the cummulative sum, so this plot is indicating that there are 5 (= 10 /2) dominant modes. As with the error plot, the *svs* give an indication of the number of modes contributing to the impulse responses and their use should be treated with care.

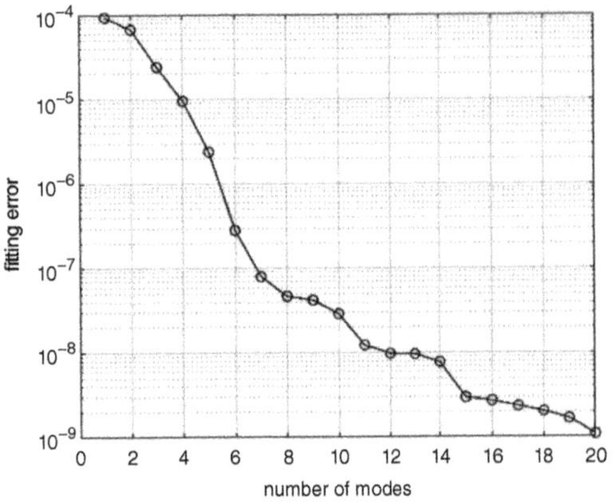

Figure 4.41 (a) Least-squares error.

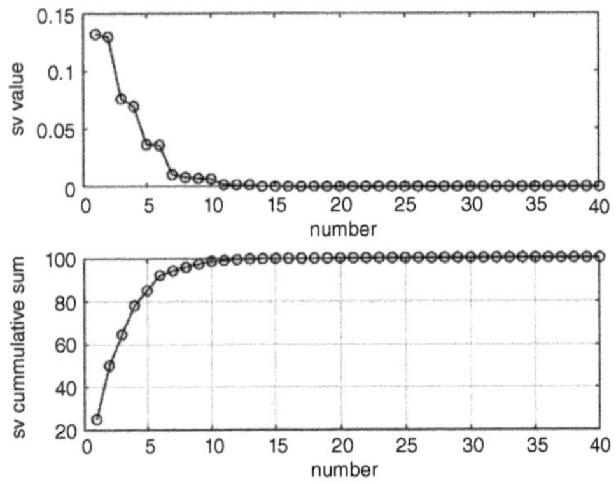

Figure 4.41 (b) Singular values plots.

Step 3 – Determination of the Mode Shapes

Once the natural frequencies and damping ratios have been estimated, via the poles λ_r, the amplitude terms C_r can be found in two ways.

i. *Time domain approach*

Considering the exponential form (4.95) for the decaying response $y_j = \sum_{k=1}^{2N} C_k e^{\lambda_k j\Delta t}$ which can be written into a matrix form for each time point giving

$$\begin{Bmatrix} y_1 \\ y_2 \\ \vdots \\ y_L \end{Bmatrix} = \begin{bmatrix} e^{\lambda_1 \Delta t} & e^{\lambda_2 \Delta t} & \cdots & e^{\lambda_{2N} \Delta t} \\ e^{\lambda_1^2 \Delta t} & e^{\lambda_2^2 \Delta t} & \cdots & e^{\lambda_{2N}^2 \Delta t} \\ \vdots & \vdots & & \vdots \\ e^{\lambda_1^L \Delta t} & e^{\lambda_2^L \Delta t} & \cdots & e^{\lambda_{2N}^L \Delta t} \end{bmatrix} \begin{Bmatrix} C_1 \\ C_2 \\ \vdots \\ C_{2N} \end{Bmatrix} \tag{4.100}$$

As one now knows the roots of the characteristic equation (4.96), these equations can be solved to find the unknown C_k amplitude terms. Improved speed for this calculation can be achieved by taking into account that the C and λ terms occur in complex conjugate pairs.

ii. *Frequency domain approach* using the Least-Squares Frequency Domain [4.24] as described in Section 4.3.4.

4.4.3 Polyreference Method

The Multiple-Input Multi-Output (*MIMO*) version of the *LSCE* method is known as Polyreference [4.29] and, for *NO* measurement stations and *NI* input cases at the kth time instant, Eq. (4.25) can be written as

$$\boldsymbol{Y}_{k+1} = \begin{bmatrix} y_{1,1} & \cdots & y_{1,NI} \\ \vdots & & \vdots \\ y_{NO,1} & \cdots & y_{NO,NI} \end{bmatrix}_k = \boldsymbol{W}\boldsymbol{D}^k\boldsymbol{V} \tag{4.101}$$

where $\boldsymbol{D} = diag\left[e^{\lambda_1 \Delta t}, e^{\lambda_2 \Delta t}, ..., e^{\lambda_{2N} \Delta t}\right]$ and $\boldsymbol{Y}_k$ is a $(NO \times NI)$ block data matrix. The use of block matrix notation enables the *SISO* formulation to be expanded directly to the *MIMO* case. $\boldsymbol{W}$ is a $(NO \times 2N)$ matrix containing the complex mode shapes and $\boldsymbol{V}$ is a $(2N \times NI)$ matrix containing the modal participation factors.

It can be shown that Eq. (4.101) is the general solution of the matrix difference equation

$$\boldsymbol{Y}_k + \boldsymbol{Y}_{k-1}\boldsymbol{A}_1 + \ldots + \boldsymbol{Y}_{k-s}\boldsymbol{A}_s = \boldsymbol{0}_{NI*NI} \tag{4.102}$$

if the matrix polynomial equation

$$\boldsymbol{D}^s\boldsymbol{V} + \boldsymbol{D}^{s-1}\boldsymbol{V}\boldsymbol{A}_1 + \ldots + \boldsymbol{D}\boldsymbol{V}\boldsymbol{A}_{s-1} + \boldsymbol{V}\boldsymbol{A}_s = \boldsymbol{0}_{N \times NI} \tag{4.103}$$

is satisfied, where $(s \times NI) = 2N$ in a similar way to (4.24), and $\boldsymbol{A}_i$ are $(NI \times NI)$ block difference equation coefficients.

Eq. (4.103) can be rewritten into a canonical form, such that

$$\begin{bmatrix} \boldsymbol{D}^{s-1}\boldsymbol{V} & \boldsymbol{D}^{s-2}\boldsymbol{V} & \ldots & \boldsymbol{V} \end{bmatrix} \begin{bmatrix} -\boldsymbol{A}_1 & \boldsymbol{I} & \boldsymbol{0} & \cdots & \boldsymbol{0} \\ -\boldsymbol{A}_2 & \boldsymbol{0} & \boldsymbol{I} & \cdots & \boldsymbol{0} \\ -\boldsymbol{A}_3 & \boldsymbol{0} & \boldsymbol{0} & & \vdots \\ \vdots & \vdots & \vdots & & \boldsymbol{I} \\ -\boldsymbol{A}_s & \boldsymbol{0} & \boldsymbol{0} & \cdots & \boldsymbol{0} \end{bmatrix} = \boldsymbol{D}\begin{bmatrix} \boldsymbol{D}^{s-1}\boldsymbol{V} & \boldsymbol{D}^{s-2}\boldsymbol{V} & \ldots & \boldsymbol{V} \end{bmatrix} \tag{4.104}$$

With this mathematical background established, a similar approach to the *LSCE* can be used to determine the modal parameters. Firstly, the block difference Eq. (4.102) is extended column-wise so that

$$\begin{bmatrix} \boldsymbol{Y}_{s+1} \\ \boldsymbol{Y}_{s+2} \\ \vdots \\ \boldsymbol{Y}_L \end{bmatrix} = \begin{bmatrix} \boldsymbol{Y}_s & \boldsymbol{Y}_{s-1} & \cdots & \boldsymbol{Y}_1 \\ \boldsymbol{Y}_{s+1} & \boldsymbol{Y}_s & \cdots & \boldsymbol{Y}_2 \\ \vdots & \vdots & & \vdots \\ \boldsymbol{Y}_{L-1} & \boldsymbol{Y}_{L-2} & \cdots & \boldsymbol{Y}_{L-s} \end{bmatrix} \begin{bmatrix} -\boldsymbol{A}_1 \\ -\boldsymbol{A}_2 \\ \vdots \\ -\boldsymbol{A}_s \end{bmatrix} \quad \rightarrow \quad \boldsymbol{Y} = \boldsymbol{\Xi}\boldsymbol{\Theta} \tag{4.105}$$

and then the Least-Squares method can be used to estimate the block $\boldsymbol{A}_i$ coefficients by minimising the sum of the squares of the error on the $\boldsymbol{Y}$ vector so that

$$\boldsymbol{\Theta} = \left(\boldsymbol{\Xi}^{\mathrm{T}}\boldsymbol{\Xi}\right)^{-1}\boldsymbol{\Xi}^{\mathrm{T}}\boldsymbol{Y} \tag{4.106}$$

The sparse Hessenberg matrix in Eq. (4.104) is formed and solving the eigenvalue problem leads give the roots of the block polynomial via the eigenvalues, giving the system frequencies and damping ratios as before. The modal participation factors are then found from the eigenvectors of the eigenproblem in Eq. (4.104).

Stability plots are then employed to determine the system modes by comparing solutions from different model orders.

The computation of the mode shapes can either be undertaken in the Least-Squares Frequency Domain method described above, or in the time domain by expanding Eq. (4.101) column-wise so that

$$\begin{bmatrix} \boldsymbol{Y}_1^T \\ \boldsymbol{Y}_2^T \\ \vdots \\ \boldsymbol{Y}_L^T \end{bmatrix} = \begin{bmatrix} \boldsymbol{V}^T \\ \boldsymbol{V}^T\boldsymbol{D} \\ \vdots \\ \boldsymbol{V}^T\boldsymbol{D}^{L-1} \end{bmatrix} \boldsymbol{W}^T \tag{4.107}$$

which can again be solved using least-squares to obtain the mode shapes.

4.4.4 Eigensystem Realisation Algorithm

The *LSCE* and Polyreference methods have the least-squares method at the heart of their computation [4.18] and consequently overspecification of the mathematical model is required in order to reduce the effects of bias and produce accurate estimates, leaving many spurious modes to be separated from the system modes. The Eigensystem Realization Algorithm (*ERA*) [4.30] was introduced to produce what is known as "minimum realization" solutions so that although a large amount of model order overspecification can be included in the formulation to reduce any potential bias, the actual number of estimated modes can be much less. Mathematically, this approach is much more efficient than using a full rank solution.

Starting in a similar way to the Polyreference method by defining a time block of impulse responses for *NO* simultaneously sampled response stations responding to *NI* input cases and such that for the $(k+1)^{th}$ time instant (see Eq. (4.26)), one has

$$\boldsymbol{Y}_{k+1} = \begin{bmatrix} y_{1,1} & \cdots & y_{1,NI} \\ \vdots & & \vdots \\ y_{NO,1} & \cdots & y_{NO,NI} \end{bmatrix}_k = \boldsymbol{C}\boldsymbol{A}^k\boldsymbol{B} \tag{4.108}$$

where $\boldsymbol{Y}$ is a $(NO \times NI)$ data matrix, $\boldsymbol{A}$ is the $(2N \times 2N)$ system matrix, $\boldsymbol{C}$ and $\boldsymbol{B}$ are $(NO \times 2N)$ and $(2N \times NI)$ measurement and input matrices respectively. Given the transient responses, the aim of the process is to identify the $\boldsymbol{A}$, $\boldsymbol{B}$ and $\boldsymbol{C}$ matrices from which the modal parameters can be determined. This formulation fits in very neatly with the state-space formulations used by control engineers [4.14].

Computing the eigen decomposition of the system matrix $\boldsymbol{A}$ such that

$$\boldsymbol{A} = \boldsymbol{\Psi}\boldsymbol{Z}\boldsymbol{\Psi}^{-1} \tag{4.109}$$

where $\boldsymbol{Z}$ is a diagonal matrix of system eigenvalues whose elements have the form $\lambda = \mathrm{e}^{\left(-\xi\omega_n \pm i\omega_n\sqrt{1-\xi^2}\right)\Delta t}$ then Eq. (4.108) becomes

$$Y_{k+1} = C\Psi Z^k \Psi^{-1} B \tag{4.110}$$

where matrix $C\Psi$ gives the system mode shapes, and matrix $\Psi^{-1}B$ contains the modal participation factors, which indicate how effective a particular input is at exciting each mode.

A block data Hankel matrix is constructed such that for the kth time increment

$$H_k = \begin{bmatrix} Y_{k+1} & Y_{k+2} & \cdots & Y_{k+\nu+1} \\ Y_{k+2} & Y_{k+3} & \cdots & Y_{k+1} \\ \vdots & \vdots & & \vdots \\ Y_{k+\mu+1} & Y_{k+\mu+2} & \cdots & Y_{k+\mu+\nu+1} \end{bmatrix} = \begin{bmatrix} C \\ CA \\ \vdots \\ CA^{\mu} \end{bmatrix} A^k \begin{bmatrix} B & AB & \dots & A^{\nu}B \end{bmatrix} = VA^kW \tag{4.111}$$

where V is a $((\mu+1)NO \times 2N)$ observability matrix, W is a $(2N \times (\nu+1)NI)$ controllability matrix, and μ and ν dictate the dimensions of the block Hankel matrix.

The *SVD* of the Hankel matrix at $k = 0$ can be written as

$$H_0 = P_F D_F Q_F^T \approx PDQ^T \tag{4.112}$$

and if H_0 has dimensions $(a \times b)$ with $b > a$, then the compact *SVD* gives the dimensions of the full orthogonal matrices P_F and Q_F^T as $(a \times a)$ and $(a \times b)$ respectively, and the diagonal singular value matrix D_F. One can now truncate the matrices, in the same manner as in Section 4.2.2 about the *SVD*, by choosing only *2N* singular values, so that the dimensions of P, D and Q^T become $(a \times 2N)$, $(2N \times 2N)$ and $(2N \times b)$ respectively.

The pseudo-inverse of H_0, denoted as H^+, is defined as

$$H_0 H^+ H_0 = H_0 \quad \text{where} \quad H^+ = QD^{-1}P^{\mathrm{T}} \tag{4.113}$$

and it can be shown that

$$\begin{gathered} Y_{k+1} = E_{NO}^T H_k E_{NI} \\ \text{with} \quad E_{NO}^T = \begin{bmatrix} I_{NO} & 0_{NO} & \dots & 0_{NO} \end{bmatrix} \quad \text{and} \quad E_{NI}^T = \begin{bmatrix} I_{NI} & 0_{NI} & \dots & 0_{NI} \end{bmatrix} \end{gathered} \tag{4.114}$$

where I_k and 0_k are $(k \times k)$ identity and null matrices respectively.

Combining all the above equations leads, after some manipulation, to the minimum realisation (so-called as the truncation of the singular values matrix reduces the size of all the computed matrices to the smallest amount that will still provide information about all of the modes):

$$Y_{k+1} = E_{NO}^T P D^{1/2} \left(D^{-1/2} P^T H_1 Q D^{-1/2} \right)^k D^{1/2} Q^T E_{NI} \tag{4.115}$$

Thus,

$$
\begin{aligned}
A &= D^{-1/2} P^T H_1 Q D^{-1/2} = \Psi Z^k \Psi^{-1} \\
B &= D^{1/2} Q^T E_{NI} \\
C &= E_{NO}^T P D^{1/2}
\end{aligned}
\tag{4.116}
$$

Following the identification of the $\boldsymbol{A}$ matrix, the natural frequencies and damping ratios can be computed from the eigenvalues of the diagonal $\boldsymbol{Z}$ matrix Determining the eigenvector matrix $\boldsymbol{\Psi}$ enables the mode shapes and modal participation factors to be computed from $\boldsymbol{C\Psi}$ and $\boldsymbol{\Psi}^{-1}\boldsymbol{B}$, respectively. If the impulse responses have been obtained via the *FRFs*, then the mode shape estimation can be undertaken using the *LSFD* method as described above.

The implementation of the *ERA* method is exactly the same as the *LSCE* and Polyreference methods, only instead of varying the model order to produce the stability plots, a large initial model size is chosen (smallest dimension of the $\boldsymbol{H}_0$ matrix) and then different minimum realisation orders are used to form the stability plots. Figure 4.42 (a) shows a typical stability plot and the equivalent damping *versus* frequency cluster plot is seen in Figure 4.42 (b). Equivalent parameter estimates to the other methods are found, and typically the stabilisation plots can be easier to interpret; there is also very little scatter in frequency and damping for the system modes.

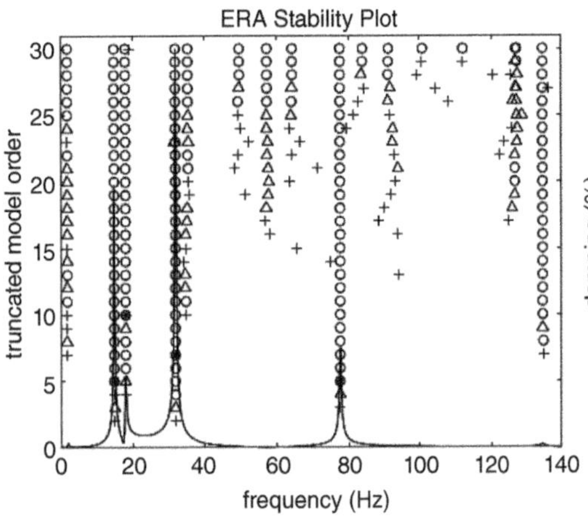

Figure 4.42 (a) *ERA* Stability Plot.
o = stabilised in freq and damping
Δ = stabilised in freq
+ = not stabilised.

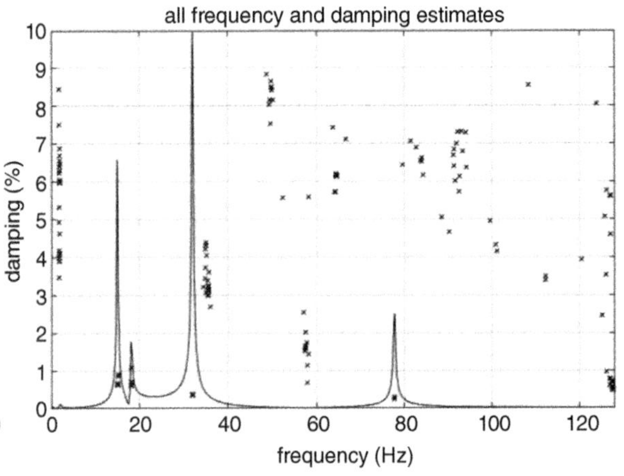

Figure 4.42 (b) *ERA* – Frequency *versus* Damping plots.

4.4.5 Reverse Data Fitting

A simple approach that can be used to improve the phase separation time domain methods that curve-fit transient decays, including the ambient response methods discussed in the next section, is to consider the data sets in the reversed direction [4.31]. In this case, the exponential envelope expands rather than contracts, and therefore the system modes have negative damping of the same magnitude as the forwards analysis, but the natural frequencies remain the same.

Having completed the analysis all the user has to do is multiply each identified damping ratio by –1. Moreover, it can be shown mathematically that when the impulse responses are analysed in the reverse direction, the spurious poles tend to lie on the opposite side of the unit circle to the system poles; so it is straightforward to remove these estimates from the stability plots as their damping values will have the opposite sign. There is no loss of accuracy on the pole estimates.

Figure 4.43 shows the *forwards* impulse response and its corresponding *reversed* time history. The corresponding stability and damping *versus* frequency cluster plots for the *reverse* data sets can be seen in Figure 4.44 and they are less "busy" than the equivalent forwards plots. The reason for this observation is shown in Figure 4.45 where there are many negative damping spurious estimates that can be eliminated from the analysis.

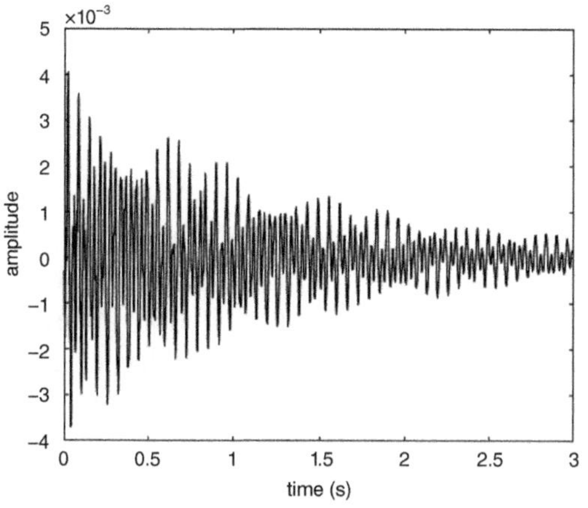

Figure 4.43 (a) Forwards Impulse Response Function.

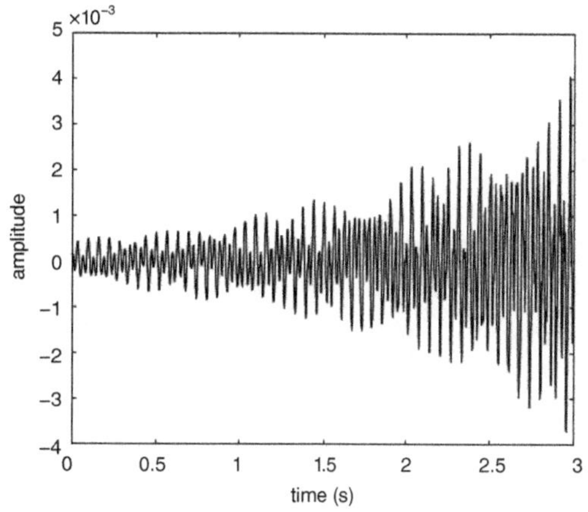

Figure 4.43(b) Reversed Impulse Response Function.

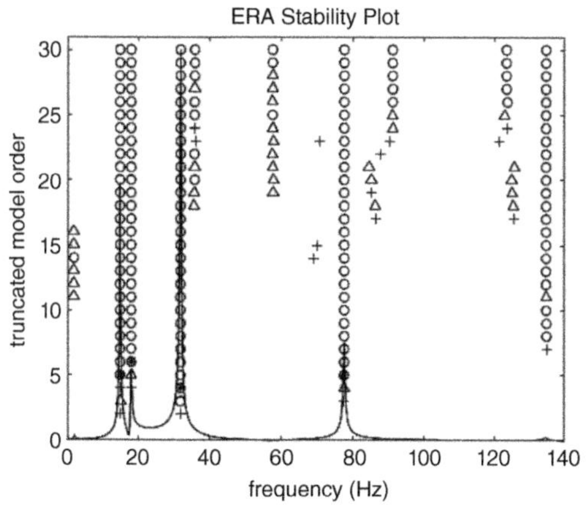

Figure 4.44 (a) Reversed *ERA* Stability Plot.
o = stabilised in freq and damping
Δ = stabilised in freq
+ = not stabilised.

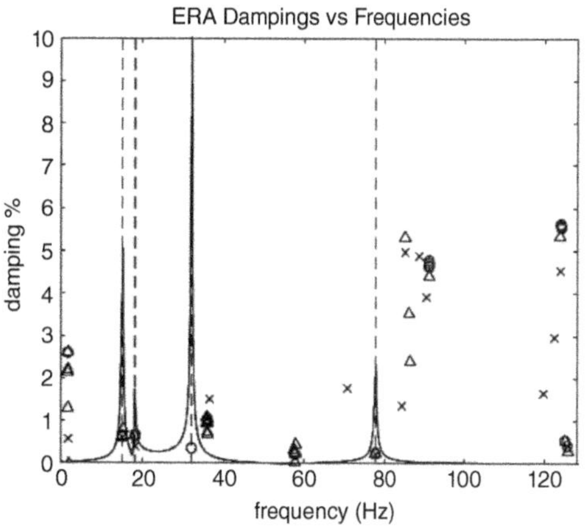

Figure 4.44 (b) Reversed *ERA* – Frequency *versus* Damping plots.

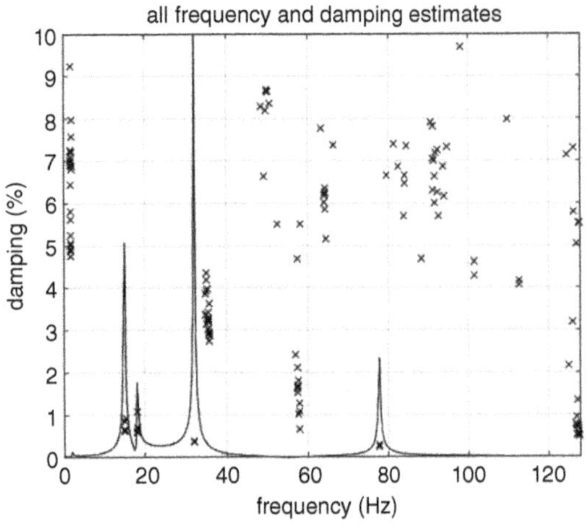

Figure 4.45 (a) *ERA*. All estimated frequencies and damping ratios.

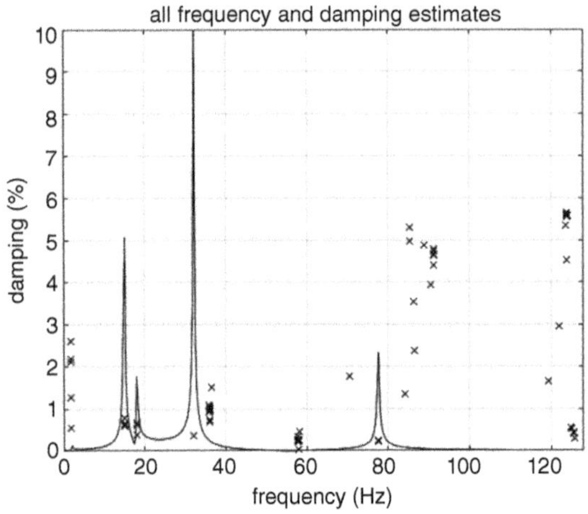

Figure 4.45 (b) Reversed *ERA*. All estimated frequencies and damping ratios.

4.5 Ambient Excitation – Operational Modal Analysis

In some situations it is common for structures to be excited in normal operating conditions, e.g. aircraft flying in turbulence, bridges subjected to moving traffic or wind excitation, buildings subjected to wind or earthquake excitations, where there is no measured excitation. A branch of modal analysis, often referred to as Operational Modal Analysis (*OMA*) [4.32], has emerged to develop methods to identify the natural frequencies, damping ratios and mode shapes from these ambient responses.

As can be seen in Figure 4.46, the key assumption for this type of technique is that the unmeasured input is "white", i.e. it has a flat spectrum across the frequency range of interest. Any spike or dip in the input frequency distribution will result in errors in the ambient analysis.

With some minor adjustments, all the identification techniques that have been previously described can be used on ambient data sets; a reasonable amount of information can be deduced from the measured time sequences, and the corresponding power and cross spectra.

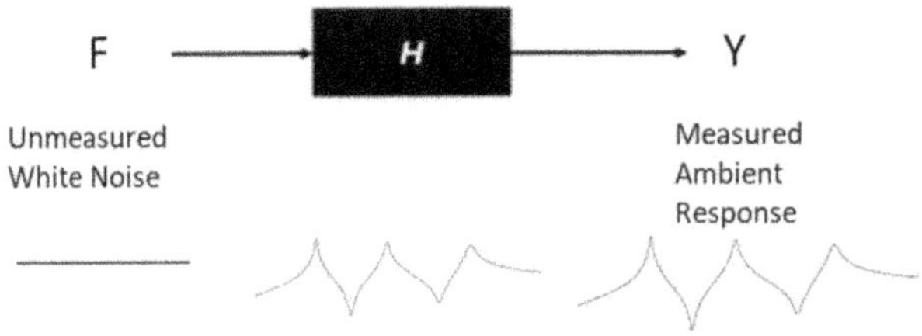

Figure 4.46 Basis of Operational Modal Analysis.

Consider the Wing-Pylon model that has previously been used as a test case, but now subjected to a random input (5 Hz – 100 Hz) resulting in the time sequences shown in Figure 4.47, with the corresponding power spectra shown in Figure 4.48. The modes that have been described before are clearly seen, with the two modes involving the engine pylons being most apparent from outputs 8 and 9. Using the auto-power spectra directly is straightforward to determine the natural frequencies and damping ratios for each time sequence using the pick picking / half-power points approach; however, this approach gives no information about the mode shapes.

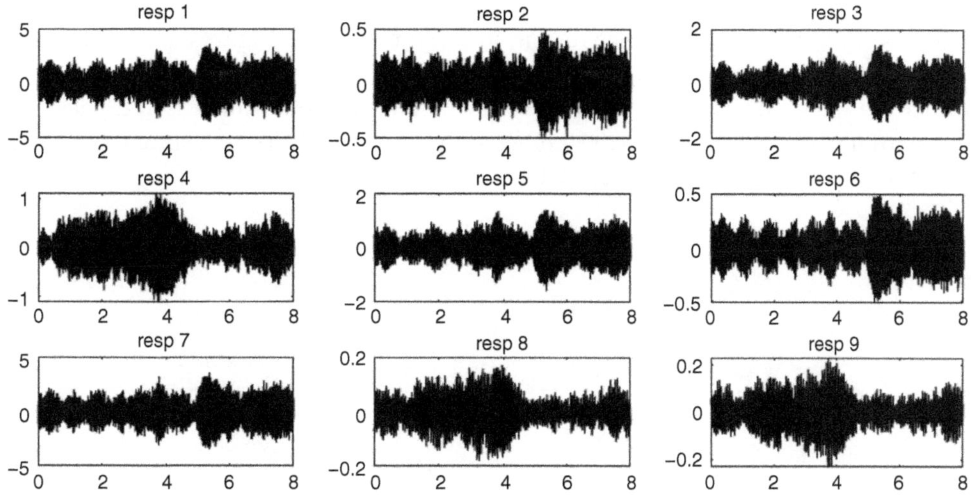

Figure 4.47 Ambient time responses for the Wing-Pylon model (Acceleration *versus* Time).

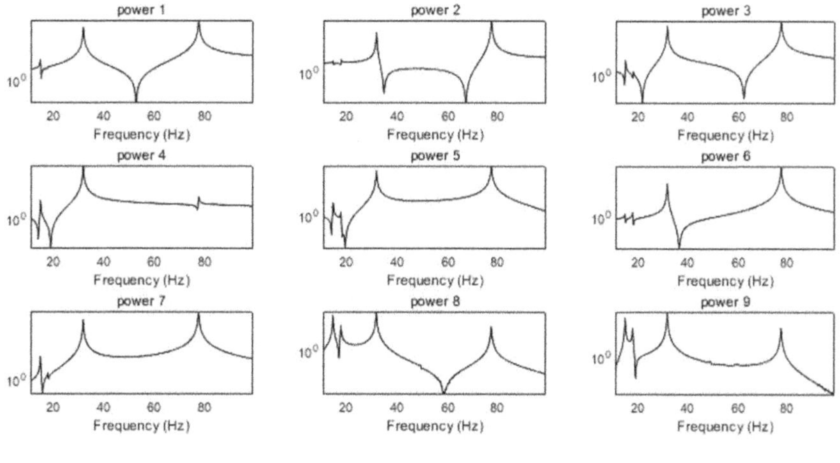

Figure 4.48 Auto-power spectra of ambient time responses for the Wing-Pylon model.

4.5.1 Ambient Analysis – Frequency Domain

The frequency domain mathematical model for the response to an unknown white spectrum has been considered before in Eq. (4.7). Information about the mode shape contribution can be determined by considering the cross-spectra for the various response channels Y_k. So, for measured time histories at station *k*, $y_k(t)$, and the corresponding Fourier Transform $Y_k(\omega)$ then the auto or power spectral density (*PSD*) and cross power spectral density (*CSD*) are defined in a similar way to Eq. (4.5):

$$S_{kk}(\omega) = Y_k(\omega)Y_k^*(\omega) \quad \text{and} \quad S_{kj}(\omega) = Y_k(\omega)Y_j^*(\omega) \tag{4.117}$$

after taking averages[6].

The cross power spectra with reference to station 1 can be seen in Figures 4.49 and 4.50. Peak Picking / Half-Power points can be used as before with the additional observation that the phase at resonance will be 0 or 180 degrees for well separated modes, which is the phase of that particular station corresponding to the reference point. The phase for measurement station 1 is always 0 degrees as this is the reference station.

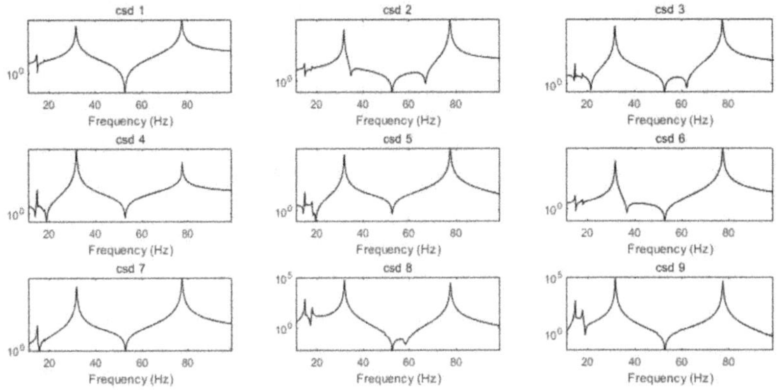

Figure 4.49 Cross Power Spectra amplitude with reference to station 1.

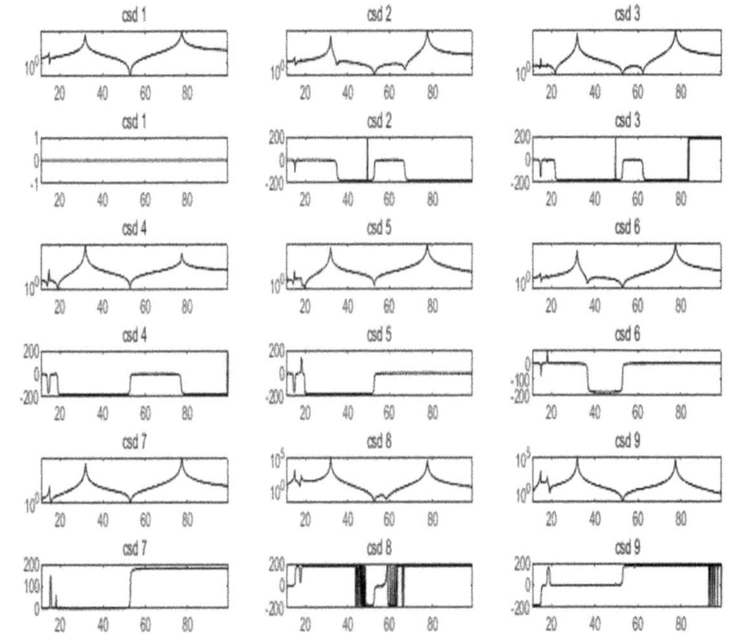

Figure 4.50 Cross Power Spectra amplitude and phase with reference to station 1.

[6] See Chapter 1 for details.

Any frequency domain system identification method can be used to curvefit the *CSDs* in exactly the same way as they are used for fitting the *FRFs*.

Frequency Domain Decomposition Method
Eq. (4.7) shows how, for an unknown "white" input, the $(NO \times NO)$ output power spectrum matrix can be written as

$$S_{YY}(\omega) = \sum_{r=1}^{N}\left(\frac{d_r \psi_r L_r^T}{i\omega - \lambda_r} + \frac{d_r^* \psi_r^* L_r^H}{i\omega - \lambda_r^*} + \frac{d_r \psi_r L_r^T}{-i\omega - \lambda_r} + \frac{d_r^* \psi_r^* L_r^H}{-i\omega - \lambda_r^*}\right) \tag{4.118}$$

For a lightly damped structure with modes that do not interact with each other, it can be assumed that at a certain frequency ω only a limited number of modes have a significant contribution to the output power spectrum.

The Frequency Domain Decomposition (*FDD*) method [4.33] takes the same approach as the *CMIF* and performs the *SVD* on the output power spectrum at each frequency line so that

$$S_{YY}(\omega_i) = U_i \Sigma_i U_i^H \tag{4.119}$$

as S_{YY} is symmetric.

The $(NO \times NO)$ U_i matrix contains the vectors $u_{i1} \ldots u_{iN}$ corresponding to each of the singular values in diagonal matrix Σ_i. Near a peak representing the kth mode, and with one dominant singular value, u_{i1} is the estimate of the kth mode shape. The accuracy of this estimate will diminish if there is another close mode at that frequency contributing to the power spectrum.

The original version of the *FDD* method computed the natural frequency by peak picking, whose accuracy is dependent upon the frequency resolution and noise level. There was no capability of computing the damping values for each mode, as the entire analysis is based at on a single frequency line for each mode.

Enhanced Frequency Domain Decomposition Method
A number of improvements have subsequently been made to the *FDD* method. The main improvement of the so-called Enhanced Frequency Domain Decomposition (*EFDD*) method [4.34] is to take the *PSD* function for a range of frequencies around an individual mode, and then to do an Inverse Fourier Transform into the time domain and to use zero crossings for a more accurate natural frequency computation, allied with a logarithmic decrement identification of the damping. Other developments consider how to remove the presence of spurious harmonic components that can appear in the *PSD*. The quality of the mode shapes obtained from the singular vectors at each modal frequency is compared using the *MAC* (Modal Assurance Criterion)[7] with other shapes obtained at frequencies around the resonance as a means for determining its quality and also the presence of other close modes.

4.5.2 Ambient Analysis – Time Domain

There can be problems in the use of the frequency domain approaches when considering ambient data sets, in particular the assumption that the unknown input is "white" and also the

[7] See Chapter 6, Section 6.3.1, for details.

data sets may not be very long, resulting in poor frequency resolution. An alternative approach is to generate Impulse Response Functions *(IRFs)* from the random responses and then to make us of some of the time domain methods that have been described earlier in this chapter. A number of routes to *IRF* generation are possible, as described in the following sub-sections.

Random Decrement Method

The Random Decrement method was introduced in the late 1960s [4.12] as a means to analyse vibrations of stochastically loaded structures. Further developments [4.35, 4.36] were made in following years and it has been primarily used on civil structures such as bridges.

Consider the ambient time response y_j at time $j\Delta t$ of some point on a structure shown in Figure 4.51. The initial formulation of the Random Decrement method for impulse response generation sets a level s and then saved the sequences from each crossing point, as seen in Figure 4.52. Assume that each of these time histories are made up of a summation of several responses:

- due to a step input of s
- due to an initial velocity
- due to random fluctuations in the response

Averaging of the time histories theoretically should result in the latter two terms cancelling out (velocities may be positive or negative and the randomness in the signal should average to zero as well) in the time sequence shown in Figure 4.53, where the response due to a step input s is found.

Further improvements to the method were obtained by setting the triggering condition to be a zero crossing with a positive slope, and also reversing the sign on a zero crossing with a negative slope. These triggered histories for the same initial sequence are shown in Figure 4.54 and the resulting random decrement time history, shown in Figure 4.55, is an impulse response.

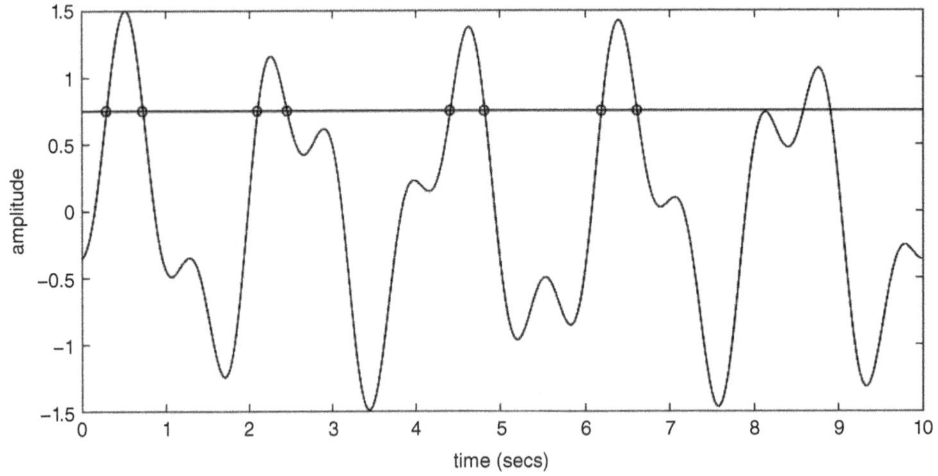

Figure 4.51 Trigger level used for Random Decrement time signatures.

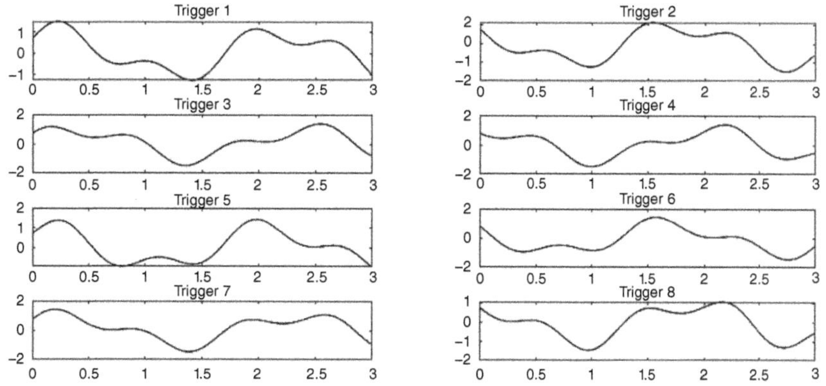

Figure 4.52 Time Histories from different trigger points – Amplitude *versus* Time.

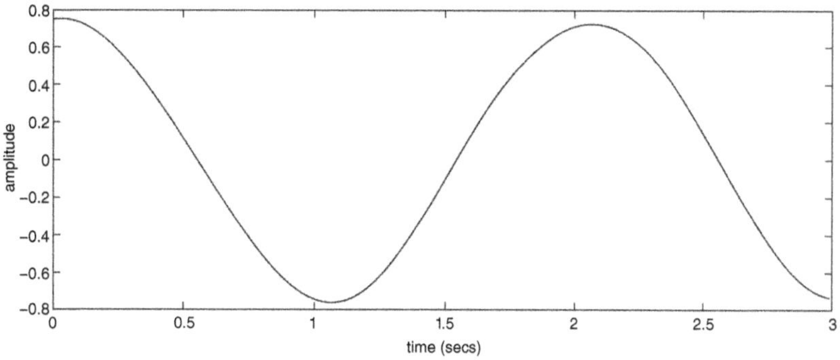

Figure 4.53 Summation of triggered responses.

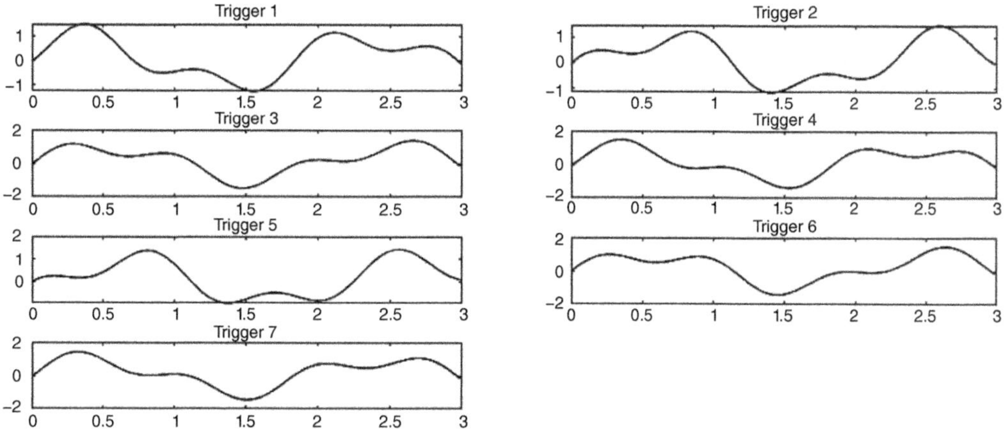

Figure 4.54 Triggered time histories for zero crossings – Amplitude *versus* Time.

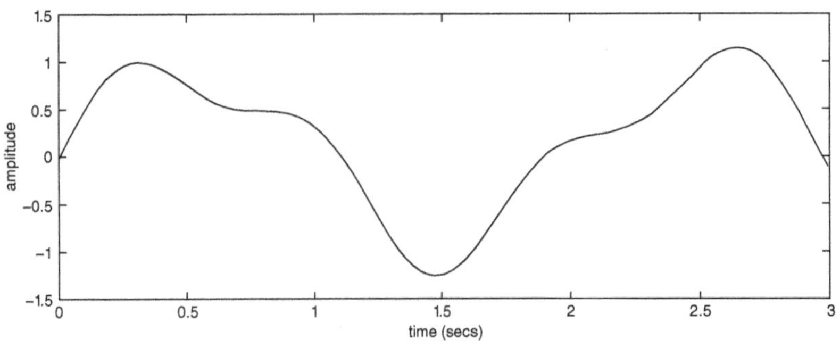

Figure 4.55 Zero Crossing Random Decrement Response.

Initially, the Random Decrement method was developed for application to individual time histories, permitting only the extraction of natural frequencies and damping ratios after application of the system identification methods. The logical extension was to enable mode shape estimation, which is achieved by using one channel, say time history x in this case, as the reference and employing these trigger times to create averaged responses from other channels, here labelled as y. *SIMO* and *MIMO* methods can then be used as before to obtain the modal parameters.

Figure 4.56 shows the zero crossings identified for the reference signal and the corresponding points on the other signal; note that in this case the trigger points are not at zero. The triggered time signals are then shown in Figure 4.57 and these are then averaged as shown in Figure 4.58. Although the frequency and damping content is the same for each channel, differences in both *IRFs* are expected due to different transducer positions experiencing different responses due to the mode shape amplitude and phases variations.

Figure 4.59 shows an example of the Random Decrement time history generated from one of the Wing-Pylon data sets and its corresponding frequency content. Although the signal is not so clean as for an *FRF* generated using an measured input, the frequency content to the first 4 modes are clearly visible.

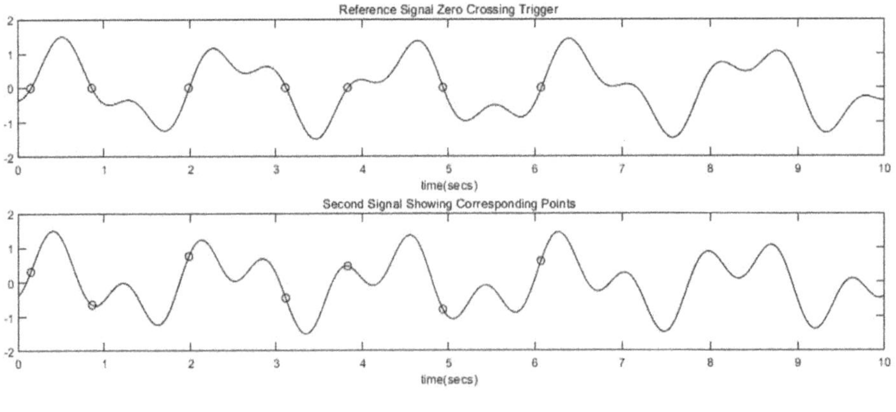

Figure 4.56 Trigger points on the reference signal (x) and corresponding points on the second signal (y).

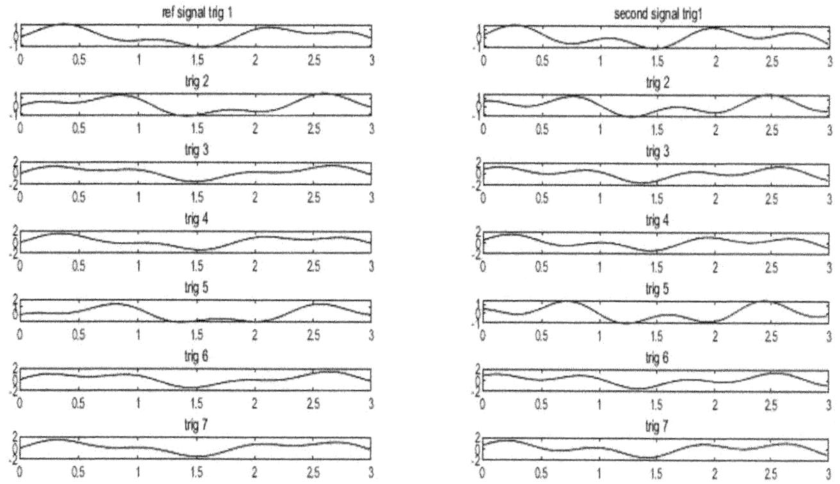

Figure 4.57 Time histories for triggered time points – both signals.

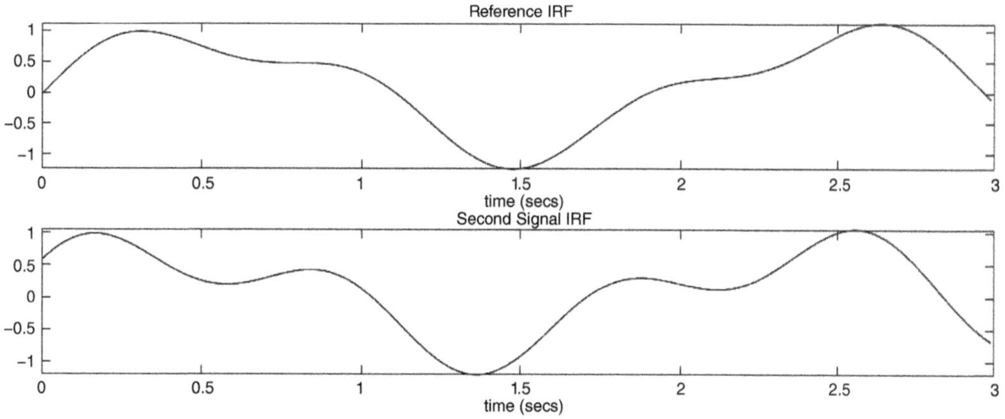

Figure 4.58 Averaged *IRFs* for the reference and the other signal.

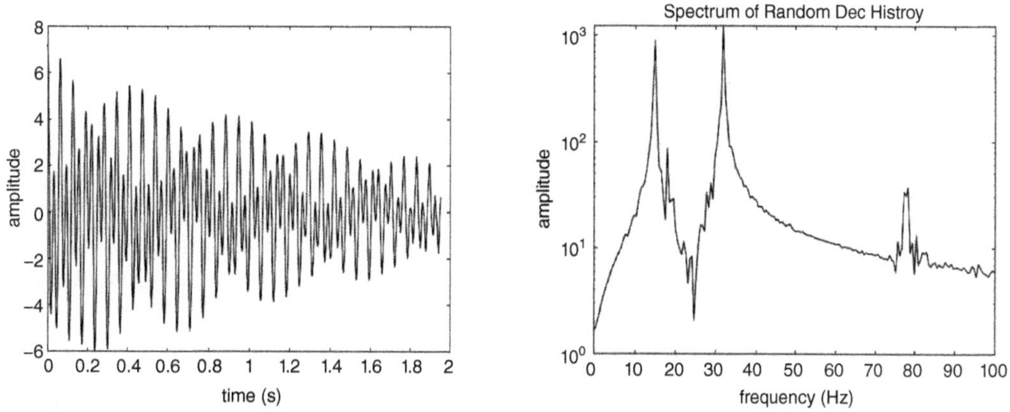

Figure 4.59 Random Decrement time history and frequency spectrum from the Wing-Pylon Data.

Generation of Impulse Response Functions from *PSDs* and *CSDs*

It is possible to produce impulse response functions by inverse Fourier Transforming the *CSDs*, shown in Figure 4.49, resulting in the time histories shown in Figure 4.60. Only the first half of the sequences needs to be considered. These time histories can then be curve-fitted using any of the time domain methods that can be applied to impulse response functions.

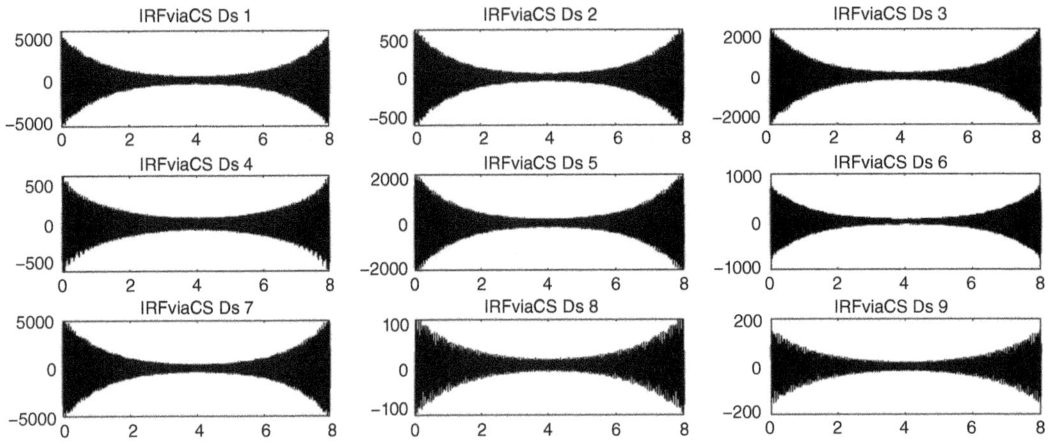

Figure 4.60 Impulse Response Functions generated from the *CSDs*. Amplitude *versus* Time.

Generation of Impulse Response Functions from Auto and Cross Correlations

Considering the time domain representation for the response at some point on the structure y_j of an *N DOF SISO* system subjected to a white excitation ε_j, then its *AR* formulation, as used previously for time domain methods based upon impulse response functions, is obtained as

$$y_j + a_1 y_{j-1} + \ \dots \ + a_{2N} y_{j-2N} = 0 \tag{4.120}$$

so multiplying by y_j and then summing between $j = 2N+1$ and L (number of data points) gives

$$\sum_{j=2N+1}^{L} y_j y_j + a_1 \sum_{j=2N+1}^{L} y_j y_{j-1} + a_2 \sum_{j=2N+1}^{L} y_j y_{j-2} + \ \dots \ + a_{2N} \sum_{j=2N+1}^{L} y_j y_{j-2N} = 0$$

or

$$R_{yy_0} + a_1 R_{yy_1} + a_2 R_{yy_2} + \ \dots \ + a_{2N} R_{yy_{2N}} = 0 \tag{4.121}$$

where $R_{yy_k} = \sum_{j=2N+1}^{L} y_j y_{j-k}$ and $k = 0, \dots 2N$ are often known as autocorrelations (or inner products) for lag k. It is usual to divide each correlation by $(L\text{-}2N)$ which imparts a very small but negligible approximation to the above. The same approach can be used to generate cross-correlations between measurements at positions x and y on the structure, such that

$$\sum_{j=2N+1}^{L} x_j y_j + a_1 \sum_{j=2N+1}^{L} x_j y_{j-1} + a_2 \sum_{j=2N+1}^{L} x_j y_{j-2} + \ \dots \ + a_{2N} \sum_{j=2N+1}^{L} x_j y_{j-2N} = 0 \tag{4.122}$$

or

$$R_{xy_0} + a_1 R_{xy_1} + a_2 R_{xy_2} + \;\ldots\; + a_{2N} R_{xy_{2N}} = 0 \qquad (4.123)$$

The same structure of the *AR* equation is maintained, and therefore the time domain approaches described previously can be used to determine the modal parameters using the auto and cross correlations rather than the measured time data. Such reasoning means that for random response data, the correlations will give decay responses akin to those obtained from the Impulse Response Function.

Consider the random time responses shown in Figure 4.47; computing the auto and cross correlation functions directly from the time histories gives the correlation functions shown in Figure 4.61, which have the corresponding power spectra shown in Figure 4.62.

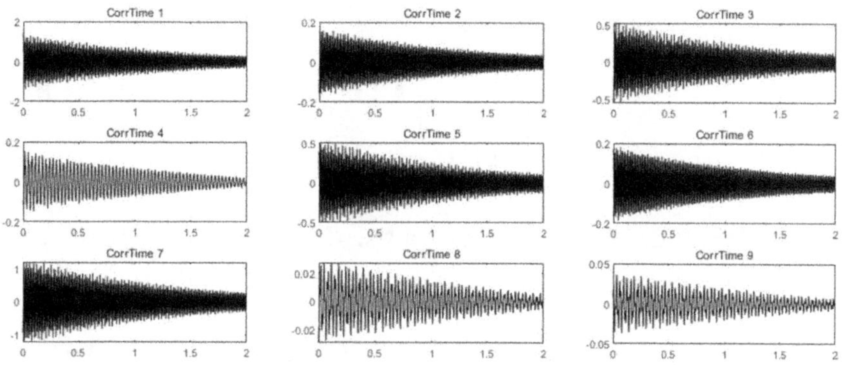

Figure 4.61 Cross-Correlation Functions computed directly from the Ambient Time histories.

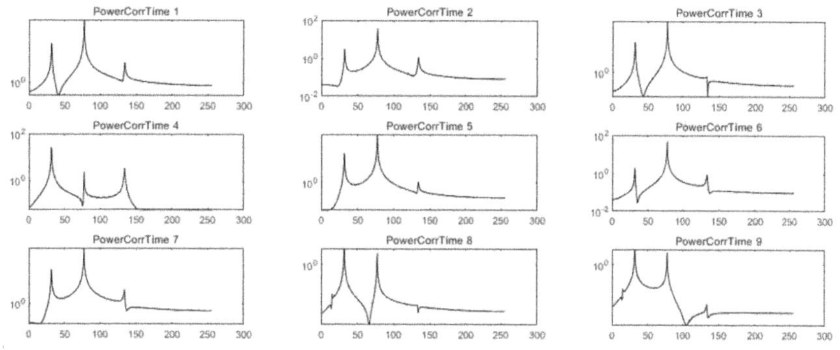

Figure 4.62 Power Spectra of Time Cross-Correlation Functions. Magnitude *versus* Frequency.

Eigensystem Realisation Algorithm using Data Correlations

The *ERA/DC* method [4.37, 4.38] is a development of the *ERA* method, but formulated with data correlations instead of the raw data sets. It can either be used for the identification of impulse response functions, where it is particularly useful for very long data sets, such as would occur in very lightly damped structures, or for ambient data sets. Starting in the same way to the *ERA* method by defining a time block of impulse or ambient responses for *NO* simultaneously sampled response stations responding to *NI* input cases and such that for the $(k+1)^{th}$ time instant

$$\boldsymbol{Y}_{k+1} = \begin{bmatrix} y_{1,1} & \cdots & y_{1,NI} \\ \vdots & & \vdots \\ y_{NO,1} & \cdots & y_{NO,NI} \end{bmatrix}_k = \boldsymbol{C}\boldsymbol{A}^k\boldsymbol{B} \tag{4.124}$$

where $\boldsymbol{Y}$ is a $(NO \times NI)$ data matrix, $\boldsymbol{A}$ is the $(2N \times 2N)$ system matrix, $\boldsymbol{C}$ and $\boldsymbol{B}$ are $(NO \times 2N)$ and $(2N \times NI)$ measurement and input matrices, respectively. A $((\mu+1)NO \times (\nu+1)NI)$ block data Hankel matrix $\boldsymbol{H}_k$ can be constructed such that, for the kth time increment,

$$\boldsymbol{H}_k = \begin{bmatrix} \boldsymbol{Y}_{k+1} & \boldsymbol{Y}_{k+2} & \cdots & \boldsymbol{Y}_{k+\nu+1} \\ \boldsymbol{Y}_{k+2} & \boldsymbol{Y}_{k+3} & \cdots & \boldsymbol{Y}_{k+\nu+2} \\ \vdots & \vdots & & \vdots \\ \boldsymbol{Y}_{k+\mu+1} & \boldsymbol{Y}_{k+\mu+2} & \cdots & \boldsymbol{Y}_{k+\mu+\nu+1} \end{bmatrix} = \begin{bmatrix} \boldsymbol{C} \\ \boldsymbol{C}\boldsymbol{A} \\ \vdots \\ \boldsymbol{C}\boldsymbol{A}^{\mu} \end{bmatrix} \boldsymbol{A}^k \begin{bmatrix} \boldsymbol{B} & \boldsymbol{A}\boldsymbol{B} & \ldots & \boldsymbol{A}^{\nu}\boldsymbol{B} \end{bmatrix} = \boldsymbol{V}\boldsymbol{A}^k\boldsymbol{W} \tag{4.125}$$

where $\boldsymbol{V}$ is a $((\mu+1)NO \times 2N)$ observability matrix, $\boldsymbol{W}$ is a $(2N \times (\nu+1)NI)$ controllability matrix, and μ and ν dictate the dimensions of the block Hankel matrix.

A square $((\mu+1)NO \times (\mu+1)NO)$ correlation matrix $\boldsymbol{R}_q$ is now defined such that

$$\boldsymbol{R}_q = \boldsymbol{H}_q\boldsymbol{H}_0^T = \boldsymbol{V}\boldsymbol{A}^q\boldsymbol{W}\boldsymbol{W}^T\boldsymbol{V} = \boldsymbol{V}\boldsymbol{A}^q\boldsymbol{W}_c \tag{4.126}$$

which contains approximate auto and cross-correlations for a range of different lag values and can be used as the basis of the $\boldsymbol{U}_q$ block correlation matrix such that

$$\boldsymbol{U}_q = \begin{bmatrix} \boldsymbol{R}_q & \boldsymbol{R}_{q+r} & \cdots & \boldsymbol{R}_{q+\beta r} \\ \boldsymbol{R}_{q+r} & \boldsymbol{R}_{q+2r} & \cdots & \boldsymbol{R}_{q+(\beta+1)r} \\ \vdots & \vdots & & \vdots \\ \boldsymbol{R}_{q+\alpha r} & \boldsymbol{R}_{q+(\alpha+1)r} & \cdots & \boldsymbol{R}_{q+(\alpha+\beta)r} \end{bmatrix} = \begin{bmatrix} \boldsymbol{V} \\ \boldsymbol{V}\boldsymbol{A}^r \\ \vdots \\ \boldsymbol{V}\boldsymbol{A}^{\alpha r} \end{bmatrix} \boldsymbol{A}^q \begin{bmatrix} \boldsymbol{W}_c & \boldsymbol{A}^r\boldsymbol{W}_c & \ldots & \boldsymbol{A}^{\beta r}\boldsymbol{W}_c \end{bmatrix} = \boldsymbol{V}_{\alpha}\boldsymbol{A}^q\boldsymbol{W}_{\beta} \tag{4.127}$$

Taking the *SVD* of this block correlation matrix at $k = 0$, written as

$$\boldsymbol{U}_q = \boldsymbol{P}_F\boldsymbol{\Sigma}_F\boldsymbol{Q}_F^T \approx \boldsymbol{P}\boldsymbol{\Sigma}\boldsymbol{Q}^T \tag{4.128}$$

as before and, if $\boldsymbol{U}_q$ has dimensions $(a \times b)$ with $b > a$, then the compact *SVD* gives the dimensions of the full orthogonal matrices $\boldsymbol{P}_F$ and $\boldsymbol{Q}_F^T$ as $(a \times a)$ and $(a \times b)$ respectively, and the diagonal singular value matrix $\boldsymbol{\Sigma}_F$. Truncating the matrices, as before, by choosing only $2N$ non-zero singular values, the dimensions of $\boldsymbol{P}$, $\boldsymbol{\Sigma}$ and $\boldsymbol{Q}^T$ become $(a \times 2N)$, $(2N \times 2N)$ and

$(2N \times b)$, respectively. Following some algebraic manipulations, one arrives at the realisations for time instant *j*:

$$\begin{aligned} \boldsymbol{R}_{q+j} &= \boldsymbol{E}^T \boldsymbol{P}\boldsymbol{D}^{1/2}\left(\boldsymbol{D}^{-1/2}\boldsymbol{P}^T\boldsymbol{U}_{q+1}\boldsymbol{Q}\boldsymbol{D}^{-1/2}\right)^j \boldsymbol{D}^{1/2}\boldsymbol{Q}^T\boldsymbol{E} \\ \text{thus,}\quad \boldsymbol{A} &= \boldsymbol{D}^{-1/2}\boldsymbol{P}^T\boldsymbol{U}_{q+1}\boldsymbol{Q}\boldsymbol{D}^{-1/2} = \boldsymbol{\Psi}\boldsymbol{Z}^k\boldsymbol{\Psi}^{-1} \\ \boldsymbol{B} &= \boldsymbol{D}^{1/2}\boldsymbol{Q}^T\boldsymbol{E} \\ \boldsymbol{C} &= \boldsymbol{E}^T\boldsymbol{P}\boldsymbol{D}^{1/2} \end{aligned} \tag{4.129}$$

where $\boldsymbol{E} = [\boldsymbol{I}\ \boldsymbol{0}\ \ \ldots\ \ \boldsymbol{0}]$ is a matrix made up of identity and null matrices of the dimension $(NO \times NO)$.

So, as with the *ERA* method, upon identification of the $\boldsymbol{A}$ matrix, the natural frequencies and damping ratios can be computed from the eigenvalues on the diagonal matrix $\boldsymbol{Z}$. Determining the eigenvector matrix $\boldsymbol{\Psi}$ enables the mode shapes and modal participation factors to be computed from $\boldsymbol{C}\boldsymbol{\Psi}$ and $\boldsymbol{\Psi}^{-1}\boldsymbol{B}$, respectively. If the impulse responses have been obtained via the *FRFs*, then the mode shape estimation could also be undertaken using the *LSFD* method as described above.

4.6 Phase Resonance (Normal Modes / Force Appropriation)[8] Testing

The "*Phase Separation*" methods are implemented by exciting all the modes simultaneously and then performing the identification process to estimate the modes at the same time. An alternative approach, which has been in existence in some form since modal tests were first performed, is to use the so-called "*Phase-Resonance*" (sometimes referred to as normal modes or force appropriation) methods whereby each mode is excited one at a time at its natural frequency through tuning the excitation frequency and adjusting the amplitude and phase of each shaker. The result motion will be in the corresponding normal mode. Although much more time consuming and difficult to perform, it is arguable that such techniques can produce more accurate estimates than *Phase Separation* techniques, particularly for very close modes. A review of the available methods can be found in [4.39, 4.40].

Considering an *MDOF* system in the usual formulation

$$\boldsymbol{M}\ddot{\boldsymbol{x}} + \boldsymbol{C}\dot{\boldsymbol{x}} + \boldsymbol{K}\boldsymbol{x} = \boldsymbol{f} \tag{4.130}$$

then a harmonic input of $\boldsymbol{f}(t) = \boldsymbol{F}\sin\omega t$ results in the steady-state response of $\boldsymbol{x}(t) = \boldsymbol{X}\sin(\omega t - \alpha)$[9], leading to

$$\left[\left(\boldsymbol{K} - \omega^2\boldsymbol{M}\right)\sin(\omega t - \alpha) + \omega\boldsymbol{C}\cos(\omega t - \alpha)\right]\boldsymbol{X} = \boldsymbol{F}\sin\omega t \tag{4.131}$$

and comparing *sin* and *cos* terms, gives

[8] Sometimes also called Tuned-Sinusoidal testing.
[9] See Chapter 1 for details.

$$\left(\left(\boldsymbol{K}-\omega^2\boldsymbol{M}\right)\cos\alpha+\omega\boldsymbol{C}\sin\alpha\right)\boldsymbol{X}=\boldsymbol{F} \quad \text{and} \quad \left(-\left(\boldsymbol{K}-\omega^2\boldsymbol{M}\right)\sin\alpha+\omega\boldsymbol{C}\cos\alpha\right)\boldsymbol{X}=\boldsymbol{0} \tag{4.132}$$

Earlier in this chapter it has been seen that for well-separated modes there is a rapid change of phase in the *FRFs* around each resonance and there the phase lag is $\alpha = 90^{\circ}$. This condition corresponds to the real part of the *FRF* being zero and the 90° phase between the force and response is referred to as being in *quadrature*. For the general case of a sinusoidal excitation with force vector $\boldsymbol{F}$ applied to an *MDOF* system, Eqs. (4.132) become

$$\omega\boldsymbol{C}\boldsymbol{X}=\boldsymbol{F} \quad \text{and} \quad \left(\boldsymbol{K}-\omega^2\boldsymbol{M}\right)\boldsymbol{X}=\boldsymbol{0} \tag{4.133}$$

which give the phase resonance condition for the excitation of the undamped normal modes at the resonance frequencies, known as the appropriated force vector $\boldsymbol{F}_a$ (*NI* x *1*), and the standard normal mode eigenvalue solution. So, for the jth mode, the force vector to achieve the jth mode shape is found as

$$\boldsymbol{F}_{a_j}=\omega_j\boldsymbol{C}\boldsymbol{X}_j \tag{4.134}$$

Another way to look at this formulation is to start from the standard *FRF* relationship between the input and the output in terms of the real and imaginary parts,

$$\boldsymbol{H}(\omega)\boldsymbol{F}(\omega)=\boldsymbol{X}(\omega) \quad\rightarrow\quad \boldsymbol{HF}=\boldsymbol{X} \quad\rightarrow\quad \boldsymbol{H}_R\boldsymbol{F}+i\boldsymbol{H}_I\boldsymbol{F}=\boldsymbol{X}_R+i\boldsymbol{X}_I \tag{4.135}$$

Thus, to excite a normal mode at the resonance frequency, the appropriated force vector $\boldsymbol{F}_a$ must be such that the real part of the response is zero whilst the imaginary part corresponds to the undamped normal mode shape $\boldsymbol{\phi}$, thus

$$\boldsymbol{H}_R\boldsymbol{F}_a=\boldsymbol{X}_R=\boldsymbol{0} \quad \text{and} \quad \boldsymbol{H}_I\boldsymbol{F}_a=\boldsymbol{X}_I=\boldsymbol{\phi} \tag{4.136}$$

Alternatively, rewriting the complex amplitude of the response $\boldsymbol{X}$ as

$$\boldsymbol{X}e^{-i\alpha}=\boldsymbol{X}\left(\cos\alpha-i\sin\alpha\right)=\left(\boldsymbol{A}+i\boldsymbol{B}\right)\boldsymbol{F} \tag{4.137}$$

and separating the real and imaginary parts leads to

$$\boldsymbol{AF}=\boldsymbol{X}\cos\alpha \quad \text{and} \quad \boldsymbol{BF}=-\boldsymbol{X}\sin\alpha$$

or

$$\boldsymbol{H}_R\boldsymbol{F}=\boldsymbol{X}\cos\alpha \quad \text{and} \quad \boldsymbol{H}_I\boldsymbol{F}=-\boldsymbol{X}\sin\alpha \tag{4.138}$$

which, by eliminating $\boldsymbol{X}$, gives an expression for the appropriated force vector $\boldsymbol{F}_a$ at the resonance frequencies as

$$\left(\boldsymbol{A}\tan\alpha+\boldsymbol{B}\right)\boldsymbol{F}_a=\boldsymbol{0}$$

or

$$\left(\boldsymbol{H}_R\tan\alpha+\boldsymbol{H}_i\right)\boldsymbol{F}_a=\boldsymbol{0} \tag{4.139}$$

Note that Eq. (4.139) is in the generalised eigenvalue form

$$\boldsymbol{A}\boldsymbol{F}_{\mathrm{a}} = -\frac{1}{\tan\alpha}\boldsymbol{B}\boldsymbol{F}_{\mathrm{a}} = -\mu\boldsymbol{B}\boldsymbol{F}_{\mathrm{a}}$$

or

$$\left(\boldsymbol{H}_R + \mu\boldsymbol{H}_I\right)\boldsymbol{F}_{\mathrm{a}} = \boldsymbol{0} \quad \text{with} \quad \mu = \frac{1}{\tan\alpha} \tag{4.140}$$

There are a range of approaches that have been developed using the above formulations to determine the appropriated force vectors; these are summarised in Table 4.12. Their use is demonstrated upon the simple 3 *DOF* system used in Section 4.1.4.

Table 4.12 Phase Resonance methods considered here.

Size of the *FRF* matrices	**Methods**
Square (*NI* = *NO*)	Asher, Modified Asher and Traill-Nash
Rectangular (*NO* > *NI*)	Extended Asher and *MMIF*
Rank Reduction of rectangular matrices	Modified *MMIF* and *SVD* method

4.6.1 Square *FRF* Matrices

When the number of inputs *NI* is equal to the number of outputs *NO*, the exact non-trivial solution to the homogeneous Eqs. (4.136) can only occur when the determinant of $\boldsymbol{H}_R$ is zero, thus

$$det\left(\boldsymbol{H}_R\right) = 0 \tag{4.141}$$

and there are several approaches to obtaining a solution to (4.141).

Asher's Method

The Asher method [4.41] uses the zero crossing of the determinant in (4.141), which gives the resonant frequencies, as seen in Figure 4.63 and then the values of the $\boldsymbol{H}_R$ matrix at these frequencies can be used to determine the forcing vector $\boldsymbol{F}$. However, the implementation of this simple approach is fraught with difficulty as if the real part of the *FRF* is even slightly non-singular, as would occur if the quadrature condition is not perfectly met, then the resulting force vector is trivial (a column of zeros).

Modified Asher's Method

A more robust approach, known as the Modified Asher's Method [4.42], is to rewrite Eq. (4.136) into an equivalent eigenvalue form such that

$$\boldsymbol{H}_R\boldsymbol{F} = \lambda\boldsymbol{F} \tag{4.142}$$

and solving for the eigenvalues of the $(NI \times NI)$ $(= NO \times NO)$ $\boldsymbol{H}_R$ matrix, then the zero values of the characteristic equation indicate the natural frequencies; the corresponding eigenvector at each natural frequency defines the appropriated force vector $\boldsymbol{F}_a$ (Figure 4.64).

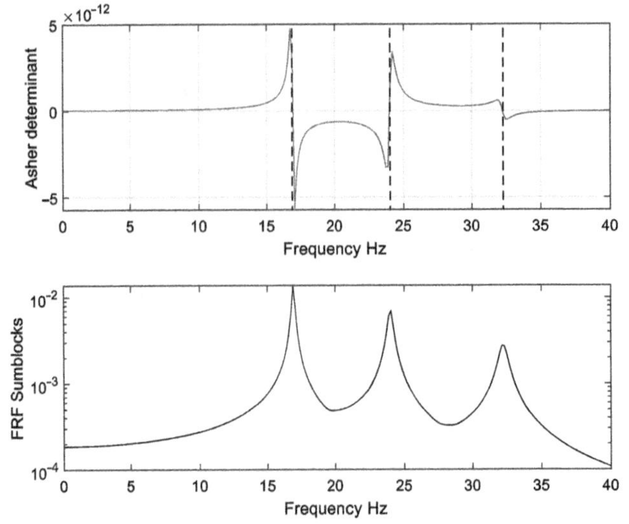

Figure 4.63 Asher determinant values – zero determinant values indicate the resonant condition.

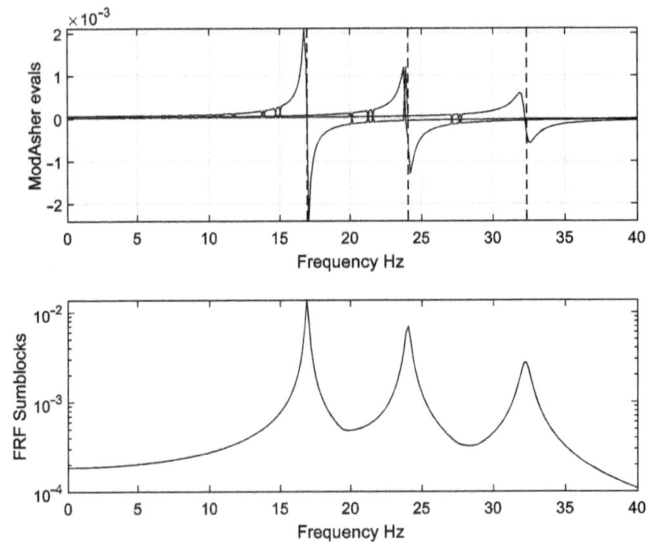

Figure 4.64 Modified Asher eigenvalues – zero values indicate the resonant condition.

Traill-Nash Method

The final square *FRF* matrix normal mode testing approach to be considered, known as the Traill-Nash method [4.43], makes use of the solution of the generalised eigenvalue problem shown in Eq. (4.140). Figure 4.65 shows how the zero crossings of the eigenvalues indicate the natural frequencies, and the corresponding eigenvectors at these frequencies give the appropriated force vectors.

4.6.2 Rectangular *FRF* Matrices

It is usual to have many more measurement stations than excitors, and therefore it is logical to extend the square *FRF* matrix methods to include these extra signals, resulting in a rectangular *FRF* matrix at each frequency.

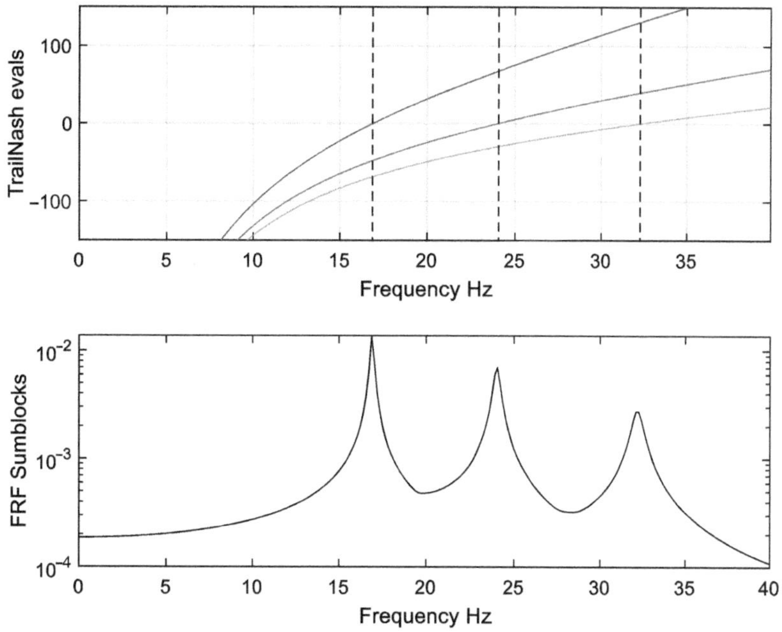

Figure 4.65 Traill-Nash eigenvalues.

Extended Asher's Method

The Extended Asher's Method [4.44] is formulated to employ the Asher technique for rectangular *FRF* matrices; this implies a rewriting of Eq. (4.142) for the quadrature condition into an equivalent eigenvalue form of the square matrix $\boldsymbol{H}_R^T\boldsymbol{H}_R$, such that

$$\boldsymbol{H}_R^T\boldsymbol{H}_R\boldsymbol{F} = \lambda\boldsymbol{F} \tag{4.143}$$

Solving for the eigenvalues of the $(NI \times NI)$ $\boldsymbol{H}_R^T\boldsymbol{H}_R$ matrix, then the zero values indicate the natural frequencies and the corresponding eigenvector at each of these frequencies defines the appropriated force vector $\boldsymbol{F}_a$.

As can be seen in Figure 4.66, it is not straightforward to determine where the minimum values of the eigenvalues occur. A similar plot is found when the singular values of $\boldsymbol{H}_R$ are plotted (which is to be expected, as they are simply the square roots of the other terms).

Multivariate Mode Indicator Function

The Multivariate Mode Indicator Function (*MMIF*) method [4.4] is a further approach that considers the rectangular *FRF* matrix case; it has been shown to work very well in a robust manner. This approach has already been met previously in Section 4.1.4 as a means of detecting how many modes there are in a set of *FRF* data.

Defining the 2 norm of the real and total responses as

$$\|\boldsymbol{X}_R\|^2 = \boldsymbol{X}_R^T\boldsymbol{M}\boldsymbol{X}_R \quad \text{and} \quad \|\boldsymbol{X}_R + i\boldsymbol{X}_I\|^2 = \boldsymbol{X}_R^T\boldsymbol{M}\boldsymbol{X}_R + \boldsymbol{X}_I^T\boldsymbol{M}\boldsymbol{X}_I \tag{4.144}$$

where $\boldsymbol{M}$ is the mass matrix (or some estimate of the local mass distribution).

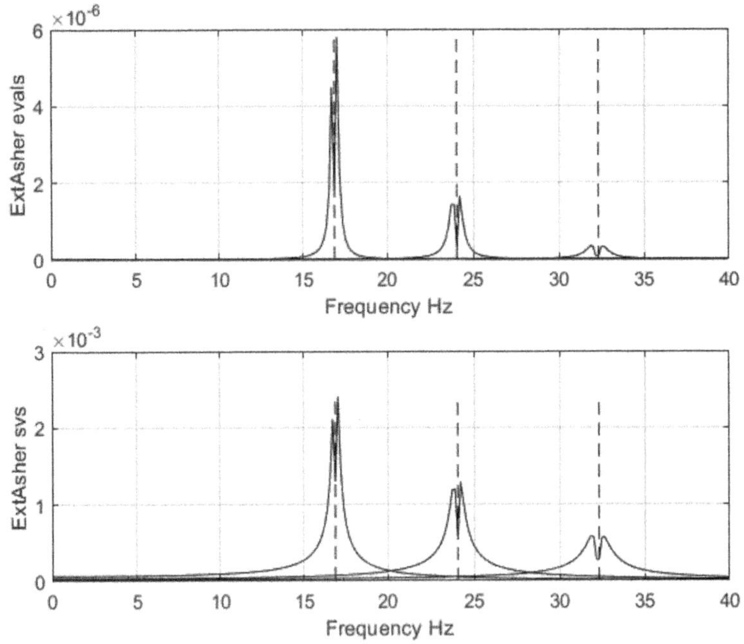

Figure 4.66 Extended Asher eigenvalues and singular values.

If the mass matrix is not available, then simply using the identity matrix is a reasonable alternative. The *MMIF* can then be found, after some algebraic manipulation, by minimising the function

$$\alpha = \frac{\|\boldsymbol{X}_R\|^2}{\|\boldsymbol{X}_R + i\boldsymbol{X}_I\|^2} = \frac{\boldsymbol{F}^T \boldsymbol{A}\boldsymbol{F}}{\boldsymbol{F}^T(\boldsymbol{A}+\boldsymbol{B})\boldsymbol{F}} \tag{4.145}$$

w.r.t. $\boldsymbol{F}$, where $\boldsymbol{A} = \boldsymbol{H}_R^T \boldsymbol{M}\boldsymbol{H}_R$ and $\boldsymbol{B} = \boldsymbol{H}_I^T \boldsymbol{M}\boldsymbol{H}_I$, $\boldsymbol{A}$ and $\boldsymbol{B}$ being both $(NI \times NI)$ matrices and $\boldsymbol{F}$ is an $(NI \times 1)$ vector.

This minimisation problem, which is equivalent to minimising the real part compared to the total function, can be shown to be the equivalent of finding, for measured $\boldsymbol{A}$ and $\boldsymbol{B}$ terms, the smallest eigenvalue λ of the eigenproblem (assuming $\boldsymbol{M} = \boldsymbol{I}$) leading to the expression

$$\boldsymbol{A}\boldsymbol{F} = (\boldsymbol{A}+\boldsymbol{B})\boldsymbol{F}\lambda \;\rightarrow\; \boldsymbol{H}_R^T \boldsymbol{H}_R \boldsymbol{F} = \lambda\left(\boldsymbol{H}_R^T \boldsymbol{H}_R + \boldsymbol{H}_I^T \boldsymbol{H}_I\right)\boldsymbol{F} \tag{4.146}$$

and the eigenproblem is of size $(NI \times NI)$ for *NI* inputs. As shown previously, plotting out the *MMIF* eigenvalues *versus* frequency, as in Figure 4.67, drops in the smallest values corresponds to a mode. Should more than one of the *MMIF* eigenvalues tend to zero at the same frequency, it indicates that there is more than one coincident mode and multiple shakers are required to excite them.

The computed eigenvectors that correspond to a dip close to zero in the *MMIF* eigenvalues are the appropriated force vectors $\boldsymbol{F}_a$. If more than one *MMIF* dips down, then the first and second sets of eigenvectors indicate the appropriated force vectors for two modes.

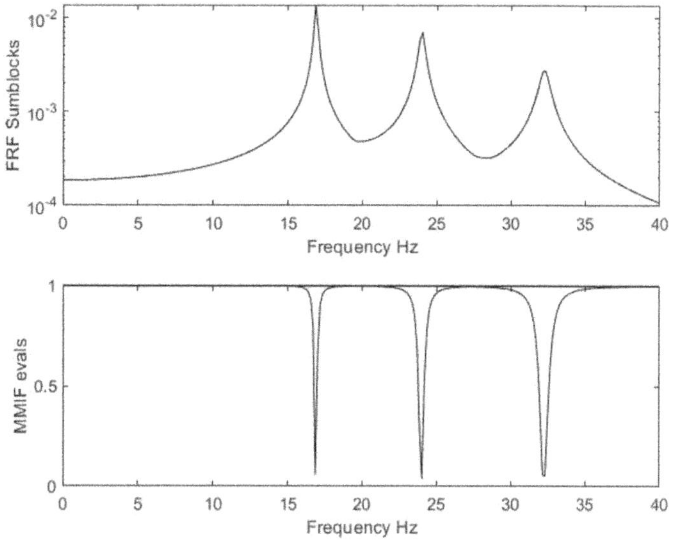

Figure 4.67 Sum of *FRFs* and *MMIF* eigenvalues.

Normal Mode Purity Function

The Normal Mode Purity Function (*NMPF*) [4.45] can be used to determine how well a normal mode has been appropriated. Remembering that

$$\boldsymbol{HF} = \boldsymbol{X} \qquad \rightarrow \quad \boldsymbol{H}_R\boldsymbol{F} + i\boldsymbol{H}_I\boldsymbol{F} = \boldsymbol{X}_R + i\boldsymbol{X}_I \tag{4.147}$$

and that at the phase resonance condition the real part of the response is required to be as close to zero as possible, then the *NMPF* can be defined as

$$\Delta = 1 - \frac{\sum_{j=1}^{NO} \left|\boldsymbol{X}_{R_j}\right| \left|\boldsymbol{X}_j\right|}{\sum_{j=1}^{NO} \left|\boldsymbol{X}_j\right|^2} \tag{4.148}$$

which has a value between 0 and 1 where X_j is the jth term of the output vector, corresponding to the jth transducer. If $\Delta = 1$, then a perfect undamped normal mode has been isolated, as the real part is zero and all the responses are monophase and in quadrature with the applied excitation. However, in practice, a value of $\Delta > 0.85$ indicates an adequately appropriated mode, $\Delta > 0.90$ a well appropriated mode, and $\Delta > 0.95$ indicates a very good appropriation.

Application to Wing-Pylon Data

The computed eigenvectors corresponding to the natural frequencies of the wing engine test case are shown in Table 4.13 along with the computed *NMPFs*. Note that the negative signs indicate application of the sinewave 180° out of phase from a positive term. When these computed force vectors are applied to the system, the appropriated modes will be excited. This process can be achieved numerically, an approach that is called *soft tuning,* by substituting the computed eigenvectors $\boldsymbol{F}$ (2×1 in this case) at each resonant frequency into the *FRF* expression $\boldsymbol{H}(\omega)\boldsymbol{F}(\omega)=\boldsymbol{Y}(\omega)$ and the resulting (9 x 1) output vector $\boldsymbol{Y}$ shows the amplitude and phase of the appropriated response. For a perfectly appropriated mode the *phase scatter plots* in Figure 4.68 should indicate responses with zero real part; however, it can be seen in practice that there is a little deviation from this ideal, but in general the appropriation methodology has worked well here. Modes 1 and 3 have the least amount of real contribution in them and have been appropriated most successfully, a finding that is confirmed by the *NMPF* values.

Table 4.13 Computed *MMIF* eigenvectors and Normal Mode Purity Function values

	Mode 1	Mode 2	Mode 3	Mode 4	Mode 5
Input #3	-0.8003	-1.1878	0.7684	0.3265	-0.0751
Input #5	-0.7906	2.0406	-0.9254	0.3985	-0.2761
NMPF	0.9540	0.8110	0.9440	0.8640	0.8450

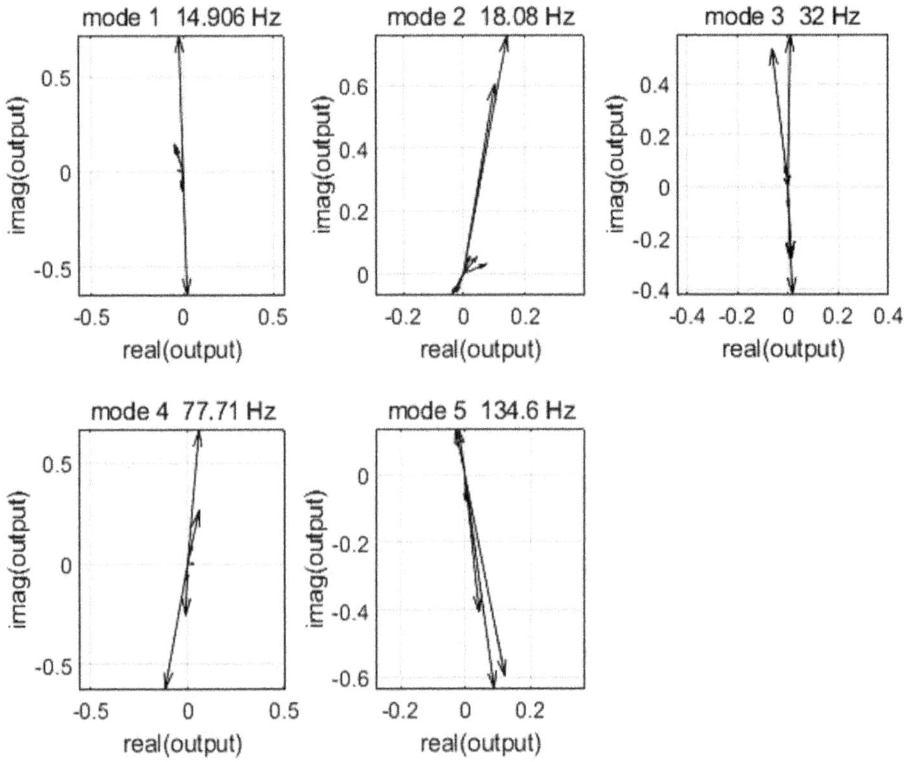

Figure 4.68 Phase Scatter plots of the soft tuned modes for the Wing-Pylon model using the *MMIF* method.

This *soft tuning* approach can be used for all the phase resonance methods that are discussed in this section. Further techniques exist to experimentally adjust the undamped resonant frequency and force vector slightly in order to achieve the phase resonance condition more closely [4.39].

4.6.3 Rank Reduction Force Appropriation Methods

If the number of exciters (*NI*) is greater than the number of effective degrees of freedom (n^*) at a particular frequency, then it is possible for the above normal mode methods to fail due to rank deficiencies in the *FRF* matrices, causing problems with the mathematical operations. Rank reduction techniques overcome this difficulty.

Modified Multivariate Mode Indicator Function
The Modified Multivariate Mode Indicator Function (*ModMMIF*) was introduced [4.46] to remedy the rank deficiency problem through the use of a rank reduction approach.

Performing the eigenvalue decomposition (in a similar way to the *SVD* rank reduction approach) of the $\boldsymbol{B}$ matrix (remembering that $\boldsymbol{A} = \boldsymbol{H}_R^T \boldsymbol{M} \boldsymbol{H}_R$ and $\boldsymbol{B} = \boldsymbol{H}_I^T \boldsymbol{M} \boldsymbol{H}_I$) and partitioning the matrices to relate to non-zero and zero eigenvalues such that

$$\boldsymbol{B} = \boldsymbol{Q}\boldsymbol{\Lambda}\boldsymbol{Q}^T = \begin{bmatrix} \boldsymbol{Q}_1 & \boldsymbol{Q}_0 \end{bmatrix} \begin{bmatrix} \boldsymbol{\Lambda}_1 & 0 \\ 0 & \boldsymbol{\Lambda}_0 \end{bmatrix} \begin{bmatrix} \boldsymbol{Q}_1^T \\ \boldsymbol{Q}_0^T \end{bmatrix} \approx \boldsymbol{Q}_1 \boldsymbol{\Lambda}_1 \boldsymbol{Q}_1^T \tag{4.149}$$

then the rank of the $\boldsymbol{B}$ matrix is found from the number of non-zero eigenvalues ($n^* \leq NI$) in the diagonal matrix $\boldsymbol{\Lambda}_1$, in practice the zero eigenvalues (diagonal matrix $\boldsymbol{\Lambda}_0$) will be very small, and there are ($NI - n^*$) zero eigenvalues. Each of the non-zero eigenvalues has a corresponding principal vector contained in $\boldsymbol{Q}_1$.

Now, assuming that the force vectors $\boldsymbol{F}$ ($NI \times 1$) can be expressed as a linear combination of the principal vectors $\boldsymbol{Q}_1$ ($NI \times n^*$) and a vector $\boldsymbol{a}$ ($n^* \times 1$) of unknown participation factors such that

$$\boldsymbol{F} = \boldsymbol{Q}_1 \boldsymbol{a} \tag{4.150}$$

then the *MMIF* cost function (4.145) becomes

$$\alpha = \frac{\boldsymbol{a}^T \boldsymbol{Q}_1^T \boldsymbol{A} \boldsymbol{Q}_1 \boldsymbol{a}}{\boldsymbol{a}^T \boldsymbol{Q}_1^T (\boldsymbol{A} + \boldsymbol{B}) \boldsymbol{Q}_1 \boldsymbol{a}} \tag{4.151}$$

The *MMIF* eigenvalue formulation (4.146) can then be rewritten as

$$\boldsymbol{Q}_1^T \boldsymbol{A} \boldsymbol{Q}_1 \boldsymbol{a} = \lambda \boldsymbol{Q}_1^T (\boldsymbol{A} + \boldsymbol{B}) \boldsymbol{Q}_1 \boldsymbol{a} \tag{4.152}$$

which is solved and used in the same way as with the *MMIF* method. The zero *ModMMIF* eigenvalues λ indicate the resonant frequencies for each of the normal modes and the corresponding participation vectors $\boldsymbol{a}$ can be employed in Eq. (4.150) to define the physical appropriated force vectors $\boldsymbol{F}_a$ for each of the normal modes.

***SVD* Multipoint Excitation Method**
An alternative to the *ModMMIF* approach is the so-called *SVD* multipoint method [4.47], which employs the minimisation of the real part of the response whilst simultaneously maximising the

imaginary response. The incorporation of the *SVD* enables any issues due to having too many input forces compared to effective modes at a particular frequency to be resolved.

Starting with the forcing conditions $\boldsymbol{F}_a$ for an appropriated mode $\boldsymbol{\phi}$ (Eq. (4.136)),

$$\boldsymbol{H}_R\boldsymbol{F}_a = \boldsymbol{X}_R = \boldsymbol{0} \quad \text{and} \quad \boldsymbol{H}_I\boldsymbol{F}_a = \boldsymbol{X}_I = \boldsymbol{\phi} \tag{4.153}$$

then for *NO* outputs, the minimisation of the cost function J_1, which considers the real part, can be defined as

$$min\, J_1 = min\left(\boldsymbol{F}^T\boldsymbol{H}_R^T\boldsymbol{H}_R\boldsymbol{F}\right) \tag{4.154}$$

whereas the equivalent cost function for the imaginary part, J_2, is written as

$$max\, J_2 = \sum_{k=1}^{NO}\left(max\left(\boldsymbol{F}^T\boldsymbol{H}_I^T\boldsymbol{H}_I\boldsymbol{F}\right)\right) \tag{4.155}$$

Performing the *SVD* on the real part of the *FRF* matrix gives

$$\boldsymbol{H}_R = \begin{bmatrix}\boldsymbol{P}_1 & \boldsymbol{P}_0\end{bmatrix}\begin{bmatrix}\boldsymbol{D}_1 & \boldsymbol{0}\\ \boldsymbol{0} & \boldsymbol{D}_0\end{bmatrix}\begin{bmatrix}\boldsymbol{Q}_1^T\\ \boldsymbol{Q}_0^T\end{bmatrix} \tag{4.156}$$

where the subscript "$_1$" relates to the n^* non-zero singular values whereas the subscript "$_0$" relates to the $(NI - n^*)$ zero singular values, hence

$$\boldsymbol{X}_R = \boldsymbol{H}_R\boldsymbol{F} = \left(\boldsymbol{P}_1\boldsymbol{D}_1\boldsymbol{Q}_1^T + \boldsymbol{P}_0\boldsymbol{D}_0\boldsymbol{Q}_0^T\right)\boldsymbol{F} \tag{4.157}$$

The drop in the singular values can be used to identify the undamped natural frequencies (Figure 4.69) as was also seen for the Extended Asher eigenvalues.

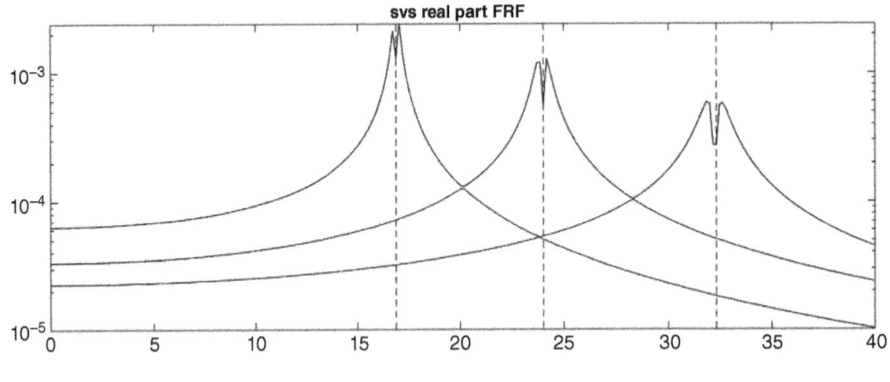

Figure 4.69 *SVD* approach: singular values of the real part of the *FRF*.

The decomposition is partitioned according to the effective rank (n^*) of $\boldsymbol{H}_R$ at a given resonance frequency. For $\boldsymbol{X}_R$ to be small, one needs to take $\boldsymbol{F}$ from the null space, so define

$$F = Q_0 c \tag{4.158}$$

where c is an unknown vector. Thus,

$$X_R = H_R F = P_1 D_1 Q_1^T Q_0 c + P_0 D_0 Q_0^T Q_0 c = P_0 D_0 c \tag{4.159}$$

Hence Eq. (4.154) becomes

$$J_1 = c^T D_0^2 c = \sum_{j=1}^{q} c_j^2 d_{0j}^2 \tag{4.160}$$

so the force vector in (4.158) leads to the minimisation of J_1; however, J_2 also needs to be maximised as follows.

Performing the *SVD* of the ($NO \times (NI - n^*)$) BQ_0 matrix, then

$$X_I = H_I F = BQ_0 c = U\Sigma V^T c \tag{4.161}$$

and assuming that the vector c is equal to a combination of the right singular vectors of BQ_0 such that

$$c = Vg \tag{4.162}$$

then the cost function J_2 becomes

$$J_2 = X_I^T X_I = \left(U\Sigma V^T c\right)^T U\Sigma V^T c = g^T \Sigma^2 g = \sum_{j=1}^{NI-n^*} g_j^2 \sigma_j^2 \tag{4.163}$$

As the singular values are always placed in descending order, then J_2 is maximised when $g^T = [1 \ \ 0 \ \ \ldots \ \ 0]$ and thus $J_2 = \sigma_1^2$ and $c = v_1$. Consequently, the appropriated force vector F_a that meets both cost functions is

$$F_a = Q_0 c = Q_0 v_1 \tag{4.164}$$

4.7 Overall Approach for Linear System Identification

To summarise the preceding sections in this chapter, a suggested approach for linear system identification is:

- explore the measured time data, auto and cross spectra, auto and cross correlations, *FRF*s and *MMIF*s (if available) and *IRF*s to determine how good the data are, considering the noise level, any nonlinear effects, the frequency resolution, the length of time data, the Nyquist frequency compared to the highest frequency of interest, etc.;
- get an overview of the system characteristics – number of modes?, are the modes close?, is the damping high or low?, are there modes that are only visible on certain transducers?;

- apply *SDOF* time or frequency methods to determine a first estimate of natural frequencies, damping ratios and mode shapes. Applying more than one approach will give confidence in the quality of the modal parameter estimates;
- if a more sophisticated analysis is needed, employ one or more of the *MDOF phase-separation* identification methods:
  - use stability plots and frequency *versus* damping plots to determine modal estimates and the scatter on them;
  - estimate the residues and mode shapes;
  - regenerate the *FRFs* using the identified parameters and assess how well the identification has been performed;
  - if the curve-fits are not satisfactory, repeat the analysis using a sub-set of the measured data sets or frequency ranges;
- If a greater accuracy is still required, perform a *phase resonance* analysis;
  - use the appropriated force vectors for each mode to determine how well the phase resonance condition has been met, employing the Normal Mode Purity Factor and also visualisation of the normal modes at each frequency. Such an assessment can be carried using a numerical "*soft tuning*" approach;
  - The damping values can be computed by *sweeping around each resonance peak*, using a stepped sine approach, with the corresponding appropriated force values, and then applying the half-power points or circle fitting methods;
  - If the appropriation has not worked very well, it may be that some form of tuning is required on the real structure to optimise the forcing vector to further reduce the real part of the appropriated mode – more sophisticated test software may be required to perform this steps;
- Remember that in all aspects of the system identification process *engineering judgement* is of upmost importance. One is recommended to always consider whether the identified modes are what is expected and whether the required accuracy and confidence in the modal parameter estimates has been achieved.

References

[4.1] Ewins, D. J., *Modal Testing - Theory, Practice and Application*, 2nd Ed., Research Studies Press, Baldock, England, 2000.

[4.2] Maia, N. M. M., Silva, J. M. M., He, J., Lieven, N. A. J., Lin, R. M., Skingle, G. W., To, W.-M., Urgueira, A. P. V., *Theoretical and Experimental Modal Analysis*, Research Studies Press Ltd., Tauton, 1997.

[4.3] Bendat, J. S., Piersol, A. G., *Random Data: Analysis and Measurement Procedures*, 4th Ed., John Wiley & Sons, 2010.

[4.4] Williams, R., Crowley, J., Vold, H., *The Multivariate Mode Indicator Function in Modal Analysis*, Proc. 4th International Modal Analysis Conference, 1986.

[4.5] Shih, C. Y. Tsuei, Y. G., Allemang, R. J., Brown. D. L., *Complex Mode Indication Function and its Applications to Spatial Domain Parameter Estimation*, Proc. 7th International Modal Analysis Conference, 1989.

[4.6] Golub, G. H., van Loan C. F., *Matrix Computations*, 3rd Ed., John Hopkins Press, 1996.

[4.7] Kay, S. M., *Fundamentals of Statistical Signal Processing*, Prentice-Hall, 1993.

[4.8] Braun, S., *Discover Signal Processing: An Interactive Guide for Engineers*, John Wiley & Sons, 2008.

[4.9] Prabhu, K. M. M., *Window Functions and their Applications in Signal Processing*, CRC Press, 2014.

[4.10] Rao, S. S., *Mechanical Vibrations*, 6th Ed., Pearson, 2016.

[4.11] Box, G. E. P., Jenkins, G. M., Reinsel, G. C., Ljung, G. M., *Time Series Analysis: Forecasting and Control*, 5th Ed., John Wiley & Sons Inc., New Jersey, pp. 712, 2015.

[4.12] Cole, H. A. Jr., On-Line *Failure Detection and Damping Measurement of Aerospace Structures by Random Decrement Signatures*, NASA CR-2205, March 1973.

[4.13] Cooper, J. E., Desforges, M. J., Wright, J. R., *Modal Parameter Identification using an Unknown Coloured Random Input*, Mechanical Systems and Signal Processing, 9(6), pp. 685–695, 1995.

[4.14] Jacques, J. F., Commandeur, S., Koopman, S. J., *An Introduction to State Space Time Series Analysis*, Oxford University Press, 2007.

[4.15] Platten, M. F., Wright, J. R., Cooper J. E., *Identification of a Wing with Discrete Nonlinear Attachments using an Extended Modal Model*, J. Aircraft, 46(5), pp. 1614–1626, 2009.

[4.16] Peeters, B., Van der Auweraer, H., Guillaume, P., Leuridan, J., *The PolyMAX Frequency Domain Method – A New Standard for Modal Parameter Estimation*, Shock and Vibration, 11, pp. 395–409, 2004.

[4.17] Guillaume, P., Verboven, P., Vanlanduit, S., Van der Auweraer, H., Peeters, B., *A Poly-Reference Implementation of the Least-Squares Complex Frequency-Domain Estimator*, Proc. 23rd International Modal Analysis Conference, 2003.

[4.18] Cooper, J. E., *Comparison of Some Time Domain System Identification Techniques using Approximate Data Correlations*, International Journal of Analytical and Experimental Modal Analysis, 4, pp. 51–57, April 1989.

[4.19] Dobson, B. J., *A Straight-Line Technique for Extracting Modal Parameters from Frequency Response Data*, Mechanical Systems and Signal Processing, 1(1), pp. 29–40, 1987.

[4.20] Ben-Israel, A., Greville, T. N. E., *Generalized Inverses: Theory and Applications*, 2nd Ed., Springer, 2003.

[4.21] Gaukroger, D. R., Skingle, C. W., Heron, K. H., *Numerical Analysis of Vector Response Loci*, Journal of Sound and Vibration, 29, pp. 341–353, 1973.

[4.22] Richardson M. H., Formenti, D. L., *Global Curvefitting of Frequency Response Measurements using the Rational Fraction Polynomial Method*, Proc. 3rd International Modal Analysis Conference, 1985.

[4.23] Wu, C., Liu, H., Qin, X., Wang, J., *Stabilization Diagrams to distinguish Physical Modes and Spurious Modes for Structural Parameter Identification*, Journal of Vibroengineering, 19(4), pp. 2777–2794, 2017.

[4.24] Brown, D. L., Allemang, R., Zimmerman, R., Mergeay, M., *Parameter Estimation Techniques for Modal Analysis*, SAE Technical Paper 790221, 1979.

[4.25] Cooper, J. E., *Extending the Logarithmic Decrement Method to Analyse Two Degree of Freedom Transient Responses*, Mechanical Systems and Signal Processing, 10(4), pp. 497–500, 1996.

[4.26] Cooper, J. E., *Experimental Validation of the Extended Logarithmic Decrement Method*, Proc. 15th International Modal Analysis Conference, 1997.

[4.27] Crittenden, J. L., Mulholland, R., Hill, J., Martinez, E., *Sampling and Data Analysis Properties of Prony's Method applied to Modal Identification*, International Journal of Systems Science, 14(5), pp. 571–584, 1983.

[4.28] Cooper, J. E., *Comparison of Modal Parameter Estimation Techniques on Aircraft Structural Data*, Mechanical Systems and Signal Processing, 4(2), pp. 157–172, 1990.

[4.29] Vold, H., Kundrat, J., Rocklin, G. T., Russell, R., *A Multi-Input Modal Estimation Algorithm for Mini-Computers*, SAE Technical Paper 820194, 1982.

[4.30] Juang, J-N, Pappa, R. S., *An Eigensystem Realisation Algorithm for Modal Parameter Identification and Model Reduction*, Journal of Guidance, Control, and Dynamics, 8(5), pp. 620–627, 1985.

[4.31] Cooper, J. E., *The Use of Backwards Models for Modal Parameter Identification*, Mechanical Systems and Signal Processing, 6(3), pp. 217–228, 1992.

[4.32] Brincker, R., Ventura, C., *Introduction to Operational Modal Analysis*, John Wiley & Sons, 2015.

[4.33] Brincker, R., Zhang, L., Anderson, P., *Modal Identification of Output-Only Systems using Frequency Domain Decomposition*, Smart Materials and Structures, 10(3), 2001.

[4.34] Jacobson, N.-J., Anderson, P., Brincker, R., *Using Enhanced Frequency Domain Decomposition as a Robust Technique to Harmonic Excitation in Operational Modal Analysis*, Proc. International Seminar on Modal Analysis, 2006.

[4.35] Rodrigues, J., Brinker, R., *Application of the Random Decrement Technique in Modal Analysis*, Proc. 1st International Operational Modal Analysis Conference, 2005.

[4.36] Ibrahim S. R.. *Random Decrement Technique for Modal Identification of Structures*, Journal of Spacecraft and Rockets. v14 n11, 1977.

[4.37] Juang, J.-N., Cooper, J. E., Wright, J. R., *An Eigensystem Realisation Algorithm using Data Correlations (ERA/DC) for Modal Parameter Identification*, Control - Theory and Advanced Technology, 4(1), 1988, pp. 5–14.

[4.38] Cooper, J. E., Wright, J. R., *Spacecraft In-Orbit Identification using Eigensystem Realisation Methods*, Journal of Guidance, Control, and Dynamics 16(2), pp. 352–359, 1992.

[4.39] Cooper, J. E., Hamilton, M. J., Wright, J. R., *Experimental Evaluation of Various Normal Mode Force Appropriation Methods*, International Journal of Analytical and Experimental Modal Analysis, 10(2), pp. 118–130, 1995.

[4.40] Wright, J. R., Cooper, J. E., Desforges, M. J., *Normal Mode Force Appropriation – Theory and Application*, Mechanical Systems and Signal Processing, 31(2), pp. 217–240, 1999.

[4.41] Asher, G. W., *A Method of Normal Mode Excitation using Admittance Measurements*, Proc. of the National Specialists Meeting on Dynamics and Aeroelasticity, Journal of Aeronautical Sciences, 6, pp. 56–96, 1958.

[4.42] Alexiou, K., Wright, J. R., *Comparison of Some Multipoint Force Appropriation Methods*, International Journal of Analytical and Experimental Modal Analysis, 8(2), pp. 119–136, 1993.

[4.43] Traill-Nash, R. W., *On the Excitation of Pure Natural Modes in Aircraft Resonance Testing*, Journal of Aerospace Sciences, 25(12), 1958.

[4.44] Ibañez, P., *Force Appropriation by Extended Asher's Method*, SAE Paper No. 760873, SAE Aerospace Engineering and Manufacturing Meeting, 1976.

[4.45] Breitbach, E., *A Semi Automatic Modal Survey Test Technique for Complex Aircraft and Spacecraft Structures*, Proc. 3rd ESRO Testing Symposium, pp 519–528, 1973

[4.46] Nash, M., *Use of Principal Force Vectors with the Multivariate Mode Indicator Function*, Proc. 9th International Modal Analysis Conference, 1991.

[4.47] Juang, J-N, Wright, J. R., *A Multi-Point Force Appropriation Method based upon a Singular Value Decomposition Approach*, Trans. ASME Journal of Vibration and Acoustics, 113, pp. 176–181, 1991.

Chapter 5 Nonlinearity in Engineering Dynamics

5.1 The Significance of Nonlinearity

All of the dynamics considered here so far has been *linear*; all of the equations of motion have been linear. This chapter will explore the ways and means of characterising and understanding *nonlinearity* in some detail. For now, it will suffice to say that an equation is linear if it contains no powers of the variables of interest higher than the first. In the context of dynamics, 'variables of interest' means the inputs and outputs for the system of interest and any of their derivatives.

Nonlinearity must be addressed because it is *pervasive*; all physical systems are nonlinear to some extent or other. Sometimes it is a good approximation to assume a linear representation, but often this leads to serious errors. One might go further and argue that *all* interesting systems are nonlinear. The core of this argument would be that nonlinearity is a signature of *interactions*, and that systems are important or interesting because they contain or express interactions. Although it will prove to be a short digression from engineering dynamics, it is worth looking at how widespread and important the effects of nonlinearity actually are.

5.1.1 Nonlinearity in Fundamental Physics

At its most fundamental level, physics is concerned with the interactions of particles. Like structures in engineering, particles in physics are governed by equations of motion and interact by exerting forces on each other. Although there are many types of particles and forces one could consider, the discussion here will be restricted (initially) to the electromagnetic forces between electrons. As electrons are all negatively charged, they will exert a repulsive force on each other by way of interaction. The electron e^- will be represented by a state variable ψ (to maintain consistency in notation) which is governed by the *Dirac equation* [5.1]. The variable ψ has four components; it is not actually a vector, it is an object called a *spinor*, but the distinction will not be important here. The equation of motion has the form,

$$i\left(\gamma^t \frac{\partial}{\partial t} + \gamma^x \frac{\partial}{\partial x} + \gamma^y \frac{\partial}{\partial y} + \gamma^z \frac{\partial}{\partial z}\right)\psi = m\psi \tag{5.1}$$

in a standard Cartesian coordinate (x, y, z) system, with time variable t. The objects γ^i are constant 4×4 matrices composed of $0, \pm1, \pm i$, and m is the mass of the electron. One can think of this as a state-space representation and it is thus first order. The equation is otherwise a little unusual by structural dynamics standards because it is complex, but the quantum world is an unusual place. Most importantly, for the discussion here, the equation is *linear*. The critical observation here is that this equation represents a single electron moving freely in space,

Structural Dynamics in Engineering Design, First Edition, Nuno M. M. Maia, Dario Di Maio, Alex Carrella, Francesco Marulo, Chaoping Zang, Jonathan E. Cooper, Keith Worden, and Tiago A. N. Silva.

uninfluenced by any other particles. The question is then: how does one introduce the idea of forces between electrons in this picture?

In the quantum world, particles experience forces by *exchanging particles*; the mathematics of the processes involved is complicated, but particle physicists have developed a pictorial representation of interactions which helps a great deal, called *Feynman diagrams*. This representation was invented by the theoretical physicist Richard Feynman in the first half of the twentieth century [5.1]. Figure 5.1 shows the simplest possible interaction between two electrons e_1^- and e_2^-.

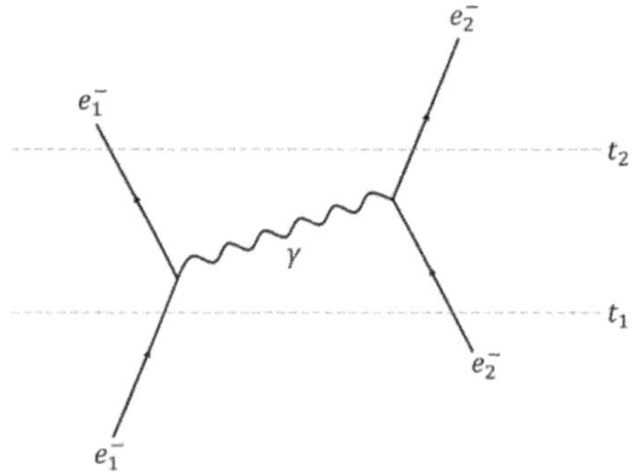

Figure 5.1 Feynman diagram for tree-level interaction between two electrons e_1^- and e_2^-.

Feynman diagrams are *space-time* representations of processes; they must be understood to show time t, increasing from bottom to top, and the spatial variable x, increasing from left to right. The lines represent paths of particles. With these conventions, one can see from Figure 5.1 that, at the time t_1, the electrons are moving towards each other, while at the later time t_2, they are moving apart. This change of direction indicates that something has happened such that the two electrons have 'repelled' each other.

The critical 'something' which as generated the apparent repulsive force, is that the electrons have exchanged a particle. Shortly after time t_1, the electron e_1^- emits a particle γ, in the positive x-direction. The particle γ has momentum, so e_1^- experiences a recoil and changes direction. The electron e_2^- absorbs the new particle, picks up its momentum, and changes direction also. An observer simply looking at the picture before and after the event would not see the temporary particle γ, and would conclude that the electrons repelled each other when they became close (it is easy to think of a classical analogy here: imagine two people on skateboards throwing some heavy object back and forth between them; the continuous momentum exchange would cause them to repel each other. Generating an *attractive* force with this picture is another matter).

This seems to be an odd way to think about forces between particles, but it turns out to be the right way. The full quantum theory of forces between electrons (or charged particles generally) – *quantum electrodynamics* – is the most accurate theory known to humanity, and makes predictions that agree with experiment to 11-12 significant figures.

However, what is the mysterious 'exchange' particle γ? It turns out that γ is a *photon* – a particle of light. This observation means that, in a full mathematical theory, the photon is going to need an equation of motion also. It transpires that one can represent γ by a vector also – usually denoted $\boldsymbol{A}$ (and it really is a vector this time). The equation of motion is interesting, it is,

$$\frac{\partial A_\mu}{\partial x_\nu} - \frac{\partial A_\nu}{\partial x_\mu} = 0 \tag{5.2}$$

where the A_μ are the components of $\boldsymbol{A} = \left(A_t, A_x, A_y, A_z\right)$, and ∂_μ is shorthand for $\partial/\partial t$, $\partial/\partial x$, etc.

As in the case of the Dirac equation, this equation is manifestly linear and only applies to the photon in the case of *free* motion, when it is not interacting with the electrons. In the Feynman graph, this corresponds to the wavy line in the graph. In a similar way, the solid lines in the graph correspond to the free motion of the electrons. However, the interactions are the critical parts of the graph which generate the forces; the vertices at which the photon is emitted and reabsorbed. In fact, emission and absorption are basically the same physics (just imagine time running forwards and backwards respectively), so the key interaction in the Feynman diagram occurs at the vertex V in Figure 5.2.

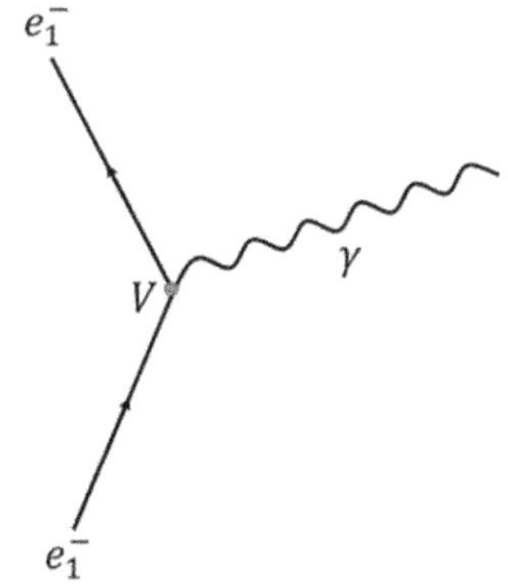

Figure 5.2 Feynman diagram showing the vertex for emission of a photon.

To make the particles interact, one needs to add the state $\boldsymbol{A}$ to the Dirac equation, and the state ψ to the photon equation. The vertex explains exactly how to do this; at a specific point in space and time, two copies of an electron line and a photon line are meeting, suggesting a *nonlinear* interaction term of the form $\psi^2 A$. In fact, this is basically correct; however there are a couple of subtleties. In the first case, the interaction term is actually added to an overall Lagrangian for the theory, where it naturally makes its way into both field equations; the end result is the same. Secondly, the term contains a complicated sum over products of components ψ_μ and A_μ; the details of the theory can be found in [5.1].

This result is quite dramatic; without nonlinearity there are no electromagnetic forces. Without nonlinearity, electrons would not bind to nuclei and atoms could not exist; the universe would be a soup of freely-moving, non-interacting particles: there would be no atoms, no molecules, no life.

Quantum electrodynamics is an intricate theory with vector and spinor fields and a complicated interaction. As is usual in such cases, mathematicians and physicists wanted a simpler 'toy' theory with which to practice; a theory with the main features of reality – like interactions – but not the full complexity. The main simplification made was to work with a *scalar* field $\phi(t,x,y,z)$, i.e. one with a single component. In this case, the free equation of motion is called the *Klein-Gordon* equation, and is given by,

$$\left(-\frac{\partial^2}{\partial t^2}+\frac{\partial^2}{\partial x^2}+\frac{\partial^2}{\partial y^2}+\frac{\partial^2}{\partial z^2}+m^2\right)\phi=0 \tag{5.3}$$

where m is the mass of the ϕ particle.

This equation is again linear and has no interactions; it is basically a wave equation with mass. To make it interesting, one needs to add interactions again, via the Lagrangian. Adding ϕ^2 leaves the equation of motion linear, and adding ϕ^3 causes some technical problems, so the simplest satisfactory interesting term is ϕ^4; such theories are called $\lambda\phi^4$ theories.

This time going from the interaction term to the vertex, one arrives at the picture in Figure 5.3 where four ϕ lines meet at a vertex.

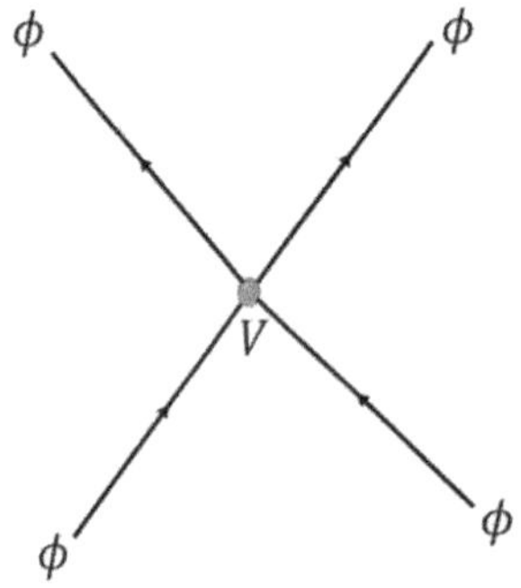

Figure 5.3 Feynman diagram for a $\lambda\phi^4$ interaction.

Individual Feynman diagrams are not the whole picture; they actually represent terms in a perturbation expansion (more of that later), and must be summed together to get overall results. The lines in the diagram represent free movement (and the linear equations) and the vertices represent the nonlinearity. The simplest diagram showing how ϕ particles exert a force on each other, that one can build from lines and vertices, is shown in Figure 5.4.

In fact, quantum field theories suffer from a severe technical problem, in the sense that some diagrams, or the overall sum, can diverge to infinity. In 'well-behaved' theories, this problem is solved by a process called *renormalisation* [5.1]. The infinities are often associated with diagrams with loops; Figure 5.5 shows an interesting example of a loop diagram, which

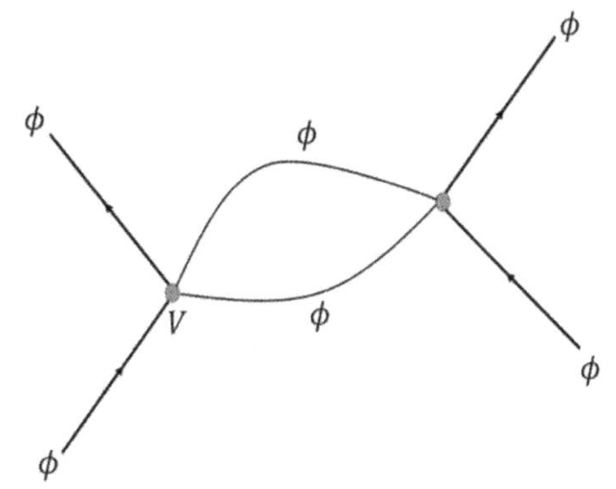

Figure 5.4 Feynman diagram for two ϕ particles scattering from each other in $\lambda\phi^4$ theory.

contributes to *self-interaction* for a particle. Clearly this diagram is only possible because of the specific nature of the interaction vertex. A final observation on $\lambda\phi^4$ theory, is that the interaction produces a ϕ^3 term in the actual equation of motion for ϕ.

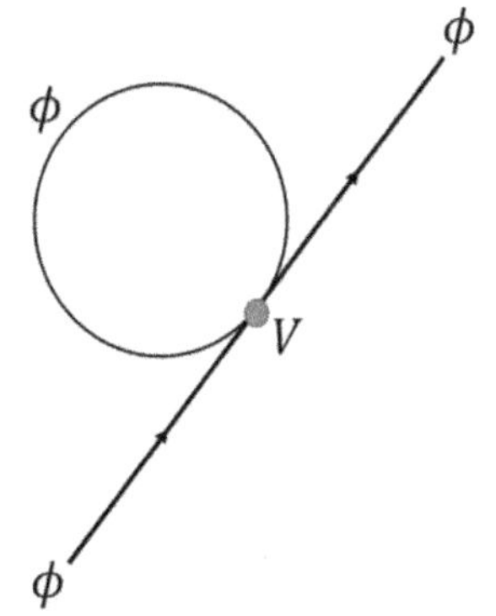

Figure 5.5 Feynman diagram showing self-interaction for a ϕ particle.

At this point, the alert reader will protest and point out that structural dynamicists quite happily include linear interactions by coupling masses together with springs, and this is a valid point; the masses are manifestly interacting. However, as earlier chapters in this book have shown, there is a change of coordinates or variables (via modal analysis) which gives a representation in which the masses are *SDOF* systems and are *not interacting*. For this reason, it would seem reasonable to call such interactions *removable*. One might argue that nonlinear interactions might prove to be removable also, and this is the basis of an approach by Poincaré called *normal form* analysis [5.2]; however, this is an approximate theory and there are no exact results to show that nonlinear interactions are completely removeable.

5.1.2 Nonlinearity in Epidemiology

Mathematical epidemiology aims to characterise and quantify the transmission of infectious diseases across populations [5.3]. These are dynamic processes, again determined by interactions; an infection cannot occur without interaction. The main models in use are differential equations, and as in the examples from physics discussed earlier, the interactions produce nonlinear equations.

For amusement, the epidemic discussed here will be a fictitious zombie outbreak, although the principles applied and the resulting equations will be typical of infectious diseases generally.

Rather than consider 'microscopic' interactions, the variables chosen will reflect sub-population statistics. Suppose one begins with some initial population; this must include at least one zombie. The human fraction of the population will be denoted by H, the zombie fraction by Z.

Consider the rate of change of the human population under natural circumstances; the population will increase because of new births, and a constant rate Π will be assumed here for simplicity. The population will decrease because of natural deaths, and the rate will depend on the size of the population, so the basic equation will be,

$$\dot{H} = \Pi - k_d H \tag{5.4}$$

where the constant k_d encodes the natural death rate. From this point onwards and as in the previous chapters, derivatives *w.r.t.* time will be denoted by overdots as the equations will largely be ordinary differential equations.

The human population will also decrease because of interactions with zombies; clearly the rate of interactions (attacks) will be higher if there are more people or more zombies, so a sensible and simple postulate reflecting this is,

$$\dot{H} = \Pi - k_h H - \alpha HZ \tag{5.5}$$

which is now nonlinear.

The zombie population also has dynamics; assuming that there is some natural decay (pun intended) of the population, it is clear that,

$$\dot{Z} = -k_z H + \alpha HZ \tag{5.6}$$

because each attack that removes a human, creates a zombie.

As in the physics example, the system is *only* interesting because of the interactions, and the nonlinearity is therefore essential.

These equations have actually been studied for some time; In fact they are a version of the Volterra-Lotke *predator-prey* equations, which arise when considering a general predator population interacting with its food source [5.4]. The most commonly-discussed example of the dynamics considers foxes and rabbits. Despite the simplicity of the system, it generates interesting dynamics. Consider a situation in which the initial condition has a few foxes and many rabbits. In those circumstances, with a steady food supply, the foxes breed and increase their populations, while the rabbits decrease accordingly. This pattern continues until the rabbits have been hunted almost to extinction, at which point the food supply cannot support a large fox population, and the number of foxes starts to decrease. With a small number of foxes present, the rabbits been to thrive again and their population increases, and so on. What entails here is a pattern of oscillations, with (in this picture) the fox oscillations lagging in phase behind the rabbit oscillations. The dynamics is interesting because the oscillations arise naturally *without* any external driving mechanism; this cannot arise in any naturally damped linear system, and all population dynamics problems are damped because of natural death rates.

So the basis of mathematical epidemiology is something like a predator-prey system; however, this is not quite realistic enough. The important point missed in the equations is that people can recover from diseases: i.e. the transition from uninfected to infected is reversible (although this is not the case in the zombie scenario). Furthermore, people can derive immunity from their illness and are not susceptible to further infection. In order to allow for these possibilities, a more general situation is considered in modern epidemiology, where three populations are involved [5.3]:

- **Susceptibles** S : This is the fraction of the population at risk of infection: i.e. not immune; this is the human population in the context of the zombie outbreak;

- **Infected** I : This is the fraction of the population infected at a given time: i.e. the zombies in the outbreak scenario;

- **Removed** R : This is the new addition, it is the fraction of the population that is not infected or at risk of infection (because of immunity).

Continuing with the zombie outbreak picture, the removed R, comprises two groups of entities. The first group is those humans who have died natural deaths, but do not resurrect as zombies; this will generate a linear decay rate on the human population and a corresponding increase in the removed. More interesting is the possibility that zombies can also be removed. In most examples of the zombie genre, the correct way to remove a zombie is to decapitate it, or otherwise incapacitate the brain. Clearly this will be the result of further interactions between humans and zombies, and will require a term of the form βHZ , where β is a different constant to α . In all cases here, the removed group comprises the dead, so the notation D will be used in the outbreak context.

Some context-specific assumptions here will be: (i) zombies do not attack other zombies and incapacitate them; (ii) the removed/dead cannot interact further. Different parts of the zombie genre have different methods for generating new zombies: for example in the 'Walking Dead' picture [5.5], any human who dies will resurrect as a zombie; this will not be assumed here. With all these points taken into account, the full HZD (SIR) equations become [5.6],

$$\dot{H} = \Pi - k_h H - \alpha HZ \tag{5.7}$$

$$\dot{Z} = -k_z Z + \alpha HZ - \beta HZ \tag{5.8}$$

$$\dot{D} = k_h H + k_z Z + \beta HZ \tag{5.9}$$

The processes at work are summarised in Figure 5.6.

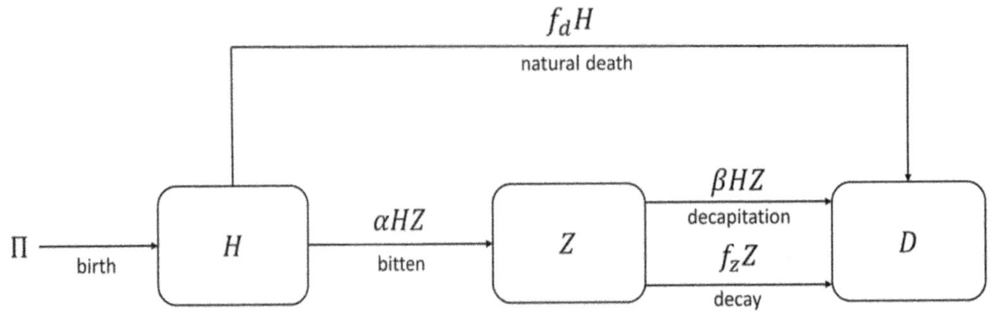

Figure 5.6 Schematic of the main processes at work in the zombie outbreak model.

Clearly this is a quite versatile framework for modelling zombie outbreaks, various features from the genre can be incorporated like the resurrection of all the dead ('The Walking Dead' [5.5]) or even a zombie birth rate ('Girl with all the Gifts' [5.7]). The most 'realistic' addition to the model, consistent with many of the fictions, would be a period of latency, e.g. a period in which a human remains human after being bitten. All these modifications are left as exercises for the reader. A simplification would be to assume that the outbreak occurs over such a short period of time that natural deaths and births could be neglected: i.e. $f_h = f_z = \Pi = 0$.

5.1.3 Nonlinearity in Meteorology

As a model of interacting systems, the *SIR* or zombie model is remarkably similar to those for other systems of interest. In particular, a system of great historical interest is the *Lorenz* system [5.8], which models a highly-simplified version of atmospheric fluid mechanics, i.e. convective movement in atmospheric layers. The reader is referred to the original reference for the details, but the system of three equations is given by,

$$\dot{x} = \sigma(y - x) \tag{5.10}$$

$$\dot{y} = \rho x - y + xz \tag{5.11}$$

$$\dot{z} = -\beta z + xy \tag{5.12}$$

which is clearly very similar to the *SIR* equations; however, in this case there are two distinct nonlinear interactions: xy and xz.

The equations are of great historical importance because they represent one of the first situations in which the phenomenon of *chaos* was observed; again, more of that later.

5.1.4 Nonlinearity in Structural Dynamics

All of the previous discussion is fascinating, and is clear evidence of the significance of nonlinearity; however, the subject of this chapter is nonlinearity in *engineering dynamics*, so it is important to see where this might arise.

In general, structural dynamicists do not think of nonlinearity as arising from interactions; rather, nonlinearity is considered to arise from geometrical sources (large deflections) or from non-Hookian material behaviour where nonlinear force-displacement behaviour is seen. The example given here will demonstrate a source of geometrical nonlinearity, i.e. for large displacements of an encastré beam – that is, a beam with built-in/clamped boundary conditions at both ends; no deflection, no slope. In this case, 'large' displacements will mean flexural deformations greater than the thickness of the beam itself (Figure 5.7).

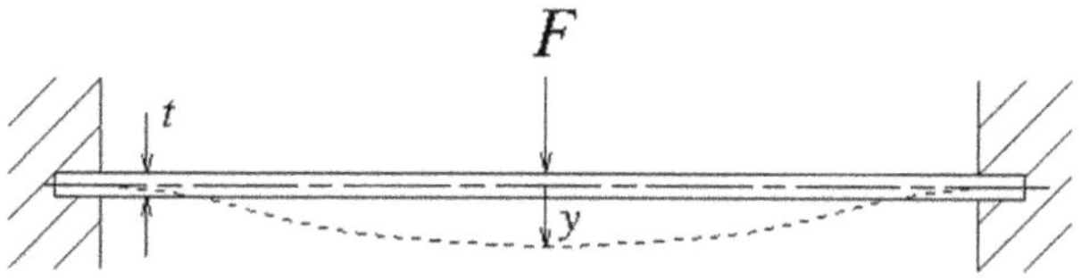

Figure 5.7 Deflection of the centre point of an encastré beam under central transverse loading.

It will be sufficient here to consider static behaviour of the beam and demonstrate a nonlinear force-stiffness relationship. Consider the encastré beam under a centrally-applied static-load (Figure 5.7). The deflection shape, with the coordinates located at the mid-point of the beam, can be assumed to be a polynomial which satisfies all the boundary conditions and the eigenvalue problem, i.e. an admissible function,

$$y(x) = Y\left(1 - \frac{ax^2}{\left(\frac{L}{2}\right)^2} + \frac{bx^4}{\left(\frac{L}{2}\right)^4} - \frac{cx^6}{\left(\frac{L}{2}\right)^6} + \ldots\right) \tag{5.13}$$

where the coefficients have been given a convenient dimensional representation, and Y has the dimensions of displacement.

Using this assumed shape and by deriving the axial and flexural strain energies, an expression for the lateral stiffness at the centre of the beam can be found. If only the first three terms in the series are used with the appropriate values for the constants (the details of the eigenfunction analysis are omitted here as they are not particularly relevant, and would require quite a large digression), the expression for the deflection is,

$$y(x) = Y\left(1 - 2.15\frac{x^2}{\left(\frac{L}{2}\right)^2} + 1.30\frac{x^4}{\left(\frac{L}{2}\right)^4} - 0.15\frac{x^6}{\left(\frac{L}{2}\right)^6} + \ldots\right) \tag{5.14}$$

and the flexural strain energy V_F is found from,

$$V_F = \int_{-L/2}^{L/2} \frac{M^2}{2EI} dx = \frac{EI}{2} \int_{-L/2}^{L/2} \left(\frac{d^2y}{dx^2}\right)^2 dx \tag{5.15}$$

to be,

$$V_F = \frac{EIY^2}{\left(\frac{L}{2}\right)^{12}} \int_0^{L/2} \left(-4.3\left(\frac{L}{2}\right)^4 + 15.6\left(\frac{L}{2}\right)^2 x^2 - 4.5x^4 \right)^2 dx \tag{5.16}$$

so finally,

$$V_F = 98.9EI\frac{Y^2}{L^3} \tag{5.17}$$

The flexural strain energy is going to result in a simple linear stiffness function, if substituted into a Lagrangian; the nonlinearity is going to arise from the fact that, at large displacements, there will be an overall change in the length of the beam which will generate non-negligible *axial strains*.

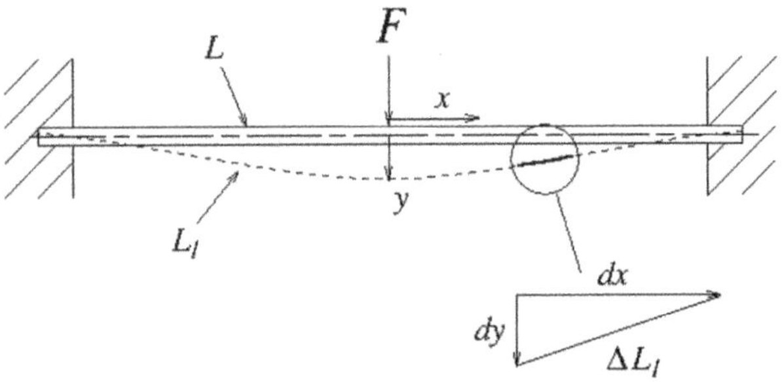

Figure 5.8 An encastré (clamped – clamped) beam under a centrally-applied static load undergoes a change in length from L to L_1. The elemental length represents the axial extension.

The strain energy due to the in-plane axial load is found from the expression governing the axial extension (see Figure 5.8),

$$\Delta L_1 = \left(dx^2 + dy^2\right)^{1/2} = dx\left(1 + \left(\frac{dy}{dx}\right)^2\right)^{1/2} \tag{5.18}$$

where the first equality is basically just Pythagoras' theorem. It follows from a simple application of the binomial theorem (assuming that the slope dy/dx is small) that,

$$\Delta L_1 = dx\left(1 + \frac{1}{2}\left(\frac{dy}{dx}\right)^2 - \frac{1}{8}\left(\frac{dy}{dx}\right)^4 + \ldots\right) \approx dx\left(1 + \frac{1}{2}\left(\frac{dy}{dx}\right)^2\right) \tag{5.19}$$

Therefore,

$$L_1 = \int_{-L/2}^{L/2} \left(1 + \frac{1}{2}\left(\frac{dy}{dx}\right)^2\right) dx = L + \frac{1}{2}\int_{-L/2}^{L/2} \left(\frac{dy}{dx}\right)^2 dx \tag{5.20}$$

and ΔL, the change in axial length of the beam is given by,

$$L_1 - L = \frac{1}{2}\int_{-L/2}^{L/2}\left(\frac{dy}{dx}\right)^2 dx \tag{5.21}$$

Substituting for $y(x)$ from Eq. (5.13) gives,

$$\Delta L = 2.44\frac{Y^2}{L} \tag{5.22}$$

Thus, the axial strain energy is,

$$V_A = \frac{1}{2}\frac{EA}{L}(\Delta L)^2 = 2.98\,EA\frac{Y^4}{L^3} \tag{5.23}$$

From Lagrange's equations, the stiffness terms are given by,

$$\frac{\partial}{\partial Y}(V_F + V_A) = 197.8\frac{EIY}{L^3} + 11.92\frac{EAY^3}{L^3} \tag{5.24}$$

i.e. the linear elastic stiffness term is $k_1 = 197.8EI/L^3$ and the nonlinear stiffness term is $k_3 = 11.92EAY^2/L^3$. (Note that the linear elastic stiffness term k_1 should be $192EI/L^3$ from simple bending theory. The small error is due to limiting the assumed deflection polynomial to only three terms.)

Now, if one assumes that the central point of the beam has an associated mass m (which could be calculated via a Rayleigh approximation if needed, but is not particularly relevant here), the end result of using the Euler-Lagrange equations in the dynamic case would be,

$$m\ddot{Y} + k_1 Y + k_3 Y^3 = F \tag{5.25}$$

which is the undamped version of an equation called *Duffing's equation* [5.9].

In many respects, Duffing's equation is the simplest possible nonlinear dynamic equation. One might argue that an equation with a quadratic nonlinearity would be simpler, but this is physically not the case. In order to generate a quadratic term from the analysis above, one would have to pre-stress the beam, i.e. create an initial offset in one direction or other, and this would break the symmetry of the physics, but still yield a cubic term from the axial strain. Furthermore, a quadratic nonlinearity introduces stability issues that are not present in the cubic equation.

This result is quite pleasing in another respect: Duffing's equation is the analogue of $\lambda\phi^4$ theory in theoretical physics; however, the corresponding model in structural dynamics is actually physically relevant, rather than just a 'toy' model. Duffing's equation is an important benchmark for almost all of the methods of nonlinear structural dynamics, and will recur many times in the following discussion.

Because of its importance as a benchmark, various versions of Duffing's equation are going to appear in the coming pages. As discussed, the simplest version considered will be the *unforced (free) undamped symmetric* equation,

$$m\ddot{y}+ky+k_3y^3=0 \tag{5.26}$$

where the term 'symmetric' refers to the fact that the restoring stiffness $ky+k_3y^3$ is a purely odd function. By adding a quadratic term to the stiffness function, one obtains the *unforced (free) undamped asymmetric* equation,

$$m\ddot{y}+ky+k_2y^2+k_3y^3=0 \tag{5.27}$$

Introducing a force term x on the *RHS* of the equation, yields a *forced* Duffing oscillator,

$$m\ddot{y}+ky+k_2y^2+k_3y^3=x(t) \tag{5.28}$$

Finally, adding a viscous damping term gives a *damped* oscillator,

$$m\ddot{y}+c\dot{y}+ky+k_2y^2+k_3y^3=x(t) \tag{5.29}$$

and this last expression is the most comprehensive form of the equation.

5.2 Solution of Nonlinear Equations of Motion

Having established that nonlinearity is pervasive, and that one is likely to encounter nonlinear systems in physics and engineering, it is important to look at the options for solving such equations. In terms of exact mathematical solutions, it will become clear that this is a limited option; this means that one must usually resort to approximate or numerical analysis. This section will discuss all the main possibilities for looking at solutions, mainly concentrating on variants of Duffing's equation, by way of illustration.

5.2.1 *Exact Solutions

This section is one that could be passed over at a first reading of this text. It is not that the mathematics is particularly sophisticated, but rather that the analysis depends on an understanding of some special functions which are not usually covered in undergraduate engineering or physics – the *elliptic* functions. The basic definitions are provided in Appendix B for completeness. The analysis here will follow very closely that in [5.10].

The equation considered here will be the unforced (free) Duffing equation,

$$m\ddot{y}+k_1y+k_3y^3=0 \tag{5.30}$$

which can be written without loss of generality as,

$$\ddot{y}+ay+by^3=0 \tag{5.31}$$

where $a=k_1/m$ and $b=k_3/m$. The initial conditions are chosen to be $y(0)=Y$ and $\dot{y}(0)=0$.

The first step is to calculate a *first integral* for the equation; this is done by writing the equation in terms of the velocity $v = \dot{y}$. The function of a function rule from basic calculus then gives $\ddot{y} = v\,dv/dy$. The equation of motion becomes,

$$v\frac{dv}{dy} = -ay - by^3 \tag{5.32}$$

and the first integral is simply,

$$\int_0^V v\,dv = -\int_Y^y \left(ay + by^3\right)dy \tag{5.33}$$

which is just,

$$v^2 = a\left(Y^2 - y^2\right) + \frac{b}{2}\left(Y^4 - y^4\right) \tag{5.34}$$

where (as will be shown later) this is basically the equation of conservation of energy of the system.

Choosing the negative root of v^2, one finds after a little algebra,

$$v = \frac{dy}{dt} = -\sqrt{a}\left(Y^2 - y^2\right)^{1/2}\left(1 + \frac{b}{2a}\left(Y^2 + y^2\right)\right)^{1/2} \tag{5.35}$$

which integrates 'directly' to,

$$\int_0^t dt = -\int_Y^y \frac{1}{\sqrt{a}\left(Y^2 - y^2\right)^{1/2}\left(1 + \frac{b}{2a}\left(Y^2 + y^2\right)\right)^{1/2}}\,dy \tag{5.36}$$

Now, this may look intractable, but in fact the *RHS* is a known function. One now makes the change of variable $y = Y\cos\theta \Rightarrow dy = -Y\sin\theta\,d\theta$, and the integral becomes,

$$t = \int_0^{\varphi} \frac{1}{\sqrt{a}\left(1 + \frac{bY^2}{2a}\left(1 + \cos^2\theta\right)\right)^{1/2}}\,d\theta \tag{5.37}$$

where the upper limit has been changed from θ to φ, as one should not mix the integral limits with the integrand variable. Naturally, $\cos\varphi = y/Y$, so $\varphi = \cos^{-1}\left(y/Y\right)$. A little more algebra places this equation in the form,

$$t = \frac{1}{\left(a + bY^2\right)^{1/2}}\int_0^{\varphi} \frac{1}{\left(1 - \lambda^2\sin^2\theta\right)^{1/2}}\,d\theta \tag{5.38}$$

where,

$$\lambda^2 = \frac{bY^2}{2\left(a + bY^2\right)} \tag{5.39}$$

Now the second term on the *RHS* of (5.38) is a (perhaps unfamiliar) standard integral,

$$F(\lambda, \varphi) = \int_0^{\varphi} \frac{1}{\left(1 - \lambda^2 sin^2\theta\right)^{1/2}} d\theta \tag{5.40}$$

called the *incomplete elliptic integral of the first kind with modulus* λ [5.11].

This is an exact solution of the problem of interest. Actually, it is an exact solution of the inverse problem; what is required is $y(t)$, but for the moment, what one has is $t(y)$,

$$t = \frac{1}{\left(a + bY^2\right)^{1/2}} F(\lambda, \varphi) \tag{5.41}$$

or,

$$t = \frac{1}{\left(a + bY^2\right)^{1/2}} F\left(\lambda, cos^{-1}\frac{y}{Y}\right) \tag{5.42}$$

Despite being an inverse function, this equation is still immediately useful. Consider the physics of the problem concerned. The system starts at the position $y = Y$, and is driven towards the origin by the cubic restoring force. Because the spring force is symmetrical about the origin, and the system is undamped, the displacement will then decrease to $y = -Y$, and the motion will reverse. The solution $y(t)$ will thus clearly be periodic, with some period τ. By the symmetry of the situation, the time taken for the system to initially go from $y = Y$ $(\varphi = 0)$ to $y = 0$ $(\varphi = \pi/2)$ will be $\tau/4$; it therefore follows that,

$$\tau = \frac{4}{\left(a + bY^2\right)} F\left(\lambda, \frac{\pi}{2}\right) \tag{5.43}$$

which is an *exact* expression for the period of the motion.

Consider the case when Y is small, and terms in Y^2 can be neglected; in this case,

$$\tau = \frac{4}{\sqrt{a}} F\left(0, \frac{\pi}{2}\right) \tag{5.44}$$

Now, from the definition in Eq. (5.40), it is clear that $F(0, \varphi) = \varphi$, so the expression above reduces to,

$$\tau = \frac{2\pi}{\sqrt{a}} = 2\pi\sqrt{\frac{m}{k_1}} \tag{5.45}$$

which is the period of the underlying linear system[1]. This result is entirely to be expected, as for small values of y, the nonlinear term in the equation of motion can be neglected.

Returning to the main problem; for simplicity of notation, let,

$$\alpha = \frac{1}{\left(a+bY^2\right)^{1/2}} \tag{5.46}$$

so that Eq. (5.38) becomes

$$t = \alpha \int_0^{\varphi} \frac{1}{\left(1-\lambda^2 sin^2\theta\right)^{1/2}} d\theta \tag{5.47}$$

Now, making the change of variable $x = \sin\theta$, the equation becomes,

$$\frac{t}{\alpha} = \int_0^{\sin\varphi} \frac{1}{(1-x^2)^{1/2}(1-\lambda^2 x^2)^{1/2}} dx \tag{5.48}$$

Again, this looks unpleasant; however, the *RHS* is basically the definition of the *first Jacobian elliptic function sn* (see Appendix B), which is sometimes called the 'elliptic sine'. It follows that,

$$\frac{t}{\alpha} = sn^{-1}\left(sin\,\varphi\right) \tag{5.49}$$

and inverting this equation, gives,

$$\varphi = sin^{-1}\left(sn\left(\frac{t}{\alpha}\right)\right) \tag{5.50}$$

and recalling that $\varphi = cos^{-1}\left(y/Y\right)$, one obtains,

$$y = Y cos\left(sin^{-1}\left(sn\left(\frac{t}{\alpha}\right)\right)\right) \tag{5.51}$$

which is still looking pretty messy; however, things can be simplified a lot further. Consider the function $z = cos\left(sin^{-1}x\right)$. Now, let $x = sin\,\phi$; it follows that $z = cos\,\phi$. Now, because $cos^2\phi + sin^2\phi = 1$, one has that $z = \sqrt{1-sin^2\phi} = \sqrt{1-x^2}$. This argument establishes an identity,

$$cos\left(sin^{-1}x\right) = \sqrt{1-x^2} \tag{5.52}$$

---

[1] Throughout this chapter, the *underlying linear system* for a given nonlinear system is simply that obtained by deleting all nonlinear terms. In general, this system will be independent of the forcing amplitude as distinct from the *linearised* systems discussed later, which will only be defined with respect to a fixed level of excitation.

On using this identity, Eq. (5.51) reduces to,

$$y = Y\sqrt{1 - sn^2\left(\frac{t}{\alpha}\right)} \tag{5.53}$$

Now, there is a reason why *sn* is called the elliptic sine; one of the consequences is that it possesses a sister function *cn* – the *elliptic cosine*. Like the ordinary trigonometric functions, these functions obey an identity $cn^2(x) + sn^2(x) = 1$ (see Appendix B). With the help of the identity, Eq. (5.53) attains its final simple form,

$$y = Y cn\left(\frac{t}{\alpha}\right) \tag{5.54}$$

or, in terms of the original constants,

$$y = Y cn\left(\sqrt{\frac{k_1 + k_3 Y^2}{m}}\, t\right) \tag{5.55}$$

Now, *cn* – like the cosine – is a periodic function, and this solution is thus oscillatory with frequency given by,

$$\omega^2 = \frac{k_1 + k_3 Y^2}{m} \tag{5.56}$$

and although this solution is exact, this type of frequency behaviour is, interestingly, going to turn up later in *approximations*.

At this point, it is interesting to look at the behaviour for small Y again. The easiest way to do this is to assume again that Y^2, and thus λ^2, are negligible. With this assumption, Eq. (5.47) becomes,

$$t = \alpha \int_0^{\varphi} d\theta = \varphi \tag{5.57}$$

Now, using once more $y/Y = cos\,\varphi$, and approximating α by $1/\sqrt{a}$, it follows that,

$$y = Y cos\left(\sqrt{a}\, t\right) \tag{5.58}$$

or, on restoring the original constants,

$$y = Y cos\left(\sqrt{\frac{k_1}{m}}\, t\right) \tag{5.59}$$

which one expects to see for small amplitude oscillations.

This completes the solution for the unforced (free) undamped Duffing oscillator. It is immediately clear that the solution is considerably more demanding than for a linear system. Furthermore, it is also evident that the 'first integral' – if it exists – will be dependent on the type of nonlinearity; elliptic functions will not always work, and it may not be possible to express even the first integral in closed form. In fact, the solution above is one of very few exact results for nonlinear systems, so other means of analysis are needed; the main approaches will be covered in the next sections.

5.2.2 Approximate Solutions: Perturbation Theory

There are many different approaches to approximating the solution of nonlinear differential equations; however, arguably the most common is to use *perturbation theory*. The core idea is to assume that, if the effect of the nonlinearity is small, the solution will be *close* to the linear system. Even with perturbation theory, there are many methods; however, only the simplest will be considered here. The reader is referred to the classic texts [5.12, 5.13] for further information; a powerful method for further study is the method of *multiple scales*. As in the previous section, the equation of interest will be the free undamped Duffing equation. In a simplified form (which will be explained later), the equation can be written as,

$$\ddot{y} + y + \varepsilon y^3 = 0 \tag{5.60}$$

In this case, one might expect the effects of the nonlinearity to be small if the coefficient ε is small: i.e. very much less than unity. Now, if one assumes that the form of the solution is fairly well-behaved, one would expect that it can be expressed as $y(t,\varepsilon)$, and one could expand this function in a Taylor series,

$$y(t,\varepsilon) = y_0(t) + \varepsilon y_1(t) + \varepsilon^2 y_2(t) + \ldots \tag{5.61}$$

For notational simplicity, the explicit dependence on t will be removed in the following, so at the risk of repetition,

$$y = y_0 + \varepsilon y_1 + \varepsilon^2 y_2 + \ldots \tag{5.62}$$

Now, if one assumes that ε is so small that terms of order ε^2 and above can be neglected, one has the first-order approximation,

$$y = y_0 + \varepsilon y_1 \tag{5.63}$$

Substituting this trial solution into Eq. (5.60) yields,

$$\frac{d^2}{dt^2}(y_0 + \varepsilon y_1) + y_0 + \varepsilon y_1 + \varepsilon(y_0 + \varepsilon y_1)^3 = 0 \tag{5.64}$$

If one expands this equation and discards all terms of order ε^2 and above, the result is,

$$\ddot{y}_0 + \varepsilon \ddot{y}_1 + y_0 + \varepsilon y_1 + \varepsilon y_0^3 = 0 \tag{5.65}$$

Now, as this equation must be true for *all possible values* of ε, it must be true for each power of ε separately, and this leads to the system,

$$\ddot{y}_0 + y_0 = 0 \tag{5.66}$$

$$\ddot{y}_1 + y_1 = -y_0^3 \tag{5.67}$$

with similar equations for higher-order terms, if the expansion were allowed to continue to higher powers of ε.

Exercise 1. Derive the equations for a perturbation expansion of Eq. (5.60) up to order ε^3.

Something rather remarkable has happened, Eq. (5.66) is linear, and determines y_0; once y_0 is fixed, Eq. (5.67) is *also linear*.

Consider (5.66), and choose the initial conditions $y_0(0) = Y$, $\dot{y}_0(0) = 0$; the solution is of course,

$$y_0(t) = Y cos\, t \tag{5.68}$$

One can move to Eq. (5.67) now, but need to compute y_0^3. As a reminder, one uses De Moivre's theorem as follows,

$$y_0^3 = Y^3 cos^3 t = \left(\frac{Y}{2}\left(e^{it} + e^{-it}\right)\right)^3$$

$$= \frac{Y^3}{8}\left(e^{3it} + 3e^{it} + 3e^{-it} + e^{-3it}\right)$$

$$= \frac{Y^3}{4}\left(\frac{3}{2}\left(e^{it} + e^{-it}\right) + \frac{1}{2}\left(e^{3it} + e^{-3it}\right)\right)$$

and finally,

$$y_0^3 = (Y cos\, t)^3 = \frac{Y^3}{4}(3 cos\, t + cos\, 3t) \tag{5.69}$$

Eq. (5.67) now becomes,

$$\ddot{y}_1 + y_1 = -\frac{Y^3}{4}(3 cos\, t + cos\, 3t) \tag{5.70}$$

As observed above, this equation is linear, but is *inhomogeneous*, the *RHS* is not zero; this means that the full solution is composed of a *complementary function* and a *particular integral* [5.4]. The complementary function is just the solution of the homogeneous equation ($RHS = 0$), so is simply $Y\cos t$ again. The particular integral is a little more complicated to find, it must satisfy the inhomogeneous equation. The usual approach to this problem is to try some trial solutions and adjust them until they fit.

The first observation is that a $\cos t$ term is needed on the *RHS*; however, this cannot be obtained from a simple $\cos t$ in the solution, as this will produce zero on the *RHS*. The next simplest possibility is $t\cos t$. Because a term of the form $\cos 3t$ is also needed on the *RHS*, a trial solution is chosen of the form,

$$y_1 = Ct\cos t + D\cos 3t \tag{5.71}$$

where C and D are constants which can be adjusted. After a little algebra, one obtains,

$$\ddot{y}_1 + y_1 = -2C\sin t - 8D\cos 3t \tag{5.72}$$

and this needs to equate in the *RHS* to (5.70). Because of the $\sin t$ term, this is not possible with the trial solution chosen; however, the answer is fairly clear; one uses a trial solution,

$$y_1 = Ct\cos(t+\phi) + D\cos 3t \tag{5.73}$$

with the additional free parameter ϕ. With this form, the equation becomes,

$$\ddot{y}_1 + y_1 = -2C\sin(t+\phi) - 8D\cos 3t \tag{5.74}$$

and choosing $\phi = \pi/2$ converts the sine term into the required cosine term. The rest of the analysis is straightforward, and yields $C = 3Y^3/8$ and $D = Y^3/32$. The final result is,

$$y_1 = Y\cos t + \frac{3Y^3}{8}t\cos\left(t+\frac{\pi}{2}\right) + \frac{Y^3}{32}\cos 3t$$

or,

$$y_1 = Y\cos t - \frac{3Y^3}{8}t\sin t + \frac{Y^3}{32}\cos 3t \tag{5.75}$$

The full perturbation solution to order ε is thus,

$$y = Y\cos t + \varepsilon\left(Y\cos t - \frac{3Y^3}{8}t\sin t + \frac{Y^3}{32}\cos 3t\right) \tag{5.76}$$

Now, there is a subtle error here and this equation can be corrected and simplified. One needs to 'share out' the initial condition among the $y_i(0)$, in order that $y(0) = Y$; the current equation has $y(0) = (1+\varepsilon)Y$ (+ higher-order terms). If one assigns the entire initial condition to the y_0 equation, the perturbative solution is,

$$y = Y\cos t + \varepsilon\left(-\frac{3Y^3}{8}t\sin t + \frac{Y^3}{32}\cos 3t\right) \tag{5.77}$$

However, this presents an immediate problem. Under the circumstances, one would expect to see a periodic solution. In fact, in this case, the exact periodic solution was demonstrated in the previous section. The issue here is that the term $t\sin t$ is *not* periodic; it actually grows without bound as t increases. This type of term is called a *secular term*, and it is an artefact of the perturbation method. The problem is that the series expansion has been truncated; were it not truncated, the secular term would effectively cancel with a term or terms later in the series. However, one is *forced* by pragmatism to truncate the series, so ideally one would wish a valid solution at any given level of the expansion. It turns out that the solution to the problem actually kills two birds with one stone.

It is known from the exact solution for the free undamped Duffing oscillator, that the 'natural frequency' actually increases with Y; however, as things stand above, this behaviour is not captured by the perturbative solution. This situation can be accommodated by replacing Eq. (5.60) by,

$$\ddot{y} + \omega^2(\varepsilon)y + \varepsilon y^3 = 0 \tag{5.78}$$

with $\omega^2(0) = 1$, and defining a further Taylor expansion,

$$\omega^2(\varepsilon) = 1 + \varepsilon\alpha_1 + \varepsilon^2\alpha_2 \ldots \tag{5.79}$$

With this extra flexibility, one expands the equation of motion as before; while the y_0 equation remains the same, the equation for y_1 becomes,

$$\ddot{y}_1 + y_1 = -\alpha_1 y_0 - y_0^3 \tag{5.80}$$

and one can select the appropriate value of α_1 to cancel the secular term. As a bonus, one recovers amplitude dependence of the frequency of oscillation as,

$$\omega^2 = 1 + \varepsilon\frac{3Y^2}{8} \tag{5.81}$$

Exercise 2. Verify Eq. (5.81).

There now remains a small piece of tidying up. At the beginning of the section, Duffing's equation was presented in the simplified form (5.60). It is fairly straightforward to derive this, and the calculation is worth carrying out for the full damped forced equation,

$$m\ddot{y} + c\dot{y} + ky + k_3 y^3 = x(t) \tag{5.82}$$

One observes that there is a freedom to *rescale* the two variables x and y *and* the independent variable t,

$$u = \alpha y, \quad p = \beta x, \quad \tau = \gamma t \tag{5.83}$$

It follows from the time rescaling that,

$$\frac{d}{d\tau} = \frac{1}{\gamma}\frac{d}{dt}$$

Now, using dashes to denote differentiation with respect to τ, substituting in all the scaled variables leads to,

$$\frac{m\gamma^2}{\alpha}u'' + \frac{c\gamma}{\alpha}u' + \frac{k}{\alpha}u + \frac{k_3}{\alpha^3}u^3 = \frac{1}{\beta}p$$

The first of the constants is easy to remove, one simply sets $\beta = 1$, and then,

$$\frac{m\gamma^2}{\alpha}u'' + \frac{c\gamma}{\alpha}u' + \frac{k}{\alpha}u + \frac{k_3}{\alpha^3}u^3 = p$$

The next constant is removed by setting $\alpha = k$, so that,

$$\frac{m\gamma^2}{k}u'' + \frac{c\gamma}{k}u' + u + \frac{k_3}{k^3}u^3 = p$$

Finally, one removes the 'inertia' constant by setting $\gamma = \sqrt{k/m} = \omega_n$, leaving,

$$u'' + \frac{c\omega_n}{k}u' + u + \frac{k_3}{k^3}u^3 = p$$

Now, note that $\frac{c\omega_n}{k} = \frac{c}{k}\sqrt{\frac{k}{m}} = \frac{c}{\sqrt{km}} = 2\xi$, so finally one has,

$$u'' + 2\xi u' + u + \varepsilon u^3 = p \tag{5.84}$$

where $\varepsilon = k_3/k^3$. In the final form, only the damping ratio and coefficient of the nonlinear term are visible, all other quantities have been removed by the rescaling.

In this context, 'small' nonlinearity means that k_3 is small compared to k^3.

5.2.3 Numerical Solutions: Simulation

So far, things have not looked particularly hopeful in terms of understanding nonlinear systems via solution of their equations of motion. In terms of exact solutions, there is very little hope; very few systems are amenable to such attacks, even the Duffing system is only exactly solvable in the free undamped case. Approximate solutions offer a way out, but as the last section will have suggested, accuracy is very much dependent on the order of approximation and the calculations at higher order can become unwieldy very quickly. Fortunately, there is a universal approach available in the form of *numerical analysis*, which offers a branch specifically attuned to the solution of differential equations. Numerical solutions can usually be found quickly and with acceptable accuracy without too much difficulty, as long as the number of degrees of freedom is moderate; *SDOF* problems rarely present a significant challenge. The branch of numerical mathematics concerned is said to deal with *initial-value problems*. There are many excellent text books on the subject, personal recommendations of the current authors are [5.14, 5.15]. A real stand-out in the literature is [5.16], which not only explains the algorithms, but provides working code (in a variety of languages); in a pre-*MATLAB* world, this book was an absolute Godsend (clearly, only the 'senior' author here recalls these Jurassic times).

The first stage in numerical simulation is almost always to convert the system of interest into first-order *state-space* form,

$$\dot{\mathbf{y}} = f(t, \mathbf{y}) \tag{5.85}$$

with corresponding initial conditions,

$$\mathbf{y}(0) = \mathbf{y}_0 \tag{5.86}$$

This form is quite general, and allows explicit time dependence as well as through the states. The conversion to first-order form is straighforward; consider the forced linear system,

$$m\ddot{y} + c\dot{y} + ky = x(t) \tag{5.87}$$

One simply defines $y_1 = y$ and $y_2 = \dot{y}$; the velocities are then $\dot{y}_1 = y_2$, which is an identity, and $\dot{y}_2 = \left(x(t) - cy_2 - ky_1\right)/m$ from the equation of motion. The first-order form is thus,

$$\begin{Bmatrix} \dot{y}_1 \\ \dot{y}_2 \end{Bmatrix} = \begin{Bmatrix} y_2 \\ \dfrac{1}{m}\left(x(t) - cy_2 - ky_1\right) \end{Bmatrix} = f(t, \mathbf{y}) \tag{5.88}$$

Now, the *initial-value problem* is to integrate the equation of motion from time $t = 0$ to $t = T$. The strategy will be to divide the interval into small steps Δt and predict forward, one step at a time until the goal is reached. Because Δt is assumed small, the prediction steps are usefully expressed in terms of a Taylor series,

$$\mathbf{y}(t+\Delta t) = \mathbf{y}(t) + \Delta t\,\dot{\mathbf{y}}(t) + \frac{1}{2}\Delta t^2\,\ddot{\mathbf{y}}(t) + O\left(\Delta t^3\right) \tag{5.89}$$

where a term $O\left(\Delta t^n\right)$ indicates that the lowest-order missing term is of the order Δt^n in the expansion parameter.

Rearranging this equation gives,

$$\dot{y}(t) = \frac{y(t+\Delta t) - y(t)}{\Delta t} + O(\Delta t) \tag{5.90}$$

which shows that the *forward-difference* approximation to the derivative is accurate to $O(\Delta t)$: i.e. one would expect errors in the approximation to decrease with Δt.

The Euler Method

The Taylor expansion approach is a quite general means of producing numerical approximation schemes; the more terms that are retained, the more accurate (potentially) the scheme will be. Clearly, the simplest possible approach is to truncate at first-order, and this produces the so-called *Euler method*. Substituting Eq. (5.85) into the first-order approximation gives,

$$\mathbf{y}(t+\Delta t) = \mathbf{y}(t) + \Delta t\, f(t, \mathbf{y}) + O(\Delta t^2) \tag{5.91}$$

As discussed, one would normally expect the numerical solution to improve uniformly as Δt decreases, and this is generally the case; however, the *error term* at order $O(\Delta t^n)$ can sometimes behave badly. The behaviour of the error term is discussed in most numerical texts, but good behaviour will be assumed on faith here.

In fact, because this is a dynamics text, it is useful to regard Eq. (5.91) as specifying a *discrete dynamical system*,

$$\mathbf{y}_{n+1} = \mathbf{y}_n + \Delta t\, f(t_n, \mathbf{y}_n) + O(\Delta t^2) \tag{5.92}$$

With the linear system of Eq. (5.88) discretised as,

$$\begin{Bmatrix} y_{1,n+1} \\ y_{2,n+1} \end{Bmatrix} = \begin{Bmatrix} y_{1,n} \\ y_{2,n} \end{Bmatrix} + \begin{Bmatrix} y_{2,n} \\ \frac{1}{m}\left(x_n - c y_{2,n} - k y_{1,n}\right) \end{Bmatrix} \tag{5.93}$$

It will prove useful to have the *FRF* of this discrete system, and this is easily accomplished by using *harmonic probing* [5.17, 5.9]. The approach exploits the fact that, if the input to a linear system is the fictitious harmonic input $x_n = e^{i\omega t_n}$, of angular frequency ω, the output will be $y_n = H(\omega) e^{i\omega t_n}$. It follows then that $x_{n+j} = e^{ij\omega\Delta t} x_n$ and $y_{n+j} = H(\omega) e^{ij\omega\Delta t} y_n$. With these observations, the linear system of Eq. (5.93), can be written in the frequency domain as,

$$\begin{Bmatrix} H_1(\omega) \\ H_2(\omega) \end{Bmatrix} e^{i\omega\Delta t} = \begin{Bmatrix} H_1(\omega) \\ H_2(\omega) \end{Bmatrix} + \frac{\Delta t}{m} \begin{Bmatrix} mH_2(\omega) \\ 1 - cH_2(\omega) - kH_1(\omega) \end{Bmatrix} \tag{5.94}$$

where $H_1(\omega)$ is the *FRF* for the discrete process $x_n \to y_n$, and $H_2(\omega)$ is the corresponding quantity for $x_n \to \dot{y}_n$.

While this expression might look a little messy, it is really just a pair of simultaneous equations for $H_1(\omega)$ and $H_2(\omega)$; simplifying the first row gives,

$$H_2(\omega) = \frac{\left(e^{i\omega\Delta t} - 1\right)}{\Delta t} H_1(\omega) \tag{5.95}$$

and expanding this as a Taylor series yields,

$$H_2(\omega) = i\omega H_1(\omega) + O(\Delta t) \tag{5.96}$$

which reproduces the correct relationship between the *FRF*s for displacement and velocity.

Solving the simultaneous equations (left as an exercise to the reader), leads to the desired result,

$$H_1(\omega) = \frac{\Delta t^2}{m\left(e^{i\omega\Delta t} - 1\right)^2 + c\,\Delta t\left(e^{i\omega\Delta t} - 1\right) + k\Delta t^2} \tag{5.97}$$

Exercise 3. Show that, in the continuous system limit $\Delta t \to 0$, Eq. (5.97) gives,

$$H_1(\omega) = \frac{1}{-m\omega^2 + ic\omega + k} \tag{5.98}$$

as expected.

This computation is not a pointless digression. Under normal circumstances, when discussing numerical methods, it is right and proper to benchmark them on known problems to establish their efficacy (or not). In this case for example, one could assume a sinusoidal input $x(t) = X\,sin(\Omega t)$ to the linear system and compare the numerical predictions with the exact output – which will also be a sinusoid. However, because of the emphasis on engineering dynamics here, it is much more interesting to take a holistic *frequency domain* viewpoint. The numerical integration scheme needs to be accurate at *all* frequencies of interest to the dynamicist, and this is a much stronger constraint. In frequency domain terms, it is useful to think of the time-step Δt as a sampling interval, with an associated sampling frequency $f_s = 1/\Delta t$. It is also important to realise that there will be an associated *Nyquist frequency* $f_N = f_s/2$: i.e. the integration scheme can only attempt to represent the real system up to the Nyquist frequency, any predictions beyond that frequency will be corrupted by aliasing [5.18].

This reasoning means that the dynamicist must choose the time-step Δt, with all the frequencies of interest in mind, and certainly not below the Nyquist limit for any frequencies present in the response. As an illustration, consider the linear system with the specific parameter values: $m = 1$, $c = 20$, $k = 10^4$. In this case, the natural frequency $\omega = \sqrt{k/m}$ is 100 rad/s, which is approximately 16 Hz. Suppose that, to capture the interesting dynamics, the Nyquist frequency is taken as 32 Hz; in this case, the predicted *FRF* from the Euler method is as shown in Figure 5.9.

It is clear that the Euler method is very poor, except at low frequencies, where the time-step is short enough to capture the dynamics of interest.

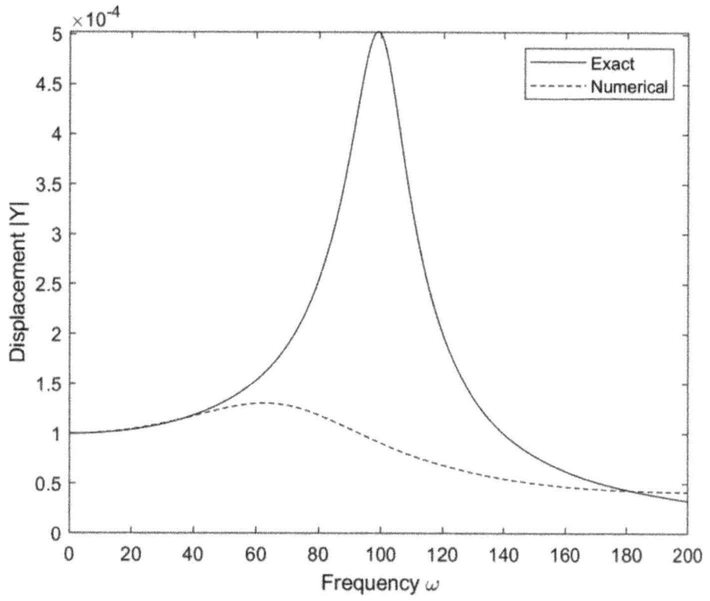

Figure 5.9 Predicted *FRF* for linear system using Euler's method with $f_N = 32$ Hz, compared with exact *FRF*.

In fact, because of the frequency domain ideas available here, it is possible to compute a very useful benchmark, in the form of the *Hamming FRF* [5.18, 5.9]; this is defined by,

$$H_a(\omega) = \frac{FRF/spectrum\ of\ predicted\ result}{FRF/spectrum\ of\ exact\ result} \tag{5.99}$$

For the simulation in Figure 5.9, the Hamming *FRF* is as shown in Figure 5.10.

The Hamming *FRF* has some interesting structure associated with the *FRF*; however, this is not the most important message. If the integration method is to be appropriately accurate for dynamics purposes, the function $H_a(\omega)$ needs to be *unity* up to the highest frequency of interest. This is clearly not the case, for the current simulation, which means that the time-step is not high enough: i.e. a higher Nyquist frequency is needed. Figure 5.11 shows the predicted *FRF*s for Nyquist frequencies of $f_N =$ 32, 320 and 32000 Hz; it is clear that only the very highest of these frequencies gives an accurate prediction for the exact *FRF*.

Exercise 4. Show that the Hamming *FRF* $H_a(\omega)$ for a linear system and the Euler method, tends to unity at low frequencies. Estimate the first frequency for parameters $m = 1$, $c = 20$, $k = 10^4$, at which there is a 10% error.

At this point, the Hamming *FRF* comes into its own. Consider Figure 5.12, which shows the function for the highest Nyquist frequency considered here ($f_N = 32000$ Hz). The greater accuracy at the lower frequencies has now washed out the structure close to resonance, and the main properties of the algorithm are revealed. Using the Hamming *FRF*, one can now choose

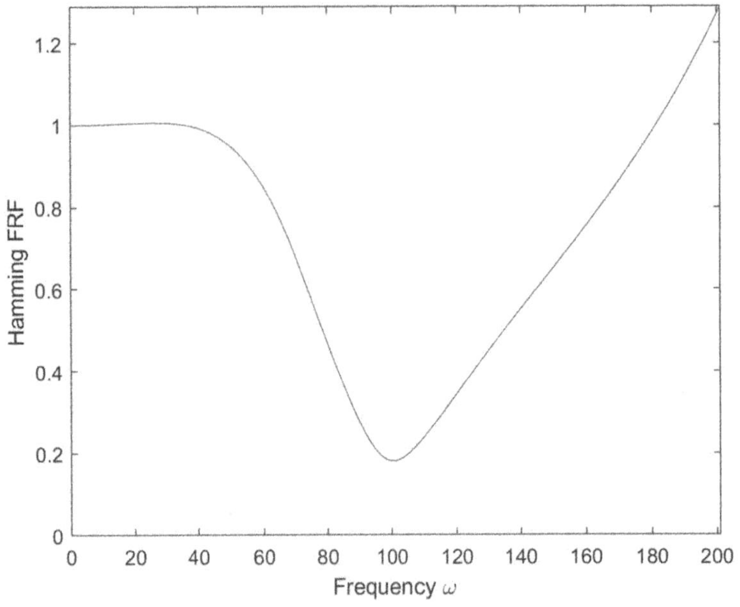

Figure 5.10 Hamming *FRF* for linear system using Euler's method with $f_N = 32$ Hz.

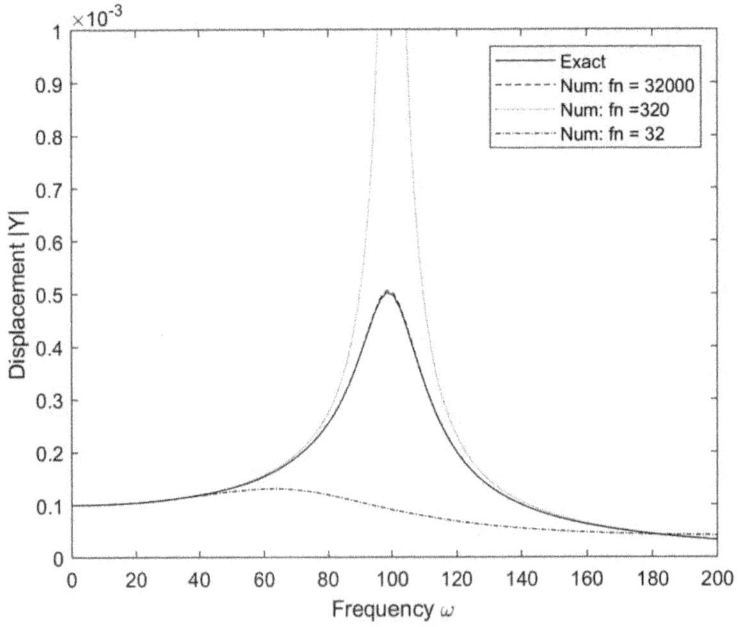

Figure 5.11 Predicted *FRF* for linear system using Euler's method with $f_N = 32$ Hz, $f_N = 320$ Hz and $f_N = 32000$ Hz; compared with exact *FRF*.

an appropriate Nyquist frequency, and hence time-step, to ensure accuracy over the frequency range of interest. Furthermore, the Hamming *FRF* shows clearly that the Euler method *overestimates* the response amplitude at all frequencies. It is important to realise that the *FRF* is *not* the *FRF* of the Euler *method*, it is specific to the Euler method for the linear system; however, once the Nyquist frequency is high enough that any resonant structures are accurately represented, the Hamming *FRF* can provide a general guide to choosing time-steps.

The overall message here is that even a very simple method like the Euler method can produce acceptably accurate simulations, if the time-step is small enough. However, as one might imagine, one can do much better than a first-order Taylor expansion in generating numerical methods. In fact, by generating higher-order Taylor expansions, one can have error terms to any order of Δt that one wishes. Apart from the additional complexity of the analysis, there can also be issues with the exact forms of the error terms, which mean that some compromise is usually chosen.

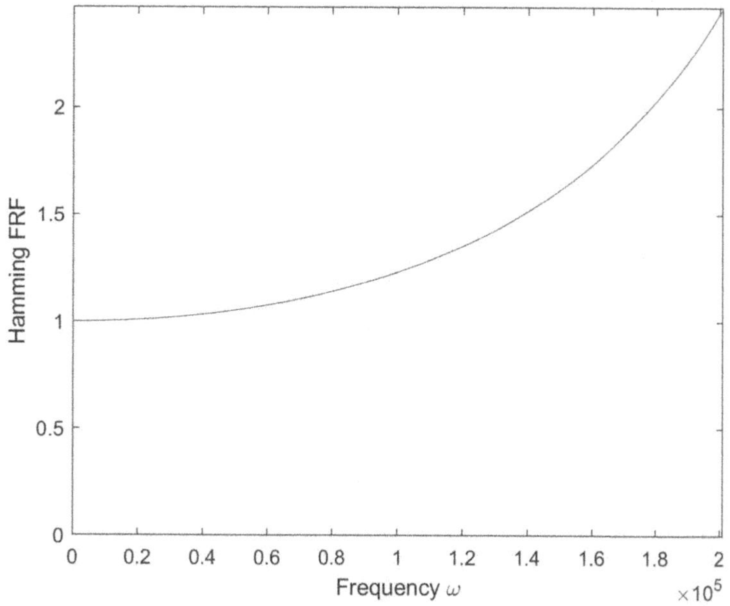

Figure 5.12 Hamming *FRF* for linear system using Euler's method with $f_N = 32000$ Hz.

Runge-Kutta's Methods

An important family of higher-order methods is the *Runge-Kutta* (*RK*) class of algorithms. It would be too much of a digression to derive these methods beyond what has already been presented in Chapter 3, Section 3.2.2; it suffices to point the reader at standard references [5.14, 5.15, 5.16]. What distinguishes the *RK* algorithms, is that they evaluate the functions of interest *in between* time steps; this is one of the features that allow them greater accuracy. By far the most historically-important *RK* algorithm is the *fourth-order Runge-Kutta*, which will be referred to here as '*RK4*'. *RK4* requires evaluations of the system equations at the mid-points of each time interval; the detailed equations of the algorithm are, in going from step t_i to $t_{i+1} = t_i + \Delta t$,

$$\boldsymbol{k}_1 = \Delta t f\left(t_i, \boldsymbol{y}_i\right) \tag{5.100}$$

$$k_2 = \Delta t f\left(t_i + \frac{\Delta t}{2}, y_i + \frac{1}{2}k_1\right) \tag{5.101}$$

$$k_3 = \Delta t f\left(t_i + \frac{\Delta t}{2}, y_i + \frac{1}{2}k_2\right) \tag{5.102}$$

$$k_4 = \Delta t f\left(t_i + \Delta t, y_i + k_3\right) \tag{5.103}$$

$$y_{i+1} = y_i + \frac{1}{6}\left(k_1 + 2k_2 + 2k_3 + k_4\right) \tag{5.104}$$

Because the *RK4* algorithm is so widely-used, a *MATLAB* implementation is included here for the use of the reader in the form of a function *simulate.m* (Appendix C). The function does not use any special toolboxes, and has been used to generate all the examples later in this section.

The function *simulate* has a couple of specific features:

1. It is often the case that a dynamicist wishes to model a system in order to compare the results with experimental data; for example while conducting *system identification* [5.9]. In this case, the excitation force is not available as a time-dependent function, but as an array of sampled values $\{x_i : i = 1,\ldots,N\}$. For this reason, *simulate* has been written to predict the response for a discrete array of force samples; the user supplies this array together with the sampling time Δt. If one wishes to simulate the response to some deterministic function, one simply generates that function in the main part of *simulate*. The implementation shown here contains lines for both a sinusoidal input and a Gaussian white noise sequence; for other options, one simply comments in the function of interest. Simulating an impulse is also simple; one specifies a zero force array and an initial condition $y(0) = 0$, $\dot{y}(0) = V$ – this is equivalent to a Dirac impulse.

2. Because of the above feature, the values of the functions of interest are *not available* at the mid-points of sampling intervals; however, they are *required* by the *RK4* algorithm. This problem is circumvented by linearly interpolating between samples of the input force. This strategy works perfectly well as long as the sampling frequency is high enough.

Before showing some explicit examples of using *RK4*, it is important to consider how the algorithm might be benchmarked. In fact, the *RK4* equations still represent a discrete dynamical system, so it is possible to compute the *FRF* of the discrete *RK4* system and compare to the exact result, exactly as for the Euler algorithm; the caveat is that the calculation is considerably more complicated than for the Euler algorithm. Furthermore, it is possible to calculate the Hamming *FRF* for *RK4* simulating a linear system. Only the results will be presented here.

First of all, the simulation of the linear system above with $f_N = 32$ Hz is repeated. Figure 5.13 shows the predicted and exact *FRF*s for this simulation; while the result is far better than that for the Euler method in Figure 5.9, it is still not acceptable. However, this is where the power of the *RK4* algorithm comes into play; the error term for *RK4* is $O(\Delta t^4)$. Figure 5.14 shows the predicted and exact *FRF*s for an *RK4* simulation with $f_N = 64$ Hz; the results are excellent, comparable with the Euler method with $f_N = 32000$, even though the time-step is 500 times as large.

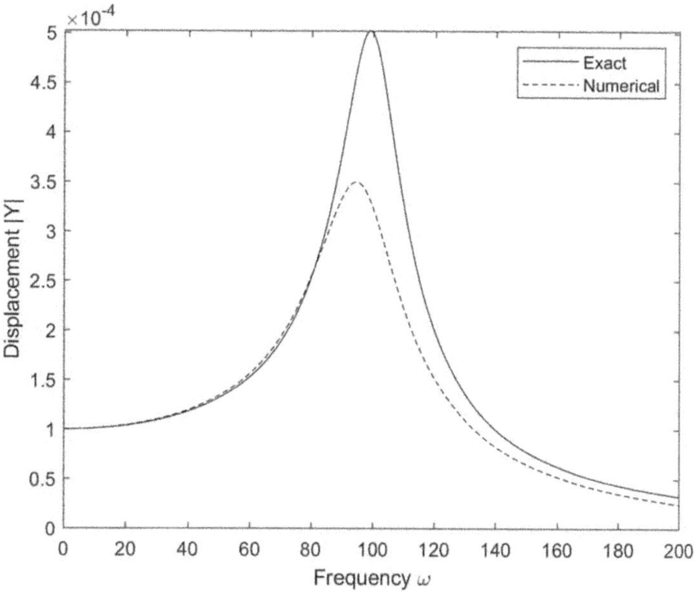

Figure 5.13 Predicted *FRF* for linear system using *RK4* method with $f_N = 32$ Hz, compared with exact *FRF*.

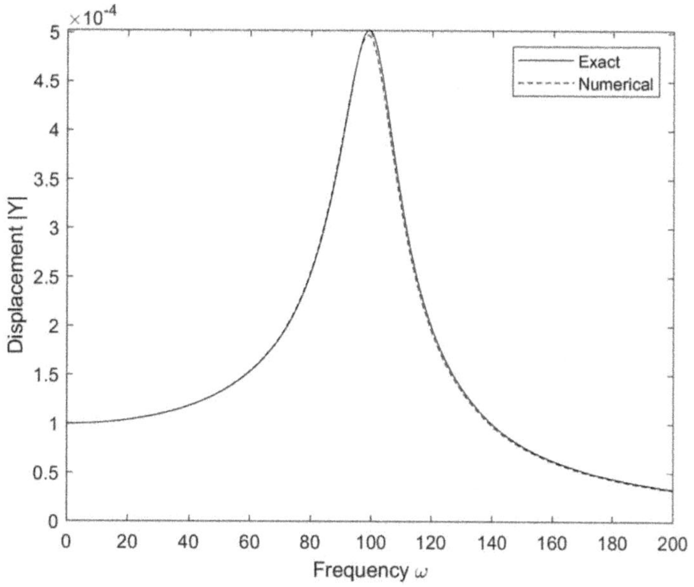

Figure 5.14 Predicted *FRF* for linear system using *RK4* method with $f_N = 64$ Hz, compared with exact *FRF*.

Figure 5.15 shows the Hamming *FRF* for the linear system with $f_N = 64$ Hz. The *FRF* is almost indistinguishable from unity up to $\omega = 200$ rad/s, and still excellent up to $\omega = 400$ rad/s, which is 2/5 of the Nyquist frequency. In this example, excellent results are thus obtained by taking the sampling frequency at ten times the resonance frequency. This is quite a good rule of thumb in terms of choosing a sampling frequency, and thus time-step, for simulations using *RK4*. However, it will soon become clear that this rule must be applied carefully if the system of interest is *nonlinear*. Finally, it is worth noting that, unlike the Euler algorithm, the *RK4* algorithm *underestimates* all frequencies up to Nyquist.

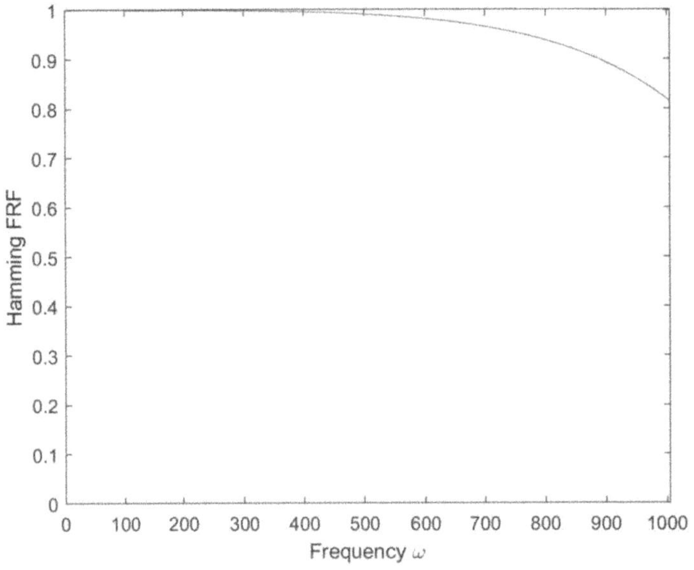

Figure 5.15 Hamming *FRF* for linear system using *RK4* method with $f_N = 160$ Hz.

Simulating Nonlinear Systems

Having developed the necessary simulation machinery, it is time to look at the response of some nonlinear systems. The first system of interest will be the forced, damped symmetric Duffing oscillator,

$$m\ddot{y} + c\dot{y} + ky + k_3 y^3 = x(t) \tag{5.105}$$

Simulation is critical here, because this equation does not have an exact solution. The values of the constants chosen here are: $m = 1$, $c = 20$, $k = 10^4$. $k_3 = 5 \times 10^9$. A sinusoidal excitation has been chosen: $x(t) = 20\,sin(100t)$, and the sampling frequency is 1000 Hz, corresponding to a time-step of 0.001 seconds. This excitation has been chosen high enough to generate nonlinear features in the response; for low levels of forcing, the response will be indistinguishable to that from the underlying linear system. Exactly as in the case of linear systems, the response will contain a starting transient corresponding to the complementary function for the system. In order to avoid recording the transient, the simulation runs for 16 cycles of the excitation before recording any data; it then saves 16 cycles. The simulated response is given in Figure 5.16.

Despite the comparatively high level of forcing, the response is not significantly different from sinusoidal; the waveforms are slightly 'sharper'. In fact, waveform distortions are more visible in the velocity response (the reason for this will be explained later) as shown in Figure 5.17.

Another effective way of showing nonlinear distortions in time data, is to plot responses in the $(y(t), \dot{y}(t))$ plane – the *phase-plane*.

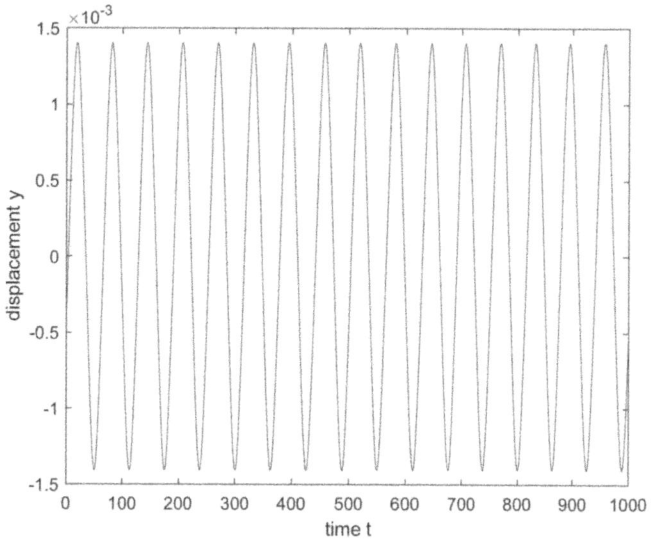

Figure 5.16 Simulated displacement response for Duffing system using *RK4* method with $f_N = 1000$ Hz.

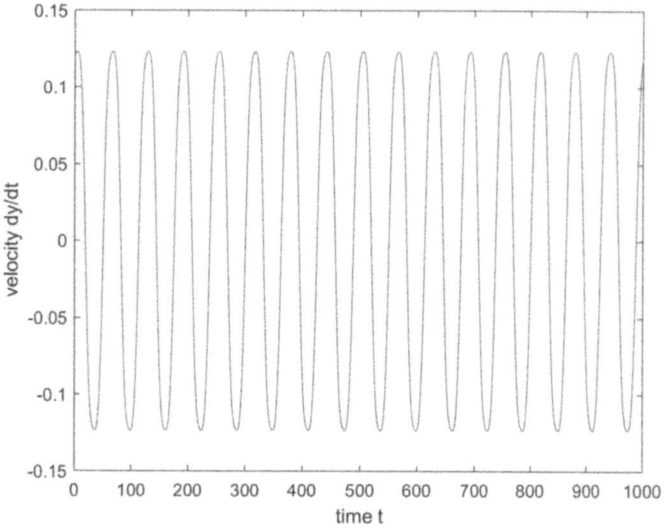

Figure 5.17 Simulated velocity response for Duffing system using *RK4* method with $f_N = 1000$ Hz.

Figure 5.18 shows the result for a collection of ten responses, generated using excitation amplitudes from $X = 2.0$ to $X = 20.0$. Each of these curves is called a *phase trajectory*; it should be clear that periodic solutions of an equation will appear in the phase plane as *closed curves*. For linear systems, sinusoidal forcing results in phase trajectories which are ellipses; this is visible in Figure 5.18 as the curves become more like ellipses as they approach the origin. It is fairly simple to show that linear system phase trajectories are perfect ellipses; however, the demonstration is postponed until the next section.

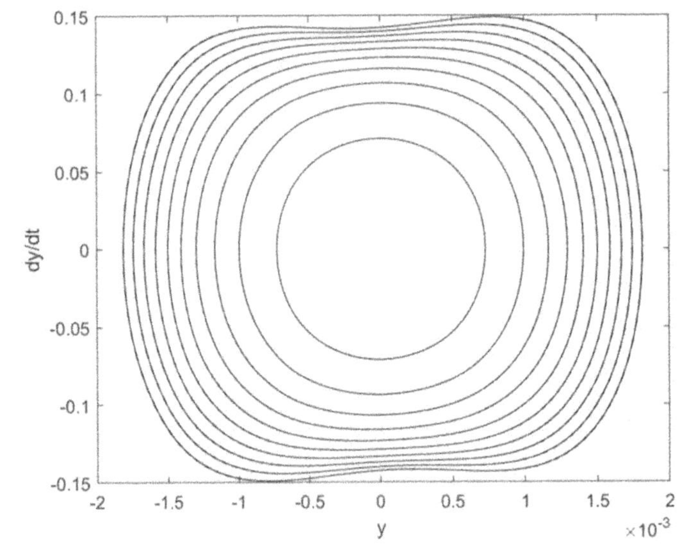

Figure 5.18 Simulated phase trajectories for Duffing system using *RK4* method with $f_N = 1000$ Hz; ten trajectories for excitations $X = 2.0$ to $X = 20.0$.

The distortions in the waveforms, i.e. the departures from a sinusoidal form, are the result of extra frequencies in the response because of the nonlinearity. The most common occurrence of 'extra' frequencies, is as *harmonics* of a fundamental frequency. For example, if a force is applied at a single frequency Ω, the response of a nonlinear system will often contain components at frequencies 2Ω, 3Ω, 4Ω etc. These multiples of the fundamental are called *superharmonics*, and their amplitude will usually decay with frequency. However, this phenomenon has an important consequence for the simulation of nonlinear systems. As dicussed above, a good rule of thumb for choosing a time-step is to ensure that the associated sampling frequency is at ten times the highest resonance or otherwise highest frequency of interest. In simulating nonlinear systems, one should take care to sample at ten times the frequency of the highest expected harmonic.

The next system simulated will be the free *Van der Pol* oscillator, which has equation of motion,

$$\ddot{y} + 0.2\left(y^2 - 1\right)\ddot{y} + y = 0 \tag{5.106}$$

For this system, the nonlinearity is in the damping, and this has a very interesting form. For small values of displacement $y^2 < 1$, the damping is negative, so the phase trajectory is driven away from the origin. For larger values of displacement $y^2 >> 1$, the damping is positive and the dynamics drive the trajectory towards the origin. The result is a compromise stable periodic solution based around $y^2 = 1$, when there is no damping at all.

This behaviour is like that mentioned for the predator-prey equations; even though there is damping but no forcing, the system can sustain oscillations – the periodic response in this case is an example of a *limit cycle*. The simulated response for the Van der Pol oscillator is shown in Figure 5.19; the simulation is started at a very small value of y; one can see that the phase trajectory spirals away from the origin onto the limit cycle; this reflects the fact that the origin is an *unstable equilibirum* for the system, a matter which will be discussed in the next section.

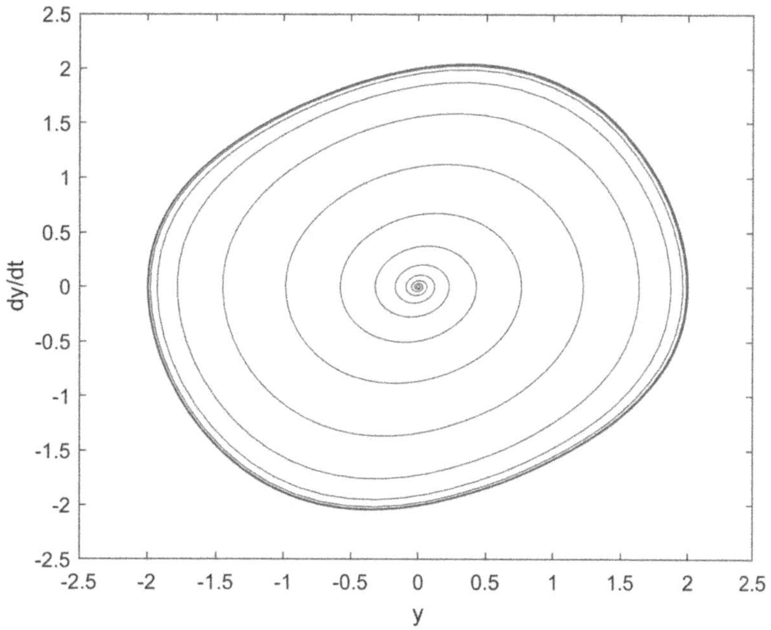

Figure 5.19 Simulated limit cycle for Van der Pol using *RK4* method with $f_N = 10000$ Hz. A high sample rate has been used to get good resolution of the trajectory.

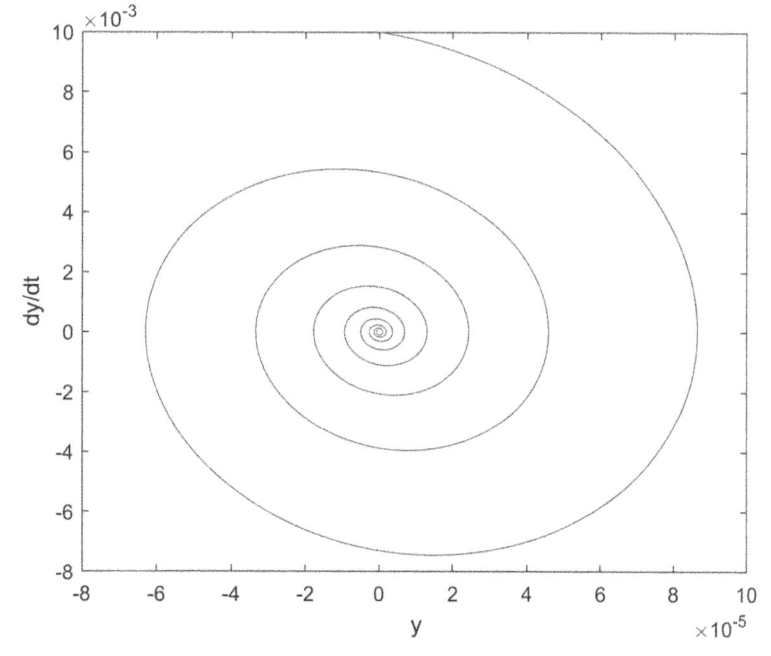

Figure 5.20 Simulated impulse response for linear system using *RK4* method with $f_N = 10000$ Hz. A high sample rate has been used to get good resolution of the trajectory.

The next simulation shows an impulse response. As it is quite difficult to distinguish nonlinear behaviour in such a transient anyway, the simulation is presented for the linear system. The sampling frequency of the simulation was taken as 10000 Hz, simply to capture details throughout the short duration of the transient. Figure 5.20 shows the response in the phase plane;

in this case, the trajectory spirals in to the origin, which is a *stable* equilibrium for positive damping.

The final simulation presented here corresponds to a situation of great practical importance, when the input force is *random*. This is important because the system is probed at all frequencies of interest at the same time, and this reduces the number of tests needed in many cases. The function *simulate* allows this case simply by sampling the input force from a Gaussian (normal) distribution. In abstract terms, such a strategy would lead to all possible frequencies being present with equal weight, so the excitation is called *white noise* by analogy with the spectrum of visible light.

True white noise excitation presents technical difficulties for both simulation and mathematical analysis. True white noise would have infinite power (variance) and so is not physically realisable; real excitations will always have some upper frequency limit.

In terms of analysis, differential equations driven by white noise are technically quite different to deterministic equations; they are called *stochastic differential equations* (*SDEs*) and have a quite different theory associated with them [5.19]. One significant difference in the theory has an impact on simulations and deterministic algorithms are subject to modification; for example, the Euler method is modified to the *Euler-Maruyama* method [5.19].

The problems of *SDEs* are avoided by the function *simulate* for two reasons. In the first place, only a single realisation of the random process is used for excitation; this means that the excitation is effectively deterministic. The second point relates to the frequency cut-off of the force; even if the x_n array is generated by sampling from the Gaussian distribution, there is an upper limit on its frequency content forced by the Nyquist criterion. Furthermore, the linear interpolation between sampling points does not allow higher frequencies to be introduced via the *RK* algorithm.

As in the case of the impulse response, it is difficult – if not impossible – to visibly discern signatures of nonlinearity in a random time signal; for this reason, the simulation presented here is for the linear system. A time-step of 0.001 seconds has been chosen, corresponding to a Nyquist frequency of 500 Hz. Figure 5.21 shows the input force sequence generated by sampling from the Gaussian distribution; it is almost impossible to see any structure to the signal. The simulated response for the system is shown in Figure 5.22. The linear system has acted as a band-pass filter restricting response frequencies to those close to the resonance frequency (16 Hz in this case).

5.2.4 Qualitative Solutions: The Phase Plane

Some of the simulations in the previous section have already made it clear that phase trajectories provide valuable information about nonlinear behaviour – e.g. closed trajectories correspond to periodic solutions.

In fact a lot of features of a pictorial characterisation – a *phase portrait* – can be deduced *without* solving the equations of motion; even a *qualitative* analysis in the phase plane can shed a great deal of light on how a nonlinear system might behave.

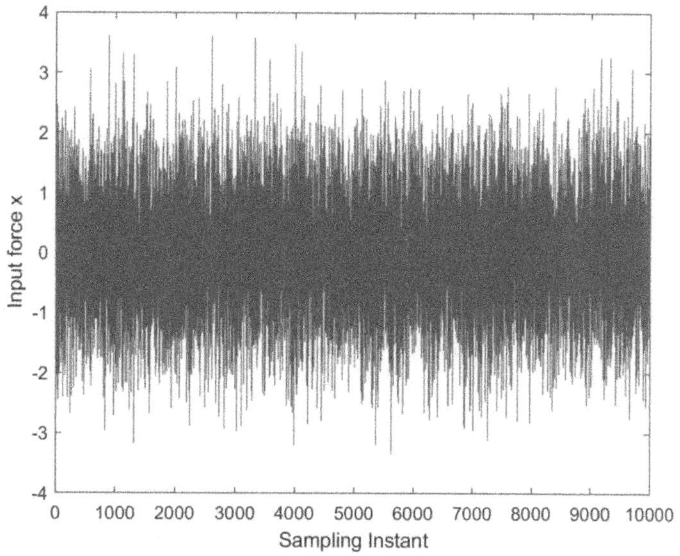

Figure 5.21 Simulated input force for randomly-excited linear system using *RK4* method with $f_N = 1000$ Hz, so Δt is 1 milli-second.

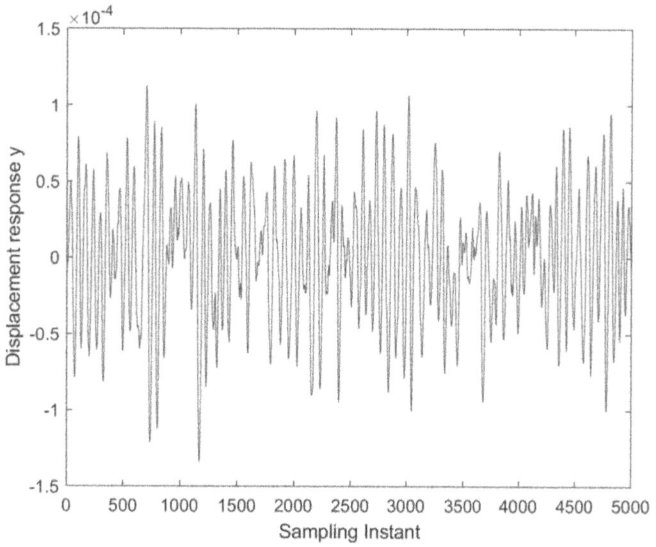

Figure 5.22 Simulated displacement response for randomly-excited linear system using *RK4* method with $f_N = 1000$ Hz.

As in initial-value problems, the first step in phase-plane analysis is always to put the equations of interest into first-order form $\dot{\mathbf{y}} = f(t, \mathbf{y})$. For the free (unforced) linear system, one has,

$$\begin{Bmatrix} \dot{y}_1 \\ \dot{y}_2 \end{Bmatrix} = \begin{Bmatrix} y_2 \\ \frac{1}{m}\left(-c y_2 - k y_1\right) \end{Bmatrix} \tag{5.107}$$

and for the free symmetric Duffing system,

$$\begin{Bmatrix} \dot{y}_1 \\ \dot{y}_2 \end{Bmatrix} = \begin{Bmatrix} y_2 \\ \frac{1}{m}\left(-cy_2 - ky_1 - k^3 y_1^3\right) \end{Bmatrix} \tag{5.108}$$

It is assumed in this section that the systems of interest are *SDOF*: i.e. the dynamics are for a single point mass m.

The first features that one wishes to identify in a phase portrait are *equilibria*, otherwise known as *fixed points*; these are points at which the system stays stationary for all time, unless disturbed by an external force – they are *positions of rest.*

Now, at such a point, it is clear that both the velocity and acceleration of the system must be zero, otherwise the mass would simply move away. Thus the conditions for an equilibrium are $\dot{y} = y_2 = 0$ and $\ddot{y} = \dot{y}_2 = 0$; in total, $\dot{\boldsymbol{y}} = \boldsymbol{0}$. From the general equation of motion, it follows that,

$$f(t, \boldsymbol{y}) = \boldsymbol{0} \tag{5.109}$$

is the condition for a fixed point. Focussing on the linear equation in (5.107), the condition is simply,

$$\begin{Bmatrix} 0 \\ 0 \end{Bmatrix} = \begin{Bmatrix} y_2 \\ \frac{1}{m}\left(-cy_2 - ky_1\right) \end{Bmatrix} \tag{5.110}$$

The first element in the equation is not interesting; it simply reasserts that the velocity vanishes at a fixed point. It is the second element which is system specific. In this case, one has,

$$\frac{1}{m}\left(-cy_2 - ky_1\right) = 0 \tag{5.111}$$

and as $y_2 = 0$, it immediately follows that $y_1 = 0$. This argument shows that the only equilibrium for the linear system is $\boldsymbol{y} = \{0, 0\}^T$, which is the origin.

For *SDOF* nonlinear systems, one can simplify the notation here a little; suppose the original second-order equation is,

$$m\ddot{y} + g(y, \dot{y}) = 0 \tag{5.112}$$

with a quite general nonlinearity in displacement and velocity. In this case, the first-order form is simply,

$$\begin{Bmatrix} \dot{y}_1 \\ \dot{y}_2 \end{Bmatrix} = \begin{Bmatrix} y_2 \\ -\frac{1}{m} g(y_1, y_2) \end{Bmatrix} \tag{5.113}$$

and it follows that the equilibrium condition is simply,

$$g(y_1, 0) = 0 \tag{5.114}$$

and with $y_2 = 0$, this completes the specification.

Returning to the linear system; armed with the fact that there is a single equilibrium, one can derive the rest of the phase portrait. First of all, it is simplest to consider a system with no damping; such a system is called *conservative* because energy is conserved (and not because it has dubious politics).

It was shown in the last section that, with no damping, and an initial condition $y(0) = Y$, $\dot{y}(0) = 0$ (i.e. $\mathbf{y}(0) = \{Y, 0\}^T$), the phase trajectories are periodic solutions $y(t) = Y\cos(\omega t)$, where $\omega^2 = k/m$, which appear as closed curves – in this case, ellipses – in the phase plane. Considering all possible initial displacements $0 \le Y \le \infty$, generates a set of ellipses which fills the phase plane, and is therefore the entire phase portrait.

Now, there are powerful theorems [5.20], which show that the solutions to differential equations are *unique*, under quite general circumstances (and when they *exist* at all). In geometrical terms, this fact means that phase trajectories *cannot ever cross*. Suppose that two phase trajectories exist that meet at some point (y_{10}, y_{20}). If one were to choose this point as the initial condition for a trajectory, it would appear to have two directions in which to move (remember there is no driving force), and this would violate uniqueness of solutions.

Uniqueness means that each point in the phase plane is on one phase trajectory *only* – one ellipse – and so the set of ellipses is said to *foliate* the plane. The phase portrait is simply the set of ellipses through the points $(Y, 0)$ for $0 \le Y \le \infty$, with the origin as a special ellipse of zero extent.

If damping is introduced $c \ne 0$, the solution for $y(t)$ is an exponentially-damped harmonic; which appears in the phase plane as a spiral moving inwards towards the origin; which is the only equilibrium. Again, the spirals foliate the plane; although this situation is a little harder to visualise. Again, no point can be on more than one spiral, to be completely clear. If the linear system is disturbed from the equilibrium, it will either oscillate close to it (no damping), or spiral back in to the origin (damped).

One can now imagine a system with *negative damping*: i.e. $c < 0$. From an initial condition $(Y, 0)$, the solution is an *exponentially growing* harmonic; the phase trajectory spirals out from the initial point. For such a system, the origin is an *unstable* equilibrium. A mass point starting at $(0, 0)$ will indeed stay there for all time; however, if it is disturbed by even the tiniest amount, it will spiral away to infinity. The set of spirals still foliate the phase plane; however, if one were to indicate the direction of motion by arrows on the trajectories, one would see the arrows reversed in the two (stable/unstable) cases. (Actually, it is possible to imagine that the dynamics is just a function of the direction of time.)

Even with positive, or no damping, equilibria can be stable (i.e. attract trajectories), or unstable (i.e. repel trajectories). To see this, it is simplest to consider conservative systems for now. In this case, the nonlinear function is simply $g(y)$, and the equation of free motion is,

$$m\ddot{y} = -g(y) \tag{5.115}$$

Now, multiplying by $\dot{y}$ gives,

$$m\dot{y}\ddot{y} = -\dot{y}g(y) \tag{5.116}$$

One can integrate this equation with respect to time trivially, to give,

$$\frac{1}{2}m\dot{y}^2 = -\int g(y)\frac{dy}{dt}dt + A = -\int g(y)dy + a \tag{5.117}$$

where this is simply a *first integral* as discussed in the last section, and A is just a constant of integration. In fact, a trivial rearrangement shows that,

$$\frac{1}{2}m\dot{y}^2 + \int g(y)dy = A \tag{5.118}$$

Now, the first term on the *LHS* is simply the kinetic energy of the particle of interest. As this system is conservative, it is immediately clear that the second term on the *LHS* must be the *potential energy* $V(y)$ of the system, and the constant A is just the total (conserved) energy E, i.e.

$$\frac{1}{2}m\dot{y}^2 + V(y) = E \tag{5.119}$$

and,

$$\int g(y)dy = V(y) \tag{5.120}$$

which means that the original equation can be interpreted as,

$$m\ddot{y} = -g(y) = -\frac{dV}{dy} \tag{5.121}$$

All of this is rather standard classical mechanics; however, in phase plane-terms, something rather wonderful has happened. Consider the linear system again, in which case $g(y) = ky$, and Eq. (5.119) becomes,

$$\frac{1}{2}m\dot{y}^2 + \frac{1}{2}ky^2 = E \tag{5.122}$$

Furthermore, suppose that the initial condition is $(Y,0)$, so that the initial energy is only potential, $V_0 = kY^2/2$; the equation of interest becomes,

$$\frac{1}{2}m\dot{y}^2 + \frac{1}{2}ky^2 = \frac{1}{2}kY^2 \tag{5.123}$$

and this is the equation of the phase trajectory starting at $(Y,0)$, and is the equation of an *ellipse*. The wonderful thing alluded to earlier is that the first integral has sufficed to give an exact equation for the phase trajectories. This is not so remarkable for the linear system, as it is straightforward to solve the equation of motion anyway; however, this is a general result for *nonlinear* systems, as long as they are conservative.

Before looking at nonlinear systems, it is worth spending a little more time with the linear system; there is quite a lot more to learn from the potential energy function. For the undamped linear oscillator, one has simply,

$$V(y) = \frac{1}{2}ky^2 \tag{5.124}$$

which represents a simple quadratic *potential well* as shown in Figure 5.23.

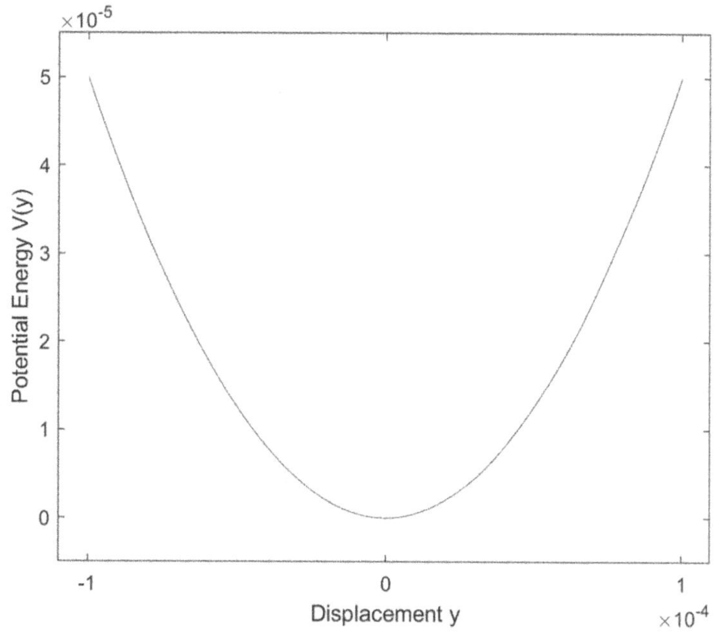

Figure 5.23 Potential energy function for a linear system with $k > 0$.

The figure is a *well* because $k > 0$. In this situation, the spring force always pushes back towards the origin if the system is disturbed; this makes the equilibrium at the origin *stable*. Actually, the potential energy curve explains the free dynamics of the system completely. Suppose the mass point starts at rest at some position $y = Y$. If $Y \neq 0$, the mass will move because it is not at an equilibrium. The mass cannot move upwards because it will increase its potential energy while also increasing its kinetic energy by virtue of the motion; this violates conservation of energy. It is clear now that, considering the mass as moving on its potential energy curve, it must always move *downwards*. When the point reaches the origin, it will have picked up kinetic energy and cannot stay at the origin; it will thus move *up* the curve at negative y, until the kinetic energy has all been converted into potential energy again, and as the system is symmetrical and undamped, this will occur at $y = -Y$. Again, because of symmetry, the process will repeat in the opposite direction and return the point to its initial position. The whole cycle repeats and one arrives at a periodic motion.

In the opposite situation where $k < 0$, the potential energy curve becomes a hill, as shown in Figure 5.24. If the particle is at rest at $y = 0$, it will stay there; the origin is an equilibrium. However, if the point is displaced this time, the negative spring pushes it further away. In fact, the point still has to travel downhill to conserve energy; however, it will accelerate without limit, conserving energy by falling without limit and allowing the potential energy to tend to $-\infty$. The origin is an *unstable* equilibrium in this case.

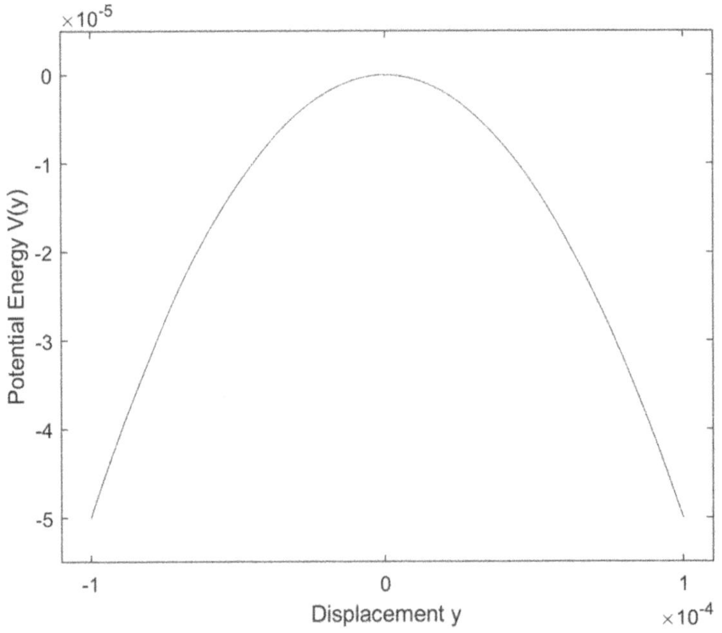

Figure 5.24 Potential energy function for a linear system with $k < 0$.

Now, one can see a general principle here; a point of zero gradient in the potential energy diagram is an equilibrium. In fact, this follows directly from Eq. (5.121); equilibria are zeroes of $g(y)$ and thus zeroes of dV/dy; this means that, by simple calculus, equilibria correspond to maxima and minima of the potential energy function for a system. Furthermore, simple geometry and calculus now shows that minima of the potential energy function are stable equilibria, while maxima are unstable equilibria. For conservative systems, all this turns out to be all one needs in order to establish stability of equilibria.

At this point, enough machinery has amassed to examine some interesting nonlinear systems. In the first case, consider the free undamped Duffing equation, which has $g(y) = ky + k_3 y^3$; the potential energy function for this system is,

$$V(y) = \frac{1}{2}ky^2 + \frac{1}{4}k_3 y^4 \tag{5.125}$$

First, consider the case $k > 0,\ k_3 > 0$; the equilibria are the solutions of,

$$g(y) = ky\left(1 + \frac{k_3}{k}y^2\right) = 0 \tag{5.126}$$

and the only real solution is at $y = 0$; so in the phase plane, the only equilibrium is at the origin again, like the linear system. There is no surprise here; if one plots the potential energy curve, it has a single minimum at the origin. The only difference for Duffing's equation in this case is that the sides of the well are steeper because the well is quartic. This observation leads again to the conclusion that the solutions for the system are periodic; however, because the well is not quadratic, one does not expect elliptic phase trajectories in this case. In fact, looking at the first integral, conservation of energy gives the equation of the phase trajectories for an initial condition $(Y, 0)$ as closed curves,

$$\frac{1}{2}m\dot{y}^2 + \frac{1}{2}ky^2 + \frac{1}{4}k_3y^4 = \frac{1}{2}kY^2 + \frac{1}{4}k_3Y^4 = Const \tag{5.127}$$

and a family of these non-elliptic trajectories has already appeared here in Figure 5.18.

A more interesting situation occurs when $k < 0$ and $k_3 > 0$. Despite the odd sign for the linear stiffness, this is actually a physically interesting system and corresponds to an axially-loaded beam with the potential to buckle.

If one writes $k = -|k|$, the equilibria are given by,

$$g(y) = ky\left(1 - \frac{k_3}{|k|}y^2\right) = 0 \tag{5.128}$$

and this equation has *three* real solutions. The potential energy curve is the quartic,

$$V(y) = -\frac{1}{2}|k|y^2 + \frac{1}{4}k_3y^4 \tag{5.129}$$

and is illustrated in Figure 5.25 for the specific values $k = 10^4$ and $k_3 = 5\times10^9$.

As the figure shows, there are three extrema in the potential energy function; an unstable equilibrium at the origin (indicated by the black circle), and two stable equilibria at the points $y = \pm\sqrt{|k|/k_3}$ (indicated by the black squares). For obvious reasons, systems like this are referred to as *twin-well* oscillators.

Qualitatively, one can begin to derive the phase portrait for such a system. Because the system is free (unforced), all the trajectories are generated by considering all possible initial conditions. Qualitatively, there are only three types of positions at which a particle can start:

1. at the origin;
2. in one of the wells;
3. further up the outer slope of one of the wells.

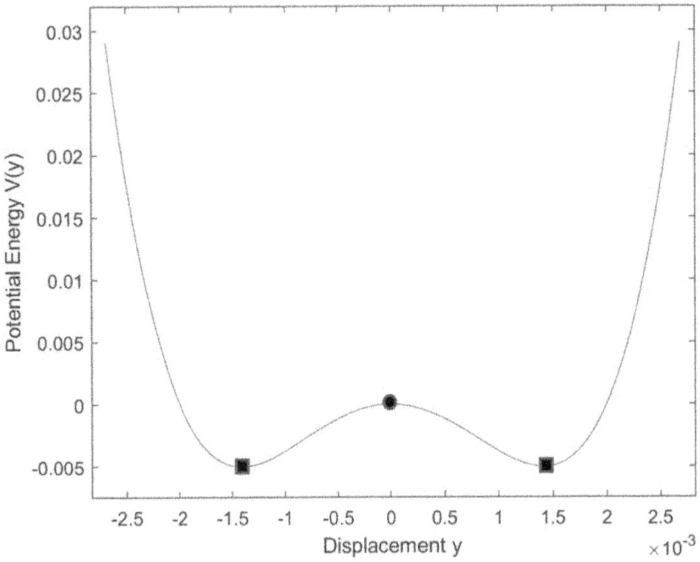

Figure 5.25 Potential energy function for a 'twin-well' Duffing system with $k < 0$.

In the case of (1), the trajectories will move away from the origin if disturbed a little, because it is unstable, and will oscillate in one of the wells. For (2), e.g. particles starting in a well, they will stay in the well and generate a periodic solution around the stable equilibrium at the bottom of the well; phase trajectories will be closed curves within the well. In the case of (3), something a little more interesting occurs. Such a particle will fall into the well and rise toward the origin; the particle will have non-zero kinetic energy at the origin, and will cross into the opposite well, where it will climb to its original height and then reverse its motion. Such trajectories are periodic solutions – but closed curves in the phase plane containing *both* wells.

An interesting limiting condition occurs when $V = 0$ and $|y| > \sqrt{|k|/k_3}$. In this case, the particle falls into the well and rises towards the origin; because it has enough initial potential energy, it will eventually reach the origin, but will become slower and slower on its approach as its kinetic energy is used up. It can be shown that the particle will take an *infinite* time to reach the origin. Such a phase trajectory is called a *homoclinic* orbit. The curve is also interesting for another reason; it separates the phase plane into two regions; one containing single-well oscillations, and the other showing double-well oscillations (Figure 5.26).

This idea of dynamics as masses moving on potential energy curves has been beautifully realised experimentally by Virgin and Todd, and described in the book [5.21]. In those experiments, the masses were formed from small trucks moving on rails formed into the shape of the potential wells.

As a final case study, which is partly left as an exercise to the reader, one can consider the equation of the simple pendulum,

$$\ddot{\theta} + \frac{l}{g}\sin\theta = 0 \tag{5.130}$$

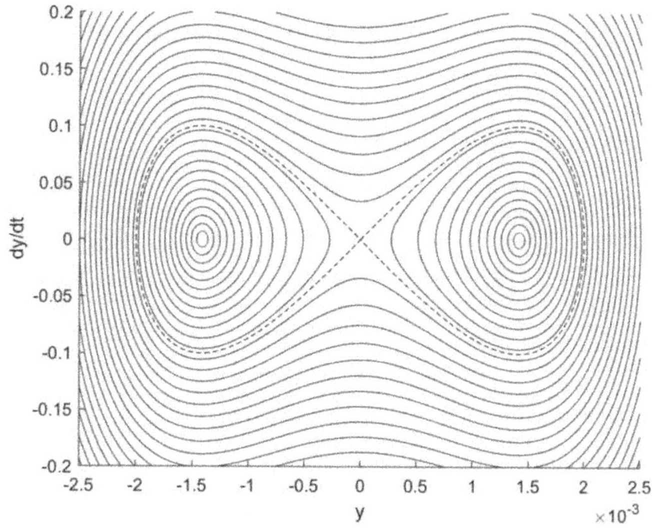

Figure 5.26 Phase portrait for a 'twin-well' Duffing system with $k < 0$. The dashed is the homoclinic orbit which separates single- and twin-well oscillations.

where θ represents the angular displacement from the vertical, l is the length of the pendulum and g is the acceleration due to gravity. In this case, the equilibria are the solutions of $g(\theta) = \sin\theta = 0$, and there are an *infinite number*, at $\theta = n\pi$, where n is any integer. The stability of the equilibria are found from the potential energy function, which is,

$$V(\theta) = -\frac{l}{g}cos\theta \tag{5.131}$$

It follows that the equilibria at $\theta = 0$ and $n\pi$ with n even are stable, while those at $n\pi$ with n odd are unstable.

Exercise 5. Deduce and label the phase portrait for the simple pendulum, as shown in Figure 5.27. (Hint: in this case, it is useful to think in terms of trajectories in terms of their *initial velocity*.)

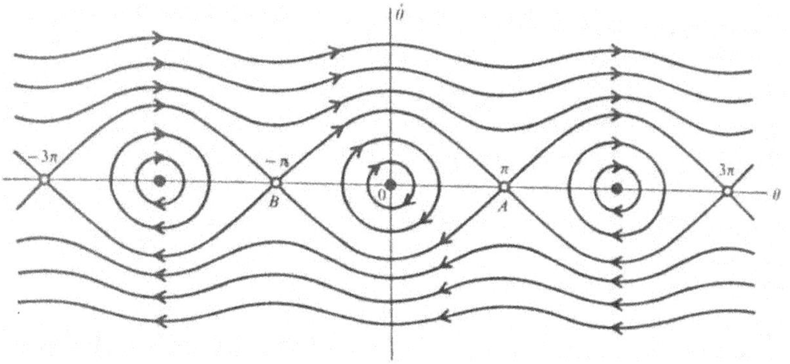

Figure 5.27 Phase portrait for the simple pendulum.

Most of the material so far has been concerned with the mathematics of nonlinearity; however, engineering must concern itself with the practical issues of physical reality. The following sections will look at how nonlinearity expresses itself in the behaviour of engineering structures.

5.3 Signatures of Nonlinearity

It is important to be able to recognise if an engineering structure or system is, in fact, nonlinear. As many of the properties of linear systems do not hold for nonlinear ones, observing these properties (or their absence), can show signatures of the nonlinearity and can lead to tests for the condition.

5.3.1 Definition of Linearity: the Principle of Superposition

The principle of superposition is more than just a *property* of linear systems; in mathematical terms one can actually regard it as *defining* what is linear and what is not. The principle, when it holds, applies to both the static and dynamic behaviour of structures; it simply states that the total response of a linear structure to a set of simultaneous inputs can be broken down into several experiments where each input is applied individually and the output to each of these separate inputs can be summed to give the total response.

A precise mathematical definition is as follows. If a system in an initial condition $S_1 = \{y_1(0), \dot{y}_1(0)\}$ responds to an input $x_1(t)$ with an output $y_1(t)$ and in a separate test an input $x_2(t)$ to the system initially in state $S_2 = \{y_2(0), \dot{y}_2(0)\}$ produces an output $y_2(t)$ then superposition holds if and only if the input $\alpha x_1(t) + \beta x_2(t)$ to the system in initial state $S_3 = \{\alpha y_1(0) + \beta y_2(0), \alpha \dot{y}_1(0) + \beta \dot{y}_2(0)\}$ results in the output $\alpha y_1(t) + \beta y_2(t)$ for all constants α, β, and all pairs of inputs $x_1(t), x_2(t)$.

Despite its fundamental status, the principle has limited use as a *test* of linearity. The reason being: in order to establish linearity beyond doubt, an infinity of tests would be required spanning all α, β, $x_1(t)$ and $x_2(t)$, and this is clearly impossible. However, to show nonlinearity without doubt, only one set of $\alpha, \beta, x_1(t), x_2(t)$ which violates superposition is needed. In general practice it may be more or less straightforward to establish such a set. Many of the practical tests, some of which are described later, rely on breakdown of superposition.

Figure 5.28 shows an example of the static application of the principle of superposition to an encastré beam like the one discussed earlier in Section 5.1.4, subjected to transverse static loading at its centre. It can be seen that superposition holds to a high degree of approximation when the static deflections are small, i.e. less than the thickness of the beam. However, as the applied load is increased, producing deflections greater than the beam thickness, the principle of superposition is violated since the applied loads $F_1 + F_2$ do not result in the sum of the deflections $y_1 + y_2$. The observed nonlinearity is called a *hardening-stiffness* which occurs because the boundary conditions restrict the axial straining of the middle surface (the neutral axis) of the beam as the lateral amplitude is increased. The symmetry of the situation dictates that, if the load direction is reversed, the deflection characteristic will follow the same pattern resulting in an *odd* nonlinear stiffness characteristic as shown in Figure 5.29. (The defining property of an odd function is that $F(-y) = -F(y)$).

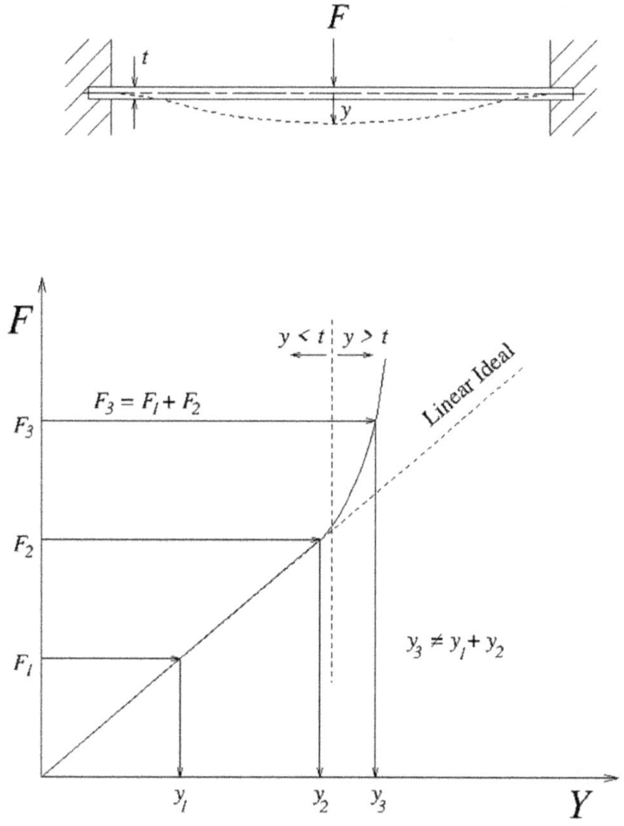

Figure 5.28 Schematic showing violation of superposition for an encastré beam.

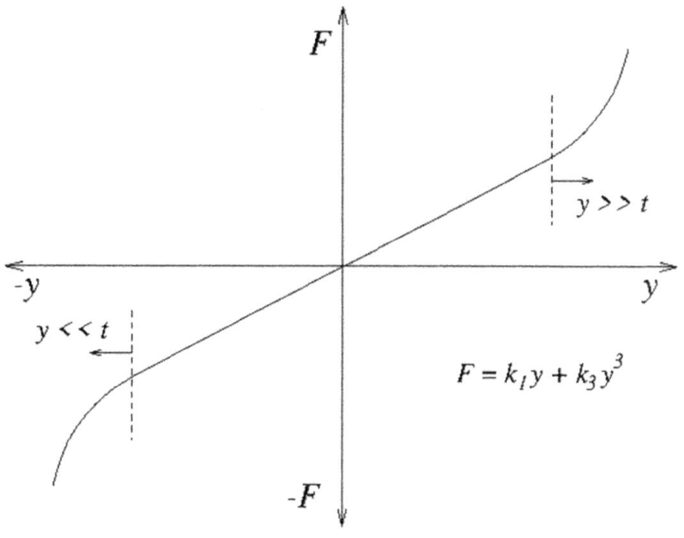

Figure 5.29 Schematic showing static stiffness characteristic for an encastré beam.

If the beam were pre-loaded, the static equilibrium point would not be centred at $(0,0)$ as in Figure 5.29, and the resulting force-deflection characteristic would become a general function lacking symmetry as shown in Figure 5.30. When these various stiffness functions are added to equations of motion, one obtains the range of variants of Duffing's equation discussed at the end of Section 5.1.4.

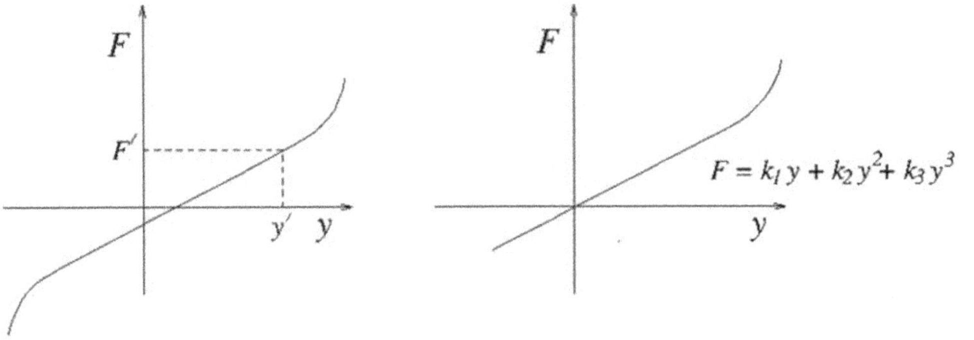

Figure 5.30 Asymmetric stiffness characteristic for a preloaded encastré beam.

In practice, because it is not possible to fully implement the principle of superposition, i.e. spanning all the possibilities of inputs, simpler procedures are employed. Since best practice in dynamic testing should always include some check of linearity, it is important that easy-to-use procedures for detecting nonlinearity are available. The most commonly-used procedures are based on harmonic distortion, homogeneity, reciprocity and the coherence function, and these will be discussed in some detail in the following sections.

5.3.2 Harmonic Distortion

Harmonic or *waveform distortion* is one of the simplest and clearest indicators of the presence of nonlinearity. It is a straightforward consequence of the principle of superposition. If the excitation to a linear system is a monoharmonic signal, i.e. a sine or cosine wave of frequency ω, the response will be monoharmonic at the same frequency (after any transients have decayed away). The proof is elementary, and proceeds as follows.

Suppose $x(t) = sin(\omega t)$ is the input to a linear system. First of all, it is observed that $x(t) \to y(t)$ implies that $\dot{x}(t) \to \dot{y}(t)$ and $\ddot{x}(t) \to \ddot{y}(t)$; this is because superposition demands that,

$$\frac{x(t+\Delta t)-x(t)}{\Delta t} \to \frac{y(t+\Delta t)-y(t)}{\Delta t} \tag{5.132}$$

and $\dot{x}(t) \to \dot{y}(t)$ follows in the limit as $\Delta t \to 0$. There is an implicit assumption of *time invariance* here, namely that $x(t) \to y(t)$ implies $x(t+\tau) \to y(t+\tau)$ for any τ. Again, by superposition,

$$x_1(t) + \omega^2 x_2(t) \to y_1(t) + \omega^2 y_2(t) \tag{5.133}$$

so taking $x_1(t) = \ddot{x}(t)$ and $x_2(t) = x(t)$ gives,

$$\ddot{x}(t) + \omega^2 x(t) \rightarrow \ddot{y}(t) + \omega^2 y(t) \tag{5.134}$$

Now, as $x(t) = sin(\omega t)$,

$$\ddot{x}(t) + \omega^2 x(t) = 0 \tag{5.135}$$

In the steady-state, a zero input to a linear system results in a zero output. It therefore follows from Eq. (5.134) that,

$$\ddot{y}(t) + \omega^2 y(t) = 0 \tag{5.136}$$

and the general solution of this differential equation is,

$$y(t) = Asin(\omega t - \phi) \tag{5.137}$$

and this establishes the result. The proof *only* uses the fact that $x(t)$ satisfies a homogeneous linear differential equation to prove the result; the implication is that any such function will not suffer distortion in passing through a linear system.

It *does not follow* from the result above, that a sine wave input to a nonlinear system will *not* generally produce a sine wave output; however, this is usually the case and this is the basis of a simple and powerful test for nonlinearity, as sine waves are simple signals to generate in practice. It will be shown later that the distortion occurs because higher harmonics appear in the response, such as $sin(3\omega t)$, $sin(5\omega t)$ etc.

Distortion can be easily detected on an oscilloscope by observing the input and output time response signals. Figures 5.31 and 5.32 show examples of harmonic waveform distortion where a sinusoidal excitation signal is warped due to nonlinearity.

In Figure 5.31 the output response from a nonlinear system is shown in terms of the displacement, velocity and acceleration. Just as in the examples presented in Section 5.2.3, the velocity is more distorted than the displacement, and the acceleration suffers most of all; this effect is actually easily explained. Let $x(t) = sin(\omega t)$ be the input to the nonlinear system. As stated above, the output will generally (at least for weak nonlinear systems; i.e. there is no subharmonic generation or chaos [5.9]) be represented as a Fourier series composed of superharmonics,

$$y(t) = A_1 sin(\omega t + \phi_1) + A_2 sin(2\omega t + \phi_2) + A_3 sin(3\omega t + \phi_3) + \ldots \tag{5.138}$$

and the corresponding acceleration is,

$$\ddot{y}(t) = -\omega^2 A_1 sin(\omega t + \phi_1) - 4\omega^2 A_2 sin(2\omega t + \phi_2) - 9\omega^2 A_3 sin(3\omega t + \phi_3) - \ldots \tag{5.139}$$

Thus, the nth output acceleration term is weighted by the factor n^2 compared to the fundamental.

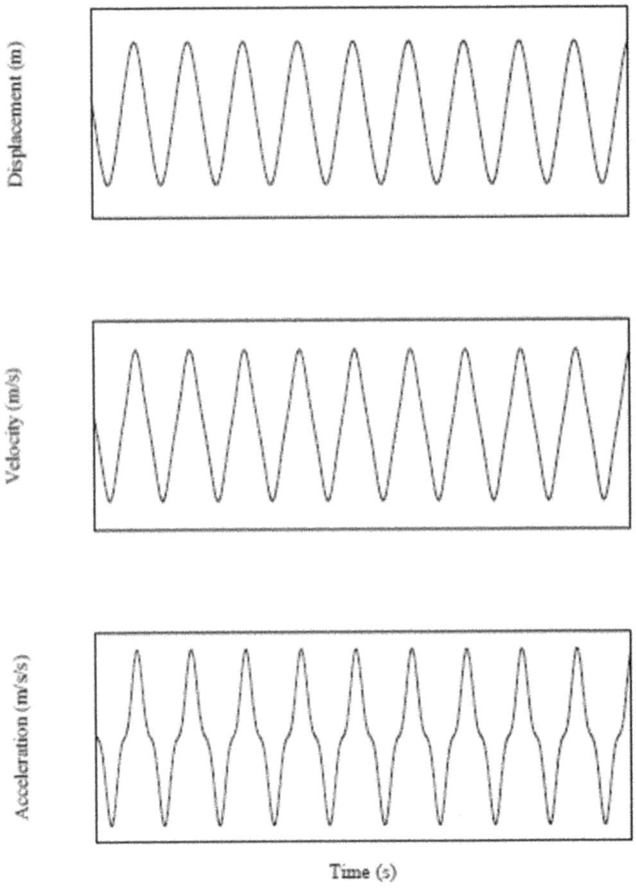

Figure 5.31 Output responses from a nonlinear system showing varying degrees of harmonic/waveform distortion: displacment (upper), velocity (centre), acceleration (lower).

As a practical example, Figure 5.32 shows the output of a force transducer during a modal test. The distortion is due to shaker misalignment resulting in friction between the armature of the shaker and the internal magnet – a nonlinearity.

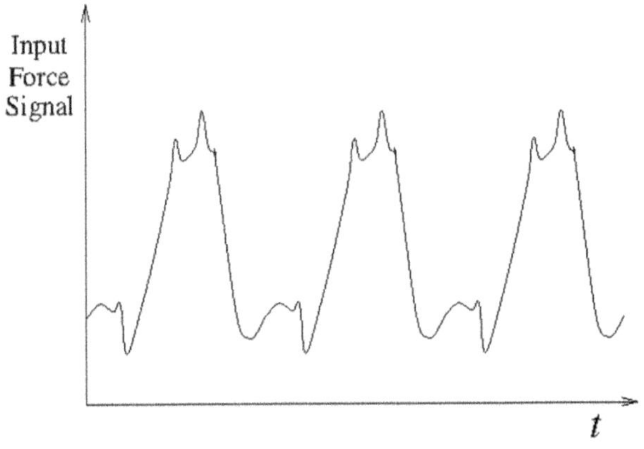

Figure 5.32 Measured force from shaker during a modal test; although the shaker receives a sinusoid from the amplifier, a friction nonlinearity produces waveform distortion.

If non-sinusoidal waveforms are used, such as band-limited random signals, waveform distortion is generally impossible to detect visually and additional procedures are employed, such as the coherence function described later in this section.

5.3.3 Homogeneity and *FRF* Distortion

This test represents an application of a restricted form of the principle of superposition. It is probably the most common method for detecting nonlinearity in dynamic testing. A *homogeneity* condition is said to hold if $x(t) \rightarrow y(t)$ implies $\alpha x(t) \rightarrow \alpha y(t)$ for all α. In essence, homogeneity is an indicator of the insensitivity of the system to the magnitude of the input signal. For example, if an input $\alpha x_1(t)$ always produces an output $\alpha y_1(t)$, the ratio of output to input is independent of the constant α. The most striking consequence of this is in the frequency domain.

First, note that $\alpha x(t) \rightarrow \alpha y(t)$ implies $\alpha X(\omega) \rightarrow \alpha Y(\omega)$; this means that if $x(t) \rightarrow \alpha x(t)$,

$$H(\omega) = \frac{Y(\omega)}{X(\omega)} \rightarrow \frac{\alpha Y(\omega)}{\alpha X(\omega)} = H(\omega) \tag{5.140}$$

and the *FRF* is invariant under changes of α, or effectively of the level of excitation.

The homogeneity test is usually applied in dynamic testing to *FRF*s where the input levels are usually mapped over a range covering typical operating levels. If the *FRF*s for different levels overlay, linearity is inferred. The test is not infallible as there are some systems which are nonlinear which nonetheless show homogeneity (e.g. some bilinear systems, as discussed in the next section); homogeneity is a *weaker* condition than superposition.

An example of a homogeneity test is shown in Figure 5.33. In this case, band-limited random excitation has been used, but in principle any type of excitation may be used. The structure of interest here is a bridge; there are distinct differences between the *FRF*s (low and high excitation) at the higher frequencies, and this infers the presence of nonlinearity. In fact, this test was carried out as part of a *Structural Health Monitoring* (*SHM*) campaign. The reason why nonlinearity is important for *SHM* is that structures which are designed and manufactured to be linear, can become *nonlinear* due to the introduction of damage; in such cases, detection of nonlinearity thus amounts to detection of damage [5.22]. Although a visual check is often sufficient to observe significant differences between *FRF*s, numerical metrics like mean-square difference can also be used. The exact form of the distortion in the *FRF* as the input amplitude changes depends on the type of the nonlinearity; some common types of *FRF* distortion will be discussed in the following sections.

5.3.4 Reciprocity

Reciprocity is another important property of linear system *FRF*s which, if violated, can be used to infer the presence of nonlinearity. To deduce linearity, reciprocity is a *necessary but not a sufficient* condition since some symmetrical nonlinear systems may exhibit reciprocity but will not satisfy the principle of superposition.

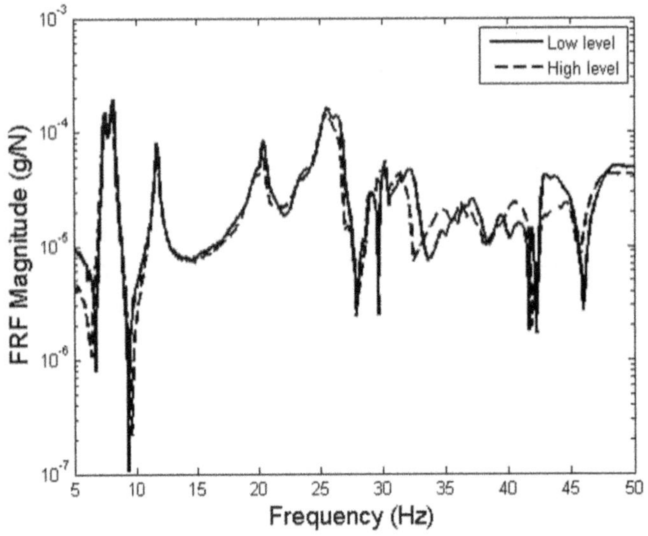

Figure 5.33 Example of a homogeneity test on a structural *FRF*.

Reciprocity is said to hold if an output y_B at a point B due to an input x_A at a point A, gives a ratio y_B/x_A numerically equal to that when the input and output points are reversed giving y_A/x_B. It follows that if this condition holds, the *FRF*s for the processes $x_A \rightarrow y_B$ and $x_B \rightarrow y_A$ are equal, and this is the usual basis of the experimental test.

Figure 5.34 shows the results of a reciprocity test on a bridge structure using band-limited random excitation and the frequency response functions between two different points, A and B. As in the homogeneity test, the difference is usually assessed by eye; in this case, the agreement is good enough that one would accept that the structure is linear.

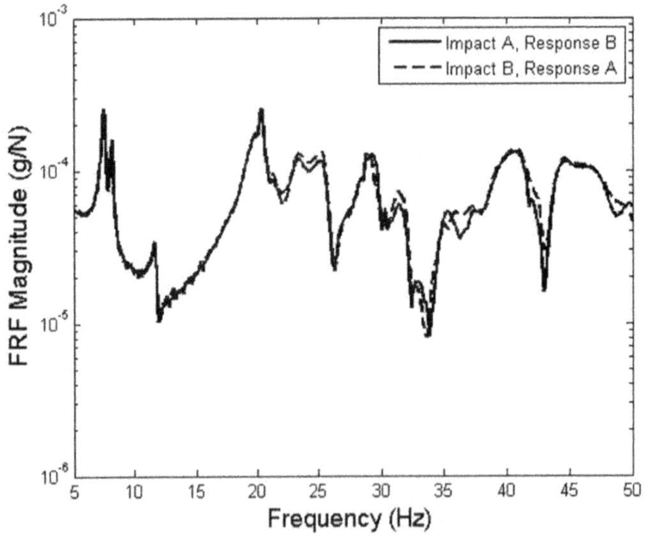

Figure 5.34 Example of a reciprocity test on structural *FRF*s measured on the Alamosa Canyon Bridge as part of a Structural Health Monitoring campaign.

When testing for reciprocity, it is important to note that all the response parameters must be the same (e.g. displacements or accelerations) and all the inputs must be forces. Furthermore a completely credible reciprocity test requires that the measurement chain is switched for the tests. If reciprocity holds, then by definition, the stiffness matrix of a structure will be symmetric as will be the frequency response function matrix.

5.3.5 The Coherence Function

The coherence function (already defined in Chapter 1 and briefly discussed in Chapter 2) is a spectrum (i.e. a function of frequency) and is therefore usually used with random or impulse excitation; it can provide a quick visual confirmation of the quality of an associated *FRF* and, in many cases, is a rapid indicator of the presence of nonlinearity in specific frequency bands or resonance regions. It is arguably the most commonly-used test of nonlinearity, by virtue of the fact that almost all commercial spectrum analysers have a built-in function allowing its calculation.

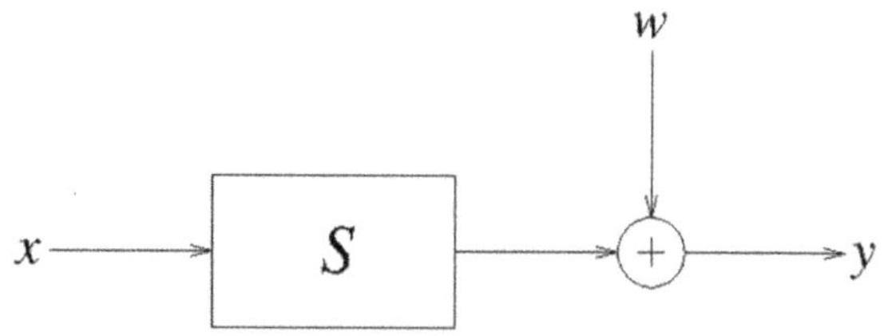

Figure 5.35 Schematic showing input-output system with measurement noise on the output.

The coherence function actually detects deviations from linear energy transport through a structure or system, and is actually sensitive to both nonlinearity *and* the presence of (input or output) measurement noise. Before discussing nonlinearity, it is convenient to derive the coherence function for *linear* systems subject to measurement noise on the output (Figure 5.35); such systems have time-domain equations of motion,

$$y(t) = S\left[x(t)\right] + m(t) \tag{5.141}$$

where $m(t)$ is the measurement noise. In the frequency domain, because the system is linear one has,

$$Y(\omega) = H(\omega)X(\omega) + M(\omega) \tag{5.142}$$

Multiplying this equation by its complex conjugate yields,

$$YY^* = HXH^*X^* + HXM^* + H^*X^*M + MM^* \tag{5.143}$$

and taking expectations gives, in terms of auto-spectra and cross-spectra[2],

[2] See Chapter 1 for the definitions of auto-spectra and cross-spectra, when explaining the response to a random excitation.

$$S_{yy}(\omega) = |H(\omega)|^2 S_{xx}(\omega) + H(\omega) S_{xm}(\omega) + H^*(\omega) S_{mx}(\omega) + S_{mm}(\omega) \tag{5.144}$$

Now, if x and m are uncorrelated signals (unpredictable from each other), then $S_{mx}(\omega) = S_{xm}(\omega) = 0$ and Eq. (5.144) reduces to,

$$S_{yy}(\omega) = |H(\omega)|^2 S_{xx}(\omega) + S_{mm}(\omega) \tag{5.145}$$

and a simple rearrangement gives,

$$\frac{|H(\omega)|^2 S_{xx}(\omega)}{S_{yy}(\omega)} = 1 - \frac{S_{mm}(\omega)}{S_{yy}(\omega)} \tag{5.146}$$

The quantity on the *RHS* is the fraction of the output power, which *can be linearly correlated with the input.* It is called the *coherence* function and denoted $\gamma^2(\omega)$. Now, as $\gamma^2(\omega)$ and $S_{mm}(\omega)/S_{yy}(\omega)$ are both positive quantities, if follows that,

$$0 \le \gamma^2 \le 1 \tag{5.147}$$

with $\gamma^2 = 1$ only if $S_{mm}(\omega) = 0$, i.e. if there is no measurement noise. The coherence function therefore detects if there is noise in the output. In fact, by very similar arguments to those above, it is straightforward to show that $\gamma^2 < 1$ if there is noise *anywhere* in the measurement chain.

Exercise 6. Show that if a linear system has measurement noise on the input, the coherence will be less than unity.

If the coherence is plotted as a function of ω, any departures from unity will be readily identifiable and easily associated with frequency ranges. The coherence is usually expressed as (see Chapter 1),

$$\gamma^2(\omega) = \frac{|S_{yx}(\omega)|^2}{S_{yy}(\omega) S_{xx}(\omega)} \tag{5.148}$$

All the quantities in the last expression are easily computed by commercial spectrum analysers designed to estimate $H(\omega)$; this is why coherence estimation is so readily available in standard instrumentation.

The coherence function also detects nonlinearity as promised above. The relationship between input and output spectra for nonlinear systems will be assumed to have the form,

$$Y(\omega) = H(\omega) X(\omega) + F[X(\omega)] \tag{5.149}$$

where F is a rather complicated function, dependent on the nonlinearity. In fact, this form follows from the Volterra expansion for nonlinear systems which is applicable to any weakly nonlinear system [5.9].

Multiplying Eq. (5.149) by Y^* and taking expectations gives,

$$S_{yy}(\omega) = |H(\omega)|^2 S_{xx}(\omega) + H(\omega) S_{xf}(\omega) + H^*(\omega) S_{fx}(\omega) + S_{ff}(\omega) \quad (5.150)$$

where this time the cross-spectra S_{fx} and S_{xf} will not necessarily vanish. In terms of the coherence,

$$\gamma^2(\omega) = 1 - 2\Re\left(H(\omega) \frac{S_{xf}(\omega)}{S_{yy}(\omega)} \right) - \frac{S_{ff}(\omega)}{S_{yy}(\omega)} \quad (5.151)$$

and the coherence will *generally* only be unity if $f = F = 0$, i.e. the system is linear.

By way of illustration, consider the Duffing oscillator of Eq. (5.105). If the level of excitation is low, the response y will be small and y^3 will be negligible in comparison. In this regime, the system will behave as a linear system and the coherence function for input and output will be unity (Figure 5.36). As the excitation is increased, the nonlinear terms will begin to play a

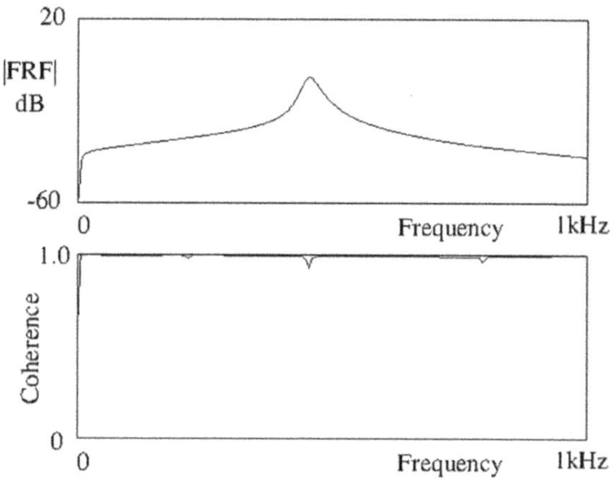

Figure 5.36 Coherence for randomly-excited Duffing oscillator system: low excitation.

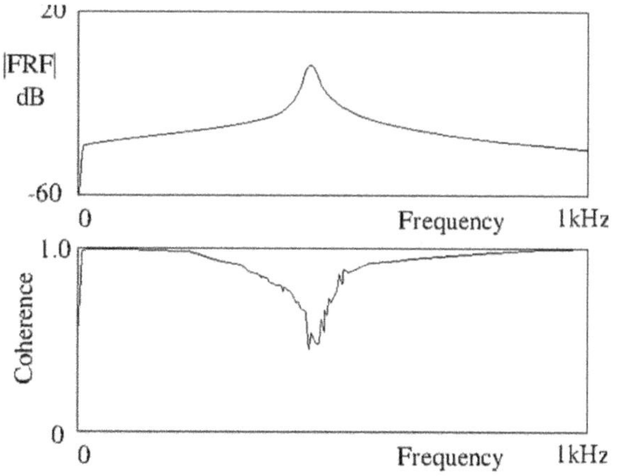

Figure 5.37 Coherence for randomly-excited Duffing oscillator system: high excitation.

part and the coherence will drop (Figure 5.37). This type of situation will occur for all polynomial nonlinearities. However, if one considers Coulomb friction (see next section), the opposite occurs. At high excitation, the friction breaks out and a nominally-linear response will be obtained and hence unit coherence.

Note that the coherence is *only* meaningful if averages are taken. For a one-shot measurement, a value of unity will always occur i.e.

$$\gamma^2 = \frac{YX^*XY^*}{YY^*XX^*} = 1 \tag{5.152}$$

Finally it is important to stress again that in order to use the coherence function for detecting nonlinearity it is necessary to realise that a reduction in the level of coherency can be caused by a range of problems, such as noise on the output and/or input signals which may in turn be due to incorrect gain settings on amplifiers, or roll-off from instrumentation. Such obvious causes should be checked before structural nonlinearity is suspected. As an illustration, consider Figure 5.38, which shows the coherences associated with the reciprocity test in Figure 5.34; the coherences are good except at low frequency, and this is probably because the accelerometers used did not have good signal-to-noise at low frequencies; there is another dip in coherence at around 34 Hz, and this is associated with low signal-to-noise at the anti-resonance in the *FRF*s. One should take care to interpret coherences as far as possible before deducing nonlinearity.

This section has touched on how noise and nonlinearity can be introduced at different points in the measurement chain; this discussion is expanded in the next section.

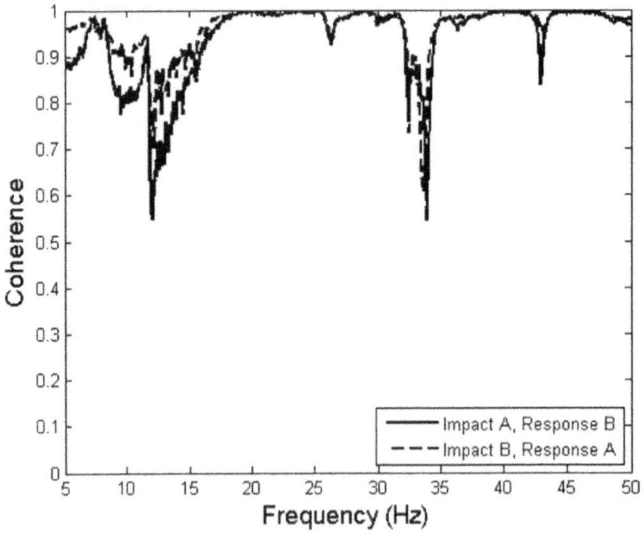

Figure 5.38 Coherences for structural *FRF*s measured on the Alamosa Canyon Bridge, as shown in Figure 5.34.

5.3.6 Nonlinearity in the Measurement Chain

It is not uncommon for nonlinearity to be unintentionally introduced in a test programme because of insufficient checks on the test set-up and/or the instrumentation used. There are several common sources of nonlinearity whose effects can (and should) be minimised at the outset of a test programme and consideration should be given to simple visual and acoustic inspection procedures (listening for rattles etc.) before the full set of tests commences.

The principal sources of nonlinearity arising from insufficient care in the test set-up are:

- Misalignment
- Exciter problems
- Looseness
- Pre-loads
- Cable rattle
- Overloads/Offset loads
- Temperature effects
- Impedance mismatching
- Poor transducer mounting

Most of these problems are easily detectable in the sense that they nearly all cause waveform distortion of some form or other. Unless one observes the actual input and output signals periodically during testing it is *impossible* to know whether or not any problems are occurring. Although tests frequently involve the measurement of *FRF*s or spectra, it is strongly recommended that a visual check is maintained of the individual drive/excitation and response voltage signals. This can be done very simply by the use of an oscilloscope. By way of illustration, Figure 5.32 earlier, showed waveform distortion which was the result of a friction nonlinearity in a shaker caused by misalignment.

In modal testing it is usual to use a force transducer (or transducers in the case of multi-point testing) as the reference input signal. In such circumstances, it is highly recommended that this signal is continuously (or at least periodically) monitored on an oscilloscope. This action is particularly important as harmonic distortion of the force excitation signal is not uncommon, often due to shaker misalignment, as discussed above, or 'force drop-out' at resonance [5.23]. Distortion can create errors in the measured *FRF* which may not be immediately apparent, and it is very important to ensure that the force input signal is not distorted.

Usually in dynamic testing one may have the choice of observing the waveform in terms of displacement, velocity or acceleration. For a linear system in which no distortion of the signal occurs, it makes little difference which variable is used. However, when nonlinearity is present this generally results in harmonic distortion. As discussed earlier in this section, under sinusoidal excitation, harmonic distortion is much easier to observe when acceleration is measured. Thus it is recommended that during testing with a sine wave, a simple test of the quality of the output waveform is to observe it on an oscilloscope in terms of the acceleration response. Any distortion or noise present will be more easily visible.

Due to the nature of random signals, waveform distortion is much more difficult to observe using an oscilloscope than with a sine wave input. However, it is still recommended that such signals are observed on an oscilloscope during testing since the effect of extreme nonlinearities – like clipping of waveforms – can easily be seen.

5.4 Common Types of Nonlinearity

The main type of nonlinearity discussed so far has been the cubic stiffness nonlinearity found in Duffing's equation. However, as one might expect, there are many types of nonlinearity to be found in structures, and it will be useful to list some of the main ones here. This section will also serve another important purpose, in introducing and discussing another important signature of nonlinearity – *FRF distortion*. This phenomenon is associated with the fact that *FRF*s measured from nonlinear systems will usually be functions of the level of excitation. For example, if one considers a polynomial nonlinearity; at low excitations, the linear terms will dominate and one will obtain an *FRF* indistinguishable from that of the underlying linear system. However, at higher levels of excitation, the nonlinearity will affect the behaviour and the *FRF* will deviate or distort from the linear form. For example, for the hardening cubic nonlinearity, the effective stiffness will be higher than the linear k suggests, and because $\omega = \sqrt{k/m}$, the peak in the *FRF* corresponding to the 'resonance' frequency will shift upwards.

The most common types of nonlinearity encountered in dynamic testing are those resulting from: polynomial stiffness and damping, clearances, impacts, friction and saturation effects. As one would expect, because of the complex physics involved, these nonlinearities are usually amplitude, velocity and frequency dependent. However, it is usual to simplify and idealise these in order that they can be incorporated into analysis, simulation and prediction capabilities. As the greatest simplification imaginable, one can onsider an *SDOF* oscillator with separate nonlinear damping and stiffness terms,

$$m\ddot{y} + f_d(\dot{y}) + f_s(y) = x(t) \tag{5.153}$$

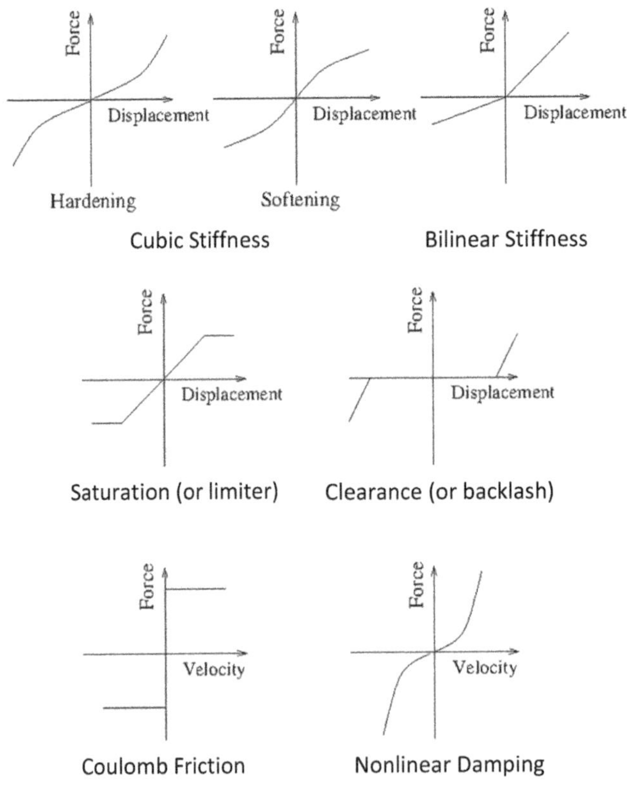

Figure 5.39 The main types of nonlinearity commonly encountered in structural dynamic tests.

Figure 5.39 summarises the most common types of stiffness and damping nonlinearities in terms of their idealised force-against-displacement or force-against-velocity characteristics. If a structure incorporates actuators, bearings, linkages or elastomeric elements, these can act as localised nonlinearities whose characteristics may be represented by one or more of those shown in Figure 5.39.

Although the assumption of an *SDOF* system might be considered an oversimplification, the effects of nonlinearities on the *FRF*s of *MDOF* systems will be similar for individual modes reasonably well separated from neighbouring modes. In this section, *FRF* distortions will be illustrated in both the *Bode* and *Nyquist* (or *Argand*) representations. The Bode representation of an *FRF* is the plot of the magnitude and phase separately as a function of frequency – only the magnitude plots will be given here. The Nyquist/Argand representation of an *FRF* plots the real part against the imaginary part, yielding a *locus* in the complex plane as the frequency varies. For a linear system, the Nyquist plot for a mobility (velocity) *FRF* gives an ellipse.

Figure 5.40 shows the characteristic *FRF* distortions for the various types of nonlinearity discussed here. In the context of modal analysis, where system parameters are often obtained by curve-fitting to *FRF*s [5.24, 5.23], nonlinear distortions – if not accounted for correctly – may lead to badly-estimated parameters. It is instructive to consider each nonlinearity in a little more detail in turn.

5.4.1 Cubic Stiffness

In this case – Duffing's equation – the force displacement characteristic has the form,

$$f_s(y) = ky + k_3 y^3 \tag{5.154}$$

and k_3 may be positive or negative. If $k_3 > 0$, one can see that at high levels of excitation the restoring force will be greater than that expected from the linear term alone. The extent of this excess will increase as the forcing level increases and for this reason such systems are referred to as *hardening*. Examples of such systems are clamped plates and beams as discussed earlier. If $k_3 < 0$, the effective stiffness decreases as the level of excitation increases and such systems are referred to as *softening*. Truly softening cubic systems are *unphysical* in the sense that the restoring force changes sign at a certain distance from equilibrium and begins to drive the system towards infinity. Systems with such characteristics are always found to have higher-order polynomial terms in the stiffness function with positive coefficients which dominate at high levels and restore stability. Systems which appear to show softening cubic behaviour over limited ranges include buckling beams.

The *FRF* distortions characteristic of these systems are shown in Figures 5.40 (a) and 5.40 (b). The most important observation is that the resonance frequency shifts up for the hardening system as the level of excitation is raised; this is consistent with the increase in effective stiffness. In contrast, the resonance frequency for the softening system shifts down.

5.4.2 Bilinear Stiffness or Damping

In this case, the stiffness characteristic has the form,

$$f_s(y) = \begin{cases} k_1 y & y > 0 \\ k_2 y & y < 0 \end{cases} \tag{5.155}$$

with a similar definition for bilinear damping.

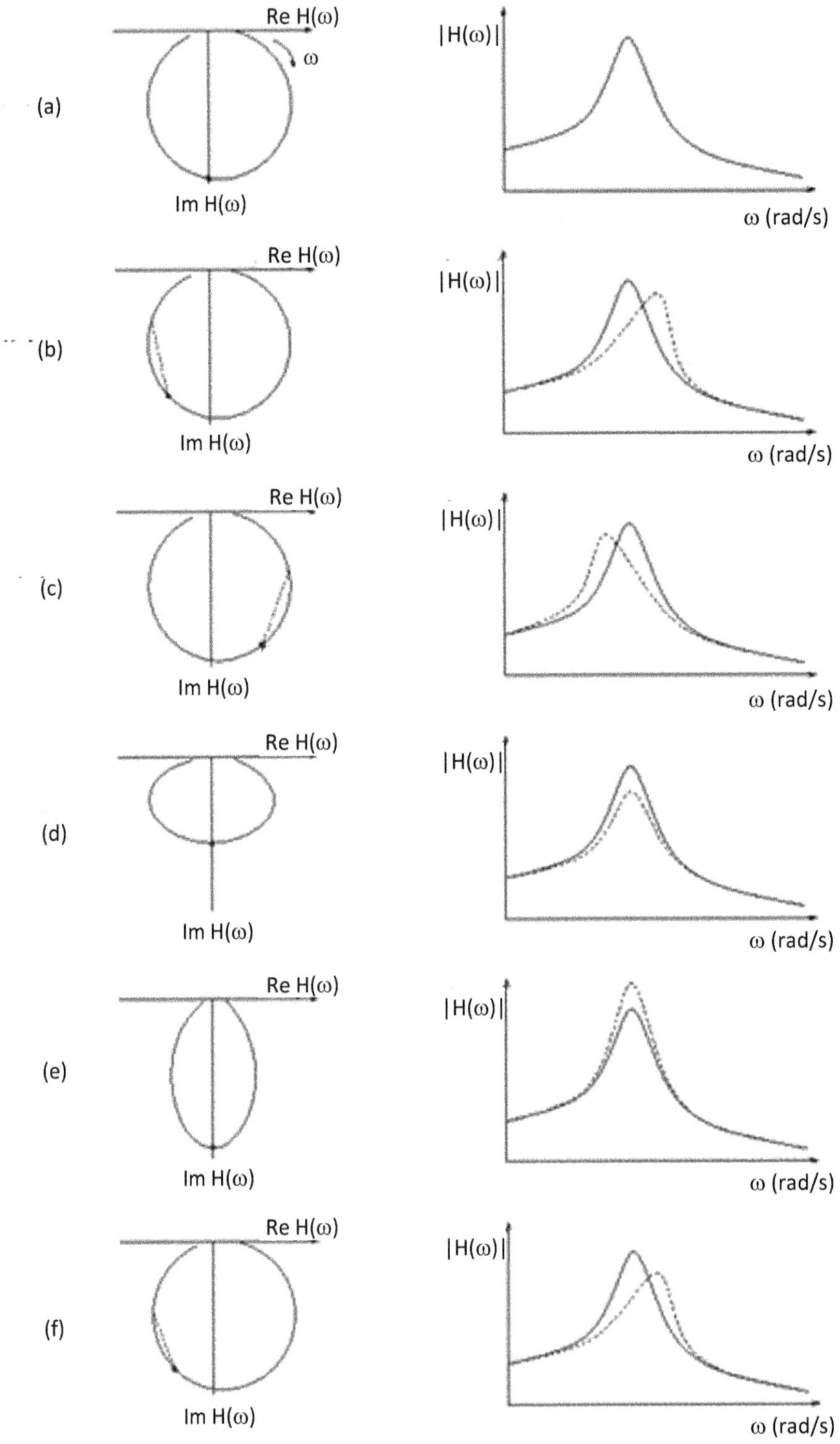

Figure 5.40 Characteristic *FRF* distortions observed for different types of linear and nonlinear systems (a) linear, (b) hardening cubic stiffness, (c) softening cubic stiffness, (d) polynomial damping, (e) Coulomb friction (damping), (f) piecewise-linear stiffness.

The most extreme example of a bilinear system is the impact oscillator for which $k_1 = 0$ and $k_2 = \infty$, this corresponds to a ball bouncing against a hard wall. Such systems can display extremely complex behaviour indeed (see chapter 15 of [5.25]). One system which approximates to a bilinear damping system is the standard automotive damper or shock absorber which is designed to have different damping constants in compression and rebound.

Figure 5.40 does not show the *FRF* distortions characteristic of this system because it is one of the rare nonlinear systems which display homogeneity. (This last remark is only true if the position of the change in stiffness is at the origin, if it is offset by any degree, the system will fail to show homogeneity if the level of excitation is taken sufficiently high.)

5.4.3 Nonlinear Damping

The most common form of polynomial damping is quadratic,

$$f_d(\dot{y}) = c_2 \dot{y}|\dot{y}| \tag{5.156}$$

(where the absolute value term is to assure that the force is always opposed to the velocity). This type of damping occurs when fluid flows through an orifice or around a slender member. The former situation is common in automotive dampers and hydromounts, the latter occurs in the fluid loading of offshore structures; the fundamental equation there is Morison's equation [5.26],

$$F(t) = c_1 \dot{u}(t) + c_2 u(t)|u(t)| \tag{5.157}$$

where F is the force on the member and u is the velocity of the flow.

The effect of increasing excitation level is to increase the effective damping as shown in Figure 5.40 (d).

5.4.4 Coulomb Friction

This type of damping has characteristic,

$$f_d(\dot{y}) = c_F \, sgn(\dot{y}) \tag{5.158}$$

as shown in Figure 5.39. This type of nonlinearity is common in any situation with relative motion of interfaces in contact. It is particularly prevalent in demountable structures, such as grandstands. The conditions of constant assembly and disassembly are suitable for creating interfaces which allow relative motion. In this sort of structure, friction will often occur in tandem with clearance nonlinearities. It is unusual here in the sense that it is most evident at low levels of excitation, where in extreme cases, *stick-slip* motion can occur. At higher levels of excitation, the friction 'breaks out' and the system will behave nominally linearly. The characteristic *FRF* distortion (Figure 5.40 (d)) is the reverse of the quadratic damping case, with the higher damping at lower excitations.

5.4.5 Piecewise Linear Stiffness

The form of the stiffness function in this case is,

$$f_s(y) = \begin{cases} k_2 y + (k_1 - k_2)d & y > d \\ k_1 y & |y| < d \\ k_2 y - (k_1 - k_2)d & y < -d \end{cases} \tag{5.159}$$

Two of the nonlinearities in Figure 5.39 are special cases of this form. The *saturation* or *limiter* nonlinearity has $k_2 = 0$ and the *clearance* or *backlash* nonlinearity has $k_1 = 0$. Typical *FRF* distortion is shown in Figure 5.40 (f) for a hardening characteristic ($k_2 > k_1$).

5.5 Linearisation: Effective *FRFs* for Nonlinear Systems

In structural/engineering dynamics, the most commonly-used method of visualising the input-output properties of a system is to construct the *Frequency Response Function* (*FRF*). So important is the technique, that it is usually the first step in any vibration test and almost all commercially-available spectrum analysers provide *FRF* capability. The *FRF* encodes most of the information necessary to specify the dynamics of a structure: resonances, anti-resonances, modal density and phase are directly visible. If *FRF*s are available for a number of response points, the system mode shapes can also be constructed. In addition, as discussed in the last section, the *FRF* can rapidly indicate whether a system is linear or nonlinear; one simply constructs the *FRF*s for a number of different excitation levels and searches for changes in the frequency or magnitude of the resonance peaks. Alternatively, in numerical simulations, the *FRF*s are invaluable for benchmarking algorithms, structural modification studies and updating numerical models.

This section considers how *FRF*s are defined and constructed for nonlinear systems. The interpretation of the *FRF*s is discussed and it is shown that they provide a representation of the system as it is *linearised* about a particular operating point. *FRF* distortions are used to provide information about nonlinearity. *FRF*s can be defined for many different excitation types, and for linear systems, all the functions obtained will coincide; however, this is not the case for nonlinear systems. Different *FRF*s, corresponding to different excitation types, will induce different linearisations. The first case considered here will correspond to sinusoidal excitations; the most direct linearisation technique in this case is referred to as *harmonic balance*.

5.5.1 Harmonic Balance

For the discipline of engineering in general, the role of applied mathematics is to describe, elucidate and support experiment; mathematical analysis should yield information in a form which is readily comparable with observation. The method of harmonic balance conforms to this principle beautifully as a means of approximating the *FRF*s of nonlinear systems.

The simplest definition of an *FRF* is based on harmonic/sinusoidal excitation; if a signal $X sin(\omega t)$ is input to a system and evokes a response $Y sin(\omega t + \phi)$, the *FRF* is,

$$H(\omega) = \left| \frac{Y}{X}(\omega) \right| e^{i\phi(\omega)} \tag{5.160}$$

This quantity is very straightforward to obtain experimentally. Over a range of frequencies $[\omega_{min}, \omega_{max}]$ at a fixed frequency increment $\Delta\omega$, sinusoids $X sin(\omega t)$ are injected sequentially into the system of interest. At each frequency, the time histories of the input and response signals are recorded after transients have died out, and Fourier transformed. The ratio of the (complex) response spectrum to the input spectrum yields the *FRF* value at the frequency of interest. In the case of a linear system, the response to a sinusoid is always a sinusoid at the same frequency and the *FRF* in Eq. (5.160) summarises the input/output process in its entirety, and does not depend on the amplitude of excitation X ; such an *FRF* will be referred to here as *pure*.

In the case of nonlinear systems, sinusoidal forcing results in response components at frequencies other than the excitation frequency and the distribution of energy amongst these frequencies depends on the level of excitation X, so the measurement process described above will also lead to a quantity which depends on X. However, because the process is so simple, it is often carried out experimentally in an unadulterated fashion for nonlinear systems. The *FRF* resulting from such a test will be referred to as *composite*, (the justification for this term can be found in [5.9]), and denoted by $\Lambda_s(\omega)$. $\Lambda_s(\omega)$ is often called a *describing function*, particularly in the literature of Electrical and Control Engineering [5.27]. The form of the composite *FRF* also depends on the type of excitation as discussed above. If white noise of constant power spectral density P is used and the *FRF* is obtained by taking the ratio of the cross- and auto-spectral densities,

$$\Lambda_r(\omega, P) = \frac{S_{yx}(\omega)}{S_{xx}(\omega)} = \frac{S_{yx}(\omega)}{P} \tag{5.161}$$

The function $\Lambda_r(\omega, P)$ is distinct from the $\Lambda_s(\omega, X)$ obtained from a stepped-sine test. However, for linear systems the forms (5.160) and (5.161) coincide. In the following, the Λ subscripts will be omitted if the excitation type is clear from the context.

Harmonic balance is the analytical analogue of the stepped-sine test. It is only one of a number of basic techniques for approximating the response of nonlinear systems. However, it is presented here in detail as it provides the neatest means of deriving the *FRF*.

The system considered here will be the symmetric Duffing equation; at the risk of repetition,

$$m\ddot{y} + c\dot{y} + ky + k_3 y^3 = x(t) \tag{5.162}$$

Harmonic balance mimics the spectrum analyser in simply assuming that the response to a sinusoidal excitation is a sinusoid at the same frequency. A trial solution $Y sin(\omega t)$ is thus substituted in the equation of motion; in the case of the symmetric Duffing oscillator one has,

$$m\ddot{y} + c\dot{y} + ky + k_3 y^3 = X sin(\omega t - \phi) \tag{5.163}$$

To simplify matters, the phase has been transferred onto the input to allow Y to be taken as real. Substituting the trial solution yields,

$$-m\omega^2 Y sin(\omega t) + c\omega Y cos(\omega t) + kY sin(\omega t) + k_3 Y^3 sin^3(\omega t) = X sin(\omega t - \phi) \quad (5.164)$$

and after a little elementary trigonometry this becomes,

$$-m\omega^2 Y sin(\omega t) + c\omega Y cos(\omega t) + kY sin(\omega t) + k_3 Y^3\left(\frac{3}{4} sin(\omega t) - \frac{1}{4} sin(3\omega t)\right)$$
$$= X sin(\omega t) cos\phi - X cos(\omega t) sin\phi \quad (5.165)$$

Equating the coefficients of $sin(\omega t)$ and $cos(\omega t)$ (the *fundamental* components) yields the equations,

$$\left(-m\omega^2 Y + kY + \frac{3}{4} k_3 Y^3\right) = X cos\phi \quad (5.166)$$

$$c\omega Y = -X sin\phi \quad (5.167)$$

Squaring and adding these equations yields,

$$X^2 = Y^2\left(\left(-m\omega^2 + k + \frac{3}{4} k_3 Y^2\right)^2 + c^2\omega^2\right) \quad (5.168)$$

which gives an expression for the gain or *modulus* of the system,

$$\left|\frac{Y}{X}\right| = \frac{1}{\left(\left(-m\omega^2 + k + \frac{3}{4} k_3 Y^2\right)^2 + c^2\omega^2\right)^{1/2}} \quad (5.169)$$

The phase is obtained from the ratio of equations (5.166) and (5.167) as,

$$\phi = tan^{-1}\frac{-c\omega}{-m\omega^2 + k + \frac{3}{4} k_3 Y^2} \quad (5.170)$$

These expressions can be combined into the complex composite *FRF*,

$$\Lambda(\omega) = \frac{1}{k + \frac{3}{4} k_3 Y^2 - m\omega^2 + ic\omega} \quad (5.171)$$

One can regard this object as the *FRF* of a linear system,

$$m\ddot{y} + c\dot{y} + k_{eq} y = X sin(\omega t - \phi) \quad (5.172)$$

where the *effective* or *equivalent stiffness* is amplitude dependent,

$$k_{eq} = k + \frac{3}{4}k_3 Y^2 \tag{5.173}$$

At a fixed level of excitation, the effective *FRF* has natural frequency,

$$\omega_n = \sqrt{\frac{k + \frac{3}{4}k_3 Y^2}{m}} \tag{5.174}$$

which depends on Y and thus indirectly on X. If $k_3 > 0$, the natural frequency increases with X – a *hardening* system. If $k_3 < 0$ the system is *softening*; the natural frequency decreases with increasing X. Note that the expression (5.174) is in terms of Y rather than X, this leads to a subtlety which has so far been ignored. Although the apparent resonance frequency changes with X in the manner described above, the *form* of the *FRF* is not that of a linear system. For given X and ω, the displacement response Y is obtained by solving the cubic Eq. (5.168). (This expression is essentially cubic in Y as one can disregard negative amplitude solutions). As complex roots occur in conjugate pairs, (5.168) will either have one or three real solutions – the complex solutions are ignored here as unphysical.

At low levels of excitation, the *FRF* is a barely-distorted version of that for the underlying linear system as the k term will dominate for $Y \ll 1$. A unique response amplitude (a single real root of (5.168) is obtained for all ω. As X increases, the *FRF* becomes more distorted, i.e. departs from the linear form; however, a unique response is still obtained for all ω. This continues until X reaches a critical value X_{crit} where the *FRF* has a vertical tangent. Beyond this point a range of ω values, $\left[\omega_{low}, \omega_{high}\right]$, is obtained over which there are three real solutions for the response.

This is an example of a *bifurcation point* of the parameter X; although X varies continuously, the number and stability types of the solutions changes abruptly. As the test or simulation steps past the point ω_{low}, two new responses become possible and persist until ω_{high} is reached and two solutions disappear. The graph of the response looks like Figure 5.41. In the interval $\left[\omega_{low}, \omega_{high}\right]$, the solutions $Y^{(1)}$, $Y^{(2)}$ and $Y^{(3)}$ are possible with $Y^{(1)} > Y^{(2)} > Y^{(3)}$. However, it can be shown that the solution $Y^{(2)}$ is unstable and will therefore never be observed in practice.

The corresponding experimental situation occurs in a stepped-sine or sine-dwell test. Consider an upward sweep. A unique response exists up to $\omega = \omega_{low}$. However, beyond this point, the response stays on branch $Y^{(1)}$ essentially by continuity. This persists until, at frequency ω_{high}, $Y^{(1)}$ ceases to exist and the only solution is $Y^{(3)}$, a jump to this solution occurs giving a discontinuity in the *FRF*. Beyond ω_{high} the solution stays on the continuation of $Y^{(3)}$, which is the unique solution in this range. The type of *FRF* from such a test is shown in Figure 5.42.

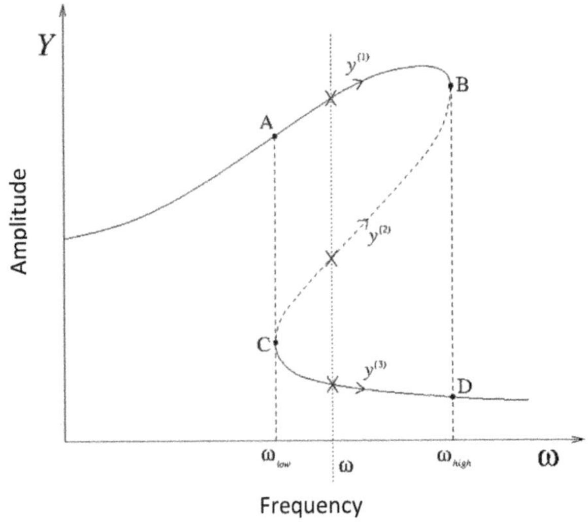

Figure 5.41 Amplitudes of analytically-allowed responses for Duffing system with a sine input.

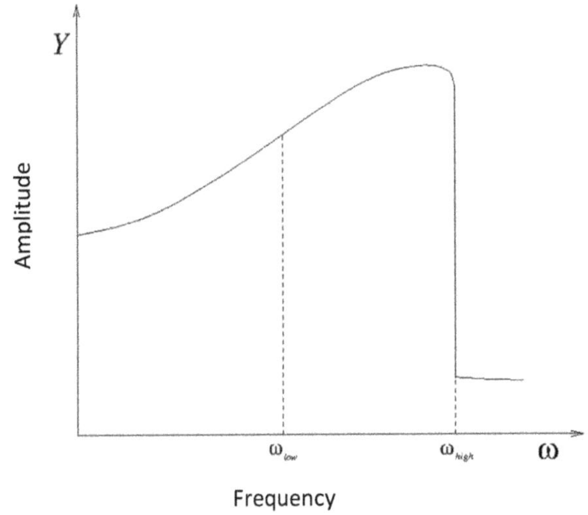

Figure 5.42 Amplitudes of actual responses for Duffing system with a stepped-sine input of increasing frequency.

The downward sweep is very similar. When $\omega > \omega_{high}$, a unique response is obtained. In the multi-valued region, branch $Y^{(3)}$ is obtained by continuity and this persists until ω_{low} when it ceases to exist and the response jumps to $Y^{(1)}$ and thereafter remains on the continuation of that branch (Figure 5.43).

If $k_3 > 0$, the resonance peak moves to higher frequencies and the jumps occur on the right-hand side of the peak as described above. If $k_3 < 0$, the jumps occur on the left of the peak and the resonance shifts downward in frequency. These discontinuities are frequently observed in experimental *FRF*s when high levels of excitation are used. The discontinuities also occur in the phase of the response, as one might expect. The position of the discontinuities can be computed analytically [5.9].

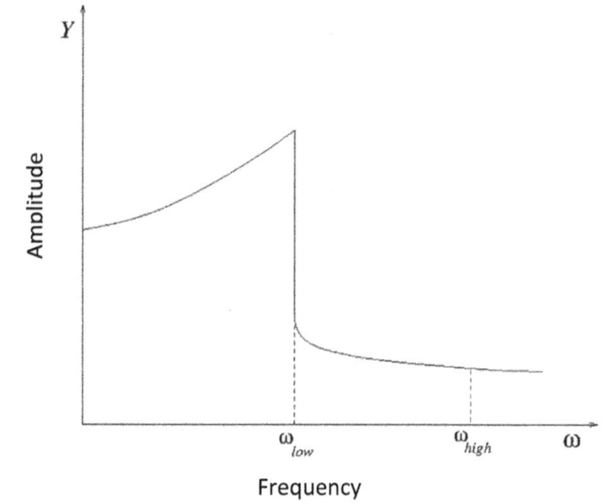

Figure 5.43 Amplitudes of actual responses for Duffing system with a stepped-sine input of decreasing frequency.

Before continuing with the approximation of *FRF*s within the harmonic balance method it is important to recognise that nonlinear systems *do not* respond to a monoharmonic signal with a monoharmonic at the same frequency. The next two sections discuss how departures from this condition arise.

5.5.2 Harmonic Generation in Nonlinear Systems

The harmonic balance analysis just described is *not* the whole story. Eq. (5.163) is *not* solved by equating coefficients of the fundamental components; a term $\frac{1}{4}k_3Y^2 sin(3\omega t)$ is *not* balanced. Setting this term equal to zero leads to the conclusion that k_3 or Y is zero, which is clearly incorrect. The reason for the problem is of course that $y(t) = Y sin(\omega t)$ is an unacceptable solution to Eq. (5.163); things are much more complicated for nonlinear systems. An apparent immediate fix is to add a term proportional to $sin(3\omega t)$ to the trial solution to potentially allow balance; this yields,

$$y(t) = Y_1 sin(\omega t + \phi_1) + Y_3 sin(3\omega t + \phi_3) \tag{5.175}$$

(with the phases explicitly represented). This *ansatz* is substituted in the phase-adjusted version of (5.163),

$$m\ddot{y} + c\dot{y} + ky + k_3y^3 = X sin(\omega t) \tag{5.176}$$

and projecting out the coefficients of $sin(\omega t)$, $cos(\omega t)$, $sin(3\omega t)$ and $cos(3\omega t)$ leads to the following system of equations,

$$-m\omega^2 Y_1 \cos\phi_1 - c\,\omega Y_1 sin\,\phi_1 + kY_1 \cos\phi_1 +$$
$$\frac{3}{4}k_3Y_1^3 \cos\phi_1 + \frac{3}{2}k_3Y_1Y_3^2 \cos\phi_1 - \frac{3}{4}k_3Y_1^2 y_3 \cos\phi_3 \cos 2\phi_1 = X \tag{5.177}$$

$$-m\omega^2 Y_1 \sin\phi_1 - c\omega Y_1 \cos\phi_1 + k Y_1 \sin\phi_1 + \frac{3}{4}k_3 Y_1^3 \sin\phi_1 + \frac{3}{2}k_3 Y_1 Y_3^2 \sin\phi_1 - \frac{3}{4}k_3 Y_1^2 y_3 \sin\phi_3 \cos 2\phi_1 = 0 \quad (5.178)$$

$$-9m\omega^2 Y_3 \cos\phi_3 - 3c\omega Y_3 \sin\phi_3 + kY_3 \cos\phi_3 - \frac{1}{4}k_3 Y_1^3 \cos^3\phi_1 + \frac{3}{4}k_3 Y_3^3 \cos\phi_3 - \frac{3}{4}k_3 Y_1^3 \cos\phi_1 \sin^2\phi_1 + \frac{3}{2}k_3 Y_1^2 Y_3 \cos\phi_3 = 0 \quad (5.179)$$

$$-9m\omega^2 Y_3 \sin\phi_3 + 3c\omega Y_3 \cos\phi_3 + kY_3 \sin\phi_3 + \frac{1}{4}k_3 Y_1^3 \sin^3\phi_1 + \frac{3}{4}k_3 Y_3^3 \sin\phi_3 - \frac{3}{4}k_3 Y_1^3 \cos^2\phi_1 \sin\phi_1 + \frac{3}{2}k_3 Y_1^2 Y_3 \sin\phi_3 = 0 \quad (5.180)$$

Solving this system of equations gives a better approximation to the *FRF*. However, now the cubic term generates terms with $\sin^3(\omega t)$, $\sin^2(\omega t)\sin(3\omega t)$, $\sin(\omega t)\sin^2(3\omega t)$ and $\sin^3(3\omega t)$ which decompose to give harmonics at $5\omega t$, $7\omega t$ and $9\omega t$. Equating coefficients up to third-order now leaves these components uncancelled. In order to properly deal with them, a trial solution of the form,

$$y(t) = Y_1 \sin(\omega t + \phi_1) + Y_3 \sin(3\omega t + \phi_3) + Y_5 \sin(5\omega t + \phi_5) + Y_7 \sin(7\omega t + \phi_7) + Y_9 \sin(9\omega t + \phi_9) \quad (5.181)$$

is required; however, this in turn will generate higher-order harmonics and one is led to the conclusion that the only way to obtain consistency is to include *all* odd harmonics in the trial solution, so,

$$y(t) = \sum_{i=1}^{\infty} Y_{2i+1} \sin([2i+1]\omega t + \phi_{2i+1}) \quad (5.182)$$

is the necessary expression. This expansion explains the appearance of harmonic components in nonlinear systems as described in Section 5.3.2. The fact that only odd harmonics are present is a consequence of the stiffness function $ky + k_3 y^3$, being odd. If the function were even or generic, *all* harmonics would be present. To see this, consider the system,

$$m\ddot{y} + c\dot{y} + ky + k_2 y^2 = X \sin(\omega t - \phi) \quad (5.183)$$

and assume a sinusoidal trial solution $y(t) = Y \sin(\omega t)$. Substituting this into Eq. (5.183) generates a term $Y^2 \sin^2(\omega t)$ which decomposes to give $\frac{1}{2}Y^2 - \frac{1}{2}Y^2 \cos(2\omega t)$, so *DC* and second harmonic appear. This change requires an amendment to the trial solution as before, so $y(t) = Y_0 + Y_1 \sin(\omega t) + Y_2 \sin(2\omega t)$ (neglecting phases). It is clear that iterating this procedure will ultimately generate all harmonics – both odd and even – and also a *DC* term. Figure 5.44 shows the pattern of harmonics in the response of the system (note the log scale),

$$\ddot{y} + 20\dot{y} + 10^4 y + 5\times 10^9 y^3 = 4 \sin(30t) \quad (5.184)$$

The relative size of the harmonics can be determined analytically by probing the equation of motion with an appropriately high-order trial solution. This action results in a horrendous set of coupled nonlinear equations. A more direct route to the information is possible using the Volterra series [5.9], but this is beyond the scope of the current discussion.

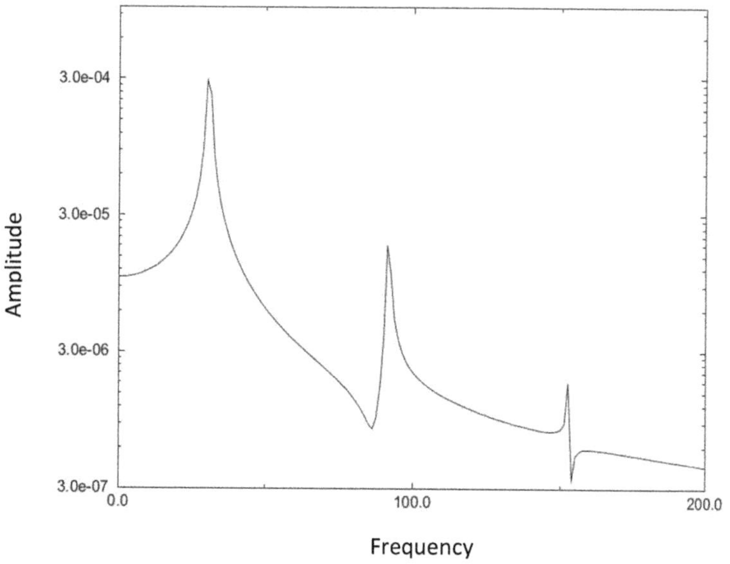

Figure 5.44 Amplitudes of harmonics in response of symmetric Duffing oscillator to sinusoidal input.

5.5.3 Sum and Difference Frequencies

It has been shown above that nonlinear systems can respond at multiples of the forcing frequency if the excitation is a pure sinusoid. The situation becomes a little more complicated if the excitation is not a pure tone. For simplicy, consider the quadratic nonlinear Eq. (5.183), with the forcing function a sum of two sinusoids or a *two-tone* signal,

$$x(t) = X_1 sin(\omega_1 t) + X_2 sin(\omega_2 t) \tag{5.185}$$

It is clear that any trial solution must at least have the form,

$$y(t) = Y_1 sin(\omega_1 t) + Y_2 sin(\omega_2 t) \tag{5.186}$$

with Y_1 and Y_2 allowed to be complex, to encode phase. The nonlinear stiffness gives a term,

$$k_2 \left(Y_1 sin(\omega_1 t) + Y_2 sin(\omega_2 t)\right)^2 = k_2 \left(Y_1^2 sin^2(\omega_1 t) + 2Y_1 Y_2 sin(\omega_1 t) sin(\omega_2 t) + Y_2^2 sin^2(\omega_2 t)\right) \tag{5.187}$$

which can be decomposed into harmonics using elementary trigonometry; the result is,

$$k_2 \left(\frac{1}{2} Y_1^2 \left(1 - cos(2\omega_1 t)\right) + Y_1 Y_2 cos\left((\omega_1 - \omega_2)t\right) - Y_1 Y_2 cos\left((\omega_1 + \omega_2)t\right) + \frac{1}{2} Y_2^2 \left(1 - cos(2\omega_2 t)\right)\right) \tag{5.188}$$

This means that balancing the coefficients of *sines* and *cosines* in Eq. (5.183) requires a trial solution,

$$y(t) = Y_0 + Y_1 sin(\omega_1 t) + Y_2 sin(\omega_2 t) + Y_{11}^+ sin(2\omega_1 t) + Y_{22}^+ sin(2\omega_2 t) + Y_{12}^+ cos([\omega_1 + \omega_2]t) + Y_{12}^- cos([\omega_1 - \omega_2]t) \quad (5.189)$$

where $Y_{ij}^{\pm}$ is simply the component of the response at the frequency $\omega_i \pm \omega_j$.

If this is substituted into (5.183), one again begins a sequence of iterations, which ultimately results in a trial solution containing all frequencies,

$$\pm p\omega_1 \pm q\omega_2 \quad (5.190)$$

with p and q integers. If the exercise above is repeated for the symmetric Duffing oscillator ($k_2 = 0$), the same result is obtained except that p and q are only allowed to sum to odd values. To lowest nonlinear order, this means that the frequencies $3\omega_1$, $2\omega_1 \pm \omega_2$, $\omega_1 \pm 2\omega_2$ and $3\omega_2$ will be present.

The *FRF* cannot encode information about sum and difference frequencies, it only makes sense for single-input single-tone systems; a full analysis again requires the Volterra series [5.9].

The theory above provides other instances of a nonlinear system violating the principle of superposition. If excitations $X_1 sin(\omega_1 t)$ and $X_2 sin(\omega_2 t)$ are presented to the asymmetric Duffing oscillator separately, each case results only in multiples of the relevant frequency in the response. If the excitations are presented together, the new response contains novel frequencies of the form (5.190); novel anyway as long as ω_1 is not an integer multiple of ω_2.

5.5.4 Harmonic Balance revisited

The analysis given in Section 5.5.1, although effective, was not very systematic. Fortunately, there is a more general formulation, leading to a simple formula for the effective stiffness, given the form of the nonlinear restoring force. Restricting for now to the case of nonlinear stiffness, consider the equation of motion,

$$m\ddot{y} + c\dot{y} + f_s(y) = x(t) \quad (5.191)$$

What is needed, is a general means to obtain,

$$f_s(y) = k_{eq} y \quad (5.192)$$

for a given operating condition. If the excitation is a phase-shifted sinusoid, $X sin(\omega t - \phi)$, substituting the harmonic balance trial solution $Y sin(\omega t)$ yields the nonlinear form $f_s(Y sin(\omega t))$. This function can be expanded as a Fourier series (see also Section 1.4.5, under "Response to a periodic excitation"),

$$f_s\left(Y sin\left(\omega t\right)\right)=\frac{a_0}{2}+\sum_{n=1}^{\infty}a_n cos\left(n\omega t\right)+\sum_{n=1}^{\infty}b_n sin\left(n\omega t\right) \tag{5.193}$$

and this will be a finite sum, if f_s is a polynomial. For the purposes of harmonic balance, the only important parts of this expansion are the fundamental terms. Elementary Fourier analysis applies and,

$$a_0=\frac{1}{\pi}\int_0^{2\pi}f_s\left(Y sin\left(\omega t\right)\right)d\left(\omega t\right) \tag{5.194}$$

$$a_1=\frac{1}{\pi}\int_0^{2\pi}f_s\left(Y sin\left(\omega t\right)\right)cos\left(\omega t\right)d\left(\omega t\right) \tag{5.195}$$

$$b_1=\frac{1}{\pi}\int_0^{2\pi}f_s\left(Y sin\left(\omega t\right)\right)sin\left(\omega t\right)d\left(\omega t\right) \tag{5.196}$$

or, in a more convenient notation,

$$a_0=\frac{1}{\pi}\int_0^{2\pi}f_s\left(Y sin\theta\right)d\theta \tag{5.197}$$

$$a_1=\frac{1}{\pi}\int_0^{2\pi}f_s\left(Y sin\theta\right)cos\theta\, d\theta \tag{5.198}$$

$$b_1=\frac{1}{\pi}\int_0^{2\pi}f_s\left(Y sin\theta\right)sin\theta\, d\theta \tag{5.199}$$

It is immediately obvious from Eq. (5.197), that the response will always contain a *DC* component if the stiffness function has an even component. In fact, if the stiffness function is purely odd, i.e. $f_s\left(-y\right)=-f_s\left(y\right)$, then it follows straightforwardly that $a_0=a_1=0$. Now, considering terms up to the fundamental in this case, Eq. (5.192) becomes,

$$f_s\left(Y sin\left(\omega t\right)\right)=b_1 sin\left(\omega t\right)=k_{eq}Y sin\left(\omega t\right) \tag{5.200}$$

which then gives,

$$k_{eq}=\frac{b_1}{Y}=\frac{1}{\pi Y}\int_0^{2\pi}f_s\left(Y sin\theta\right)sin\theta\, d\theta \tag{5.201}$$

so the *FRF* takes the form,

$$\Lambda\left(\omega\right)=\frac{1}{k_{eq}-m\omega^2+ic\omega} \tag{5.202}$$

(combining both amplitude and phase). It is straightforward to derive the expressions (5.201) and (5.202) for the case of the symmetric Duffing oscillator. The stiffness function is $f_s\left(y\right)=ky+k_3y^3$, so substituting into (5.201) yields,

$$k_{eq} = \frac{k}{\pi Y}\int_0^{2\pi} Y \sin\theta \sin\theta d\theta + \frac{k_3}{\pi Y}\int_0^{2\pi} Y^3 \sin^3\theta \sin\theta d\theta \tag{5.203}$$

The first integral trivially gives the linear part k; the contribution from the nonlinear stiffness is,

$$\frac{k_3}{\pi Y}\int_0^{2\pi} Y^3 \sin^4\theta d\theta = \frac{k_3 Y^2}{\pi}\int_0^{2\pi} \frac{1}{8}(3 - 4\cos 2\theta + \cos 4\theta)d\theta = \frac{3}{4}k_3 Y^2 \tag{5.204}$$

so,

$$k_{eq} = k + \frac{3}{4}k_3 Y^2 \tag{5.205}$$

in agreement with Eq. (5.173).

As described above, this process represents a naive replacement of the nonlinear system with an *effective* linear system. This action begs the question: *what is the significance of the linear system?*; this is quite simple to answer and fortunately the answer agrees with intuition.

A measure of how well the linear system represents the nonlinear system is given by the error function,

$$E = \lim_{T\to\infty}\frac{1}{T}\int_0^T \left(y(t) - y_{lin}(t)\right)^2 dt \tag{5.206}$$

A system which minimises E is called an *optimal quasi-linearisation*. It can be shown [5.27], that a linear system minimises E if and only if,

$$R_{xy}(\tau) = R_{xy_{lin}}(\tau) \tag{5.207}$$

where R is the cross-correlation function,

$$R_{pq}(\tau) = \lim_{T\to\infty}\frac{1}{T}\int_0^T p(t)q(t+\tau)dt \tag{5.208}$$

(This is quite a remarkable result, no higher-order statistics are needed.)

It is fairly straightforward to verify that Eq. (5.207) is satisfied by the system with harmonic balance relations (5.198) and (5.199), for the particular reference signal used[3]. It suffices to show that if,

$$f(t) = \frac{a_0}{2} + \sum_{n=1}^{\infty} a_n \cos(n\omega t) + \sum_{n=1}^{\infty} b_n \sin(n\omega t) \tag{5.209}$$

and,

$$f_{lin}(t) = a_1 \cos(\omega t) + b_1 \sin(\omega t) \tag{5.210}$$

[3] Note that linearisations exist for *all* types of reference signal, there is no restriction to harmonic signals.

then,

$$R_{xf}(\tau) = R_{xf_{lin}}(\tau) \tag{5.211}$$

with $x(t) = X sin(\omega t + \phi)$. This means that the linear system predicted by harmonic balance is indeed an optimal quasi-linearisation.

The physical content of Eq. (5.201) is easy to discern; it simply represents the average value of the restoring force over one cycle of excitation, divided by the value of displacement. This expression gives the mean value of the stiffness experienced by the system over a cycle. For this reason, harmonic balance, to this level of approximation, is sometimes referred to as an *averaging method.* Use of such methods dates back to the seminal work of Krylov and Boguliubov in the first half of the Twentieth Century. So strongly is this approach associated with these pioneers, that it is sometimes referred to as the method of Krylov and Boguliubov [5.28].

5.5.5 Nonlinear Damping

The formulæ presented for harmonic balance so far have been restricted to the case of nonlinear stiffness. However, the method, in principle, has no restrictions on the form of the nonlinearity examined and it is a simple matter to extend the theory to nonlinear damping. Consider the system,

$$m\ddot{y} + f_d(\dot{y}) + ky = X sin(\omega t - \phi) \tag{5.212}$$

Choosing a trial output $y(t) = Y sin(\omega t)$ yields a nonlinear function,

$$f_d(\omega Y cos(\omega t)) \tag{5.213}$$

Now, truncating the Fourier expansion at the fundamental as before gives,

$$f_d(\omega Y cos(\omega t)) = a_0 + a_1 cos(\omega t) + b_1 sin(\omega t) \tag{5.214}$$

and further, restricting f_d to be an odd function yields, $a_0 = a_1 = 0$ and,

$$b_1 = \frac{1}{\pi}\int_0^{2\pi} f_d(\omega Y cos\,\theta) sin\,\theta d\theta \tag{5.215}$$

Defining the equivalent (effective, linearised) damping from,

$$f_d(\dot{y}) = c_{eq}\dot{y} \tag{5.216}$$

one then obtains,

$$f_d(\omega Y cos(\omega t)) = c_{eq}\omega Y cos(\omega t) = b_1 sin(\omega t) \tag{5.217}$$

giving finally,

$$c_{eq} = \frac{b_1}{\omega Y} = \frac{1}{\pi \omega Y} \int_0^{2\pi} f_d(\omega Y \cos\theta) \sin\theta d\theta \tag{5.218}$$

with a corresponding effective, or linearised *FRF*,

$$\Lambda(\omega) = \frac{1}{k - m\omega^2 + ic_{eq}\omega} \tag{5.219}$$

As discussed in Section 5.4.3, an interesting physical example of nonlinear damping is given by,

$$f_d(\dot{y}) = c_2 \dot{y}|\dot{y}| \tag{5.220}$$

which corresponds to the drag force experienced by bodies moving at high velocities in viscous fluids. The equivalent damping is given by,

$$c_{eq} = \frac{c_2}{\pi \omega Y} \int_0^{2\pi} \omega Y \cos\theta |\omega Y \cos\theta| \cos\theta d\theta = \frac{c_2 \omega Y}{\pi} \int_0^{2\pi} \cos^2\theta |\cos\theta| d\theta \tag{5.221}$$

and in order to evaluate this, it is necessary to split the integral to account for the $|\ |$ function, so that,

$$c_{eq} = \frac{2c_2\omega Y}{\pi} \int_0^{\pi/2} \cos^3\theta\, d\theta - \frac{c_2\omega Y}{\pi} \int_{\pi/2}^{3\pi/2} \cos^3\theta\, d\theta$$

$$= \frac{c_2\omega Y}{2\pi} \int_0^{\pi/2} (\cos 3\theta + 3\cos\theta) d\theta - \frac{c_2\omega Y}{4\pi} \int_{\pi/2}^{3\pi/2} (\cos 3\theta + 3\cos\theta) d\theta \tag{5.222}$$

After a little algebraic manipulation, this expression becomes,

$$c_{eq} = \frac{8c_2\omega Y}{3\pi} \tag{5.223}$$

Exercise 7. Evaluate the integrals needed to establish Eq. (5.223).

The analysis shows that the *FRF* for a simple oscillator with this type of quadratic damping is,

$$\Lambda(\omega) = \frac{1}{k - m\omega^2 + i\frac{8c_2\omega Y}{3\pi}\omega} \tag{5.224}$$

which appears to be the *FRF* of an undamped linear system,

$$\Lambda(\omega) = \frac{1}{k - m_{eq}\omega^2} \tag{5.225}$$

with complex mass,

$$m_{eq} = m + i\frac{8c_2Y}{3\pi} \tag{5.226}$$

This is an interesting phenomenon and a similar effect is exploited in the definition of *hysteretic damping* in the contexts of linear structural dynamics and modal analysis [5.23, 5.24]. Damping always manifests itself as the imaginary part of the *FRF* denominator. Depending on the frequency dependence of the term, it can sometimes be absorbed in a redefinition of one of the other parameters. If the damping has no dependence on frequency, a complex stiffness can be defined $k^* = k(i + i\eta)$ (where η is called the *loss factor*); this is hysteretic damping[4]. Polymers and viscoelastic materials appear to have damping with quite complicated frequency dependence and this makes their modelling quite demanding at times [5.29].

The analysis of systems with mixed nonlinear damping and stiffness presents no new difficulties. In fact in the case where the nonlinearity is *additively separable*, i.e.

$$m\ddot{y} + f_d(\dot{y}) + f_s(y) = X\sin(\omega t - \phi) \tag{5.227}$$

equations (5.201) and (5.218) still apply and the linearised *FRF* is,

$$\Lambda(\omega) = \frac{1}{k_{eq} - m\omega^2 + ic_{eq}\omega} \tag{5.228}$$

5.5.6 Two Systems of Particular Interest

In this section, two dynamic systems of some importance are studied, whose analysis by harmonic balance presents interesting subtleties.

Quadratic Stiffness

Consider the system specified by the equation of motion,

$$m\ddot{y} + c\dot{y} + ky + k_2y^2 = X\sin(\omega t - \phi) \tag{5.229}$$

If one naively follows the harmonic balance procedure in this case and substitutes the trial solution $y(t) = Y\sin(\omega t)$, one obtains,

$$-m\omega^2Y_1\sin(\omega t) + c\omega Y_1\cos(\omega t) + kY_1\sin(\omega t) + \frac{1}{2}k_2Y_1^2 - \frac{1}{2}k_2Y_1^2\cos(2\omega t) = X\sin(\omega t - \phi) \tag{5.230}$$

and equating the coefficients of the fundamentals leads to the *FRF* of the underlying linear system. The problem here is that the trial solution not only requires a higher-harmonic component for balance, it also needs a lower-order part – a *DC* term. If the trial solution $y(t) = Y_0 + Y_1\sin(\omega t)$ is adopted, one obtains after substitution in Eq. (5.229),

[4] See Chapter 1.

$$-m\omega^2 Y_1 sin(\omega t) + c\omega Y_1 cos(\omega t) + kY_0 + kY_1 sin(\omega t) +$$
$$k_2 Y_0^2 + 2k_2 Y_0 Y_1 sin(\omega t) + \frac{1}{2} k_2 Y_1^2 - \frac{1}{2} k_2 Y_1^2 cos(2\omega t) = X sin(\omega t - \phi) \tag{5.231}$$

Equating coefficients of sin and cos yields the *FRF*,

$$\Lambda(\omega) = \frac{1}{k + 2k_2 Y_0 - m\omega^2 + ic\omega} \tag{5.232}$$

so the effective undamped natural frequency is,

$$\omega_n = \sqrt{\frac{k + 2k_2 Y_0}{m}} \tag{5.233}$$

and a little more effort is needed in order to interpret this correctly.

Recalling the energy analysis discussed in Section 5.2.4, consider the potential energy function $V(y)$, corresponding to the stiffness $f_s(y) = ky + k_2 y^2$. As the restoring force is given by,

$$f_s = \frac{\partial V}{\partial y} \tag{5.234}$$

then,

$$V(y) = \int f_s(y)\, dy = \frac{1}{2} k y^2 + \frac{1}{3} k_2 y^3 \tag{5.235}$$

Now, if $k_2 > 0$, a function is obtained like that shown in Figure 5.45. Note that if the forcing places the system beyond point A on this curve, the system falls into an infinitely deep potential well, i.e. escapes to $-\infty$. For this reason, the system must be considered unstable

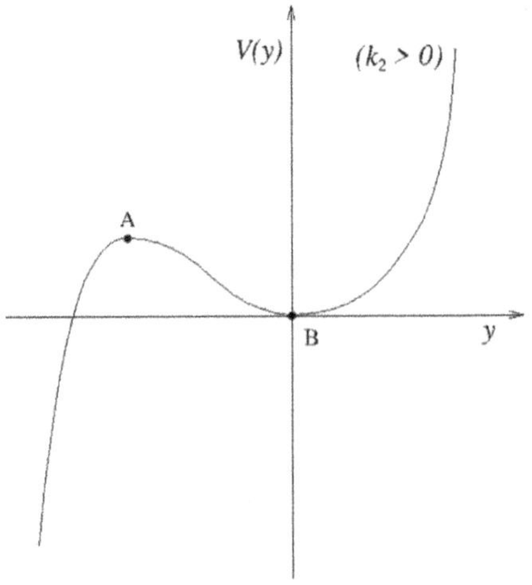

Figure 5.45 Potential energy function for a quadratic stiffness nonlinearity with $k_2 > 0$.

except at low amplitudes where the linear term dominates and always returns the system to the stable equilibrium at B. In any case, if the motion remains bounded, less energy is required to maintain negative displacements, so the mean operating point $Y_0 < 0$. This offset means that the product $k_2 Y_0 < 0$. Alternatively, if $k_2 < 0$, a potential curve as in Figure 5.46, arises. The system is again unstable for high enough excitation, with escape this time to ∞. However, in this case, $Y_0 > 0$; so $k_2 Y_0 < 0$ again.

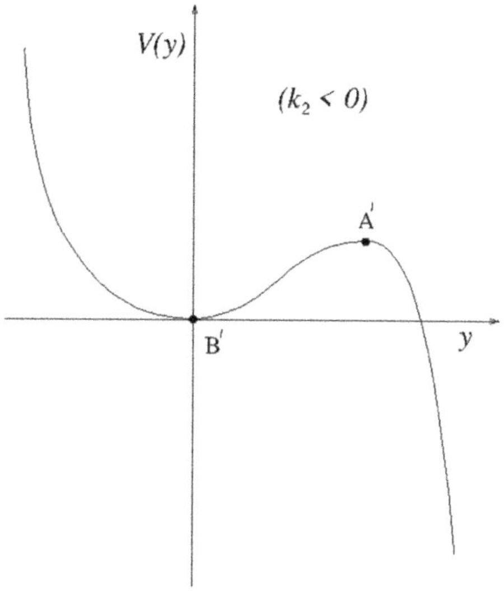

Figure 5.46 Potential energy function for a quadratic stiffness nonlinearity with $k_2 < 0$.

This result indicates that the effective natural frequency for this system (given in (5.233)) *always decreases* with increasing excitation; i.e. the system is always softening, independently of the sign of k_2. This is in contrast to the situation for cubic systems.

Although one cannot infer jumps from the *FRF* at this level of approximation, they are found to occur, always below the linear natural frequency as shown in Figure 5.47 which is computed from a simulation – the numerical equivalent of a stepped-sine test. The equation of motion for the simulation was (5.229) with parameter values $m = 1$, $c = 20$, $k = 10^4$ and $k_2 = 10^7$.

Because of the unstable nature of the pure quadratic, 'Second-order' behaviour is sometimes modelled with a term of the form $k_2 y|y|$. The *FRF* for a system with this nonlinearity is given by,

$$\Lambda(\omega) = \frac{1}{k + \frac{8k_2 Y}{3\pi} - m\omega^2 + ic\omega} \tag{5.236}$$

and the bifurcation analysis is similar to that in the cubic case, but a little more complicated as the equation for the response amplitude is a quartic,

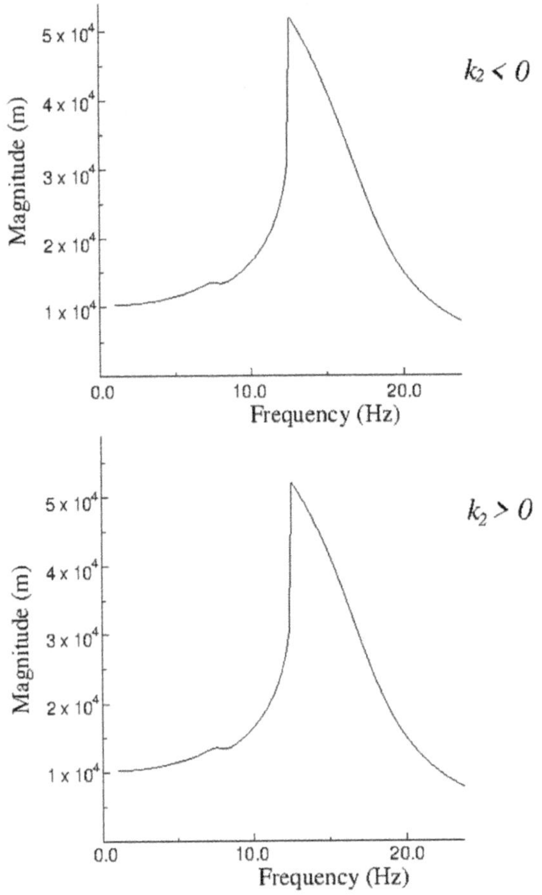

Figure 5.47 Harmonic balance *FRF*s for systems with quadratic stiffness showing jump phenomena.

$$X^2 = Y^2\left(\left[k + \frac{8k_2Y}{3\pi} - m\omega^2\right]^2 + c^2\omega^2\right) \tag{5.237}$$

Bilinear Stiffness

Another system which is of physical interest is that with bilinear stiffness function of the form (Figure 5.48),

$$f_s(y) = \begin{cases} k \quad if \quad y < y_c \\ k'y + (k - k')y_c \quad if \quad y \geq y_c \end{cases} \tag{5.238}$$

Without loss of generality, one can specify that $y_c > 0$. Although the equivalent stiffness is given by Eq. (5.201), there is again a slight subtlety here, the integrand changes when the displacement $Y sin(\omega t)$ exceeds y_c. This corresponds to a point in the cycle $\theta_c = \omega t_c$ where,

$$\theta_c = sin^{-1}\left(\frac{y_c}{Y}\right) \tag{5.239}$$

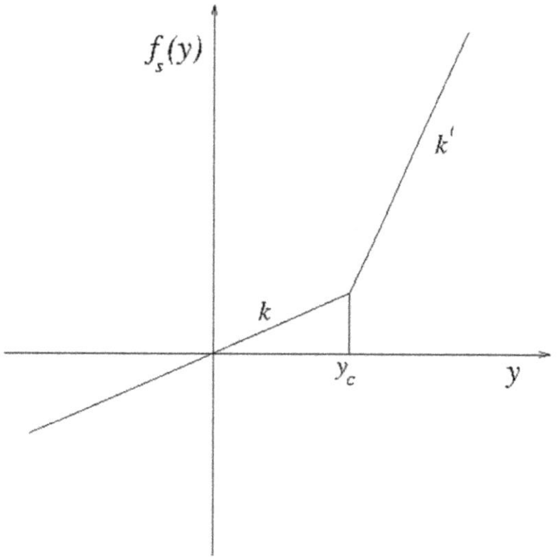

Figure 5.48 Bilinear stiffness characteristic with offset discontinuity.

The integrand switches back when $\theta = \pi - \theta_c$. A little thought shows that the equivalent stiffness must have the form,

$$k_{eq} = k + \frac{(k' - k)}{\pi} \int_{\theta_c}^{\pi-\theta_c} \sin\theta \left(\sin\theta - \frac{y_c}{Y} \right) d\theta \tag{5.240}$$

so that, after a little algebra,

$$k_{eq} = k + \frac{(k' - k)}{2\pi} \left(\pi - 2\theta_c + \sin 2\theta_c - \frac{4y_c}{Y} \cos\theta_c \right) \tag{5.241}$$

or,

$$k_{eq} = k + \frac{(k' - k)}{2\pi} \left(\pi - 2\sin^{-1}\left(\frac{y_c}{Y}\right) + \sin\left(2\sin^{-1}\left(\frac{y_c}{Y}\right)\right) - \frac{4y_c}{Y} \cos\left(\sin^{-1}\left(\frac{y_c}{Y}\right)\right) \right) \tag{5.242}$$

As a check, substituting $k = k'$ or $Y = y_c$ yields $k_{eq} = k$ as necessary. The effective or linearised *FRF* has the form,

$$\Lambda(\omega) = \frac{1}{k + \frac{(k' - k)}{2\pi} \left(\pi - 2\sin^{-1}\left(\frac{y_c}{Y}\right) - \frac{2y_c}{Y}\sqrt{Y^2 - y_c^2} \right) - m\omega^2 + ic\omega} \tag{5.243}$$

Now, let $y_c = 0$ (Figure 5.49). The expression (5.242) collapses to,

$$k_{eq} = \frac{1}{2}(k + k') \tag{5.244}$$

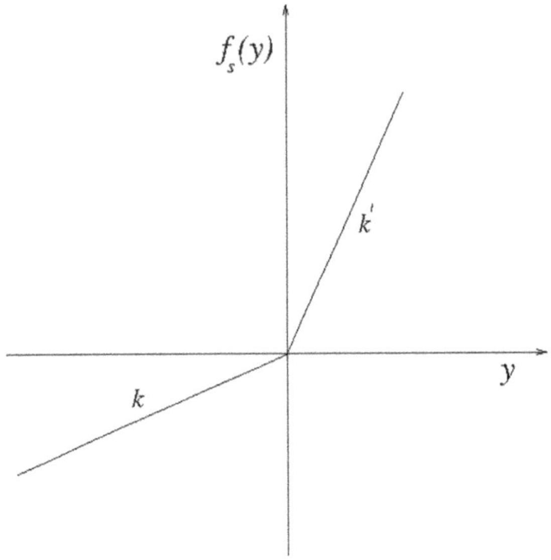

Figure 5.49 Bilinear stiffness characteristic with discontinuity at the origin.

which is simply the average stiffness. So the system has an effective natural frequency and *FRF*, independent of the size of Y and therefore, independent of X. The system is thus *homogeneous* as described in Section 5.3.3; the homogeneity test would fail to detect that this system is nonlinear. That it is nonlinear is manifest; the Fourier expansion of $f_s(y)$ (Figure 5.49) contains *all* harmonics so the response of the system to a sinusoid will also contain all harmonics. The homogeneity of this system is a consequence of the fact that the stiffness function looks the same at all length scales. This analysis is only first-order; however, Figure 5.50 shows *FRF*s for different levels of excitation for the simulated system,

$$\ddot{y}+20\dot{y}+10^4 y+4\times10^4 y\Theta(y)=X sin(30t) \tag{5.245}$$

where Θ is the Heaviside or step function.

The curves overlay and this demonstrates why homogeneity is a *necessary but not sufficient* condition for linearity.

5.5.7 *Statistical Linearisation

In summary, the idea of linearisation – in the case of the *SDOF* systems discussed so far – is to find an *equivalent* linear *FRF*,

$$H_{eq}(\omega)=\frac{1}{-m_{eq}\omega^2+ic_{eq}\omega+k_{eq}} \tag{5.246}$$

which approximates most closely (in some sense) that of the nonlinear system or structure of interest. In the time-domain this implies a best linear model of the form,

$$m_{eq}\ddot{y}+c_{eq}\dot{y}+k_{eq}y=x(t) \tag{5.247}$$

and such a model is again a *linearisation*.

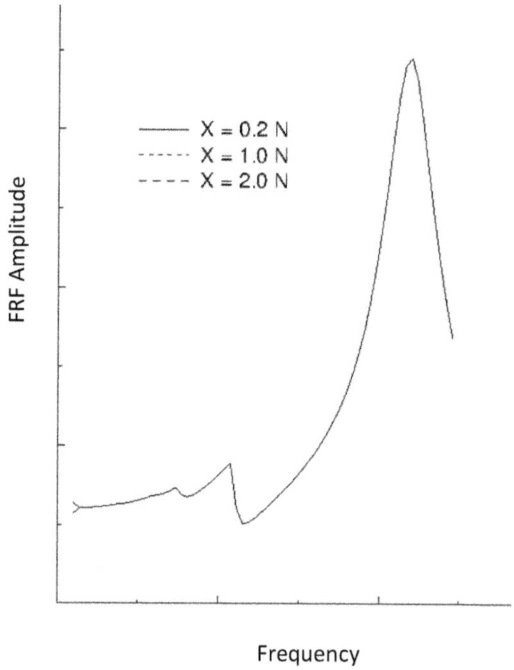

Figure 5.50 Harmonic balance *FRF*s for homogeneous bilinear system at different forcing levels.

As the nonlinear system *FRF* will usually change its shape as the level of excitation is changed, any linearisation is only valid for a given level. Also, because the form of the *FRF* is a function of the *type* of excitation as discussed in the previous section, different forcing types of nominally the same amplitude will require different linearisations. These are clear limitations.

So far, all of the linearisations calculated have been for harmonic excitations; however, other types of excitation may prove more convenient for a given test, in particular, random excitation has many things to recommend it. For this reason, in this section, linearisations based on random excitation will be discussed. This section is starred because it demands a high degree of familiarity with probability theory, which is not required elsewhere in this chapter.

Linearisations from random vibration are arguably more fundamental because it is possible to show that random excitation is the only type which generates nonlinear system *FRFs* which *look* like linear system *FRFs*.

Theory

The basic theory presented here does not proceed via the *FRFs*; one operates directly on the equations of motion. The technique – *equivalent* or more accurately *statistical linearisation* – dates back to the fundamental work of Caughey [5.30]. The following discussion is limited to *SDOF* systems; however, this is not a fundamental restriction of the method.

Given a general *SDOF* nonlinear system,

$$m\ddot{y}+f(y,\dot{y})=x(t) \tag{5.248}$$

one seeks an equivalent linear system of the form (5.247). As the excitation is random, an apparently sensible strategy would be to minimise the average difference between the nonlinear restoring force and the linear system (it will be assumed that the apparent mass is unchanged, i.e. $=m$), i.e. find the c_{eq} and k_{eq} which minimise,

$$J_1\left(y, c_{eq}, k_{eq}\right)=E\left[f(y,\dot{y})-c_{eq}\dot{y}-k_{eq}y\right] \tag{5.249}$$

In fact this strategy is not sensible as the differences will generally be a mixture of negative and positive and could still average to zero for a wildly-inappropriate system. The correct strategy is to minimise the expectation (mean) of the squared differences i.e.

$$J_2\left(y, c_{eq}, k_{eq}\right)=E\left[\left(f(y,\dot{y})-c_{eq}\dot{y}-k_{eq}y\right)^2\right] \tag{5.250}$$

or,

$$J_2\left(y, c_{eq}, k_{eq}\right)=E\left[f(y,\dot{y})^2+c_{eq}^2\dot{y}^2+k_{eq}^2y^2-2f(y,\dot{y})c_{eq}\dot{y}-2f(y,\dot{y})k_{eq}y+2c_{eq}k_{eq}y\dot{y}\right] \tag{5.251}$$

so this is a type of *least-squared error* approach.

Now, using elementary calculus, the values of c_{eq} and k_{eq} which minimise Eq. (5.251) are those which satisfy the equations,

$$\frac{\partial J_2}{\partial c_{eq}}=\frac{\partial J_2}{\partial k_{eq}}=0 \tag{5.252}$$

The first of these equations yields,

$$E\left[c_{eq}\dot{y}^2-\dot{y}f(y,\dot{y})+k_{eq}y\dot{y}\right]=c_{eq}E\left[\dot{y}^2\right]-E\left[\dot{y}f(y,\dot{y})\right]+k_{eq}E\left[y\dot{y}\right]=0 \tag{5.253}$$

and the second,

$$E\left[k_{eq}y^2-yf(y,\dot{y})+c_{eq}y\dot{y}\right]=k_{eq}E\left[y^2\right]-E\left[yf(y,\dot{y})\right]+c_{eq}E\left[y\dot{y}\right]=0 \tag{5.254}$$

after using the linearity of the expectation operator.

Now, it is a basic theorem of stochastic processes that $E[y\dot{y}]=0$ for a wide range of processes[5]. With this assumption, equations (5.253) and (5.254) become,

$$c_{eq}=\frac{E\left[\dot{y}f(y,\dot{y})\right]}{E\left[\dot{y}^2\right]} \tag{5.255}$$

---

[5] The proof is elementary and depends on the processes being *stationary*, i.e. that the statistical moments of $x(t)$, mean, variance, etc. do not vary with time. With this assumption,

$$\frac{d\sigma_y^2}{dt}=0=\frac{d}{dt}E\left[y^2\right]=E\left[\frac{dy^2}{dt}\right]=2E\left[y\dot{y}\right]$$

and,

$$k_{eq} = \frac{E\left[yf(y,\dot{y})\right]}{E\left[y^2\right]} \tag{5.256}$$

Now, all that remains is to evaluate the expectations. Unfortunately this turns out to be nontrivial. The expectation of a function of random variables like $f(y,\dot{y})$ is given by,

$$E\left[f(y,\dot{y})\right] = \int_{-\infty}^{\infty}\int_{-\infty}^{\infty} p(y,\dot{y}) f(y,\dot{y})\, dy d\dot{y} \tag{5.257}$$

where $p(y,\dot{y})$ is the Probability Density Function (*PDF*) for the processes y and $\dot{y}$. The problem is that the *PDF* of the response is not known for general nonlinear systems, estimating it presents formidable problems of its own.

The solution to this problem is to approximate $p(y,\dot{y})$ by $p_{eq}(y,\dot{y})$ – the *PDF* of the equivalent linear system (5.247); however, even this still requires a little thought. The fact that comes to the rescue is a basic theorem of random vibrations of *linear* systems [5.31], namely: if the excitation to a linear system is a zero-mean Gaussian signal, then so is the response. To say that $x(t)$ is Gaussian zero-mean is to say that it has the *PDF*,

$$p(x) = \frac{1}{\sqrt{2\pi}\sigma_x}\exp\left(-\frac{x^2}{2\sigma_x^2}\right) \tag{5.258}$$

where σ_x^2 is the variance of the process $x(t)$.

The theorem states that the *PDFs* of the responses are Gaussian also, so,

$$p_{eq}\left(y_{eq}\right) = \frac{1}{\sqrt{2\pi}\sigma_{y_{eq}}}\exp\left(-\frac{y_{eq}^2}{2\sigma_{y_{eq}}^2}\right) \tag{5.259}$$

and,

$$p_{eq}\left(\dot{y}_{eq}\right) = \frac{1}{\sqrt{2\pi}\sigma_{\dot{y}_{eq}}}\exp\left(-\frac{\dot{y}_{eq}^2}{2\sigma_{\dot{y}_{eq}}^2}\right) \tag{5.260}$$

so the joint *PDF* is,

$$p_{eq}\left(y_{eq},\dot{y}_{eq}\right) = p_{eq}\left(y_{eq}\right)p_{eq}\left(\dot{y}_{eq}\right) = \frac{1}{\sqrt{2\pi}\,\sigma_{y_{eq}}\sigma_{\dot{y}_{eq}}}\exp\left(-\frac{y_{eq}^2}{2\sigma_{y_{eq}}^2} - \frac{\dot{y}_{eq}^2}{2\sigma_{\dot{y}_{eq}}^2}\right) \tag{5.261}$$

In order to make use of these results it will be assumed from now on that $x(t)$ is zero-mean Gaussian.

Matters can be simplified further by assuming that the nonlinearity is separable, i.e. the equation of motion takes the form,

$$m\ddot{y}+c\dot{y}+ky+\phi(\dot{y})+\psi(y)=x(t) \tag{5.262}$$

in this case, $f(y,\dot{y})=c\dot{y}+ky+\phi(\dot{y})+\psi(y)$.

Eq. (5.255) becomes,

$$c_{eq}=\frac{E\left[\dot{y}\left(c\dot{y}+ky+\phi(\dot{y})+\psi(y)\right)\right]}{E\left[\dot{y}^2\right]} \tag{5.263}$$

or, using the linearity of E,

$$c_{eq}=\frac{cE\left[\dot{y}^2\right]+kE\left[\dot{y}y\right]+E\left[\dot{y}\phi(\dot{y})\right]+E\left[\dot{y}\psi(y)\right]}{E\left[\dot{y}^2\right]} \tag{5.264}$$

which reduces to,

$$c_{eq}=c+\frac{E\left[\dot{y}\phi(\dot{y})\right]+E\left[\dot{y}\psi(y)\right]}{E\left[\dot{y}^2\right]} \tag{5.265}$$

and a similar analysis based on (5.256) gives,

$$k_{eq}=k+\frac{E\left[y\phi(\dot{y})\right]+E\left[y\psi(y)\right]}{E\left[y^2\right]} \tag{5.266}$$

Now, consider the term $E\left[y\phi(\dot{y})\right]$ in Eq. (5.266); this is given by,

$$E\left[y\phi(\dot{y})\right]=\int_{-\infty}^{\infty}\int_{-\infty}^{\infty}p_{eq}(y,\dot{y})\,y\phi(\dot{y})\,dy\,d\dot{y} \tag{5.267}$$

and because the *PDF* factors, i.e. $p_{eq}(y_{eq},\dot{y}_{eq})=p_{eq}(y_{eq})p_{eq}(\dot{y}_{eq})$, so does the integral, hence,

$$E\left[y\phi(\dot{y})\right]=\left(\int_{-\infty}^{\infty}p_{eq}(y)\,y\,dy\right)\left(\int_{-\infty}^{\infty}p_{eq}(\dot{y})\phi(\dot{y})\,d\dot{y}\right)=E[y]E\left[\phi(\dot{y})\right] \tag{5.268}$$

but the response is zero-mean Gaussian by the theorem cited earlier and therefore $E[y]=0$. It follows that $E\left[y\phi(\dot{y})\right]=0$ and therefore Eq. (5.266) becomes,

$$k_{eq}=k+\frac{E\left[y\psi(y)\right]}{E\left[y^2\right]} \tag{5.269}$$

and a similar analysis for Eq. (5.265) yields,

$$c_{eq} = c + \frac{E\left[\dot{y}\phi(\dot{y})\right]}{E\left[\dot{y}^2\right]} \tag{5.270}$$

Now, assuming that the expectations are taken with respect to the linear system *PDFs* from equations (5.259) and (5.260), Eq. (5.270) becomes,

$$c_{eq} = c + \frac{1}{\sqrt{2\pi\sigma_{\dot{y}_{eq}}^3}} \int_{-\infty}^{\infty} \dot{y}\phi(\dot{y}) \exp\left(-\frac{\dot{y}^2}{2\sigma_{\dot{y}_{eq}}^2}\right) d\dot{y} \tag{5.271}$$

and Eq. (5.269) becomes,

$$k_{eq} = k + \frac{1}{\sqrt{2\pi\sigma_{y_{eq}}^3}} \int_{-\infty}^{\infty} y\psi(y) \exp\left(-\frac{y^2}{2\sigma_{y_{eq}}^2}\right) dy \tag{5.272}$$

which are the final forms required. Although it may now appear that the problem has been reduced to the evaluation of integrals, unfortunately things are still not quite that simple. It remains to estimate the variances in the integrals. Now, standard theory (see [5.32]) gives,

$$\sigma_{y_{eq}}^2 = \int_{-\infty}^{\infty} \left|H_{eq}(\omega)\right|^2 S_{xx}(\omega)\, d\omega = \int_{-\infty}^{\infty} \frac{S_{xx}(\omega)}{\left(k_{eq} - m\omega^2\right)^2 + c_{eq}^2\omega^2} d\omega \tag{5.273}$$

and

$$\sigma_{\dot{y}_{eq}}^2 = \int_{-\infty}^{\infty} \frac{\omega^2 S_{xx}(\omega)}{\left(k_{eq} - m\omega^2\right)^2 + c_{eq}^2\omega^2} d\omega \tag{5.274}$$

and here lies the final problem. Eq. (5.272) expresses k_{eq} in terms of the variance $\sigma_{y_{eq}}^2$ and (5.273) expresses $\sigma_{y_{eq}}^2$ in terms of k_{eq}; the result is a rather nasty pair of coupled nonlinear algebraic equations which must be solved for k_{eq}. The same is of course true of c_{eq}. In order to see how progress can be made, it is useful to move to a concrete example.

Application to Duffing's Equation

The equation of interest is going to be the symmetric Duffing equation, so as usual,

$$\psi(y) = k_3 y^3 \tag{5.275}$$

and the expression for the effective stiffness, from Eq. (5.272) is,

$$k_{eq} = k + \frac{k_3}{\sqrt{2\pi\sigma_{y_{eq}}^3}} \int_{-\infty}^{\infty} y^4 \exp\left(-\frac{y^2}{2\sigma_{y_{eq}}^2}\right) dy \tag{5.276}$$

In order to obtain a tractable expression for the variance from Eq. (5.273) it will be assumed that $x(t)$ is a *white* zero-mean Gaussian signal, i.e. $S_{xx}(\omega)=P$ a constant; it is then a standard result that [5.32],

$$\sigma_{y_{eq}}^2 = P\int_{-\infty}^{\infty} \frac{1}{\left(k_{eq}-m\omega^2\right)^2 + c_{eq}^2\omega^2}\,d\omega = \frac{\pi P}{c\,k_{eq}} \tag{5.277}$$

This expression then results in,

$$k_{eq} = k + \frac{k_3}{\sqrt{2\pi}\left(\dfrac{\pi P}{ck_{eq}}\right)^{3/2}}\int_{-\infty}^{\infty} y^4 \exp\left(-\frac{ck_{eq}y^2}{2\pi P}\right)dy \tag{5.278}$$

Now, making use of the standard integral result[6],

$$\int_{-\infty}^{\infty} y^4 \exp\left(-ay^2\right)dy = \frac{3\pi^{1/2}}{4a^{5/2}} \tag{5.279}$$

gives,

$$k_{eq} = k + \frac{3\pi k_3 P}{ck_{eq}} \tag{5.280}$$

and the required k_{eq} satisfies the quadratic equation,

$$ck_{eq}^2 - ckk_{eq} - 3\pi k_3 P = 0 \tag{5.281}$$

The desired root is (after a little algebra),

$$k_{eq} = \frac{k}{2} + \frac{k}{2}\sqrt{1+\frac{12\pi k_3 P}{ck^2}} \tag{5.282}$$

which shows the expected behaviour, i.e. k_{eq} increases if P or k_3 increase. If k_3P is small, the binomial approximation gives,

$$k_{eq} = k + \frac{3\pi k_3 P}{ck} + O\left(k_3^2 P^2\right) \tag{5.283}$$

To illustrate the use of Eq. (5.282), the parameters $m=1$, $c=20$, $k=10^4$ and $k_3=5\times10^9$ were chosen for the Duffing oscillator. Time data under random excitation were simulated using

---

[6] Integrals of the type $\int_{-\infty}^{\infty} y^n \exp\left(-ay^2\right)dy$ occur fairly often in the equivalent linearisation of polynomial nonlinearities. Fortunately, they are fairly straightforward to evaluate using recursion formulae [5.9].

the fourth-order Runge-Kutta and the *FRFs* were estimated by standard spectral methods. Figure 5.51 shows the linearised *FRF* with k_{eq} given by (5.282) with levels of excitation corresponding to $P = 0$, 0.01 and 0.02. The values of k_{eq} found are respectively 10000.0, 11968.6 and 13492.5, giving equivalent natural frequencies of $\omega_n =$ 100.0, 109.4 and 116.2.

In order to validate this result, the linearised *FRF* for $P = 0.02$ is compared to the *FRF* estimated from the full nonlinear system in Figure 5.52.

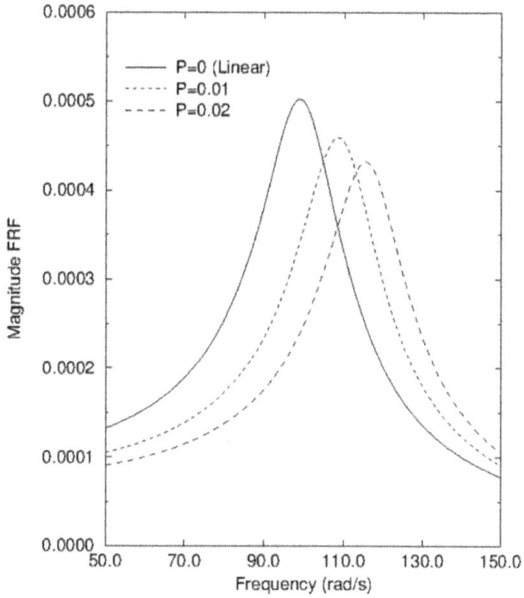

Figure 5.51 *FRFs* from equivalent linearisation of Duffing oscillator for different white noise excitation levels.

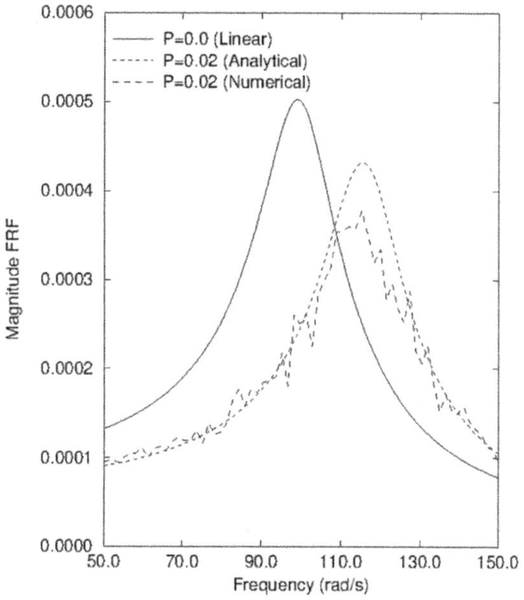

Figure 5.52 *FRF* from equivalent linearisation of Duffing oscillator for white noise excitation level $P = 0.02$; comparison with numerical simulation.

5.6 Chaos

The most spectacular manifestation of nonlinearity is *chaos*. This phenomenon could quite happily have been discussed in Section 5.3, as it is very much a signature of nonlinearity – chaos cannot occur in linear systems. However, it is more convenient to end with the subject as the discussion touches on various matters covered earlier. The phenomenon was discovered in various contexts across the twentieth century and is covered by a huge literature, which includes many authoritative textbooks. A good introductory text is [5.33] and a much deeper treatment can be found in [5.20]. There are many characteristics of chaos, some of which are more concerned with deep mathematics and some which are pertinent for Engineering, including apparent random response of systems to periodic forcing and attracting sets of fractal dimension [5.34]. Perhaps the most fundamental aspect of chaos, and the one considered first here is *sensitive dependence on initial conditions*. This phenomenon will be illustrated here using the chaotic oscillator whose properties were explored in [5.35]. The system is a variant of Duffing's equation,

$$m\ddot{y}+c\dot{y}+k_3y^3 = X\cos(\omega t) \tag{5.284}$$

but note the absence of the linear term.

Figure 5.53 shows a segment of a displacement time-history for this system. The parameter values for the simulation were $m=1$, $c=0.05$, $k_3=1$, $X=7.5$ and $\omega=1$. The system was integrated forward from initial conditions $y_0=3$, $\dot{y}_0=4$ using a fourth-order Runge-Kutta scheme with a sampling frequency set to give 64 points per cycle at $\omega=1$.

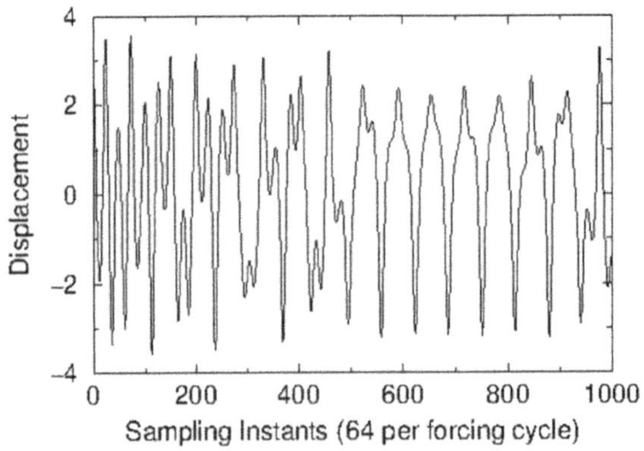

Figure 5.53 Segment of the forced response from the Ueda oscillator.

As mentioned above, a signature of chaos is that a harmonic forcing produces an apparently random response; in this case, the aperiodic nature of the response is clear. The chaotic nature is a little more difficult to establish because one might simply be seeing a complicated pattern of harmonics; however, one can make use of a useful tool here – the *Poincaré map*. This type of map is basically a phase portrait *stroboscopically sampled* at the forcing frequency. The map is constructed simply by sampling from the response *once per forcing cycle* and plotting in the phase plane. Now, for a linear system (after transients have decayed), the response will be periodic with the same period as the forcing, so the Poincaré map will be a *single point*. For a nonlinear system where the response has superharmonics, the Poincaré map will still be a single point. In fact, interesting structure will only appear if the system of interest is strongly nonlinear. Consider what will happen if a subharmonic of order two is present; in this case the period of the response will be twice that of the forcing and sampling at the forcing frequency will produce two distinct points in the Poincaré map. In general, an nth-order subharmonic will produce a Poincaré map with n distinct points. Now, suppose that the response to the periodic force could be *apparently random*; in this case the samples for the Poincaré map would never repeat and a situation like that in Figure 5.54 would occur.

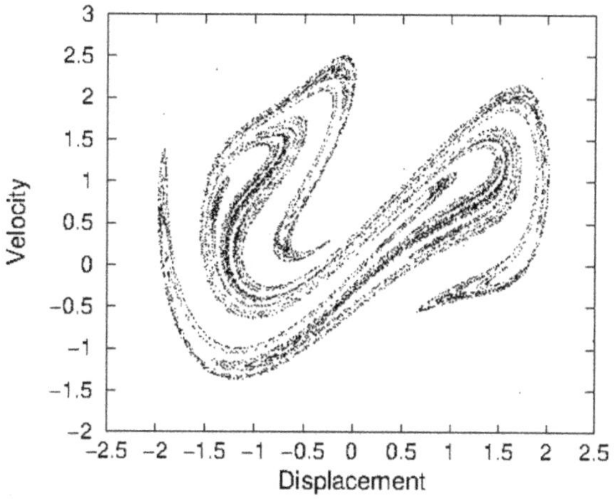

Figure 5.54 Poincaré's map from the Ueda oscillator in Eq. (5.284).

Figure 5.54 shows the Poincaré map for the Ueda oscillator. There is no repetition, so the response is apparently random; however, there does appear to be regularity – there is a *pattern* – which one would not expect to see if the sampling were truly random. To discuss this issue in detail would be too much of a digression, but a brief explanation is as follows. When a system is excited with a periodic force, one expects a pattern in the response; one expects to see the same frequencies, or harmonics. This is related to the idea of an *attractor*, which is the expected pattern in the phase plane. For a linear system, one would always see an ellipse in the forced response, or an equilibrium in the free motion. The point is that, whatever the initial conditions, after the transients have died away, the phase trajectory is *attracted* to a specific set of points; this set is called the attractor for the dynamics. In both cases discussed here – periodic motion or fixed point – the attractor as viewed in the Poincaré map will be a single point. This idea also applies to limit cycles for nonlinear systems, as discussed in Section 5.2. 3 for the Van der Pol oscillator; a limit cycle is another type of attractor. Now, all of these attractors have well-defined *dmensions*, considered as sets; in the phase plane, periodic motions appear as lines (dimension one) and equilibria appear as points (dimension zero); in the Poincaré map, both attractors appear as points. A peculiar aspect of the chaotic motions is that they generate attractors which do not have integer dimensions! Such sets are commonly called *fractals* (sets of fractional dimension); in this case the attractors are called *strange*. Figure 5.54 shows the strange attractor for the Ueda oscillator. A good introductory treatment of fractals and chaos is [5.36]. One of the ways to detect chaos is to construct the attractor and establish that it has fractional dimension; this is tricky in practice, but can be done [5.37].

In practice, it will not be possible to construct the attractor from a measured time-series, unless one can simultaneously measure displacement and velocity or very accurately estimate one from the other. However, it is possible to construct something which is representative of its structure by the method of *embedding* [5.37]. Instead of plotting the phase portrait from the variables and their derivatives, *delayed* values of the variables are used. The actual process of determining the dimension of the embedding and the order of delay requires a little care [5.36], but Figure 5.55 shows a naive embedding derived by taking the delay for the Ueda system as one forcing cycle. Only the first two delay coordinates are shown for visualisation purposes. The attractor appears folded; however, the fractal nature is again visible. The representation shown can be proved to be topologically equivalent to that given in Figure 5.54, i.e. the fractal dimension is the same.

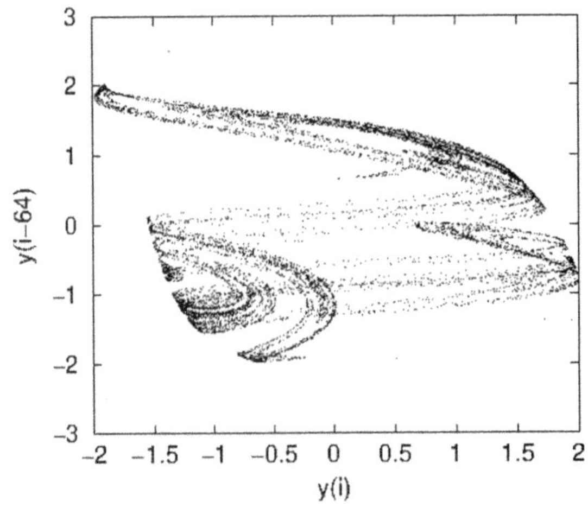

Figure 5.55 Reconstruction of Ueda's attractor by embedding.

Now, suppose the initial conditions for the simulation here are perturbed to $y_0 = 3.003$, $\dot{y}_0 = 4.004$. This corresponds to the minimal error in recording to ten-bit accuracy. One might imagine an experiment in system identification which gave perfect estimates of the parameters, but slightly flawed initial conditions. The resulting displacement time-history is shown in Figure 5.56 (dotted) with the original time-series (solid) for comparison. The response is sensitive to the initial conditions to the extent that the two signals begin to diverge after approximately 400 samples. This divergence is actually *exponential*, so all agreement with the measured (first) data is quickly lost. Although the model is perfect, a mean-square error metric would indicate complete disagreement of the time-series. However, if one considers the Poincaré map (Figure 5.57), there is excellent agreement between the perturbed and unperturbed systems; this is because the system response is always captured by the attractor, independent of the initial conditions. The time-series diverge because the responses are moving on different parts of the attractor. This raises the possibility of using the attractor itself, or topological invariants such as the dimension in order to validate models – this would be a practical use of the theory of chaotic dynamics. A single number – the dimension – would be too coarse; however, there are other invariants of the motion such as the spectrum of *Lyapunov exponents* which have higher dimension and could be used [5.36, 5.38]. Lyapunov exponents are basically estimates of the rate of exponential divergence of two phase trajectories; if a system is not chaotic, the two trajectories will *converge* on a simple attractor, so the exponent will be negative; if a system is chaotic, the exponents will be positive. As one might expect, the reconstruction of the attractor using the delay coordinates also looks identical to the original, so it is not shown here.

In case the reader is starting to suspect that the only interesting dynamical system is the Duffing oscillator, it is reassuring to look at some other systems which exhibit chaos and have strange attractors. As a first example, one can consider the Henon discrete map [5.39]; this example shows nicely that quite simple *discrete* dynamical systems can also show very complex behaviour. The equations of motion considered here (choice of parameters), correspond to,

$$\begin{aligned} x_{i+1} &= 1.4 + 0.3y_i - x_i^2 \\ y_{i+1} &= x_i \end{aligned} \tag{5.285}$$

and 200 points of the series are given in Figure 5.58. The response is clearly not periodic. For a discrete system, the natural way to plot the Poincaré map is to use the sampling interval as the stroboscope; plotting the states x_n and y_n against each other results in the strange attractor shown in Figure 5.59.

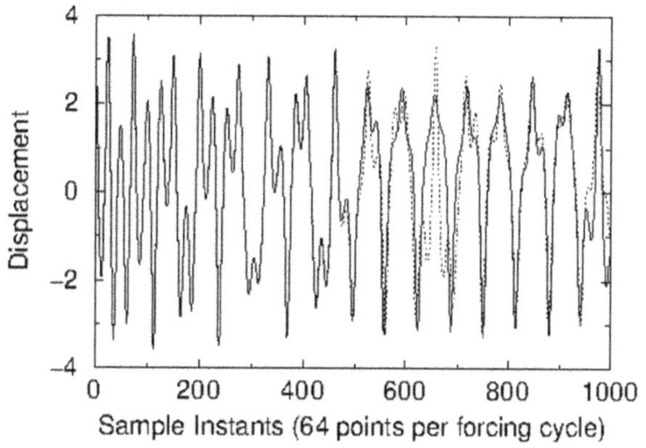

Figure 5.56 Comparison of time-series from Ueda's system and identical system with perturbed intial conditions.

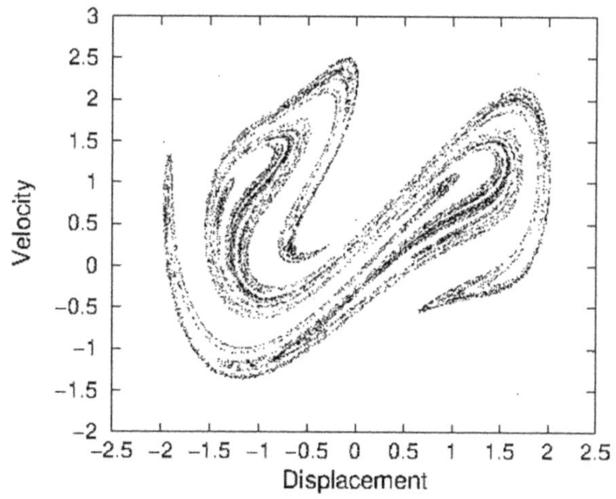

Figure 5.57 Poincaré's map from the slightly perturbed Ueda oscillator.

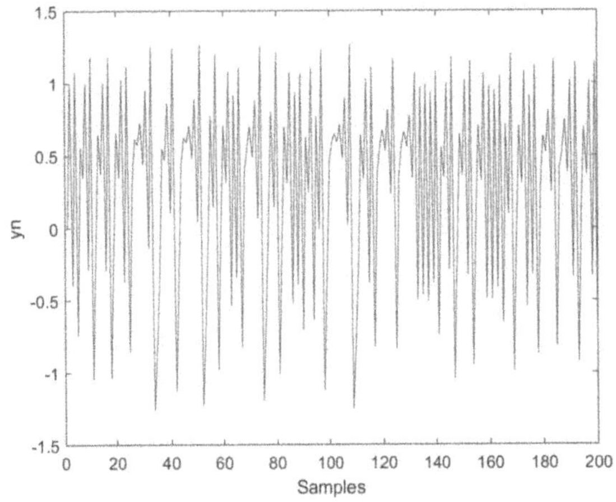

Figure 5.58 Samples from Henon system response.

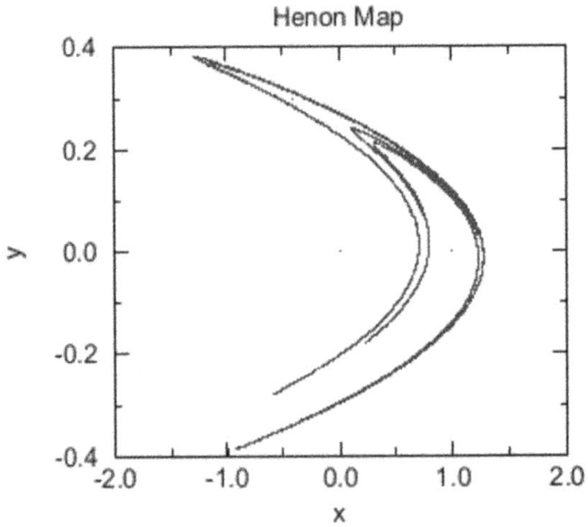

Figure 5.59 Strange attractor for the Henon system.

The final system discussed here will be the Lorenz system, discussed in the introduction to this chapter, but with parameters taken from [5.38],

$$\begin{aligned}\dot{x} &= 16.0(y+z)\\ \dot{y} &= x(45.92-z)-y\\ \dot{z} &= xy-4.0z\end{aligned} \tag{5.286}$$

In this case the attractor is plotted in the (x, y, z) phase-space, as shown in Figure 5.60; the result is quite pretty and the attractor is sometimes referred to as the 'Lorenz butterfly'.

Exercise 8. Using the simulate function from Appendix C, or otherwise, reproduce the figures for all the attractors in this section.

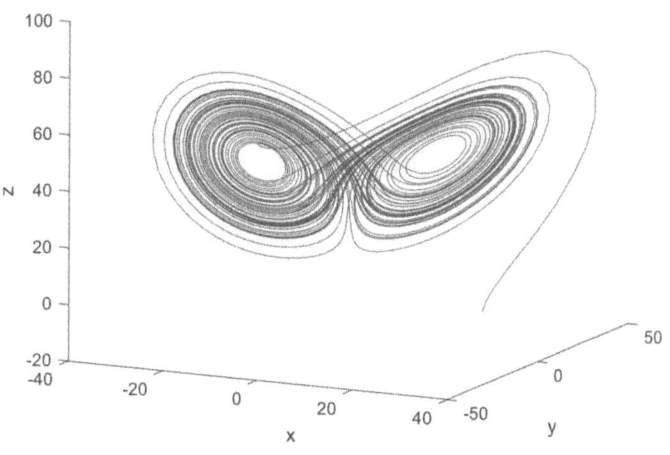

Figure 5.60 Attractor for the Lorenz system.

References

[5.1] Ryder, L. H., *Quantum Field Theory*, Cambridge University Press, 1985.

[5.2] Wagg, D., Neild, S., *Nonlinear Vibration with Control: For Flexible and Adaptive Structures*, Springer, 2nd Ed., 2014.

[5.3] Krämer, A., Kretzschmar, M., Krickeberg, K., *Modern Infectious Disease Epidemiology: Concepts, Methods, Mathematical Models, and Public Health*, Springer, 2010.

[5.4] Simmons, G. F., *Differential Equations with Applications and Historical Notes*, CRC Press, 3rd Ed., 2017.

[5.5] Kirkman, R., Moore, T., Adlard, C., *The Walking Dead*, Image Comics, 2009 – 2019.

[5.6] Munz, P., Hudea, I., Imad, J., Smith, R. J., *When Zombies Attack!: Mathematical Modelling of an Outbreak of Zombie Infection*, CRC Press, 3rd Ed., 2017.

[5.7] Carey, M. C., *The Girl with all the Gifts*, Orbit, 2014.

[5.8] Lorenz, E. N., *Deterministic Nonperiodic Flow*, Journal of the Atmospheric Sciences, 20, pp. 130–141, 1963.

[5.9] Worden, K., Tomlinson, G. R., *Nonlinearity in Structural Dynamics: Detection, Modelling and Identification*, Institute of Physics Press, 2001.

[5.10] McLachlan, N. W., *Ordinary Non-Linear Differential Equations in Engineering and Physical Sciences*, Oxford University Press, 1950.

[5.11] Greenhill, A. G., *The Applications of Elliptic Functions*, London, 1893.

[5.12] Nayfeh, A. H., *Perturbation Methods*, Wiley, 1973.

[5.13] Nayfeh, A. H., Mook, D. T., *Nonlinear Oscillations*, Wiley Interscience, 1979.

[5.14] Fox, L., *An Introduction to Numerical Linear Algebra*, Monographs on Numerical Analysis, Clarendon Press, Oxford, 1964.

[5.15] Acton, F. S., *Numerical Methods that Work*, The Mathematical Association of America, 1997.

[5.16] Press, W. H., Flannery, B. P., Teukolsky, S. A., Vetterling, W. T., *Numerical Recipes – The Art of Scientific Computing*, Cambridge University Press, 1986.

[5.17] Bedrosian, E., Rice, S. O., *The Output Properties of Volterra Systems Driven by Harmonic and Gaussian Inputs*, Proceedings IEEE, 59(12), pp. 1688–1707, 1971.

[5.18] Hamming, R. W., *Digital Filters*, Prentice Hall, 3rd Ed., 1989.

[5.19] Särkkä, A., Solin, A., *Applied Stochastic Differential Equations*, Cambridge University Press, 2019.

[5.20] Guckerheimer, J., Holmes, P., *Nonlinear Oscillators, Dynamical Systems and Bifurcations of Vector Fields*, Springer-Verlag, 1983.

[5.21] Virgin, L. N., *Introduction to Experimental Nonlinear Dynamics: A Case Study In Mechanical Vibration*, Cambridge University Press, 2000.

[5.22] Farrar, C. R., Worden, K., *Structural Health Monitoring: a Machine Learning Perspective*, John Wiley & Sons, 2012.

[5.23] Avitabile, P., *Modal Testing: A Practitioner's Guide*, Wiley, 2018.

[5.24] Ewins, D. J., *Modal Testing: Theory and Practice*, Research Studies Press, 1984.

[5.25] Thompson, J. M. T., Stewart, H. B., *Nonlinear Dynamics and Chaos*, John Wiley & Sons, 1986.

[5.26] Morison, J. R., O'Brien, M. P., Johnson, J. W., Schaaf, S. A., *The Force Exerted by Surface Waves on Piles*, Petroleum Transactions, 189, pp. 149–154, 1950.

[5.27] Vidyasagar, M., *Nonlinear Systems Analysis*, Prentice Hall, 1993.

[5.28] Krylov, N. M., Bogoliubov, N. N., *Introduction to Nonlinear Mechanics*, Princeton University Press, 1947.

[5.29] Ferry, J. D., *Viscoelastic Properties of Polymers*, John Wiley & Sons, 1961.

[5.30] Caughey, T. K., *Equivalent Linearisation Techniques*, Journal of the Acoustical Society of America, 35(11), pp. 1706–1711, 1963.

[5.31] Dinca, F., Teodosiu, C., *Nonlinear and Random Vibrations*, Academic Press, 1973.

[5.32] Newland, D. E., *An Introduction to Random Vibrations, Spectral and Wavelet Analysis*, Longman, 1979.

[5.33] Baker, G. L., Gollub, J. P., *Chaotic Dynamics: An Introduction*, Cambridge University Press, 2nd Ed., 2008.

[5.34] Moon, F. C., *Chaotic Vibrations: An Introduction for Applied Scientists and Engineers*, Wiley, 1987.

[5.35] Ueda, Y., *Random Phenomena Resulting from Nonlinearity in the System Described by Duffing's Equation*, International Journal of Nonlinear Mechanics, 20(5-6), pp. 481–491, 1985.

[5.36] Addison, P. S., *Fractals and Chaos: An Illustrated Course*, Institute of Physics Press, 1997.

[5.37] Packard, N. H., Crutchfield, J. P., Farmer, J. D., Shaw, R. S., *Geometry from Time-Series*, Physical Review Letters, 45(9), pp. 712-716, 1980.

[5.38] Wolf, A., Swift, J. B., Swinney, H. L., Vastano, J. A., *Determining Lyapunov Exponents from a Time Series*, Physica D, 16(3), pp. 285–317, 1985.

[5.39] Kaplan, D., Glass, L., *Understanding Nonlinear Dynamics*, Springer-Verlag, 1995.

[5.40] Du Val, P., *Elliptic Functions and Elliptic Curves*, London Mathematical Society Lecture Notes no. 9, Cambridge University Press, 1973.

Chapter 6 Updating of Numerical Models

6.1 Introduction

Assuming that modal testing and analysis have been carefully carried out, measured data are the most reliable information on the dynamic behaviour of a structure and usually considered as correct. Therefore, they are taken as the reference or target data. On the other hand, structural modelling based on Finite Element (*FE*) analyses became a standard on most industrial and research applications. As it is widely known, there are several sources of error at the modelling stage, roughly ranging from modelling assumptions to uncertainty in the modelling parameters [6.1]. As the initial estimates on the analytical or numerical model are often incorrect, there is a difference between the measured responses and the ones predicted using the model. Hence, it is necessary to reconcile or to ensure the agreement of the two data sets. This task is made possible by correcting the model regarding a set of reference responses, and it is referred to as model updating. Model updating is also referred to as model calibration by some authors.

Over the years, a huge range of model updating methods has been proposed and implemented in the most diverse fields. The focus of this chapter is concerned with the updating of linear models; this subject has been reviewed in detail by Mottershead and Friswell [6.2, 6.3]. In [6.1], one can find a brief, although detailed, overview of several implementations of model updating strategies, from direct to iterative methods, including methods based on robust optimisation and meta-modelling. A comprehensive review on the use of computational intelligence techniques applied to model updating can be found in [6.4]. A historical overview of model updating (1960–1990) was given by Hemez and Farrar [6.5].

The updating of nonlinear models is by far less developed than the linear model updating and a review is given by Kerschen et al. [6.6]. da Silva et al. presented a comparison among some of the most prominent available methods [6.7].

The traditional approach to *FE* model updating is usually deterministic, as it only accounts for the minimisation of a deterministic error function based on a set of deterministic modelling parameters. In the context of a model updating process, it is fundamental to ensure previously that the model and reference responses to be compared with are compatible in terms of size and number of response variables (scalar or vectorial ones). This task is referred to as model matching. Regarding the comparison between the compatible model and the reference responses, it is required to establish the correlation metrics used to assess their similarity. In other words, the quality of the updated model responses. Moreover, if one collects a set of dynamic responses, a set of observations can be built for each reference response and the set of modelling parameters may be defined as random variables. In such a case, the model updating approach becomes stochastic, enabling for a more informative process that may be solved using probabilistic or non-probabilistic approaches.

In this chapter, the fundamentals of model matching and correlation towards model updating are covered, and details on selected deterministic and stochastic strategies are given.

Structural Dynamics in Engineering Design, First Edition, Nuno M. M. Maia, Dario Di Maio, Alex Carrella, Francesco Marulo, Chaoping Zang, Jonathan E. Cooper, Keith Worden, and Tiago A. N. Silva.

6.2 Model Matching

As the experimental data are usually collected from few coordinates and in a limited frequency range, when the comparison of numerical and experimental data is needed, the problem of model incompleteness arises. Hence, it is inevitable to deal with model incompleteness. Several methods have been proposed and are being used in order to reduce the theoretical model to the test points or to expand the experimental data over the degrees-of-freedom (*DOF*) of the numerical model, as the model updating methods usually require compatible data sets, enabling a one-to-one comparison between the *DOFs* of the model and the experimental ones.

6.2.1 Model Reduction[1]

Recalling the dynamic equilibrium equation for an *N DOF* system with viscous damping,

$$\boldsymbol{M}\ddot{\boldsymbol{x}}(t)+\boldsymbol{C}\dot{\boldsymbol{x}}(t)+\boldsymbol{K}\boldsymbol{x}(t)=\boldsymbol{f}(t) \tag{6.1}$$

from the *DOFs* of an *FE* model, one may define two sets. The set of primary (or master) *p DOFs* is formed by all the active *DOFs*, encompassing *DOFs* of interest, i.e., *DOFs* where excitations are or might be applied and *DOFs* at measurement locations. The secondary (or slave) *s DOFs* are the remaining ones. Once the sets of primary and secondary *DOFs* have been defined, and neglecting now the damping, Eq. (6.1) can be recast as

$$\begin{bmatrix} \boldsymbol{M}_{pp} & \boldsymbol{M}_{ps} \\ \boldsymbol{M}_{sp} & \boldsymbol{M}_{ss} \end{bmatrix} \begin{Bmatrix} \ddot{\boldsymbol{x}}_p \\ \ddot{\boldsymbol{x}}_s \end{Bmatrix} + \begin{bmatrix} \boldsymbol{K}_{pp} & \boldsymbol{K}_{ps} \\ \boldsymbol{K}_{sp} & \boldsymbol{K}_{ss} \end{bmatrix} \begin{Bmatrix} \boldsymbol{x}_p \\ \boldsymbol{x}_s \end{Bmatrix} = \begin{Bmatrix} \boldsymbol{f}_p \\ \boldsymbol{0} \end{Bmatrix} \tag{6.2}$$

where, to make things simpler, the time dependence has been omitted.

Note that forces can only act at the primary *DOFs* and that this subset is composed by a given number of *DOFs* to be retained after a model reduction. The selection of the retained *DOFs* can be challenging and it must be done in such a way that the modal properties of the reduced system can be described in the best possible way. The objective of all reduction methods is to obtain a reduced system given by

$$\boldsymbol{M}_R\ddot{\boldsymbol{x}}_p+\boldsymbol{K}_R\boldsymbol{x}_p=\boldsymbol{f}_p \tag{6.3}$$

where $\boldsymbol{M}_R$ and $\boldsymbol{K}_R$ are the reduced mass and stiffness matrices, respectively.

The following subsections present some of the most referred reduction methods, with special emphasis on the transformation matrix proposed by each one. Detailed reviews on model reduction techniques can be found in [6.8 - 6.10].

Guyan's Reduction

Introduced by Guyan [6.11], the Guyan or static reduction technique can be found in several commercial computational applications, although its accuracy is limited to the lower order modal properties, as it neglects the inertial terms of Eq. (6.2). Hence, one can state that

[1] Also known as Model Condensation.

$$\boldsymbol{K}_{sp}\boldsymbol{x}_p + \boldsymbol{K}_{ss}\boldsymbol{x}_s = \boldsymbol{0} \tag{6.4}$$

from which

$$\boldsymbol{x}_s = -\left(\boldsymbol{K}_{ss}^{-1}\boldsymbol{K}_{sp}\right)\boldsymbol{x}_p \tag{6.5}$$

Therefore,

$$\begin{Bmatrix} \boldsymbol{x}_p \\ \boldsymbol{x}_s \end{Bmatrix} = \begin{bmatrix} \boldsymbol{I} \\ -\boldsymbol{K}_{ss}^{-1}\boldsymbol{K}_{sp} \end{bmatrix} \boldsymbol{x}_p \tag{6.6}$$

or

$$\begin{Bmatrix} \boldsymbol{x}_p \\ \boldsymbol{x}_s \end{Bmatrix} = \boldsymbol{T}_S\boldsymbol{x}_p \tag{6.7}$$

where $\boldsymbol{T}_S = \begin{bmatrix} \boldsymbol{I} \\ -\boldsymbol{K}_{ss}^{-1}\boldsymbol{K}_{sp} \end{bmatrix}$ is the transformation matrix of the Guyan reduction, relating the spatial model coordinates to the primary ones. Consequently, the reduced model matrices are given by

$$\boldsymbol{M}_R = \boldsymbol{T}_S^T\boldsymbol{M}\boldsymbol{T}_S \quad \text{and} \quad \boldsymbol{K}_R = \boldsymbol{T}_S^T\boldsymbol{K}\boldsymbol{T}_S \tag{6.8}$$

Dynamic Reduction

As the inertial terms have been neglected in the formulation of the transformation matrix $\boldsymbol{T}_S$, the dynamic response of the reduced model is obviously exact only in static conditions. Moreover, as the excitation frequency increases, the accuracy of the reduced model responses tends to degrade. Thus, the Guyan reduction technique can be extended in order to reproduce the exact response of a structure at any given frequency ω [6.12]. This extension is easily accomplished by substituting in the transformation matrix $\boldsymbol{T}_S$ the static stiffness $\boldsymbol{K}$ by the dynamic stiffness $\boldsymbol{Z} = \boldsymbol{K} - \omega^2\boldsymbol{M}$. Therefore, the transformation matrix of the dynamic reduction technique is given by

$$\boldsymbol{T}_D = \begin{bmatrix} \boldsymbol{I} \\ -\left(\boldsymbol{K}_{ss} - \omega^2\boldsymbol{M}_{ss}\right)^{-1}\left(\boldsymbol{K}_{sp} - \omega^2\boldsymbol{M}_{sp}\right) \end{bmatrix} \tag{6.9}$$

and, consequently, the reduced model matrices are obtained similarly to Eq. (6.8):

$$\boldsymbol{M}_R = \boldsymbol{T}_D^T\boldsymbol{M}\boldsymbol{T}_D \quad \text{and} \quad \boldsymbol{K}_R = \boldsymbol{T}_D^T\boldsymbol{K}\boldsymbol{T}_D \tag{6.10}$$

Qu and Fu [6.13] presented a dynamic reduction technique based on the subspace iteration method in the eigenproblem. The transformation matrix of this iterative technique is derived from the ith approximation of the first m mass-normalised eigenvectors $\boldsymbol{\Phi}_m^{(i)}$, such as

$$\boldsymbol{\Phi}_m^{(i)} = \begin{bmatrix} \boldsymbol{\Phi}_{pm}^{(i)} \\ \boldsymbol{\Phi}_{sm}^{(i)} \end{bmatrix} = \begin{bmatrix} \boldsymbol{I} \\ \boldsymbol{\Phi}_{sm}^{(i)}\left(\boldsymbol{\Phi}_{pm}^{(i)}\right)^{-1} \end{bmatrix} \boldsymbol{\Phi}_{pm}^{(i)} \tag{6.11}$$

If one considers the subspace approximation $\boldsymbol{X}_m^{(i+1)}$ of $\boldsymbol{\Phi}_m^{(i+1)}$, defined as

$$\boldsymbol{X}_m^{(i+1)} = \boldsymbol{K}^{-1}\boldsymbol{M}\boldsymbol{\Phi}_m^{(i)} \tag{6.12}$$

the reduced system matrices are given by

$$\boldsymbol{M}_R = \left(\boldsymbol{X}_m^{(i+1)}\right)^T \boldsymbol{M}\boldsymbol{X}_m^{(i+1)} \quad \text{and} \quad \boldsymbol{K}_R = \left(\boldsymbol{X}_m^{(i+1)}\right)^T \boldsymbol{K}\boldsymbol{X}_m^{(i+1)} \tag{6.13}$$

Improved Reduction System

The Improved Reduction System (*IRS*) technique, proposed by O'Callahan [6.14], develops the Guyan technique to include the inertial terms of the secondary coordinates as pseudo-static forces, which generates the following transformation matrix,

$$\boldsymbol{T}_I = \boldsymbol{T}_S + \boldsymbol{S}\boldsymbol{M}\boldsymbol{T}_S\boldsymbol{M}_R^{-1}\boldsymbol{K}_R \tag{6.14}$$

where $\boldsymbol{T}_S$ is the transformation matrix of the Guyan reduction, $\boldsymbol{M}_R$ and $\boldsymbol{K}_R$ are the statically reduced mass and stiffness matrices, respectively, and

$$\boldsymbol{S} = \begin{bmatrix} \boldsymbol{0} & \boldsymbol{0} \\ \boldsymbol{0} & \boldsymbol{K}_{ss}^{-1} \end{bmatrix} \tag{6.15}$$

Hence, the reduced model matrices are generated as in Eq. (6.8), considering $\boldsymbol{T}_I$ instead of $\boldsymbol{T}_S$.

The transformation given by Eq. (6.14) is a perturbation on Guyan's reduction technique that includes the inertia effect and therefore it is more accurate than the former. Friswell et al. [6.15] defined the *IRS* technique based on a perturbation on the dynamic reduction, where the transformation is exact for a selected non-zero frequency. This reduction technique is the Dynamic *IRS* (*DIRS*) and its transformation matrix is,

$$\boldsymbol{T}_{DI} = \boldsymbol{T}_D + \boldsymbol{S}_D\boldsymbol{M}\boldsymbol{T}_D\boldsymbol{M}_R^{-1}\boldsymbol{K}_R \tag{6.16}$$

where

$$\boldsymbol{S}_D = \begin{bmatrix} \boldsymbol{0} & \boldsymbol{0} \\ \boldsymbol{0} & \left(\boldsymbol{K}_{ss} - \omega^2\boldsymbol{M}_{ss}\right)^{-1} \end{bmatrix} \tag{6.17}$$

The *IRS* technique was also modified by Friswell et al. [6.15] in order to improve its accuracy and is known as the Iterated *IRS* (*IIRS*) technique, where the transformation matrix $\boldsymbol{T}_I$ is replaced by,

$$\boldsymbol{T}_{II}^{(k)} = \boldsymbol{T}_S + \boldsymbol{S}\boldsymbol{M}\boldsymbol{T}_{II}^{(k-1)}\left(\boldsymbol{M}_R^{(k-1)}\right)^{-1}\boldsymbol{K}_R^{(k-1)} \tag{6.18}$$

where the superscript $^{(k)}$ denotes the kth iteration, for $k > 1$, with $\boldsymbol{T}_{II}^{(0)} = \boldsymbol{T}_S$ and $\boldsymbol{T}_{II}^{(1)} = \boldsymbol{T}_I$ given in Eq. (6.14). The reduced model matrices are given by,

$$\boldsymbol{M}_R^{(k)} = \left(\boldsymbol{T}_{II}^{(k)}\right)^T \boldsymbol{M}\,\boldsymbol{T}_{II}^{(k)} \quad \text{and} \quad \boldsymbol{K}_R^{(k)} = \left(\boldsymbol{T}_{II}^{(k)}\right)^T \boldsymbol{K}\,\boldsymbol{T}_{II}^{(k)} \tag{6.19}$$

Further details on this technique are given in [6.15, 6.16].

Xia and Lin [6.17] proposed a modification to the *IIRS* technique, which leads to the following transformation matrix:

$$\boldsymbol{T}_{III}^{(k)} = \boldsymbol{T}_S + \boldsymbol{S}\boldsymbol{M}\boldsymbol{T}_{III}^{(k-1)}\left(\boldsymbol{M}_D^{(k)}\right)^{-1}\boldsymbol{K}_S \tag{6.20}$$

with

$$\boldsymbol{M}_D^{(k)} = \boldsymbol{T}_S^T \boldsymbol{M}\,\boldsymbol{T}_{III}^{(k-1)} \quad \text{and} \quad \boldsymbol{K}_S = \boldsymbol{T}_S^T \boldsymbol{K}\,\boldsymbol{T}_S \tag{6.21}$$

One should note that $\boldsymbol{T}_{III}^{(0)} = \boldsymbol{T}_S$ and $\boldsymbol{T}_{III}^{(1)} = \boldsymbol{T}_I$, as for the *IIRS*, and the reduced model matrices are obtained similarly to Eq. (6.19). This technique proved to be computationally more efficient than the basic *IIRS* [6.17].

System Equivalent Reduction Expansion Process

The System Equivalent Reduction Expansion Process (*SEREP*) was introduced by O'Callahan et al. [6.18], although a similar technique had been proposed by Kammer [6.19]. This technique has the ability of being suitable for both model reduction and modal data expansion purposes, with the same level of accuracy. *SEREP* requires the solution of the generalised eigenproblem in order to obtain the numerical modal model, defined by the natural frequencies, the mode shapes and the modal damping ratios [6.1]. The computed eigenvectors must then be partitioned, so that

$$\boldsymbol{\psi}^{(r)} = \begin{Bmatrix} \boldsymbol{\psi}_p \\ \boldsymbol{\psi}_s \end{Bmatrix}^{(r)} \tag{6.22}$$

where the superscript $^{(r)}$ refers to the rth numerical mode shape.

Considering the analytical counterpart of the k identified mode shapes, an $m \times k$ matrix $\boldsymbol{\Psi}_p$ is formed. The Moore-Penrose pseudo-inverse of $\boldsymbol{\Psi}_p$, denoted by $\boldsymbol{\Psi}_p^+$, is used to expand the mode shapes at the primary *DOFs* by means of the transformation matrix,

$$\boldsymbol{T}_{SE} = \begin{Bmatrix} \boldsymbol{\Psi}_p \\ \boldsymbol{\Psi}_s \end{Bmatrix}\left(\boldsymbol{\Psi}_p^T\boldsymbol{\Psi}_p\right)^{-1}\boldsymbol{\Psi}_p^T = \boldsymbol{\Psi}\boldsymbol{\Psi}_p^+ \tag{6.23}$$

The reduced model matrices are obtained as in Eq. (6.8). If the eigenvectors are mass-normalised, the reduced model matrices can be efficiently computed, regarding the orthogonality conditions. Recalling that these are given by[2]

$$\boldsymbol{\Phi}^T\boldsymbol{M}\boldsymbol{\Phi} = \boldsymbol{I} \quad \text{and} \quad \boldsymbol{\Phi}^T\boldsymbol{K}\boldsymbol{\Phi} = \boldsymbol{\Lambda} \tag{6.24}$$

---

[2] See Chapter 1, Section 1.5.2, under 'Orthogonality Properties and Normalisation of the Mode Shapes', for details.

where $\boldsymbol{\Phi}$ is the mass-normalised modal matrix and $\boldsymbol{\Lambda}$ is a diagonal matrix of the eigenvalues, it follows that

$$\boldsymbol{M}_R = \left(\boldsymbol{\Phi}_p^+\right)^T \boldsymbol{\Phi}_p^+ \quad \text{and} \quad \boldsymbol{K}_R = \left(\boldsymbol{\Phi}_p^+\right)^T \boldsymbol{\Lambda}\boldsymbol{\Phi}_p^+ \tag{6.25}$$

Due to rank deficiency issues, *SEREP* can only be applied if the number of primary *DOFs* is greater or equal to the number of eigenvectors. However, as it can be derived from the *SEREP* formulation, the reduced model has exactly the same eigenfrequencies and eigenmodes as the full model for the modes used to form $\boldsymbol{T}_{SE}$. The *IIRS* technique, proposed by Friswell et al. [6.15], converges exactly to the same solution of *SEREP*, though avoiding the extraction of modal properties.

Kammer [6.20] proposed a hybrid reduction technique based on modal reduction of a set of target mode shapes and a more accurate compensation of the residual ones. The technique overcomes sensitivity issues related to differences between test and model mode shapes and its transformation matrix is given by

$$\boldsymbol{T}_H = \boldsymbol{T}_S + \left(\boldsymbol{T}_{T_m} - \boldsymbol{T}_S\right)\boldsymbol{P}_T \tag{6.26}$$

where $\boldsymbol{T}_S$ is the transformation matrix of the Guyan reduction, $\boldsymbol{T}_{T_m}$ is a modal reduction transformation matrix that provides an exact representation of the target mode shapes at the primary *DOFs* and $\boldsymbol{P}_T$ is an orthogonal projector matrix used to quantify the contribution of the target mode shapes to the responses at the primary *DOFs*, thus $\boldsymbol{P}_T$ is close to the transformation matrix of *SEREP*.

Modal Truncation

Considering the modal matrix $\boldsymbol{\Psi}$, one can derive the system matrices in modal coordinates, known as the modal mass and modal stiffness matrices, respectively:

$$\mathcal{M} = \boldsymbol{\Psi}^T \boldsymbol{M}\boldsymbol{\Psi} \quad \text{and} \quad \mathcal{K} = \boldsymbol{\Psi}^T \boldsymbol{K}\boldsymbol{\Psi} \tag{6.27}$$

If the mass-normalised modal matrix is used, the orthogonality conditions of Eq. (6.24) are verified.

In the modal domain, the equations of motion are completely decoupled and therefore constitute a set of linearly independent equilibrium equations related to each eigenvalue. Hence, one can obtain a reduced model just by truncation of the modal system matrices at a given mode of interest. Thus, for a truncated modal system defined by $\boldsymbol{\Phi}_R$ and $\boldsymbol{\Lambda}_R$, one can obtain the following reduced model matrices:

$$\boldsymbol{M}_R = \left(\boldsymbol{\Phi}_R^T\right)^+ \boldsymbol{\Phi}_R^+ \quad \text{and} \quad \boldsymbol{K}_R = \left(\boldsymbol{\Phi}_R^T\right)^+ \boldsymbol{\Lambda}_R\boldsymbol{\Phi}_R^+ \tag{6.28}$$

As it can be observed, the reduced mass and stiffness matrices can be directly obtained from the experimental modal data, after normalisation. If the damping is not neglected, a second order modal truncation must be considered, as described in [6.9].

The use of modal truncation is the basis of an efficient approach to the assembly of substructures or super-elements, the modal-based assembly (*MBA*) technique.

Component Mode Synthesis

Developed as a modal coupling technique, one of the most accurate, although expensive, model reduction techniques is given by the Component Mode Synthesis (*CMS*) methods. These methods attain a reduced order model for the substructures of a given assembled structure by reducing the number of mode shapes used to describe the dynamics of each substructure, while preserving all the physical *DOFs*. A general *CMS* algorithm can be framed into three main categories: the fixed-interface, the free-interface, and the residual-flexible free-interface [6.1, 6.9, 6.21, 6.22]. The choice for a specific *CMS* is problem dependent, although in structural mechanics the fixed-interface method is usually recommended, as it accurately retains the lower to mid frequency eigenvalues. Among several *CMS* hypothesis, the Craig-Bampton (*CB*) method [6.23] is perhaps the most used one and therefore it is here presented. Furthermore, it is implemented in several commercial *FE* software packages.

Regarding a fixed-interface *CMS* method, Eq. (6.2) can be re-written in terms of boundary or connection *c* and interior *i DOFs*, as

$$\begin{bmatrix} \boldsymbol{M}_{cc} & \boldsymbol{M}_{ci} \\ \boldsymbol{M}_{ic} & \boldsymbol{M}_{ii} \end{bmatrix} \begin{Bmatrix} \ddot{\boldsymbol{x}}_c \\ \ddot{\boldsymbol{x}}_i \end{Bmatrix} + \begin{bmatrix} \boldsymbol{K}_{cc} & \boldsymbol{K}_{ci} \\ \boldsymbol{K}_{ic} & \boldsymbol{K}_{ii} \end{bmatrix} \begin{Bmatrix} \boldsymbol{x}_c \\ \boldsymbol{x}_i \end{Bmatrix} = \begin{Bmatrix} \boldsymbol{f}_c \\ \boldsymbol{f}_i \end{Bmatrix} \tag{6.29}$$

where $\boldsymbol{x}_c$ and $\boldsymbol{f}_c$ are the displacement and force vectors associated to the connection *DOFs* between substructures, respectively; and $\boldsymbol{x}_i$ and $\boldsymbol{f}_i$ are the ones related to the interior *DOFs*. The connection *DOFs* are actually the primary ones, in the sense of the previous notation.

The fixed-interface *CMS* algorithm is based on the observation that the secondary *DOFs* can be expressed in terms of both elastic and static mode shapes, considering that the connection *DOFs* are fixed ($\boldsymbol{x}_c = \boldsymbol{0}$) and no forces are acting at the interior *DOFs* ($\boldsymbol{f}_i = \boldsymbol{0}$). These assumptions lead to the approximation of the vector of interior displacements $\boldsymbol{x}_i$ by a linear combination of the elastic fixed-interface modes or *CB* modes $\boldsymbol{\Phi}_{CB}$, given by

$$\boldsymbol{x}_i = \boldsymbol{\Phi}_{CB}\, \boldsymbol{p}_m \tag{6.30}$$

where $\boldsymbol{p}_m$ is the displacement vector of the interior *DOFs* in generalised coordinates, with *m* interior *DOFs*. The modal matrix $\boldsymbol{\Phi}_{CB}$ can be reduced to a given number of modes to be retained or, from a different perspective, $\boldsymbol{\Phi}_{CB}$ can be reduced in order to retain as much modes as the number of primary *DOFs* (*k*).

On the other hand, the constraint or static modes are also required to approximate the displacements at the interior *DOFs*. In the *CB* method, these static modes are in fact given by the Guyan reduction, as it can be inferred from the manipulation of Eq. (6.29). Hence, one can relate the displacements at the interior coordinates to the connection ones:

$$\boldsymbol{x}_i = -\left(\boldsymbol{K}_{ii}^{-1}\boldsymbol{K}_{ic}\right)\boldsymbol{x}_c = \boldsymbol{T}_{S_{CB}}\boldsymbol{x}_c \tag{6.31}$$

Thus, connection and interior *DOFs* can be related by the transformation matrix of the *CB* method $\boldsymbol{T}_{CB}$, derived from Eqs. (6.30) and (6.31), as follows:

$$\begin{Bmatrix} \boldsymbol{x}_i \\ \boldsymbol{x}_c \end{Bmatrix} = \begin{bmatrix} \boldsymbol{T}_{S_{CB}} & \boldsymbol{\Phi}_{CB} \\ \boldsymbol{I} & \boldsymbol{0} \end{bmatrix} \begin{Bmatrix} \boldsymbol{x}_c \\ \boldsymbol{p}_k \end{Bmatrix} = \boldsymbol{T}_{CB} \begin{Bmatrix} \boldsymbol{x}_c \\ \boldsymbol{p}_k \end{Bmatrix} \tag{6.32}$$

Consequently, the reduced model matrices, for the *k DOFs*, are defined as

$$\boldsymbol{M}_R = \boldsymbol{T}_{CB}^T \boldsymbol{M} \boldsymbol{T}_{CB} \quad \text{and} \quad \boldsymbol{K}_R = \boldsymbol{T}_{CB}^T \boldsymbol{K} \boldsymbol{T}_{CB} \tag{6.33}$$

Note that the *CB* method can be improved by considering a more accurate approximation of the static modes.

The *CB* method has been modified by several authors. Rixen [6.24] proposed a dual *CB* method which is based on a hybrid free-interface and residual-flexible free-interface components, using the connection forces in order to assemble the substructures. The modification introduced by this method is concerned with the inclusion of the rigid body modes in the description of the static modes, given in Eq. (6.31). Further improvements on the dual *CB* method are reported by Rixen in [6.25]. Aiming at eliminating the spurious modes that can be found in the reduced system obtained using the dual *CB* method, Rixen proposed to improve the reduction basis of the basic method by using a set of correction modes computed via modal truncation.

Sum of Weighted Accelerations Technique

Carne et al. [6.26] presented the Sum of Weighted Accelerations Technique (*SWAT*), a force identification technique based on the concept of a modal filter. This technique can be used as a reduction technique in the sense that the model can be replaced by its local time dependent response [6.27].

SWAT uses the rigid body mode shapes in order to derive a weighting matrix that separates the rigid body accelerations from the elastic response within a frequency band of interest. Consider the transformation of generalised coordinates into physical ones, given by

$$\boldsymbol{x} = \boldsymbol{\Psi} \boldsymbol{q} \tag{6.34}$$

and the acceleration vector approximated by a sum of modal contributions, as

$$\ddot{\boldsymbol{x}} = \boldsymbol{\Psi} \ddot{\boldsymbol{q}} \tag{6.35}$$

the referred weighting matrix $\boldsymbol{W}$ is derived such as a vector of rigid body accelerations $\ddot{\boldsymbol{x}}_{RB}$, also known as *SWAT DOFs*, can be extracted from the acceleration vector defined in Eq. (6.35), so that,

$$\ddot{\boldsymbol{x}}_{RB} = \boldsymbol{W} \ddot{\boldsymbol{x}} \tag{6.36}$$

Rewriting Eq. (6.35) in terms of the contribution of the rigid body modes $\boldsymbol{\Phi}_{RB}$ and the flexible ones $\boldsymbol{\Phi}$, one has

$$\ddot{\boldsymbol{x}} = [\boldsymbol{\Phi}_{RB} \quad \boldsymbol{\Phi}] \begin{Bmatrix} \ddot{\boldsymbol{x}}_{RB} \\ \ddot{\boldsymbol{q}} \end{Bmatrix} \tag{6.37}$$

Using Eqs. (6.36) and (6.37), it is possible to define the weighting matrix W that extracts only the rigid body accelerations from the measured ones, as described by Allen and Carne [6.28]:

$$W = \left[\left[\boldsymbol{\Phi}_{RB} \quad \boldsymbol{\Phi} \right]^T \right]^+ \begin{bmatrix} I \\ 0 \end{bmatrix} \tag{6.38}$$

If one truncates the elastic mode shapes to a given number of modes of interest (k), defined as the number of primary *DOFs* to retain, as addressed by the *CB* method, one can relate the spatial model coordinates to the primary k ones based on Eq. (6.37), as

$$\boldsymbol{x} = \begin{Bmatrix} \boldsymbol{x}_p \\ \boldsymbol{x}_s \end{Bmatrix} = \left[\boldsymbol{\Phi}_{RB} \quad \boldsymbol{\Phi}_k \right] \begin{Bmatrix} \boldsymbol{x}_{RB} \\ \boldsymbol{x}_k \end{Bmatrix} = \boldsymbol{T}_{SW} \begin{Bmatrix} \boldsymbol{x}_{RB} \\ \boldsymbol{x}_k \end{Bmatrix} \tag{6.39}$$

Thus, the reduced system matrices are given similarly to Eq. (6.8) and their dimensions are given by the number of rigid body modes in addition to the k modes of interest.

Reduction of Damped Models

Jeong et al. [6.29] derived a transformation matrix equivalent to the *IIRS* technique in the state-space formulation to be used for damped systems. The equation of motion for non-proportional damping must be reformulated in the state-space form to derive the quadratic eigenvalue problem, given by

$$\boldsymbol{A}\dot{\boldsymbol{y}} + \boldsymbol{B}\boldsymbol{y} = \boldsymbol{b} \tag{6.40}$$

with

$$\boldsymbol{A} = \begin{bmatrix} \boldsymbol{C} & \boldsymbol{M} \\ \boldsymbol{M} & \boldsymbol{0} \end{bmatrix}, \quad \boldsymbol{B} = \begin{bmatrix} \boldsymbol{K} & \boldsymbol{0} \\ \boldsymbol{0} & -\boldsymbol{M} \end{bmatrix}, \quad \boldsymbol{y} = \begin{Bmatrix} \boldsymbol{x} \\ \dot{\boldsymbol{x}} \end{Bmatrix} \quad \text{and} \quad \boldsymbol{b} = \begin{Bmatrix} \boldsymbol{f} \\ \boldsymbol{0} \end{Bmatrix} \tag{6.41}$$

From Eq. (6.40), one can write the generalised eigenvalue problem, as

$$\boldsymbol{B}\tilde{\boldsymbol{\Psi}} = \boldsymbol{A}\tilde{\boldsymbol{\Psi}}\tilde{\boldsymbol{\Lambda}} \tag{6.42}$$

whose eigensolutions are real or exist in complex conjugate pairs given by

$$\tilde{\boldsymbol{\Psi}} = \begin{bmatrix} \boldsymbol{\Psi} & \boldsymbol{\Psi}^* \\ \boldsymbol{\Lambda}\boldsymbol{\Psi} & \boldsymbol{\Lambda}^*\boldsymbol{\Psi}^* \end{bmatrix} \quad \text{and} \quad \tilde{\boldsymbol{\Lambda}} = \begin{bmatrix} \boldsymbol{\Lambda} & \boldsymbol{0} \\ \boldsymbol{0} & \boldsymbol{\Lambda}^* \end{bmatrix} \tag{6.43}$$

where $\tilde{\boldsymbol{\Psi}}$ and $\tilde{\boldsymbol{\Lambda}}$ are matrices of complex conjugate pairs of eigenvectors and eigenvalues.

Rewriting Eq. (6.42) in a partitioned form, one has

$$\begin{bmatrix} \boldsymbol{B}_{pp} & \boldsymbol{B}_{ps} \\ \boldsymbol{B}_{sp} & \boldsymbol{B}_{ss} \end{bmatrix} \begin{bmatrix} \tilde{\boldsymbol{\Psi}}_p \\ \tilde{\boldsymbol{\Psi}}_s \end{bmatrix} = \begin{bmatrix} \boldsymbol{A}_{pp} & \boldsymbol{A}_{ps} \\ \boldsymbol{A}_{sp} & \boldsymbol{A}_{ss} \end{bmatrix} \begin{bmatrix} \tilde{\boldsymbol{\Psi}}_p \\ \tilde{\boldsymbol{\Psi}}_s \end{bmatrix} \tilde{\boldsymbol{\Lambda}} \tag{6.44}$$

from which one derives a transformation matrix $\boldsymbol{T}_{SS}$ equivalent to the *IIRS* transformation given in Eq. (6.18), such as

$$T_{SS}^{(k)} = \begin{bmatrix} I \\ -B_{ss}^{-1} B_{sp} \end{bmatrix} + \begin{bmatrix} 0 & 0 \\ 0 & B_{ss}^{-1} \end{bmatrix} B T_{SS}^{(k-1)} \left(B_R^{(k-1)} \right)^{-1} A_R^{(k-1)} \tag{6.45}$$

where the superscript $^{(k)}$ denotes the kth iteration, for $k > 1$, and $T_{SS}^{(0)} = \begin{bmatrix} I \\ -B_{ss}^{-1} B_{sp} \end{bmatrix}$, that is equivalent to the static reduction, and the reduced model matrices are given by

$$A_R^{(k)} = \left(T_{SS}^{(k)} \right)^T M T_{SS}^{(k)} \quad \text{and} \quad B_R^{(k)} = \left(T_{SS}^{(k)} \right)^T K T_{SS}^{(k)} \tag{6.46}$$

Das and Dutt [6.30] proposed a modified *SEREP* formulated in state-space, including the gyroscopic effect and both internal and external damping. This technique is thought to reduce finite element models of rotor systems, although it can be applied to any linear system of equations.

6.2.2 Expansion of Measured Data

As the referred reduction techniques point out, the great majority of the available techniques were developed either in the spatial or the modal domain and they can be directly used to expand the modal data over the secondary coordinates. Such an approach can be viewed as an inverted reduction, where the mode shape expansion is achieved by using the transformation matrices of the aforementioned reduction techniques [6.31].

Hence, one can obtain the complete r^{th} mode shape $\psi^{(r)}$ considering the expansion of $\psi_p^{(r)}$, using one of the several transformation matrices developed for model reduction purposes, such as,

$$\psi^{(r)} = \begin{Bmatrix} \psi_p \\ \psi_s \end{Bmatrix}^{(r)} = T \psi_p^{(r)} \tag{6.47}$$

If one considers not just one particular mode shape but a set of mode shapes, Eq. (6.47) becomes

$$\Psi = \begin{bmatrix} \Psi_p \\ \Psi_s \end{bmatrix} = T \Psi_p \tag{6.48}$$

where Ψ is a rectangular modal matrix constituted by a set of measured mode shapes Ψ_p expanded over the set of secondary coordinates.

Kidder's Method

Kidder [6.32] proposed a spatial transformation using the analytical mass and stiffness matrices and the solution of the generalised eigenproblem, given by

$$\left[\begin{bmatrix} K_{pp} & K_{ps} \\ K_{sp} & K_{ss} \end{bmatrix} - \omega_r^2 \begin{bmatrix} M_{pp} & M_{ps} \\ M_{sp} & M_{ss} \end{bmatrix} \right] \begin{Bmatrix} \psi_p \\ \psi_s \end{Bmatrix}^{(r)} = \begin{Bmatrix} 0 \\ 0 \end{Bmatrix} \tag{6.49}$$

where ω_r is the rth natural frequency.

From Eq. (6.49), the expanded mode shapes are given by

$$\boldsymbol{\psi}^{(r)} = \begin{Bmatrix} \boldsymbol{\psi}_p \\ \boldsymbol{\psi}_s \end{Bmatrix}^{(r)} = \begin{bmatrix} \boldsymbol{I} \\ -\left(\boldsymbol{K}_{ss} - \omega_r^2 \boldsymbol{M}_{ss}\right)^{-1}\left(\boldsymbol{K}_{sp} - \omega_r^2 \boldsymbol{M}_{sp}\right) \end{bmatrix} \boldsymbol{\psi}_p^{(r)} = \boldsymbol{T}_K \boldsymbol{\psi}_p^{(r)} \tag{6.50}$$

where $\boldsymbol{T}_K$ is the transformation matrix of Kidder's method for the rth mode shape.

One is reminded that the Kidder's expansion process is conducted mode by mode, and that the experimental mode shapes can be expanded over the entire set of model coordinates, although the out-of-range modes are neglected.

Expansion using Analytical Modes

Perhaps the simplest approach to expand measured data is the one where the data at unmeasured *DOFs* are directly replaced by their analytical counterpart, although numerical discontinuities may arise from this kind of procedure [6.1]. However, in this context, a feasible technique was introduced by Lipkins and Vandeurzen [6.33]. This expansion technique relies on the assumption that each mode can be obtained by a linear combination of the analytical modes. Considering the analytical modal matrix partitioned in terms of primary and secondary coordinates, and selected/retained and discarded mode shapes, one has,

$$\boldsymbol{\Psi}_A = \begin{bmatrix} \boldsymbol{\Psi}_{pk} & \boldsymbol{\Psi}_{pd} \\ \boldsymbol{\Psi}_{sk} & \boldsymbol{\Psi}_{sd} \end{bmatrix} \tag{6.51}$$

where the submatrix $\begin{bmatrix} \boldsymbol{\Psi}_{pk} & \boldsymbol{\Psi}_{sk} \end{bmatrix}^T$ is related to the k selected mode shapes, while the submatrix $\begin{bmatrix} \boldsymbol{\Psi}_{pd} & \boldsymbol{\Psi}_{sd} \end{bmatrix}^T$ is related to the d discarded ones. Hence, the expanded experimental mode shapes are generated from the following linear combination,

$$\boldsymbol{\psi}_X = \begin{Bmatrix} \boldsymbol{\psi}_p \\ \boldsymbol{\psi}_s \end{Bmatrix}^{(X)} = \begin{bmatrix} \boldsymbol{\Psi}_{pk} \\ \boldsymbol{\Psi}_{sk} \end{bmatrix} \boldsymbol{v} \tag{6.52}$$

where $\boldsymbol{v}$ is a vector of unknown coefficients obtained in a least-squares sense, as

$$\boldsymbol{v} = \boldsymbol{\Psi}_{pk}^{+} \boldsymbol{\psi}_p^{(X)} \tag{6.53}$$

and therefore the mode shapes at the secondary coordinates are computed by

$$\boldsymbol{\psi}_s^{(X)} = \boldsymbol{\Psi}_{sk} \boldsymbol{v} = \boldsymbol{\Psi}_{sk} \boldsymbol{\Psi}_{pk}^{+} \boldsymbol{\psi}_p^{(X)} \tag{6.54}$$

Expansion of Frequency Response Functions

Rather than expanding the mode shapes, the need to expand a set of measured *FRFs* has been addressed by several authors.

This subject has arisen interest in the field of structural coupling, where several methods require the rotational *FRFs* [6.34]. Furthermore, if one can obtain the complete *FRF* matrix, the model updating task is solved directly, as will be discussed in Section 6.4.2.

Modified Kidder's Method

To expand a set of measured *FRFs* $\boldsymbol{H}_{pj}(\omega)$ over the secondary model coordinates, one may use a modified version of Kidder's expansion method, considering that the experimental responses are due to a set of q excitations applied individually at j locations. Instead of being based on the solution of the eigenproblem, this technique is based on the notion of dynamic stiffness and it is derived from the equilibrium equation for a particular frequency, as

$$\left[\begin{bmatrix} \boldsymbol{K}_{pp} & \boldsymbol{K}_{ps} \\ \boldsymbol{K}_{sp} & \boldsymbol{K}_{ss} \end{bmatrix} - \omega^2 \begin{bmatrix} \boldsymbol{M}_{pp} & \boldsymbol{M}_{ps} \\ \boldsymbol{M}_{sp} & \boldsymbol{M}_{ss} \end{bmatrix}\right] \begin{Bmatrix} \boldsymbol{H}_{pj}(\omega) \\ \boldsymbol{H}_{sj}(\omega) \end{Bmatrix} = \begin{Bmatrix} \boldsymbol{I}_j \\ \boldsymbol{0} \end{Bmatrix} \tag{6.55}$$

where $\boldsymbol{I}_j$ is a vector identifying each excitation location j.

From Eq. (6.55), a complete *FRF* vector is given by,

$$\boldsymbol{H}_j(\omega) = \begin{Bmatrix} \boldsymbol{H}_{pj}(\omega) \\ \boldsymbol{H}_{sj}(\omega) \end{Bmatrix} = \begin{bmatrix} \boldsymbol{I} \\ -\left(\boldsymbol{K}_{ss} - \omega^2 \boldsymbol{M}_{ss}\right)^{-1}\left(\boldsymbol{K}_{sp} - \omega^2 \boldsymbol{M}_{sp}\right) \end{bmatrix} \boldsymbol{H}_{pj}(\omega) = \boldsymbol{T}_M \boldsymbol{H}_{pj}(\omega) \tag{6.56}$$

which can also be extended to be used for damped systems.

Regarding the modified Kidder's method, the expansion process is conducted spectral line by spectral line for a given experimental frequency range of interest, for each *FRF* vector.

A Complete Matrix of Frequency Response Functions

Avitabile and O'Callahan [6.35] proposed an approach using *SEREP* to obtain a complete *FRF* matrix from a set of measured *FRFs*. This expansion technique recalls the *FRF* partial fraction form, written to include the effect of all the modes of the system. This method can be implemented in one of two ways: i) one needs to synthesise the *FRFs* at translational *DOFs* from the measured modal quantities through its partial fraction form and to expand them over the secondary *DOFs* (both translational and rotational ones) using *SEREP*; additionally, one includes the residual effect of the unmeasured modes and the expansion of it using the dynamic expansion method; and ii) where the synthesised *FRFs* are expanded over the secondary *DOFs* using *SEREP* and the residual effect is related to the difference of the *SEREP* expansion and the dynamic one at a particular frequency.

However, the described approach is based on modal identification. To avoid modal identification, one can address the straightforward expansion of the experimental *FRF* matrix using the transformation matrix of the modified Kidder's method (Eq. (6.56)) and a specific classification of each element of the *FRF* matrix [6.36].

Considering the *FRF* matrix partitioned in terms of translational t and rotational θ response *DOFs* and *DOFs* where forces f and moments τ may be applied, and omitting the frequency dependency notation for the sake of simplicity, one has,

$$\boldsymbol{H} = \begin{bmatrix} \boldsymbol{H}_{tf} & \boldsymbol{H}_{t\tau} \\ \boldsymbol{H}_{\theta f} & \boldsymbol{H}_{\theta\tau} \end{bmatrix} \tag{6.57}$$

Experimentally, it is common to measure only part of the translational *DOFs* and to apply excitations on a subset of f. Thus, only part of the submatrix $\boldsymbol{H}_{tf}$ is obtained and it often results in a rectangular submatrix, as the number of measured *DOFs* is usually different from the number of the excited ones. As it is very difficult to measure rotational *DOFs* or excite them by applying pure moments, one shall assume that this kind of measurements are unavailable. However, if these *DOFs* are accessible, the correspondent *FRFs* may be included in the measured set. Hence, if one chooses to organize the subsets of t and f types of *DOFs*, by their nature, one can define the four types of *DOFs,* respectively: A for measured/forced *DOFs*; B for measured/unforced *DOFs*; E for unmeasured/forced *DOFs*; and U for unmeasured/unforced *DOFs* [6.36]. Regarding this classification, one may rewrite Eq. (6.57) in a more detailed way, as

$$\boldsymbol{H} = \begin{bmatrix} \boldsymbol{H}_{AA} & \boldsymbol{H}_{AB} & \boldsymbol{H}_{AE} & \boldsymbol{H}_{AU} & \boldsymbol{H}_{A\tau} \\ \boldsymbol{H}_{BA} & \boldsymbol{H}_{BB} & \boldsymbol{H}_{BE} & \boldsymbol{H}_{BU} & \boldsymbol{H}_{B\tau} \\ \boldsymbol{H}_{EA} & \boldsymbol{H}_{EB} & \boldsymbol{H}_{EE} & \boldsymbol{H}_{EU} & \boldsymbol{H}_{E\tau} \\ \boldsymbol{H}_{UA} & \boldsymbol{H}_{UB} & \boldsymbol{H}_{UE} & \boldsymbol{H}_{UU} & \boldsymbol{H}_{U\tau} \\ \boldsymbol{H}_{\theta A} & \boldsymbol{H}_{\theta B} & \boldsymbol{H}_{\theta E} & \boldsymbol{H}_{\theta U} & \boldsymbol{H}_{\theta\tau} \end{bmatrix} \tag{6.58}$$

It is convenient to assemble the set of measured *FRFs* on the left upper position of the matrix, before starting the expansion process. Therefore Eq. (6.58) may be recast as

$$\boldsymbol{H} = \begin{bmatrix} \boldsymbol{H}_{AA} & \boldsymbol{H}_{AE} & \boldsymbol{H}_{AB} & \boldsymbol{H}_{AU} & \boldsymbol{H}_{A\tau} \\ \boldsymbol{H}_{BA} & \boldsymbol{H}_{BE} & \boldsymbol{H}_{BB} & \boldsymbol{H}_{BU} & \boldsymbol{H}_{B\tau} \\ \boldsymbol{H}_{EA} & \boldsymbol{H}_{EE} & \boldsymbol{H}_{EB} & \boldsymbol{H}_{EU} & \boldsymbol{H}_{E\tau} \\ \boldsymbol{H}_{UA} & \boldsymbol{H}_{UE} & \boldsymbol{H}_{UB} & \boldsymbol{H}_{UU} & \boldsymbol{H}_{U\tau} \\ \boldsymbol{H}_{\theta A} & \boldsymbol{H}_{\theta E} & \boldsymbol{H}_{\theta B} & \boldsymbol{H}_{\theta U} & \boldsymbol{H}_{\theta\tau} \end{bmatrix} \tag{6.59}$$

Now, the submatrices concerning measured (A and B) and excited (A and E), highlighted in grey, are used to estimate *FRFs* at all the excited, but unmeasured, *DOFs* (response sets E, U and θ). Based on the principle of reciprocity, the process follows to a second expansion step to estimate *FRFs* at unexcited *DOFs* (B, U and τ). Eventually, the full *FRF* matrix is obtained. Details on the application of this method can be found in [6.36].

6.3 Model Correlation

Ensuring that the model incompleteness issue is solved, one should establish a procedure to assess the agreement between the experimental and predicted responses. A first comparison can be made using the reference and predicted eigenfrequencies, which is achieved without the need of model matching. However, to determine the similitude between the experimental and theoretical models, one must address the use of correlation techniques capable of comparing response vectors, either in the modal or in the frequency domain. In what follows, the most commonly used model correlation metrics are presented.

6.3.1 Modal Domain

The Modal Assurance Criterion (*MAC*) is probably the most used correlation criterion, proposed by Allemang and Brown [6.38] to assess the degree of correlation between two modal vectors.

Recalling the orthogonality conditions of Eq. (6.24), if one considers two mass-normalised modal vectors, respectively, one measured mode $\boldsymbol{\phi}_{X_i}$ and one predicted mode $\boldsymbol{\phi}_{A_j}$, the *MAC* is defined as

$$MAC\left(\boldsymbol{\phi}_{X_i},\boldsymbol{\phi}_{A_j}\right)=\frac{\left|\boldsymbol{\phi}_{X_i}^H\boldsymbol{\phi}_{A_j}\right|^2}{\left(\boldsymbol{\phi}_{X_i}^H\boldsymbol{\phi}_{X_i}\right)\left(\boldsymbol{\phi}_{A_j}^H\boldsymbol{\phi}_{A_j}\right)} \tag{6.60}$$

This correlation criterion ranges from 0 to 1, meaning, respectively, uncorrelated and perfectly correlated mode shapes. Due to its simple implementation and interpretation, the *MAC* has become the most used correlation criterion in modal analysis, being reinterpreted and extended by many authors [6.38].

A matrix form is often used to represent the *MAC*, where a set of experimental mode shapes is compared to its predicted counterpart. An ideally perfect match leads to the identity matrix. As it is widely known, it is not realistic to assume that all the predicted eigenfrequencies can be identified from the experimental data set, as there can be some unexcited modes. Thus, the off-diagonal terms of *MAC*, with a significant value, are related to unpaired mode shapes. When the calculation is made between two mode shapes from the same set, it is known as *AutoMAC*. The Alternated Search Modal Assurance Criterion (*ASMAC*) was proposed to assess the eigenfrequencies pairing, which can be used to improve the analysis of the *MAC* values [6.39].

A well-known variant of *MAC* is the Coordinate Modal Assurance Criterion (*COMAC*), given by Lieven and Ewins [6.40]. This criterion attempts to evaluate which *DOF* i has a poor contribution to the *MAC*, regarding all the mode shapes within a frequency range of interest, and is defined as

$$COMAC(i)=\frac{\left(\sum_{j=1}^{L}\left|\phi_A(i,j)\phi_X^*(i,j)\right|\right)^2}{\sum_{j=1}^{L}\left|\phi_A(i,j)\right|^2\sum_{j=1}^{L}\left|\phi_X(i,j)\right|^2} \tag{6.61}$$

where L is the number of correlated mode pairs. The normalisation of mode shapes must be performed consistently due to the sensitivity of *COMAC* to differences in scale regarding analytical and experimental modal amplitudes. Mass-normalisation is an option, although it is not the only one.

Several variations and extensions have been proposed over the years [6.38], although *MAC* and *COMAC* are still the most used correlation techniques in the modal domain.

6.3.2 Frequency Domain

Regarding the correlation techniques developed in the frequency domain, the parallel with the ones deduced in the modal domain can be established [6.38].

Pascual et al. introduced the Frequency Domain Assurance Criterion (*FDAC*) [6.41], which quantifies the correlation between measured and simulated *FRFs* for a given excitation location q and a spectral range, which can be different for each *FRF* set. This indicator ranges between 0 and 1 and is given by

$$FDAC(\omega_i,\omega_j)_q = \frac{\left(\boldsymbol{H}_X(\omega_i)_q^H \boldsymbol{H}_A(\omega_j)_q\right)^2}{\left(\boldsymbol{H}_X(\omega_i)_q^H \boldsymbol{H}_X(\omega_i)_q\right)\left(\boldsymbol{H}_A(\omega_j)_q^H \boldsymbol{H}_A(\omega_j)_q\right)} \tag{6.62}$$

However, the *FDAC* computed using Eq. (6.62) is insensitive to phase differences in the considered *FRFs*. Therefore, an improved *FDAC*, considering the cosine between *FRFs* to be compared, was proposed [6.42]. The improved *FDAC*, also mentioned in [6.43], is given by

$$FDAC(\omega_i,\omega_j)_q = \frac{\boldsymbol{H}_X(\omega_i)_q^H \boldsymbol{H}_A(\omega_j)_q}{\left|\boldsymbol{H}_X(\omega_i)_q\right|\left|\boldsymbol{H}_A(\omega_j)_q\right|} \tag{6.63}$$

and ranges between -1 and 1, accounting for the possibility of an *FRF* pair being in phase opposition (*FDAC* close to -1).

Pascual et al. also proposed the Frequency Response Scale Factor (*FRSF*) [6.42], which is analogue to the Modal Scale Factor (*MSF*) [6.1] and measures the difference in magnitude between a measured and a predicted *FRF*.

The Response Vector Assurance Criterion (*RVAC*) is a particular case of *FDAC* [6.43], where the same spectral range for both measured and predicted *FRFs* is assumed:

$$RVAC(\omega_i)_q = \frac{\left(\boldsymbol{H}_X(\omega_i)_q^H \boldsymbol{H}_A(\omega_i)_q\right)^2}{\left(\boldsymbol{H}_X(\omega_i)_q^H \boldsymbol{H}_X(\omega_i)_q\right)\left(\boldsymbol{H}_A(\omega_i)_q^H \boldsymbol{H}_A(\omega_i)_q\right)} \tag{6.64}$$

As *FDAC*-like techniques are computed for a given spectral range, its computational effort can be demanding.

A local assessment criterion in the frequency domain, the Frequency Response Assurance Criterion (*FRAC*), is defined regarding the correlation of pairs of *FRF* vectors for a given response at *DOF* j due to an input at location q, within a certain frequency range. This correlation technique was proposed by Heylen and Lammens [6.44] and it is defined as

$$FRAC(j,q) = \frac{\left|\boldsymbol{H}_X(\omega)_{jq}^H \boldsymbol{H}_A(\omega)_{jq}\right|^2}{\left(\boldsymbol{H}_X(\omega)_{jq}^H \boldsymbol{H}_X(\omega)_{jq}\right)\left(\boldsymbol{H}_A(\omega)_{jq}^H \boldsymbol{H}_A(\omega)_{jq}\right)} \tag{6.65}$$

where $\boldsymbol{H}_{\cdot}(\omega)_{jq}$ is an *FRF* vector in a given spectral interval.

FRAC assesses the shape correlation of an *FRF* set, being more sensitive to changes in mass and stiffness. As *FRAC* compares $\boldsymbol{H}_{\cdot}(\omega)_{jq}$ at successive frequencies, its value can be difficult to interpret due to phase and order of magnitude differences between the *FRF* pair at *DOF* j due to an excitation at *DOF* q [6.45]. Thus, it can give false indications and should be used in conjunction with another correlation indicator.

Given an *FRF* pair, another significant measure of discrepancy between the pair is its amplitude difference. Hence, the Frequency Amplitude Assurance Criterion (*FAAC*) was introduced [6.43]. The *FAAC* is defined as

$$FAAC(j,q)=\frac{2\left|\boldsymbol{H}_X(\omega)_{jq}^H\,\boldsymbol{H}_A(\omega)_{jq}\right|}{\boldsymbol{H}_X(\omega)_{jq}^H\,\boldsymbol{H}_X(\omega)_{jq}+\boldsymbol{H}_A(\omega)_{jq}^H\,\boldsymbol{H}_A(\omega)_{jq}} \tag{6.66}$$

As it quantifies the amplitude discrepancies between *FRF* pairs, it is sensitive to errors in the modelling of damping.

Despite the type of information, both the *FRAC* and *FAAC* provide correlation information in a spatial domain, whereas the *FDAC* provides information in the frequency domain [6.46]. Moreover, it can be seen, from the formulation point of view, that the *FDAC* is equivalent to the *MAC*, although in the frequency domain, whereas the *FRAC* is equivalent to the *COMAC*.

In order to combine the amplitude and shape correlation information as a function of frequency, Grafe proposed the Global Shape Criterion (*GSC*) and the Global Amplitude Criterion (*GAC*) [6.47], while Dascotte and Strobbe introduced the Cross Signature Correlation (*CSC*) [6.48]. All these criteria can be classified as global due to the fact that they gather information about all the measured *DOFs*.

The *CSC* is composed by two criteria, the Cross Signature Assurance Criterion (*CSAC*) and the Cross Signature Scale Factor (*CSF*). These two criteria are the same as the ones given in [6.47], being the *CSAC* similar to the *GSC* and the *CSF* similar to the *GAC*. For simplicity, here only the *CSC* formulation is presented. The *CSC* is fully implemented in commercial software for *FE* model updating, not only for model correlation, but also as a metric to be used in an updating scheme based on the sensitivity of the *CSC* to the modelling parameters.

The *CSAC* is computed like the *MAC*, as it considers *FRF* vectors defined by all the measured *DOFs* j at each excited *DOF* q and frequency line ω_i, $\boldsymbol{H}(\omega_i)_q$. The *CSAC* assesses the frequency shift or shape correlation, being sensitive to mass and stiffness changes, and is given by

$$CSAC(\omega_i)_q=\frac{\left|\boldsymbol{H}_X(\omega_i)_q^H\,\boldsymbol{H}_A(\omega_i)_q\right|^2}{\left(\boldsymbol{H}_X(\omega_i)_q^H\,\boldsymbol{H}_X(\omega_i)_q\right)\left(\boldsymbol{H}_A(\omega_i)_q^H\,\boldsymbol{H}_A(\omega_i)_q\right)} \tag{6.67}$$

Due to the insensitivity of the *CSAC* to amplitude discrepancies, it is not able to distinguish FRFs with similar shapes but with different amplitudes. Furthermore, if only one measured *DOF* is considered in the *FRF* pair, the terms in Eq. (6.67) become scalars, leading to $CSAC = 1$ for the entire frequency range, even for uncorrelated *FRFs*. Hence, the need for a correlation metric concerning amplitude discrepancies arises. The *CSF* assesses the amplitude differences, being sensitive to damping changes, and it is computed as

$$CSF(\omega_i)_q=\frac{2\left|\boldsymbol{H}_X(\omega_i)_q^H\,\boldsymbol{H}_A(\omega_i)_q\right|}{\boldsymbol{H}_X(\omega_i)_q^H\,\boldsymbol{H}_X(\omega_i)_q+\boldsymbol{H}_A(\omega_i)_q^H\,\boldsymbol{H}_A(\omega_i)_q} \tag{6.68}$$

Both the *CSC* criteria values range between 0 and 1 for each frequency line ω_i, so it is useful to plot these criteria along with an *FRF* pair. However, if one desires to have only one value to look at, one can consider the mean values of the *CSAC* and *CSF* over a given frequency range. Note that both the *CSAC* and *CSF* allow to assess the correlation between *FRFs* in specific frequency regions.

Complementary, one may need to assess the correlation locally, concerning the measured *DOFs*. Therefore, Zang et al. proposed the Local Amplitude Criterion (*LAC*) [6.46], a variant of the *GAC*, defined by

$$LAC(\omega_i)_{jq} = \frac{2\left|H_X(\omega_i)^*_{jq} H_A(\omega_i)_{jq}\right|}{H_X(\omega_i)^*_{jq} H_X(\omega_i)_{jq} + H_A(\omega_i)^*_{jq} H_A(\omega_i)_{jq}} \tag{6.69}$$

which is obtained using Eq. (6.68) with elements of the *FRFs* instead of vectors. So, $H_{\cdot}(\omega_i)_{jq}$ is an *FRF* element taken at the measured *DOF j*, excitation *DOF q* and frequency ω_i. *LAC* values also range between 0 and 1 for each frequency line ω_i and a mean *LAC* can also be computed.

Some of these correlation techniques and derivations have also been used in the context of damage detection [6.49, 6.50].

6.3.3 A Brief Note on Model Validation

Often, the notion of model validation is misused. It is usual to address indifferently the notions of model validation and model correlation. In Section 6.3, several model correlation techniques or metrics have been presented. These techniques are tools that allow for the assessment of the quality of the updated model, by the correlation of its responses with the reference ones.

Model validation is defined in [6.51], as *the process of determining the degree to which a model is an accurate representation of the real world from the perspective of the intended uses of the model.*

In the context of deterministic model updating, all that one can assess is the agreement between the sets of model responses and their experimental counterparts, by the use of deterministic metrics. The referred accordance is supported by the considered correlation metrics and therefore, in this case, model correlation can be interpreted as model validation. However, within a stochastic model updating framework[3], the comparison of responses cannot be made response by response. One must consider instead the scatters or distributions of each response, as the prediction errors must take into account the existence of uncertainties [6.52]. Therefore, model validation is intended as a more refined concept than model correlation, where besides the agreement of a particular response correlation, one should evaluate the agreement of the model responses within the validation domain, with a certain degree of confidence based on statistical evidence [6.5, 6.53]. Further details on *Model Validation and Uncertainty*

[3] See Section 6.5.

Quantification can also be found in the courseware given by Cogan and Foltête [6.54] or Hemez [6.55].

6.4 Deterministic Model Updating

Addressing only the updating techniques applied to linear models, given their nature, it can be difficult to classify them properly, although it is commonly accepted that those methods can be classified as direct or iterative. Hemez and Farrar introduce three major categories, (i) optimum matrix updating methods, (ii) small perturbation methods, and (iii) iterative sensitivity-based methods [6.5]. Some of the most popular approaches are here reviewed.

6.4.1 Direct Optimum Matrix Updating Method

The optimum matrix updating methods are based on global corrections to the mass and stiffness matrices. These methods consider that the system matrices can be found if the initial estimated matrices are optimally corrected:

$$\boldsymbol{M} = \boldsymbol{M}_0 + \Delta\boldsymbol{M} \quad \text{and} \quad \boldsymbol{K} = \boldsymbol{K}_0 + \Delta\boldsymbol{K} \tag{6.70}$$

Baruch proposed one of the first attempts of model updating [6.56]. Assuming that the mass matrix is correct ($\Delta\boldsymbol{M} = \boldsymbol{0}$), a constraint objective problem can be formulated through the following Euclidean norm:

$$min\frac{1}{2}\left\|\boldsymbol{M}^{-1/2}\left(\boldsymbol{K} - \boldsymbol{K}_0\right)\boldsymbol{M}^{-1/2}\right\|_2^2 \tag{6.71}$$

subjected to symmetry and orthogonality conditions:

$$\boldsymbol{K} = \boldsymbol{K}^T \quad \text{and} \quad \boldsymbol{\Phi}^T\boldsymbol{K}\boldsymbol{\Phi} = \boldsymbol{\Lambda} \tag{6.72}$$

where $\boldsymbol{\Phi}$ and $\boldsymbol{\Lambda}$ are the experimental matrices of the mode shapes and squared natural frequencies, respectively. The constraint optimisation problem is solved using Lagrange multipliers and an optimal correction to the stiffness matrix is given by

$$\Delta\boldsymbol{K} = -\boldsymbol{K}_0\boldsymbol{\Phi}\boldsymbol{\Phi}^T\boldsymbol{M} - \boldsymbol{M}\boldsymbol{\Phi}\boldsymbol{\Phi}^T\boldsymbol{K}_0 + \boldsymbol{M}\boldsymbol{\Phi}\boldsymbol{\Phi}^T\boldsymbol{K}_0\boldsymbol{\Phi}\boldsymbol{\Phi}^T\boldsymbol{M} + \boldsymbol{M}\boldsymbol{\Phi}\boldsymbol{\Lambda}\boldsymbol{\Phi}^T\boldsymbol{M} \tag{6.73}$$

Berman and Nagy adopted a similar approach in order to compute an optimal correction to both mass and stiffness matrices by solving two subsequent constraint optimisation problems [6.57]. Hence, an optimal correction of the mass matrix is defined, solving

$$min\frac{1}{2}\left\|\boldsymbol{M}^{-1/2}\left(\boldsymbol{M} - \boldsymbol{M}_0\right)\boldsymbol{M}^{-1/2}\right\|_2^2 \tag{6.74}$$

subjected to the orthogonality condition $\boldsymbol{\Phi}^T\boldsymbol{M}\boldsymbol{\Phi} = \boldsymbol{I}$, using Lagrange multipliers. The optimal correction of the mass matrix is given by

$$\Delta\boldsymbol{M} = -\boldsymbol{M}_0\boldsymbol{\Phi}\bar{\boldsymbol{M}}_0^{-1}\left(\boldsymbol{I} - \bar{\boldsymbol{M}}_0\right)\bar{\boldsymbol{M}}_0^{-1}\boldsymbol{\Phi}^T\boldsymbol{M}_0 \tag{6.75}$$

with $\bar{\boldsymbol{M}}_0 = \boldsymbol{\Phi}^T \boldsymbol{M}_0 \boldsymbol{\Phi}$. With the updated mass matrix, one can compute the optimal correction to the stiffness matrix given in Eq. (6.73).

Several developments and adaptations to these methods are reviewed in [6.3]. Friswell et al. proposed an augmented constraint optimisation problem to simultaneously update the stiffness and damping matrices, considering an *a priori* correct mass matrix [6.58]. In this sense, Eq. (6.71) is augmented to include a norm concerning the damping error, such that the objective function J to be minimised is

$$J = \frac{1}{2}\left\| \boldsymbol{M}^{-1/2}\left(\boldsymbol{K}-\boldsymbol{K}_0\right)\boldsymbol{M}^{-1/2}\right\|_2^2 + \frac{1}{2}\gamma\left\| \boldsymbol{M}^{-1/2}\left(\boldsymbol{C}-\boldsymbol{C}_0\right)\boldsymbol{M}^{-1/2}\right\|_2^2 \tag{6.76}$$

subjected to the symmetry and orthogonality conditions

$$\boldsymbol{K} = \boldsymbol{K}^T, \quad \boldsymbol{C} = \boldsymbol{C}^T \quad \text{and} \quad \boldsymbol{\Phi}^T \boldsymbol{M} \boldsymbol{\Phi} = \boldsymbol{I} \tag{6.77}$$

In (6.76), γ is a weighting coefficient, enabling the improvement of the damping matrix estimation. This is a regularisation type of coefficient that can be determined by trial and error or using the corner of the L-curve, as described by Ahmadian et al. [6.59]. Extensive reviews on damping identification methods and modelling can be found in [6.60, 6.61].

6.4.2 *FRF*-based Direct Updating Method

In this section, a direct updating method in the frequency domain, capable of identifying the mass, stiffness and damping matrices, is presented. In an updating process, one must minimise the difference between the responses obtained numerically (e.g. using a spatial model) and the ones obtained experimentally. Usually, the relation between numerical and response models can be stated starting with the dynamic equilibrium equation that, in steady-state conditions, for a harmonic excitation ($\boldsymbol{f}(t) = \boldsymbol{F}\mathrm{e}^{i\omega t}$), leads to:

$$\left(\boldsymbol{K} - \omega^2 \boldsymbol{M} + i\omega \boldsymbol{C}\right)\boldsymbol{X} = \boldsymbol{F} \quad \Rightarrow \quad \boldsymbol{Z}\boldsymbol{X} = \boldsymbol{F} \tag{6.78}$$

One should bear in mind that $\boldsymbol{Z}$ is a function of the frequency ω, although this dependence is omitted for the sake of simplicity.

Considering the separation of the complex $\boldsymbol{Z}$ into its real and imaginary parts, $\boldsymbol{Z}_R$ and $\boldsymbol{Z}_I$, respectively, one has

$$\left(\boldsymbol{Z}_R + i\omega \boldsymbol{C}\right)\boldsymbol{X} = \boldsymbol{F} \tag{6.79}$$

Often, all one has from the numerical model is the undamped dynamic stiffness matrix $\boldsymbol{Z}_R$. As the corresponding response model is given by $\boldsymbol{X} = \boldsymbol{H}\boldsymbol{F}$, where $\boldsymbol{H}$ is the receptance matrix, it follows that

$$\boldsymbol{Z}^{-1} = \boldsymbol{H} \tag{6.80}$$

As $\boldsymbol{H}$ is obtained experimentally, it is complex. Therefore, $\boldsymbol{H}$ can be written as $\boldsymbol{H}_R + i\boldsymbol{H}_I$ and Eq. (6.80) can be recast as:

$$\left(\boldsymbol{I} + i\omega \boldsymbol{Z}_R^{-1}\boldsymbol{C}\right)^{-1}\boldsymbol{Z}_R^{-1} = \boldsymbol{H}_R + i\boldsymbol{H}_I \tag{6.81}$$

or

$$\boldsymbol{Z}_R^{-1} = \left(\boldsymbol{I} + i\omega \boldsymbol{Z}_R^{-1}\boldsymbol{C}\right)\left(\boldsymbol{H}_R + i\boldsymbol{H}_I\right) \tag{6.82}$$

Equating the real and imaginary parts, it follows that:

$$\boldsymbol{Z}_R^{-1} = \boldsymbol{H}_R - \omega \boldsymbol{Z}_R^{-1}\boldsymbol{C}\boldsymbol{H}_I \tag{6.83}$$

$$\boldsymbol{0} = \omega \boldsymbol{Z}_R^{-1}\boldsymbol{C}\boldsymbol{H}_R + \boldsymbol{H}_I \tag{6.84}$$

Solving Eqs. (6.83) and (6.84) for $\boldsymbol{Z}_R$ and $\boldsymbol{C}$, one finally obtains:

$$\boldsymbol{Z}_R = \left(\boldsymbol{H}_R + \boldsymbol{H}_I\boldsymbol{H}_R^{-1}\boldsymbol{H}_I\right)^{-1} \tag{6.85}$$

and

$$\boldsymbol{C} = -\frac{1}{\omega}\left(\boldsymbol{H}_I + \boldsymbol{H}_R\boldsymbol{H}_I^{-1}\boldsymbol{H}_R\right)^{-1} \tag{6.86}$$

If one had chosen the hysteretic damping model, the term $i\omega\boldsymbol{C}$ in Eq. (6.78) would be $i\boldsymbol{D}$, where $\boldsymbol{D}$ is the hysteretic damping matrix. In that case the damping matrix would be computed as:

$$\boldsymbol{D} = -\left(\boldsymbol{H}_I + \boldsymbol{H}_R\boldsymbol{H}_I^{-1}\boldsymbol{H}_R\right)^{-1} \tag{6.87}$$

This set of equations (Eqs. (6.85) and (6.86)) lead to damping matrices that are function of the frequency, and therefore some kind of average should be undertaken (see Eq. (6.94)).

In principle, the aforementioned set of equations allows for a direct identification of the system in terms of its spatial properties, without the need for any special initial concerns on damping. However, in practice, issues related to model matching arise, as the response model is always very limited in size, demanding either a condensation of the numerical model or an expansion of the response one. Moreover, one can take advantage of the initial undamped numerical model, which may already be considerably close to the experimental results, looking only for relatively small adjustments $\Delta\boldsymbol{Z}_R$ such that the numerical prediction of $\boldsymbol{Z}_R$ is defined as $\boldsymbol{Z}_R = \boldsymbol{Z}_R^0 + \Delta\boldsymbol{Z}_R$, considering the initial undamped model $\boldsymbol{Z}_R^0$. Hence, one has,

$$\Delta\boldsymbol{Z}_R = \left(\boldsymbol{H}_R + \boldsymbol{H}_I\boldsymbol{H}_R^{-1}\boldsymbol{H}_I\right)^{-1} - \boldsymbol{Z}_R^0 \tag{6.88}$$

If the damping matrix is defined as $\boldsymbol{C} = \boldsymbol{C}_0 + \Delta\boldsymbol{C}$ (or $\boldsymbol{D} = \boldsymbol{D}_0 + \Delta\boldsymbol{D}$) and the initial estimate for the damping $\boldsymbol{C}_0$ (or $\boldsymbol{D}_0$) is taken as zero, one can use directly Eq. (6.86) (or Eq. (6.87)) to estimate the damping matrix for each frequency line. Otherwise, one can chose to estimate the damping matrix iteratively, therefore $\Delta\boldsymbol{C}$ is defined as:

$$\Delta\boldsymbol{C} = -\frac{1}{\omega}\left(\boldsymbol{H}_I + \boldsymbol{H}_R\boldsymbol{H}_I^{-1}\boldsymbol{H}_R\right)^{-1} - \boldsymbol{C}_0 \tag{6.89}$$

or $\Delta\boldsymbol{D}$ as:

$$\Delta\boldsymbol{D} = -\left(\boldsymbol{H}_I + \boldsymbol{H}_R\boldsymbol{H}_I^{-1}\boldsymbol{H}_R\right)^{-1} - \boldsymbol{D}_0 \tag{6.90}$$

It can be observed that Eqs. (6.89) and (6.90) also lead to damping matrices that are function of the frequency, and therefore some kind of average should be undertaken.

The model updating technique itself is a direct procedure, if a complete *FRF* matrix is available. So, this method is based on the idealistic hypothesis of experimentally obtaining a full set of measured *FRFs* at all *DOFs*. If the full experimental response model is attained, the model updating problem is linear and can be solved in one iteration with a perfect correlation between test and model data. However, in practice, it is not possible to measure or to apply excitation onto all the *DOFs* considered in the *FE* model. Hence, it is inevitable to deal with model incompleteness, as discussed in Section 6.2. The modified Kidder's method may be used with advantage in this context.

The identification of the damping matrix is explicitly defined in Eqs. (6.86) and (6.89) (or Eqs. (6.87) and (6.90)). On the other hand, the identification of mass and stiffness matrices is performed recurring to the real part of the dynamic stiffness matrix, updated by Eq. (6.88), based on a least squares approximation of the referred system matrices; care must be taken in the selection of the frequency lines used for identification, as it can lead to a degradation of the updated results.

The coefficients of the mass and stiffness matrices are identified using the Moore-Penrose pseudo-inverse, for each element of matrix $\boldsymbol{Z}$, for all the N_ω frequency lines considered in the identification process, as

$$z_{ij}\left(\omega_k\right) = k_{ij} - \omega_k^2 m_{ij} \qquad k = 1,\ldots,N_\omega \tag{6.91}$$

Rearranging Eq. (6.91), one has

$$\begin{Bmatrix} z_{ij}\left(\omega_1\right) \\ z_{ij}\left(\omega_2\right) \\ \vdots \\ z_{ij}\left(\omega_{N_\omega}\right) \end{Bmatrix} = \begin{bmatrix} 1 & -\omega_1^2 \\ 1 & -\omega_2^2 \\ \vdots & \vdots \\ 1 & -\omega_{N_\omega}^2 \end{bmatrix} \begin{Bmatrix} k_{ij} \\ m_{ij} \end{Bmatrix} = \boldsymbol{\Omega} \begin{Bmatrix} k_{ij} \\ m_{ij} \end{Bmatrix} \tag{6.92}$$

From Eq. (6.92), one extracts the $\boldsymbol{K}$ and $\boldsymbol{M}$ elements, reassembling the identified system matrices, considering

$$\begin{Bmatrix} k_{ij} \\ m_{ij} \end{Bmatrix} = \boldsymbol{\Omega}^+ \boldsymbol{z}_{ij} \tag{6.93}$$

Regarding the identification of an updated damping matrix averaged over a given frequency range of interest, one can reformulated Eq. (6.86) as

$$\boldsymbol{C} = -\frac{1}{N_\omega}\sum_{i=1}^{N_\omega}\omega_i^{-1}\left(\boldsymbol{H}(\omega_i)_I + \boldsymbol{H}(\omega_i)_R\boldsymbol{H}(\omega_i)_I^{-1}\boldsymbol{H}(\omega_i)_R\right)^{-1} \tag{6.94}$$

The process used to identify modelling properties is sensitive to the frequency, so care must be taken when selecting a frequency range or different sets of frequency lines to perform the identification of $\boldsymbol{C}$, $\boldsymbol{K}$ and $\boldsymbol{M}$. This topic has been addressed by Kwon and Lin [6.62]. With respect to $\boldsymbol{K}$ and $\boldsymbol{M}$ it is good practice to select frequencies avoiding resonant peaks within the frequency range considered for updating. However, in what damping is concerned, better estimates should be obtained with frequencies near the resonant peaks, where damping dominates the response. For details on system identification, the reader is directed to [6.1].

Example 6.4.2 Application of the *FRF*-based direct model updating

A planar clamped-clamped Bernoulli-Euler steel beam ($E = 200$ GPa, $\rho = 7800$ kg/m^3), modelled by the *FE* method[4] with 30 elements (Figure 6.1), is considered to illustrate the *FRF*-based direct updating procedure. The modelled beam is 910 mm long and it has a rectangular cross-section with 50 mm in width by 5 mm in thickness (h).

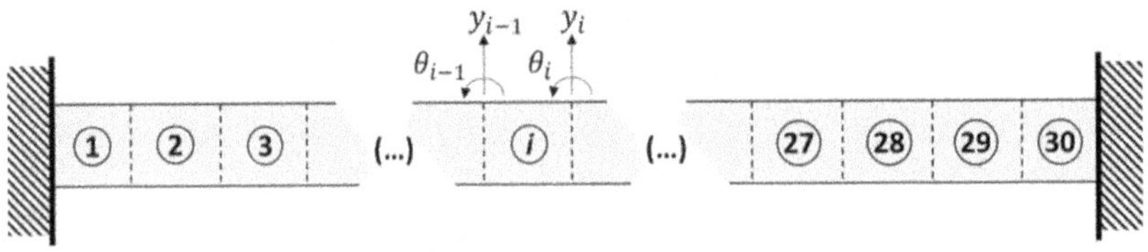

Figure 6.1 A clamped-clamped Bernoulli-Euler beam with 30 *FE* and the definition of the *DOFs* at element i.

The simulated experimental *FRFs* (reference model) are generated from the referred *FE* model, considering proportional damping ($\boldsymbol{C} = 1\times10^{-5}\boldsymbol{K}$), in the frequency range of [0,800] Hz with 1200 frequency points. As test reference, common to all examples, one has chosen the $FRF_{9,9}$ (a *driving point* translation/force *FRF*) and the $FRF_{16,9}$ (a *cross* rotation/force *FRF*).

As an initial estimate for updating, one has considered an undamped *FE* model based on a perturbation of the reference model, such as a global increase of 10% in the thickness of all the elements, except for the local perturbations on 6 of them, where the thickness is defined as $h_i = [1.20, 1.40, 1.25, 1.40, 1.30, 1.30]h$ for $i = 3, 5, 11, 16, 25, 29$. The result before updating is shown in Figure 6.2, for $FRF_{9,9}$, where the discrepancy between the reference and the initial model responses is supported by the *CSC* criteria (Section 6.3.2).

[4] See Section 3.5.4.

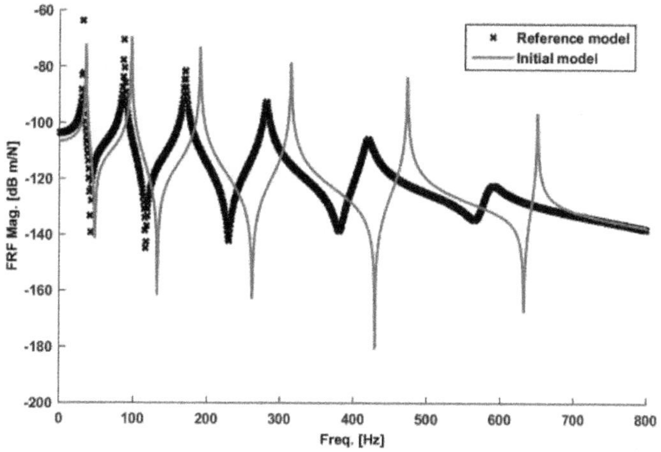

(a) Magnitude of $FRF_{9,9}$

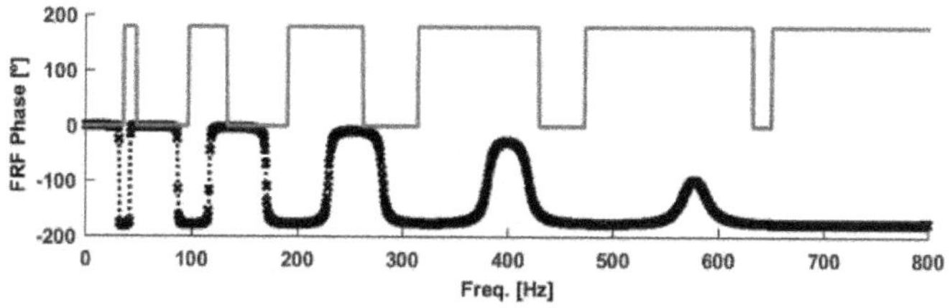

(b) Phase of $FRF_{9,9}$

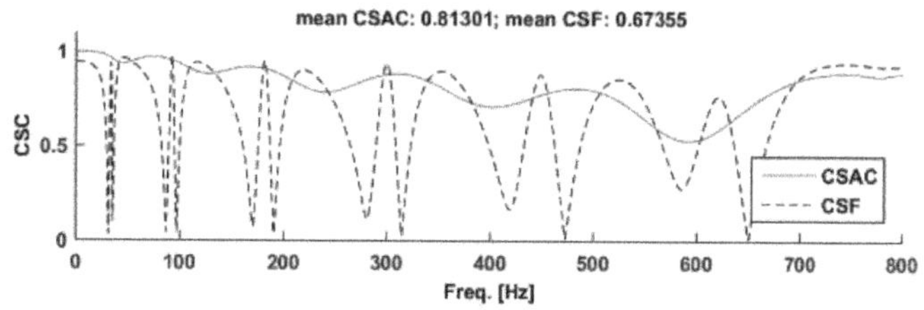

(c) *CSC* of $FRF_{9,9}$

Figure 6.2 Comparison of $FRF_{9,9}$ for the reference model and for the model before updating, and the corresponding *CSC*, for noise-free data.

This example starts by considering that it is possible to measure the complete noise-free *FRF* matrix. The example is far from the experimental reality; however, it is important to verify the theoretical validity of the updating method. Moreover, by this idealistic case, one can conclude that if a perfect expanded *FRF* matrix is available, the model is perfectly updated. Figure 6.3 shows a perfect agreement between the reference and the updated *FRFs*.

Moreover, as the *FRF* matrix is reconstructed using the identified model matrices, the model matrices themselves are updated as well. In terms of the physical connectivity, direct updating methods usually produce updated system matrices that have lost their physical connectivity to

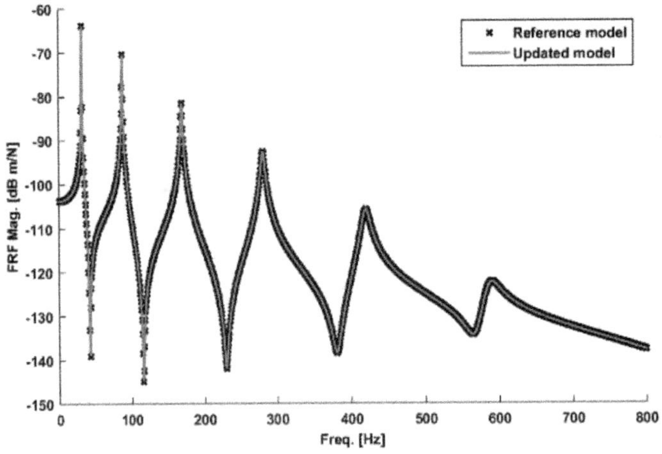

(a) Magnitude of $FRF_{9,9}$

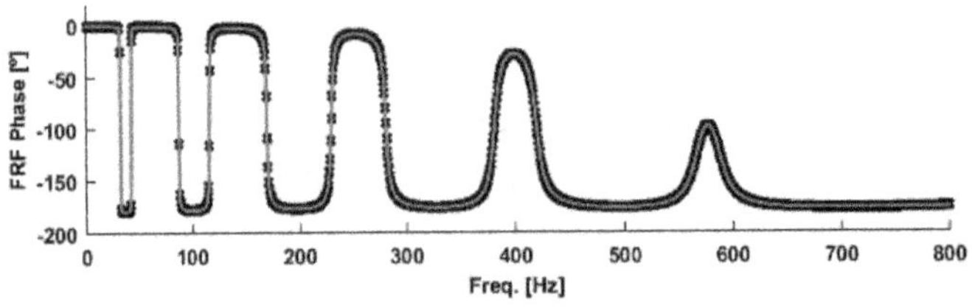

(b) Phase of $FRF_{9,9}$

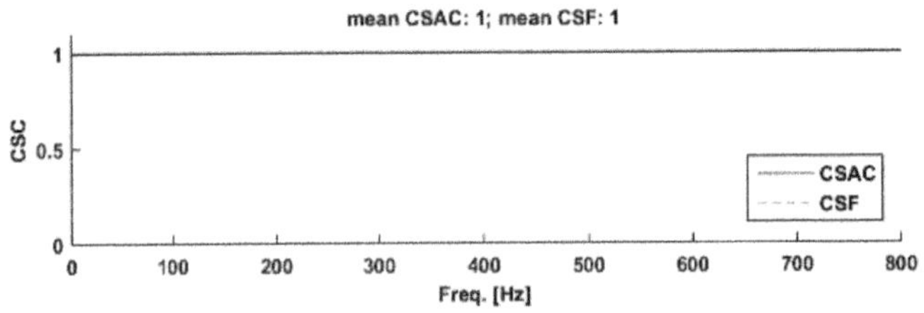

(c) CSC of $FRF_{9,9}$

Figure 6.3 Comparison of $FRF_{9,9}$ for the reference model and for the model after updating, and the corresponding CSC, for noise-free data.

the reference model, i.e., the updated matrices are populated with off-diagonal elements, while the original ones are sparse matrices concentrated around the main diagonal. However, for this ideal case, the updated matrices preserve the physical connectivity, as shown in Figure 6.4, where one can evaluate the perfect reconstruction of the reference model matrices by the value of the difference between the initial (subscript "$_0$") and the updated matrices. Here, the perturbation imposed to the reference model is notorious. Note the higher changes in the elemental matrices, related to elements 3, 5, 11, 16, 25, 29, identifiable specially in Figures 6.4 (a)-(b).

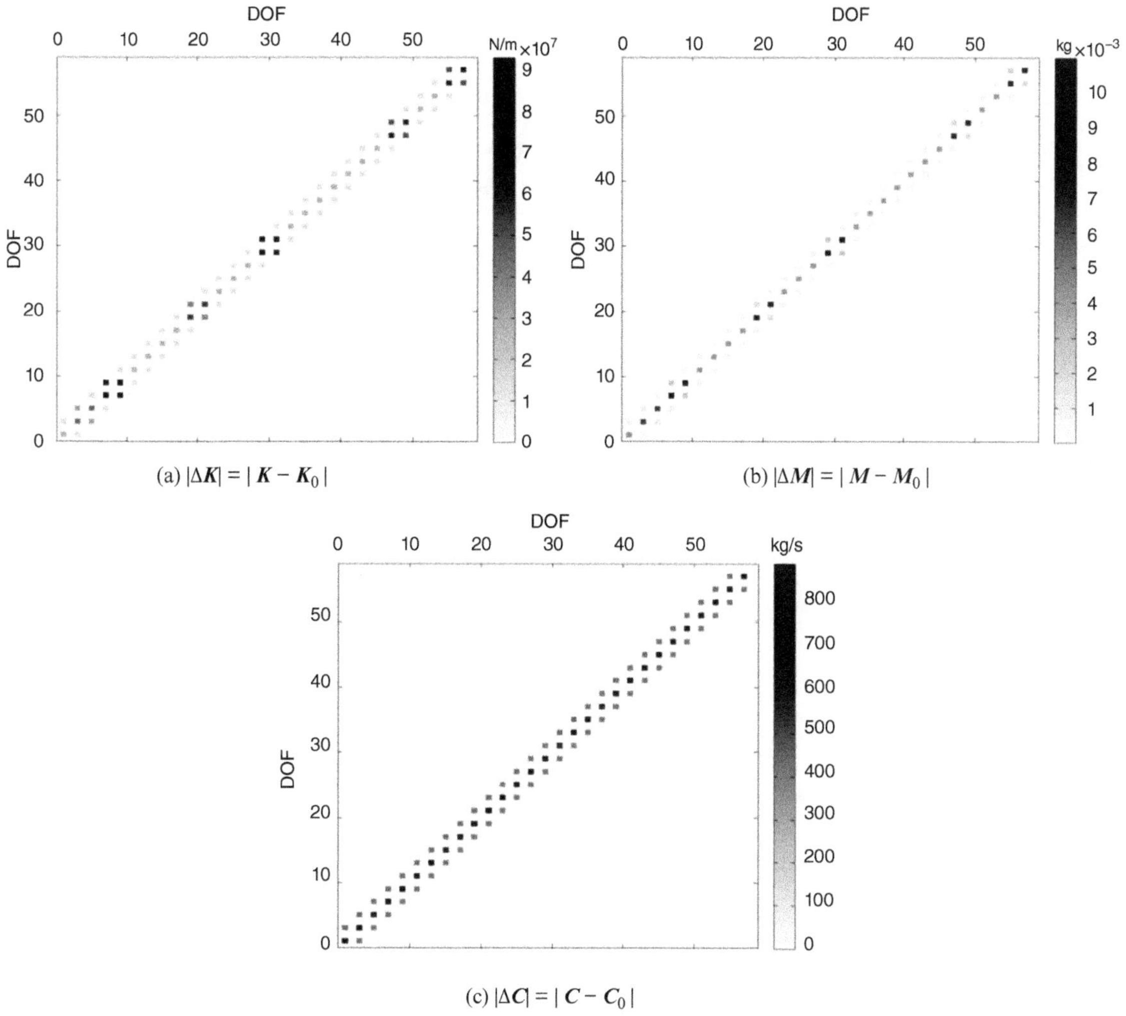

(a) $|\Delta K| = |\boldsymbol{K} - \boldsymbol{K}_0|$

(b) $|\Delta M| = |\boldsymbol{M} - \boldsymbol{M}_0|$

(c) $|\Delta C| = |\boldsymbol{C} - \boldsymbol{C}_0|$

Figure 6.4 Change in the model matrices after updating when compared to the initial estimate, for a completely measured matrix $\boldsymbol{H}$ and for noise-free data (Note that the darker the colour, the higher the change).

Regarding the conditions for this example, if one adds noise as a zero-mean uniformly distributed perturbation to the reference *FRF* data, the quality of the obtained results after updating are perturbed, as shown in Figure 6.5. However, the mean values of the *CSC* still indicate a very high-level of correlation.

The updating process is very sensitive to noise, due to the corruption of the identified model matrices, as it is evident by the changes to the system matrices shown in Figure 6.6. In the present case, the system matrices lost the physical connectivity. However, the updated *FRFs* are in good agreement with the reference ones. The performance of the method is also very sensitive to model incompleteness.

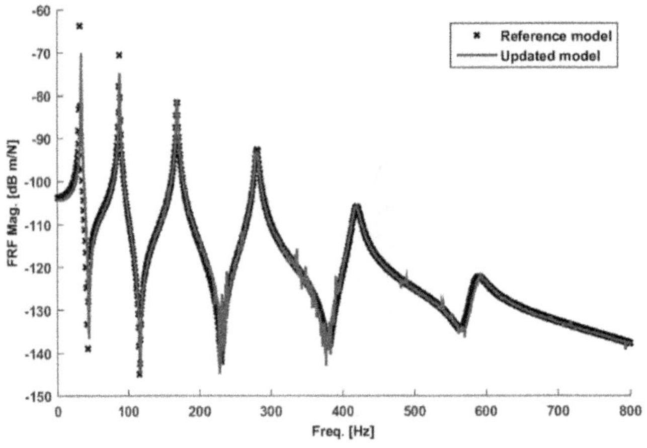

(a) Magnitude of $FRF_{9,9}$

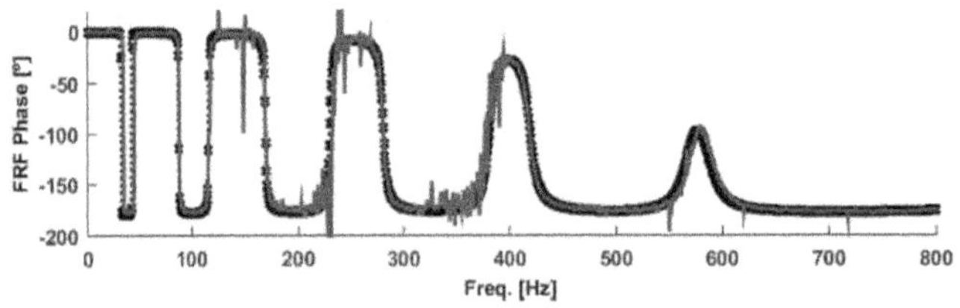

(b) Phase of $FRF_{9,9}$

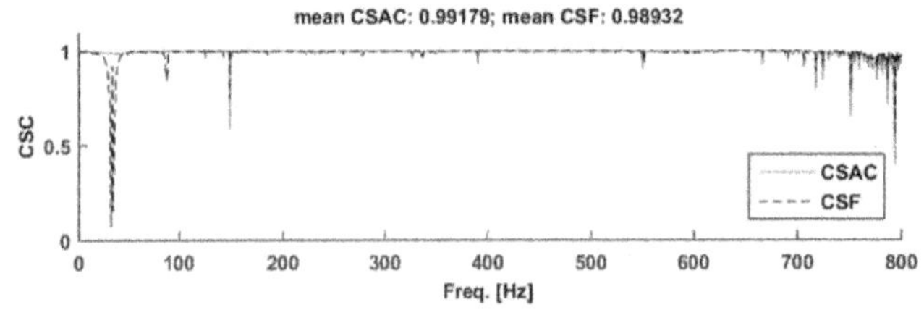

(c) CSC of $FRF_{9,9}$

Figure 6.5 Comparison of $FRF_{9,9}$ for the reference model and for the model after updating, and the corresponding CSC, for data with 1% noise.

6.4.3 Sensitivity-based Model Updating

The sensitivity-based model updating method was reviewed by Mottershead et al. [6.63]. Generally, this kind of methods consider that the experimental or reference model can be cast as a perturbation about a theoretical model approximation. By this, the error between the experimental and predicted responses is

$$\boldsymbol{\varepsilon}_{z_j} = z_m - z(\theta_j) \tag{6.95}$$

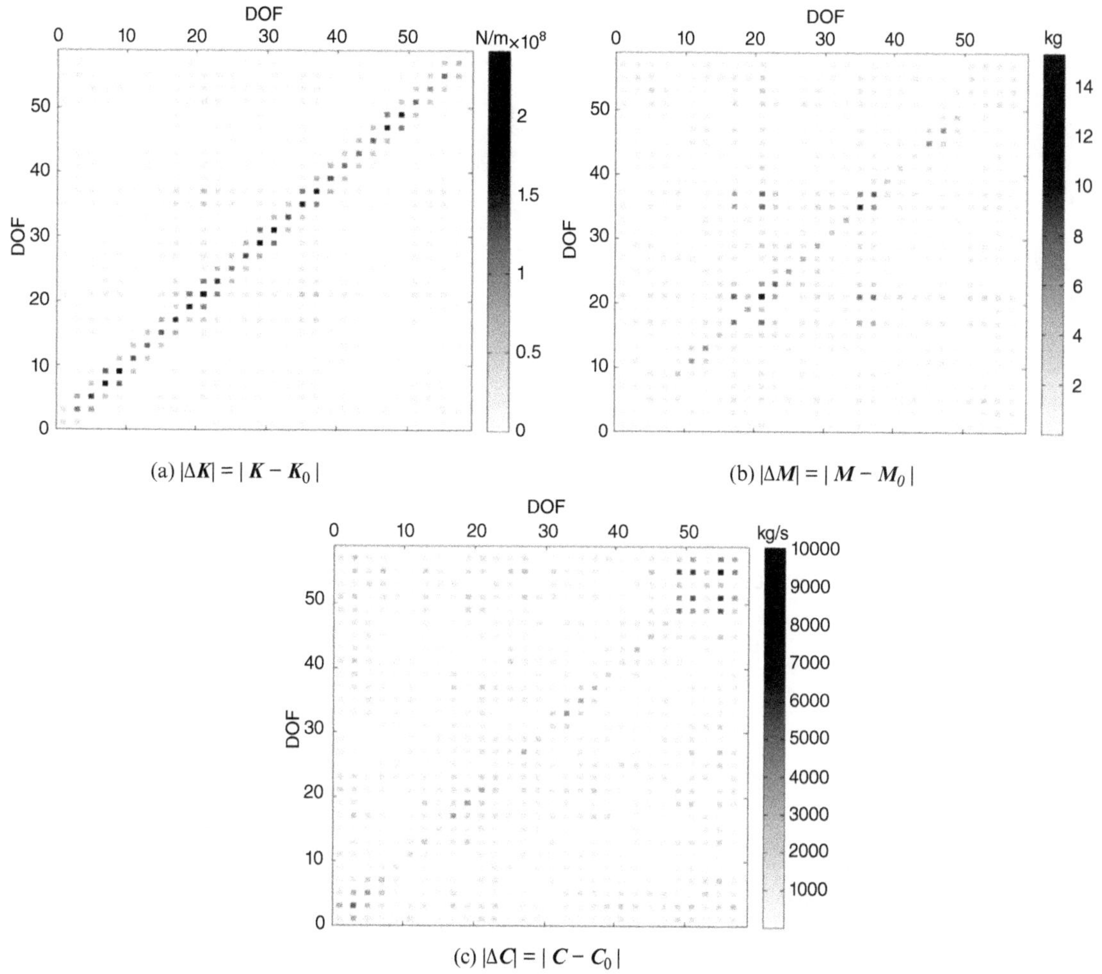

Figure 6.6 Change in the model matrices after updating when compared to the initial estimate, for a completely measured matrix $\boldsymbol{H}$ and for noise-free data (Note that the darker the colour, the higher the change).

where $\boldsymbol{z}_m$ is the vector of n experimentally measured responses and $\boldsymbol{z}(\boldsymbol{\theta}_j)$ is the vector of the corresponding counterparts predicted by the model at the jth iteration step, and therefore function of the modelling parameters vector, or – to be more precise – the updating parameters vector $\boldsymbol{\theta}_j$. Here, it is considered that the response vectors may include the eigenvalues and eigenvectors, or elements of them, such as a generic response vector given by

$$\boldsymbol{z} = \begin{bmatrix} \omega_1^2 & \omega_2^2 & \cdots & \omega_{r_\omega}^2 & \boldsymbol{\phi}_1^T & \boldsymbol{\phi}_2^T & \cdots & \boldsymbol{\phi}_{r_\phi}^T \end{bmatrix}^T \tag{6.96}$$

where r_ω and r_ϕ denote that one may consider a different number of eigenfrequencies and eigenvectors. Mass-normalised eigenvectors are here considered, although this is not mandatory.

The error given in Eq. (6.95) can be approximated by an expansion in Taylor series of $z(\boldsymbol{\theta}_j)$, as

$$\boldsymbol{\varepsilon}_{z_j} = z_m - \left(z_j + \boldsymbol{S}_j\left(\boldsymbol{\theta}_{j+1} - \boldsymbol{\theta}_j\right) + O\left(\boldsymbol{\theta}_j^2\right)\right) \approx z_m - z_j - \boldsymbol{S}_j \Delta\boldsymbol{\theta}_j \tag{6.97}$$

where $\boldsymbol{S}_j$ is the sensitivity matrix assembled as

$$\boldsymbol{S}_j = \sum_{l=1}^{k} \left.\frac{\partial z_j}{\partial \theta_l}\right|_{\theta=\theta_j} \tag{6.98}$$

for $i = 1, \ldots, n$ experimentally measured outputs and $l = 1, \ldots, k$ modelling parameters at the jth iteration.

As $\boldsymbol{\varepsilon}_{z_j} \to \boldsymbol{0}$, Eq. (6.97) can be re-written as

$$z_m - z_j = \boldsymbol{S}_j\left(\boldsymbol{\theta}_{j+1} - \boldsymbol{\theta}_j\right) \tag{6.99}$$

or

$$\Delta z_j = \boldsymbol{S}_j \Delta\boldsymbol{\theta}_j \tag{6.100}$$

Eigensensitivity Approach
The exact analytical solution of Eq. (6.98), in terms of sensitivity of the eigensolutions was given by Fox and Kapoor [6.64]. The sensitivity of the predicted eigenvalues is found by the partial derivative (for undamped systems) *w.r.t.* the updating parameters, as

$$\frac{\partial \omega_i^2}{\partial \theta_l} = \boldsymbol{\phi}_i^T \left(\frac{\partial \boldsymbol{K}}{\partial \theta_l} - \omega_i^2 \frac{\partial \boldsymbol{M}}{\partial \theta_l} \right) \boldsymbol{\phi}_i \tag{6.101}$$

As the eigenvectors are linearly independent, the sensitivity of the predicted eigenvectors is found by a weighted linear combination of the $M \leq N$ eigenvectors of interest, defined as

$$\frac{\partial \boldsymbol{\phi}_i}{\partial \theta_l} = \sum_{r=1}^{M} a_{ir}^{(l)} \boldsymbol{\phi}_r \tag{6.102}$$

which can be proved to result in the following expression [6.1],

$$\frac{\partial \boldsymbol{\phi}_i}{\partial \theta_l} = \sum_{r=1, r\neq i}^{M} \left(\frac{\boldsymbol{\phi}_r \boldsymbol{\phi}_r^T}{\omega_i^2 - \omega_r^2} \left(\frac{\partial \boldsymbol{K}}{\partial \theta_l} - \omega_i^2 \frac{\partial \boldsymbol{M}}{\partial \theta_l} \right) \boldsymbol{\phi}_i \right) - \frac{1}{2} \boldsymbol{\phi}_i \boldsymbol{\phi}_i^T \frac{\partial \boldsymbol{M}}{\partial \theta_l} \boldsymbol{\phi}_i \tag{6.103}$$

Knowing the sensitivity matrix, the updated vector of modelling parameters is then given by

$$\boldsymbol{\theta}_{j+1} = \boldsymbol{\theta}_j + \boldsymbol{T}_j \left(\boldsymbol{z}_m - \boldsymbol{z}_j \right) \tag{6.104}$$

where the transformation matrix $\boldsymbol{T}_j$ is generally the weighted pseudo-inverse of the sensitivity matrix $\boldsymbol{S}_j$,

$$\boldsymbol{T}_j = \left(\boldsymbol{S}_j^T \boldsymbol{W}_\varepsilon \boldsymbol{S}_j + \boldsymbol{W}_\vartheta \right)^{-1} \boldsymbol{S}_j^T \boldsymbol{W}_\varepsilon \tag{6.105}$$

and $\boldsymbol{W}_\varepsilon$ and $\boldsymbol{W}_\vartheta$ are scaling and regularisation matrices, respectively, to allow for the regularisation of ill-posed sensitivity equations [6.59, 6.65].

The vector of updated parameters of Eq. (6.104) can be computed using a formulation based on a constraint optimisation problem, as detailed for instance by Link [6.66]. The referred optimisation problem consists of the minimisation of the following objective function:

$$J(\boldsymbol{\theta}) = \boldsymbol{\varepsilon}_z^T \boldsymbol{W}_\varepsilon \boldsymbol{\varepsilon}_z + \Delta\boldsymbol{\theta}_j^T \boldsymbol{W}_\vartheta \Delta\boldsymbol{\theta}_j \tag{6.106}$$

leading to

$$\Delta\boldsymbol{\theta}_j = \left(\boldsymbol{S}_j^T \boldsymbol{W}_\varepsilon \boldsymbol{S}_j + \boldsymbol{W}_\vartheta \right)^{-1} \boldsymbol{S}_j^T \boldsymbol{W}_\varepsilon \Delta \boldsymbol{z}_j \tag{6.107}$$

which is a weighted least-squares approximation to the solution and, obviously, this is the result of Eq. (6.104).

Usually, for well-posed problems, one takes $\boldsymbol{W}_\varepsilon = \boldsymbol{I}$ and $\boldsymbol{W}_\vartheta = \boldsymbol{0}$, so that $\boldsymbol{T}_j$ is the pseudo-inverse of the sensitivity matrix. In general, the regularisation matrix can be given by $\boldsymbol{W}_\vartheta = \lambda^2 \boldsymbol{I}$, where λ is the regularisation parameter, which is able to control the step size during the updating process, in terms of the correction to the parameters. So, the higher the value of λ, the slower is the convergence of the iterative process. Regarding regularisation, a detailed discussion is given in [6.65]. There, two different main approaches to establish $\boldsymbol{W}_\vartheta$ are presented, one of them due to Link [6.66], where the regularisation matrix is formed using the inverse of the squared weighted sensitivity matrix, and another one due to Ahmadian et al. [6.59], in which the regularisation parameter corresponds to the corner of a *L*-curve, which results from the representation of the norm of the constraint deemed to regularise the problem $\Delta\boldsymbol{\theta}_j^T \boldsymbol{W}_\vartheta \Delta\boldsymbol{\theta}_j$ against the norm of the weighted residual $\boldsymbol{\varepsilon}_z^T \boldsymbol{W}_\varepsilon \boldsymbol{\varepsilon}_z$.

The sensitivity method has been extended to lead with stochastic model updating as given in [6.67-6.69] and reviewed in [6.70]. This issue is treated in detail in Section 6.5.

***FRF* Sensitivity-based Approach**

The least-squares approximation can also be applied to updating problems stated in the frequency domain, which is the case of the Response Function Method (*RFM*), proposed by Lin and Ewins [6.71]. This method is based on the existent difference between experimental and predicted response functions and it is deduced for a single measured *FRF* vector $\boldsymbol{H}_m$ as follows.

Using the dynamic stiffness concept, one can match the responses of the models, as

$$\boldsymbol{Z}_{j+1}\boldsymbol{H}_m = \boldsymbol{Z}_j\boldsymbol{H}_j \tag{6.108}$$

where $\boldsymbol{H}_j$ is the predicted *FRF* vector and $\boldsymbol{Z}_j$ is the dynamic stiffness matrix of the model at the jth iteration. Therefore, $\boldsymbol{Z}_{j+1}$ is the dynamic stiffness matrix of the updated model.

Rearranging Eq. (6.108), one obtains

$$\left(\boldsymbol{Z}_{j+1} - \boldsymbol{Z}_j\right)\boldsymbol{H}_m = \boldsymbol{Z}_j\left(\boldsymbol{H}_j - \boldsymbol{H}_m\right) \tag{6.109}$$

pre-multiplying both sides of Eq. (6.109) by the predicted *FRF* matrix $\boldsymbol{H}_j$ leads to

$$\boldsymbol{H}_j\Delta\boldsymbol{Z}_j\boldsymbol{H}_m = \left(\boldsymbol{H}_j - \boldsymbol{H}_m\right) = \Delta\boldsymbol{H}_j \tag{6.110}$$

The residual $\Delta\boldsymbol{Z}_j$ can be expanded in Taylor series, as

$$\boldsymbol{Z}_{j+1} = \boldsymbol{Z}_j + \sum_{l=1}^{k}\frac{\partial \boldsymbol{Z}_j}{\partial\theta_l}\Delta\theta_l\bigg|_{\theta=\theta_j} + O\left(\theta_j^2\right) \tag{6.111}$$

where θ_l is referred to the lth parameter of the parameter vector $\boldsymbol{\theta}_j$ at the jth iteration.

Considering the truncation of the series after the first-order terms, one has

$$\Delta\boldsymbol{Z}_j = \sum_{l=1}^{k}\frac{\partial \boldsymbol{Z}_j}{\partial\theta_l}\Delta\theta_l\bigg|_{\theta=\theta_j} \tag{6.112}$$

Substituting Eq. (6.112) in Eq. (6.110) leads to

$$\Delta\boldsymbol{H}_j = \sum_{l=1}^{k}\left(\boldsymbol{H}_j\frac{\partial \boldsymbol{Z}_j}{\partial\theta_l}\boldsymbol{H}_m\right)\Delta\theta_l \tag{6.113}$$

that can be recast as an expression similar to Eq. (6.100):

$$\Delta\boldsymbol{H}_j = \boldsymbol{S}_j\Delta\boldsymbol{\theta}_j \tag{6.114}$$

from which the vector of updated parameters $\boldsymbol{\theta}_{j+1}$ is computed using a least-squares approximation, such as described in the case of the eigensensitivities,

$$\boldsymbol{\theta}_{j+1} = \boldsymbol{\theta}_j + \boldsymbol{S}_j^{+}\Delta\boldsymbol{H}_j \tag{6.115}$$

where $\boldsymbol{S}_j^{+}$ is the pseudo-inverse of $\boldsymbol{S}_j$, the sensitivity matrix of an *FRF* to the set of updating parameters, given by

$$\boldsymbol{S}_j = \boldsymbol{H}_j \begin{bmatrix} \dfrac{\partial \boldsymbol{Z}_j}{\partial \theta_1} & \dfrac{\partial \boldsymbol{Z}_j}{\partial \theta_2} & \cdots & \dfrac{\partial \boldsymbol{Z}_j}{\partial \theta_k} \end{bmatrix} \boldsymbol{H}_m \tag{6.116}$$

In the beginning of the development, one has stated that the formulation is derived for a single measured *FRF* vector, although the sensitivity matrix can be extended to contain information of more than one measured response vector; thus,

$$\boldsymbol{S}_j = \boldsymbol{H}_j \begin{bmatrix} \begin{bmatrix} \dfrac{\partial \boldsymbol{Z}_j}{\partial \theta_1} & \dfrac{\partial \boldsymbol{Z}_j}{\partial \theta_2} & \cdots & \dfrac{\partial \boldsymbol{Z}_j}{\partial \theta_k} \end{bmatrix} \boldsymbol{H}_{m_1} \\ \begin{bmatrix} \dfrac{\partial \boldsymbol{Z}_j}{\partial \theta_1} & \dfrac{\partial \boldsymbol{Z}_j}{\partial \theta_2} & \cdots & \dfrac{\partial \boldsymbol{Z}_j}{\partial \theta_k} \end{bmatrix} \boldsymbol{H}_{m_2} \\ \vdots \\ \begin{bmatrix} \dfrac{\partial \boldsymbol{Z}_j}{\partial \theta_1} & \dfrac{\partial \boldsymbol{Z}_j}{\partial \theta_2} & \cdots & \dfrac{\partial \boldsymbol{Z}_j}{\partial \theta_k} \end{bmatrix} \boldsymbol{H}_{m_n} \end{bmatrix} \tag{6.117}$$

Lin and Ewins [6.71] developed the *RFM* for undamped systems and therefore

$$\frac{\partial \boldsymbol{Z}_j}{\partial \theta_l} = \frac{\partial \boldsymbol{K}_j}{\partial \theta_l} - \omega^2 \frac{\partial \boldsymbol{M}_j}{\partial \theta_l} \tag{6.118}$$

please recall that $l = 1, \ldots, k$ are modelling parameters at the jth iteration.

An extension of the *RFM* to deal with damped systems was proposed by Lin and Zhu [6.72]. As the responses of damped system are complex, Eq. (6.114) is partitioned into its real and imaginary parts:

$$\begin{Bmatrix} \Delta \boldsymbol{H}_j^R \\ \Delta \boldsymbol{H}_j^I \end{Bmatrix} = \begin{bmatrix} \boldsymbol{S}_j^R \\ \boldsymbol{S}_j^I \end{bmatrix} \Delta \boldsymbol{\theta}_j \tag{6.119}$$

where the superscripts ^{R} and ^{I} denote – respectively – the real and imaginary parts and $\Delta \boldsymbol{\theta}_j$ contains parameters related to the mass and stiffness, identified from the real part of Eq. (6.119), and parameters related to damping, identified from the imaginary part of Eq. (6. 119).

The sensitivity of *FRF* correlation metrics (see Section 6.3.2) to the updating parameters has also been used by several authors. These updating methods have in common the use of the sensitivity of the correlation criteria to improve the level of correlation by adjusting a vector of updating parameters. The structure of the updating problem is similar to the ones of Eqs. (6.100) and (6.114), although the residual to be minimised is defined as $1-\varsigma(\omega)$, where $\varsigma(\omega)$ is a generic correlation metric in the frequency domain, ranging from 0 to 1.

Example 6.4.3 Application of the deterministic eigensensitivity-based model updating

The deterministic application of the described eigensensitivity-based model updating method, where the vector of updated parameters is given in Eq. (6.104), is illustrated by the following example. This updating approach aims at minimising the dissimilarity between reference and model responses in a deterministic least-squares iterative sense.

Consider the 3 degree of freedom mass-spring system shown in Figure 6.7. The nominal values of the reference model, the set of parameters of the "experimental" system, are: $m_i = 1.0\ kg$ $(i = 1, 2, 3)$, $k_i = 1.0$ N/m $(i = 1, 2, \ldots 5)$ and $k_6 = 3.0$ N/m.

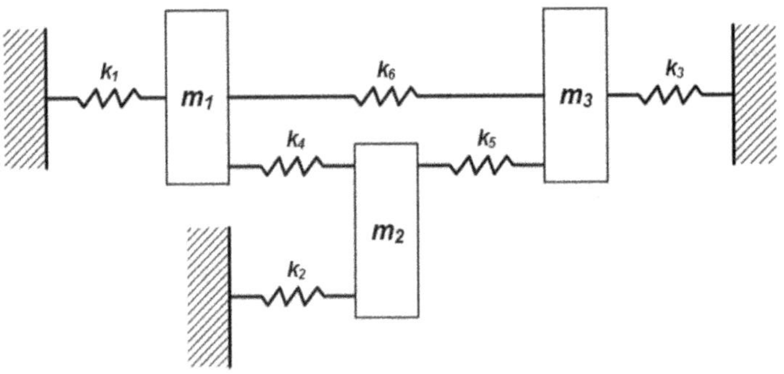

Figure 6.7 Three degree of freedom undamped system.

As an example, assume that the only unknown/uncertain parameters are k_1, k_2 and k_5. Consequently, the values of all the remaining parameters (mass and stiffness ones) are set to be known with values equal to the ones of the reference model. To start the updating procedure, the initial guess for the uncertain parameters (k_1, k_2 and k_5) is set as 2.0 N/m to all three (100% higher than their reference values). In this example, the vector of parameters to be updated (θ_j) is composed by k_1, k_2 and k_5 and a set comprising the three natural frequencies of the system is the reference response set (z_m). The summary of the updating procedure is given in Table 6.1, where reference, initial and updated values, after convergence, for both parameters and responses (natural frequencies) are given. Figure 6.8 presents the convergence of the results during the iterative updating process. The adopted convergence criterion was that the deviation of the predicted eigenfrequencies *w.r.t.* the experimental ones should be less than a specified tolerance.

Table 6.1 Parameter and response values, before and after deterministic eigensensitivity-based updating.

	Reference	**Initial**	**Updated** (error %)
k_1 [N/m]	1.0	2.0	1.01 (-0.03)
k_2 [N/m]	1.0	2.0	0.97 (0.06)
k_3 [N/m]	1.0	2.0	1.01 (-0.02)
f_1 [Hz]	0.1586	0.2030 (28.02)	0.1586 (-0.00)
f_2 [Hz]	0.3180	0.3960 (24.54)	0.3180 (-0.00)
f_3 [Hz]	0.4505	0.4823 (7.06)	0.4505 (-0.00)
# Iterations	-	-	6

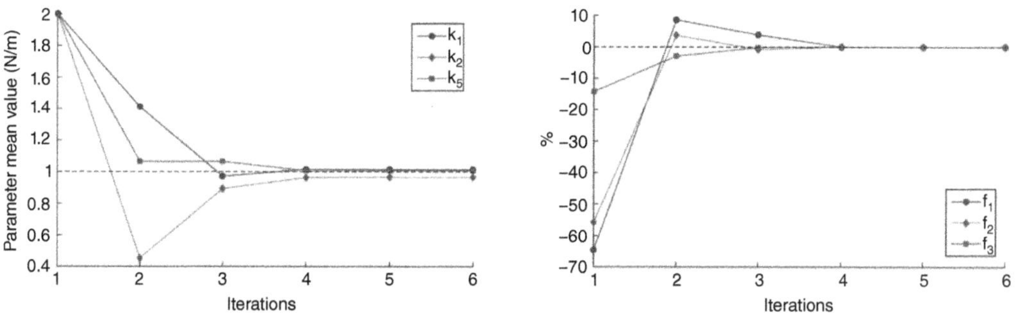

(a) Parameter values.

(b) Eigenfrequencies deviation *w.r.t.* the reference values.

Figure 6.8 Convergence plots (deterministic eigensensitivity-based updating) (dashed line: reference values).

6.4.4 On the Localisation of Modelling Errors

The model updating process may be separated into a two steps approach: a localisation step and a correction one, bearing in mind a potentially significant reduction on the number of updating parameters involved in the correction step. In such an approach, the localisation step is used to identify parts of the structure deemed to promote the dissimilarity between model and reference responses (e.g., poor modelling of joints).

Ladevèze [6.73] proposed a method for the localisation of modelling errors with application to model updating and damage identification [6.74]. This method is formulated as a continuous mechanics problem and it is based on the notion of error in the material constitutive relations. By identifying the erroneous elements in the structure, this concept avoids exploiting the entire space of variables during the correction or updating step, as the updating is performed at the element or group of elements for which the error has exceeded a given threshold. Hence, one can judiciously implement an updating method of one's choice to locally minimise the modelling errors, basically element by element until that minimum threshold is reached. An explicit formulation of the method in terms of frequency response data *FRFs* has been developed and can be found in [6.75].

6.5 Stochastic Model Updating

As stated in the introductory section of this chapter, the uncertainty quantification is required in order to generate valid models. Thus, the model updating implementation demands a formulation where the variability of measurements and the uncertainty of modelling parameters can be included. This is the case of stochastic model updating.

In the context of stochastic model updating, the approaches may be framed into two main categories: probabilistic and non-probabilistic approaches. Here, some aspects of both approaches are considered, namely through model updating techniques based on the perturbation method and on the interval theory. It is important to start by giving a preliminary explanation on some fundamentals of probability and statistics, aiming at directing the reader to the used concepts and tools.

6.5.1 Fundamentals of Probability and Statistics

Random Variables

For a given set of outcomes of some experiment, the sampled space, the concept of random variable is related to the generation of a function that allows to transform the elements in the set of outcomes into real numbers. If the sampled space is a countable set (e.g. the number of faults observed in an equipment during a given interval in time), the random variable is discrete. On the other hand, if the sample space is uncountable (e.g. measured temperature or natural frequencies changes over time), the random variable is continuous. In this chapter, the nature of the treated variables is continuous, thus one focus our attention on the description of that type of random variables.

Continuous models usually give good approximations of the real-world events and they can be described by their Probability Density Function (*PDF*). *PDF* gives the curve that describes the probability distribution of some continuous random variable; it is a non-negative function in $\mathbb{R}$ and its integral equals 1 (Figure 6.9 (a)). More formally, for a continuous random variable X, the *PDF* is a function $f(x)$, such that, for any a and b ($a \le b$), the probability of occurrence of such an event is given by $P(a \le X \le b) = \int_a^b f(x)dx$, which represents the area under the function $f(x)$ between a and b. It is also possible to describe a continuous random variable using the Cumulative Distribution Function (*CDF*), denoted by $F(x)$, obtained by the integration of the *PDF* between $-\infty$ and x. Note that x is a specific value of the continuous random variable X. More formally, the *CDF* of X is defined as $F(x) = P(X \le x) = \int_{-\infty}^{x} f(\xi)d\xi$. This function is continuous, non-decreasing and ranges from 0 to 1 (Figure 6.9 (b)).

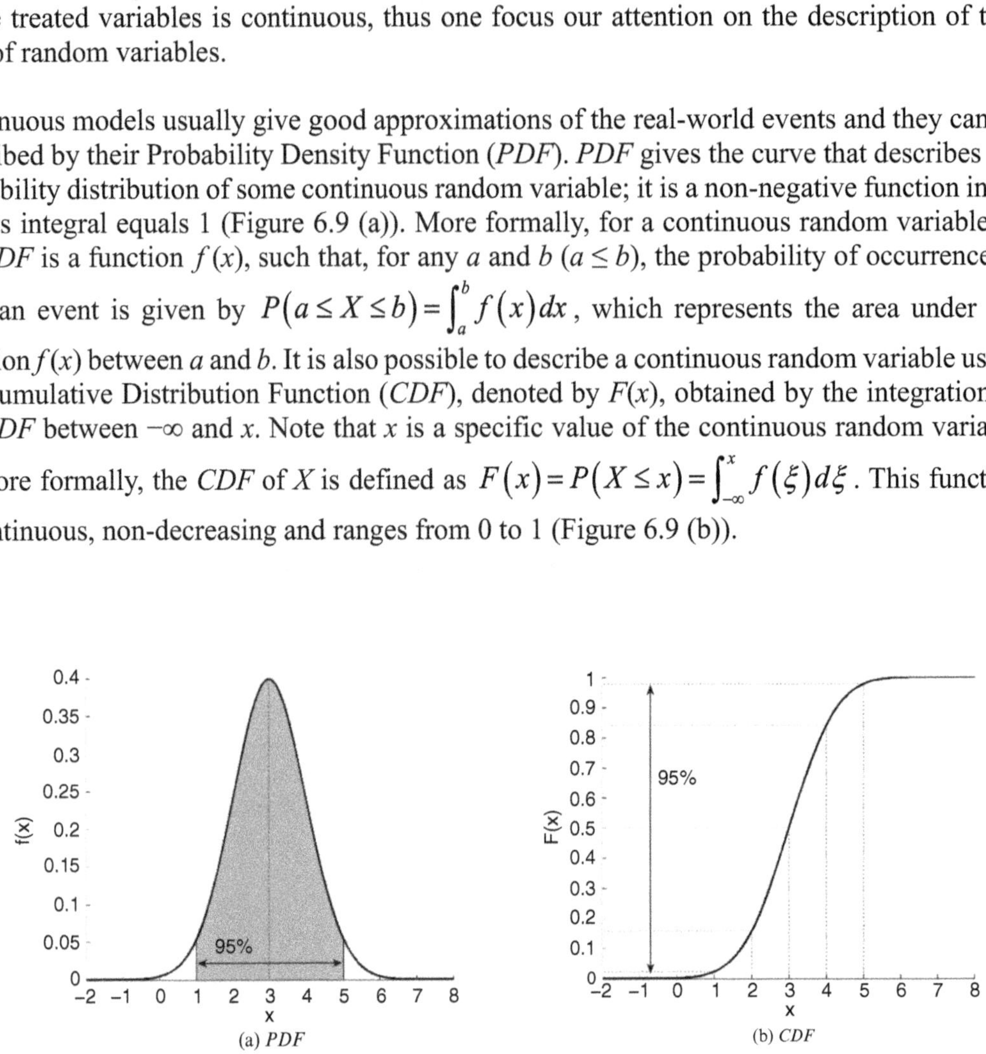

Figure 6.9 Example of a Gaussian model ($\mu = 3$ and $\sigma = 1$).

There are several important parameters that are useful to characterize a random variable. In terms of location, the expected or mean value of the population is given by $\mu = E(X) = \int_{\mathbb{R}} x f(x)dx$. This is the most used parameter and it is by intuition the expected value of a given experiment that locates the centre of gravity of the *PDF*. The mean value is a location parameter that gives an idea of the magnitude of the random variable, but does not describe its variability. To measure the variability associated to a random variable, one can use the expected value of the squared differences from the mean value, which is the variance of the population, defined as $var(X) = \sigma^2 = E\left[(X-\mu)^2\right] = E(X)^2 - \mu^2 = \int_{\mathbb{R}} x^2 f(x)dx - \mu^2$. However, it is common to use the standard deviation of the population, $\sigma = \sqrt{var(X)}$. In several situations, it is useful to compare the spread of different random variables. In those cases, it is better to use the Coefficient of Variation, given by $CoV = \frac{\sigma}{\mu}$, which measures the variability in relation to the mean value of the population.

Random variables can be modelled by different probability functions. Due to its nature, in model updating, the Gaussian or normal model is perhaps the most used one. However, care must be taken to limit the sampling of physically non-meaningful observations, and a way to ensure this is to use a truncated normal model or a lognormal one.

The normal model is symmetric around the mean value of the population μ and the spread is defined by σ, the standard deviation of the population, the *PDF* being given by

$$f(x) = \frac{1}{\sqrt{2\pi}\sigma} e^{-\frac{(x-\mu)^2}{2\sigma^2}} \quad x, \mu \in \mathbb{R} \quad \text{and} \quad \sigma > 0 \qquad (6.120)$$

Hence, a random variable x modelled by a normal model can be denoted as $X \sim N(\mu, \sigma)$, i.e., X is distributed as a normal random variable with μ and σ.

One of the important features of the normal distribution is the tolerance intervals based on the standard deviation. Thus, about 68% of the values drawn from a normal distribution stay within the interval $[\mu - \sigma, \mu + \sigma]$, one standard deviation away from the mean value. Therefore, if one considers two (three) standard deviations, the percentage of values of the population lying within that interval is 95.45% (99.73%). The area represented in Figure 6.9 (a) is the area corresponding to the tolerance interval for 2σ.

Another important continuous probability model is the uniform one, commonly used if one desires to assign an equal probability value for the entire interval, defined between a and b, such as

$$f(x) = \begin{cases} \frac{1}{\overline{x} - \underline{x}} & \Leftarrow \underline{x} < x < \overline{x} \\ 0 & \Leftarrow otherwise \end{cases}, \quad \underline{x} < \overline{x} \qquad (6.121)$$

In the case of the uniform model, one can define a random variable x uniformly distributed in $[\underline{x}, \overline{x}]$, as $X \sim U(\underline{x}, \overline{x})$.

Usually, there is more than one random variable, leading to study the relationship between pairs of variables, denoted by (X, Y). To measure the strength of the correlation between two variables, one can use the covariance between them, as $cov(X,Y) = E\left[(X-\mu_X)(Y-\mu_Y)\right] = E[X,Y] - \mu_X\mu_Y$. For a pair of uncorrelated variables, the covariance is zero. Otherwise, the covariance is positive (negative) if Y tends to increase (decrease) as X

increases. Independent random variables are uncorrelated, but the reverse is not necessarily true. Often, the independence between random variables is assumed in order to facilitate the approach to a problem, although this assumption may be unreasonable, leading to a biased solution of the given problem. Figure 6.10 illustrates the properties of a bivariate random variable, where the effect of the variance of each variable and the correlation between the pair can be evaluated. The dashed isodensity lines, in Figures 6.10 (d) to 6.10 (f), mark the tolerance interval for 2σ.

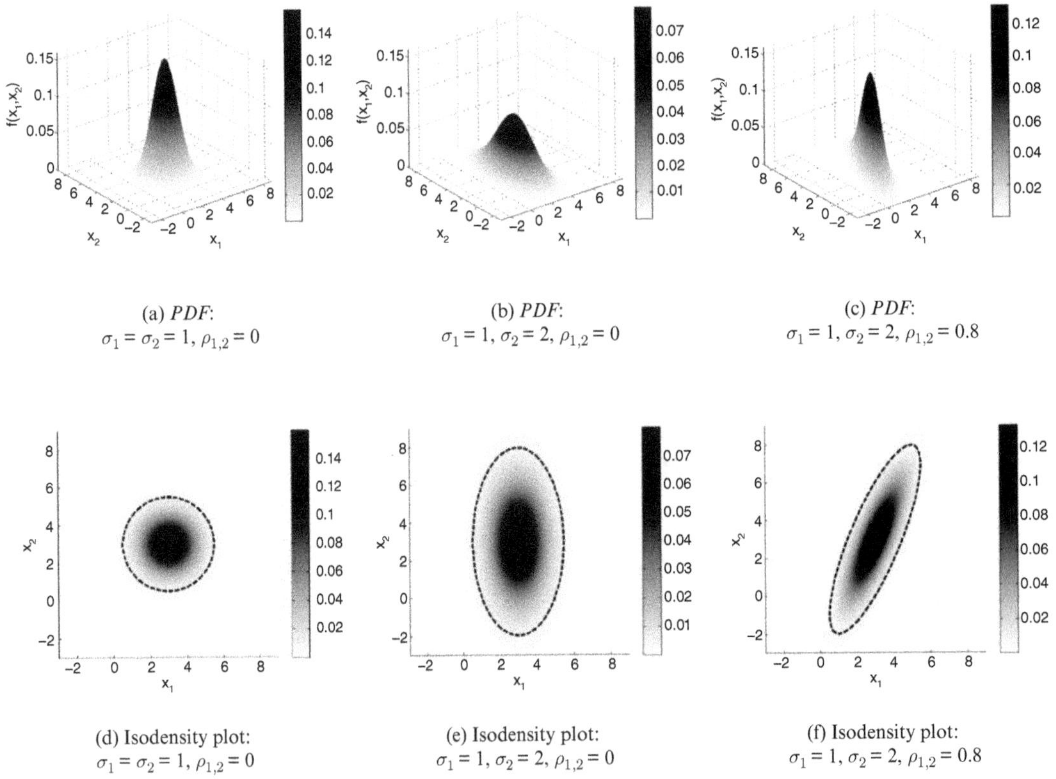

(a) *PDF*: $\sigma_1 = \sigma_2 = 1$, $\rho_{1,2} = 0$

(b) *PDF*: $\sigma_1 = 1$, $\sigma_2 = 2$, $\rho_{1,2} = 0$

(c) *PDF*: $\sigma_1 = 1$, $\sigma_2 = 2$, $\rho_{1,2} = 0.8$

(d) Isodensity plot: $\sigma_1 = \sigma_2 = 1$, $\rho_{1,2} = 0$

(e) Isodensity plot: $\sigma_1 = 1$, $\sigma_2 = 2$, $\rho_{1,2} = 0$

(f) Isodensity plot: $\sigma_1 = 1$, $\sigma_2 = 2$, $\rho_{1,2} = 0.8$

Figure 6.10 Example of bivariate Gaussian distributions: *PDFs* and Isodensity plots for $\mu_1 = \mu_2 = 3$ and different covariance matrices.

The covariance between two variables can be normalised *w.r.t.* their standard deviations, leading to the definition of the correlation coefficient, $\rho_{X,Y} = cov(X,Y)/(\sigma_X \sigma_Y)$. The interpretation of this measure is analogous to the covariance, though in the range $[-1,1]$.

Using sampling techniques, one can generate a sample of observed values from any continuous model. The Latin Hypercube Sampling (*LHS*) is here considered with the capability of imposing correlation between sampled variables [6.76]. The description of a particular sample can be performed using the histogram or the Empirical *CDF* (*eCDF*). Figure 6.11 illustrates these two types of graphical representations.

Depending on the problem, one can define a multivariate random variable $\boldsymbol{X}$ with k dimensions, such as $\boldsymbol{X} = \{X_1 \cdots X_k\}$. Sampling n times this random variable, the previous measures can be

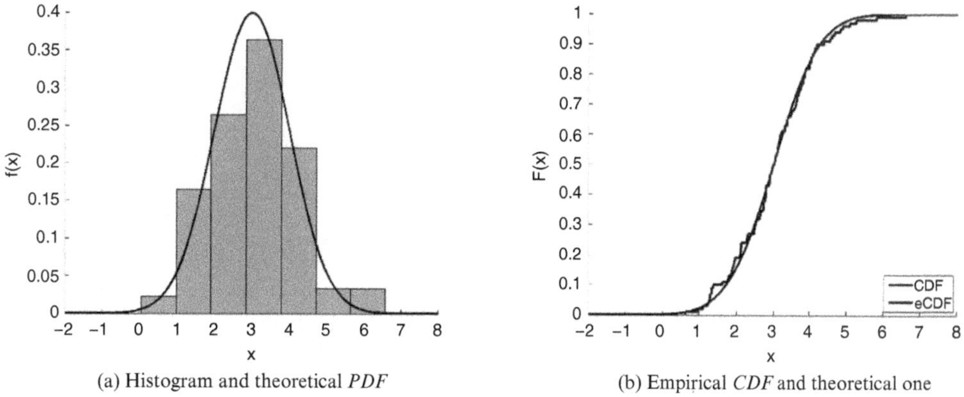

(a) Histogram and theoretical *PDF* (b) Empirical *CDF* and theoretical one

Figure 6.11 Simulation of $n = 100$ observations of x from a Gaussian model ($X \sim N(3,1)$).

discretised and a vector of mean values is defined as

$$\tilde{\boldsymbol{x}} = \{\tilde{x}_1 \cdots \tilde{x}_k\}^T \tag{6.122}$$

where $\tilde{x}_i = \frac{1}{n}\sum_{ob=1}^{n} x_{i_{ob}}$ is the mean value of a sample taken from the ith dimension of the multivariate random variable $\boldsymbol{X}$, denoted as $\boldsymbol{x}_i = \{x_{i_1} \cdots x_{i_n}\}^T$. The mean value of a sample taken from a random variable is here defined as $\tilde{x}_i$ to distinguish it from the mean value of the population, μ, and from the upper bound of a random variable defined by its interval, given by $\bar{x}_i$.

For a generic discretised multivariate random variable $\boldsymbol{X} = [\boldsymbol{x}_1 \cdots \boldsymbol{x}_k]$, the covariance matrix is given by

$$cov(\boldsymbol{X},\boldsymbol{X}) = \begin{bmatrix} s_1^2 & cov(\boldsymbol{x}_1,\boldsymbol{x}_2) & \cdots & cov(\boldsymbol{x}_1,\boldsymbol{x}_k) \\ cov(\boldsymbol{x}_2,\boldsymbol{x}_1) & s_2^2 & \cdots & cov(\boldsymbol{x}_2,\boldsymbol{x}_k) \\ \vdots & \vdots & \ddots & \vdots \\ cov(\boldsymbol{x}_k,\boldsymbol{x}_1) & cov(\boldsymbol{x}_k,\boldsymbol{x}_2) & \cdots & s_k^2 \end{bmatrix} \tag{6.123}$$

where $s_i^2 = \frac{1}{n-1}\sum_{ob=1}^{n}\left(x_{i_{ob}} - \tilde{x}_i\right)^2$ and $cov(\boldsymbol{x}_i,\boldsymbol{x}_j) = \frac{1}{n-1}\sum_{ob=1}^{n}\left(x_{i_{ob}} - \tilde{x}_i\right)\left(x_{j_{ob}} - \tilde{x}_j\right)$. Note that s_i is the standard deviation of a sample taken from from the ith dimension of the multivariate random variable $\boldsymbol{X}$. Thus, s_i denotes the standard deviation of a sample instead of the one concerned to the population, σ.

Statistical Hypothesis Tests

Hypothesis tests are used to test the validity of an assumption made about a given population (e.g., which type of distribution is followed by a sample), which can be classified as parametric or non-parametric ones. Any hypothesis test attempts to decide between the rejection of an assumption, the null hypothesis denoted by H_0, and the rejection of the alternative hypothesis H_1 (see Eq. (6.124)). The test is made assuming that the null hypothesis is true and the decision is to reject or to fail to reject H_0. When the null hypothesis is not rejected, it does not mean that H_0 is true; instead, there was not sufficient evidence to reject H_0. On the other hand, when the null hypothesis is rejected, the alternative one is consequently accepted. The acceptance/rejection decision can be based on a significance level (usually 5%), fixed *a priori*, which defines the critical regions of rejection and non-rejection and is compared with the test statistic. Another approach is to compute the probability of observing an event even rarer than the one given by the test statistic. This remarkably rare probability is known as *p*-value and it is compared to a given significance level in order to reject, or not, the null hypothesis. Therefore, the *p*-value weighs the strength of the evidence against the null hypothesis. Most authors refer to *statistically significant* if the *p*-value < 5% and *statistically highly significant* if the *p*-value < 0.1%, less than one in a thousand chances of being wrong.

A goodness-of-fit test is a non-parametric hypothesis test used to assess the degree of agreement between an observed sample and some specific theoretical distribution. There are several possibilities based either on the *PDF* or on the *CDF* [6.77].

One of the most popular hypothesis test is the *Chi-square* test. Simple to implement, it is not well-suited for data sets with few observations. As an alternative, the Kolmogorov-Smirnov goodness-of-fit test (*KS-test*) can be considered. The *KS-test* is based on the maximum difference between an empirical and a theoretical *CDF* [6.78].

The *KS-test* may be used to find the theoretical *PDF* that fits better the sampled data from a set of feasible probabilistic models. Regarding the *KS-test*, let $F_0(x)$ be the theoretical *CDF* of a continuous variable *X* under the null hypothesis and $F(x)$ the *eCDF* of a given sample. This type of *KS-test* is also known as Kolmogorov-Smirnov goodness-of-fit test for one sample. Thus, the test formulation is given by

$$H_0:\ F(x)=F_0(x) \quad vs \quad H_1:\ F(x)\neq F_0(x) \tag{6.124}$$

Therefore, if H_0 is true, it is expected that the difference between $F_0(x)$ and $F(x)$ is close to zero and within the limits of a random error. The Kolmogorov's statistic is defined as $D=max\left|F_0(x_i)-F(x_i)\right|$, for i = 1, 2, ..., n, where $F_0(x_i)=P(X\leq x_i)$ and $F(x_i)=i/n$, the *eCDF*. *D* is called the maximum deviation and it is compared with critical probability values [6.78]. Hence, if the *p*-value is less than the significance level of the test, there is no statistical evidence to reject H_0, and $F(x)$ is considered to fit well the data. Note for instance the example given in Figure 6.11 (b), using the *KS-test* to compare an *eCDF* obtained from a data set with the *CDF* for the normal model $(X \sim N(3,1))$, the computed *p*-value (≈ 94%) is clearly higher than the significance level of 5%, which means that there is no evidence to reject the null hypothesis, i.e. there is no evidence to reject the normal model as the one from which the data were sampled. The *KS-test* is a non-parametric test, requiring the complete definition of a theoretical model beforehand. If it is necessary to estimate the model parameters, one must consider the *Lilliefors* test.

Additionally, the test for two *eCDFs* is addressed by the Kolmogorov-Smirnov goodnessof-fit test for two samples, which is defined as the *KS*-test, but considering two *eCDFs* instead of testing an *eCDF* against a theoretical one. This test may be used as a comparison metric between distributions, namely a reference distribution and an updated one.

Two popular distances on the comparison between distributions are the Mahalanobis and the Battacharrya distances. The Mahalanobis distance can be presented as an Euclidean distance weighted by the inverse of the covariance matrix (the precision matrix), computed for each observation in a sample against a reference sample. On the other hand, the Battacharrya distance presents the ability of measuring the similarity between two samples, as it measures the superposition of two distributions, that may be empirical ones [6.79].

6.5.2 Updating the Parameter Covariance Matrix of a Model

Probably the simplest approach, conceptually, to stochastic model updating would be to carry out deterministic model updating multiple times. In this way, multiple sets of updated parameters could be obtained, each one corresponding to a different set of measurements to take account of manufacturing variability, difference in test conditions, different signal processing techniques, environmental and operational variabilities, etc. Finally, after carrying out a very large number of computations it would be possible to say something about the distributions on the parameters deemed to be responsible for the observed variability in the data. Such an approach is in principle capable of delivering very accurate updated stochastic models, but it depends upon the availability of large volumes of data, which are rarely available in industry and exceedingly expensive to acquire. An alternative, that might be practical, would be to obtain an updated stochastic model simply in terms of the parameter mean values and covariances.

In this section, two stochastic model updating methods are shown to be equivalent and a simplified equation is presented. The simplified equation can be applied without the use of expensive forward propagation of the parameters through the model, the so-called Monte-Carlo Simulations (*MCS*), to determine the output covariance matrix. It is demonstrated that the choice of updating parameters is critical to this process. If the correct parameters are chosen, then the output covariance matrix is reconstructed faithfully. However, this is generally not the case when updating parameters are chosen wrongly, even though the output means may be accurately reconstructed. It is shown that the scaled output covariance matrix may be decomposed, using Singular Value Decomposition (*SVD*)[5], to allow the contributions of each candidate parameter to be assessed. Use of the classical linearised sensitivity permits the assessment to be carried out efficiently, as some numerical examples show.

Recalling the deterministic model updating formulation, given in Section 6.4.3, the updated vector of parameters was given in Eq. (6.104), here repeated, so that

$$\boldsymbol{\theta}_{j+1} = \boldsymbol{\theta}_j + \boldsymbol{T}_j \left(z_m - z_j \right)$$

where the transformation matrix $\boldsymbol{T}_j$ is generally the weighted pseudo-inverse of the sensitivity matrix $\boldsymbol{S}_j$.

[5] See Chapter 3, Section 3.4.7 and Chapter 4 (end of Section 4.2.2).

In the following analysis one considers a "cloud" of data points in a multi-dimensional space so that each point represents a separate test. There is a corresponding cloud of predicted points and the objective is to adjust the statistics of the parameters so that the position and orientation of the cloud, determined according to the mean values and covariances of the predictions, is made to agree with the data cloud.

The Perturbation Method

Equivalently to the updated vector of parameters in the deterministic context (Eq. (6.104)), the stochastic model updating equation may be written as

$$\tilde{\boldsymbol{\theta}}_{j+1}+\Delta\boldsymbol{\theta}_{j+1}=\tilde{\boldsymbol{\theta}}_j+\Delta\boldsymbol{\theta}_j+\left(\tilde{\boldsymbol{T}}_j+\Delta\boldsymbol{T}_j\right)\left(\tilde{\boldsymbol{z}}_m+\Delta\boldsymbol{z}_m-\tilde{\boldsymbol{z}}_j-\Delta\boldsymbol{z}_j\right) \tag{6.125}$$

where $\tilde{\bullet}$ and Δ denote the mean and perturbation on the mean, respectively. Then, by separating the zeroth-order and first-order terms, one obtains the following two expressions [6.67]:

$$O\left(\Delta^0\right):\quad \tilde{\boldsymbol{\theta}}_{j+1}=\tilde{\boldsymbol{\theta}}_j+\tilde{\boldsymbol{T}}_j\left(\tilde{\boldsymbol{z}}_m-\tilde{\boldsymbol{z}}_j\right) \tag{6.126}$$

$$O\left(\Delta^1\right):\quad \Delta\boldsymbol{\theta}_{j+1}=\Delta\boldsymbol{\theta}_j+\tilde{\boldsymbol{T}}_j\left(\Delta\boldsymbol{z}_m-\Delta\boldsymbol{z}_j\right)+\sum_{i=1}^{n}\frac{\partial\tilde{\boldsymbol{T}}_j}{\partial z_m^{(i)}}\Delta z_m^{(i)}\left(\tilde{\boldsymbol{z}}_m-\tilde{\boldsymbol{z}}_j\right) \tag{6.127}$$

Here, the third term on the right-hand-side of Eq. (6.127) will be neglected on the grounds that $\left(\tilde{\boldsymbol{z}}_m-\tilde{\boldsymbol{z}}_j\right)$ is itself a small quantity of $O(\Delta^1)$. Eq. (6.126) is the updating equation for the parameters mean values (equivalent to Eq. (6.104)), exactly the same result as given by Govers and Link [6.68].

Eq. (6.127) is used to form a covariance matrix as

$$cov\left(\Delta\boldsymbol{\theta}_{j+1},\Delta\boldsymbol{\theta}_{j+1}\right)=cov\left(\Delta\boldsymbol{\theta}_j+\tilde{\boldsymbol{T}}_j\left(\Delta\boldsymbol{z}_m-\Delta\boldsymbol{z}_j\right)+\boldsymbol{A}_j\Delta\boldsymbol{z}_m,\Delta\boldsymbol{\theta}_j+\tilde{\boldsymbol{T}}_j\left(\Delta\boldsymbol{z}_m-\Delta\boldsymbol{z}_j\right)+\boldsymbol{A}_j\Delta\boldsymbol{z}_m\right) \tag{6.128}$$

where

$$\boldsymbol{A}_j=\left[\left.\frac{\partial\tilde{\boldsymbol{T}}_j}{\partial z_m^{(1)}}\right|_{z_m^{(1)}=\tilde{z}_m^{(1)}}\left(\tilde{\boldsymbol{z}}_m-\tilde{\boldsymbol{z}}_j\right)\quad \left.\frac{\partial\tilde{\boldsymbol{T}}_j}{\partial z_m^{(2)}}\right|_{z_m^{(2)}=\tilde{z}_m^{(2)}}\left(\tilde{\boldsymbol{z}}_m-\tilde{\boldsymbol{z}}_j\right)\quad\cdots\quad \left.\frac{\partial\tilde{\boldsymbol{T}}_j}{\partial z_m^{(n)}}\right|_{z_m^{(n)}=\tilde{z}_m^{(n)}}\left(\tilde{\boldsymbol{z}}_m-\tilde{\boldsymbol{z}}_j\right)\right] \tag{6.129}$$

A complete expression without further approximations was developed, but had the disadvantage of introducing computationally expensive second-order sensitivity terms.

Here, the covariance matrix is developed using just the first two right-hand terms of Eq. (6.127). Thus,

$$cov\left(\Delta\boldsymbol{\theta}_{j+1},\Delta\boldsymbol{\theta}_{j+1}\right)=cov\left(\Delta\boldsymbol{\theta}_j+\tilde{\boldsymbol{T}}_j\left(\Delta\boldsymbol{z}_m-\Delta\boldsymbol{z}_j\right),\Delta\boldsymbol{\theta}_j+\tilde{\boldsymbol{T}}_j\left(\Delta\boldsymbol{z}_m-\Delta\boldsymbol{z}_j\right)\right) \tag{6.130}$$

Expanding this expression leads to

$$\begin{aligned}cov\left(\Delta\boldsymbol{\theta}_{j+1},\Delta\boldsymbol{\theta}_{j+1}\right)&=cov\left(\Delta\boldsymbol{\theta}_j,\Delta\boldsymbol{\theta}_j\right)-cov\left(\Delta\boldsymbol{\theta}_j,\Delta\boldsymbol{z}_j\right)\tilde{\boldsymbol{T}}_j^T-\tilde{\boldsymbol{T}}_j cov\left(\Delta\boldsymbol{z}_j,\Delta\boldsymbol{\theta}_j\right)\\&\quad+\tilde{\boldsymbol{T}}_j cov\left(\Delta\boldsymbol{z}_j,\Delta\boldsymbol{z}_j\right)\tilde{\boldsymbol{T}}_j^T+\tilde{\boldsymbol{T}}_j cov\left(\Delta\boldsymbol{z}_m,\Delta\boldsymbol{z}_m\right)\tilde{\boldsymbol{T}}_j^T\end{aligned} \tag{6.131}$$

where the updated parameters $\Delta\boldsymbol{\theta}_j$ and hence the predictions $\Delta\boldsymbol{z}_j$ are assumed to be statistically independent of the measurements $\Delta\boldsymbol{z}_m$. Khodaparast et al. [6.67] showed that every term of the matrix $\boldsymbol{A}_j$ contained derivatives of the parameter means with respect to the data. Thus, $\boldsymbol{A}_j$ was found to disappear under the above assumption, thereby confirming the correct omission of the third right-hand-side term in Eq. (6.127). The assumption is strictly correct only for the first iteration since the initial analytical prediction is completely unrelated to the data. On subsequent iterations the assumption is invalid, but long experience, dating back to the work of Collins et al. [6.80], has shown that reliable engineering solutions are obtainable under such an approximation. Also, Khodaparast et al. [6.67] found almost no difference between the covariances developed from Eq. (6.128) and those obtained from Eq. (6.131).

Similarly, Govers and Link [6.68] addressed the adjustment of the parameter covariance matrix by the identification of the parameter covariance matrix changes, given by the residual

$$\boldsymbol{R}_{\Delta\theta_j} = cov\left(\Delta\boldsymbol{\theta}_{j+1}, \Delta\boldsymbol{\theta}_{j+1}\right) - cov\left(\Delta\boldsymbol{\theta}_j, \Delta\boldsymbol{\theta}_j\right) \tag{6.132}$$

which minimises the Frobenius norm of the weighted difference between the covariance matrix of the measured data and the one of the model responses, described by the residual matrix $\boldsymbol{R}_{\Delta j} = cov\left(\Delta\boldsymbol{z}_m, \Delta\boldsymbol{z}_m\right) - cov\left(\Delta\boldsymbol{z}_j, \Delta\boldsymbol{z}_j\right)$. So, the problem may be expressed as

$$\min_{\boldsymbol{R}_{\Delta\theta_j}} \frac{1}{2} \left\| \boldsymbol{W}_R \boldsymbol{R}_{\Delta j} \boldsymbol{W}_R^T \right\|_F^2 \tag{6.133}$$

where $\boldsymbol{W}_R$ is a weighting diagonal matrix.

Solving Eq. (6.133) and assuming the model responses to be statistically independent from the model parameters, the residual matrix $\boldsymbol{R}_{\Delta j}$ can be cast as

$$\boldsymbol{R}_{\Delta j} = cov\left(\Delta\boldsymbol{z}_m, \Delta\boldsymbol{z}_m\right) - cov\left(\Delta\boldsymbol{z}_j, \Delta\boldsymbol{z}_j\right) - \tilde{\boldsymbol{S}}_j \boldsymbol{R}_{\Delta\theta_j} \tilde{\boldsymbol{S}}_j^T \tag{6.134}$$

Considering Eq. (6.134), the problem given in Eq. (6.133) leads to the updated parameter covariance matrix, given by

$$\boldsymbol{R}_{\Delta\theta_j} = \tilde{\boldsymbol{T}}_j \left(cov\left(\Delta\boldsymbol{z}_m, \Delta\boldsymbol{z}_m\right) - cov\left(\Delta\boldsymbol{z}_j, \Delta\boldsymbol{z}_j\right) \right) \tilde{\boldsymbol{T}}_j^T \tag{6.135}$$

Expanding Eq. (6.135), one has

$$cov\left(\Delta\boldsymbol{\theta}_{j+1}, \Delta\boldsymbol{\theta}_{j+1}\right) = cov\left(\Delta\boldsymbol{\theta}_j, \Delta\boldsymbol{\theta}_j\right) + \tilde{\boldsymbol{T}}_j \left(cov\left(\Delta\boldsymbol{z}_m, \Delta\boldsymbol{z}_m\right) - cov\left(\Delta\boldsymbol{z}_j, \Delta\boldsymbol{z}_j\right) \right) \tilde{\boldsymbol{T}}_j^T \tag{6.136}$$

The covariances $cov\left(\Delta\boldsymbol{z}_m, \Delta\boldsymbol{z}_m\right)$ and $cov\left(\Delta\boldsymbol{z}_j, \Delta\boldsymbol{z}_j\right)$ are readily available from the data and from forward propagation using the distribution with mean $\tilde{\boldsymbol{\theta}}_j$ and $cov\left(\Delta\boldsymbol{z}_j, \Delta\boldsymbol{z}_j\right)$ determined at the previous updating iteration. Khodaparast et al. [6.67] also determined $cov\left(\Delta\boldsymbol{\theta}_j, \Delta\boldsymbol{z}_j\right)$ by forward propagation. However, a further simplification is available, as will be introduced next.

Small Perturbation about the Mean

In the case of small parameter variability, θ_j may be approximated by the perturbation on the mean, as

$$\theta_j = \tilde{\theta}_j + T(\tilde{\theta}_j)(z_j - z(\tilde{\theta}_j)) \tag{6.137}$$

or

$$\Delta\theta_j = T(\tilde{\theta}_j)(z_j - z(\tilde{\theta}_j)) \tag{6.138}$$

When it is assumed that

$$\tilde{T}_j = T(\tilde{\theta}_j) \quad \text{and} \quad \tilde{z}_j = z(\tilde{\theta}_j) \tag{6.139}$$

then

$$\Delta\theta_j = \tilde{T}_j(z_j - \tilde{z}_j) = \tilde{T}_j \Delta z_j \tag{6.140}$$

and thus the terms of Eq. (6.131) can be re-written as

$$cov(\Delta\theta_j, \Delta z_j)\tilde{T}_j^T = \tilde{T}_j cov(\Delta z_j, \Delta z_j)\tilde{T}_j^T \tag{6.141}$$

and

$$\tilde{T}_j cov(\Delta z_j, \Delta\theta_j) = \tilde{T}_j cov(\Delta z_j, \Delta z_j)\tilde{T}_j^T \tag{6.142}$$

Substitution of Eqs. (6.141) and (6.142) into eq. (6.131) leads immediately to

$$cov(\Delta\theta_{j+1}, \Delta\theta_{j+1}) = cov(\Delta\theta_j, \Delta\theta_j) + \tilde{T}_j(cov(\Delta z_m, \Delta z_m) - cov(\Delta z_j, \Delta z_j))\tilde{T}_j^T \tag{6.143}$$

which is equal to Eq. (6.136).

Using the same assumptions, and the relationship given in Eq. (6.140), one has

$$\tilde{T}_j cov(\Delta z_j, \Delta z_j)\tilde{T}_j^T = cov(\Delta\theta_j, \Delta\theta_j) \tag{6.144}$$

$$cov(\Delta\theta_j, \Delta z_j)\tilde{T}_j^T = cov(\Delta\theta_j, \Delta\theta_j) \tag{6.145}$$

$$\tilde{T}_j cov(\Delta z_j, \Delta\theta_j) = cov(\Delta\theta_j, \Delta\theta_j) \tag{6.146}$$

Substitution of Eqs. (6.144) to (6.146) into Eq. (6.131), or Eq. (6.144) into Eq. (6.143), leads to a very simple expression [6.81],

$$cov(\Delta\theta_{j+1}, \Delta\theta_{j+1}) = \tilde{T}_j cov(\Delta z_m, \Delta z_m)\tilde{T}_j^T \tag{6.147}$$

Eq. (6. 147) allows the computation of the outputs covariances using the transformation matrix $\tilde{T}_j$, obtained at the final step of deterministic updating of the mean values using Eq. (6.126), without the need for forward propagation, which reduces significantly the computational effort of the perturbation method.

An Equivalent Formulation

Alternatively, if one considers the truncated form of Eq. (6.127), such as

$$\Delta\boldsymbol{\theta}_{j+1} - \Delta\boldsymbol{\theta}_j = \tilde{\boldsymbol{T}}_j\left(\Delta\boldsymbol{z}_m - \Delta\boldsymbol{z}_j\right) \tag{6.148}$$

the stochastic model updating problem may be expressed as

$$\left(\boldsymbol{z}_m - \tilde{\boldsymbol{z}}_m\right) = \tilde{\boldsymbol{S}}_j\left(\boldsymbol{\theta}_{j+1} - \tilde{\boldsymbol{\theta}}_{j+1}\right) + \boldsymbol{\varepsilon}_{j+1} \tag{6.149}$$

by the assumption of small perturbation about the mean, as described before, $\Delta\boldsymbol{z}_m = \boldsymbol{z}_m - \tilde{\boldsymbol{z}}_m$ and $\Delta\boldsymbol{\theta} = \boldsymbol{\theta} - \tilde{\boldsymbol{\theta}}$. It should be stressed that $\boldsymbol{\varepsilon}_{j+1}$ represents errors introduced from various sources including inaccuracy of the model and measurement imprecision ($\boldsymbol{\varepsilon}_m$), and it may be defined as

$$\boldsymbol{\varepsilon}_{j+1} = \left(\boldsymbol{z}_j - \tilde{\boldsymbol{z}}_j\right) - \tilde{\boldsymbol{S}}_j\left(\boldsymbol{\theta}_j - \tilde{\boldsymbol{\theta}}_j\right) + \boldsymbol{\varepsilon}_m \tag{6.150}$$

From Eq. (6.149), one can write that the matrix of measured output covariances is given by

$$cov\left(\Delta\boldsymbol{z}_m, \Delta\boldsymbol{z}_m\right) = \tilde{\boldsymbol{S}}_j cov\left(\Delta\boldsymbol{\theta}_{j+1}, \Delta\boldsymbol{\theta}_{j+1}\right)\tilde{\boldsymbol{S}}_j^T + cov\left(\boldsymbol{\varepsilon}_{j+1}, \boldsymbol{\varepsilon}_{j+1}\right) \tag{6.151}$$

Then, if the covariance of the error is considered as small, an estimate of the parameter covariances may be obtained by inversion of Eq. (6.151), leading to

$$cov\left(\Delta\boldsymbol{\theta}_{j+1}, \Delta\boldsymbol{\theta}_{j+1}\right) = \tilde{\boldsymbol{T}}_j cov\left(\Delta\boldsymbol{z}_m, \Delta\boldsymbol{z}_m\right)\tilde{\boldsymbol{T}}_j^T \tag{6.152}$$

Eq. (6.152) was already given by Eq. (6.147), although the starting point for their development is rather different, as it can be seen. Thus, Eq. (6.152) also allows for the computation of the updated covariance of the parameters $cov\left(\Delta\boldsymbol{\theta}_{j+1}, \Delta\boldsymbol{\theta}_{j+1}\right)$ using only the transformation matrix $\tilde{\boldsymbol{T}}_j$, obtained at the final step of deterministic updating of the means, and the measured output covariance matrix. Moreover, it avoids expensive forward propagation of uncertain parameters through the model, required by alternative approaches, as already referred.

Example 6.5.2 Application of the stochastic sensitivity-based model updating

Recalling the deterministic eigensensitivity-based model updating example given in Section 6.4.3, this example extends Example 6.4.3 to the stochastic eigensensitivity-based model updating by the perturbation method.

As in Example 6.4.3, the nominal values of the reference model, the set of parameters of the "experimental" system, are: $m_i = 1.0\ kg$ ($i = 1, 2, 3$), $k_i = 1.0$ N/m (i = 1, 2, … 5) and $k_6 = 3.0$ N/m. However, in this example, the uncertain modelling parameters k_1, k_2 and k_5 are random variables assumed to have Gaussian distributions with mean values, $\mu_{k_1} = \mu_{k_2} = \mu_{k_5} = 2.0$ N/m and standard deviations $\sigma_{k_1} = \sigma_{k_2} = \sigma_{k_5} = 0.3$ N/m. The true mean values are the nominal values, given above, with standard deviations $\sigma_{k_1} = \sigma_{k_2} = \sigma_{k_5} = 0.2$ N/m (20% of the true mean values). Parameters k_1, k_2 and k_5 are assumed to be independent, thus their covariances are null.

Considering a set of updating parameters that includes all the modelling parameters deemed to be responsible for the observed response variability, one has a consistent set of updating parameters. In this example, k_1, k_2 and k_5 are the randomised variables. Thus, the variability of the three natural frequencies of the system is due to the uncertainty on k_1, k_2 and k_5.

Eqs. (6.143) and (6.147) above were applied and the initial cloud of predicted natural frequencies was made to converge upon the cloud of "measured" natural frequencies, as shown in Figure 6.12 for the hyper-plane of the first two natural frequencies (f_1 and f_2). The measured data consisted of 30 separate measurement points (30 points in the 3 dimensional space of the natural frequencies) and the predictions are represented by 1000 points, obtained by forward propagation of the sampled parameters θ_j in Eq. (6.143), in order to determine Δz_j from $\Delta\boldsymbol{\theta}_j$. One should bear in mind that the parameters were sampled from a normal distribution by *LHS* with imposed correlation [6.76], to insure independence. Furthermore, it is important to retain that Eq. (6.147) does not require this forward propagation and it is therefore very fast in updating the parameter covariance matrix.

Figure 6.12 shows the results produced by the two equations, where it is apparent that the updated covariance ellipses from the two solutions are almost indistinguishable from each other or from the covariance ellipse of the "measured" data. The covariance ellipses in the scatter plots encompass 95% of the data (2σ ellipses). Figure 6.13 enables an easier comparison between the scatter of reference and updated natural frequencies. Note the updated correlation value given in Figure 6.13, obtained using Eq. (6.147).

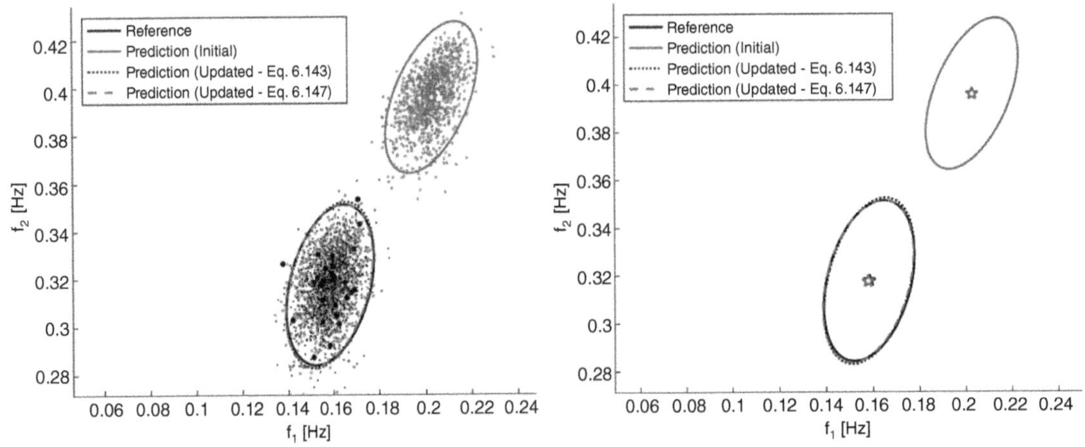

(a) Scatter plot

(b) Detailed view of the 2σ ellipses

Figure 6.12 Scatter plot of the first two natural frequencies, before and after updating, using Eqs. (6.143) and (6.147): scatter, 2σ ellipses and their centres.

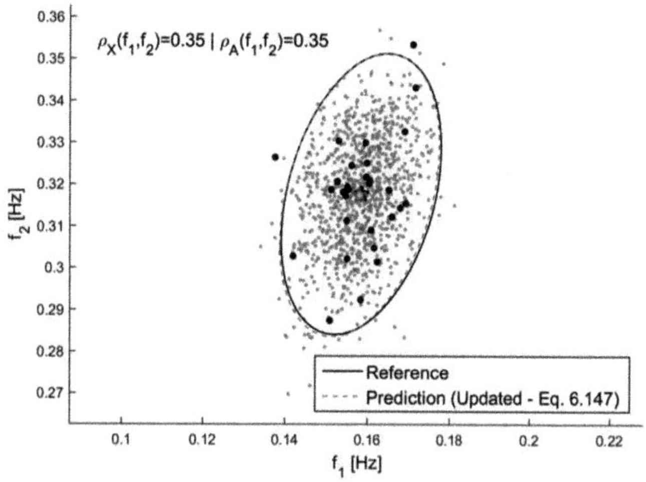

Figure 6.13 Scatter plot of the first two natural frequencies, after updating, using Eq. (6.147).

Typical convergence characteristics are shown in Figure 6.14 and the updated parameter values are given in Table 6.2. The adopted convergence criterion was again based on the minimisation of the deviation between the predicted eigenfrequencies *w.r.t.* the experimental ones, for a specified tolerance.

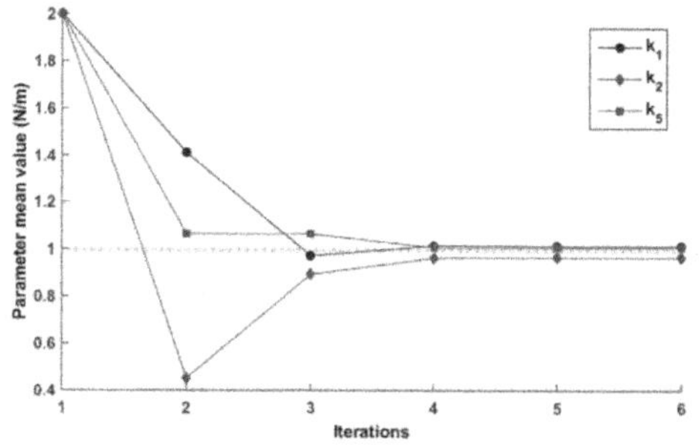

(a) Mean values of the updating parameters.

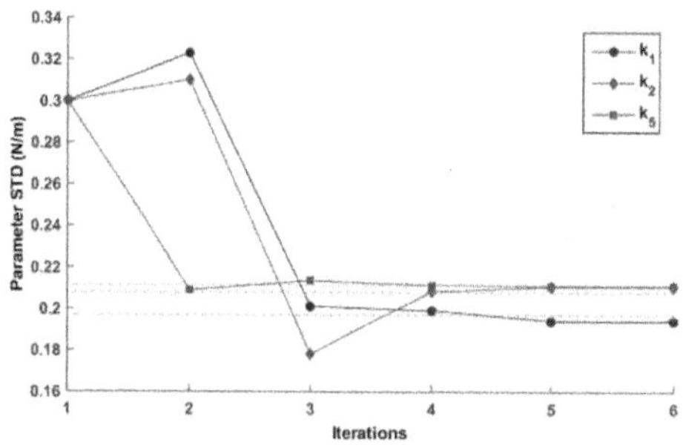

(b) Standard deviation of the updating parameters.

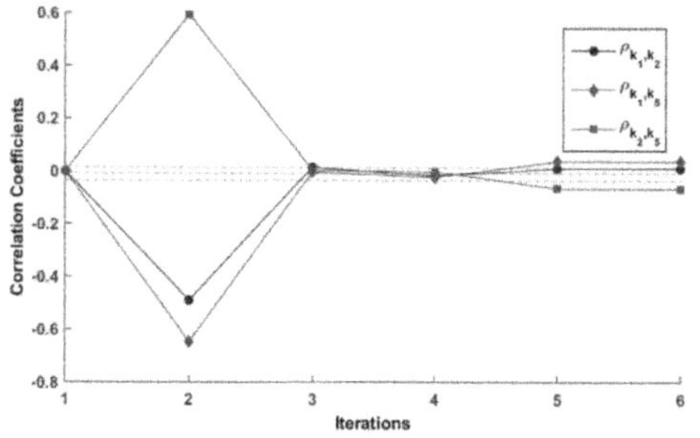

(a) Correlation of the updating parameters.

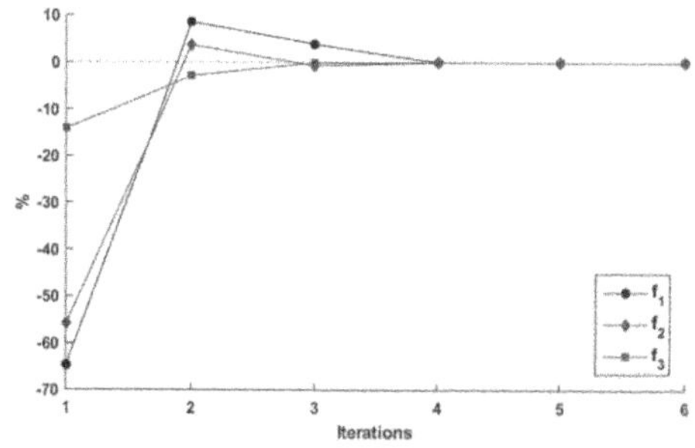

(b) Estimated mean eigenfrequencies deviation *w.r.t.* the reference values.

Figure 6.14 Convergence plots (stochastic eigensensitivity-based updating: Eq. (6.147)) (dashed lines: reference values).

Table 6.2 Parameters and eigenfrequencies values, before and after updating, using Eqs. (6.143) and (6.147).

	Reference 30 obs.	**Initial** (error %)	**Updated** (error %) **Eq. (6.143)** - 1000 obs.	**Updated** (error %) **Eq. (6.147)**
$\tilde{k}_1$ [N/m]	1.001	2.0 (99,73)	1.001 (-0.03)	1.014 (1.26)
$\tilde{k}_2$ [N/m]	0.992	2.0 (101.55)	0.993 (0.06)	0.966 (-2.68)
$\tilde{k}_5$ [N/m]	1.001	2.0 (99.84)	1.001 (-0.02)	1.008 (0.69)
s_{k_1} [N/m]	0.197	0.3 (52.59)	0.194 (-1.59)	0.194 (-1.36)
s_{k_2} [N/m]	0.208	0.3 (44.58)	0.213 (2.35)	0.211 (1.40)
s_{k_5} [N/m]	0.211	0.3 (41.94)	0.211 (-0.05)	0.211 (-0.11)
f_1 [Hz]	0.1586	0.2030 (28.02)	0.1586 (-0.00)	0.1586 (-0.00)
f_2 [Hz]	0.3180	0.3960 (24.54)	0.3180 (-0.00)	0.3180 (-0.00)
f_3 [Hz]	0.4505	0.4823 (7.06)	0.4505 (-0.00)	0.4505 (0.00)
# Iterations	-	-	9	6
CPU time ratio	-	-	300	1

The *CPU* time ratios shown in Table 6.2 are determined *w.r.t.* the updating solution using Eq. (6.147). It is seen that for this particular 3 *DOF* problem, the calculation of the parameter covariance matrix is approximately 300 times faster by Eq. (6.147) than by Eq. (6.143).

The reported results support the applicability of Eq. (6.147) as an efficient way to update the parameter covariance matrix. Hence, in the following, the updating is carried out using Eq. (6.147) only.

The 3 *DOF* system has five stiffness elements, so it is not expected that one can identify beforehand the modelling parameters considered as responsible for the observed response variability. Hence, the updating parameter set is now extended to include the stiffness parameters $k_1 - k_5$. By this, one has a consistent set of updating parameters, although some parameters are not randomised in the simulation. The updating parameter set is now $k_1 - k_5$ and the inclusion of more responses is required to make the sensitivity matrix overdetermined. So, one has now included the mode shapes in the reference response set.

Here, the unknown parameters are assumed to have Gaussian distributions, as previously stated, but now with mean values $\mu_{k_1} = \mu_{k_2} = \mu_{k_5} = 2.0\,N/m$ and $\mu_{k_3} = \mu_{k_4} = 0.5\,N/m$ and standard deviations $\sigma_{k_1} = \sigma_{k_2} = \sigma_{k_5} = 0.3\,N/m$ and $\sigma_{k_3} = \sigma_{k_4} = 0.1\,N/m$. The random properties of the first group of parameters (k_1, k_2 and k_5) have not changed, except that now k_2 and k_5 are considered to be correlated, with $\rho_{k_2,k_5} = -0.9$. The reference data set, with 30 observations, is the same as before.

Figure 6.15 shows a very good agreement between the reference and updated scatter ellipses, using the reference sample with 30 observations and an updated *LHS* of 1000 observations, generated with the updated mean values and covariance matrix.

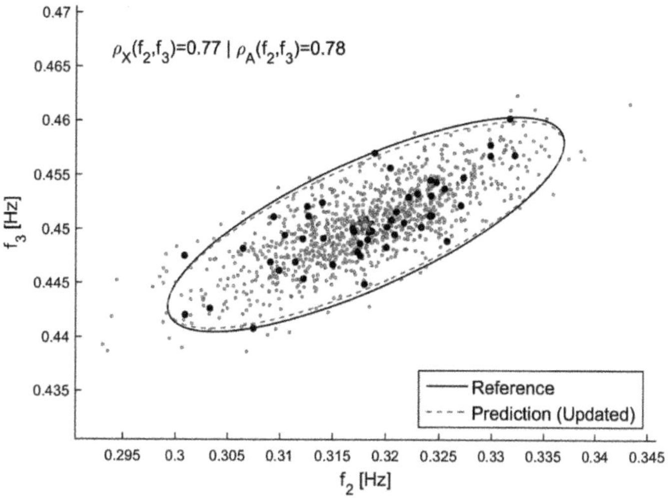

Figure 6.15 Scatter plot of the first two natural frequencies, after updating, using Eq. (6.147).

Typical convergence characteristics are shown in Figure 6.16, where the standard deviation of the randomised stiffnesses (k_1, k_2 and k_5 with $\sigma_i = 0.2\,N/m$) is approximated, but not obtained perfectly, as both standard deviations and correlation coefficients of the parameters are not fully recovered, regarding their reference values (Figures 6.16 (b) and 6.16 (c)). This is because the non-randomised stiffnesses k_3 and k_4 become random variables after updating, as the updated covariance matrix has also components related to these initially non-randomised parameters.

The updating is successful for consistent sets of updating parameters, but what if one is not able to include in the updating parameter set all the parameters deemed to be responsible for the observed response variability? This is a frequent situation, namely due to modelling simplifications and will be discussed in what follows.

For this case, the reference data set, with 30 observations, is the same as before. Thus, the reference data set was produced with randomised k_1, k_2 and k_5, while the updating parameter set is now composed of k_1, k_2 and k_6, i.e., the uncertain k_5 is not included in the updating parameter set. This is the case of an inconsistent updating parameter set. The unknown random parameters are assumed to have Gaussian distributions with mean values $\mu_{k_1} = \mu_{k_2} = \mu_{k_6} = 2.0\,N/m$ and standard deviations $\sigma_{k_1} = \sigma_{k_2} = \sigma_{k_6} = 0.3\,N/m$. Remember that k_6 is not a randomised parameter in the reference data set and that its nominal value is $k_6 = 3.0\,N/m$. In this case regularisation of the sensitivity matrix is considered, with $\boldsymbol{W}_\varepsilon = \boldsymbol{I}$ and $\boldsymbol{W}_\vartheta = 0.1 \times \boldsymbol{I}$.

Figures 6.17 and 6.18 show the results of the updating process. The scatter plots of Figure 6.17 show that the output means are reconstructed faithfully but the choice of an inconsistent set of updating parameters has resulted in large errors in the reconstructed covariance ellipses. The updating parameters k_1, k_2 and k_6 have fully converged after 30 iterations to incorrect values, in terms of the parameters variability (Figures 6.18 (b) and 6.18 (c)), while the parameter means are correctly updated (Figure 6.18 (a)). This result demonstrates that the selection of updating parameters based on the reconstruction of the output means is not sufficient to ensure that the output covariances will be well reconstructed. This may support the more informed analysis based on stochastic approaches against deterministic ones.

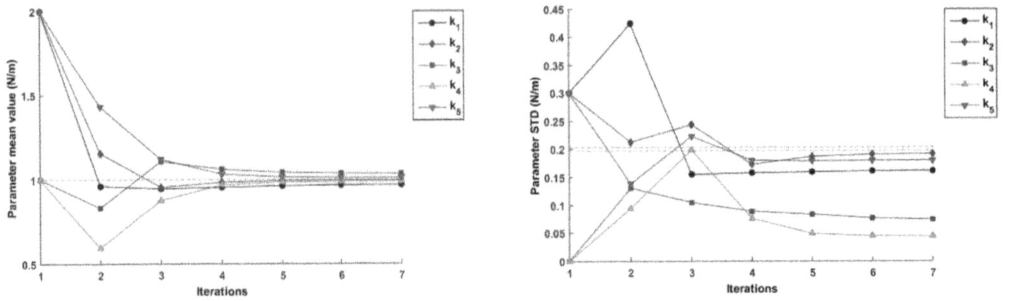

(a) Mean values of the updating parameters.

(b) Standard deviation of the updating parameters.

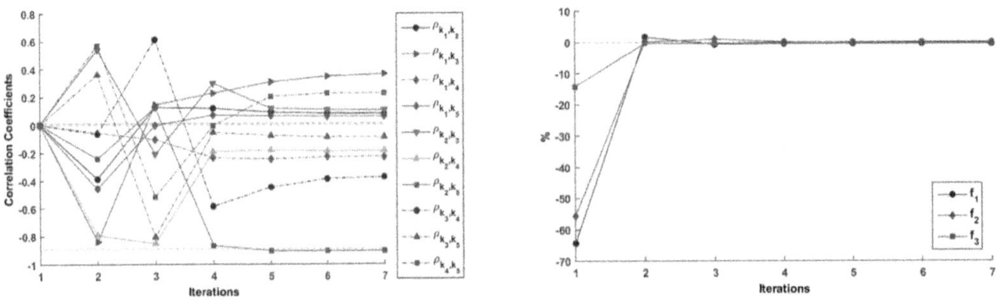

(a) Correlation of the updating parameters.

(b) Estimated mean eigenfrequencies deviation *w.r.t.* the reference values.

Figure 6.16 Convergence plots (stochastic eigensensitivity-based updating: Eq. (6.147)) (dashed lines: reference values).

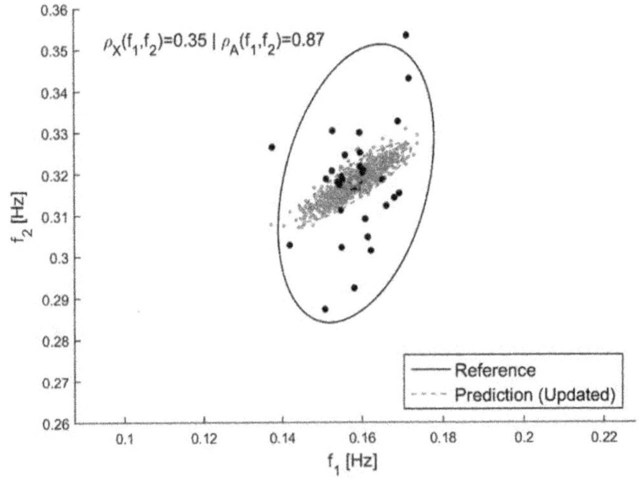

Figure 6.17 Scatter plot of the first two natural frequencies, after updating, using Eq. (6.147), for an inconsistent updating parameter set.

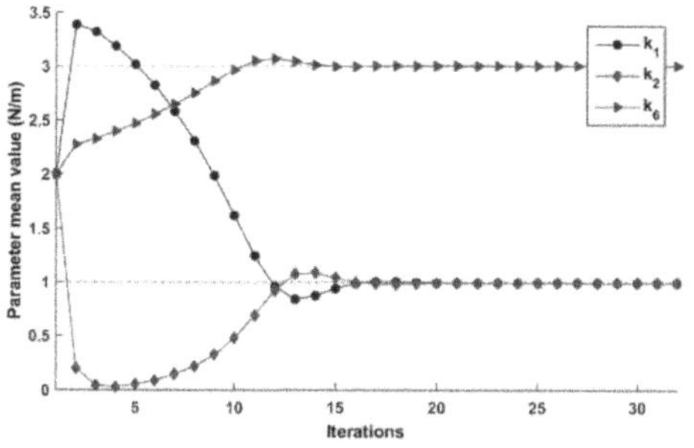

(a) Mean values of the updating parameters.

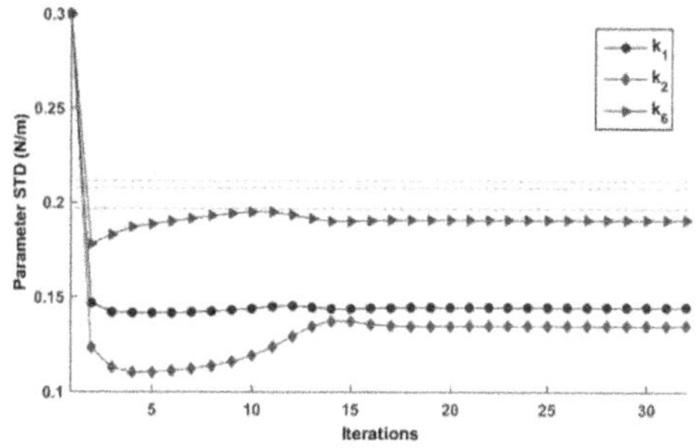

(b) Standard deviation of the updating parameters.

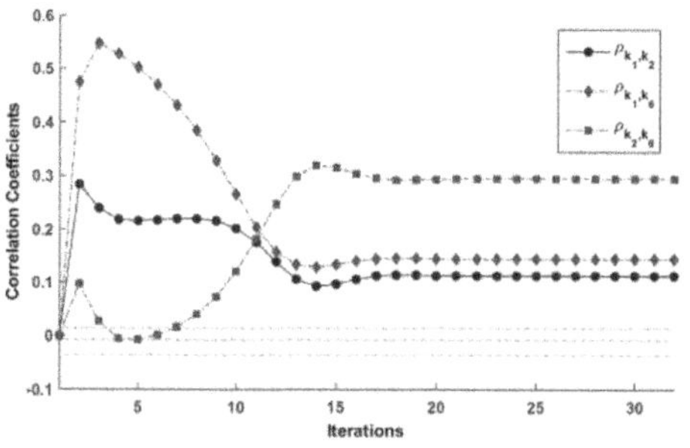

(a) Correlation of the updating parameters.

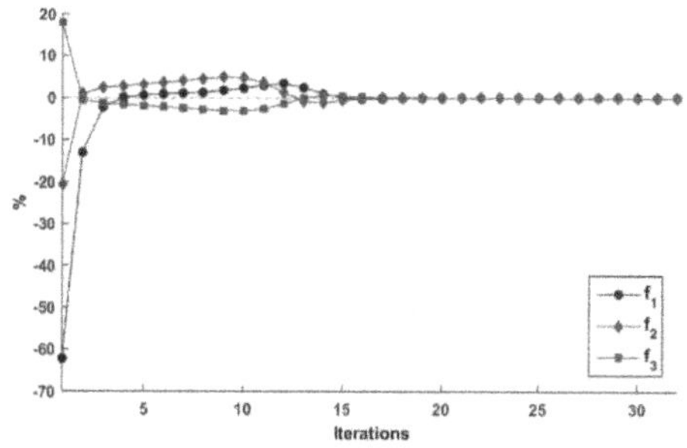

(b) Estimated mean eigenfrequencies deviation *w.r.t.* the reference values.

Figure 6.18 Convergence plots (stochastic eigensensitivity-based updating: Eq. (6.147)), for an inconsistent updating parameter set (dashed lines: reference values).

The problem to be addressed in the following section is, therefore, to define a procedure to identify the parameters supposed to be responsible for the response variability, ensuring that the output covariances as well as the mean values are reconstructed accurately.

6.5.3 Selection of Parameters for Stochastic Model Updating

In this section, two approaches for the selection of updating parameters in the scope of stochastic model updating are addressed. The first approach is based on the Principal Component Analysis (*PCA*)[6]. The second approach is based on the orthogonal projection of scaled sensitivity vectors. These approaches aim at assessing the contribution of each modelling parameter to the response covariance matrix, thereby enabling the selection of updating parameters. Examples 6.5.3-1 and 6.5.3-2 use the 3 *DOFs* system of Examples 6.4.3 and 6.5.2, while Example 6.5.3-3 explores the selection of parameters for updating concerning a pin-jointed truss.

[6] See Chapter 3, Section 3.4.8.

Towards Updating Parameters Selection using *PCA*

In Example 6.5.2, an extended, although consistent, updating parameter set was considered; therefore, it is interesting to assess how the variability of the updating parameters are related to the variability of the output responses. Hence, the use of *PCA* is addressed in order to perform this evaluation. *PCA* is based on the *SVD* of the covariance matrix of the observed response set, whose eigenvalues are function of the sampled parameters and the principal components or directions of maximal variance are given by the eigenvectors [6.82]. The number of principal components necessary to describe a certain portion of the total variance can be found using distinct rules, namely the Kaiser-Meyer-Olkin (*KMO*) criterion [6.83] or the Horn's Parallel Analysis (*PA*) [6.84]. These criteria allow for an indication on the number of principal components that should be retained, in order to reduce the number of components without much loss in the description of the total variance.

This approach to the selection of updating parameters is able to identify the set of parameters that are responsible for the response variability, i.e. the set of randomised or uncertain modelling parameters, although the procedure is performed after a previous updating stage.

Example 6.5.3-1 Selection of updating parameters using *PCA*

Recalling the deterministic eigensensitivity-based model updating example given in Section 6.4.3, this example deals with the selection or identification of updating parameters from the set of modelling ones.

Figure 6.19 shows the scree plots obtained by *PCA* of the extended updating parameter set sampled after updating, in this particular case for the noise-free reference data. The scree plot of Figure 6.19 (a) has a threshold line at 90% of the total variance, which is the threshold value for a very high quality description of the total variance, according to the *KMO* criterion, and therefore gives an indication on the number of principal components to be retained if a reduced problem is required. As it can be observed, around 93% of the total variance is described by the two first principal components. Moreover, the results of the *PA* (Figure 6.19 (b)) reinforce the conclusion that the two first principal components are the most significant in the description of the total variance and therefore they should be retained for interpretation (Figure 6.20). According to the Horn's *PA* criterion, the principal components to be retained are the ones with the eigenvalues higher than the eigenvalues of the 95th percentile of the random data of *PA*.

Knowing the number of significant principal components, it is possible to analyse the weight or score of each parameter in the structure of each principal component and by this to find the parameter set responsible for the variability of the model responses. Thus, from Figure 6.20, it is obvious which are the parameters responsible for the definition of the two first principal components, the ones with the highest scores, k_1, k_2 and k_5. This result is expected as the reference data was generated with randomised k_1, k_2 and k_5. It must be reminded that all these parameters were included in the updating parameter set and that the obtained updating results were quite good (see Example 6.5.2).

However, what happens to the *PCA* if the updating parameter set is inconsistent? Recalling the scatter plots after updating (Figure 6.17), the correlation between the reference and predicted responses is unacceptable, although the mean values are quite well updated. As the covariance matrix of the parameters is a transformation of the covariance matrix of the outputs, a sensitivity analysis of the updated model responses *w.r.t.* all the modelling parameters, or an extended set

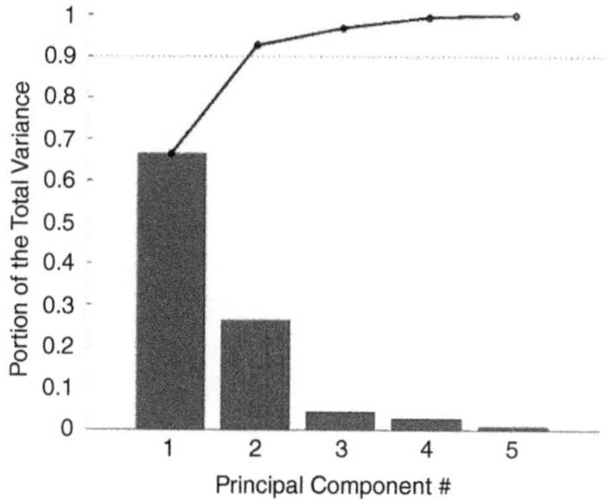

(a) Portion of the total variance described by each principal component along with its cumulation and the *KMO* criterion threshold.

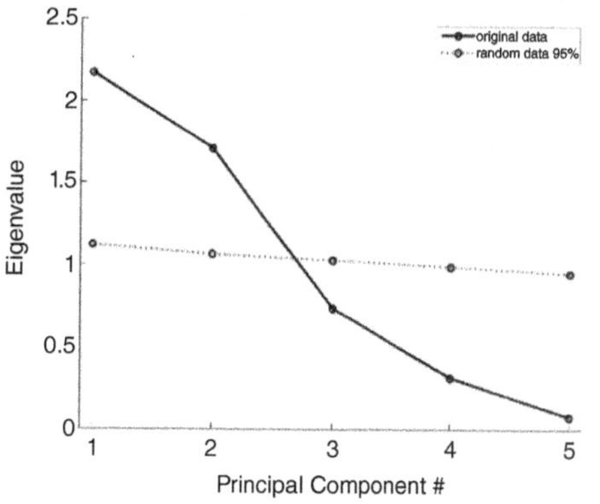

(b) Scree plot of both *PCA* and *PA* eigenvalues.

Figure 6.19 Decomposition of the total variance for the extended, and consistent, updating parameter set (see Example 6.5.2).

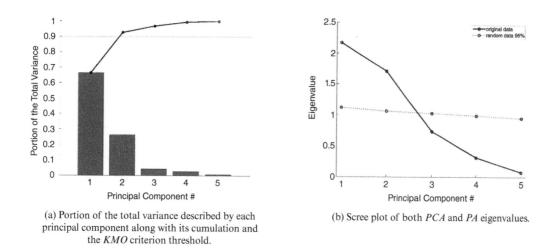

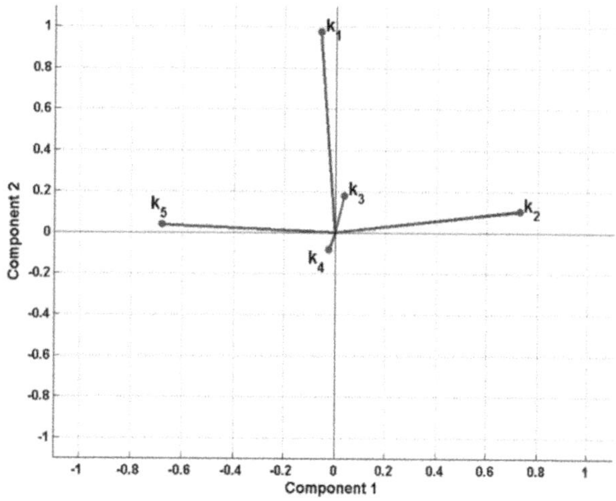

Figure 6.20 Projection of each parameter score onto the plane defined by the first two principal components for the extended, and consistent, updating parameter set (see Example 6.5.2).

of parameters susceptible to be uncertain, should be performed before applying the *PCA*. With this new and extended sensitivity matrix, one can compute the covariance matrix of the parameters, using Eq. (6.147), and generate a new sample of modelling parameters, which are used in the *PCA*. As the covariance matrix of the parameters is a transformation of the covariance matrix of the measured responses, their variabilities are closely related. Thus, it is expected that only the parameters deemed to be uncertain in the reference data set are significant in the description of the variance and therefore *PCA* will be able to identify those parameters. After identification, only this reduced set of parameters will be propagated through the model to generate a new set of model responses.

Figure 6.21 shows the scree plots obtained for the inconsistent updating parameter set sampled after updating, considering all the modelling parameters $k_1 - k_6$. The scree plots of Figures 6.19 and 6.21 are quite similar, which is somehow expectable as the reference data is the same in both cases. Hence, as in the previous case, around 93% of the total variance is described by the first two principal components and again the scree plot of Figure 6.21 (b) reinforces that only those principal components should be retained for analysis.

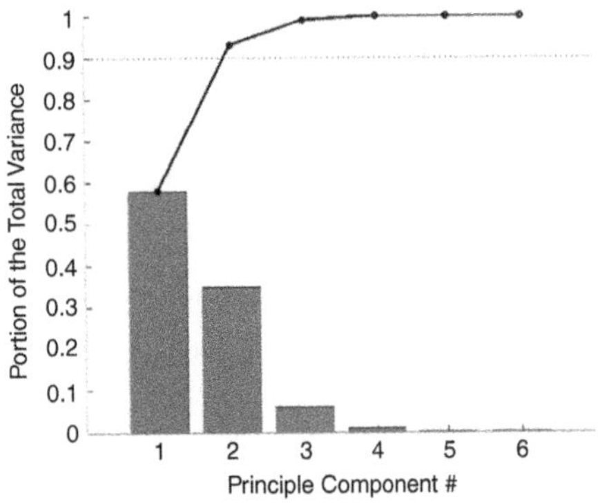

(a) Portion of the total variance described by each principal component along with its cumulation and the *KMO* criterion threshold.

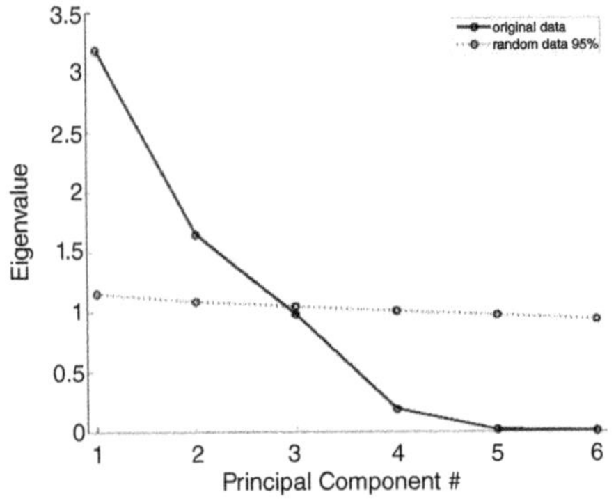

(b) Scree plot of both *PCA* and *PA* eigenvalues.

Figure 6.21 Decomposition of the total variance for the extended, and inconsistent, updating parameter set (see Example 6.5.2).

Figure 6.22 shows the plane defined by the two first principal components, from which it is evident that these principal components are mainly described by the scores related to k_1, k_2 and k_5. This result is very interesting because it shows that the inclusion of k_6 as an uncertain parameter does not contribute to the description of the total variance of the data set. Moreover, if one only uses the significant parameters, in terms of the description of variance, it is possible to update their means and – without performing a new model updating process – the updated scatter of the responses may be obtained, as the updating solution is readily available after the

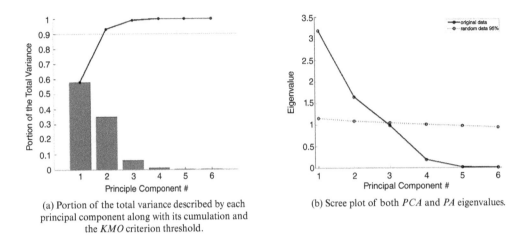

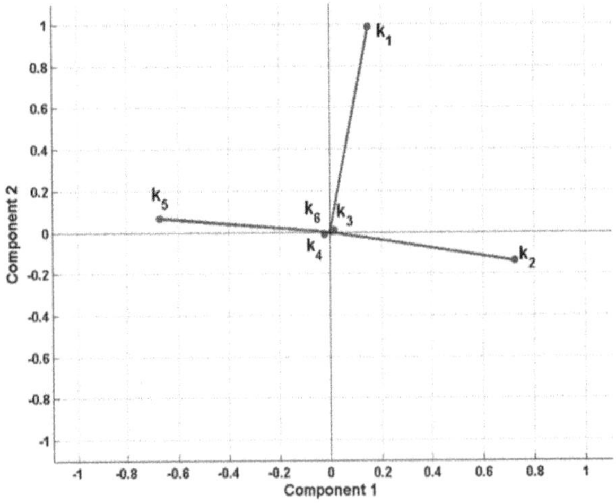

Figure 6.22 Projection of each parameter score onto the plane defined by the first two principal components for the extended, and inconsistent, updating parameter set (see Example 6.5.2).

mean values are updated. Hence, one should only repeat the forward propagation of the correct randomised parameters through the already updated model, using the updated mean values and sensitivity with Eq. (6.147). This shows that it is possible to correct the predicted model variability after updating the parameter means.

The described use of *PCA* to identify the correct random parameters requires the parameters mean values to be updated to their exact values. The need for an exact updating of all the parameters mean values is a huge limitation of this approach to parameter selection, as in practice is nearly impossible to include all the exact structural variables in the model.

Consider a new inconsistent set of updating parameters. This new set is simulated considering that the reference data set, given in Example 6.5.2, also contains an unknown uncertain element $k_6 = 3.5\,N/m$, which was not localised and therefore not included in the updating parameter set. However, the updating parameter set includes all the randomised parameters k_1, k_2 and k_5, which were used to generate the reference data.

Figures 6.23 and 6.24 show the results of the updating process for such case. The scatter plots of Figure 6.24 (d) show that the output means are reconstructed faithfully and that the agreement between reference and updated response scatter ellipses is quite good (Figure 6.23). However, this remarkable reconstruction of the measured responses is achieved by an updated parameter set that converged to incorrect values (Figures 6.24 (a) and (c)), due to the non-localised uncertain k_6.

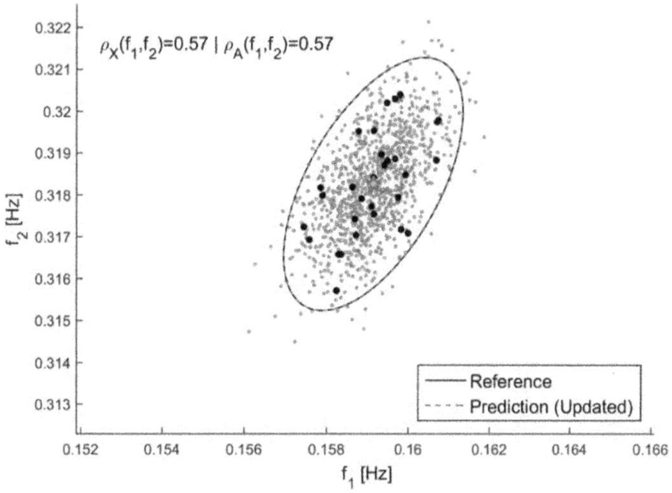

Figure 6.23 Scatter plot of the first two natural frequencies, after updating, using Eq. (6.147), for a new inconsistent updating parameter set.

In such conditions, the result of the *PCA* is inconclusive, as the plane defined by the first two principal components is populated by all the possible parameters (Figure 6.25).

This result demonstrates that the identification of the randomised parameters is only possible if the mean values are previously and correctly updated. As referred, this is a strong limitation and therefore the problem to be addressed in the following subsection is to define a procedure for parameter selection, prior to model updating, that ensures that the output covariances as well as the mean values are reconstructed accurately.

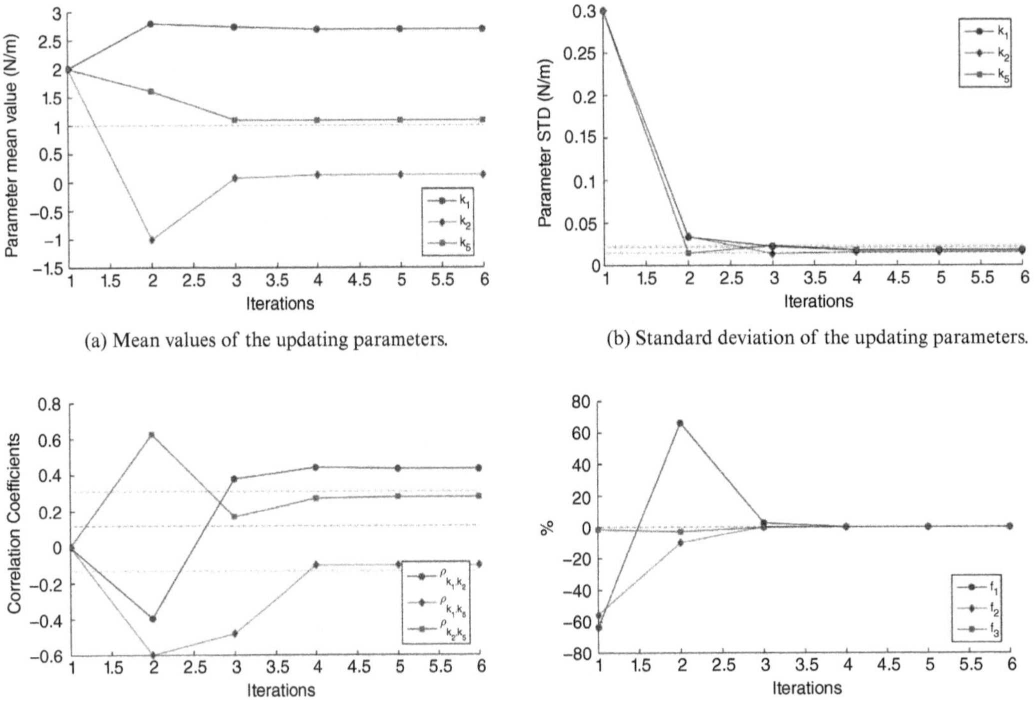

(a) Mean values of the updating parameters.

(b) Standard deviation of the updating parameters.

(a) Correlation of the updating parameters.

(b) Estimated mean eigenfrequencies deviation *w.r.t.* the reference values.

Figure 6.24 Convergence plots (stochastic eigensensitivity-based updating: Eq. (6.147)), for a new inconsistent updating parameter set (dashed lines: reference values).

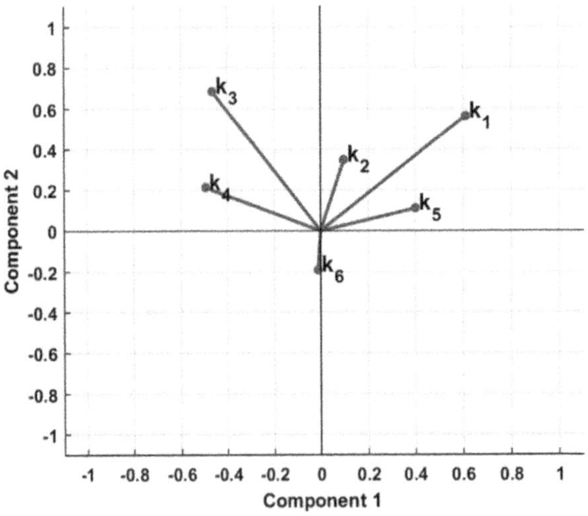

Figure 6.25 Projection of each parameter score onto the plane defined by the two first principal components (for the case with a new inconsistent set of updating parameters).

Selection of Parameters using Orthogonal Projections

Recalling Eq. (6.149), if one assumes an initial parameter estimate for parameter selection purposes, it can be recast as

$$\left(z_m - \tilde{z}_m\right) = \tilde{\boldsymbol{S}}_0\left(\boldsymbol{\theta}_0 - \tilde{\boldsymbol{\theta}}_0\right) + \boldsymbol{\varepsilon} \tag{6.153}$$

Considering n experimentally measured outputs and k modelling parameters, Eq. (6.153) may be written in scaled form using the output and parameter standard deviations, such that

$$\begin{Bmatrix} \dfrac{z_1 - \tilde{z}_1}{\sigma_{z_1}} \\ \dfrac{z_2 - \tilde{z}_2}{\sigma_{z_2}} \\ \vdots \\ \dfrac{z_n - \tilde{z}_n}{\sigma_{z_n}} \end{Bmatrix} = \begin{bmatrix} \dfrac{\sigma_{\theta_1}}{\sigma_{z_1}}\dfrac{\partial \tilde{z}_1}{\partial \tilde{\theta}_1} & \dfrac{\sigma_{\theta_2}}{\sigma_{z_1}}\dfrac{\partial \tilde{z}_1}{\partial \tilde{\theta}_2} & \cdots & \dfrac{\sigma_{\theta_k}}{\sigma_{z_1}}\dfrac{\partial \tilde{z}_1}{\partial \tilde{\theta}_k} \\ \dfrac{\sigma_{\theta_1}}{\sigma_{z_2}}\dfrac{\partial \tilde{z}_2}{\partial \tilde{\theta}_1} & \dfrac{\sigma_{\theta_2}}{\sigma_{z_2}}\dfrac{\partial \tilde{z}_2}{\partial \tilde{\theta}_2} & \cdots & \dfrac{\sigma_{\theta_k}}{\sigma_{z_2}}\dfrac{\partial \tilde{z}_2}{\partial \tilde{\theta}_k} \\ \vdots & \vdots & \ddots & \vdots \\ \dfrac{\sigma_{\theta_1}}{\sigma_{z_n}}\dfrac{\partial \tilde{z}_n}{\partial \tilde{\theta}_1} & \dfrac{\sigma_{\theta_2}}{\sigma_{z_n}}\dfrac{\partial \tilde{z}_n}{\partial \tilde{\theta}_2} & \cdots & \dfrac{\sigma_{\theta_k}}{\sigma_{z_n}}\dfrac{\partial \tilde{z}_n}{\partial \tilde{\theta}_k} \end{bmatrix} \begin{Bmatrix} \dfrac{\theta_1 - \tilde{\theta}_1}{\sigma_{\theta_1}} \\ \dfrac{\theta_2 - \tilde{\theta}_2}{\sigma_{\theta_2}} \\ \vdots \\ \dfrac{\theta_k - \tilde{\theta}_k}{\sigma_{\theta_k}} \end{Bmatrix} + \underset{\sim}{\varepsilon} \tag{6.154}$$

where the subscripts "m" and "0" are now omitted, for the sake of simplicity, and $\underset{\sim}{\varepsilon}$ denotes the scaled error vector. Thus,

$$\underset{\sim}{z} = \underset{\sim}{\boldsymbol{S}}\,\underset{\sim}{\boldsymbol{\theta}} + \underset{\sim}{\varepsilon} \tag{6.155}$$

where $\underset{\sim}{z}$, $\underset{\sim}{\boldsymbol{\theta}}$ and $\underset{\sim}{\boldsymbol{S}}$ are the scaled vectors of responses and parameters and the scaled sensitivity matrix, respectively.

The covariance matrix of the normalised parameters is given by the correlation matrix. If the chosen parameters are independent, then the covariance matrix is given by the identity matrix,

$$cov\left(\underset{\sim}{\boldsymbol{\theta}}, \underset{\sim}{\boldsymbol{\theta}}\right) = \boldsymbol{I} \tag{6.156}$$

Assuming the error $\underset{\sim}{\varepsilon}$ in Eq. (6.154) to be independent of the parameters, then the output covariance matrix may be expressed as

$$cov\left(\underset{\sim}{z}, \underset{\sim}{z}\right) = \underset{\sim}{\boldsymbol{S}}\,\underset{\sim}{\boldsymbol{S}}^T + cov\left(\underset{\sim}{\varepsilon}, \underset{\sim}{\varepsilon}\right) \tag{6.157}$$

Eq. (6. 157) may be expanded so that,

$$cov\left(\underset{\sim}{z}, \underset{\sim}{z}\right) = \boldsymbol{s}_{\theta_1}\boldsymbol{s}_{\theta_1}^T + \boldsymbol{s}_{\theta_2}\boldsymbol{s}_{\theta_2}^T + \ldots + \boldsymbol{s}_{\theta_l}\boldsymbol{s}_{\theta_l}^T + cov\left(\underset{\sim}{\varepsilon}, \underset{\sim}{\varepsilon}\right) \tag{6.158}$$

where $\boldsymbol{s}_{\theta_l}$ denotes the lth column of the scaled sensitivity matrix $\underset{\sim}{\boldsymbol{S}}$. The term $\boldsymbol{s}_{\theta_l}\boldsymbol{s}_{\theta_l}^T$ on the right-hand-side of Eq. (6.158) therefore represents the contribution of the lth parameter to the scaled output covariance matrix. Hence, for parameter selection purposes, one would like to select those parameters that make the most significant contributions [6.81].

The covariance matrix of measured outputs may be expressed by its *SVD*, as

$$\boldsymbol{A} = cov\left(\underset{\sim}{z}, \underset{\sim}{z}\right) = \boldsymbol{U\Sigma V}^H \tag{6.159}$$

As A is a square matrix, the right-singular vectors V are equal to the left ones, so that

$$A = U \Sigma U^H \tag{6.160}$$

The left-singular vectors corresponding to the non-zero singular values of A span its range. Therefore, the rank of A equals the number of non-zero singular values of Σ. Hence, the range of A is spanned by the columns of U corresponding to the non-zero singular values, so

$$range(A) = span(U_{\Sigma \neq 0}) \tag{6.161}$$

Any vector $x \in range(A)$ can be decomposed into its orthogonal components as

$$x = U_{\Sigma \neq 0} \alpha + U_{\Sigma = 0} \beta \tag{6.162}$$

where $U_{\Sigma \neq 0} \alpha$ is the projection of x onto the $range(A)$ and $U_{\Sigma = 0} \beta$ is the projection of x onto the $range(A)^{\perp} = null(A^T)$. Since $U_{\Sigma \neq 0}^T U_{\Sigma \neq 0} = I$ and $U_{\Sigma \neq 0}^T U_{\Sigma = 0} = 0$, pre-multiplying Eq. (6.162) by $U_{\Sigma \neq 0}^T$ leads to

$$\alpha = U_{\Sigma \neq 0}^T x \tag{6.163}$$

and therefore the projection of x onto the $range(A)$ is given by

$$x' = U_{\Sigma \neq 0} U_{\Sigma \neq 0}^T x \tag{6.164}$$

where $U_{\Sigma \neq 0} U_{\Sigma \neq 0}^T$ is the orthogonal projector onto the $range(A)$.

From the right-hand-sides of Eqs. (6.158) and (6.160), it is clear that the number of parameters that contribute to A must be equal to the number of non-zero singular values.

Thus, the projection onto $U_{\Sigma \neq 0}$ of the contribution of each parameter θ_l to A, i.e. each term on the right-hand-side of Eq. (6. 158), is then given by

$$s'_{\theta_l} = U_{\Sigma \neq 0} U_{\Sigma \neq 0}^T s_{\theta_l} \tag{6.165}$$

Ideally, if a parameter θ_l makes a non-zero contribution, then s'_{θ_l} must be given exactly by a linear combination of the columns of $U_{\Sigma \neq 0}$, so that s_{θ_l} and s'_{θ_l} are identical. In practice, they will be different and the cosine distance may be used to assess the closeness between s_{θ_l} and s'_{θ_l}:

$$1 - cos \Psi_l = 1 - \frac{\left| s_{\theta_l}^T s'_{\theta_l} \right|}{\left\| s_{\theta_l} \right\| \left\| s'_{\theta_l} \right\|} \tag{6.166}$$

where Ψ_l denotes the angle between s_{θ_l} and s'_{θ_l}. The cosine distance takes a value between zero and unity; in practice, if it is less than a chosen threshold,

$$1 - cos\,\Psi_l < \varepsilon_\Psi \tag{6.167}$$

then θ_l may be considered to be a contributing parameter.

The test for parameter θ_l in Eq. (6.166) requires that $cov(\underset{\sim}{z}, \underset{\sim}{z})$ must be less than full rank, so there are columns of $\boldsymbol{U}$ corresponding to small (theoretically zero) singular values, i.e., $\boldsymbol{U}_{\Sigma=0} \neq 0$. Otherwise, it is not possible to recognise wrongly selected parameters. This means that there must be more outputs than significant parameters.

This approach for the selection of updating parameters can identify the set of parameters deemed to be responsible for the response variability based on an initial parameter estimate before updating. This is a relevant result, as it is convenient to define the set of updating parameters, before starting the updating process.

Example 6.5.3-2 Selection of updating parameters using orthogonal projections

To assess the applicability of the method to select updating parameters from the set of modelling ones based on the orthogonal projection of scaled sensitivity vectors, the 3 *DOFs* structure shown in Figure 6.7 is here considered. Therefore, the nominal values for the system parameters are the ones defined and used in Example 6.4.3.

In each of the parameter selection exercises shown in Figures 6.26 to 6.28 the data are produced using a limited set of randomised parameters. However, all six stiffness terms are tested for significance with initial mean estimates of twice their nominal values (100% error) and standard deviations of half their mean values (50% error).

Considering a set of just two randomised modelling parameters, the results of parameter selection are given in Figures 6.26 and 6.27. The result of using only the sensitivity of the natural frequencies is shown in Figures 6.26 (a) and 6.27 (a), whereas the result of using both the sensitivity of natural frequencies and mode shapes is shown in Figures 6.26 (b) and 6.27 (b).

It is clear that in every case the correct parameters responsible for variability in the outputs are identified, as their corresponding cosine distance is below the specified threshold (0.05), marked by the dashed line. In Figures 6.26 and 6.27 not only are the correct parameters selected, but incorrect ones are also found. This is due to the symmetry of the model, where parameters k_1 and k_3 have the same effect on the sensitivity of the model, as well as parameters k_4 and k_5. In the cases where these parameters are randomised, the inclusion of the sensitivity of the mode shapes improves the procedure, potentially allowing for the selection of only the correct parameters.

If one had considered that three modelling parameters were randomised, it would be necessary to use both the sensitivity of the natural frequencies and mode shapes, so that the number of outputs were greater than the number of parameters. Again, the correct parameters responsible

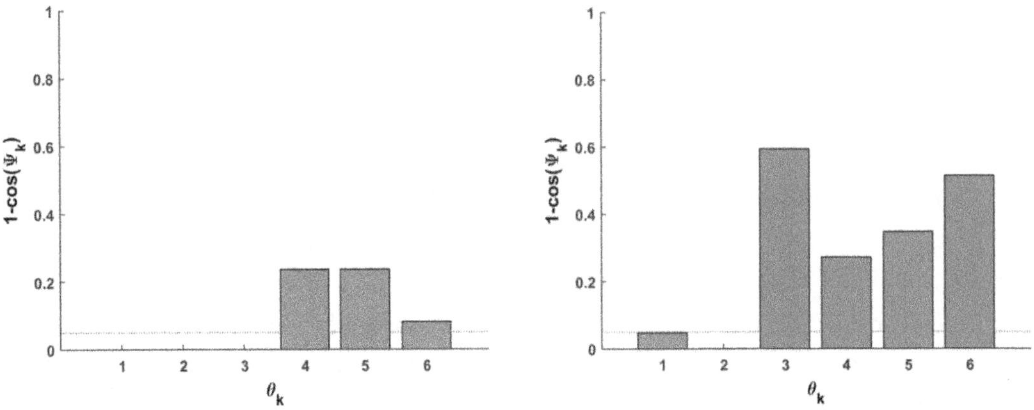

(a) Computed sensitivities: 3 natural frequencies.

(b) Computed sensitivities: 3 natural frequencies and 3 mode shapes.

Figure 6.26 Cosine distance: Randomised parameters k_1 and k_2 .

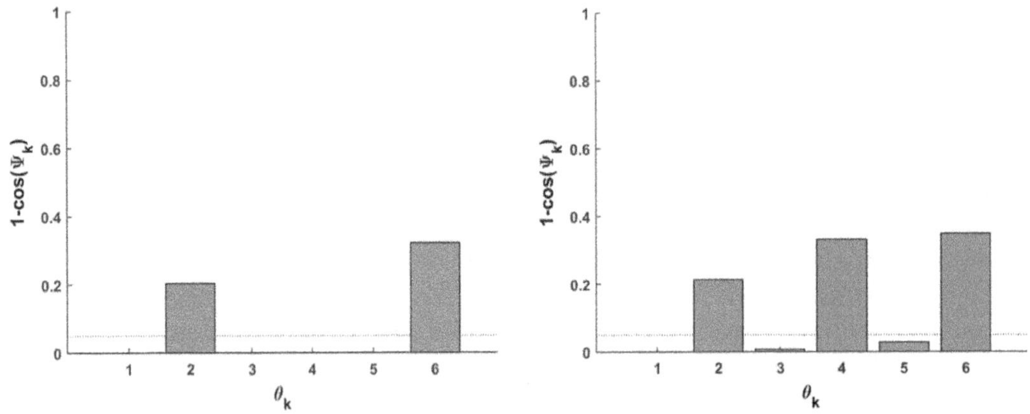

(a) Computed sensitivities: 3 natural frequencies.

(b) Computed sensitivities: 3 natural frequencies and 3 mode shapes.

Figure 6.27 Cosine distance: Randomised parameters k_1 and k_5 .

for output variability are identified in Figure 6.28, although cosine distances of additional parameters fall below the threshold. This is not a problem, since it was shown in Example 6.5.2 that stochastic updating performs well with more than the necessary number of parameters, provided that the correct ones are included in the set of updating parameters.

Example 6.5.3-3 Selection of updating parameters of a truss using orthogonal projections

Using the selection approach based on orthogonal projections, as in Example 6.5.3-2, a more realistic example is here presented. The pin-jointed truss shown in Figure 6.29 has overall dimensions 5m×1m and is composed of 21 bar elements in total, each with a stiffness matrix given by,

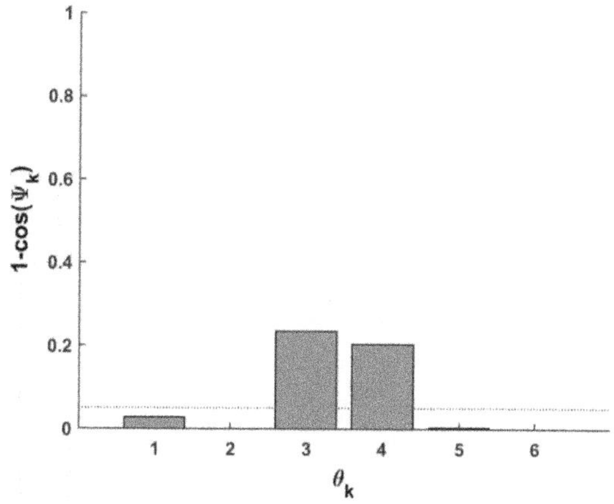

Figure 6.28 Cosine distance: Randomised parameters k_1, k_2 and k_6. Computed sensitivities: 3 natural frequencies and 3 mode shapes.

$$\boldsymbol{K}^{(i)} = k_i \begin{bmatrix} 1 & -1 \\ -1 & 1 \end{bmatrix} \tag{6.156}$$

for $i = 1, 2, ..., 21$ bar elements with a generic stiffness k_i. The elastic modulus, mass density and cross-sectional area are assumed to take the values E = 70 GPa, ρ = 2700 kg/m^3 and A = 0.03 m^2, respectively.

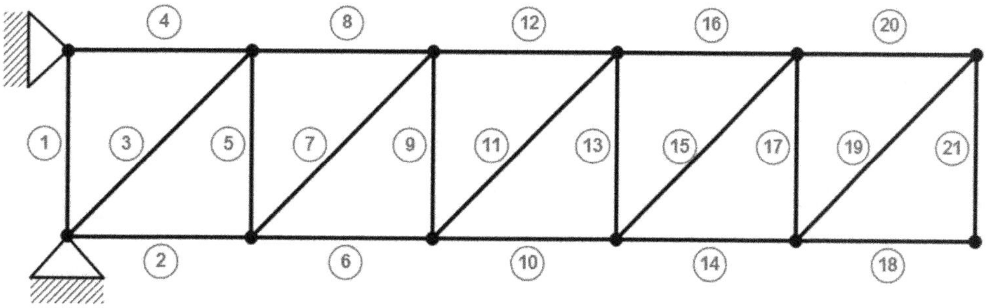

Figure 6.29 Pin-jointed truss model.

The five diagonal bars of nominal stiffness $(EA/L)_j = 1.485 \times 10^8\ N/m$ are each randomised to generate the experimental data set for updating. The true mean value of each is equal to the nominal stiffness and the standard deviations are given by $\sigma_{k_j} = 0.135\mu_{k_j}$, for $j = 3, 7, 11, 15, 19$, where 0.135 is the coefficient of variation $CoV(k_j)$. For the purposes of parameter selection, the initial estimates of all the mean stiffnesses, k_i, $i = 1, 2, ..., 21$, are considered to be 70% of the reference values and the standard deviations are given by $\sigma_{k_j} = 0.27\mu_{k_j}$.

Parameter selection results are shown in Figure 6.30. It is seen that the correct parameters for updating are recognised correctly in each case of different sensitivity analysis.

It can be seen from Figure 6.30 that the first bar element k_1 has zero cosine distance. This happens because the boundary condition prevents any extension or compression of element 1, so that all the outputs are insensitive to it. When the constraints are removed, so the truss is in free-free conditions, the cosine distance corresponding to parameter k_1 becomes finite and exceeds the threshold of 5%, as shown in Figure 6.31. It can be concluded that k_1 is not a randomised updating parameter.

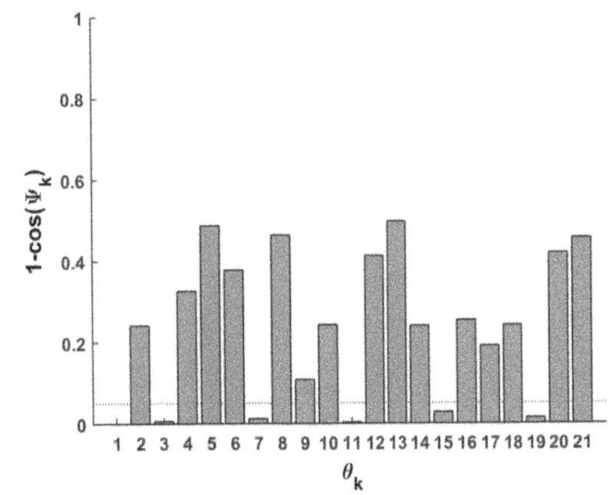

(a) Computed sensitivities: first 10 natural frequencies.

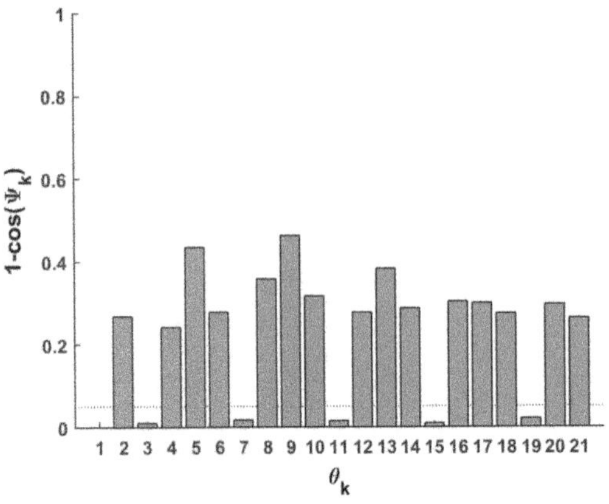

(b) Computed sensitivities: first 15 natural frequencies.

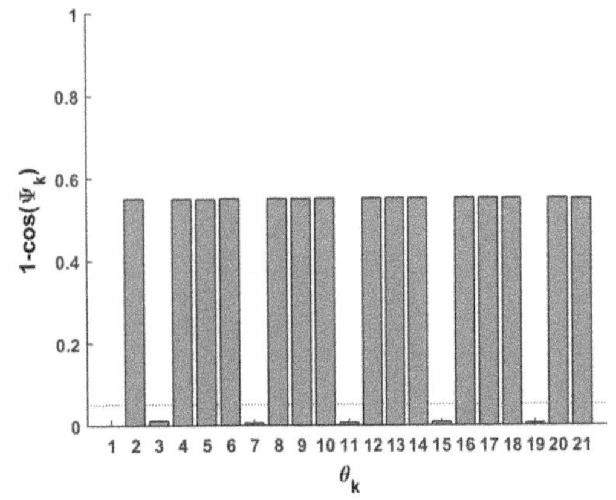

(c) Computed sensitivities: all natural frequencies.

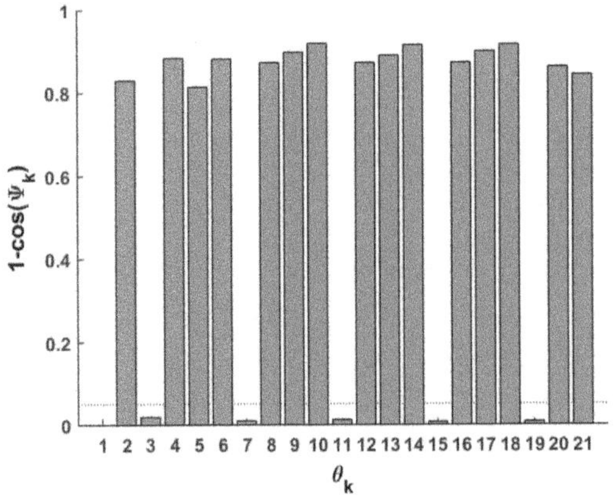

(d) Computed sensitivities: all natural frequencies and all mode shapes.

Figure 6.30 Cosine distance (Pin-jointed truss).

Having correctly identified the randomised updating parameters, it is necessary to carry out the stochastic model updating. The initial values of the updating parameters are set to:

$k_3 = 0.70\mu_{k_3}$, $k_7 = 1.20\mu_{k_7}$, $k_{11} = 0.90\mu_{k_{11}}$, $k_{15} = 0.80\mu_{k_{15}}$, $k_{19} = 1.15\mu_{k_{19}}$ with $CoV\left(k_j\right) = 2\dfrac{\sigma_{k_j}}{\mu_{k_j}}$.

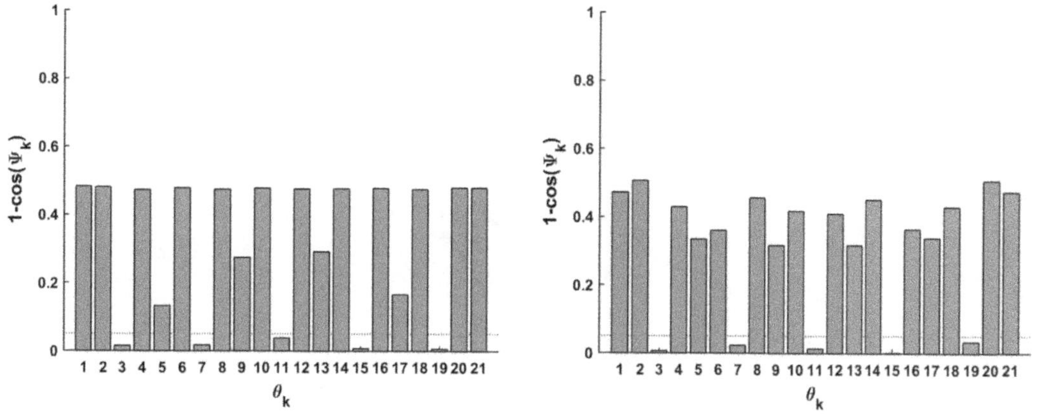

(a) Computed sensitivities: first 10 natural frequencies. (b) Computed sensitivities: all natural frequencies.

Figure 6.31 Cosine distance (Pin-jointed truss in free-free condition).

The updating results are shown in Figures 6.32 and 6.33, where one can observe that, when all the 20 natural frequencies are used in model updating, the updated parameter means and covariances are in very good agreement with the values used to generate the data (Figure 6.32). Also, the output covariances are reconstructed almost exactly (Figure 6.33).

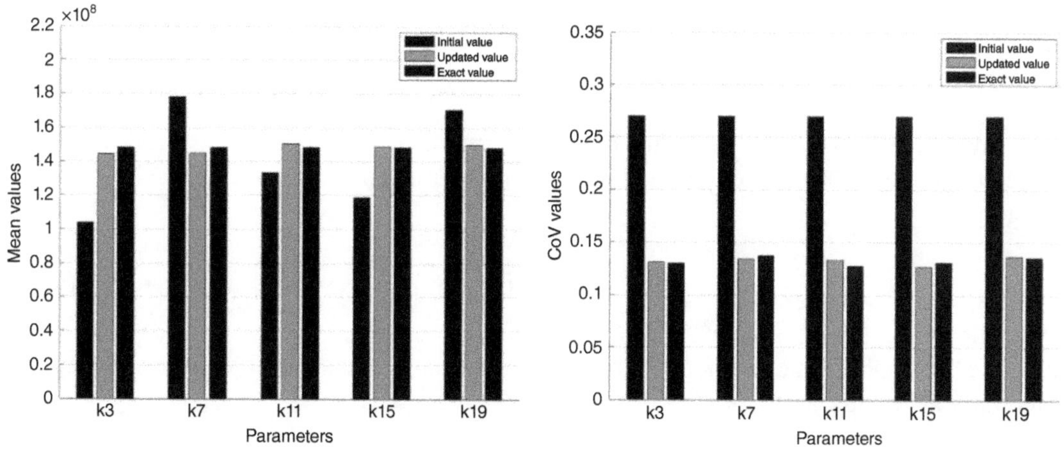

(a) Computed sensitivities: first 10 natural frequencies. (b) Computed sensitivities: all natural frequencies.

Figure 6.32 Identified parameters (Pin-jointed Truss - Eq. (6.147): using all natural frequencies).

6.5.4 A Non-Probabilistic Approach to Model Updating

The aforementioned uncertainty quantification approaches are based on Gaussian assumptions, although assumptions on the probabilistic structure or model of both the data and modelling parameters are often not clear due to lack-of-knowledge. Hence, rather than considering a probabilistic approach to stochastic model updating, one may consider a non-probabilistic approach to the problem, such as the one based on interval arithmetic [6.85, 6.86]. It is known that the techniques based on interval arithmetic tend to overestimate the parameters bounds, as it is assumed that the parameters are uniformly distributed over the entire design space, so one

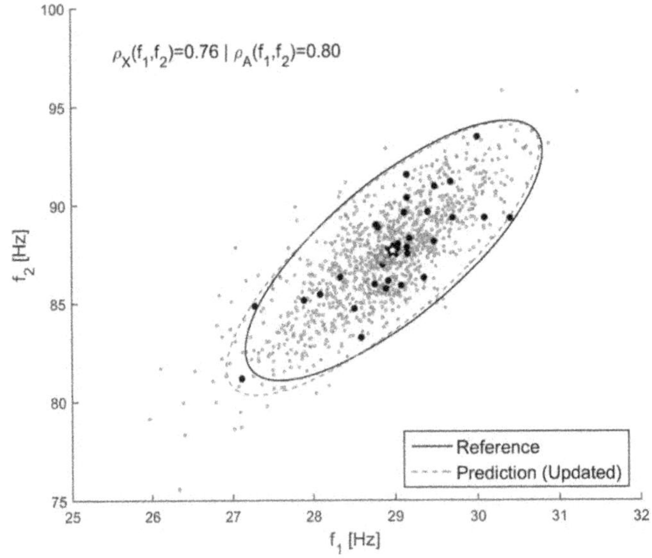

Figure 6.33 Scatter plot of the first two natural frequencies, after updating (Pin-jointed Truss - Eq. (6.147): using all natural frequencies).

cannot either distinguish or weigh regions inside the response intervals, and consequently inside the parameter intervals. The use of fuzzy logic may be addressed to enable the weighting of regions inside the referred intervals [6.85], although the basis of such an approach is still a successive application of stochastic model updating based upon interval arithmetic at different fuzzy membership levels [6.87].

Retrieving Eq. (6.104), the sensitivity-based deterministic model updating process may be adapted to compute the updated interval modelling parameters vector, in a least-squares sense, giving

$$\hat{\boldsymbol{\theta}}_{j+1} = \hat{\boldsymbol{\theta}}_j + \hat{\boldsymbol{T}}_j \left(\hat{\boldsymbol{z}}_m - \hat{\boldsymbol{z}}_j \right) \tag{6.168}$$

where the hat symbol denotes interval variables, with each interval variable defined generically by its interval bounds, as

$$\hat{x} = \left[\underline{x}, \overline{x} \right] \tag{6.169}$$

where $\underline{x}$ and $\overline{x}$ are, respectively, the lower and upper bounds of the interval variable $\hat{x}$.

The solution of the interval model updating method is often obtained by a parameter vertex solution, although this approach assumes a monotonic relation between the inputs and outputs of the system. To avoid this restriction, one may consider an interval model updating solution that uses the sensitivity of a surrogate model. Such an approach presents the advantage of not being limited to monotonic or quasi-monotonic system relations, and it is computationally efficient. It should be stressed that the initial interval on the parameters space must lead to a solution hyper-space that encompasses the entire hyper-space of the reference responses. In the context of fuzzy variables, with a feasible initial hyper-interval for the parameter set, the interval model updating is carried on successively at each fuzzy membership level, from the lowest to the highest level.

Regarding computational efficiency, if one defines the interval variables by the perturbation method, each interval variable may be defined by two different parameters: the interval centre and its radius (half of the interval amplitude). Thus, the generic interval variable given in Eq. (6.169) may be reformulated as

$$\hat{x} = \tilde{x} \pm \delta x \tag{6.170}$$

where $\tilde{x}$ denotes the interval centre and δx is the interval radius, where $\delta x = (\overline{x} - \underline{x})/2$.

With the interval variables defined by the perturbation method, it is possible to reduce the interval model updating process to the deterministic updating of the interval centres, or mean values, of all the modelling parameters, followed by the estimation of each interval radius. One should bear in mind that the variance of a generic continuous random variable, uniformly distributed in a given interval, can be obtained from the interval amplitude. Hence, one might gain advantage from the use of the sensitivity-based covariance updating given in Eq. (6.147) to compute the updated variances of each modelling parameter, from which one may approximate the interval radii of the updating parameters right after updating their centres, as the variance of the updating parameters is available from the diagonal of the updated covariance matrix.

Example 6.5.4 Application of the interval model updating

The 3 *DOFs* that has been used in the examples given in this chapter will also be considered in the following.

In this example, the uncertain parameters are defined as interval variables. Thus, the vector of updating parameters is $\hat{\theta} = \left[\hat{\theta}_1\ \hat{\theta}_2\ \hat{\theta}_3\right]^T = \left[\hat{k}_1\ \hat{k}_2\ \hat{k}_5\right]^T$ and the reference values for all the updating parameters are $\hat{\boldsymbol{\theta}}_i = \left[\underline{\theta}_i\ \overline{\theta}_i\right] = \left[0.8\ 1.2\right]$ for i = 1, 2 and 3. The reference sample was generated using *LHS* from a normal distribution with 30 observations, as in Example 6.5.2. Here, the reference response set is $\hat{z}_m = \left[\hat{z}_1 \cdots \hat{z}_4\right]^T = \left[f_1\ f_2\ f_3 \left|\phi_{1,1}\right|\right]^T$, i.e. the first three natural frequencies and the modulus of the first element of the first mode shape. Recall that the reference response set contains 30 sampled values from the interval space of $\hat{z}_m$. This is considered to be close to the data set used in Example 6.5.2, aiming at simulating an experimental data set obtained from a series of tests on identical structures.

The interval updating process starts with an initial guess on the parameters interval that must lead to model response intervals that encompass the reference, though unknown, ones. In this example, the initial guess on the parameters interval is set as $\hat{\boldsymbol{\theta}}_i = \left[\underline{\theta}_i\ \overline{\theta}_i\right] = \left[0.5\ 1.5\right]$ for i = 1, 2 and 3.

The presented results, given in Table 6.3 and Figure 6.34, summarise the outcome of the interval updating process.

Table 6.3 Initial and updated values for the parameters interval and the relative error after interval model updating.

	$\underline{k}_1$	$\overline{k}_1$	$\underline{k}_2$	$\overline{k}_2$	$\underline{k}_5$	$\overline{k}_5$
Initial	0.5	1.5	0.5	1.5	0.5	1.5
Reference	0.837	1.176	0.805	1.197	0.839	1.165
Updated	0.837	1.177	0.807	1.190	0.838	1.167
error %	-0.03	0.07	0.26	-0.64	-0.12	0.10

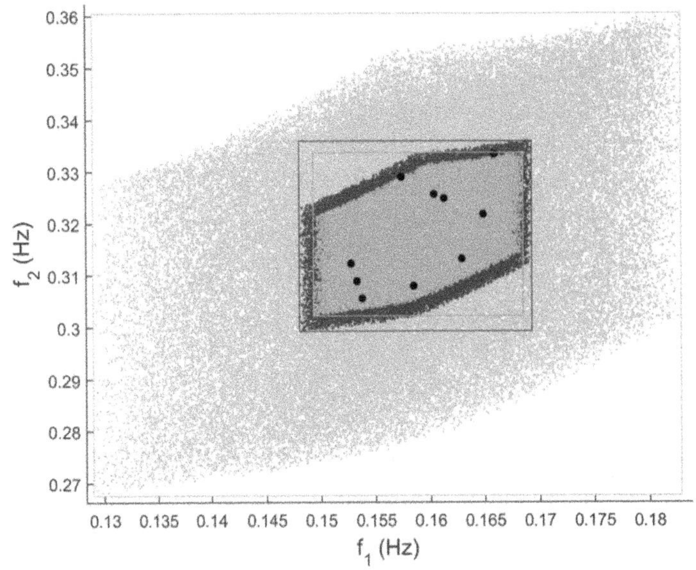

Figure 6.34 Scatter plot, with rectangular intervals, of the first two natural frequencies: initial, true (from where the reference data set is sampled) and updated interval spaces.

From Table 6.3 and Figure 6.34, one can observe that the initial parameters interval space leads to an initial interval on the response space that clearly encompasses the reference data set, and that the updated response interval falls inside the true ones. This is due to the interval updating based on the reference data set alone. As it can be observed, the reference observations are sampled from the true interval space.

The presented updating results show a very good agreement with the reference values in terms of both the parameter and the response intervals.

If Eq. (6.170) is used, one can solve the interval model updating problem based on the updating of the centre of the interval only, followed by the transformation of the covariance matrix of the reference data to estimate each interval radius.

6.5.5 On the Bayesian Approach to Model Updating

Frequentist approaches to model updating, as covered in this chapter, rely on specific or set of specific standard probability distributions to describe the uncertain modelling parameters or to frame the set of experimental responses. In contrast, the Bayesian approach to model updating is far more flexible, as probability is used to fill the lack of knowledge on uncertain parameters of different sources, assuming only that the reference data, responses, are the base to build to updating process. By the frequentist eye, the probability of an event to occur demands for a considerable number of trials to comply with the law of large numbers. In other words, to ensure the statistical significance of the outcome based on the assumption of a predefined probability distribution. In a Bayesian model updating framework, parameters related to both epistemic and aleatoric sources of uncertainty may be handled in one global updating problem using the probability theory, meaning that one may include modelling parameters as well as parameters related to the description of experimental noise. The Bayesian framework is performed using Bayes' theorem to learn on the parameters based on data evidence, usually the set of reference responses or metrics of it. Hence, the Bayesian updating process results on the estimation of a joint probability distribution of the updating parameters, instead of an estimation of parameters of a given standard probability distribution, upon prior knowledge on the parameters to be updated, and on the statistical structure of an error function to be minimised to solve a stochastic model updating problem [6.88, 6.89].

To set a proper Bayesian framework, one should set up a probabilistic model, by the definition a joint *PDF*, that encompasses information of and relations between all the relevant quantities in the underlying updating problem, including the data collection process. One must ensure that the inferred posterior *PDF* is conditioned by the observed data, meaning that it relates to the conditional *PDF* of the parameters to be updated, given the reference data. As aforementioned, generically, an updating process aims at minimising the discrepancy between experimental z_m and predicted $z(\theta_j)$ responses, being the latter a function of the uncertain modelling parameters $\boldsymbol{\theta}_j$. On the other hand, one may consider that the experimental responses, although taken as reference, are usually corrupted by experimental noise at some extent.

Bayesian inference draws conclusions, in terms of probability statements, about the set of updating parameters given a set of reference responses, defining the probabilistic model as a joint *PDF* given by the product

$$p(\boldsymbol{\theta}, z_m) = p(\boldsymbol{\theta})\, p(z_m \mid \boldsymbol{\theta}) \tag{6.171}$$

conditioning on the known value of the reference data $p(z_m)$, using Bayes' theorem, the posterior *PDF* $p(\boldsymbol{\theta} \mid z_m)$ is

$$p(\boldsymbol{\theta} \mid z_m) = \frac{p(\boldsymbol{\theta})\, p(z_m \mid \boldsymbol{\theta})}{p(z_m)} \tag{6.172}$$

where $p(\boldsymbol{\theta})$ reflects the prior knowledge on the updating parameters and $p(z_m \mid \boldsymbol{\theta})$ is, usually, the likelihood function. Sampling methods are used to estimate the posterior distributions for high dimensional models. Different implementations based on Markov Chain Monte Carlo (*MCMC*) algorithms have been proposed [6.88].

Hierarchical models have been introduced, allowing to refine the probabilistic model by the inclusion of multilevel probability specifications. Consider, as an illustrative example, the case of a joint posterior *PDF* of a set of updating parameters conditioned by the observed reference data, where the set of updating parameters includes the distribution of the updating parameters, the hyper-parameters of those distributions, estimated responses and parameters related to experimental noise.

The Bayesian approach to model updating is quite computationally demanding, though its flexibility and potential is undeniable.

References

[6.1] Maia, N. M. M., Silva, J. M. M., *Theoretical and Experimental Modal Analysis*, Research Studies Press, 1997.

[6.2] Mottershead, J. E., Friswell, M. I., *Model Updating in Structural Dynamics: A Survey*, Journal of Sound and Vibration, 167 (2), pp. 347–375, 1993.

[6.3] Friswell, M. I., Mottershead, J. E., *Finite Element Model Updating in Structural Dynamics*, Kluwer Academic Publishers, 1995.

[6.4] Marwala, T., *Finite-Element-Model Updating using Computational Intelligence Techniques: Applications to Structural Dynamics*, Springer-Verlag, 2010.

[6.5] Hemez, F. M., Farrar, C. R., *A Brief History of 30 Years of Model Updating in Structural Dynamics*, Proceedings of the 32nd International Modal Analysis Conference (IMAC XXXII), Springer, 6, pp. 53–71, 2014.

[6.6] Kerschen, G., Worden, K., Vakakis, A. F., Golinval, J.-C., *Past, Present and Future of Nonlinear System Identification in Structural Dynamics*, Mechanical Systems and Signal Processing, 20 (3), pp. 505–592, 2006.

[6.7] da Silva, S., Cogan, S., Foltête, E., Buffe, F., *Metrics for Nonlinear Model Updating in Structural Dynamics*, Journal of the Brazilian Society of Mechanical Sciences and Engineering., 31 (1), pp. 27–34, 2009.

[6.8] Qu, Z.-Q., *Model Order Reduction Techniques with Applications in Finite Element Analysis*. Springer, 2004.

[6.9] Koutsovasilis, P., *Model Order Reduction in Structural Mechanics: Coupling the Rigid and Elastic Multi Body Dynamics*, Ph.D. thesis, Technische Universität Dresden, 2009.

[6.10] Besselink, B., Tabak, U., Lutowska, A., van de Wouw, N., Nijmeijer, H., Rixen, D. J., Hochstenbach, M., Schilders, W., *A Comparison of Model Reduction Techniques from Structural Dynamics, Numerical Mathematics and Systems and Control*, Journal of Sound and Vibration, 332 (19), pp. 4403–4422, 2013.

[6.11] Guyan, R. J., *Reduction of Stiffness and Mass Matrices*, AIAA Journal, 3 (2), pp. 961–962, 1965.

[6.12] Paz, M., *Dynamic Condensation Method*, AIAA Journal, 22 (5), pp. 724–727, 1984.

[6.13] Qu, Z.-Q., Fu, Z.-F., *An Iterative Method for Dynamic Condensation of Structural Matrices*, Mechanical Systems and Signal Processing, 14 (4), pp. 667–678, 2000.

[6.14] O'Callahan, J., *A Procedure for an Improved Reduced System (IRS) Model*, Proceedings of the 7th International Modal Analysis Conference (IMAC VII), pp. 17–21, 1989.

[6.15] Friswell, M. I., Garvey, S., Penny, J., *Model Reduction using Dynamic and Iterated IRS Techniques,* Journal of Sound and Vibration, 186 (2), pp. 311–323, 1995.

[6.16] Friswell, M. I., Garvey, S., Penny, J., *The Convergence of the Iterated IRS Method*, Journal of Sound and Vibration, 211 (1), pp. 123–132, 1998.

[6.17] Xia, Y., Lin, R.-M., *Improvement on the Iterated IRS Method for Structural Eigensolutions*, Journal of Sound and Vibration, 270 (4-5), pp. 713–727, 2004.

[6.18] O'Callahan, J., Avitabile, P., Riemer, R., *System Equivalent Reduction Expansion Process (SEREP)*, Proceedings of the 7th International Modal Analysis Conference (IMAC VII), pp. 29–37, 1989.

[6.19] Kammer, D. C., *Test-Analysis Model Development using an Exact Modal Reduction*, International Journal of Analytical and Experimental Modal Analysis, 2 (3), pp. 174–179, 1987.

[6.20] Kammer, D. C., *A Hybrid Approach to Test Analysis Model Development for Large Space Structures*, Journal of Vibration and Acoustics, 113 (3), pp. 325–332, 1991.

[6.21] Craig, Jr., R. R., *Coupling of Substructures for Dynamic Analyses: An Overview*, AIAA Dynamics Specialists Conference, 2000.

[6.22] Craig, Jr., R. R., Kurdila, A. J., *Fundamentals of Structural Dynamics*, 2nd Edition, John Wiley & Sons, 2006.

[6.23] Craig, Jr., R. R., Bampton, M. C. C., *Coupling of Substructures for Dynamic Analyses*, AIAA Journal, 6 (7), pp. 1313–1319, 1968.

[6.24] Rixen, D. J., *A Dual Craig-Bampton Method for Dynamic Substructuring*, Journal of Computational and Applied Mathematics, 168 (1-2), pp. 383–391, 2004.

[6.25] Rixen, D. J., *Dual Craig-Bampton with Enrichment to Avoid Spurious Modes*, Proceedings of the 27th International Modal Analysis Conference (IMAC XXVII), 2009.

[6.26] Carne, T. G., Bateman, V. I., Mayes, R. L., *Force Reconstruction Using a Sum of Weighted Accelerations Technique*, Proceedings of the 10th International Modal Analysis Conference (IMAC X), 1992.

[6.27] Minnicino II, M., Hopkins, D., *Overview of Reduction Methods and their Implementation into Finite Element Local-to-Global Techniques*, Tech. rep., U.S. Army Research Laboratory: Aberdeen Proving Ground, 2004.

[6.28] Allen, M., Carne, T., *Comparison of Inverse Structural Filter (ISF) and Sum of Weighted Accelerations (SWAT) Time Domain Force Identification Methods* Proceedings of 47th AIAA-ASME-ASCE-AHS-ASC Structures, Structural Dynamics, and Materials Conference, AIAA, 2006.

[6.29] Jeong, J., Baek, S., Cho, M., *Dynamic Condensation in a Damped System through Rational Selection of Primary Degrees of Freedom*, Journal of Sound and Vibration, 331 (7), pp. 1655–1668, 2012.

[6.30] Das, A., Dutt, J., *Reduced Model of a Rotor-Shaft System using Modified SEREP*, Mechanics Research Communications, 35 (6), pp. 398–407, 2008.

[6.31] Balmès, E., *Sensors, Degrees of Freedom, and Generalized Modeshape Expansion Methods*, Proceedings of the 17th International Modal Analysis Conference (IMAC XVII), 1999.

[6.32] Kidder, R. L., *Reduction of Structural Frequency Equations*, AIAA Journal, 11 (6), p. 892, 1973.

[6.33] Lipkins, J., Vandeurzen, U., *The Use of Smoothing Techniques for Structural Modification Applications*, Proceedings of the 12th International Conference on Noise and Vibration Engineering (ISMA), S1–3, 1987.

[6.34] Batista, F., Maia, N. M. M., *Estimation of Unmeasured Frequency Response Functions*, Proceedings of the International Congress on Sound and Vibration (ICSV19), 2012.

[6.35] Avitabile, P., O'Callahan, J., *Frequency Response Function Expansion for Unmeasured Translation and Rotation DOFs for Impedance Modeling Applications*, Mechanical Systems and Signal Processing, 17 (4), pp. 723–745, 2003.

[6.36] Maia, N. M. M., Silva, T. A. N., *An Expansion Technique for the Estimation of Unmeasured Rotational Frequency Response Functions*, Mechanical Systems and Signal Processing, 156, 107634, 2021.

[6.37] Allemang, R. J., Brown, D. L., *A Correlation Coefficient for Modal Vector Analysis*. Proceedings of the 1st International Modal Analysis Conference (IMAC I), pp. 110–116, 1982.

[6.38] Allemang, R. J., *The Modal Assurance Criterion - Twenty Years of Use and Abuse*. Sound and Vibration, pp. 14–21, 2003.

[6.39] Meireles, J., Ambrósio, J., Montalvão e Silva, J., Pinho, A., *Structural Dynamic Analysis by Finite Element Models Experimentally Identified: An Approach Using Modal Data*, Proceedings of the International Conference on Experimental Vibration Analysis for Civil Engineering Structures (EVACES), 2007.

[6.40] Lieven, N. A. J., Ewins, D. J., *Spatial Correlation of Modeshapes: The Coordinate Modal Assurance Criterion (COMAC)*, Proceedings of the 6th International Modal Analysis Conference (IMAC VI), pp. 690–695, 1988.

[6.41] Pascual, R., Golinval, J.-C., Razeto, M., *Testing of FRF Based Model Updating Methods using a General Finite Element Program*, Proceedings of 21st International Conference on Noise and Vibration Engineering (ISMA), pp. 1933–1945, 1996.

[6.42] Pascual, R., Golinval, J.-C., Razeto, M., *A Frequency Domain Correlation Technique for Model Correlation and Updating*, Proceedings of the 15th International Modal Analysis Conference (IMAC XV), 1997.

[6.43] Heylen, W., Lammens, S., Sas, P., *Modal Analysis: Theory and Testing*. Katholieke Universiteit Leuven, 2007.

[6.44] Heylen, W., Lammens, S., *FRAC: A Consistent Way of Comparing Frequency Response Functions*, Proceedings of the International Conference on Identification in Engineering Systems, pp. 48–57, 1996.

[6.45] Heylen, W., Avitabile, P., *Correlation Considerations: Part 5 - Degree of Freedom Correlation Techniques*, Proceedings of the 16th International Modal Analysis Conference (IMAC XVI), pp. 207–214, 1998.

[6.46] Zang, C., Grafe, H., Imregun, M., *Frequency-Domain Criteria for Correlating and Updating Dynamic Finite Element Models*. Mechanical Systems and Signal Processing, 15 (1), pp. 139–155, 2001.

[6.47] Grafe, H., *Model Updating of Large Structural Dynamics Models Using Measured Response Functions*. Ph.D. thesis, Imperial College of Science, Technology and Medicine, 1998.

[6.48] Dascotte, E., Strobbe, J., *Updating Finite Element Models using FRF Correlation Functions*, Proceedings of the 17th International Modal Analysis Conference (IMAC XVII). pp. 1169–1174, 1999.

[6.49] Zang, C., Friswell, M. I., Imregun, M., *Structural Health Monitoring and Damage Assessment using Frequency Response Correlation Criteria*, Journal of Engineering Mechanics, 133 (9), pp. 981–993, 2007.

[6.50] Maia, N. M. M., Almeida, R. A. B., Urgueira, A. P. V., Sampaio, R. P. C., *Damage Detection and Quantification using Transmissibility*, Mechanical Systems and Signal Processing, 25, pp. 2475–2483, 2011.

[6.51] Schwer, L. E., *An Overview of the PTC 60/V&V 10: Guide for Verification and Validation in Computational Solid Mechanics*. Engineering with Computers, 23 (4), pp. 245–252, 2007.

[6.52] Roy, C. J., Oberkampf, W. L., *A Comprehensive Framework for Verification, Validation, and Uncertainty Quantification in Scientific Computing*. Computer Methods in Applied Mechanics and Engineering, 200 (25-28), pp. 2131–2144, 2011.

[6.53] Hemez, F. M., Doebling, S. W., *Model Validation and Uncertainty Quantification*, Proceedings of the 19th International Modal Analysis Conference (IMAC XIX), 7955, pp. 1153–1158, 2001.

[6.54] Cogan, S., Foltête, E., *Model Updating for Validation: Theory and Practice*, International Modal Analysis Conference - IMAC XXVIII (pre-conf. short-course), 2010.

[6.55] Hemez, F. M., *Verification and Validation of Structural Dynamics Models (V&VSDM)*, International Conference on Noise and Vibration Engineering - ISMA2012 (post-conf. short-course), 2012.

[6.56] Baruch, M., *Optimization Procedure to Correct Stiffness and Flexibility Matrices using Vibration Tests*, AIAA Journal, 16 (11), pp. 1208–1210, 1978.

[6.57] Berman, A., Nagy, E. J., *Improvement of a Large Analytical Model using Test Data*, AIAA Journal, 21 (8), pp. 1168–1173, 1983.

[6.58] Friswell, M. I., Inman, D. J., Pilkey, D. F., *Direct Updating of Damping and Stiffness Matrices*, AIAA Journal, 36 (3), pp. 491–493, 1998.

[6.59] Ahmadian, H., Mottershead, J. E., Friswell, M. I., *Regularisation Methods for Finite Element Model Updating*. Mechanical Systems and Signal Processing, 12 (1), pp. 47–64, 1998.

[6.60] Pilkey, D. F., *Computation of Damping Matrix for Finite Element Model Updating*, Ph.D. thesis, Virginia Polytechnic Institute and State University, 1998.

[6.61] Adhikari, S., *Damping Models for Structural Vibration*. Ph.D. thesis, University of Cambridge, 2000.

[6.62] Kwon, K.-S., Lin, R.-M., *Frequency Selection Method for FRF-based Model Updating*, Journal of Sound and Vibration, 278 (1-2), pp. 285–306, 2004.

[6.63] Mottershead, J. E., Link, M., Friswell, M. I., *The Sensitivity Method in Finite Element Model Updating: A Tutorial*, Mechanical Systems and Signal Processing, 25 (7), pp. 2275–2296, 2011.

[6.64] Fox, R. L., Kapoor, M. P., *Rates of Change of Eigenvalues and Eigenvectors*, AIAA Journal, 6 (12), pp. 2426–2429, 1968.

[6.65] Friswell, M. I., Mottershead, J. E., Ahmadian, H., *Finite-Element Model Updating using Experimental Test Data: Parametrization and Regularization*. Philosophical Transactions of the Royal Society A: Mathematical, Physical and Engineering Sciences, 359, pp. 169–186, 2001.

[6.66] Link, M., *Updating of Analytical Models - Basic Procedures and Extensions*, Modal Analysis and Testing: NATO Science Series, Kluwer, pp. 281–304, 1999.

[6.67] Khodaparast, H. H., Mottershead, J. E., Friswell, M. I., *Perturbation Methods for the Estimation of Parameter Variability in Stochastic Model Updating*, Mechanical Systems and Signal Processing, 22 (8), pp. 1751–1773, 2008.

[6.68] Govers, Y., Link, M., *Stochastic Model Updating - Covariance Matrix Adjustment from Uncertain Experimental Modal Data*, Mechanical Systems and Signal Processing, 24 (3), pp. 696–706, 2010.

[6.69] Khodaparast, H. H., Mottershead, J. E., Badcock, K. J., *Interval Model Updating with Irreducible Uncertainty using the Kriging Predictor*. Mechanical Systems and Signal Processing, 25 (4), pp. 1204–1226, 2011.

[6.70] Mottershead, J. E., Link, M., Silva, T. A. N., Govers, Y., Khodaparast, H. H., *The Sensitivity Method in Stochastic Model Updating*, Vibration Engineering and Technology of Machinery: Proceedings of VETOMAC X, 23, pp. 65–77, Springer, 2015.

[6.71] Lin, R.-M., Ewins, D. J., *Model Updating using FRF Data*, Proceedings of the 15th International Seminar on Modal Analysis Noise and Vibration Engineering (ISMA), pp. 141–162, 1990.

[6.72] Lin, R.-M., Zhu, J., *Model Updating of Damped Structures using FRF Data*, Mechanical Systems and Signal Processing, 20 (8), pp. 2200–2218, 2006.

[6.73] Ladevèze, P., *Recalage de Modélisations des Structures Complexes*, Note technique: 33.11.01.4, Tech. rep., Aerospatiale, Les Mureaux, 1983.
[6.74] Maia, N. M. M., Reynier, M., Ladevèze, P., *Error Localisation for Updating Finite Element Models using Frequency-Response-Functions*, Proceedings of the 12th International Modal Analysis Conference (IMAC XII), pp. 1299–1308, 1994.
[6.75] Silva, T. A. N., Maia, N. M. M., *Detection and Localisation of Structural Damage based on the Error in the Constitutive Relations in Dynamics*, Applied Mathematical Modelling, 46, pp. 736–749, 2017.
[6.76] Iman, R. L., Conover, W. J., *A Distribution-Free Approach to Inducing Rank Correlation among Input Variables*. Communications in Statistics - Simulation and Computation, 11 (3), pp. 311–334, 1982.
[6.77] Karian, Z. A., Dudewicz, E. J., *Fitting Statistical Distributions: The Generalized Lambda Distribution and Generalized Bootstrap Methods*. Chapman and Hall/CRC, 2000.
[6.78] Massey Jr., F. J., *Kolmogorov-Smirnov Test for Goodness of Fit*, Journal of the American Statistical Association, 46 (253), pp. 68–78, 1951.
[6.79] Bi, S., Prabhu, S., Cogan, S., Atamturktur, S., *Uncertainty Quantification Metrics with Varying Statistical Information in Model Calibration and Validation*, AIAA Journal, 55 (10), pp. 3570–3583, 2017.
[6.80] Collins, J., Hart, G., Hasselman, T., Kennedy, B., *Statistical Identification of Structures,* AIAA Journal, 12 (2), pp. 185–1980, 1974.
[6.81] Silva, T. A. N., Maia, N. M. M., Link, M., Mottershead, J. E., *Parameter Selection and Covariance Updating*, Mechanical Systems and Signal Processing, 70–71, pp. 269–283, 2016.
[6.82] Everitt, B., Dunn, G., *Applied Multivariate Data Analysis*, John Wiley and Sons, 2nd Ed., 2001.
[6.83] Kaiser, H. F., *A Second Generation Little Jiffy*, Psychometrika, 35 (4), pp. 401–415, 1970.
[6.84] Horn, J. L., *A Rationale and Test for the Number of Factors in Factor Analysis*, Psychometrika, 30 (2), pp. 179–185, 1965.
[6.85] Moens, D., Hanss, M., *Non-Probabilistic Finite Element Analysis for Parametric Uncertainty Treatment in Applied Mechanics: Recent Advances*, Finite Elements in Analysis and Design, 47 (1), pp.4–16, 2011.
[6.86] Khodaparast, H. H., Mottershead, J. E., Badcock, K. J., *Interval Model Updating with Irreducible Uncertainty using the Kriging Predictor*, Mechanical Systems and Signal Processing, 25 (4), pp. 1204–1226, 2011.
[6.87] Hanss, M., *Applied Fuzzy Arithmetic: An Introduction with Engineering Applications*, Springer, 2005.
[6.88] Gelman, A., Carlin, J., Stern, H., Dunson, D., Vehtari, A., Rubin, D., *Bayesian Data Analysis*, 3rd Ed., Chapman and Hall/CRC, 2013.
[6.89] Yuen, K.-V., *Bayesian Methods for Structural Dynamics and Civil Engineering*, John Wiley and Sons, 2010.

Chapter 7 Industrial Case Studies

7.1 General Introduction

The previous six chapters have provided a focused overview of the fundamental pillars of structural dynamics or else said it has been shown how to derive and study the dynamic behaviour of a system using its analytical, numerical and experimental models; at the same time, the disciplines that derive from the interactions of these representations have been exposed (identification, simulation and validation). In this last chapter, some real-life engineering examples will be discussed to provide a glimpse of how engineers apply the concepts presented so far to their structural analysis jobs.

It should be clear by now that there is no industrial application for which structural dynamics is not a concern. Vibrations are as natural as sound and omnipresent: in most cases, these are undesired phenomena; thus, understanding their cause and effect is pivotal to better design for more efficient and safer structures. Most likely, the reader has already a structural analysis issue at hand, as there are many possible situations: for example, precision cutting machines (milling machine or laser-plasma cutter) might suffer from tool chatter resulting in poor quality of the surface finish and higher costs. New washing machines are designed to be quieter and quieter, which poses a vibration challenge during the spinning phase; and also electronic boards, wind turbines, production cars, turbojet engines, pretty much everything is related to dynamics. Understanding the problem and accurately designing machines is the only way to produce better products and, most notably for the industry, to reduce/eliminate a costly redesign or troubleshooting campaign.

In order to provide a complete overview of some practical cases and given the authors' practical experience acquired over several years, this chapter refers to the application of structural dynamics in the aerospace industry. First, one shall present a general overview of the Ground Vibration Testing (*GVT*) and how a structural dynamicist proceeds throughout the experimental model validation. Then, two practical examples of experimental validation of linear Finite Element (*FE*) models will be given, both of which will illustrate the validation and updating procedure with several tests and numerical simulations. Next, one shall carry on through the issue of nonlinear vibrations caused by jointed interfaces. Why joints? The challenge of interface loads (often nonlinear) in the dynamics of the system is one of the causes of inaccurate dynamic simulations. Therefore, the last part of this chapter presents the nonlinear behaviour of joints in (i) bolted flanges, investigated in two aerospace structures and (ii) in a helicopter rotor. Finally, the chapter briefly extends to the challenge of modelling and validating a nonlinear *FE* model with experimental data.

7.2 An Engineering Application: the Ground Vibration Test

In the aeronautical industry, the Ground Vibration Test (*GVT*) provides the pinnacle of a dynamic study, as all the subjects discussed so far come together. It is a milestone for

Structural Dynamics in Engineering Design, First Edition, Nuno M. M. Maia, Dario Di Maio, Alex Carrella, Francesco Marulo, Chaoping Zang, Jonathan E. Cooper, Keith Worden, and Tiago A. N. Silva.

developing new aircraft (both with fixed or rotary wings) and for any program for which a structural change is made (e.g. commercial to cargo conversion or a new weapon bay). Such activity is critical to a program and financially important: it will take the effort of several engineers, technicians and most likely some external contractors over several days to plan, execute and provide the results of such a test campaign. In other words, the *GVT* is very expensive to execute (in the order of thousands of euros) and presents a financial risk because an aerostructure might or might not be certified to fly. Therefore, performing a *GVT* is both essential and expensive, and as such, it needs to be optimised to succeed.

7.2.1 Definition of the Objective(s)

As seen throughout Chapters 2 to 6, any testing and simulation activity must be made with an objective or goal. What should be measured? What should be simulated? Which test data should be used in the models? How can a simulation help to prepare the test? For example, if the various substructures that compose the aircraft have been the object of their own tests, updating, verification and validation before the actual Ground Vibration Test, then any discrepancies between test and simulation are most certainly due to the poor modelling of the joints.

Therefore, a test campaign must start with a kick-off in which all the different stakeholders (usually, program managers, stress analysts and test engineers) set out the scope of the *GVT* campaign. Here are some examples of 'objectives' discussed at such an initial meeting:

- How and where to measure on a structure the vibration responses, which will be then processed in a *response model*, e.g. in the form of Frequency Response Functions (*FRFs*);
- How to identify the modal parameters (natural frequencies, damping factors and mode shapes) and their uncertainty from those measured *FRFs*, which will allow generating a *modal model*;
- How to validate the numerical model of the aircraft, using appropriate indicators;
- How to investigate the sensitivity of the modal parameters of the *FE* model, for instance to structural modifications, such as a change in the configuration of the aircraft from lightweight to heavyweight by adding X kg of weight at position A;
- How to detect, locate, characterise and quantify sources of nonlinearities in a specific region of the aircraft;
- …

This kick-off is necessary to plan, allocate resources and define the test matrix (e.g. the type of excitation, the number of exciters, etc.).

7.2.2 Perform Pre-Test Analysis

At this stage of the program, there should be an *FE* model of the aircraft available (and if it does not exist, maybe it is not yet the time to think of running a *GVT*). If a test is performed to validate an *FE* model, the initial *FE* model can also help planning a test, as it may give, with the due approximation, a qualitative idea of the dynamics of the structure. The synergy between the test department working for the engineering team and *vice-versa* is a powerful tool to be exploited.

From the study of the previous chapters, the reader must now be aware that one must avoid placing sensors or exciting the structure close to its nodes. In the same way, it could be possible to define the best locations for exciters and sensors to measure all the modes of interest. For simple structures such as beams or plates, it should be possible also to define the good number of sensors/exciters (in addition to their location). Unfortunately, this is not the case for complex structures, such as aircrafts. So, some questions arise: "Where should the structure be excited?", "How many accelerometers should be used?" "Where should they be placed?" Answers can be found in *pre-test analysis* techniques, which are now part of the *GVT* process.

Starting with the Global Finite Element Model (*GFEM*), which is the "sum" of all the models making an aircraft, the modal characteristics of the structure, such as mode shapes and natural frequencies, can be retrieved from both the eigenvalue and eigenvector analyses of the spatial model (that is the *FE* model). As this numerical exercise will be used to simulate the *GVT*, the boundary conditions (e.g. soft suspension/deflated tyres, bungee cords…) need to be modelled as accurately as possible. In order to answer the questions of where one excites and/or where one measures, one must also be sure on which modes to capture: a practical criterion might be to select the target modes based on energy considerations, e.g. on effective modal mass or on kinetic energy [7.1]. At this point, the test engineer, based on his experience in terms of accessibility, geometry, cost, etc., defines all the possible candidate locations for the accelerometers and exciters. The best (sub)set of locations will eventually be found out of that cloud of points. There are several methods to make this selection. Some metrics base their selection on the observability of target modes, using information about modal displacement or energy (normalised modal displacement, nodal kinetic energy). Other methods proceed to iteratively eliminate sensors from the set of candidates to maintain linear independence or orthogonality between mode shapes optimally. For example, the Effective Independence Method (*EI*) is based on this logic and eliminates locations of the sensors using the Modal Assurance Criterion (*MAC*)[1] or an iterative model reduction scheme, for instance the Guyan reduction[2]. As only some of the nodes of an *FE* model can be measured, the reduction process finds the most effective points (nodes) on the structure for measuring a predefined number of vibration modes; The Effective Independence method was introduced by Krammer [7.2]. The goal is to select the measurement positions that make the mode shapes of interest as linearly independent as possible, and thus avoiding redundancy of sensors. An additional criterion is to identify the locations of the sensors that are capable of measuring sufficient information about the modal response, i.e. to maximise the signal-to-noise ratio of the sensors. Furthermore, independence of the target mode partitions is required such that the test data can be used in test-analysis correlation. The process starts from an *FE* model which is used to select the mode shapes of interest, a partition mode shape matrix $[m \times N]$, involved in the correlation exercise. In the case of the *EI* method, the sensor placement procedure starts by selecting a large set of candidate sensor locations, which is iteratively reduced to the smallest and most effective number of locations. The more sensors, the more time-consuming is the test set-up, which might not lead to any additional benefit for the modal analysis.

Hence, the sensor placement can be cast in the form of an estimation problem based on the Fisher Information (*FI*) matrix. The *FI* matrix is created by using the selected partition mode

---

[1] See Chapter 6, Section 6.3.1.

[2] See Chapter 6, Section 6.2.1.

shape matrix and multiplying the transposed matrix $[m \times N]^T$ by itself $[m \times N]$. The *FI* matrix is then used to generate the so-called Prediction Matrix, which is the product of the partition mode shape matrix $[m \times N]$ by the inverse of the *FI* matrix $[N \times N]$ and by the transpose of the partition mode shape matrix $[m \times N]^T$. Every diagonal element of the prediction matrix, represents the contribution of its corresponding degree-of-freedom to the global rank of the truncated mode shape matrix. The diagonal element which has the smallest value in the prediction matrix determines the *DOF* that has the smallest contribution to the representation of the mode shape matrix. That *DOF* is removed and the prediction matrix is calculated again in order to check the rank of the now-reduced mode shape matrix. This iterative process of removing the least-contributing *DOF* at every iteration is continued until the rank of the mode shape matrix ends up being equal to the number of modes in the selected partition matrix. Once the process is finished and the optimal locations defined, the software yields the locations of the sensors and a reduced geometry can then be imported into the test software for the *GVT* itself, as shown in Figure 7.1 [7.3]. Note that in Figure 7.1 (a) both the aerostructure and the supporting frame are modelled and used for the best sensor location. One should never assume that supporting frames are infinitely rigid. They can be dynamically coupled to the test article. Therefore, one should also add some additional sensors, for instance, onto a supporting frame, or even on the ground, to identify which mode shapes are from the test article and which are from the surrounding environment (Fig. 7.1 (b)). Experienced practitioners know that vibration energy from the surrounding environment can find its transfer path onto the test article, and spurious resonances might appear in the drive-point *FRF*(s). Hence, it is a good practice to augment the observability of the vibration responses.

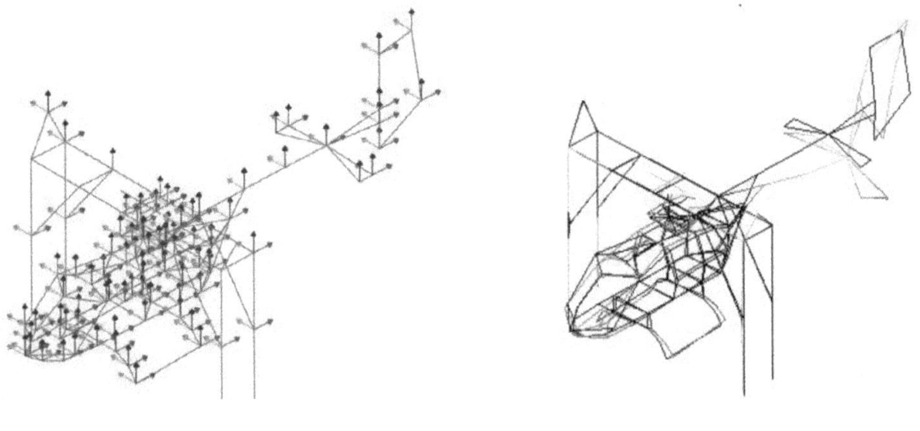

(a) (b)

Figure 7.1 Example of sensor location placement on wireframe imported into the test software[3].

7.2.3 Testing Activities

Up to this point in the campaign, the activity has been the preparation for the test. *GVT* requires a rather large team and a complex logistic: securing the test space (or hangar), making sure the fire unit is ready for when the aircraft has to be fuelled (for configuration tests), ensuring that all the sensors are appropriately labelled and calibrated and a number of other activities that are all part of the tremendous effort required for a *GVT* campaign.

[3] Full colour scheme available in the digital version of the book.

Regarding the actual vibration test, there are two significant steps to accomplish before shaking the aircraft, and both are as tedious as they are necessary. The first one is to calibrate all the accelerometers/load-cells/sensors or update/load the sensor database into the vibration test software. The second one is to create the test geometry (or ideally load the one of the *pre-test analysis*) and couple sensors-location-geometry nodes. One can immediately understand the importance of these tasks: the correct sensitivity of the sensor will ensure reading the correct amount of vibratory response and, from a bureaucracy standpoint, the documentation obligations towards the certifying authorities on sensors calibration factsheets. For example, imagine what would happen if a 100 mV/g accelerometer is set to 10 mV/g by a typing mistake! Alternatively, what would the mode shape look like if the directions were inverted (e.g. accelerometer mounted upside down or wrong direction in the channel set up of the vibration test software)? Moreover, where there are several hundreds of accelerometers (e.g. 400 for the Bombardier C-Series, or 800+ for the Airbus 350), it is easy to understand why it can become a tedious process. Table 7.1 lists of actions to be undertaken.

Table 7.1 List of actions to undertake for the *GVT*.

Create a sensor list – or load from sensor database
Create the geometry in the vibration test software (or import from the pre-test)
Attach the sensors/accelerometers on the structure – special care in case of 3D sensors
Associate the sensor/physical point on the structure with the geometry node – attention to associate the correct directions to that physical orientation and software direction match

Of course, all the other points discussed in Chapter 2 still apply, for example, to the setting of the amplifier of the shaker (set in *Constant Current* mode as opposed to *Constant Voltage*, etc.). Test activities can be started when all these tasks have been tackled and accomplished and the test item (aircraft or helicopter) is in the hanger. The test engineer(s) will be running the activities from an ad-hoc test room (the example in the photo below in Figure 7.2 is from Bombardier during the *GVT* of the C-series aircraft).

Figure 7.2 Example of *GVT* test set-up (Courtesy of C-Series Bombardier).

7.2.4 Test Plan

Preparing a detailed test schedule is time-consuming, but it is an essential task to accomplish at the start of a *GVT*. One good reason is that a complete and detailed schedule is necessary to coordinate all the related logistic activities: this becomes even more true as the test item is more complex. For example, think of fuelling operations (to study different mass configurations) or crane/suspension operations as necessary if one needs to change the boundary conditions. Another important reason is technical: a test is a means to an end, and one must define the scope and detail under which the tests run to achieve the goal. Finally, the sequence of runs (a run is a name usually given to each particular test) defines the campaign findings. So, for example, a low amplitude level broadband random test is a good starting point as it provides a wide-range set of natural frequencies and mode shapes. Then, one could think of running the same at different amplitudes or switching the excitation point. Altogether, these would give a first initial linearity test. For nonlinear studies, one could follow up with some sinusoidal tests with varying amplitude around a particular mode. The screenshot below, see Figure 7.3, shows an actual *GVT Run log* that counts 55 runs.

RUN	Input	Excitation type	Shaker location	Shaker Direction
				Full Fuel/50P
1	SIMO	Burst Random		Vertical
2	SIMO	Sine step		Vertical
3	SIMO	Sine step		Vertical
4	SIMO	Sine step		Vertical
5	SIMO	Sine step		Vertical
6	SIMO	Sine step		Vertical
7	SIMO	Sine step		Vertical
8	SIMO	Burst Random		Vertical
9	SIMO	Burst Random		Vertical
10	SIMO	Sine step		Vertical
11	SIMO	Burst Random		
12	SIMO	Sine step		
13	MIMO	Burst Random		
14	MIMO	Sine step		
15	MIMO	Sine step		
16	MIMO	Sine step		
17	MIMO	Burst Random		
18	SIMO	Burst Random		
19	SIMO	Sine step		
20	SIMO	Sine step		
21	SIMO	Sine step		
Full Fuel/50Psi on all tyres/Dummy tail Rotor hub fitted (No blades)				
22	SIMO	Burst Random	TGB	TGB, XY Comb AFT/STBD
23	SIMO	Sine step	TGB	TGB, XY Comb AFT/STBD
24	MIMO	Burst Random	TGB/Starboard Lashing Point	TGB, XYZ Comb 45deg AFT/STBD/DOWN, STBD-Comb XYZ, 45deg AFT/UP/PORT
25	MIMO	Sine step	TGB/Starboard Lashing Point	TGB, XYZ Comb 45deg AFT/STBD/DOWN, STBD-Comb XYZ, 45deg AFT/UP/PORT
26	MIMO	Burst Random	TGB/Starboard Lashing Point	TGB, XYZ Comb 45deg FWD/STBD/DOWN, STBD-Comb XYZ, 45deg AFT/UP/PORT
27	MIMO	Burst Random	TGB/Starboard Lashing Point	TGB, XYZ Comb 45deg FWD/STBD/DOWN, STBD-Comb XYZ, 45deg AFT/UP/PORT
28	MIMO	Sine step	TGB/Starboard Lashing Point	TGB, XYZ Comb 45deg FWD/STBD/DOWN, STBD-Comb XYZ, 45deg AFT/UP/PORT
Full Fuel/50Psi on all tyres/				
29	MIMO	Sine step	TGB/Starboard Lashing Point	TGB, XYZ Comb 45deg FWD/STBD/DOWN, STBD-Comb XYZ, 45deg AFT/UP/PORT

Figure 7.3 Example of a *GVT Run log*.

Naturally, test plans are subjected to changes, as things rarely go without issues. However, the reader must have appreciated the importance of creating a comprehensive test plan that will help collecting all the necessary data (one can imagine realising that a data set is missing when the aircraft is no longer available!).

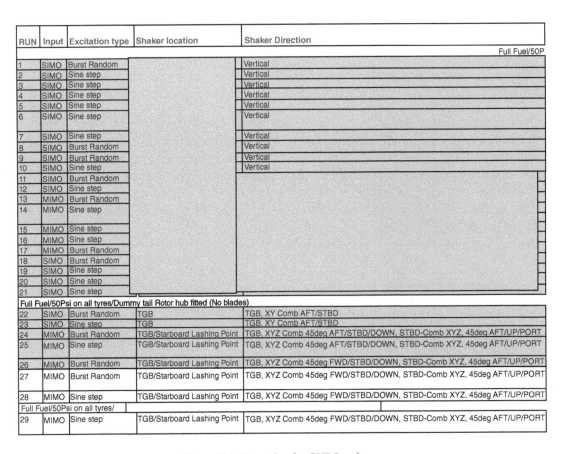

7.2.5 Data Verification

Running a *GVT* implies acquiring several gigabytes of data. Nowadays, the computational power and the storage capacity are such that one does not shy away from over-measuring. Of course, this was not the case some decades ago when each measured data point counted – given the time it took to measure it and the way it was stored. So, the first checkmark is verifying that the acquired data are usable. File corruption, missing data points, or excessively noisy data are only some of the reasons that would make data unusable. There are different strategies to transfer the data from the acquisition station to a processing one (e.g. using external storage devices like an external *HDD* or connecting multiple machines in a network). Nevertheless, the goal is clear: to have a clean and complete set of data to process during the model validation phase.

7.2.6 Modal Identification

During a test campaign, it is always important to remember the reasons for testing. Given the cost and the resources necessary to perform such an activity, it must be clear what must be measured (and especially why). As previously said, the main goal of the *GVT* is to validate the *FE* model of the structure/aircraft experimentally. This process is done by building a *response model* (i.e. measuring the *FRFs* of the aircraft) and transforming it into a *modal model* (i.e. extracting the natural frequencies, damping ratios and mode shapes of the structure), which will then be compared to the one derived from the *spatial model* (or the *FE* model). This process is called Modal Identification; in Chapter 4 some of the most used techniques have been discussed in detail. The extraction of modal parameters is based on techniques (some of which are nonlinear) which require that the vibration response is linear. The reader must refer to Chapter 5 to appreciate the difference between linear and nonlinear vibrations. The modal analysis algorithms would not distinguish between the linear and nonlinear vibrations, but all of them will still calculate the modal parameters. These parameters could be incorrect if the practitioner did not invest any time in verifying the vibration response linearity.

Some mode shapes are often not easy to identify during a modal identification process because of the high modal density. A first *AutoMAC* [4] [7.4] calculation is a good starting point to verify the data collected; visualising the mode shape associated with each pole (or natural frequency) is also a good check. Modern software allows mode shapes to be readily visualised in many formats (from complex phase plots to animated deformation). However, naming each mode shape requires some skills (and experience). At times, the engineer processing the data can (and must) request better data from the test engineer to perform a meaningful modal extraction (one can think of close modes and apply normal mode testing, see Section 4.6).

7.2.7 Model Validation and Updating

The result of the experimental modal identification allows the creation of the so-called *modal model*. Like any other model, natural frequencies, damping ratios and mode shapes derived from the test constitute a mathematical model of the tested structure. One of the exploitations of the modal model could be, for example, the simulation of *FRFs*. The extraction of these modal parameters relies, of course, on some assumptions – so, the modal model is an

[4] The *AutoMAC* measures the degree of correlation between two mode shapes from the same mode shape set; it is a particular case of the Modal Assurance Criterion (*MAC*) – see Section 6.3.1 for details.

approximation based on specific assumptions (for example, linearity, or damping model, etc.). As said, the primary scope of a test (if not the only one) is to identify the modal parameters of the structure so that a numerical/spatial model can be compared to and, if necessary, updated in order to match the experimental results.

The literature on *Model Validation* and *Updating* is ripe, and so is the availability of commercial software. Chapter 6 describes in detail the process(es) to (a) correlate a model derived from experimental data with one derived from analytical equations (a *FE* model in most cases) and (b) how to modify or "update" both the constituents of the model which are the mass and stiffness matrices. Since modal parameters depend on mass and stiffness distributions, these are the "knobs" available to improve the accuracy of the model in terms of its natural frequencies and mode shapes. Furthermore, the knowledge of the modal damping from experimental data allows for simulating forced vibrations. The reader must be reminded that the sole knowledge of natural frequencies and mode shapes is necessary but not sufficient to simulate forced responses which require damping. Without the knowledge from test data, modellers tend to distribute such damping proportionally. Unfortunately, the reality of jointed structures is that it is often non-proportionally distributed, which means specific resonant modes tend to be more attenuated than others. The challenge is to understand how to model that distribution in mechanical systems at the outset. The *GVT*, in the end, provide that piece of missing information to the *FE* model.

Some experience and engineering judgment will be necessary when dealing with cases where the modal density is high enough to have very complex data sets to correlate. An initial correlation/validation exercise can be a simple graph with the natural frequencies from the model and those obtained from the measurements, as in Figure 7.4 [7.1].

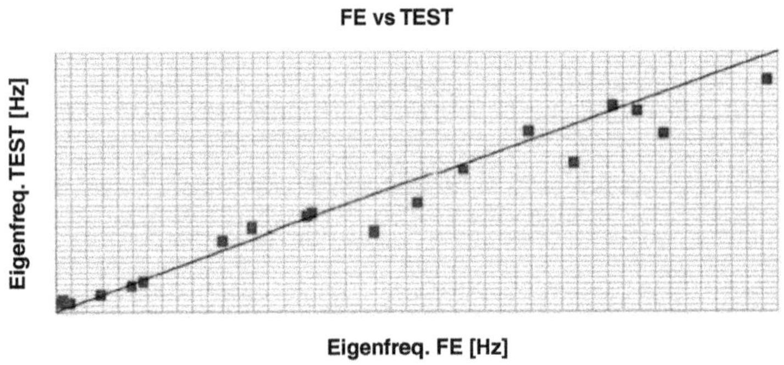

Figure 7.4 Comparison between experimental and numerical natural frequencies.

The Modal Assurance Criterion (*MAC*) is the most reliable and generally adopted indicator of mode-matching. In addition, modern algorithms have developed extensive display capabilities to give immediate visual feedback on the correlation between 2 sets of mode shapes, as shown in Figure 7.5.

In the last few years, all the engineering disciplines (from automotive to energy, aerospace to mechanical, etc.) have embraced the "digital twin" hype (this will be explained in what follows). Several definitions exist, and some vendors have their variants for marketing purposes. Moreover, of course, the complexity of digitalisation varies from object to twin: it could be only a structural component (for example, a detailed part like a clutch, a seat, or a wing)

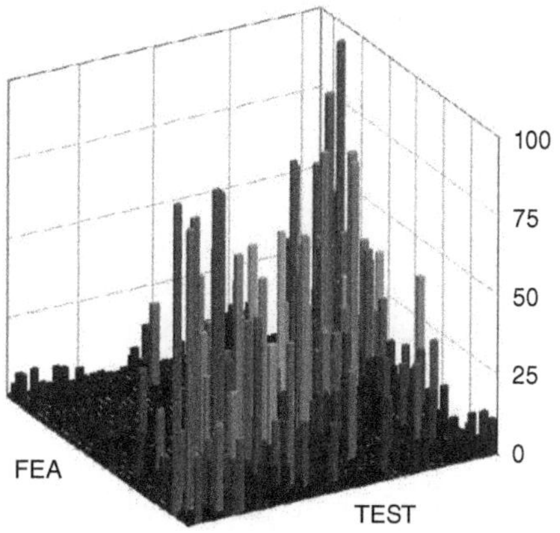

Figure 7.5 *MAC* between experimental and numerical mode shapes[5].

to be twinned. However, it could also be a mechanism (like a complete hydraulic circuit or mechatronic system behind a fly-by-wire system) that operates in a more complex structure. As one can see, validating and updating such a complex system becomes a daunting task, and the sources of errors in the final digitalised version of the physical system can hardly be counted.

Seeking a more practical definition and going back to the basic meaning of the words, one finds – according to the Cambridge Dictionary – twin: 1) *either of two children born to the same mother on the same occasion*; 2) *used to describe two similar things that are a pair.*

Consider the 2nd definition. It defines twins as "similar" and not "identical" things. This concept of the digital twin and its definition can be applied in the structural dynamics sense, too: the digital twin is a validated numerical version of a physical object, one with the similar structural dynamics "*DNA*", i.e. with similar modal parameters (natural frequencies, damping ratios and mode shapes).

If the updated numerical/*FE* model can be called the digital twin of the tested article, then the predictive simulation run using that model will have a lot more engineering value for the design of that structure and verification purposes. Nevertheless, assembling different digital twins is not a sufficient and necessary condition to create a whole digital twin of a complex structure when – for example – sources of nonlinearity are not modelled and included in the simulated vibration response.

7.3 Presentation of Test Cases

The previous sections gave the reader a general overview of a model validation procedure applied to an aerostructure. The task of experimentally validating structures subjected to dynamic loads is never easy. The *FE* models for dynamic analyses might be much coarser, as shown earlier, than the ones used for stress calculations, and so the mode shapes will never resemble the ones of continuous structures. Nevertheless, the modal analysis results can be used for planning dynamic tests, the analysis of which will be used for correlating the modal properties with numerical ones. The model updating is the last process for correcting the mass and stiffness matrices of the numerical model.

---

[5] Full colour scheme available in the digital version of the book.

To ease the comprehension of the entire process, one has produced a series of test cases focused on aerostructures. However, one shall understand that the same test and analysis methodology described hereafter can be applied to other structures, like automotive or naval, for example. The four test cases presented in this chapter are the following:

1. Experimental model validation of a composite fan blade.
2. Experimental model validation of a tower rotor test rig.
3. Experimental nonlinear model validation of an aero-engine casing with bolted flanges.
4. Experimental nonlinear modal analysis of the tail drive shaft system of a helicopter.

The test cases 1) and 2) are examples of model validation assuming a linear vibration response. Test case 3) is a more advanced one, where the validation is extended to nonlinear vibration responses. However, this test case will present the experimental nonlinear modal analysis achieved by generating linearised *FRFs* through the physical control of the vibration response amplitude. Finally, the last test case, 4), will present the nonlinear modal analysis, which will process *FRFs* measured and processed with a method called *Complex FRF Analysis*.

Before introducing the core topic of the model validation process, one shall provide the reader with some basic information about the Scanning Laser Doppler Vibrometer (*SLDV*), which is often used in this chapter alongside with the more conventional accelerometers for measuring the vibration responses. The reader must be aware that many other optical methods can be nowadays used for retrieving deflection shapes, for instance, using Digital Image Correlation (*DIC*) or Motion Magnification.

7.3.1 Introduction to the *SLDV* Measurement System

The Laser Doppler Vibrometer (*LDV*) is a system based on the physical principle known as the Doppler effect. The laser light is used to detect the frequency shift caused by a vibrating target. However, laser light cannot be directly used to measure vibration. The laser light needs signal conditioning to deliver information about the measuring target, a structure subjected to mechanical vibrations. The laser light was explored in several forms and with several different detectors for performing vibration measurements. The *LDV* system is the one that has overcome some most undesirable issues in vibration measurement, one is the quality of the Signal-to-Noise-Ratio (*SNR*), and the other one is the eye-safe laser light which allows the *LDV* to be used anywhere without stringent Health & Safety regulations. The early laser vibrometers were prone to poor *SNR*, particularly with surfaces providing little backscatter light, for instance, the dark one as composite. The recent upgrades, and updates of the laser light wavelength and better electronics for demodulation and signal conditioning of the incoming laser light, have made the *LDV* a very reliable and high-quality *SNR* measurement system. Figure 7.6 presents how a single-point *LDV* system works, describing a Mach-Zehnder interferometer and how the internal architecture of an *LDV* system is made. Without going too much into details, the interested reader is referred to further information to the following references [7.5 – 7.8]. The *LDV* laser light is split into two beams by beam splitters (*PBS1*, *PBS*, *BS2*, in Figure 7.6). One laser beam acts as the reference signal and the other one as the measurement signal; both are combined again at the photodetectors (*PD1* and *PD2*, in Figure 7.6), which will supply the *LDV* vibration signal after further signal conditioning. The Bragg Cell is fundamental for the functioning of the *LDV* because it shifts the frequency of the reference laser beam before both beams are recombined at the photo detector. The output signals from the photodetectors are amplitude modulated ones, which contain both information about the amplitude and phase of the vibration signal measured on the target. Eventually, the internal electronics of the *LDV* system take care of the signal conditioning. The *LDV* output is a voltage which must be multiplied by the *LDV* sensitivity in (meter/second). The *LDV* measures the velocity of the vibration as opposed to an accelerometer, which measures the acceleration.

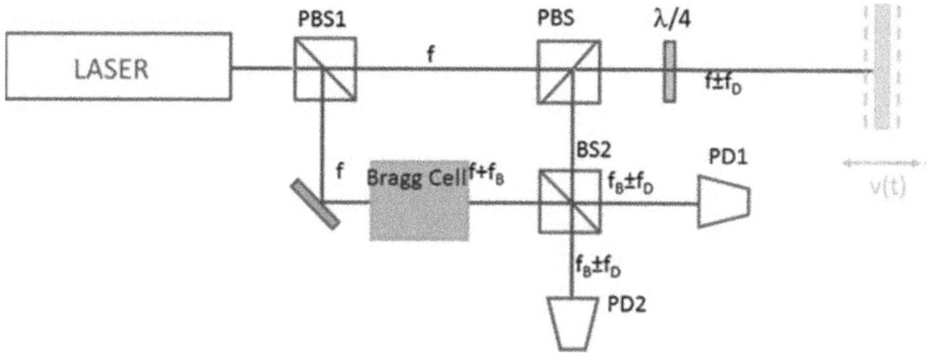

Figure 7.6 Schematic of the Mach-Zehnder interferometer [7.8].

The laser beam can be directed anywhere on a structure presenting optical access by a scanning device made of two mirrors mounted on galvanometric motors. The scanning device moves the laser beam along with the *X* and *Y* directions, thus allowing a vibrating surface to be mapped in great detail. However, optical access is also a limitation for such a system, something that contact sensors like accelerometers do not experience. Hence, an *X-Y* scanner turns an *LDV* into a *Scanning-LDV*. One should remark that the laser beam measures the vibration along its optical path. If that is not normal to the surface, it measures a component of that vibration. However, the electronics inside the galvanometric motors can measure the angle of rotation of the scanning mirrors, and the post-processing corrects for the normal-to-surface vibration component. One must also remember that curved surfaces are challenging for the *LDV*. Note that the laser beam can become tangent to the target surface and thus unable to measure. Hence, inexperienced practitioners might run into unexpected vibration amplitude scaling effects when simulated and measured vibrations are compared.

The most recent *SLDV* systems can consist of three scanning heads measuring vibration from three positions. Three-dimensional deflection shapes can be therefore achieved. *SLDV* systems can supply a variety of outputs, and the most common are frequency response functions; in general, the *LDV* output signal can be directly used as a time series too. The reader is encouraged to learn about methods and applications of *SLDV* from two journal review papers [7.6] and [7.9]. Those reviews contain many information and examples to complement the one provided in this chapter.

7.3.2 Response Model using the *SLDV* Measurement Method

A response model can be created in various ways. The accelerometers are the most used sensors due to their reliability, calibration traceability and availability/cost. Tri-axial accelerometers measure the vibrations in the three directions in space, and a set of those enables either Single- or Multiple-Input and Multiple-Output modal testing. Unquestionably, this type of sensor is very effective for standard modal testing to verify and validate an *FE* model. The shortcoming of this type of sensor is the setting up time: accelerometers are contact sensors that need to be physically attached to a test item on one end and connected to the measurement system at the other end. Nowadays, the setting up practice and the signal conditioning are well established, but significant structures such as an aero-engine casing or an aeroplane fuselage can still prove very time-consuming. The cost-benefits are accepted when standard verification/validation procedures are designed by test plans which are *FE* model-based. If these procedures are assumed valid for linear vibrations, their validity is questionable when source(s) of nonlinearity are activated by high vibration amplitudes (or in the presence of complex structures with several joints). The investigation of nonlinear responses and the source of those are challenging to plan by methods based on linear models. The experimental scoping of the mechanical vibration using accelerometers can be adequate for the time/frequency domain analysis, but might be insufficient when the highly-dense spatial information is also relevant. As dense as it can be, a network of accelerometers would never match the optical methods that can inherently (and theoretically) scan a virtually infinite number of measurement points.

As introduced in section 7.3.1, optical methods are still not seen as the best choice for performing experimental model validation. For example, a single-head *SLDV* might require an accurate set-up to avoid the vibration response being scaled incorrectly because of the angle of incidence of a laser beam or from the viewpoint of a digital camera. In fact, this uncertainty is too costly, and the industry (especially aerospace) is not prepared (yet) to integrate such optical systems for model validation applications. Nonetheless, examples of research activity might suggest the future use of non-contact optical methods in this field of operation.

7.4 Case study 1: Experimental Model Validation of a Composite Fan Blade

Composites are used for structural parts bearing both static and dynamic loads. Structures designed as such must be verified and validated during the design stage. Furthermore, composite materials are anisotropic because the material system combines matrix and fibres (for instance, unidirectional or woven type). Composite structures can be achieved by stacking a sequence of laminæ which can give *quasi*-isotropic material properties. One can start from a basic anisotropic lamina (unidirectional fibres) and finish with a *quasi*-isotropic material because of specific fibre orientation. The process of validating a model of a composite structure is paramount. On the one hand, the literature continuously builds new references on how to model and validate such structures made of novel materials. However, on the other hand, material systems change so quickly that experimental testing must be continuously adjoined to the modelling to ensure validated predictions. Hence, the process that allows predicting the forced response of a composite structure is not dissimilar from the one presented in the previous sections. However, this example will provide a few more technical details than those given earlier for the *GVT* example.

This section reports post-processed data of vibration measurements captured on a composite scaled model blade. Tests were designed so that the blade could be tested using (i) Free-Free and (ii) Fixed-Free boundary conditions. Measurements under free-free conditions were performed using two types of excitation sources which were: (i) impact (impact hammer) and (ii) white noise (loudspeaker). Measured data were post-processed by modal analysis software, and mode shapes and natural frequencies were reported. The composite blade was then clamped onto a mass block which incorporated a hydraulic piston to apply the clamping load to the blade root. Two sets of modal vibration measurements were selected using the fixed-free boundary conditions. The first set consisted of estimating natural frequencies and measurements of Operational Deflection Shapes (*ODSs*) at those frequencies. An impact excitation was produced to measure an *FRF* which was post-processed by modal analysis to recover the natural frequencies of the structure; the *Fast Scan SLDV* (*FS-SLDV*) technique was then used to capture the *ODSs* of the blade sinusoidally excited at each of the natural frequencies. The second set of measurements was to perform modal tests in a frequency range between 50 Hz and 1600 Hz, and those tests were produced using either chirp or pseudo-random excitation signals. Mode shapes and natural frequencies were obtained by modal analysis. The final modal data from the second set was used for correlating the experimental and numerical mode shapes.

7.4.1 Modal Test of a Composite Blade

This section will illustrate a series of modal testing practices recommended for inexperienced practitioners. In real life, composite blades are mounted onto a rotor hub so as to form a fan-bladed disc, and each blade can be assumed under a fixed-free boundary condition. Before exploring the dynamic response under such conditions, the best practice is to evaluate the modal properties under *free-free* boundary conditions. Unfortunately, the so-called *free-free* conditions are challenging to achieve. Nonetheless, in this section, one is assuming (here stressing the word "assumption") that a blade resting on a piece of foam can fulfil such a requirement. The next step is to evaluate the modal properties of the blade under fixed-free conditions by clamping the blade with a bespoken fixture which allows the creation of a joint as representative as possible to a real one. Finally, the experimental mode shapes and numerically calculated ones will be compared through the Modal Assurance Criterion (*MAC*) relationship.

7.4.2 Modal Test under Free-Free Boundary Conditions

A set of modal tests were first performed on the blade under *free-free* conditions, as shown in Figure 7.7. The test structure was laid on a foam cylinder, and white spots identified the positions where the *SLDV* system measured the responses. The excitation was produced by two different excitation systems: (i) by an impact hammer, the impact of which is indicated by the grey dot, and (ii) by a loudspeaker fed with a white noise waveform. Note that both the hammer and the speaker provide an excitation without a permanent bonding with the test article. The main difference between the two methods is measuring the force by hammer excitation, which cannot be achieved using the speaker. Some attempts to measure the excitation force provided by the speaker with a microphone opposite the source did not produce trustable results. Therefore, the most practical solution was to scale the *LDV* response by the voltage driving the speaker. By doing this, one knows that the mode shapes are not scaled correctly. However, it must be ensured that those mode shapes are correctly identified, as well as the natural frequencies and damping factors. The two modal testing practices are compared using the modal properties evaluated by modal analysis. One final remark is the impact location, which could not be applied opposite to a measurement point, but had to be hidden next to it for practical reasons. It is always a good practice, whenever possible, to weigh the test item because that can be compared with the one calculated from the model. Although it might sound like a trivial exercise, it gives the very first verification of your model. The weight of the composite blade was measured and found to be 443 g, and it showed a 99.9% match with the one calculated from the model.

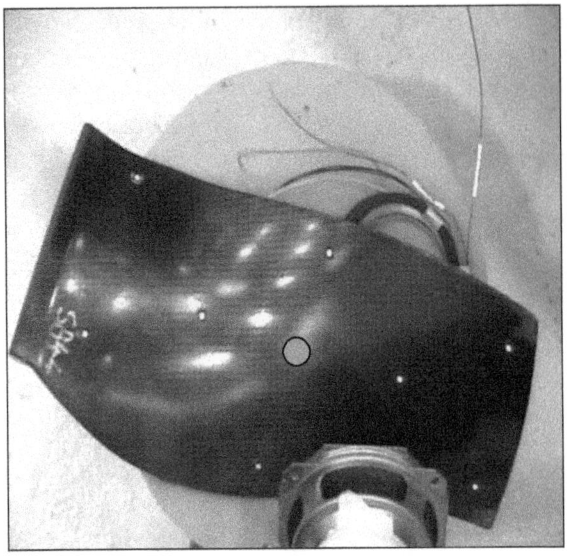

Figure 7.7 Blade set-up for the *free-free* modal test; impact produced at the grey circle.

Four modal tests were performed on the blade, three using the impact hammer and one using the loudspeaker with white noise waveform. One remark should be made about the speaker: the choice depends on the frequency range. For instance, one shall select a subwoofer for the low-frequency ranges. In the present case, the selection showed to be just as good as required for exciting the first blade resonance. The settings adopted for the tests are reported in Table 7.1.

Two examples of *FRFs* from the modal impact and speaker excitation are presented in Figure 7.8. At first glance, one notices that the *FRF* measured using the speaker is noisier than the one by modal impact, which is typically true. Both *FRFs* show several spikes for frequencies lower

Table 7.1 Measurement parameters for the modal test in free-free conditions.

	Impact hammer-settings	Loudspeaker-settings
Bandwidth:	0 Hz- 2kHz	0 Hz – 2kHz
FFT lines:	6400	6400
Sampling time:	3.2 sec	3.2 sec
Windows:	Exponential (response); Transient (force)	Rectangular (response); Rectangular (force)
Average:	3	10
Excitation signal type:	Impact	Pseudo Random
SLDV-tracking filter:	Slow	Slow
SLDV-sensitivity:	25mm/s/V	10mm/s/V
SLDV-low pass filter:	20KHz	20KHz

than 200 Hz, some of which are resonances and some others not. It will be observed that the first flexible mode of the blade is beyond 400 Hz. The lower ones are caused by the foam and blade interaction, suggesting that the blade is unconstrained but still dynamically coupled to the base foam. Those modes are rejected from the modal analysis results. These *FRFs* demonstrate that a *free-free* boundary condition is never easy to achieve (and the clamped is even more difficult!). Nevertheless, a good practitioner knows how to handle and mitigate such undesired scenarios.

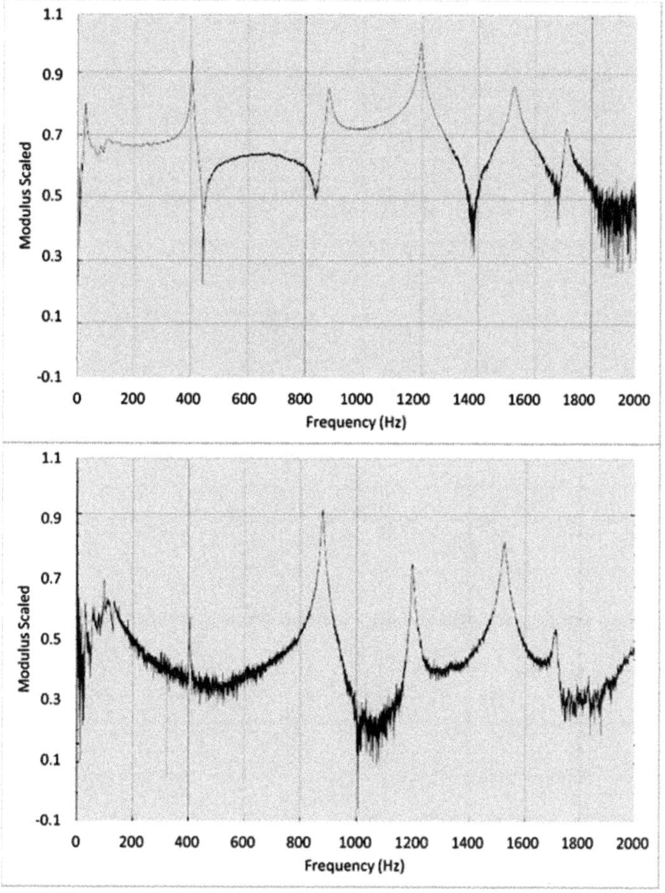

Figure 7.8 *FRFs* from modal impact (top) and loudspeaker excitation (bottom).

The data obtained from each modal test were post-processed using modal analysis software so that the natural frequencies and damping values could be compared as shown in Figures 7.9 and 7.10, while the natural frequencies values are reported in Table 7.2. Five mode shapes were extracted by modal analysis, but only two are presented in Figure 7.11, the first and the last one.

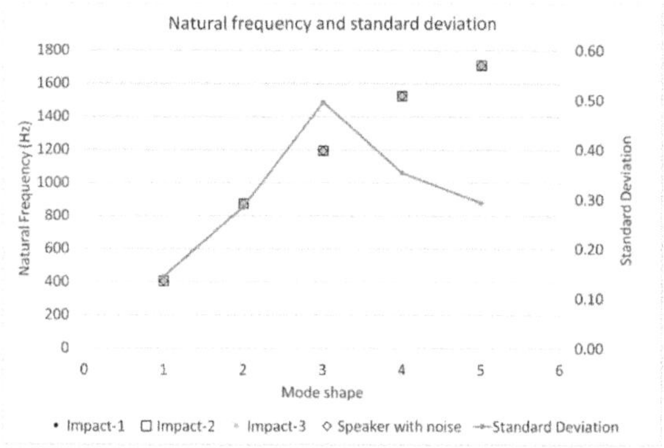

Figure 7.9 Comparison of natural frequencies obtained from modal analysis (4 tests).

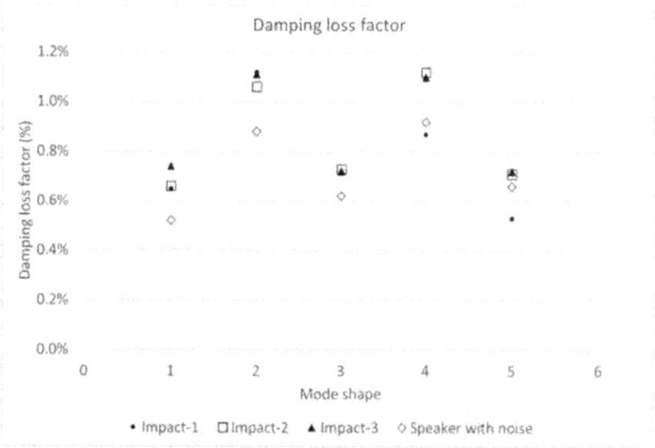

Figure 7.10 Comparison of damping loss factors obtained from modal analysis (4 tests).

Table 7.2 Natural frequencies from the four modal analysis.

	Mode-1	Mode-2	Mode-3	Mode-4	Mode-5
Test-1 (impact)	404.7 Hz	878.6 Hz	1195.4 Hz	1525.3 Hz	1711.7 Hz
Test-2 (impact)	404.5 Hz	878.0 Hz	1195.3 Hz	1525.1 Hz	1711.5 Hz
Test-3 (impact)	404.5 Hz	878.4 Hz	1195.4 Hz	1524.8 Hz	1711.0 Hz
Test-4 (speaker)	404.3 Hz	878.1 Hz	1194.4 Hz	1525.7 Hz	1711.2 Hz

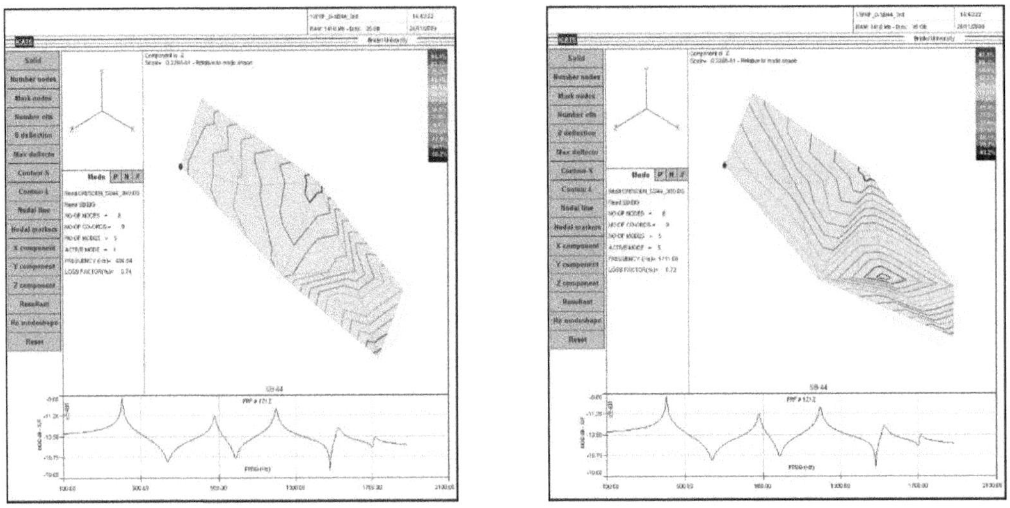

Figure 7.11 First on the left (404 Hz) and last on the right (1711 Hz) mode shapes[6].

Both excitation methods allow the correct identification of the natural frequencies, where the standard deviation is small, as shown in Figure 7.9. The damping value plot indicates a degree of variability across the four modal tests, where even mode shapes show higher damping than odd ones. It is always a good practice to create such plots to inspect the dynamic parameters against the identified mode shapes. The mode shapes of Figure 7.11, given the low number of measurement points, do not show a great visualisation of the shapes (say, segmented).

One of the first checks to assure that one has identified all the resonances present in the *FRF* is to regenerate the *FRF* from the modal parameters extracted from the modal analysis. Figure 7.12 shows that one of the resonances (around 850 Hz) was missed from the analysis, so that resonance mode is not regenerated, as visible by the dotted line in Figure 7.12. The frequency range below 200 Hz contains many spurious peaks, one of which is a resonance lookalike, a rigid body motion caused by the foam base. Having included the resonance mode at about 850 Hz, an additional check-up that must be carried out is the *AutoMAC* correlation. It consists of plotting the experimental mode shapes against themselves. The *AutoMAC* plot indicates if the mode shapes were correctly identified and if those mode shapes are unique as expected.

Figure 7.13 shows the *AutoMAC* of the modal analysis. The mode shapes are well identified, and the number of measurement points is also adequate, despite a few ones, for resolving the unicity of the mode shapes. The diagonal values (white = 100%) show good correlations. The off-diagonal values (black < 20%) show no biases between the mode shapes, which means there is no aliasing between modes given the few measurement points.

[6] Full colour scheme available in the digital version of the book.

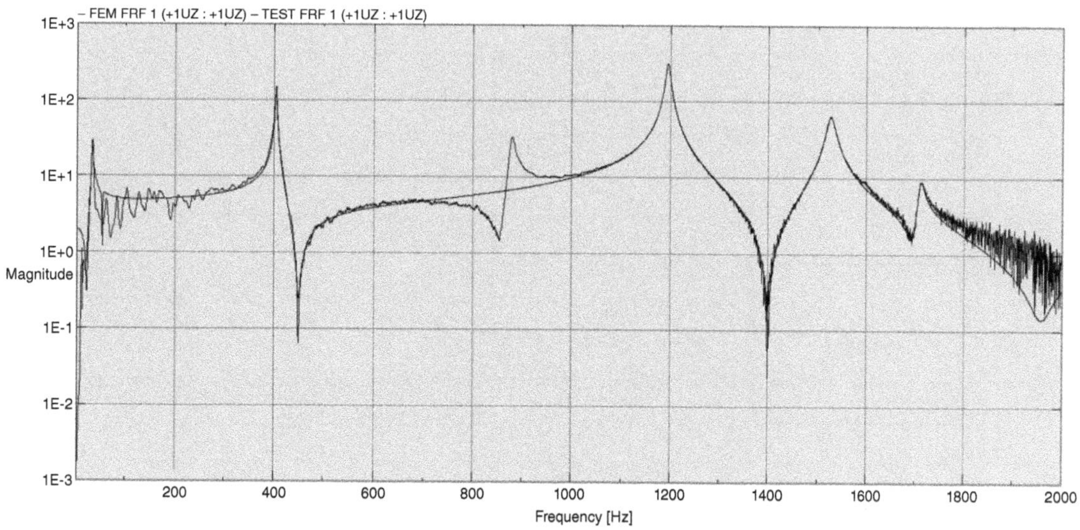

Figure 7.12 Measured (solid) and regenerated (dotted) *FRFs*.

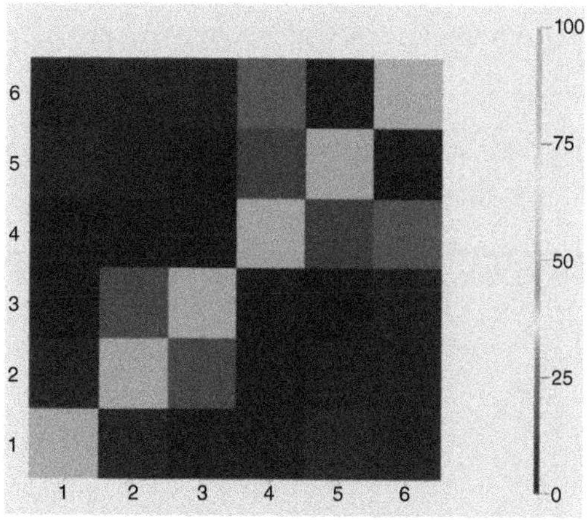

Figure 7.13 *AutoMAC*.

7.4.3 Fixed-Free Modal Test

After the modal tests were completed under free-free conditions, the blade was clamped onto a seismic clamping block (of about 100kg) to perform additional measurements under *fixed-free* boundary conditions. A sketch of the test rig is shown in Figure 7.14. A 200 N *LDS* cooled shaker was used and connected to the mass block by a driving rod. The clamping block was sat on circular metal rods (10 mm in diameter), which reduced the friction between the base of the block and the ground. The *SLDV* was positioned in front of the blade, clamped and locked by a hydraulic mechanism. Furthermore, the clamping area of the block had a matching profile of the dove-tail root of the blade. An accelerometer was mounted on the clamping block to measure the base acceleration, later used to scale the *FRFs*.

The first test to be performed using *fixed-free* conditions was the natural frequency estimation and measurement of *ODS* at each of those frequencies. It was decided to use an impact hammer to excite the blade, and the response was measured at one of the top corners, as shown in Figure

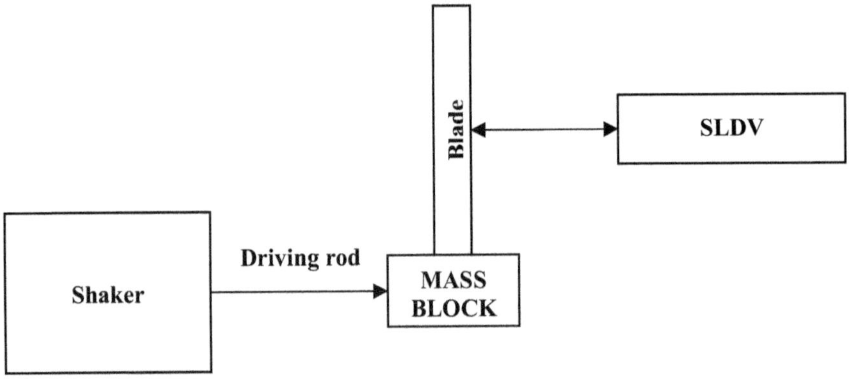

Figure 7.14 Test rig set-up scheme.

7.15. The shaker was disconnected to avoid its dynamics from being included in the measurements. The measured *FRF* was then post-processed using a Single Degree of Freedom analysis in a modal analysis programme. Natural frequencies are reported in Table 7.3.

Table 7.3 Natural frequencies form impact excitation.

	Mode 1	Mode 2	Mode 3	Mode 4	Mode 5
Natural Frequency [Hz]	162.19	404.15	740.14	1098.64	1337.18

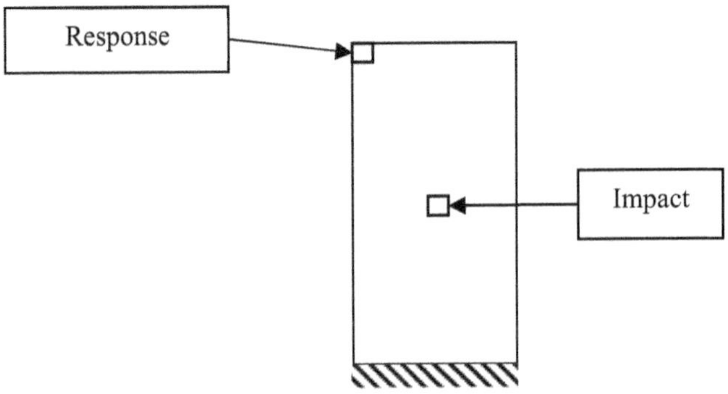

Figure 7.15 Fixed-Free natural frequency estimation.

As explained earlier, after the modal properties have been extracted, the regeneration of the measured *FRF* was performed and it showed that the modal properties obtained were accurate. The next step was to carry out *Fast Scan* (*FS-SLDV*) *ODS* measurements at the natural frequencies of the five modes, estimated by impact so that the spatial representation of the modes could be obtained. The *Fast Scan ODS* allows scanning all the measurement points in rapid succession by measuring a few number of cycles per measurement point at the driving harmonic, at one of the measured resonances. This practice of rapid testing of the deflection shapes can be considered for gathering information from a test item, let us say in a few minutes rather than in hours. For example, it can provide the rapid verification of the clamping conditions on the actual dynamics of the test article. Figure 7.16 shows the *ODSs* measured by the *SLDV*. Note that by selecting a maximum peak response (one of the *FRF* resonances) does not necessarily mean that such a peak is at the natural frequency, in particular when the frequency resolution, damping and noise are large. This is what can happen when *ODS* visualization methods are used in measurement software without modal analysis. Refer to Chapter 1 for further insight about the meaning of resonance.

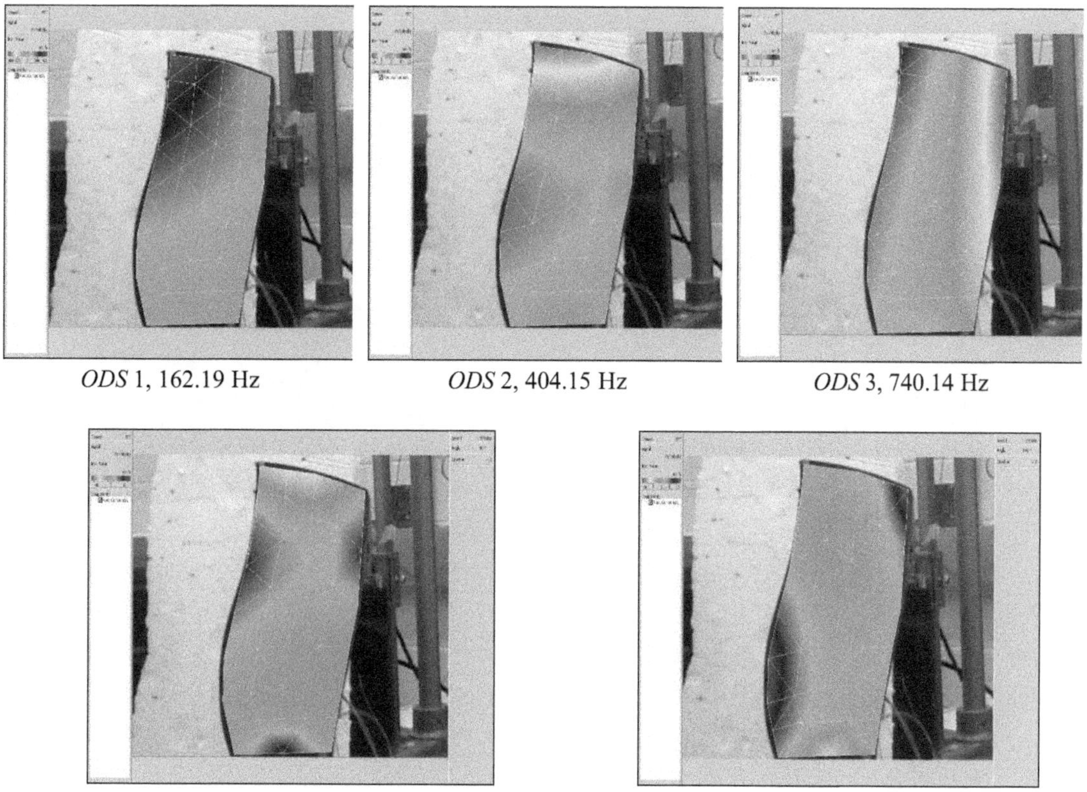

ODS 1, 162.19 Hz — *ODS* 2, 404.15 Hz — *ODS* 3, 740.14 Hz

ODS 4, 1098.18 Hz — *ODS* 5, 1337.18 Hz

Figure 7.16 *ODSs* from fixed-free measurements[7].

7.4.4 Normal Mode Shapes and Correlation with *FE* Modes

The last set of measurements consisted of modal tests to obtain normal mode shapes, which one could correlate to *FE* mode shapes. As for the *free-free* boundary conditions, two different types of excitation methods shall be presented and it will be verified that both methods deliver the same mode shapes. The correlation is achieved by the *MAC* factor obtained for selected *lock points* defined on the test structure. Those points matched the nodes of the *FE* model from which the eigenvector was extracted for the correlation. Such a manual practice is not generally necessary when post-processing *FE* programmes are available.

7.4.5 Modal Test for *MAC* Correlation

The last set of measurements was produced because the measurement points were not defined at the exact locations of the *FE* nodes. There are techniques for searching for the closest nodes-to-measurement point match, but these were not implemented here. Furthermore, the *FE* model

[7] Full colour scheme available in the digital version of the book.

was not used for generating a test plan. Hence, a set of *lock points* were identified from the *FE* model, and those were used for correlation between the experimental and analytical mode shapes. Finally, the modal displacements of the *FE* model were projected along the line of sight of the laser beam. Figure 7.17 shows the positions, listed in Table 7.4, of those seven *lock points* on the surface of the composite blade. Table 7.5 presents the natural frequencies obtained from modal analysis performed on test data captured using two different excitation systems: electromagnetic shaker and loudspeaker.

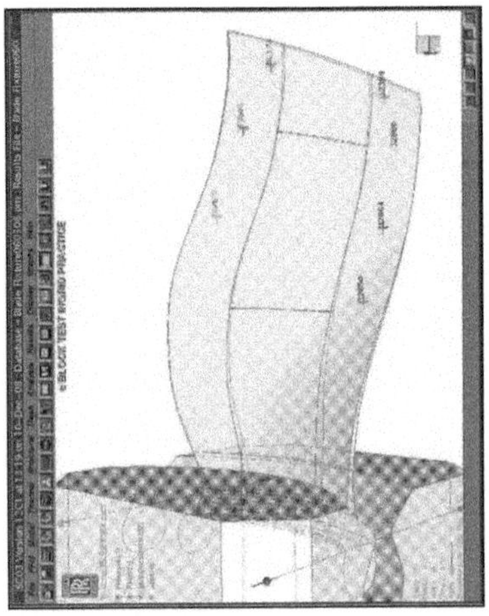

Figure 7.17 *FE* model on the left and blade with *lock points* markers on the right[8].

Table 7.4 *Lock points* positions.

Lock points:	Point 1	Point 2	Point 3	Point 4	Point 5	Point 6	Point 7
Distance (mm):	(29,15)	(119,15)	(55,15)	(115,15)	(170,15)	(120,15)	(61,15)

Table 7.5 Natural frequencies using different excitation systems.

Excitation waveform: Pseudo-Random	Mode 1	Mode 2	Mode 3	Mode 4	Mode 5
Natural Frequency [Hz] (Shaker)	161.1	402.5	734.4	1099.5	1337.9
Natural Frequency [Hz] (Loudspeaker)	162.0	402.4	740.0	1098.9	1337.4

The *MAC* relationship is now used to verify whether both experimental mode shapes are the same; hence, the experimental procedure does not influence the analysis. The mode shapes are derived from two different measurement methods: shaker (contact) and speaker (contactless). The mode shapes of one experiment are correlated against the mode shapes from the other experiments. It might read this procure similarly to *AutoMAC*, but there the same set of mode shapes are plotted against each other. Figure 7.18 shows the *MAC* correlation between the two sets of eigenvectors obtained by the two excitation systems. Note that the *MAC* compares shapes even if those shapes are not scaled by the excitation force. However, the best practice for experimental model validation is to use mode shapes which are scaled by the excitation force measured at the drive point.

[8] Full colour scheme available in the digital version of the book.

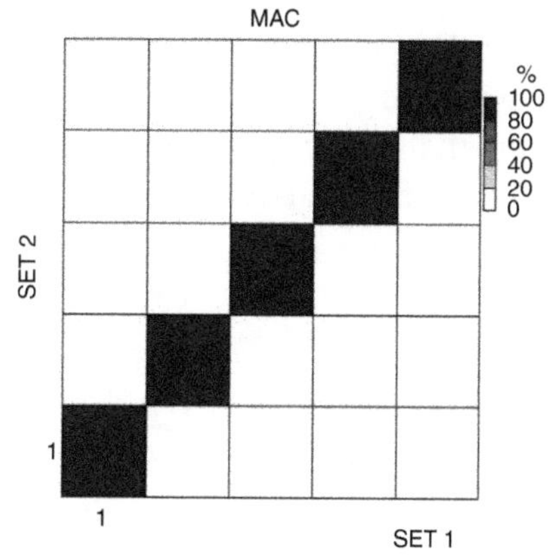

Figure 7.18 *MAC* correlation from modal data obtained using shaker and loudspeaker.

One can conclude this test case by showing the correlation between measured and simulated mode shapes, as shown in Figure 7.19. The correlation indicates that the *FE* model can predict the mode shape correctly. Ultimately, this means that mass and stiffness matrices correctly define the frequency range selected in this example. Therefore, the *FE* model does not require any updating procedure. The next test case presents an experimental model validation of a tower rotor test rig where it is shown what happens when the linear mode shapes do not fully correlate between the experimental and simulated ones.

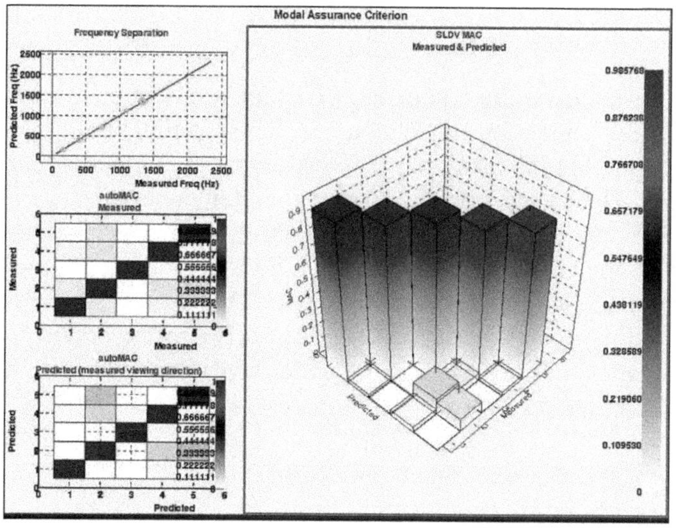

Figure 7.19 *MAC* between *FE* and experimental mode shapes[9].

7.5 Case Study 2: Experimental Model Validation of a Tower Rotor Test Rig

This second test case presents all the methods and procedures used to validate the *FE* models of a tower rotor test rig, composed of three main parts, as shown in Figure 7.20. This test rig is designed to investigate the dynamics of a tilt-rotor inside a wind tunnel facility. As specified, it is made of three sections described as follows:

[9] Full colour scheme available in the digital version of the book.

- The **Lower Rig** (or Bottom Rig), which represents the structure that lies below the wind tunnel floor section and serves as support.
- The **Upper Rig**, consisting in the aerofoil shaped tower.
- The **Top Rig**, that includes gearbox, fuselage and rotor hub.

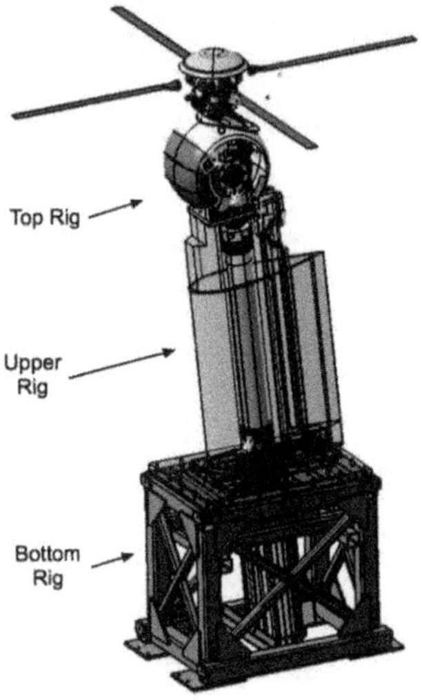

Figure 7.20 Rotor Test Rig (schematic).

This second test case focus on the Upper Rig (*UR*) and elaborates in more detail of what is explained about Model Validation in Section 7.1. This section will also include the model updating used for correcting the mass and stiffness matrices, a process that was not presented in the earlier sections. Hence, in order to obtain a complete assessment of the dynamics of the entire rig, all parts (components, separate sub-assemblies or groups of sub-assemblies) must be experimentally tested and correlated against their *FE* models. Each *FE* model is then updated in the parameters to match the experimentally measured responses. An *FE* model of a structure can be considered "valid" when it possesses the adequately similar modal quantities (i.e. natural frequencies, mode shapes) of the real structure under consideration. Each structural validation process is broken down into the following sub-processes:

1. Test Strategy
2. Pre-Test Analysis
3. Test Planning
4. Responses Measurements
5. Modal Parameters Extraction
6. *FE* model correlation
7. *FE* model updating

Points 1-3 are referred to as the "planning phase", points 4-5 constitute the "analysis phase" and points 6-7 are the "validation phase".

7.5.1 Structural Analysis of the Tower

This document section will go through the above three phases, detailing the validation process of the tower strut (Upper Rig) shown in Figure 7.21. Ideally, other components like the Bottom Rig and Top Rig were experimentally tested in-situ to provide grounds for a fully updated *FE* model of the assembly. However, these are not presented here.

The planning phase is the phase in which the strategies are defined and all the tests needed are thought out and organised. The test strategy states the objectives of the work and the tests to be performed. The pre-test analysis lays down sensors type, quantities and locations, and the test planning deals with other technical aspects like excitation, fixtures and equipment.

The final objective of this work is the validation of a *FE* model of the tower. The tower is itself composed of four elements:

a) Tower strut
b) Side cover plate
c) Leading cover plate
d) Trailing cover plate

The initial plan was to test the tower with all the cover plates (a, b and c) on and with all the cover plates off to assess the effect of the presence of the covers. However, the leading and trailing cover plates were unable to come off due to the presence of dowel pins that the lab technicians could not pull out. Therefore, the tower was only tested with the side plate on/off, leaving the other plates in place. While the plates were meant to be there for covering purposes only, it will be shown that the side cover plate plays a substantial structural role.

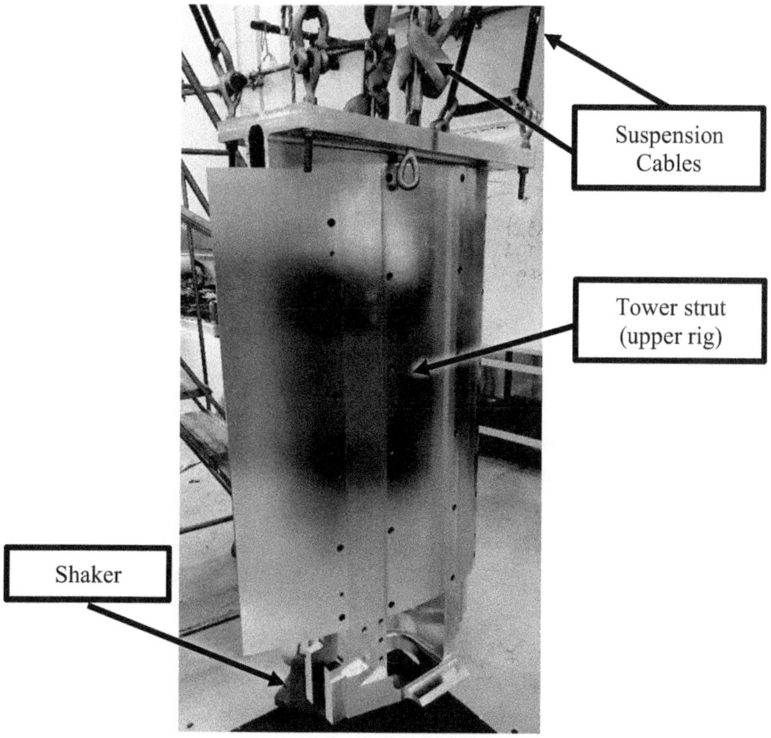

Figure 7.21 Tower strut (upper rig) suspended upside down.

Extraction of the modal parameters of the tower was executed by means of a curve fitting the measured *FRFs*. In order to obtain these *FRFs*, a random excitation signal was sent from an electromagnetic shaker to the tower and the responses were collected from an array of accelerometers. This kind of test is quick and reliable, but it needs to be modified in order to avoid leakage effects. The modification consists of the so-called "burst random" in which the driving random signal is sustained for only a percentage of the test runtime (75% in this case), while the rest is null. This minor modification permits a good *FRF* accuracy over many averages. The number of averages has been chosen to be 30. Figure 7.22 shows both the measurements of the excitation force and the acceleration response. In the dashed black box, the driving signal is null, so the force gauge measures nothing else than some background noise, which is averaged out from the *FRF* curve.

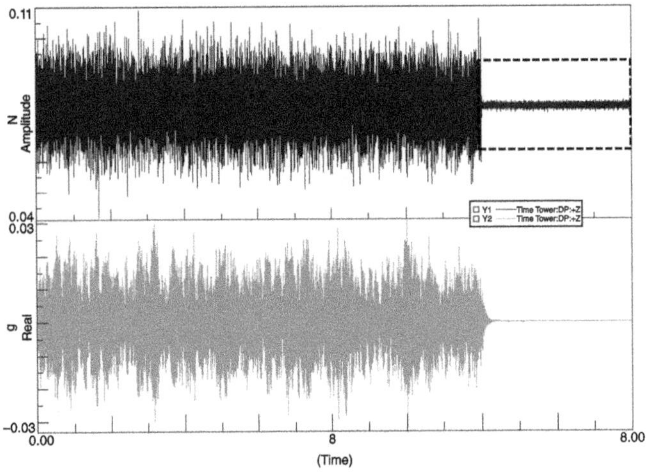

Figure 7.22 Driving point force (dark grey) and response time histories of a single test realisation.

In summary, a total of four tests were executed:

1) Side cover plate on, 75% burst random low amplitude, 30 averages
2) Side cover plate on, 75% burst random high amplitude, 30 averages
3) Side cover plate off, 75% burst random low amplitude, 30 averages
4) Side cover plate off, 75% burst random high amplitude, 30 averages

The *pre-test analysis* in a structure like this is paramount, because the choice of the type, number and locations of the sensors will make the test campaign effective. As already explained, using too few, poorly located, or the wrong type of sensors can compromise the observability of some mode shapes. The best locations for sensors are the most mobile points of the structure in terms of mode shapes, away from the nodal points (points in which the displacement is zero for a certain mode shape). In order to get a good grasp of the expected mode shapes, the preliminary (non-updated) *FE* model is analysed and the mode shapes extracted. This analysis is called a *pre-test analysis*. The sensor locations can then be selected using one of the many criteria available to the users, and nowadays those can readily be accessed in the analysis software. For instance, the Normalised Kinetic Energy method [7.3] ranks nodes by their kinetic energy, and the Normalised Modal Displacement [7.3] ranks nodes based on the average displacement of all the mode shapes in the selected frequency range. This pre-test analysis gives an array of candidate sensor locations, which can then be further narrowed down based on equipment availability, channels, and other considerations.

Figure 7.23 shows the locations of the sensors (black dots) and the measurement direction (red arrow), which is normal to the surface. Two rows of accelerometers run down in an alternated pattern along the sides of the cover plate to capture torsional effects. On the back, a single row of accelerometers accounts for the "spine" of the tower and captures the main bending modes. Inside the strut, two rows of accelerometers in an alternated pattern are placed in the fore-aft direction in order to account for the so-called "breathing modes", in which the tower expands and contracts as a ribcage. The remaining accelerometers are placed on top of the strut in all directions to observe its behaviour. The locations of the sensors (coordinates and Euler angles) can finally be exported into a text file that will later be imported into the acquisition software to create the test geometry.

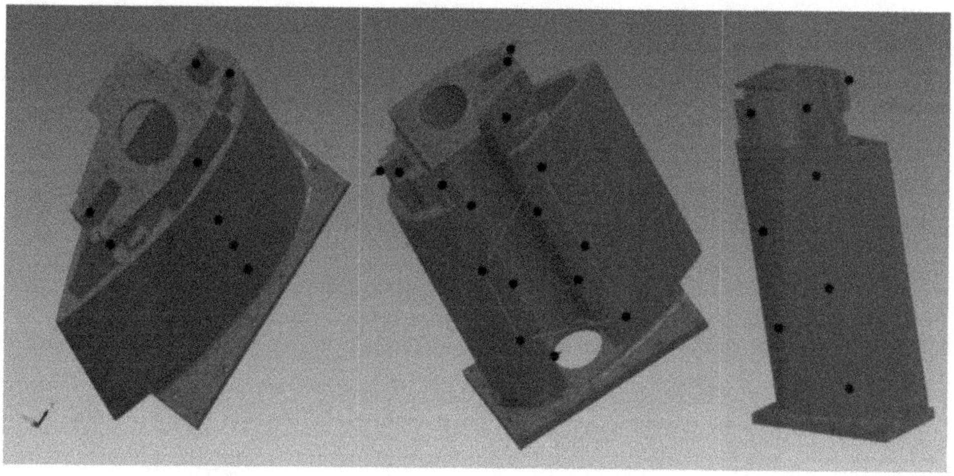

Figure 7.23 The locations of the sensors are described as black dots.

The normal-to-surface sensor direction is set in the pre-tests analysis because the surface of the test article is curved; therefore, using a global coordinate system would deal with wrong readings for misaligned sensors. Another important task to address during test planning is the fixture and excitation. Four bungee cords were used to suspend the structure and thus creating the *free-free* boundary conditions. These cords are selected by the lowest elastic stiffness, so they do not interfere with the first elastic mode of the tower. It is always advisable to check that by measuring the frequency of the rigid body motion, which should be thirty times (engineering judgment) lower than the first resonant mode of the test article.

The electromagnetic shaker used for the excitation has been suspended on four sash cords. Such an arrangement is convenient when a structure is suspended under *free-free* conditions, to minimise the interactions between the structure and the excitation source. A slender stinger is used to convey the excitation force from the shaker to the structure. The slenderness of the stinger makes sure that the force is conveyed axially along the direction of the stinger itself, thus avoiding any bending moment detrimental to the shaker functionality (and providing an extra undesired input to the structure). The positioning of the shaker is skewed with respect to the centreline of the tower, such that all the mode shapes are likely to be excited in all three spatial directions. Figure 7.24 shows the test set-up.

The analysis phase comprises the actual test and the subsequent analysis of the outcome, including the extraction of the modal parameters that fully characterise a linear structure. The *FRFs* are measured as follows: the burst random signal is sent from the signal generator to the shaker. The force and acceleration signals coming from all the sensors are then collected by the acquisition system and processed into Auto- and Cross-spectra. The *FRFs* are then estimated by the Cross- and Auto-spectra ratios, as explained in Chapters 1, 2 and 4. The results are averaged over the repetition of the random test for 30 times, as chosen during the test strategy.

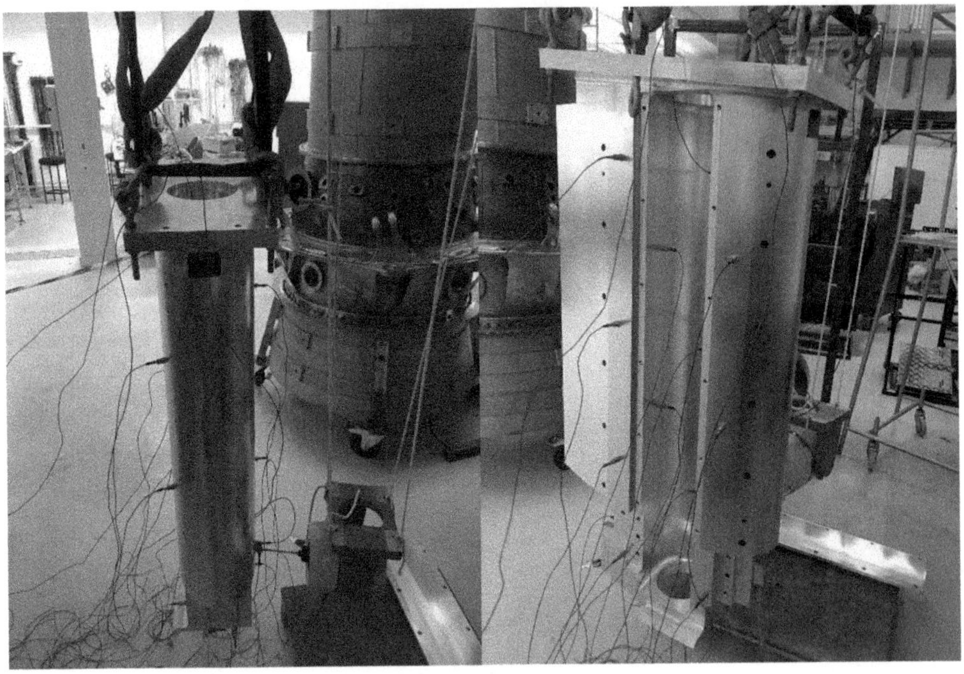

Figure 7.24 Suspension of the instrumented tower (no side plate).

The resulting *FRFs* are a complex quantity and are inspected in amplitude and phase to retrieve helpful information. The driving point *FRF* is shown in Figure 7.25. The regions of interest are the ones that show a high response amplitude normalised to the force level. In this case, the peaks at 627.63 Hz and the one at 1588.25 Hz are the most prominent ones with respective amplitude values of 11.51 [g/N] and 9.79 [g/N], while in the remaining frequency bandwidth, the *FRF* is mainly below the threshold of 0.5 [g/N]. The peaks in the lowest frequency bandwidth occur below 2 Hz and are the expression of the rigid body motion of the suspended structure.

The coherence of an *FRF* is a quantity valued between 0 and 1 that determines the correlation of the output as a result of a specific input, as discussed in Chapter 2. Figure 7.26 shows the coherence levels and the *FRF* of the driving point. A drop in coherence can be observed in the bandwidth 750-950 Hz and at the highest end of the full bandwidth. While the drop at the bandwidth edges is considered normal in an experimental test, the 750-950 Hz drop must be investigated. Coherence is a linear indicator and does not take into account nonlinear effects or the presence of noise. Since the signal-to-noise ratio (*SNR*) was above 30 for all the input channels, the coherence drop might be due to some sources of nonlinearity.

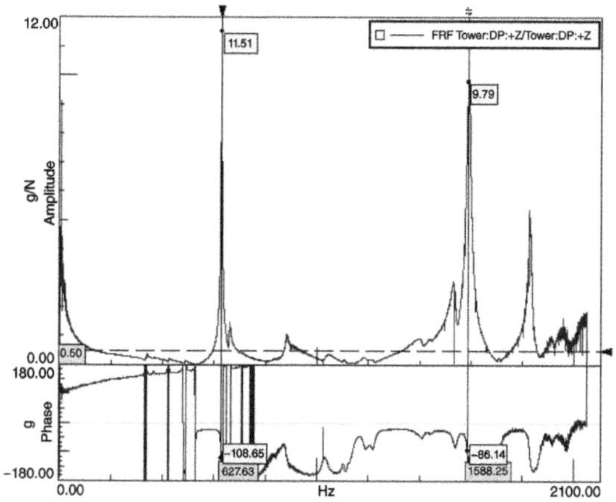

Figure 7.25 Driving Point *FRF*. Linear scale.

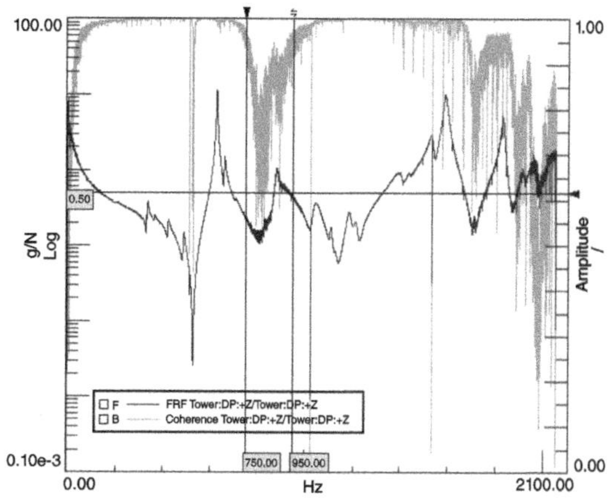

Figure 7.26 *FRF* (black) and Coherence (grey) at driving point (*FRF* in log-scale).

Since the tower comprises four components joined together, it is good practice to check for nonlinear effects arising from the joints. In order to perform these checks, a second random test was scheduled at twice the amplitude of the previous test. If the structure were behaving nonlinearly, distortions of *FRFs* and drops in coherence levels would be observed. Nonlinear effects often scale with the input levels and, therefore, a second test was repeated at two times the average input force to verify the presence of nonlinearities. The *FRF* and coherence measured at the drive point for high-vibration amplitude were superposed to the ones coming from the low-vibration amplitude test. The plot in Figure 7.27 shows the test carried out at low and high amplitudes. Focussing on the black dashed box, one might notice that the light grey FRF, generated for high amplitude, shifts to lower frequency when compared with the FRF, generated with low amplitude. By looking at the coherences, the medium grey coherence (high amplitude) drops more than the black one (low amplitude). This is a typical sign of nonlinearity, and the likely suspects for this type of behaviour are usually to be found in the joints.

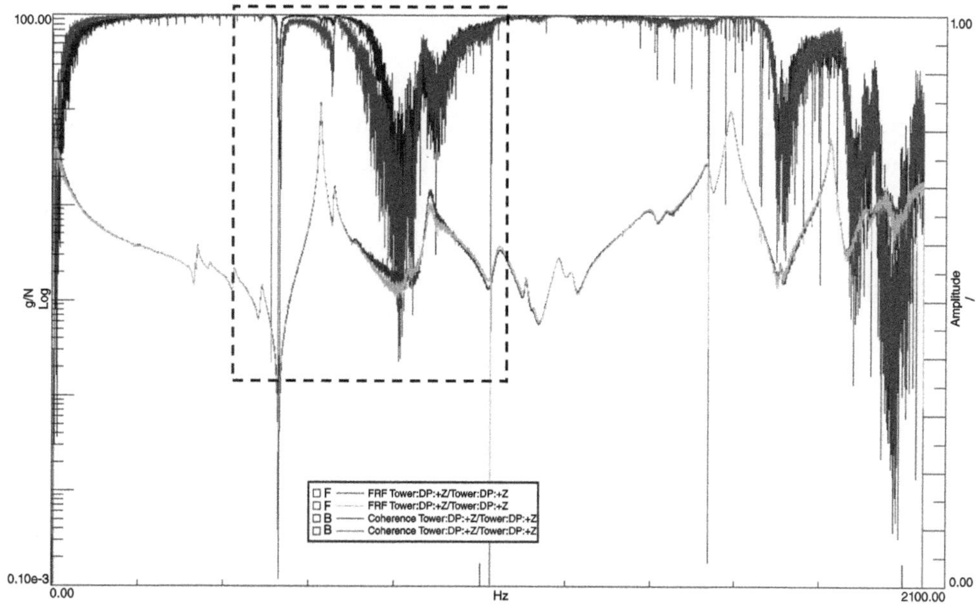

Figure 7.27 Low- and High- vibration amplitudes *FRFs* with coherences. Light grey *FRF* and medium grey coherence for high amplitude levels, while black *FRF* and black coherence for low vibration amplitudes.

The *FRF* amplitude reduction from low to high excitation force can be due to the interface slippage at the joints, increasing the damping of the system. One final test was carried out to evaluate the effect of the side panel on the vibration response. Therefore, two *FRF* measurements were carried out with and without a side panel. Figure 7.28 shows that the *FRF* is much sharper without the side cover plate, but it is also much different: while the previous peaks generally see a reduction in amplitude, a completely new array of modes appears at a lower frequency range. Those modes will be later identified as "breathing modes", and they are visible in the 250-450 Hz range. The presence of the side cover plate inhibits these modes by constraining the two sides of the tower together. Furthermore, it was noticed that all the nonlinear effects disappeared with the removal of the side cover plate. Hence, the primary source of nonlinearity is between the side plate and the tower, at the connection.

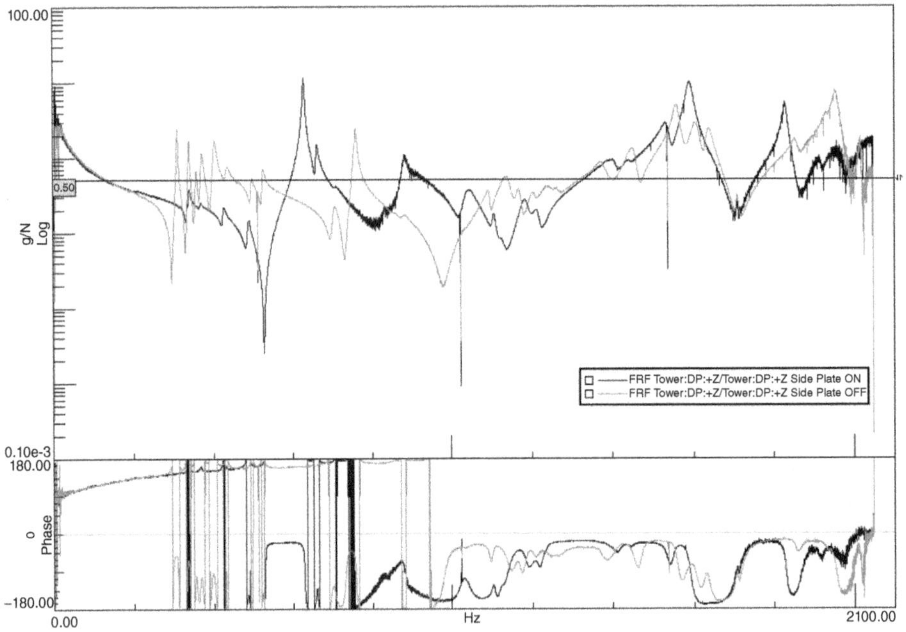

Figure 7.28 *FRFs* with side panel on (in black) and off (in light grey).

The last step of the analysis phase is the extraction of the modal parameters, which are the natural frequencies ω_r, damping ratios ξ_r and mode shapes ϕ_r. The software used to extract these parameters performs a "poly-referenced least-squares" method, as described in Chapter 4. Figure 7.29 shows, in a cropped picture, an *FRF* and the stable poles indicated by "*s*" along the light grey lines. Table 7.6 summarises the identified damping values and natural frequencies. Finally, Figure 7.30 shows an example of the first four mode shapes of the tower.

Table 7.6 Damping and frequency values for the full tower strut.

N.	Damping [%]	Frequency [Hz]	N.	Damping [%]	Frequency [Hz]
1	0.28	627.517	**7**	0.99	1390.76
2	0.40	659.767	**8**	1.48	1434.84
3	2.16	882.342	**9**	0.72	1535.6
4	1.45	1038.63	**10**	0.46	1590.52
5	0.35	1107.6	**11**	1.03	1711.02
6	0.94	1196.16			

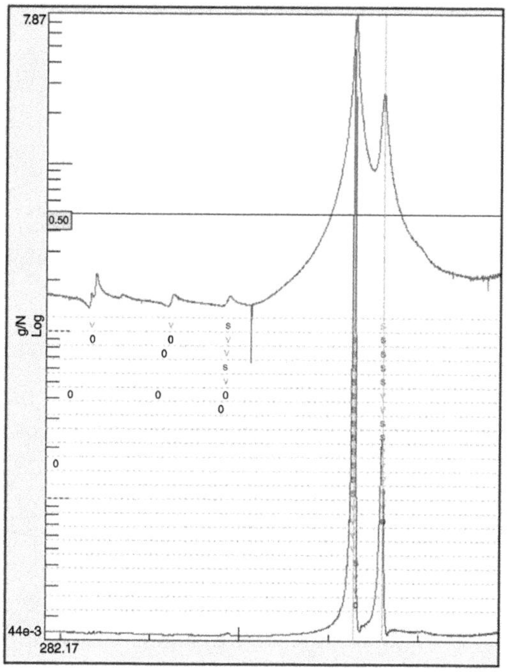

Figure 7.29 *FRF* and poles identification.

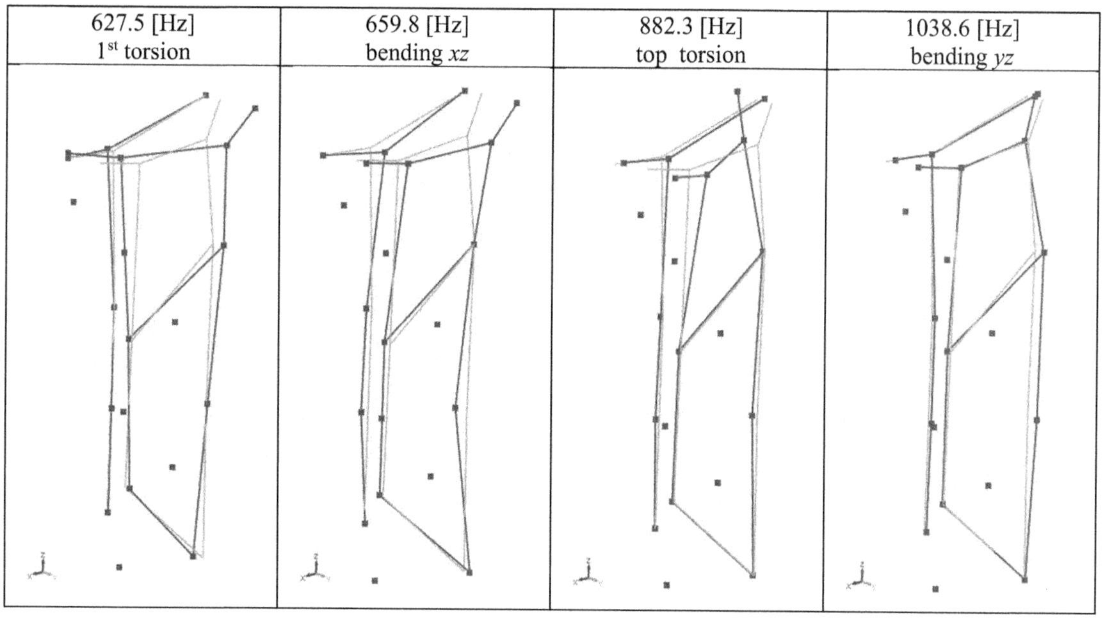

Figure 7.30 An example of the first four mode shapes of the tower.

As done in Section 7.4, the verification that the mode shapes identified by modal analysis are unique, the *AutoMAC* and complexity analyses are performed. Calculating the *AutoMAC* for each pair of mode shapes gives the *AutoMAC* matrix. Figure 7.31 shows the *AutoMAC* matrix on the top part of the figure and a table of indicators at the bottom. Looking at the *AutoMAC*, each coloured square represents the correlation between two mode shapes of the same test. The matrix contains all ones on the diagonal (the correlation of a shape with itself) and is symmetric. From the modal theory, there cannot be two identical mode shapes at different frequencies, so if a high correlation is found outside the diagonal, there might be a problem of "spatial aliasing", for which there are not enough sensors to describe the mode or the sensors are poorly located.

Looking at the bottom part, which contains many indicators, one can observe two main characteristics. The first one is the scatter indicator, indicating how accurate that mode was identified across the many measured *FRFs*. Low scatter means accurate mode identification, while high scatter means low accuracy. One of the reasons for high scatter can be due to complex mode shapes. The complexity can be inspected by looking at the phase angles of the modal vectors. If all the entries in a modal vector have the same or opposing phases, one has a "real mode" (standing wave). Otherwise, one has a "complex mode" (travelling wave).

From the modal theory, complex mode shapes arise in discrete damping elements (dashpots, some joints) that make the damping matrix non-proportional to the mass and/or stiffness ones. The most used indicator for complexity analysis is the Modal Phase Collinearity index (*MPC*) [7.10]. The *MPC* measures how the phases of the *DOF* vectors lie in the complex plane. For a real mode, the *DOF* vectors are either in-phase or out-of-phase and collinear (the *MPC* index is 100%). For complex modes, those vectors lie at different angles on the complex plane (the *MPC* index is < 100%). A threshold value of 80% is usually accepted to discriminate real modes from complex modes. Note that the high scatter matches with low *MPC* value.

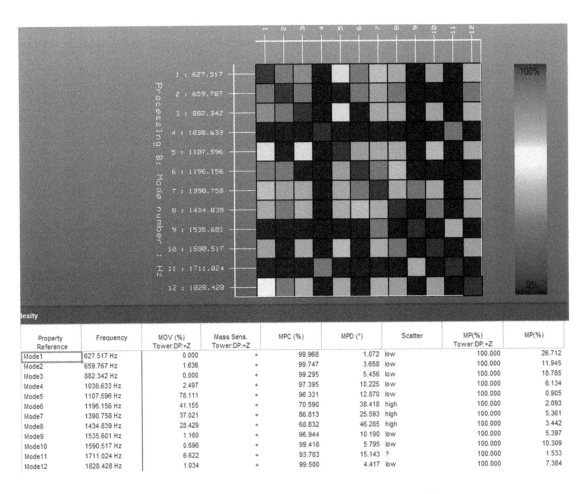

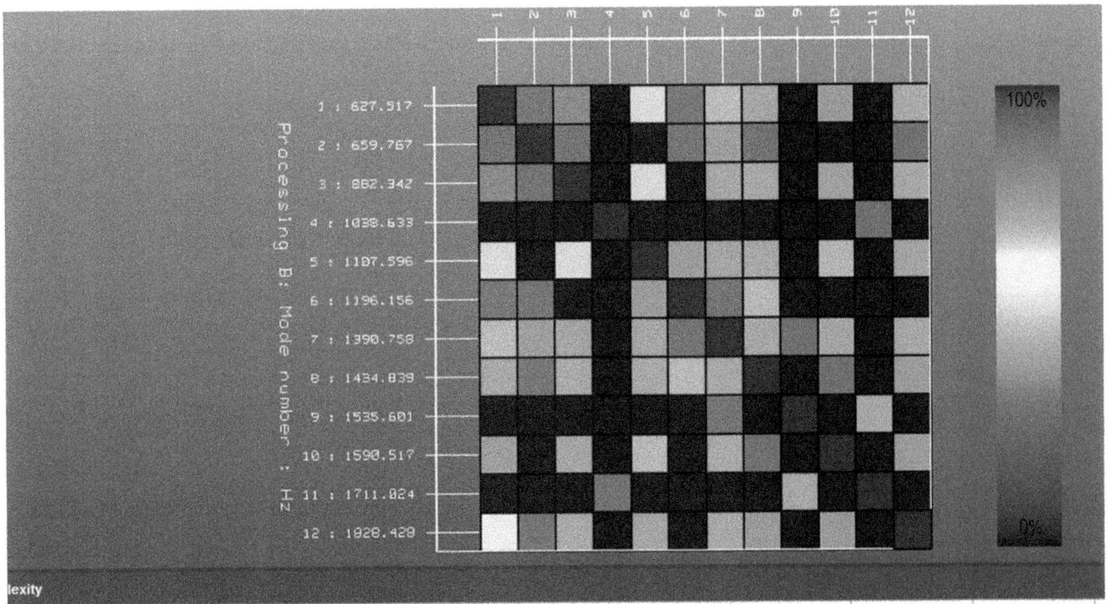

Property Reference	Frequency	MOV (%) Tower:DP:+Z	Mass Sens. Tower:DP:+Z	MPC (%)	MPD (°)	Scatter	MP(%) Tower:DP:+Z	MP(%)
Mode1	627.517 Hz	0.000	+	99.968	1.072	low	100.000	26.712
Mode2	659.767 Hz	1.636	+	99.747	3.658	low	100.000	11.945
Mode3	882.342 Hz	0.000	+	99.295	5.456	low	100.000	18.785
Mode4	1038.633 Hz	2.497	+	97.395	10.225	low	100.000	6.134
Mode5	1107.596 Hz	78.111	+	96.331	12.870	low	100.000	0.905
Mode6	1196.156 Hz	41.155	+	70.590	38.418	high	100.000	2.093
Mode7	1390.758 Hz	37.021	+	86.813	25.593	high	100.000	5.361
Mode8	1434.839 Hz	28.429	+	68.832	46.285	high	100.000	3.442
Mode9	1535.601 Hz	1.160	+	96.944	10.190	low	100.000	5.397
Mode10	1590.517 Hz	0.696	+	99.418	5.795	low	100.000	10.309
Mode11	1711.024 Hz	6.622	+	93.783	15.143	?	100.000	1.533
Mode12	1828.428 Hz	1.034	+	99.500	4.417	low	100.000	7.384

Figure 7.31 *AutoMAC* matrix and complexity analysis[10].

[10] Full colour scheme available in the digital version of the book.

7.5.2 Experimental Model Validation

The validation phase comprises the *FE* model correlation with the test results and the updating of the *FE* model parameters, in order to correct for every shortcoming of the model. One must make sure that the model contains enough parameters to be updated. Otherwise, an automatic updating procedure might result in unphysical parameters. The *FE* analysis (*FEA*) of the whole model with the cover plates on (rigidly constrained to the tower) was run and correlated with the Experimental Modal Analysis (*EMA*) parameters. The mode shapes were paired based on the following criteria:

a) *MAC* values above 60%
b) Frequency difference below 50%

These criteria are purposely kept loose to match the most significant number of mode shapes and get a general idea of the behaviour of the *FE* model. Once the responses and parameters are selected and the model updated, the pairing will be repeated with tighter thresholds. The paired mode shapes before updating are shown in tabular form (Table 7.7) and graphically (Figure 7.32 (a) and (b)).

Table 7.7 Paired mode shapes before updating.

N.	*FEA*	Hz	*EMA*	Hz	Diff. (%)	*MAC* (%)
1	1	679.33	2	659.77	2.97	92.5
2	2	765.81	1	627.52	22.04	95.5
3	3	1072.5	3	882.34	21.55	86.8
4	5	1392.9	7	1390.8	0.16	68.1

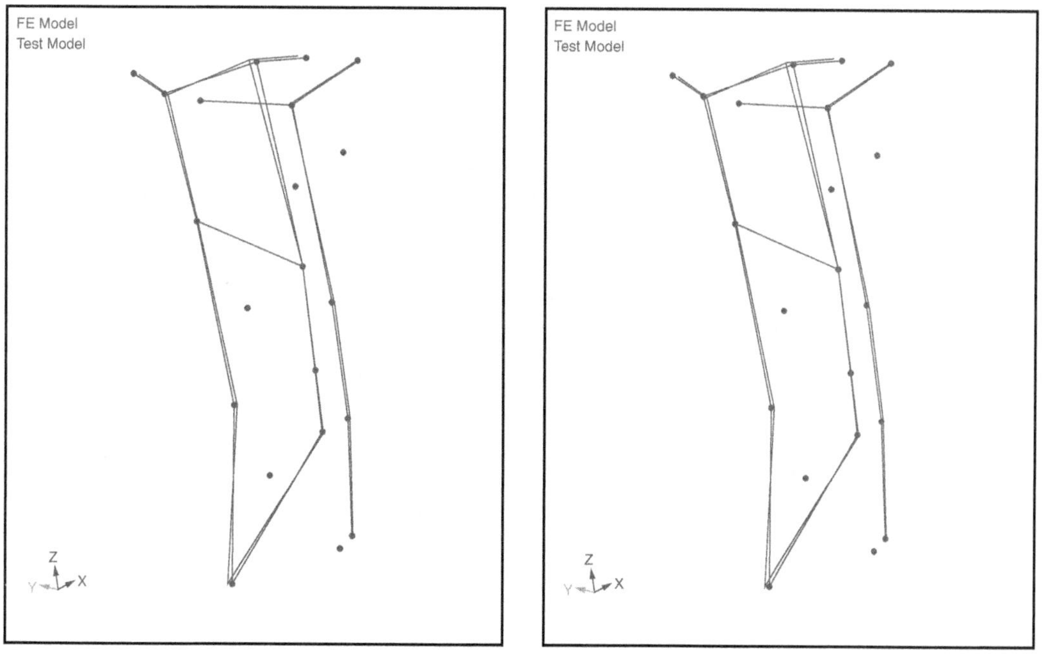

Figure 7.32 (a) Paired mode shapes 1 and 2 before updating (line & dots: *EMA*, line: *FEA*)[11].

[11] Full colour scheme available in the digital version of the book.

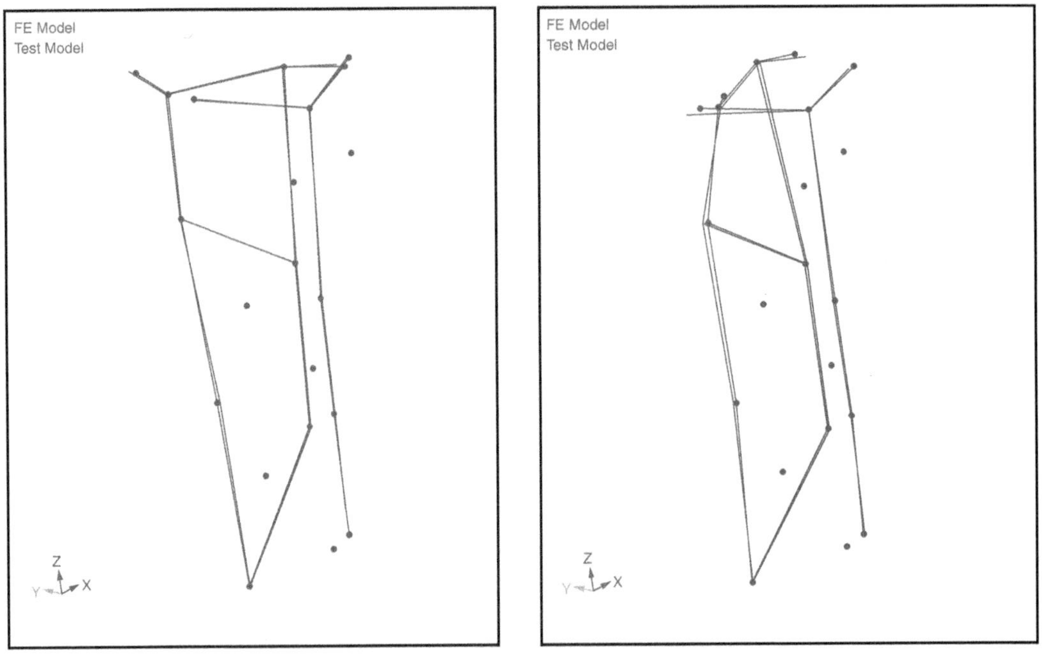

Figure 7.32 (b) Paired mode shapes 3 and 4 before updating (line & dots: *EMA*, line: *FEA*)[12].

By looking at Table 7.7, one can notice that mode shape 1 of the *FEA* (1st bending) correlates well with mode shape 2 of the *EMA* (1st bending) and *vice versa* mode shape 2 of the *FEA* (1st torsion) correlates well with mode shape 1 of the *EMA* (1st torsion). This "mode-switching" issue can promptly be recognised in the *MAC* matrix shown in Figure 7.33. A further remark on the *MAC* matrix is about the fewer mode shapes correlating above the 90%, whereas the other ones measured and simulated have little correlation. Sometimes, it is not necessary to pursue a perfect diagonal for all the mode shapes. The relevant ones (say, for the correct performance of a machine) must be correlated and then updated. This example shows exactly that situation.

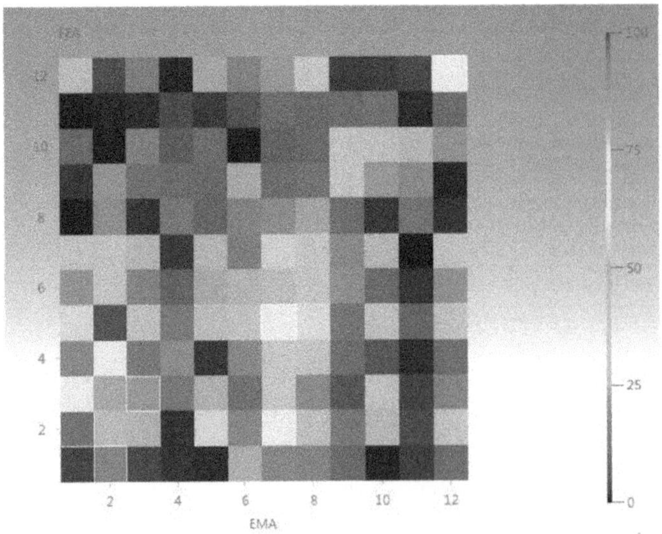

Figure 7.33 *MAC* matrix (For *FEA versus EMA*) before updating[13].

[12] Full colour scheme available in the digital version of the book.

[13] Full colour scheme available in the digital version of the book.

The model updating strategy is based on monitoring the error on the selected responses by modifying a set of selected parameters, well discussed in Chapter 6. In this case, the available responses are the frequencies, and *MAC* values of the paired modes (Table 7.7) and the available parameters are Young's modulus (*E*) and mass density (ρ) of the elements constituting the sub-components of the tower. The parameters can be changed globally (global updating) or on a per-element basis (local updating). Local updating is helpful to identify regions of the model that need refinement (e.g. some holes that have not been modelled are playing a significant role in the dynamics of some components), while global updating is used to correct significant global errors in the parameters (e.g. the wrong material was selected for the model). Three updating schemes were applied in order to get a good, updated model:

a) *FE* model with rigidly tied cover plates – global parameters change
b) *FE* model with rigidly tied cover plates – local parameters change
c) *FE* model with fastened (discrete beam elements) cover plates – global change

The updating process calculates the sensitivity of each response concerning each parameter, retrieving the so-called *sensitivity coefficients*. Sensitivity coefficients quantify the variation of a response value (e.g. resonance frequency or mass) as a result of modifying a parameter value. The coefficients obtained for all combinations of responses and parameters are stored in a *sensitivity matrix*. Analysing this matrix yields information on the sensitive and insensitive regions of the structure. Sensitivity coefficients are computed using a differential or finite difference method, depending on the parameter type and the element formulation.

The sensitivity matrix is then inverted (Moore-Penrose pseudo-inverse operation for rectangular matrices) to find a gain matrix. This gain matrix is multiplied with the difference between predicted and reference response values (error vector) to find the required parameter change to compensate for this error in the least-squares sense. This process is then repeated for a number of iterations until a selected tolerance value is reached (usually, when the norm of the error vector is below 1%). For the case of the *FE* model with rigidly tied cover plates, the global parameters change gives the following mode pairs after two iterations (Table 7.8):

Table 7.8 Paired mode shapes after global updating (Tied Cover Plates).

N.	*FEA*	Hz	*EMA*	Hz	Diff. (%)	*MAC* (%)
1	1	613.41	2	659.77	-7.03	93.5
2	2	691.49	1	627.52	10.20	96.5
3	3	968.39	3	882.34	9.75	86.8
4	5	1257.8	7	1390.8	-9.56	68.4
5	12	1903.1	12	1828.4	4.08	63.1

As shown in Table 7.8, there is one more paired shape, the overall error was reduced, but the mode-switching across modes #1 and #2 still stays on. Since this is a concerning issue, a local updating process was run to identify regions of the models that could benefit from refined modelling. Figure 7.34 shows both the *MAC* matrix (left) and the mesh with the updated elements (right).

The mode-switching issue was solved but at the expense of the stiffness of the elements around the side cover plate. This indicates that the model needs refinement in that area. In particular, it requires less stiffness (lower Young's modulus). This result is wholly expected, since the

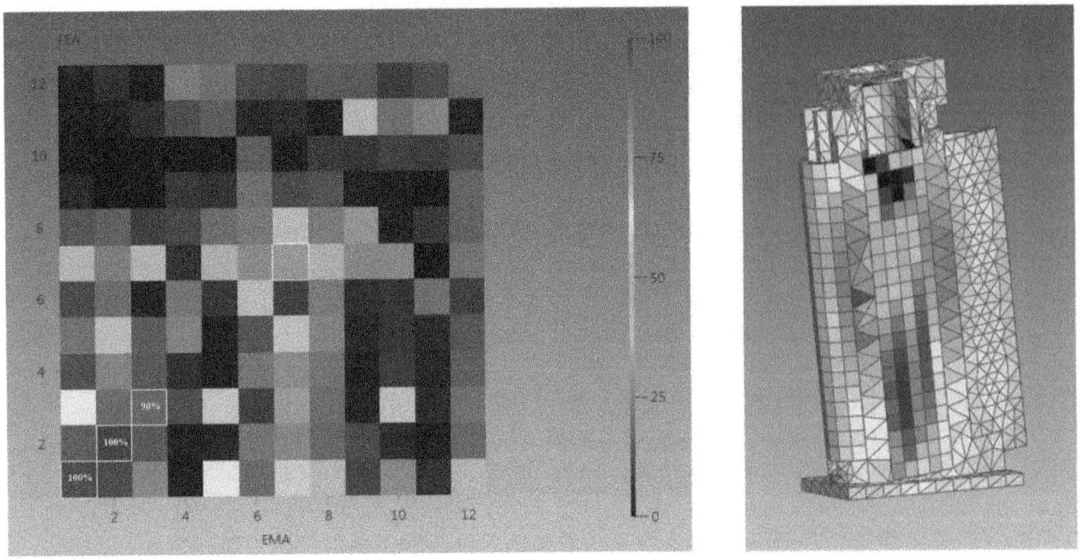

Figure 7.34 *MAC* matrix (left) and Mesh showing Young's modulus change (right) for local updating[14].

modelled cover plate is tied to the tower, leading to a stiffer interface. On the other hand, the actual cover plate is fastened with 12 bolts (6 per side), leading to a much softer overall configuration due to the non-fastened areas. A model refinement in that region is therefore deemed necessary. The model was augmented with beam-element fasteners to provide the necessary freedom at the interface to better capture the dynamics of the tower. A new model updating procedure was then started, and after eight iterations, the new paired mode shapes were retrieved, as shown in Figure 7.35, with much more minor errors and higher *MAC* values shown in Table 7.9.

Table 7.9 Paired mode shapes after updating (fastened model).

N.	*FEA*	Hz	*EMA*	Hz	Diff. (%)	*MAC* (%)
1	1	624.90	1	627.52	-0.42	96.2
2	2	649.51	2	659.77	-1.55	98.8
3	3	888.25	3	882.34	0.67	92.9
4	5	1116.4	4	1038.6	7.49	72.9
5	6	1265.4	6	1196.2	5.79	60.5
6	7	1390.1	7	1390.8	-0.04	72.1

7.6 Case Study 3: Nonlinear Behaviour of Bolted Flanges

As seen in the previous sections, dynamic testing of large structures is challenging because of the test design, the test set-up, the analysis of large datasets and their final interpretation. Most of those tasks have been, over the years, automated as much as possible. Reinstating what was already said earlier, the selection and position of sensors on large aerostructures are carried out following the analysis of *FE* models and, more rarely, by engineering judgments. Even if such statements might be generally valid for all engineering applications where *FE* models have a reasonable degree of accuracy, there are still many practical cases where the engineering judgment and "the Navigator Principle", as in Chapter 2, can be very relevant.

[14] Full colour scheme available in the digital version of the book.

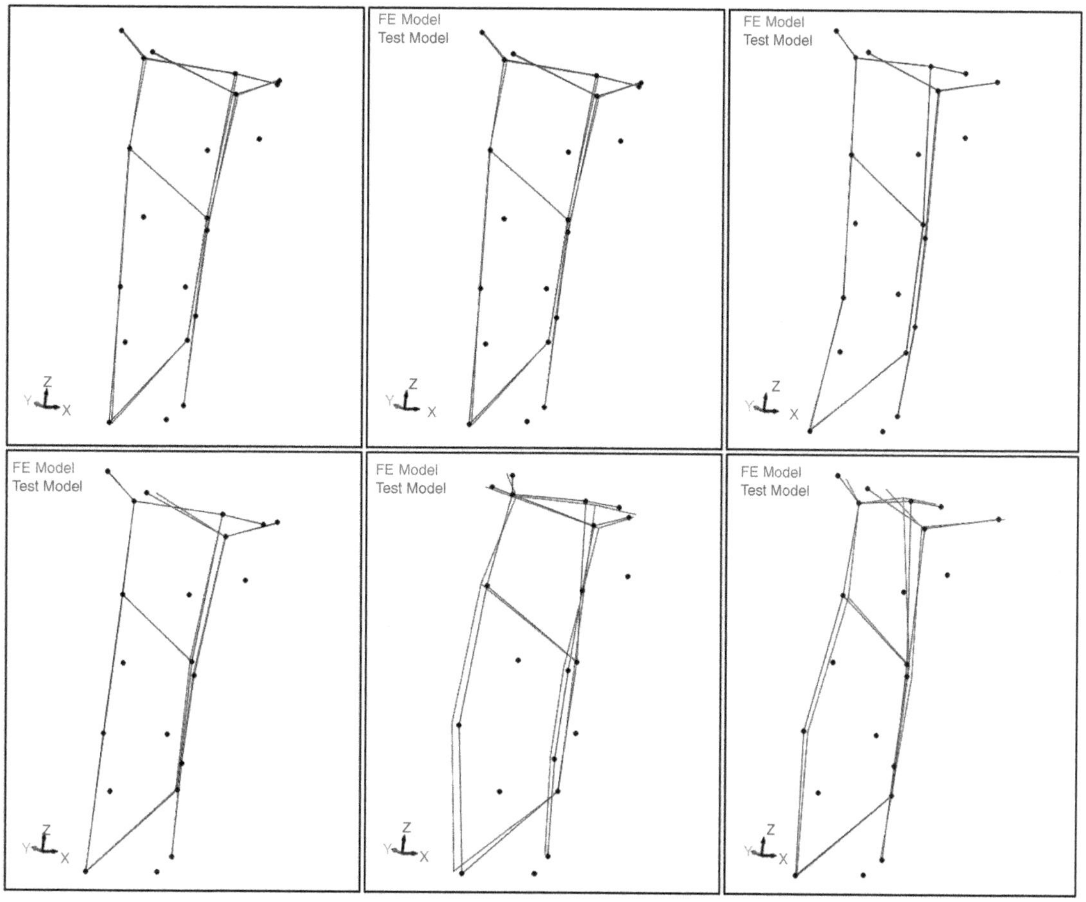

Figure 7.35 Paired mode shapes after updating (line & dots: *EMA*, line: *FEA*)

One of those cases is for dynamics that become amplitude-dependent. Mechanical systems are altogether a sum of individual components kept together by many different types of fasteners. For instance, bolted joints are a common source of amplitude-dependent responses and nonlinear behaviour. In his book [7.11], Matt Brake reports the most relevant research on that topic. Therefore, the validation of systems has become very challenging and critical in the design cycle: on the one hand, if the numerical/*FE* models one is trying to validate are linear, then the experimental activity should use low-amplitude excitation to collect data – at which point there are two linear models (one experimental, one numerical) and the validation process makes perfect sense; on the other hand, if the nonlinear behaviour should be modelled, it must also be captured and identified experimentally for a meaningful validation process.

GVTs of large aero-structures, such as aeroplanes, present challenges of nonlinearity, and some literature is available and suggested to the reader [7.12 – 7.15]. This section presents an example of how bolted joints influence the dynamics of the frequency response, and a simply bolted flange sub-system is considered.

A brief table of contents is presented to guide the reader over the following sections:

- Linear model validation using the response model from *SLDV* techniques;
- Use of *SLDV* measurements to "navigate" through the test set-up for the nonlinear modal tests;
- Technique for characterising nonlinear vibration responses;

- Example of nonlinear model validation.

Sections 7.3.1 and 7.3.2 introduced the scanning *LDV* system to the reader. One application of the measurement method to a composite blade was presented in Section 7.4. The significant benefit of using an *SLDV* system is to achieve a high-definition mode shape in a relatively short testing time. Furthermore, the laser beam did not create any mass loading on the test article, which was ideal for the composite blade. The following section will show why the *SLDV* measurement system can be advantageous, as opposed to contact accelerometers, when deflection shapes must be promptly achieved to aid the test plan when nonlinear vibrations are involved. As mentioned earlier, there are other valid optical methods, such as Motion Magnification and Digital Image Correlation (*DIC*), capable of delivering deflection shapes, but are not proposed in this chapter. The authors do not want to be exclusive with the use of one or another optical method since the practitioner must decide upon the best device for carrying out preliminary experimental investigations.

7.6.1 Rapid Validation of Fine Mesh *FEM* Axisymmetric Casings and Assemblies

This subsection presents an application case study of how Fine Mesh *FEM* (*FMFEM*) can be validated by measurements carried out by the *SLDV* system. One shall mention that an *FMFEM* is a High-Fidelity model of the test structure, meaning that every detail of the structure has been meshed. It is straightforward to understand that those *FMFEM* models can be extensive, often made of millions of *DOFs*. The objectives of this example are:

1) to show that an *FMFEM* does not need to be validated by measuring the responses at locations covering the entire assembly;
2) to show that an *SLDV* can deliver test data rapidly without time-consuming set-ups.

One starts by saying that the Whole Engine Models (*WEM*) [7.16] validation processes require that all the fundamental modes are acquired during the modal test: this is necessary to perform the modal correlation and any Computational Model Updating (*CMU*) that may be required later. Modal Test Planning and Validation of *WEM* models start with the *WEM* model itself.

The *WEM* model of an aero-engine casing is typically an axisymmetric structure, basically, a cylindrical structure. Such structures are known to present repeated roots, which means two identical, and orthogonally rotated mode shapes at the same natural frequency. Therefore, when validating large, *strongly* axisymmetric *FMFEM* models, usually only one orthogonal mode from each pair needs to be acquired on a test, since a strong correlation with one of the orthogonal modes would imply a similarly strong correlation for the other one. In practice, what was said earlier means that one of the two identical mode shapes is sufficient for the *MAC* correlation since the other one should also be correlated. For example, assume that one simulates the first orthogonal mode shape (radial deflection, not longitudinal one) of a cylinder. The *FE* solver calculates one pair of identical natural frequencies, one for each identical mode shape. In practice, those two natural frequencies can be slightly separated because of the non-perfect axis-symmetry of the physical structure. On the one hand, the achievement of a validated *FMFEM* means that such a model can become a reference because of its high-fidelity description. On the other hand, such a model becomes computationally very expensive to run. The underline idea is to use such an *FMFEM* to verify whether the process of model reduction, computationally cheaper to run, leads to a new system, the dynamics of which is incompatible with the reference one (*FMFEM*). In other words, relevant global mode shapes will not change if during the model reduction some geometrical defeaturing is applied.

This paragraph illustrates some suggestions to reduce the model validation test time required for *FMFEM* components and/or sub-assemblies from the delivery of the hardware to the test area. Modal test set-up times can still be lengthy for large casing structures despite the apparent simplicity of the *free-free* boundary conditions, which are often required. However, using the *FMFEM* model makes it possible to predict the fundamental *free-free* mode shape and mass of the test component with high confidence. Correctly marking up the test hardware prior to modal testing is critical, as it locates the test positions carefully chosen during the Test Planning onto the actual test hardware. Mistakes made during the mark-up of the test article are often challenging to evaluate tests *a posteriori* and will artificially reduce the correlation with the *FMFEM* model. A large amount of geometric detail in *FMFEMs* can be utilised to simplify the marking-up of the test hardware enormously. Many features (e.g. flange holes, casing holes, casing bosses, etc.) within the *FMFEM* can now be accurately located on the test hardware itself.

Higher initial confidence in the *FMFEM* also means that fewer test points are required for validation, i.e. additional test points that would otherwise have been specified to extract both orthogonal mode pairs or aid mode visualisation are no longer required. The following section will describe the process of correlated experimental and numerical mode shapes based on what was just explained by using one part of the *FMFEM* model rather than all of it. The use of one part of the model allows for measuring the vibrations by the *SLDV*, since that surface is optically accessible for the laser beam.

7.6.2 Sector Test Planning for Large Axisymmetric *FMFEMs*

This section illustrates how to explore a sector of the whole model to perform the test design by using an *SLDV* system. Figure 7.36 (a) shows an *FMFEM* model of a large Civil Engine casing sub-assembly. This model has over 12 million degrees of freedom, and the casing flanges have been modelled (simplistically) as rigidly fixed together. It is possible to quickly "hide" all but a sector of the whole assembly model, Figure 7.36 (b), and export the selected nodal displacements from that sector only. Figure 7.36 (c) shows how closely the geometry of the *FMFEM* sector matches the actual test structure. The first 30 *FMFEM* modes were included in the Test Planning, which is done through the *AutoMAC* analyses of the displacement patterns of the sector from those first 30 *FMFEM* modes. The *AutoMAC* of a good test plan will have low off-diagonal terms for all the modes of interest, which means that there is a sufficient number of sensors to distinguish the vibration modes without any aliasing.

One knows that the *SLDV* system can generate a dense measurement grid. However, which is the minimum number of measurement points for distinguishing all the vibration modes selected for the test plan? Figure 7.37 shows the effect on the *AutoMAC* of reducing the number of test points assumed from (a) all the 60810 points exported from the *FMFEM*, (b) from a subset of 1014 points, (c) when only 21 points have been selected. A Test Plan is usually deemed satisfactory if the off-diagonal terms of the simulated test *AutoMAC* are less than 40% for all fundamental modes. Such a threshold can be reduced even further (25%) when one deals with non-axisymmetric structures, which present well-separated mode shapes. The near-identical *AutoMACs* of Figure 7.37 (a) and (b) show that using 1014 test points would be as good as using 60810 points. The 21 points option (Figure 7.37 (c)) could be achieved using accelerometers, but the *SLDV* setting up time is definitely shorter. The more recent user interface allows to load the test plan as a measurement grid, and after some calibration between the test article and the camera field of view, the test points are exactly set as in the test plan.

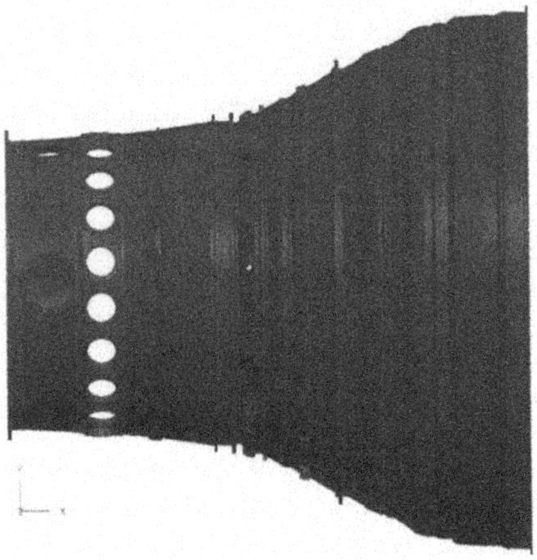

(a) *FMFEM* of the *COC/HPTIPT/LPT* casing assembly.

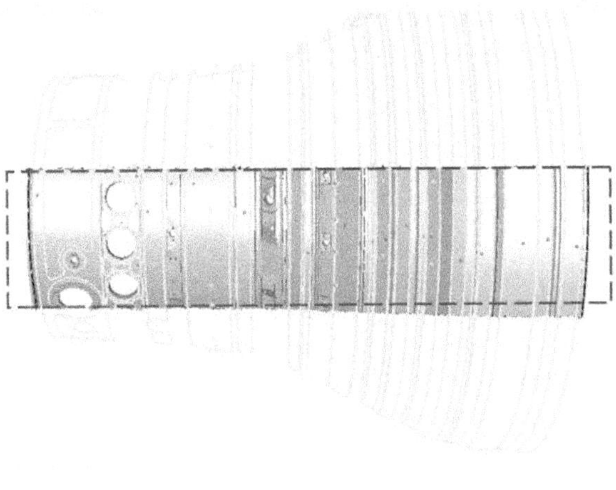

(b) Target sector used for the test planning of *COC/HPTIPT/LPT* casing assembly *FMFEM*.

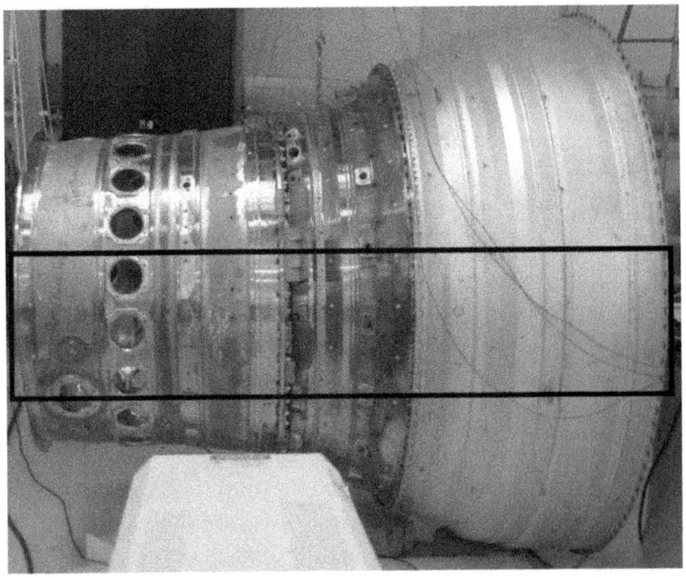

(c) Set-up showing the target sector of the *FMFEM* on the actual *CCOC/HPTIPT/LPT* casing assembly.

Figure 7.36 *FE* model (a), target sector (b) and test set-up (c).

The assumption that a smaller number of measurement points suffice for the validation activity is presented in this final paragraph. Twenty-one points on the test hardware, as presented in Figure 7.37 (c), were measured by the *SLDV* system by performing a roving hammer excitation with three averages per point. The mode shapes recovered from the *SLDV* measurement were corrected according to the laser beam angle of incidence on the target (recalling that the laser measures a component of the actual vibration). Table 7.10 compares the natural frequency difference and *MAC* value between a modal test carried out by accelerometers and roving modal hammer of the whole casing assembly and the one carried out by *SLDV* and roving modal hammer on one part of it. The comparison shows exemplary achievements for the modes in shaded grey colour. The testing time is a fraction of the one required for performing the whole casing assembly. As a final remark, a *3D SLDV* technology and features like *FEM* meshing

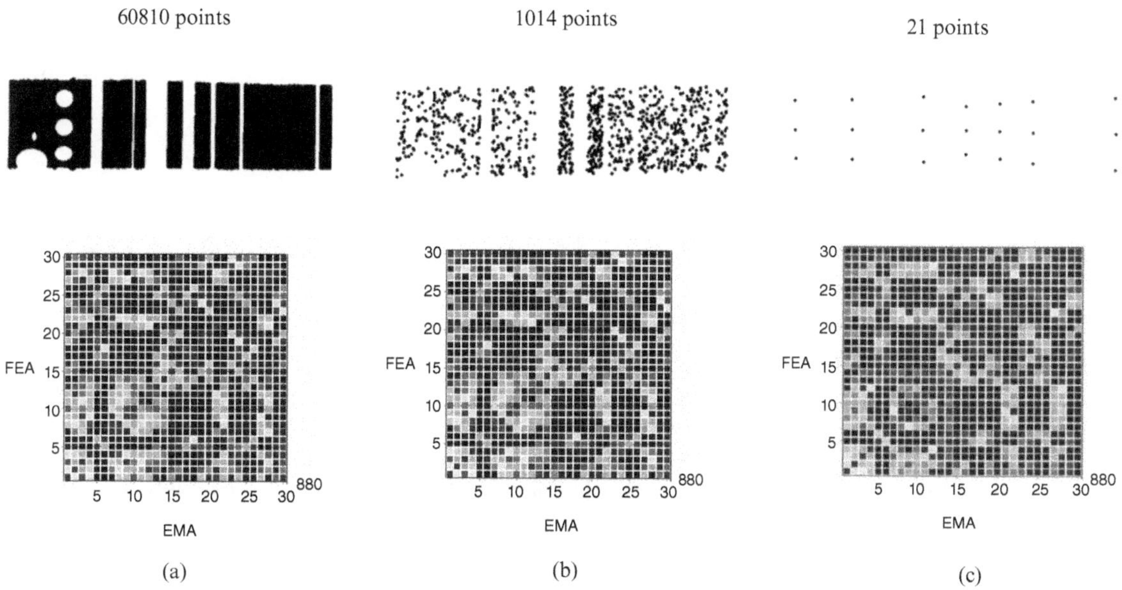

Figure 7.37 Predicted variation in the *AutoMAC* of the first 30 modes of the *FMFEM* sector due to a reduction in the test point density[15].

Table 7.10 Comparison of *MAC* values achieved by different test methods.

Mode Pair	*FMFEM* Mode	Hz	TEST Mode	Hz	Full Test Freq. Diff. (%)	MAC	Rapid Test Freq. Diff. (%)	MAC
1	1	39.3	1	39.1	0.4	97.8	**-0.9**	**91.3**
	2	40.1	2	39.7	1.1	94.4	2.5	80.9
2	3	53.7	3	53.1	1.1	99.0	**1.1**	**96.3**
	4	53.8	4	53.2	1.2	99.2	1.1	89.1
3	5	58.7	5	58.5	0.3	97.4	**-0.4**	**95.2**
	6	58.7	6	58.9	-0.3	97.1	0.4	90
4	7	84.5	7	82.3	2.7	99.5	**2.7**	**97.8**
	8	84.6	8	82.4	2.7	99.1	2.7	94.4
5	9	129.3	9	125.7	2.9	98.8	**2.8**	**96.1**
	10	129.3	10	125.8	2.8	99.0	2.9	90.6
6	11	185.5	11	180.5	2.8	97.3	**2.8**	**98.5**
	12	185.6	12	180.6	2.8	96.9	2.8	79.3
7	13	201.8	13	187.7	7.5	85.9	7.5	92.2
	14	203.0	14	189.4	7.2	96.3	**7.2**	**95.2**
8	15	206.8	15	201.1	2.9	96.9	**0.4**	**97.1**
	16	209.9	16	206.0	1.9	97.3	4.4	96.4
9	17	241.2	17	229.8	4.9	92.2	**4.9**	**94.8**
	18	242.4	18	230.3	5.3	91.6	5.3	89.5
10	19	244.8	19	232.7	5.2	98.4	**5.2**	**97.3**
	20	245.0	20	233.2	5.0	93.1	5.1	90.9
11	21	249.6	21	242.6	2.9	98.9	**2.9**	**96.6**
	22	249.7	22	243.0	2.8	99.1	2.8	93.4
12	23	290.1	23	283.5	2.3	99.4	1.9	93.5
	24	290.5	24	284.8	2.0	99.4	**2.5**	**96.2**

[15] Full colour scheme available in the digital version of the book.

mapping can make the whole process quicker and more accurate. The *3D* system can already deliver vibrations along the three Cartesian axes $\{X, Y, Z\}$, meaning that the mode shape does not need special conditioning as done by using a single scanning *LDV* head. Furthermore, the test plan can be directly imported into the *SLDV* system software, which requires some essential calibration before it can be converted into a measurement grid.

7.6.3 Test Plan for the Characterisation of Nonlinear Vibrations in Bolted Flanges

The previous sub-section provided an example of using a non-contact measurement method for linear model validation. This sub-section shows how the *SLDV* measurement method can effectively support test engineers through nonlinear modal testing. The contents are divided into two parts, one focusing on the large casing assembly (already presented in the previous sub-section) and one focused on a cut-out from the casing assembly; the latter is used for the nonlinear model validation. A simple example soon clarifies the rationale for using an optical method. Figure 7.38 shows the modulus of an *FRF* of the whole aero-engine casing. The *FRF* exhibits about fifty resonant modes in a frequency range of 800 Hz. The challenge is to explore and select vibration modes that exhibit nonlinear behaviour caused by the flanges connecting the three casings.

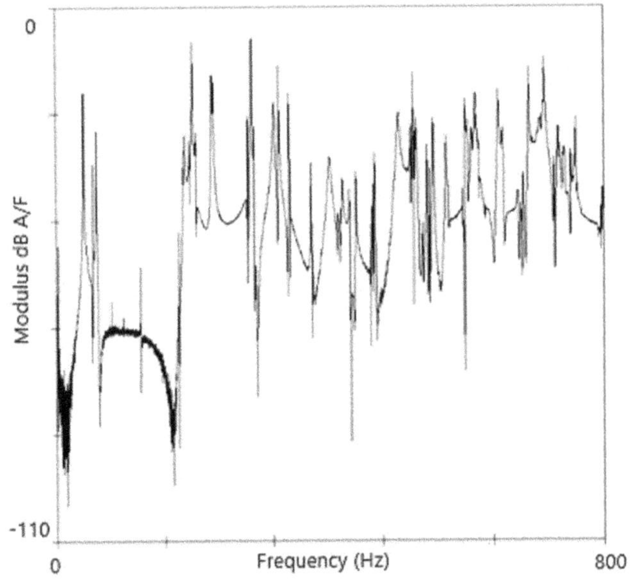

Figure 7.38 Example of the modulus of an *FRF* of the aero-engine casing.

Typically, nonlinear vibrations are detected by checking the deviation of frequency responses from low to high excitation force amplitudes. Accelerometers can carry out this detection process, perhaps already installed on the test item. Assuming that one selects a set of possible resonance peak candidates, how would one know which deflection shapes are associated with those peaks? And, furthermore, are any of those deflection shapes causing the flanges to deform up to generate nonlinear vibrations? The *SLDV* measurements can augment the capacity to carry out that resonance peak selection by providing its deflection shapes. Figures 7.39 - 7.42 show one example of *FRFs* and mode shapes, respectively. The *FRFs* measured for low and high force amplitudes show an apparent frequency deviation, resulting in a poor homogeneity correlation (note that in both plot the frequency range is 10Hz). The deflection shape at 202.9 Hz helps visualising the flanges that might be subjected to torsion and bending, which might cause nonlinearity. The following example shows in Figure 7.39 two *FRFs* at low and high excitation forces, the frequency deviation of which is not as much as seen in Figure 7.41. A minor frequency deviation, and a better homogeneity, might be due to more minor deformations in the region of the flanges, leading to less nonlinearity.

In conclusion, both Figures 7.39 and 7.41 show how much the resonance frequencies change when the excitation force is changed from low to high. As introduced, the homogeneity method, presented in Chapter 5[16] and in [7.17], can quantify the deviation and rank and decide which of the two examples needs more attention. The *SLDV* can aid that investigation further.

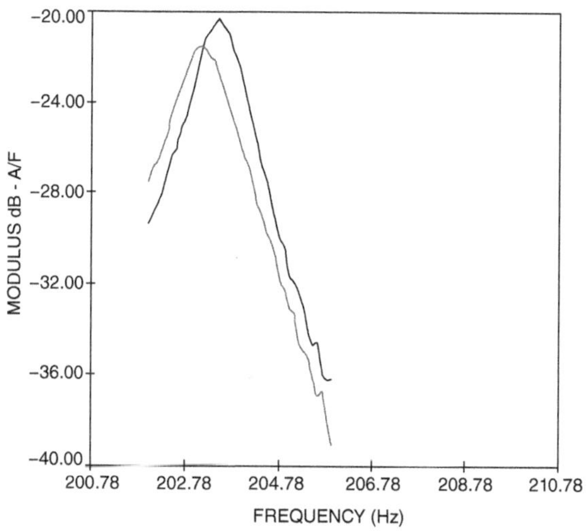

Figure 7.39 *FRFs* (zoom) around 202 Hz.

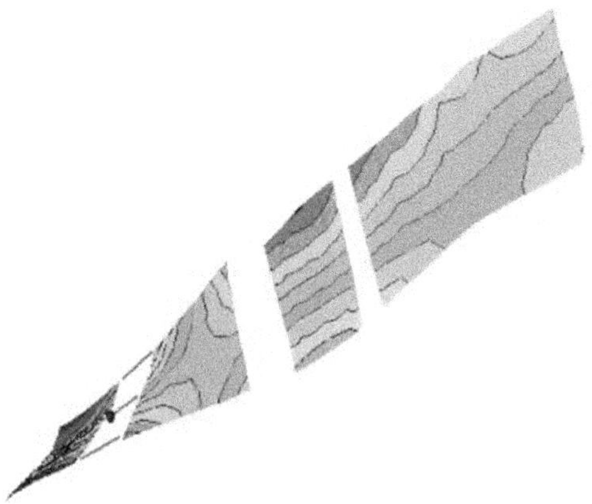

Figure 7.40 Mode shape at 202.9 Hz.

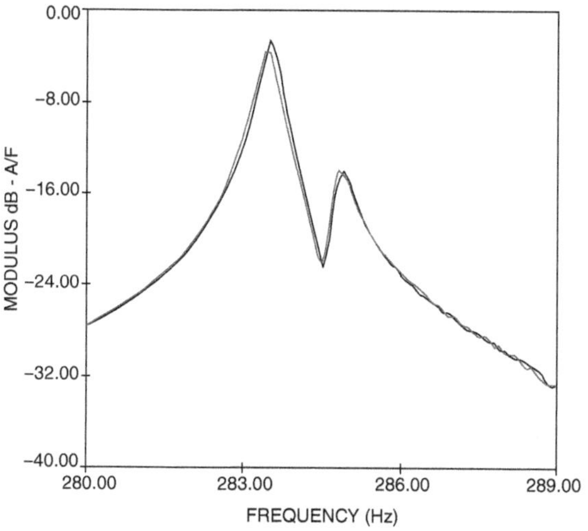

Figure 7.41 *FRFs* (zoom) around 284 Hz.

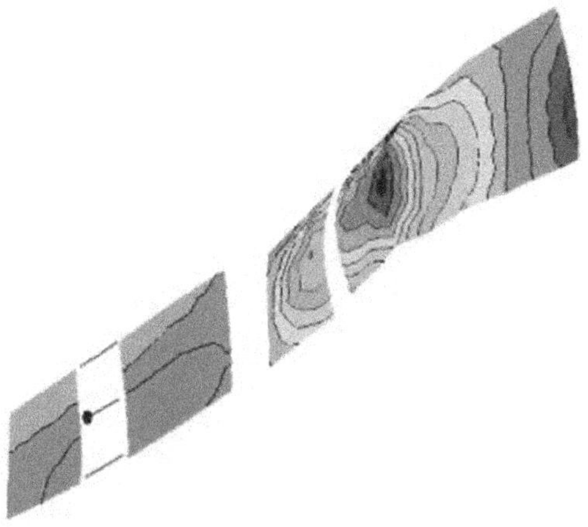

Figure 7.42 Mode shape at 284 Hz.

---

[16] Section 5.3.3.

Hence, a simple two-step approach could be (i) to inspect *FRF* measurements by homogeneity analysis and (ii) to inspect associated deflection shapes using an optical system, being that an *SLDV* or any other type of sensor.

The same procedure is applied to a sector of the casing assembly, which was used for the nonlinear model validation. Again, the *SLDV* carried out the modal survey, and Figure 7.43 shows the high modal density presented by the *FRF* measured. The set-up of the test article does not require particular effort. First, the item is suspended under *free-free* conditions and then excited by a shaker.

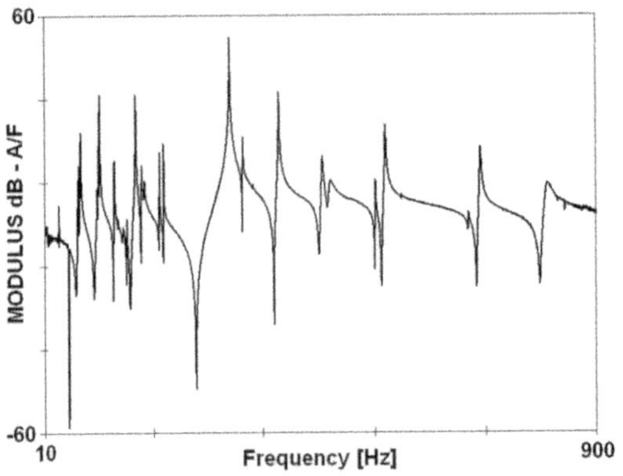

Figure 7.43 *FRF* at the driving point on the sector flange.

As explained earlier, analysing the mode shape helps selecting resonances associated with structural modes. For instance, the casing assembly is made of two parts, one of which presents a metal skin layer on the inner side, producing several local resonances. As the bolted flange and the nonlinear interface behaviour is under investigation, the local modes of the inner skin might mislead the resonance selection. Figure 7.44 shows the selection of the most relevant modes identified and selected for the nonlinear identification (note that the arrow indicates the excitation location). The criteria were adopted as follows: mode at 167 Hz presented maximum bending near the flange area, mode at 313 Hz presented two nodal lines parallel and across the flange and mode at 334 Hz showed nearly null displacements at the top part of the sector casing. The last one, mode at 462 Hz, shows a more complex deflection over the flange area.

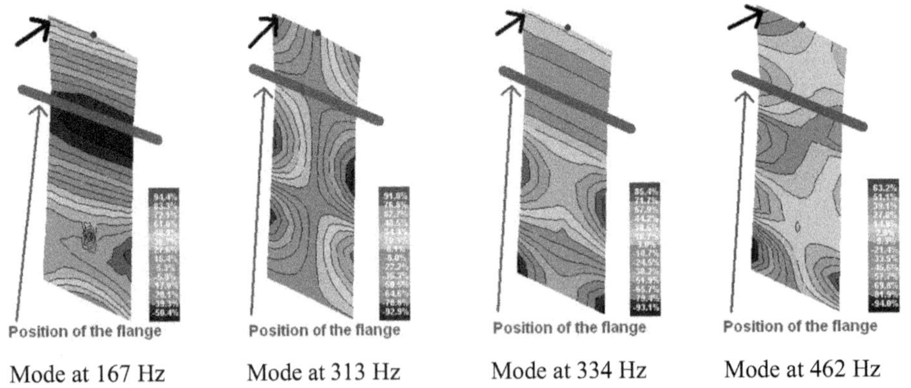

Figure 7.44 Selected mode shapes for the nonlinear identification (picture from ref. [7.25]).

Finally, one can conclude this section by saying that the optical methods, in this case the *SLDV*, are a great addition to the test engineer who can scope the dynamic response in terms of both resonances and mode shapes. The latter might help characterise areas of the structure, such as the bolted flanges, which might be sources of nonlinear vibration.

7.6.4 Test Set-ups for Characterisation of Bolted Flanges Nonlinear Vibrations

The previous sub-section discussed the inspection of deflection shapes alongside the loss of homogeneity of the frequency response functions because of the nonlinearity. The next phase, having selected the resonant modes, requires characterising the dynamic nonlinear response.

There are several ways of exerting large dynamic forces in a system, and an electromagnetic shaker is one of those. In brief, an electromagnetic shaker is made of a suspended armature oscillated by a magnetic field. The bigger the shaker, the lower the dynamic range and the higher the exerted forces. Hence, the choice of a shaker also depends on the dynamic range to investigate. There are many types of shakers. The most common ones present in most dynamic laboratories are modal shakers, called in this way because they are often employed for modal surveys. These are specified by the generated sinusoidal force, spanning from a few to a couple of thousands of Newton. The electromagnetic shaker is a versatile test device because different excitation signals can feed it. The advice is to handle them with care, avoiding the over travel of the armature (the allowable armature extension) or exposing them to uncontrolled shocks. Shakers need to be permanently attached to the test structure, and that can be a lengthy process. It is never easy, and the incorrect mounting can seriously bias the dynamic response [7.17]. Therefore, one shall check the shaker installation by running trial tests. For instance, an excitation harmonic signal must correspond to a single harmonic response (at the same point of the excitation) without noise or harmonic distortion. Chapter 2 thoroughly explored the test set-up before engaging in time-consuming tests.

For the large casing assembly, the resonances for nonlinear modal testing fell on two modes, at 194.5 Hz and 204.5 Hz. The most up-to-date test planning software can deliver the best excitation location, making most vibration modes observable. However, this might not be enough to ensure that a particular location is the best for driving the system to nonlinear vibrations. The following paragraph explains some possible shortfalls that a practitioner might face during testing.

As already said, the standard test planning achievable using *FE* models is not helpful because the algorithm has to search for the best location to make all modes observable, and that location is typically a very flexible point of the structure. The shortfall of such an approach is part of the nature of the nonlinear vibration. Large excitation forces are required, and a shaker mounted at a flexible point on the structure might experience its armature over-travelling because of a significant vibration response. The best location offers a suitable impedance to avoid over-travelling and a point able to transfer vibration energy. Several set-ups were carried out in the example of the whole casing to determine the best impedance match between the shaker and the test structure. One can appreciate the number of attempts from Figures 7.45 to 7.50. Position (6) shown in Figure 7.50 is the worst possible because the casing behaves like a bell. Think about pushing that point to exert nonlinear vibration. Since that point is very flexible and offers low resistance, the harder one pushes, the less energy one can feed into it at resonance because the armature of the shaker is almost in-phase with the excitation point. Therefore, the shaker will likely experience over-travel issues, one of which may lead to "pseudo-nonlinearities". The best position was number (3) in Figure 7.47, which provided the best results for the amplitude-controlled vibration test, which will be discussed in Section 7.3.7.

Figure 7.45 Position 1.

Figure 7.46 Position 2.

Figure 7.47 Position 3.

Figure 7.48 Position 4.

Figure 7.49 Position 5.

Figure 7.50 Position 6.

The same trial and error procedure was carried out on a smaller sub-system, shown in Figure 7.51. The shaker was mounted, and tests were attempted until a suitable position was identified, as shown in Figure 7.51. The figure also shows the test structure with the shaker and the accelerometer. Two accelerometers were set up, opposite to the excitation direction (not visible in the picture) and one at a distance from the shaker. The test article is suspended by a canvas thread to a stiff metal heavy duty frame used for lifting weights. The canvas thread did not act as a vibration energy dissipator after several experimental verifications. As a matter of advice for inexperienced test practitioners, in general it is preferable to carry out nonlinear damping measurements with fishing wires, rather than canvas treads or elastic chords, because fishing wires work well as proven in [7.18].

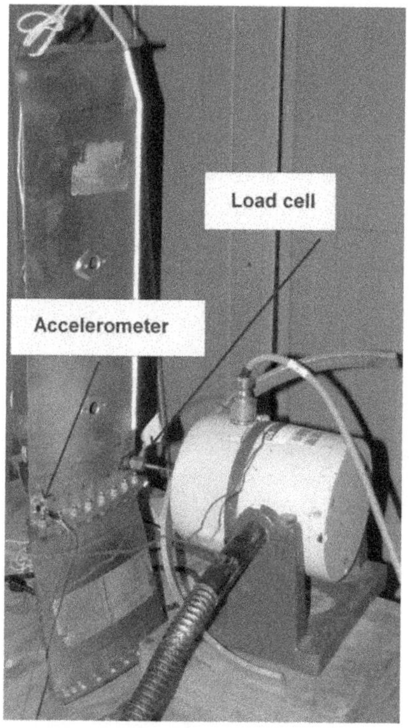

Figure 7.51 Excitation position (picture from ref. [7.25]).

One should stress again that the alignment between the shaker and the test article must be perfect, to avoid (i) pseudo-nonlinearities and (ii) more frustrating than all, incur detachment of the force sensor during tests. An exemplary set-up can be tedious and labour intensive, but the quality of the tests depends upon the correct installation. Having found an optimal position for the shaker tests, such that the excitation force generates nonlinear vibrations, the next phase is to perform nonlinear modal testing of both the whole and sector aero-engine casings.

7.6.5 Test Method

This section will discuss how the modal test is carried out on the two test structures for characterising both natural frequency and damping of the selected resonant modes. Over recent years, several new identification methods have been developed and reported in the scientific literature. This section describes a "simple" and "effective" method for characterising the nonlinear vibration response. It is described as "simple" because the method is based on linearising the frequency response. It is called "effective" because standard linear modal analysis identification tools can process the *FRF* test data. The linearisation of a frequency response function can be achieved if all the frequency points are measured at constant vibration response amplitude over the selected frequency range, as shown in Figure 7.52. An *FRF* measured as such can be analysed by linear modal analysis algorithms, such as the *Line-Fit* or Dobson's method [7.19]. This experimental approach does not require special test equipment, except for a *PID*[17] controller. In fact, the vibration test is carried at constant response amplitude and is achieved by the proportional controller designed in the acquisition system, which searches for the target amplitude defined between an upper and lower bound. The *FRF* is measured using a stepped-sine test. A downside of such a testing approach requires a shaker and the controller to work hard on a structure that must vibrate at a constant response amplitude over a range of frequencies. Think about this: near the resonance, the shaker needs to deliver a small amount of current for maintaining the vibration amplitude constant because the structure does vibrate by itself. However, aiming for large vibration amplitudes, the shaker needs more current to push the armature and thus the excitation point. Hence, the armature of the shaker is prone to over-travel near the resonance where the large vibration displacement is achieved. Controlling the vibration amplitude to be constant is a considerable challenge, as (i) away from the resonance one has to reach high amplitudes at frequencies where the structure does not vibrate much and (ii) at resonance both shaker and controller have to control the response by measuring very little force from the force transducer. It is a delicate balance that is very resonance-dependent, where practitioners might find execution difficulties.

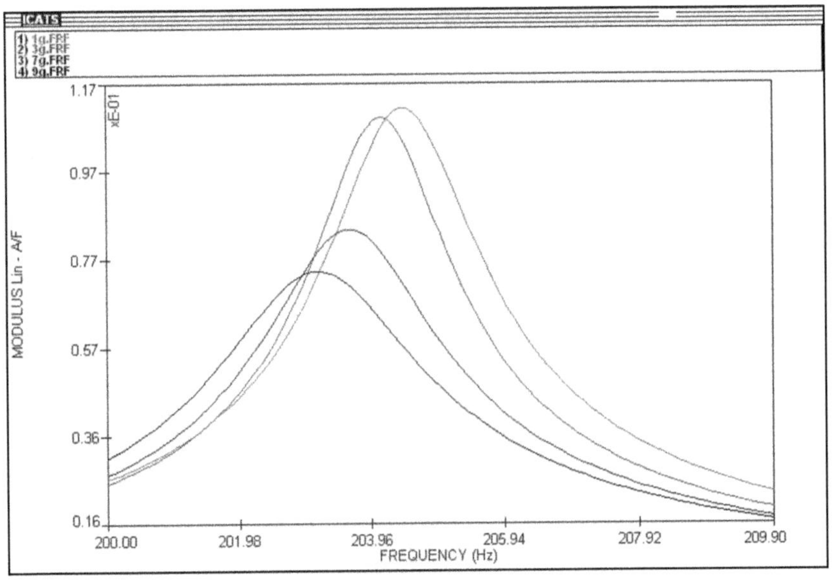

Figure 7.52 Example of linearised *FRFs*.

The correct selection of the frequency range can make testing less demanding. For instance, modal identification requires a small number of frequency lines. However, choosing fine frequency steps around the resonance and selecting a frequency range within the half-power bandwidth can make testing easier for the controller. In fact, it can step through frequency lines

---

[17] Proportional Integral Derivative (*PID*)

having to carry out smaller adjustments to achieve quicker convergence. Figure 7.52 shows an example of linearised *FRFs* measured at a different level of acceleration, which means that the acceleration amplitude stayed constant over the selected frequency range.

The literature does not report many examples of this technique, Göge [7.6] presents a *GVT* where some nonlinear testing was carried out. The primary motivation for performing the vibration test as described by Göge is the use of linear modal analysis tools to calculate the natural frequency and damping values of the *FRFs*. The measurement and analysis workflows can be achieved using commercially available equipment.

Despite the simplistic test described so far, the practitioner can yield very accessible results to structural dynamic analysts. Note that both nonlinear natural frequency and damping loss factor values are characterised by step-wise linear modal analysis carried out at various vibration amplitudes. Therefore, the more the number of levels, the better the nonlinear characterisation of the mechanical system is, although being time-consuming.

An alternative method, equivalent to the one so far described, but which does not require any type of *PID* controller for yielding linearised *FRFs*, is the one developed by Zang *et al.* [7.20 – 7.23]. The tests are carried out by the stepped-sine method, as done before. The frequency range selected for the investigation is swept as many times as the levels of the shaker drive voltage. For example, a hypothetical frequency range (as the one selected in the plot of Figure 7.52) can be swept ten times, each sweep using a different drive voltage to the shaker. Both acceleration and force spectra at the drive frequency are acquired for every shaker drive level, thus creating two surfaces, one for the acceleration and one for the force. The extraction of linearised *FRFs* from those surfaces consists of slicing the acceleration surface map at a constant acceleration level and searching the force values from the force surface. The linearised *FRF* is calculated by the ratio of pairs of acceleration and force values. Note that the process is equivalent when velocities are measured instead of accelerations. The exact process is repeated for the phase points. The method is robust and can be easily implemented with existing commercial testing software by saving the accelerations and forces for every shaker drive used within the selected frequency range. This method is a variant of the one called *complex FRF* analysis, which shall be discussed in Section 7.8 on the helicopter tail-rotor. A downside of these methods is the need to perform several *FRFs* measurements to obtain adequate natural frequency and damping functions, meaning that a single measurement is insufficient. The structural dynamics community is continuously developing much faster measurement and analysis techniques to retrieve nonlinear natural frequencies and damping from tests that are not always possible by stepped-sine testing.

7.6.6 Experimental Results

This section will review the experimental results obtained after selecting suitable resonances for the nonlinear testing by inspecting the homogeneity of the *FRFs* and the associated deflection shapes. The whole and sector engine casing were investigated primarily in terms of the damping loss factor because the natural frequencies did not show any relevant changes.

Full Assembly of the Aero-Engine Casing

The assembly did not present a strong frequency nonlinearity, opposite to a stronger damping nonlinearity caused by the bolted flanges. Figure 7.53 shows the natural frequencies obtained from *FRFs* measured between 0.1g and 5g for mode at 194.6 Hz and between 0.1g and 11g for mode at 204.7 Hz. Figure 7.54 shows the loss factor curves obtained from the *FRF* measurements.

The most significant frequency deviation is 1.6%, which could be considered aggravating or not, depending on the design choices. The damping behaviour shows a much more complex pattern for the vibration mode at 194 Hz than at 204 Hz. The former indicates that the vibration response would not be increasing with the excitation force, but follows a less predictable pattern and, therefore, more challenging for designers. For example, one could suggest that the integrity of the casing structure would be at stake because of this nonlinear behaviour. On the other hand, that behaviour might indicate an unpredictable force vibration response that might transfer vibration energy to smaller parts, like accessories mounted on the casing.

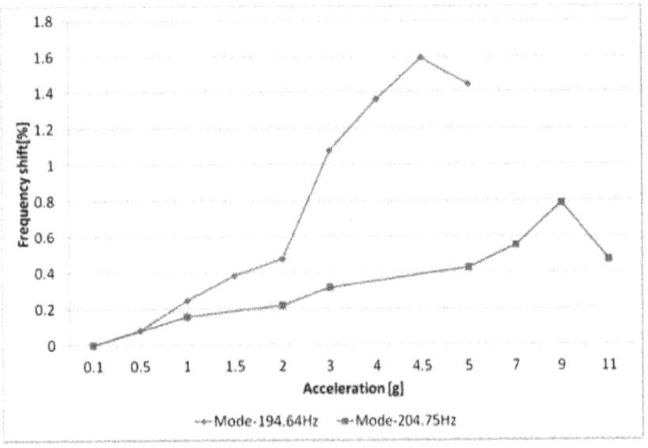

Figure 7.53 Natural frequency variation [%].

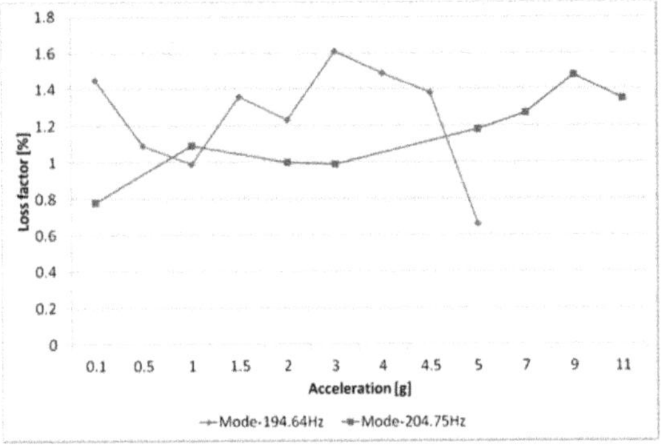

Figure 7.54 Damping Loss factor [%].

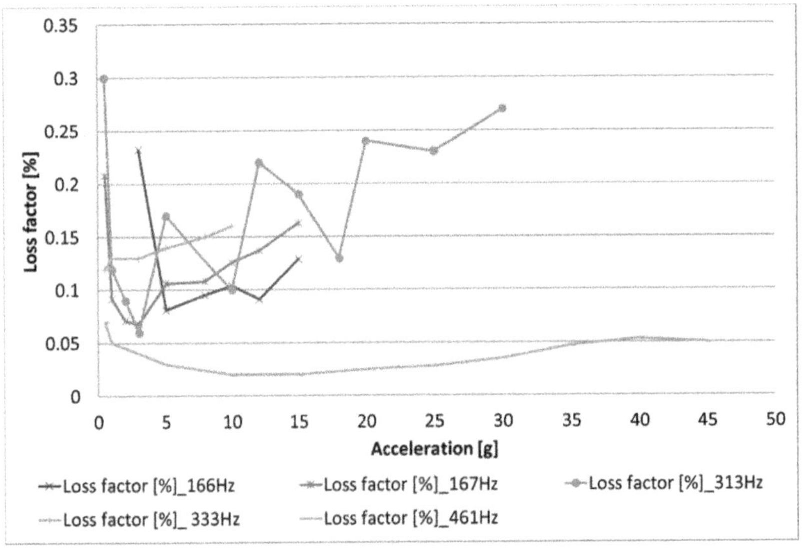

Figure 7.55 Damping factors identified from the sector flange (picture from ref. [7.25]).

Experimental Results of the Sector Flange

The frequency nonlinearity observed from the tests was negligible and not reported here. Figure 7.55 reports the damping factors for the mode shapes under investigation. Each mode shows an unusual damping behaviour over the vibration amplitudes selected for the experiments. These damping functions indicate that the contact conditions at the flange interface depend on the vibrations level, as the friction forces change. The challenge is (i) to identify and characterise such damping behaviours and (ii) to model the contact conditions and associated friction forces.

The following section, 7.6.7, presents an example of modelling such contact conditions to predict the modal damping behaviour of the test structure. More research details can be found in [7.24] and [7.25].

7.6.7 Validation of the *FE* Model

This section presents the model validation carried out on the *FE* model of the sector of the aero-engine component to predict correct responses under nonlinear vibration conditions. As introduced earlier, the entire assembly did not offer any chance of any model validation because of the computational effort involved. Therefore, this shorter part will offer a brief insight into how one can carry out model validation of a nonlinear system. Finally, the developed contact meshing tool and the suggested methods for nonlinear dynamic analysis and updating bolted flange joints presented in [7.26] and [7.27] have been applied to the aero-engine *CCOC/HPT* casing jointed sector to obtain a final validation of the suggested modelling approach.

Linear and Nonlinear Models

Based on the dimensions of the aero-engine casing, a linear *FE* model was provided (Figure 7.56 (a)). It contained a detailed and matching mesh of 20 node brick elements at the contact interface of the flange, but the casings themselves were only modelled to a moderate level of accuracy to slightly reduce the *FE* model size (3 million *DOFs*).

An initial linear model update with a rigid flange joint was carried out, resulting in a maximum frequency error of less than 5% for the first eight modes of the structure. Next, the nonlinear static stress distribution was calculated in *ABAQUS* (see Figure 7.56 (b)) using the nonlinear friction element (friction coefficient 0.66) with a penalty scheme. Thereby the preload of the bolt was applied with the thermal load, which led to similar strains to the one experienced with the bolt [7.24]. This approach allows for possible elastic deformations of the members and resulting unloading of the bolts into account. It leads to high stresses underneath the pressure cone of the bolts and relatively low compressive stress over the rest of the flange.

(a)

(b)

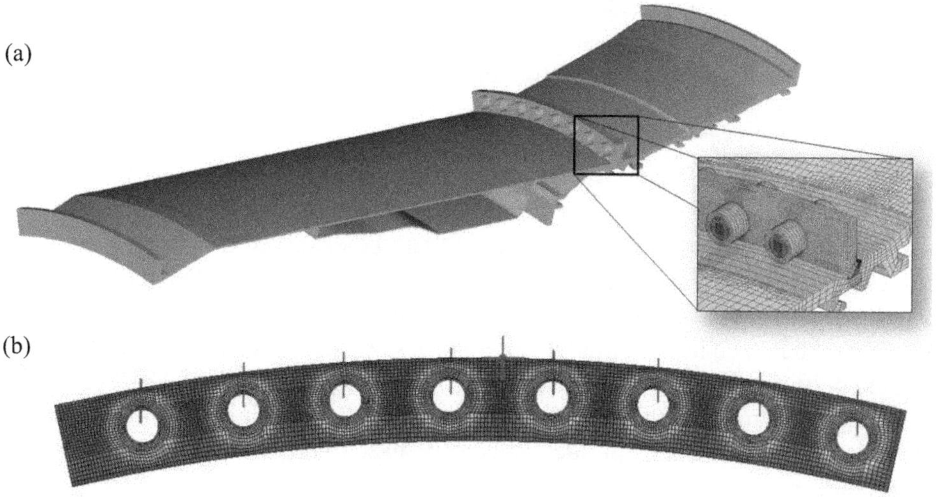

Figure 7.56 (a) Linear *FE* model of the aero-engine casing sector and (b) static normal load distribution (picture from ref. [7.25]).

Based on the linear finite element mesh and its static normal load distribution, nonlinear dynamic meshes were generated for analysis with the in-house code *FORSE* [7.27] and [7.28], which allows the calculation of the nonlinear response based on the multi-harmonic representation of the steady-state response and enables large scale realistic friction interface modelling. Full three-dimensional contact interface elements, developed in [7.29], have been used to model nonlinear interactions at the contact interfaces, allowing the representation of stick, slip, and contact separation. The interface parameters were based on measured data [7.30], leading to a friction coefficient $\mu = 0.66$, and a tangential and normal contact stiffness of 30×10^7 N/m, and the normal load of each element was obtained from the previous static analysis. An initial convergence study showed that the first two odd harmonics were enough for an accurate solution.

The nonlinear mesh was a relatively rough mesh, where each dot represents one of the 430 nonlinear elements. These were evenly spaced over the entire flange surface to provide adequate coverage of the contact interface and accurately capture the normal load distribution. The highest normal load occurred in the pressure cone area underneath the bolts. Therefore, a finer mesh with 709 nonlinear elements was incorporated. In addition, a mesh optimisation from [7.29] was used, which added more elements close to the casing-flange connection due to the previously identified higher energy dissipation in this contact zone.

The excitation point and measurement location of the nonlinear model was selected in accordance with the measurement set-up to allow a direct comparison of the results. Several nonlinear calculations were conducted for the two meshes with the nonlinear dynamic response code *FORSE*, where amplitude control was used at different excitation levels to ensure a good comparison with the measured data. The second torsional mode was the mode of interest since the measurement results indicated a nonlinear behaviour for this model.

Validated Numerical Results

The results for the nonlinear dynamic analysis of the aero-engine flange joint can be seen in Figure 7.57, where each response curve could be obtained in less than an hour due to its calculation in the frequency domain. The frequency response function at a wide range of excitation amplitudes for the model with 430 elements shows a clear nonlinear behaviour, with a drop in frequency and amplitude at higher force excitation levels. The predicted frequency value was within 5% of the measured response, which was considered adequate for this investigation that focused on the damping behaviour of the flange joint. No further updating of the nonlinear model was performed.

The predicted damping behaviour of the two nonlinear models at different response levels can be seen in Figure 7.58. Relatively low damping is present in the aero-engine casing sector up to 1g, after which a gradual increase in damping can be observed due to slip and separation in the contact. Most of the damping originates from the slip in the lower part of the flange. In contrast, the area around the bolts is stuck throughout the vibration cycle. There is a generally good agreement between the two models, more or less enveloping the measured data comparing the damping predictions to the measured damping behaviour in Figure 7.58. It can be concluded that the suggested modelling approach is capable of capturing the underlying nonlinear mechanism of a realistic flange interface and can be used to predict the forced dynamic response of bolted flange joints.

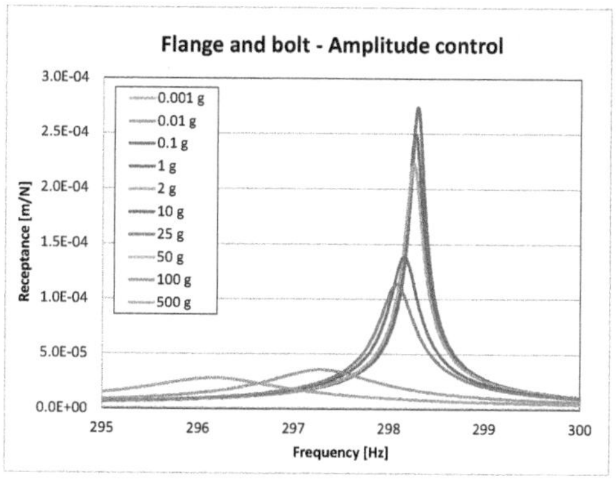

Figure 7.57 The frequency response function for 430 elements at different excitation levels (picture from ref. [7.25]).

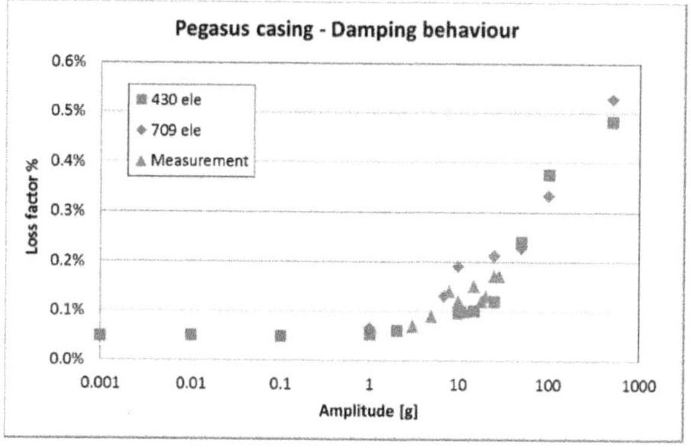

Figure 7.58 Change in damping at different response levels (picture from ref. [7.25]).

7.7 Case study 4: Experimental Nonlinear Modal Analysis of the Tail Drive Shaft System of a Helicopter

This section presents a test case about a tail transmission shaft of a helicopter. It is a complex system connected by multiple segments of the transmission shaft, multiple couplings, and multiple bearing supports. The transmission shaft adopts a thin-wall structure and is subjected to large loads and high rotational speeds; thus, it is one of the parts with a high helicopter failure rate. Furthermore, from the various faults of the tail transmission shaft system, the faults due to vibration occupy a large proportion of failure cases. The tail transmission shaft vibration may cause, in turn, a significant vibration of the engine and transmission system, affecting the regular operation of the system.

A parameter identification method based on the vibration test data is proposed to evaluate the nonlinear stiffness of a tail drive shaft system. Firstly, modal testing on the tail drive shaft system under low-amplitude excitation was carried out, and a simplified dynamic model of the tail drive shaft system was established, which was verified by testing results. Then, the stepping sine sweep testing under different excitation levels was conducted, and the relationship between natural frequency and displacement amplitude was obtained. The nonlinear modal parameters were identified by a method called *Complex FRF Analysis* [7.31]. Moreover, the finite element iterative calculation of the simplified model was carried out to achieve the relationship between the natural frequency and the equivalent stiffness of the tail drive shaft system. Finally, the relationship between the equivalent stiffness and the displacement amplitude was established, and the nonlinear stiffness of the tail drive shaft system was identified.

7.7.1 The Identification Process of Nonlinear Stiffness Parameters of the Tail Drive Shaft System

The identification process of the nonlinear stiffness parameters of the helicopter tail drive shaft system is shown in Figure 7.59, mainly including the following three steps:

(1) Linear model confirmation: conduct a mode test at a low vibration level on the tail transmission shaft system (the aim is to minimise the impact of the nonlinear behaviour, establish a simplified *FE* model of the axial system and modify the *FE* model based on the test data to obtain an updated model;

(2) Nonlinear modal identification: detect through the fast sine sweep test the presence of nonlinearity, then conduct the stepped sine sweep test at different vibration force levels and obtain the relationship between the natural frequency of the shaft at any measured displacement amplitude based on the nonlinear modal analysis method;

(3) Nonlinear stiffness parameter identification: mainly consider the nonlinear connection stiffness introduced by the diaphragm coupling connection, based on the *FE* model obtained in point (1), modify the connection stiffness, combined with the nonlinear modal parameters identified in point (2), and identify the nonlinear stiffness parameter at the axial connection, that is, the changing relationship of the connection stiffness with the displacement amplitude.

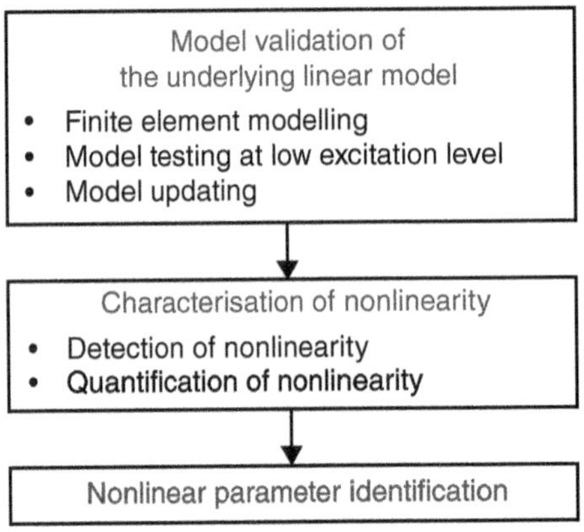

Figure 7.59 Three steps for identifying nonlinear stiffness parameters of a tail drive shaft system.

7.7.2 Identification Method of the Nonlinear Stiffness Parameters

Section 7.6.5 introduced a method for identifying the nonlinear natural frequency and damping by controlling the amplitude of the vibration response over a selected frequency range. That *SISO* method works as much with isolated as close resonance peaks. This section introduces another *SISO* method, called *Complex FRF Analysis* [7.31, 7.32], based on an *SDOF* vibration response. In practice, the *Complex FRF Analysis* method can be used when a resonance peak is well isolated, assuming that the modal contributions of the neighbour modes can be considered negligible. In all other cases, the practitioner must pay careful attention to the interpretation of the analysis results. This section explains how the *SISO* analysis method works and how the result data can be explored.

For a generic *SDOF* nonlinear system subjected to simple harmonic forces, the differential equation of motion is:

$$m\ddot{x} + id_{eq}(X)x + k_{eq}(X)x = F_0 sin(\omega t) \tag{7.1}$$

where $k_{eq}(X)$ and $d_{eq}(X)$ are the equivalent stiffness and the equivalent structural (hysteretic) damping of the system, respectively (both function of the displacement amplitude of the response, X). The receptance *FRF* is expressed as follows:

$$H(\omega)=\frac{X(\omega)}{F(\omega)}=\frac{A_r(X)+iB_r(X)}{\omega_r^2(X)-\omega^2+i\eta_r(X)\omega_r^2(X)} \tag{7.2}$$

where $A_r(X)$ and $B_r(X)$ are the real and imaginary parts of the modal constant as a function of the vibration displacement, and $\omega_r(X)$ and $\eta_r(X)$ are the natural frequency and hysteretic damping factor (loss factor), respectively, also functions of the vibration displacement. At any given modulus (see the boxed number in Figure 7.60) of the receptance, one can identify two frequencies $\omega_1 = 2\pi f_1$ and $\omega_2 = 2\pi f_2$, as shown in Figure 7.60. A pair of symmetrical points on either side of the resonance peak with the same amplitude cannot be found due to the sampling frequency steps, so an interpolation needs to establish a series of corresponding points.

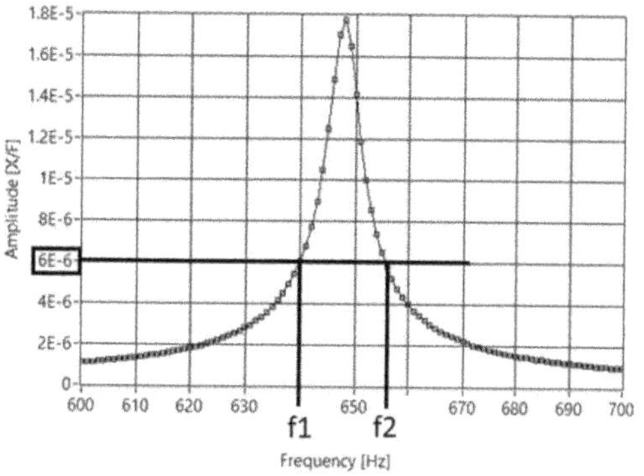

Figure 7.60 Evaluation of modal properties at equal response amplitudes.

After completing the interpolation, at each given amplitude modulus, the real and imaginary parts can be defined by a pair of symmetric displacement *FRF* points [7.31]:

$$\begin{cases} H(\omega_1)=\dfrac{A_r+iB_r}{\omega_r^2-\omega_1^2+i\eta_r\omega_r^2}=R_1+iI_1 \\ H(\omega_2)=\dfrac{A_r+iB_r}{\omega_r^2-\omega_2^2+i\eta_r\omega_r^2}=R_2+iI_2 \end{cases} \tag{7.3}$$

By equating the modulus of the right and left receptance *FRF*, one obtains the following relationship:

$$|H(\omega_1)|=\sqrt{R_1^2+I_1^2}=\sqrt{R_2^2+I_2^2}=|H(\omega_2)| \tag{7.4}$$

which can be solved both for natural frequency and damping factor at every displacement level.

Basically, for every pair of frequency points taken at the same response amplitude, one extracts the natural frequency and damping of the linear equivalent *SDOF* system. Therefore, at each excitation level, the correspondence between the natural frequency, displacement amplitude, and excitation force amplitude can be established.

$$\omega_r = f(X, F) \tag{7.5}$$

At each set of excitation forces, the natural frequency varies very little, and the intermediate value can be taken as the approximation. Then, the formula can be further reduced to

$$\omega_r = f(X) \tag{7.6}$$

The nonlinear underlying linear *FE* model of the system is established, and by changing the stiffness parameters, the frequency changes with stiffness can be established.

$$\omega_r = g(k) \tag{7.7}$$

By joining Eqs. (7.6) and (7.7), through spline value polynomials, one can finally establish the change of stiffness with the displacement amplitude, the functional characterisation form of the nonlinear stiffness parameters, as

$$k = h(X) \tag{7.8}$$

7.7.3 Validation of the Linear Model of the Tail Drive Shaft System

The modal test of the tail drive shaft system is shown in Figure 7.61. To simulate free-free boundary conditions, the two tail drive shafts are connected by a diaphragm coupling and are suspended by flexible ropes at three positions (both ends and the middle of the shaft system). The laser measurement method is adopted. After the two shafts are connected, there are a total of 32 measuring points, and the photosensitive paper is pasted on the 32 measuring points in the horizontal direction. Place the exciter on the suspension device that can be adjusted forward, backward, left, and right and connected with the pushrod. The tail drive shaft system is excited in the horizontal direction.

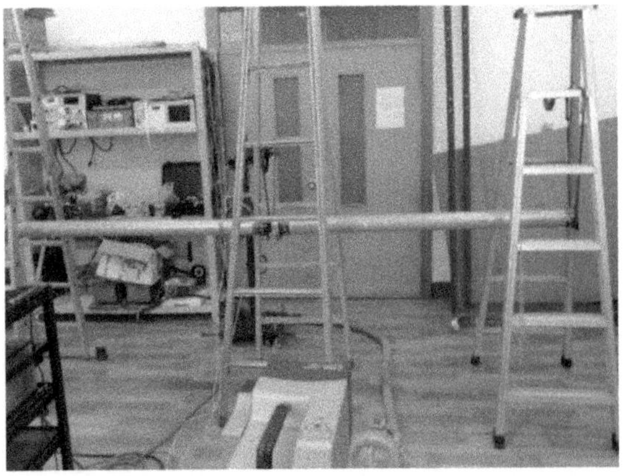

Figure 7.61 Modal test of the tail drive shaft system.

The exciter uses a fast frequency sweep signal to excite the structure, the frequency range is 0-500 Hz, and the tested number of spectrum lines is 6400. The excitation is at a position close to the middle. The laser sequentially records the speed response of the 32 measuring points to obtain the mobility *FRF*, whose amplitude is shown in Figure 7.62. The modal analysis software is used to analyse the mobility *FRF* of the tail drive shaft system. As a result, its natural frequencies and damping ratios can be obtained, as shown in Table 7.11. In the range of 0-500 Hz there are four resonant modes.

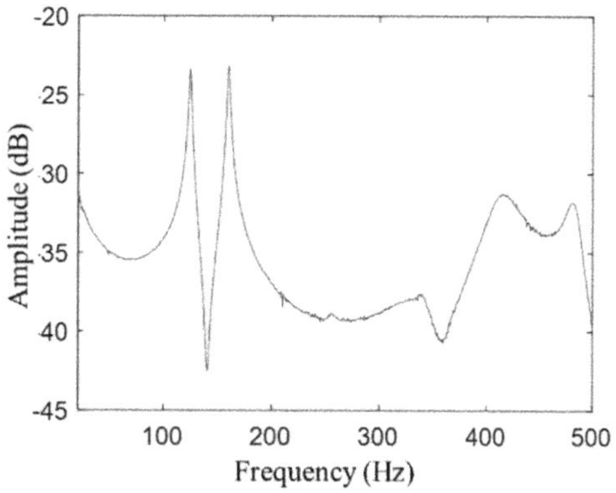

Figure 7.62 Mobility *FRF* (modulus).

Table 7.11 Modal analysis results of the tail drive shaft system.

Mode	Frequency [Hz]	Damping [%]
1	124.25	1.20
2	159.74	0.81
3	414.39	3.20
4	482.16	1.24

A simplified model of the single shaft is shown in Figure 7.63. The single shaft is divided into 15 segments and described by 16 nodes, and the software *ANSYS* is used. The *Beam188* element is used between the first 15 nodes to define the shaft body.

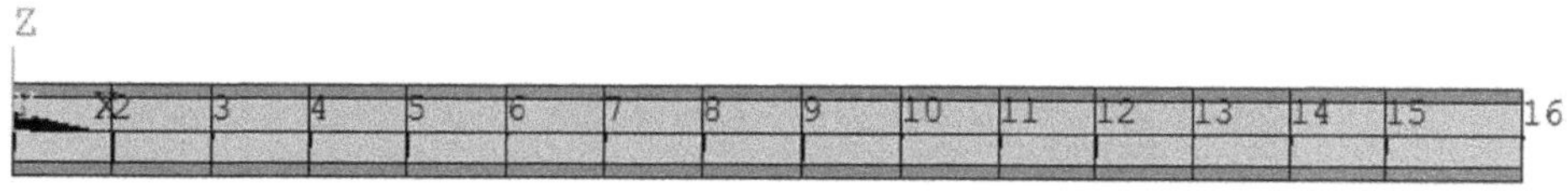

Figure 7.63 The simplified *FE* model of a single shaft.

The two shafts are joined together by a diaphragm, the mass of which, set as *Mass21* in *ANSYS*, is dived in two parts, one associated with node 16 of the first shaft and one associated with node 17 of the second shaft. The established tail drive shaft system only considers the bending mode of the shafting, not the torsional mode, so only the axial and radial rigidities of the intermediate connection need to be added. The simplified construction of the diaphragm coupling the model is shown in Figure 7.64. Three spring units are established between the two mass units of the diaphragm coupling to simulate the axial stiffness along *X*, and the radial stiffness along the *Y* and *Z* directions.

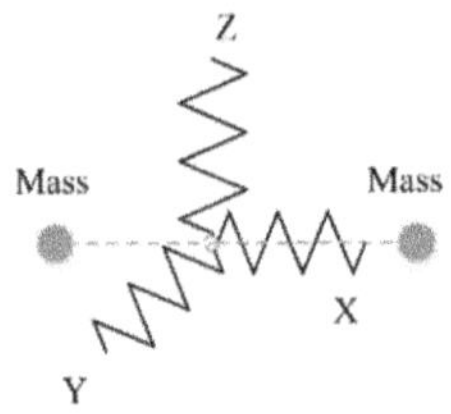

Figure 7.64 Schematic diagram of the diaphragm coupling.

Taking the axial stiffness of the diaphragm coupling along the *X* direction as a variable, after updating the parameters, the mode shapes are plotted against the axial stiffness of the connection, as shown in Figure 7.65. Similarly, the radial stiffness of the diaphragm coupling along the *Y* direction is used as a variable, and the mode shapes are plotted against the radial stiffness of the connection, as shown in Figure 7.66.

It can be seen from Figure 7.66 that with the change of the axial stiffness parameter, the simulation frequency of each order of the simplified model always remains unchanged, indicating that the axial stiffness never affects the bending mode frequency of the shaft system.

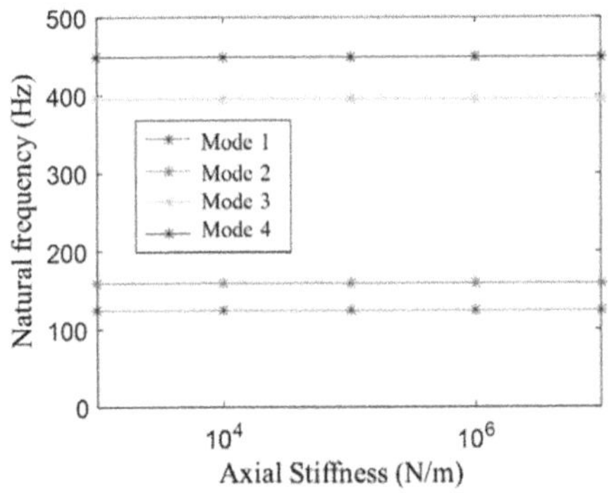

Figure 7.65 Mode shapes relationships with axial stiffness.

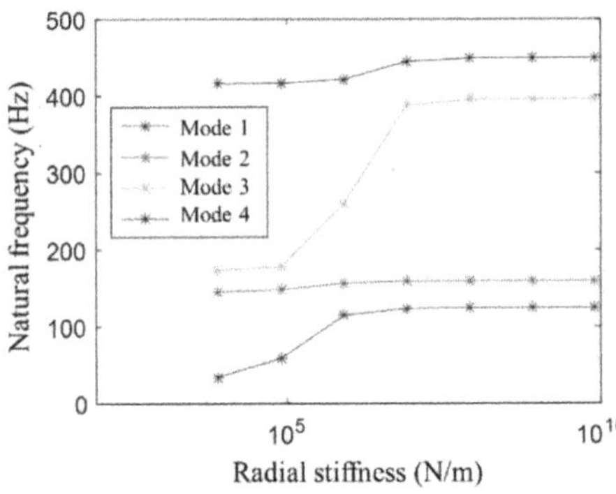

Figure 7.66 Mode shapes relationships with radial stiffness.

From Figure 7.66, as the radial stiffness parameter changes, the frequency of each order of the simplified model increases up to stiffness values of around 107 N/m, but beyond that the natural frequency stabilises. This section draws on the specific values of the literature for the axial and radial stiffness of the diaphragm coupling . The simplified model of the intermediate diaphragm coupling has the axial stiffness along the *X* direction and the radial stiffness along the *Y* and *Z* directions. The axial stiffness is selected to compare the obtained finite element simulation results and the test results in Table 7.12.

Table 7.12 Frequency comparison between test and simulation of the tail drive shaft.

Mode	Test [Hz]	Simulation [Hz]	|Error| (%)
1	124.25	124.41	0.13
2	159.74	159.29	0.28
3	414.39	395.60	4.53
4	482.16	449.08	6.86

As shown in Table 7.12, the frequency error of the first two steps does not exceed 0.3%, and the maximum frequency error is less than 7%. The simplified model of the tail drive shaft system established at this time is accurate enough, and the actual shaft system can be simulated by the established finite element simplified model, such as the dynamic characteristics of bending vibration.

7.7.4 Nonlinear Identification of the Tail Drive Shaft System

This section shows how the identification method is carried out. First of all, one should note that the frequency separation between the two modes is about 30 Hz, which is probably a borderline case. Nevertheless, it is worth attempting the procedure. In a stepped sine sweep frequency test under different excitation force levels for the tail drive shaft system, the frequency range is from 100 Hz to 150 Hz, and the frequency step is 0.25 Hz. The excitation force is again *PID* controlled to achieve a constant force amplitude across the frequency range. However, the test procedure can be even carried out without controlling the excitation force, but it is fundamental to record the force signal for every test. Figure 7.67 shows the *FRFs* measured at every force level.

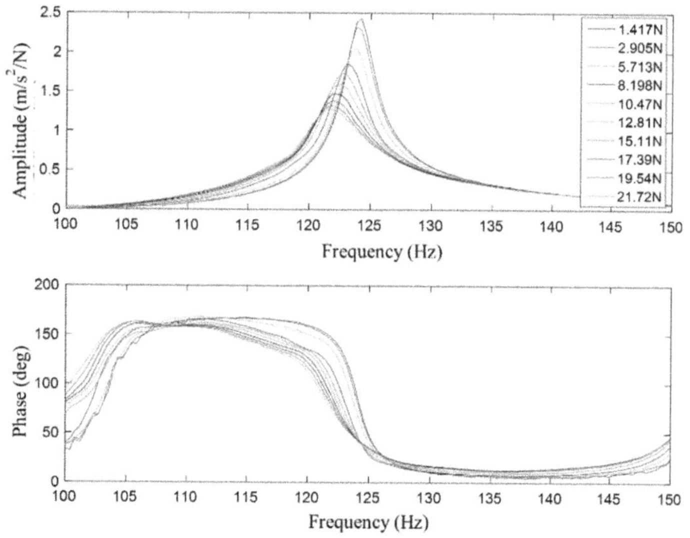

Figure 7.67 Accelerance *FRFs* under various forcing levels.

The *FRFs* thus measured are converted into receptance *FRFs*, with forces of different excitation levels. On this basis, the relationship curve between natural frequency and displacement amplitude at each excitation level is established, as shown in Figure 7.68.

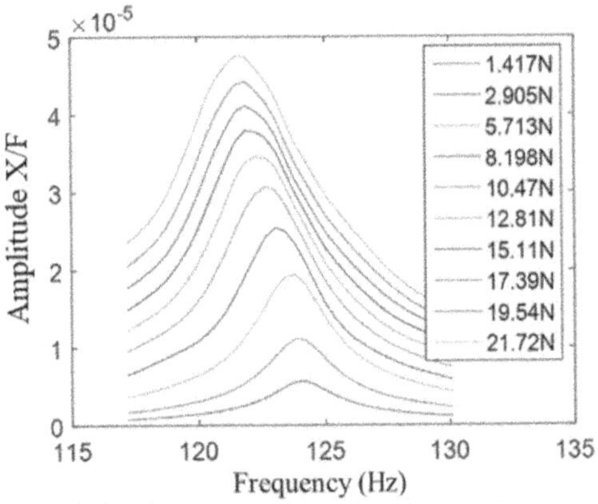

Figure 7.68 Displacement *FRFs* as a function of the forcing levels.

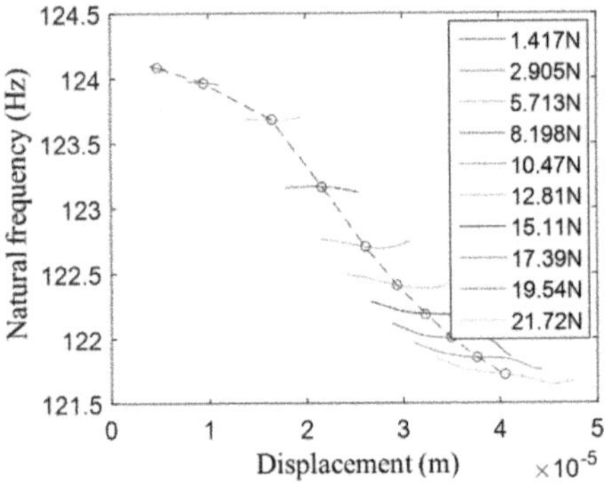

Figure 7.69 Natural frequency as a function of displacement amplitude.

In Figure 7.69, the change in natural frequency is negligible for each excitation force level, and the circle in the middle represents the average natural frequency value. With the increase of the excitation force level, the amplitude of the displacement response gradually increases, and the natural frequency decreases by about 2%, showing the characteristic of gradually softening stiffness. The polynomial fitting of the natural frequency under different excitation levels can obtain the relationship between the natural frequency and the displacement amplitude, as shown in Figure 7.70. The function of the change of the natural frequency of the tail drive shaft system with the displacement amplitude is as follows:

$$\omega_n = 9.71\times10^{13} X^3 - 7.14\times10^9 X^2 + 7.78\times10^4 X + 123.84 \tag{7.9}$$

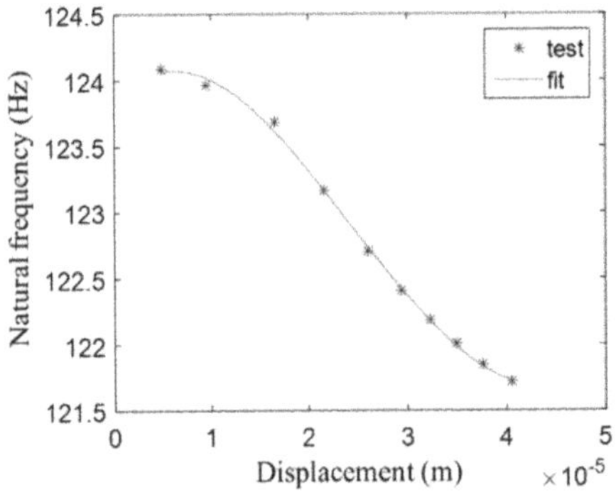

Figure 7.70 Curve-Fitting of natural frequencies.

The final challenge is the identification of nonlinear stiffness parameters of the tail drive shaft system. Based on the finite element simplified model of propeller shafting obtained in Section 0, by modifying the radial stiffness along the horizontal direction (*Y* direction) of the middle connection of the shaft. In *ANSYS*, the spring stiffness in the horizontal direction is successively used as the joint of the finite element simplified model of the shaft, and the first-order simulation frequency is obtained through modal analysis, to establish the relationship between the horizontal equivalent stiffness of the joint of the tail drive shafting and the first-order natural frequency, as shown in Figure 7.71.

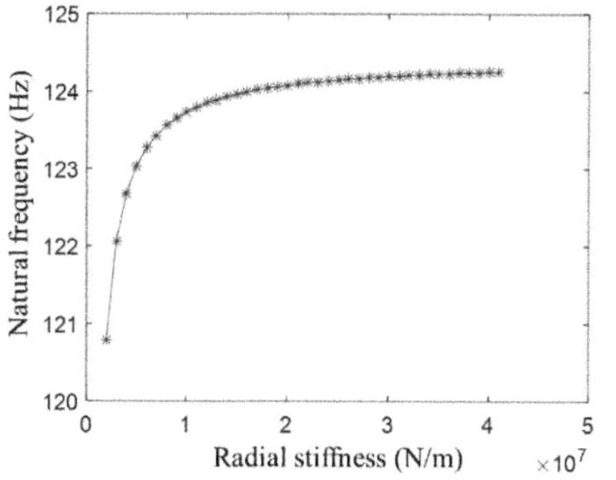

Figure 7.71 Natural frequency as a function of the radial stiffness.

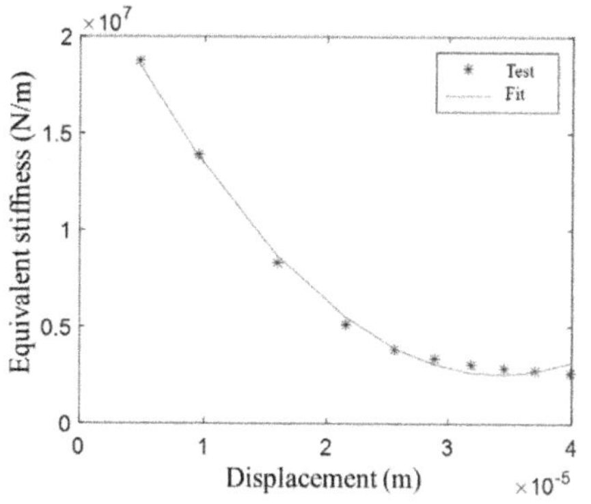

Figure 7.72 Nonlinear stiffness as a function of the displacement amplitude.

In combination with the first-order natural frequency variation with the displacement amplitude obtained from the nonlinear modal test, the relationship between the nonlinear stiffness at the connection of the tail drive shaft system and the displacement amplitude can be established through a polynomial fitting interpolation, i.e. the representation form of the nonlinear stiffness function at the connection. The curve-fitting result is shown in Figure 7.72.

As shown in Figure 7.72, with the increase in displacement amplitude, the nonlinear stiffness of the propeller shaft system decreases at first and then tends to remain constant. By polynomial fitting, the nonlinear equivalent stiffness along the horizontal direction of the connection of the tail drive shafting can be obtained:

$$k_{eq} = 1.86\times10^{16} X^2 - 1.27\times10^{12} X + 2.43\times10^7 \tag{7.10}$$

Due to the variety of nonlinearities of actual structures, it is almost impossible to use one expression to achieve all nonlinearities. One can use polynomials for curve-fitting and choose the appropriate polynomial number according to the fitting situation. If the function expression of nonlinear stiffness is established theoretically, it can be used for curve fitting. Of course, due to the limited excitation level in the test, the function expression of nonlinear stiffness fitting has its own applicable scope. In order to obtain the nonlinear stiffness expression at a higher displacement level, a vibration test with a higher excitation level is needed.

This case study integrates modal testing technology, finite element modelling theory and a nonlinear modal analysis method; it identifies nonlinear stiffness parameters of the shaft joints through standard vibration testing, providing a reference for future engineering applications. The main aspects of drawing out from this case study are as follows:

(1) The study consisted of three steps, namely, linear model confirmation, nonlinear modal identification and nonlinear stiffness parameter identification;

(2) The finite element simplified model of the propeller shaft was established, and the error between the first two orders and the test frequency was within 0.3%, and the maximum error was less than 7%;

(3) The nonlinear characteristics of the propeller shaft are accurately identified. With the increase of the excitation force, the first-order natural frequency decreased by about 2%;

(4) With the increase of the excitation force, the radial stiffness of the nonlinear connection of the propeller shaft decreased gradually at first and then tended to be stable.

7.8 Final Notes

This final chapter produced four industrial test cases showing an increasing challenge in the experimental model validation. The model validation under linear vibration responses is mature and well-established in the fabric of the product design of mechanical/aerospace assets. However, for the years to come, the major challenge is represented by the many joint interfaces, each presenting a different degree of modelling complexity. Furthermore, the simulations of forced vibrations of assemblies strongly depend on the future tools which must be developed for correcting dynamic models, both in terms of stiffness and damping, since they are the two mechanical properties sensitive to vibration amplitudes.

References

[7.1] Ciavarella, C., Priems, M., Govers, Y., Böswald, M., *An extensive Helicopter Ground Vibration Test: from Pretest Analysis to the Study of Non-linearities*, 44th European Rotorcraft Forum 2018, ERF 2018.

[7.2] Kammer, D. C., Tinker, M. L., *Optimal Placement of Triaxial Accelerometers for Modal Vibration Tests*. Mechanical systems and signal processing, 18(1), pp.29-41, 2004.

[7.3] Peeters, B., Carrella, A., Lau, J., Gatto, M., Coppotelli, G., *Advanced Shaker Excitation Signals for Aerospace Testing*, Advanced Aerospace Applications, Vol. 1, pp. 229-241, Springer, New York, 2011.

[7.4] Allemang, R. J., *The Modal Assurance Criterion–Twenty Years of Use and Abuse*, Sound and vibration, 37(8), pp. 14-23, 2003.

[7.5] Castellini, P., Martarelli, M., Tomasini, E. P., *Laser Doppler Vibrometry: Development of Advanced Solutions Answering to Technology's Needs*, Mechanical Systems and Signal Processing, 20(6), pp. 1265-1285, 2006.

[7.6] Rothberg, S. J., Allen, M. S., Castellini, P., Di Maio, D., Dirckx, J. J. J., Ewins, D. J., Halkon, B. J., Muyshondt, P., Paone, N., Ryan, T., Steger, H., *An international Review of Laser Doppler Vibrometry: Making Light Work of Vibration Measurement*, Optics and Lasers in Engineering, 99, pp. 11-22, 2017.

[7.7] Tomasini, E. P., Castellini, P., *Laser Doppler Vibrometry*, Springer, 2020.

[7.8] Tomasini, E. P., Castellini, P., *Laser Doppler Vibrometry: A Multimedia Guide to its Features and Usage*, Springer Nature, 2020.

[7.9] Di Maio, D., Castellini, P., Martarelli, M., Rothberg, S., Allen, M. S., Zhu, W. D., Ewins, D. J., *Continuous Scanning Laser Vibrometry: A Raison d'être and Applications to Vibration Measurements*, Mechanical Systems and Signal Processing, Vol. 156, 107573, 2021.

[7.10] Pappa, R. S., Elliott, K. B., Schenk, A., *Consistent-mode indicator for the eigensystem realization algorithm*, Journal of Guidance, Control, and Dynamics, 16(5), pp.852-858.

[7.11] Brake, M. R. (Ed.), *The Mechanics of Jointed Structures: Recent Research and Open Challenges for Developing Predictive Models for Structural Dynamics*, Springer, 2017.

[7.12] Göge, D., Sinapius, M., Füllekrug, U., Link, M., *Detection and Description of Non-linear Phenomena in Experimental Modal Analysis Via Linearity Plots*, International Journal of Non-Linear Mechanics, 40(1), pp. 27-48, 2005.

[7.13] Göge, D., Böswald, M., Füllekrug, U., Lubrina, P., *Ground Vibration Testing of Large Aircraft–State-of-the-Art and Future Perspectives*, Proceedings of the 25th International Modal Analysis Conference (IMAC), 2007.

[7.14] Goege, D., Füllekrug, U., Sinapius, M., Link, M., Gaul, L., *Intl–A Strategy for the Identification and Characterization of Non-linearities within Modal Survey Testing*, Proceedings of the 22nd International Modal Analysis Conference (IMAC), 2004.

[7.15] Kerschen, G., Peeters, M., Golinval, J. C., Stéphan, C., *Nonlinear Modal Analysis of a Full-Scale Aircraft*, Journal of Aircraft, 50(5), pp. 1409-1419, 2013.

[7.16] Chen, G., *Vibration Modelling and Verifications for whole Aero-Engine*, Journal of Sound and Vibration, 349, pp.163-176, 2015.

[7.17] Worden, K., Tomlinson, G. R., *Nonlinearity in Structural Dynamics: Detection, Identification and Modelling*, CRC Press, 2019.

[7.18] Ruffini, V., Schwingshackl, C. W., Green, J. S., *LDV Measurement of Local Nonlinear Contact Conditions of Flange Joint*, Topics in Nonlinear Dynamics, Vol. 1, pp. 159-168. Springer, New York, 2013.

[7.19] Dobson, B. J., *A Straight-Line Technique for Extracting Modal Properties from Frequency Response Data*, Mechanical Systems and Signal Processing, 1(1), pp. 29-40, 1987.

[7.20] Zhang, G., Zang, C., Friswell, M. I., *Parameter Identification of a Strongly Nonlinear Rotor-Bearing System based on Reconstructed Constant Response Tests*, Journal of Engineering for Gas Turbines and Power, 142(8), 2020.

[7.21] Zhang, G., Zang, C., Friswell, M. I., *Identification of Weak Nonlinearities in MDOF Systems Based on Reconstructed Constant Response Tests*, Archive of Applied Mechanics, 89(10), pp. 2053-2074, 2019.

[7.22] Zhang, G., Zang, C., Friswell, M. I., *Measurement of Multivalued Response Curves of a Strongly Nonlinear System by exploiting Exciter Dynamics*, Mechanical Systems and Signal Processing, Vol. 140, 106474, 2020.

[7.23] Zhang, G., Zang, C., Friswell, M. I., *Measurement of the Multivalued Phase Curves of a Strongly Nonlinear System by Fixed Frequency Tests*, Archive of Applied Mechanics, 90(11), pp. 2543-2560, 2020.

[7.24] Schwingshackl, C. W., Di Maio, D., Sever, I., Green, J. S., *Modeling and Validation of the Nonlinear Dynamic Behavior of Bolted Flange Joints*, Journal of Engineering for Gas Turbines and Power, 135(12), 2013.

[7.25] Di Maio, D., Schwingshackl, C., Sever, I. A., *Development of a Test Planning Methodology for performing Experimental Model Validation of Bolted Flanges*, Nonlinear Dynamics, 83(1), pp. 983-1002, 2016.

[7.26] Petrov, E. P., Ewins, D. J., *Generic Friction Models for Time-Domain Vibration Analysis of Bladed Disks*, J. Turbomach., 126(1), pp.184-192, 2004.

[7.27] Petrov, E. P., *A High-Accuracy Model Reduction for Analysis of Nonlinear Vibrations in Structures with Contact Interfaces*, Journal of Engineering for Gas Turbines and Power, 133(10), 2011.

[7.28] Petrov, E. P., Ewins, D. J., *Analytical Formulation of Friction Interface Elements for Analysis of Nonlinear Multi-Harmonic Vibrations of Bladed Disks*. J. Turbomach., 125(2), pp. 364-371, 2003.

[7.29] Schwingshackl, C. W., Petrov, E. P., Ewins, D. J., *Measured and Estimated Friction Interface Parameters in a Nonlinear Dynamic Analysis*, Mechanical Systems and Signal Processing, 28, pp. 574-584, 2012.

[7.30] Schwingshackl, C. W., Petrov, E. P., *Modeling of Flange Joints for the Nonlinear Dynamic Analysis of Gas Turbine Engine Casings*, Journal of Engineering for Gas Turbines and Power, 134(12), 2012.

[7.31] Carrella, A., Ewins, D. J., *Identifying and Quantifying Structural Nonlinearities in Engineering Applications from Measured Frequency Response Functions*, Mechanical Systems and Signal Processing, 25(3), pp. 1011-1027, 2011.

[7.32] Ewins, D. J., *Modal Testing, Theory, Practice and Application, 2nd Ed.*, Baldock, England, 2000.

Appendix A *MATLAB* Codes (Chapter 3)

Code 1: Central Difference Method (Section 3.2.1)

```
% Main function of central difference method
% Set the parameters of m, c and k
m = 1;
k = 10;
c = 2;
% Set the excitation and initial parameters
deltat = 0.01;            % Increment of the time
t = 0.01:deltat:20;       % Time vector
w = 3*2*pi;               % Excitation frequency = 3*2*pi [rad/s]
fex = 20*cos(w*t);        % Excitation amplitude = 20;
x = zeros(length(t),1);
vini = 2;
x(1) = 0;
x(2) = x(1)+vini*deltat;  % Start with the first order Taylor assumption
% Set the central difference

for(i = 3:length(t))
   x(i) = (2*deltat^2*fex(i-1)-(2*k*deltat^2-4*m)*x(i-1)-(2*m-c*deltat)*x(i-
   2))/(2*m+deltat*c);
end

% Plot result
figure
plot(t,x)
set(gca,'FontSize',14)
xlabel('Time [s]');
ylabel('Amplitude [m]');
box on
grid on
xlim([0,8])
```

Code 2: Runge-Kutta's Method (Section 3.2.2)

```
% Main code of Runge-Kutta
% Set calculation parameter
ti = 0;
tf = 20;
```

```
ui = [0;2];
% Set the solver parameter
tol = 1.0E-4;trace = 1;
% Perform the ode solver task
 [t_result,u_result] = ode45('SDOF_forced_vibration_system',[ti,tf],ui,tol,trace);
% Plot the result
figure
plot(t_result,u_result(:,1))
set(gca,'FontSize',14)
xlabel('Time [s]');
ylabel('Amplitude [m]');
xlim([0,8])
box on
grid on

function udot = SDOF_forced_vibration_system(t,u)
   % Parameter of force setting
   m = 1;
   k = 10;
   c = 2;
   w = 3*2*pi;                  % Excitation frequency = 3*2*pi [rad/s]
   fex = 20*cos(w*t);           % Excitation amplitude = 20;
   % Set the test function
   udot = zeros(2,1);
   udot(1) = u(2);
   udot(2) = (fex-c*u(2)-k*u(1))/m;
end
```

Code 3: Houbolt's Method (Section 3.2.3)

```
clear all;close all;clc;
% Set the parameters of m, c and k
m = 1;
k = 10;
c = 2;
% Set the excitation parameters
deltat = 0.001;              % Increment of the time
t = 0:deltat:20;             % Time vector
w = 3*2*pi;                  % Excitation frequency is 3*2*pi [rad/s]
f0 = 20;
f_ex = f0*cos(w*t);          % Excitation amplitude is 20
% Set initial parameters
x = zeros(length(t),1);
x(1) = 0;                    % Initial displacement is 0
v_ini = 2;                   % Initial velocity is 2
% Use the central difference method to find x1 and x2
```

```
x(2) = x(1)+v_ini*deltat;
x(3) = (2*deltat^2*f_ex(3-1)-(2*k*deltat^2-4*m)*x(3-1)-(2*m-c*deltat)*x(3-
2))/(2*m+deltat*c);
% Calculate response by Houbolt's method

for i = 4:length(t)
  x(i) = (f_ex(i)+(5*m/deltat^2+...
      3*c/deltat)*x(i-1)-...
      (4*m/deltat^2+3*c/deltat/2)*x(i-2)+...
      (m/deltat^2+c/deltat/3)*x(i-3))/...
      (2*m/deltat^2+11*c/deltat/6+k);
end

% Plot result
figure
plot(t,x);
set(gca,'FontSize',14)
xlabel('Time [s]');
ylabel('Amplitude [m]');
box on
grid on
xlim([0,8])
```

Code 4: Wilson's Method (Section 3.2.4)

```
clear all;close all;clc;
% Set the parameters of m, c and k
m = 1;
k = 10;
c = 2;
% Set the excitation parameters
deltat = 0.001;                 % Increment of the time
t = 0:deltat:20;                % Time vector
w = 3*2*pi;                     % Excitation frequency is 3*2*pi [rad/s]
f0 = 20;
f_ex = f0*cos(w*t);             % Excitation amplitude is 20
% Set initial parameters
x = zeros(length(t),1);
x(1) = 0;                       % Initial displacement is 0
v = zeros(length(t),1);
v(1) = 2;                       % Initial velocity is 2
a = zeros(length(t),1);
a(1) = (f_ex(1)-c*v(1)-k*x(1)); % Use equation of motion at initial time to find initial
                                  acceleration
% Calculate response by Wilson's method
theta = 1.4;                    % theta is 1.4(>1.37)
x_theta = zeros(length(t),1);
```

```
for i = 2:length(t)
   x_theta(i-1) = (f_ex(i-1)+theta*(f_ex(i)-f_ex(i-1))+...
          (6*m/theta^2/deltat^2+3*c/theta/deltat)*x(i-1)+...
          (6*m/theta/deltat+2*c)*v(i-1)+...
          (2*m+0.5*c*theta*deltat)*a(i-1))/...
          (6*m/deltat^2/theta^2+3*c/deltat/theta+k);
   a(i) = 6/theta^3/deltat^2*(x_theta(i-1)-x(i-1))-...
        6/theta^2/deltat*v(i-1)+(1-3/theta)*a(i-1);
   v(i) = v(i-1)+deltat/2*(a(i)+a(i-1));
   x(i) = x(i-1)+deltat*v(i-1)+deltat^2/6*(a(i)+2*a(i-1));
end

% Plot result
figure
plot(t,x);
set(gca,'FontSize',14)
xlabel('Time [s]');
ylabel('Amplitude [m]');
box on
grid on
xlim([0,8])
```

Code 5: The Newmark-β Method (Section 3.2.5)

```
% Set the parameters of m, c and k
m = 1;
k = 10;
c = 2;
% Set the excitation parameters
deltat = 0.001;                % Increment of the time
t = 0:deltat:20;               % Time vector
w = 3*2*pi;                    % Excitation frequency is 3*2*pi [rad/s]
f0 = 20;
f_ex = f0*cos(w*t);            % Excitation amplitude is 20
% Set iniitial parameters
x = zeros(length(t),1);
x(1) = 0;                      % Initial displacement is 0
v = zeros(length(t),1);
v(1) = 2;                      % Initial velocity is 2
a = zeros(length(t),1);
a(1) = (f_ex(1)-c*v(1)-k*x(1));  % Use equation of motion at initial time to find initial
                                 acceleration
% Calculate response by Newmark's method
x_newmark = zeros(length(t),1);

for i = 2:length(t)
   x_newmark(i-1) = (f_ex(i-1)+(f_ex(i)-f_ex(i-1))+...
```

```
            (6*m/deltat^2+3*c/deltat)*x(i-1)+...
            (6*m/deltat+2*c)*v(i-1)+...
            (2*m+0.5*c*deltat)*a(i-1))/...
            (6*m/deltat^2+3*c/deltat+k);
    a(i) = 6/deltat^2*(x_newmark(i-1)-x(i-1))-...
         6/deltat*v(i-1)-2*a(i-1);
    v(i) = v(i-1)+deltat/2*(a(i)+a(i-1));
    x(i) = x(i-1)+deltat*v(i-1)+deltat^2/6*(a(i)+2*a(i-1));
end

figure
plot(t,x)
set(gca,'FontSize',14)
xlabel('Time [s]');
ylabel('Amplitude [m]');
box on
grid on
xlim([0,8])
```

Code 6: Rayleigh's Method (Section 3.3.2)

```
% Main function of Rayleigh's method
% Set parameters
M = [1 0 0;0 1 0;0 0 1];
K = [20 -10 0;-10 20 -10; 0 -10 10];
% Set the assumed x
X = [1;1.7;2];
% Approximate the frequency
f = sqrt((X'*K*X)/(X'*M*X))
```

Code 7: Rayleigh Comparison (Section 3.3.2)

```
% Main function of Rayleigh method
% Set parameters
M = [1 0 0;0 1 0;0 0 1];
K = [20 -10 0;-10 20 -10; 0 -10 10];
% Set the assumed x
X = zeros(3,3);
X(:,1) = [1;0;-1];
X(:,2) = [1;-1;0];
X(:,3) = [0;-1;1];
f = zeros(3,1);
% Approximate the frequency

for(i = 1:3)
    f(i) = sqrt((X(:,i)'*K*X(:,i))/(X(:,i)'*M*X(:,i)));
```

```
end

% Plot result
figure
plot(f,'o-','color','k','LineWidth',2,'MarkerSize',5,'MarkerFaceColor','k')
set(gca,'FontSize',20)
xlabel('Number');
ylabel('Frequency [rad/s]');
box on
```

Code 8: Ritz' Method (Section 3.3.3)

```
clc;clear all;
% Main code of Ritz' method
% Set the parameter
K = [20 -10 0;-10 20 -10; 0 -10 10];
M = [1 0 0;0 1 0;0 0 1];
fai = [1 1;1.7 0;2 -1];
Ka = fai'*K*fai;
Ma = fai'*M*fai;
% Calculate the eigenproblem
U = chol(Ma);
% Set the dynamic matrix
dynamicMx = inv(U)'*Ka*inv(U);
[vector,lambda] = eig(dynamicMx);
vector = inv(U)*vector;
% Output the reuslt
modalshape1 = fai*vector(:,1);
modalshape1 = modalshape1/modalshape1(1);
modalshape2 = fai*vector(:,2);
modalshape2 = modalshape2/modalshape2(1);
modalshape = [modalshape1,modalshape2];
modalfrequency = sqrt(diag(lambda));
```

Code 9: Holzer's Method (Section 3.3.4)

```
clc;clear all;
% Set the system parameter
m = zeros(3,1);
k = zeros(2,1);

for(i = 1:3)
   m(i) = 1;
end
for(i = 1:2)
   k(i) = 100;
end
```

```
m1 = m(1);
m2 = m(2);
m3 = m(3);
k1 = k(1);
k2 = k(2);
M = diag([1,1,1]);
K = [100 -100 0;-100 200 -100;0 -100 100];
[V,D] = eig(inv(M)*K);
Modal_Fre = sqrt(diag(D));

for(i = 1:3)
   Modal_Vec(:,i) = V(:,i)/V(1,i);
end

% Set the trial frequency
Nx = 50000;
X = zeros(3,Nx);
Trial_Fre = linspace(0,3*2*pi,Nx);
Restant_Force = zeros(1,Nx);
Threshold = 0.05;
Approximated_Vector = zeros(3,3);
Approximated_Fre = zeros(3,1);
kx = 1;
% Compute the Holzer vector under trial frequency and the resultant force

for(i = 1:Nx)
   Fre = Trial_Fre(i);
   X1 = 1;
   X2 = X1-Fre^2/k1*m1*X1;
   X3 = X2-Fre^2/k2*(m1*X1+m2*X2);
   X(:,i) = [X1;X2;X3];
   Restant_Force(i) = Fre^2*(m1*X1+m2*X2+m3*X3);
end

for(i = 1:Nx-1)
      if Restant_Force(i)*Restant_Force(i+1)<=0
         if(abs(Restant_Force(i))<Threshold)
             Approximated_Fre(kx) = Trial_Fre(i);
             Approximated_Vector(:,kx) = X(:,i);
             kx = kx+1;
         end
      end
end

% Plot result
figure
hold on
plot(Trial_Fre,Restant_Force,'k','LineWidth',2.5);
```

```
x1 = linspace(0,3,3);
y1 = repmat(0,1,3);
plot(x1,y1,'--','color','k','LineWidth',2)
plot(Approximated_Fre,y1,'o','MarkerFaceColor',...
[200/255,200/255,200/255],'MarkerEdgeColor','k','MarkerSize',16)
hold off
set(gca,'FontSize',20)
xlabel('Frequency [rad/s]');
ylabel('Resultant Force [N]');
box on
% Plot result
figure
subplot(1,3,1);
y2 = [1,2,3];
plot(Modal_Vec(:,1),y2,'o-','MarkerSize',12,'MarkerEdgeColor','k','MarkerFaceColor',...
[175/255,175/255,175/255],'color','k','LineWidth',2.5);
hold on
plot(Approximated_Vector(:,1),y2,'--','color',...
[200/255,200/255,200/255],'LineWidth',2.5);
hold off
set(gca,'FontSize',20);
set(gca,'YTickLabel',[]);
ylabel('DOF');
box on
title('Mode 1');
subplot(1,3,2);
y2 = [1,2,3];
plot(Modal_Vec(:,2),y2,'o-
','MarkerSize',12,'MarkerEdgeColor','k','MarkerFaceColor',[175/255,175/255,175/255],'
color','k','LineWidth',2.5);
hold on
plot(Approximated_Vector(:,2),y2,'--
','color',[200/255,200/255,200/255],'LineWidth',2.5);
hold off
set(gca,'FontSize',20);
xlabel('Modal Shape')
set(gca,'YTickLabel',[]);
title('Mode 2');
box on
subplot(1,3,3);
y2 = [1,2,3];

plot(Modal_Vec(:,3),y2,'o-
','MarkerSize',12,'MarkerEdgeColor','k','MarkerFaceColor',[175/255,175/255,175/255],'
color','k','LineWidth',2.5);
hold on
```

```
plot(Approximated_Vector(:,3),y2,'--
','color',[200/255,200/255,200/255],'LineWidth',2.5);
hold off
set(gca,'FontSize',20);
set(gca,'YTickLabel',[]);
title('Mode 3');
box on
```

Code 10: Sturm Sequences (Section 3.4.2)

```
clc;clear all;
% Set the initial borders of the sequence
Border_left = 0;
Border_right = 40;
% Obtain the interval of the eigenvalues
[Signature_Change,C] = Get_the_turning_point(Border_left,Border_right);
% Plot the result
figure
subplot(2,1,1)
plot(C,Signature_Change,'color','k','LineWidth',2);
set(gca,'FontSize',20)
ylabel('Change of signs');
box on
ylim([0,4])
set(gca,'YTick',[0,1,2,3,4]);
subplot(2,1,2);
plot(C(1:10000),Signature_Change(1:10000)-
Signature_Change(2:10001),'color','k','LineWidth',2);
set(gca,'FontSize',20)
xlabel('C [(rad/s)^2]');
ylabel('V(cl)-V(cl+1)');
box on
ylim([0,1.2])
% Extract the precise modal frequencies from the ascertained intervals
% Set the interval of the eigenvalues
Eigenvalue_interval = zeros(3,2);
kl = 1;
% Ascertain the interval of the first eigenvalue

for(i = 1:10000)
  if((Signature_Change(i)-Signature_Change(i+1))==1)
    Eigenvalue_interval(kl,1) = C(i);
    Eigenvalue_interval(kl,2) = C(i+1);
    kl = kl+1;
  end
end
```

```
lambda = zeros(3,1);           % Set the eigenvalue
lambda_num = 1;                % The number of the eigenvalue
Precise_eigenvalue_interval = zeros(3,2);

for(i = 1:3)
  Border_left = Eigenvalue_interval(i,1);
  Border_right = Eigenvalue_interval(i,2);
  [Signature_Change,C] = Get_the_turning_point(Border_left,Border_right);

  for(i = 1:10000)
    if((Signature_Change(i)-Signature_Change(i+1))==1)
      Precise_eigenvalue_interval(lambda_num,1) = C(i);
      Precise_eigenvalue_interval(lambda_num,2) = C(i+1);
      lambda(lambda_num) = C(i+1);
      lambda_num = lambda_num+1;
      break
    end
  end

end

% Get the precise modal frequency
modal_frequency = sqrt(lambda);

function [Signature_Change,C] = Get_the_turning_point(Border_left,Border_right)
  % Set the initial borders of the system
  A = Border_left;
  B = Border_right;
  C = linspace(A,B,10001);       % Separate the border
  Sturm = zeros(4,10001);        % Set the Sturm sequence
  Signature_Change = zeros(10001,1);  % Set the change of the signatures of the
                                        adjacent elements

  for(i = 1:10001)
    % Compute the Sturm sequence
    Sturm(1,i) = C(i)^3-50*C(i)^2+600*C(i)-1000;
    Sturm(2,i) = C(i)^2-40*C(i)+300;
    Sturm(3,i) = C(i)-20;
         Sturm(4,i) = 1;
    % Exclude the element that is equal to 0
    kk = 1;
    Sturm_exclude = Sturm(:,i);
    while (kk)

      for(j = 1:length(Sturm_exclude));
        if(j==length(Sturm_exclude))
          kk = 0;
        end
        if(Sturm_exclude(j)==0)
```

```
            Sturm_exclude = [Sturm_exclude(1:j-1);Sturm_exclude(j+1:end)];
            break
         end
      end

    end

temp_signature_change = 0;

    for(j = 1:length(Sturm_exclude)-1)
       if(sign(Sturm_exclude(j))/sign(Sturm_exclude(j+1))<0)
          temp_signature_change = temp_signature_change+1;
       end
    end

  Signature_Change(i) = temp_signature_change;
  i;
  end

end
```

Code 11: ROOTS (Section 3.4.3)

```
% Main function of ROOTS method
a = [1,-50,600,-1000];
[lambda] = roots(a);
modal_frequency = sqrt(lambda);
```

Code 12: Cholesky Method (Section 3.4.4)

```
% Set parameters
M = [1 0 0;0 1 0;0 0 1];
K = [20 -10 0;-10 20 -10; 0 -10 10];
U = chol(M);
% Set the dynamic matrix
dynamicMx = inv(U)'*K*inv(U);
[vector,lambda] = eig(dynamicMx);
% Output the reuslt
modalshape = inv(U)*vector;
modalfrequency = sqrt(diag(lambda));
```

Code 13: Matrix Iteration Method (Section 3.4.5)

```
clear all;close all;clc;
n_dof = 3;
n_it = 8;
```

```
x = zeros(n_dof,n_it);
x(:,1) = ones(n_dof,1);
w = zeros(n_it,1);
m = diag(ones(n_dof,1));
D = [1,1,1;1,2,2;1,2,3];

for i = 1:n_it
   x(:,i+1) = D*x(:,i);
   w(i) = sqrt(x(1,i)/x(1,i+1));
end

n_it2 = 12;
x_n = sqrt(1/(x(:,9)'*x(:,9)))*x(:,9);
D2 = D-1/w(8)^2*x_n*x_n';
x2 = zeros(n_dof,n_it2);
x2(:,1) = ones(n_dof,1);
w2 = zeros(n_it2,1);

for i = 1:n_it2
   x2(:,i+1) = D2*x2(:,i);
   w2(i) = sqrt(x2(1,i)/x2(1,i+1));
end

n_it3 = 4;
x_n2 = sqrt(1/(x2(:,n_it2+1)'*x2(:,n_it2+1)))*x2(:,n_it2+1);
D3 = D2-1/w2(n_it2)^2*x_n2*x_n2';
x3 = zeros(n_dof,n_it3);
x3(:,1) = ones(n_dof,1);
w3 = zeros(n_it3,1);

for i = 1:n_it3
   x3(:,i+1) = D3*x3(:,i);
   w3(i) = sqrt(x3(1,i)/x3(1,i+1));
end

m1 = x(:,end)/abs(x(1,end));
m2 = x2(:,end)/abs(x2(1,end));
m3 = x3(:,end)/abs(x3(1,end));
K = 100*[2,-1,0;-1,2,-1;0,-1,1];
M = diag([1;1;1]);
[V,D] = eig(K,M);
V = V(1:3,:)./V(1,:);
figure
subplot(1,3,1);
gch = plot(m1,[1:3],V(:,1),[1:3]);
set(gca,'FontSize',14)
xlabel('Mode 1');
ylabel('DOF');
```

```
% Line properties
set(gch(1),'Color','k',...
        'LineStyle','-',...
        'LineWidth',2,...
        'Marker','o')
set(gch(2),'Color',[0.65 0.65 0.65],...
        'LineStyle','--',...
        'LineWidth',2)
set(gch(1),'DisplayName','Matrix methods')
set(gch(2),'DisplayName','Analytical Solution')
% User defined paper sizes (Variable)
set(gcf, 'PaperUnits', 'centimeters');
set(gcf, 'PaperSize', [16 6.18]);
% User defined axis sizes (Variable)
width = 2;
height = 5.82;
papersize = get(gcf, 'PaperSize');
left = (papersize(1)-6)/2;
bottom = (papersize(2))/2;
myfiguresize = [left, bottom, width, height];
set(gcf,'PaperPosition', myfiguresize);
subplot(1,3,2);
gch = plot(m2,[1:3],V(:,2),[1:3]);
set(gca,'FontSize',14)
xlabel('Mode 2');
ylabel('DOF');
% Line properties
set(gch(1),'Color','k',...
        'LineStyle','-',...
        'LineWidth',2,...
        'Marker','o')
set(gch(2),'Color',[0.65 0.65 0.65],...
        'LineStyle','--',...
        'LineWidth',2)
set(gch(1),'DisplayName','Matrix methods')
set(gch(2),'DisplayName','Analytical Solution')
% User defined axis sizes (Variable)
width = 2;
height = 5.82;
papersize = get(gcf, 'PaperSize');
left = (papersize(1)- width-6)/2;
bottom = (papersize(2)- height)/2;
myfiguresize = [left, bottom, width, height];
set(gcf,'PaperPosition', myfiguresize);
subplot(1,3,3);
gch = plot(m3,[1:3],V(:,3),[1:3]);
set(gca,'FontSize',14)
```

```
xlabel('Mode 3');
ylabel('DOF');
% Line properties
set(gch(1),'Color','k',...
        'LineStyle','-',...
        'LineWidth',2,...
        'Marker','o')
set(gch(2),'Color',[0.65 0.65 0.65],...
        'LineStyle','--',...
        'LineWidth',2)
set(gch(1),'DisplayName','Matrix methods')
set(gch(2),'DisplayName','Analytical Solution')
% User defined axis sizes (Variable)
width = 2;
height = 5.82;
papersize = get(gcf, 'PaperSize');
left = (papersize(1)- width-6)/6*5;
bottom = (papersize(2)- height)/2;
myfiguresize = [left, bottom, width, height];
set(gcf,'PaperPosition', myfiguresize);
legend show
lgd = legend(gch);
set(lgd,'FontSize',14,...
    'Location','northeastoutside');
print -dmeta myfigure
```

Code 14: Jacobi's Method (Section 3.4.6)

```
clear all;close all;clc;
n_dof = 3;
n_it = 3;
x = zeros(n_dof,n_it);
x(:,1) = ones(n_dof,1);
w = zeros(n_it,1);
m = 1;
k = 1;
D = m/k*[1,1,1;1,2,2;1,2,3];
kappa = zeros(n_dof,n_it);

for i = 1:n_it
   kappa(:,i) = [1:n_dof]';
end

R = eye(n_dof);
mode = eye(n_dof);
```

```
for j = 1:n_it
   for i = 1:n_dof
     D = R'*D*R;
     theta = theta_k(D,kappa(i,j));
     R = R_jaco(theta,kappa(i,j));
     mode = mode*R;
   end
end

w = 1./sqrt(D);

function theta = theta_k(D,kappa)
   if kappa == 1
     i = 1;j = 2;
     theta = 0.5*atan(2*D(i,j)/(D(i,i)-D(j,j)));
   end
   if kappa == 2
     i = 1;j = 3;
     theta = 0.5*atan(2*D(i,j)/(D(i,i)-D(j,j)));
   end
   if kappa == 3
     i = 2;j = 3;
     theta = 0.5*atan(2*D(i,j)/(D(i,i)-D(j,j)));
   end
end

function R = R_jaco(theta,kappa,n_dof)
   if kappa == 1
     i = 1;j = 2;
     R =  R_trig(theta,i,j,n_dof)
     R(3,3) = 1;
   end
   if kappa == 2
     i = 1;j = 3;
     R =  R_trig(theta,i,j,n_dof)
     R(2,2) = 1;
   end
   if kappa == 3
     i = 2;j = 3;
     R =  R_trig(theta,i,j,n_dof)
     R(1,1) = 1;
   end
end

function Rt = R_trig(theta,i,j,n_dof)
   Rt = zeros(n_dof);
   Rt(i,i) = cos(theta);
```

```
        Rt(i,j) = -sin(theta);
        Rt(j,i) = sin(theta);
        Rt(j,j) = cos(theta);
    End
```

Code 15: SVD (Section 3.4.7)

```
function [U,siv,V,approx] = CompactSVD(A)
  % U, V are left and right character matrix
  % Siv are the singular values,approx = U*diag(siv)*V'
  [Ui,Si,Vi] = svd(A,'econ');
  siv = diag(Si);
  cea = 0;

  for m = 1:length(siv)
    if siv(m)/siv(1)>=1e-15
       cea = cea+1;
    end
  end

  siv = siv(1:cea);X = diag(siv);U = Ui(:,1:cea);V = Vi(:,1:cea);
  approx = U*X*V';
end
```

Code 16: Natural Frequencies of a Free Bar (Section 3.5.3)

```
N_ele = 8;                          % Total number of elements
Npe = 2;                            % Number of one element nodes
dofpn = 1;Tolnod = 9;
Toldof = Tolnod*dofpn;              % Total degree of freedom
nc(1,1) = 0; nc(2,1) = 1.0; nc(3,1) = 2.0; nc(4,1) = 3.0;
nc(5,1) = 4.0;nc(6,1) = 5; nc(7,1) = 6; nc(8,1) = 7; nc(9,1) = 8;
mat(1) = 198e9;                     % Elastic modulus
mat(2) = 0.0016;                    % Cross-sectional area
mat(3) = 7850;                      % Density
n_s(1,1) = 1; n_s(1,2) = 2; n_s(2,1) = 2; n_s(2,2) = 3; n_s(3,1) = 3; n_s(3,2) = 4;
n_s(4,1) = 4; n_s(4,2) = 5; n_s(5,1) = 5;n_s(5,2) = 6;n_s(6,1) = 6;n_s(6,2) = 7;
n_s(7,1) = 7;n_s(7,2) = 8;n_s(8,1) = 8;n_s(8,2) = 9;
K = zeros(Toldof,Toldof);M = zeros (Toldof ,Toldof);
I = zeros(Npe*dofpn, 1);            % Index vector
% The mass and stiffness matrices of each element are calculated and assembled

for i = 1:N_ele
  nd(1) = n_s(i,1);nd(2) = n_s(i,2);
  x1 = nc(nd(1),1);x2 = nc(nd(2),1);
```

```
    L = (x2-x1);E = mat(1); A = mat(2);rou = mat(3);
    I = Eedof(nd,Npe,dofpn);
    ipt = 1;                      % Consistent mass matrix
    [k,m] = febar(E,L,A,rou,ipt); % Calculate element stiffness matrix
    K = Assem_e (K,k,I);          % Assembly mass matrix
    M = Assem_e (M,m,I);          % Assembly stiffness matrix
end

f2 = eig(K,M);
fre_sol = sqrt(f2)                % Natural frequencies of the structure
% Some other function will be called. These functions include Eedof.m, febar.m and
Assem_e.m.

function [I] = Eedof(nd,Npe,dofpn)
    % Calculate system dofs for the element
    k = 0;

    for i = 1:Npe
        start = (nd(i)-1)*dofpn;

        for j = 1:dofpn
            k = k+1;
            I(k) = start +j;
        end

    end

end

function [k,m] = febar(E,L,A,rou,ip)
    % Stiffness and mass matrices of the bar element
    k = (A*E/L)*[ 1 -1; -1 1];
    % Consistent mass matrix
    if ip==1
        m = (rou*A*L/6)*[ 2 1; 1 2];
        % Lumped mass matrix
    else
        m = (rou*A*L/2)*[ 1 0; 0 1];
    end
end

function [K] = Assem_e(K,k,id)
    % Element assembly
    dofe = length (id);

    for i = 1:dofe
        ii = id(i);
```

```
        for j = 1:dofe
          jj = id(j);
          K(ii,jj) = K(ii,jj)+k(i,j);
        end

      end

    end
```

Appendix B Elliptic Integrals and Functions (Chapter 5)

The objective of this appendix is simply to give the most basic definitions relating to elliptic integrals and functions in order that the notation and terminology be presented here in support of the material discussed in Section 5.2.1, specifically relating to exact solutions. The main books used for consultation were [5.11, 5.40]. The first of these references is arguably the classic text on the Jacobian elliptic functions, while the second is a fairly comprehensive, and more modern, reference on the Weierstrass elliptic function.

The subject began historically with *elliptic integrals*. These were motivated by geometry and originally arose in the determination of the arc-length of the ellipse. In this context, the *Elliptic Integral of the First Kind* arose in the form,

$$F(\varphi,k)=\int_0^{\varphi}\frac{1}{\sqrt{1-k^2 sin\,\theta}}d\theta \tag{B.1}$$

where k is referred to as the *modulus* of the elliptic integral.

This integral is referred to as the *trigonometric form*; if one makes the substitution $t = sin\,\theta$, $z = sin\,\varphi$, one arrives at the *Jacobi form* of the integral,

$$F(z,k)=\int_0^{z}\frac{1}{\sqrt{(1-t^2)(1-k^2t^2)}}dt \tag{B.2}$$

which is arguably the one most used. This integral arises in many problems in nonlinear dynamics.

The theory of *elliptic functions* began when Abel observed that the inverse function to (B.2) might be of interest. Following Abel's initial work, the main developments in the theory were largely due to Jacobi; the modern notation is his and the functions are named for him. The basic *Jacobian elliptic function* $sn(z,k)$ is defined by,

$$sn^{-1}(z,k)=\int_0^{z}\frac{1}{\sqrt{(1-t^2)(1-k^2t^2)}}dt \tag{B.3}$$

This function was quickly established to be periodic. In fact, the period $4K$ can be found from,

$$K(k)=\int_0^{1}\frac{1}{\sqrt{(1-t^2)(1-k^2t^2)}}dt \tag{B.4}$$

and this is, of course, a function of the modulus k. Much more importantly, it was discovered that, if the variable z is considered to be complex, the elliptic function sn is actually *doubly-*

periodic and also has a complex-period $2iK'$, where K' can be expressed in terms of another definite integral.

Jacobi identified nine different elliptic functions with various combinations of real and complex periods, all multiples of K or K'. Two of these functions are essentially at the same level of importance as *sn* and appear constantly throughout the theory; they are denoted *cn* and *dn* and have definitions in terms of integrals,

$$cn^{-1}(z,k) = \int_z^1 \frac{1}{\sqrt{(1-t^2)(k'^2+k^2t^2)}}\,dt \tag{B.5}$$

$$dn^{-1}(z,k) = \int_z^1 \frac{1}{\sqrt{(1-t^2)(t^2-k'^2)}}\,dt \tag{B.6}$$

where $k'^2 = 1-k^2$ is referred to as the *complementary modulus*.

There are relations between these functions that mirror the relations between the more familiar (singly-periodic) trigonometric functions. Two examples of such relations are,

$$cn^2 z + sn^2 z = 1 \tag{B.7}$$

$$k^2 sn^2 z + dn^2 z = 1 \tag{B.8}$$

The Jacobian elliptic functions have many more fascinating properties and most of them can be found in [5.11].

Although the Jacobian elliptic functions came first, it was later discovered by Weierstrass that they are all derivable from a much more fundamental construct - the *Weierstrass elliptic function* $\wp(z)$.

In order to construct the Weierstrass function, one begins with the idea of a *lattice* Ω in the complex plane. This is the set of points $\{\omega = n\,\omega_1 + m\,\omega_2\}$ where n and m span the integers. The basic periods of the lattice ω_1 and ω_2 are arbitrary complex numbers, except that they cannot be collinear; this last condition is simply enforced by asserting that their ratio cannot be a real number. Given the lattice, the Weierstrass function is defined by,

$$\wp(z) = \frac{1}{z^2} + \sum_{\omega\in\Omega}{}' \left(\frac{1}{(z-\omega)^2} - \frac{1}{\omega^2} \right) \tag{B.9}$$

where the dash on the summation indicates that one should exclude zero.

It is possible to show that the Weierstrass function satisfies a differential equation,

$$\wp'(z)^2 = 4\wp(z)^3 - g_2\wp(z) - g_3 \tag{B.10}$$

where g_2 and g_3 are constants determined by the lattice. This fact shows that the Weierstrass function also arises as the inverse of an elliptic integral,

$$\wp^{-1}(z) = \int_{-\infty}^{z} \frac{1}{\sqrt{4t^3 - g_2 t - g_3}}\,dt \tag{B.11}$$

The theory of the Weierstrass function is incredibly rich; the reader should consult [5.40] for the relevant fundamental theory.

Appendix C *RK4*: A *MATLAB* Function (Chapter 5)

RK4

```
function [x, y, dy] = simulate
    % Function SIMULATE. Runs a simulation of a nonlinear
      system of interest.
    % This version offers a choice of a linear system, a
      Duffing oscillator and
    % a Van der Pol oscillator, although it is easy to supply
      your own system
    % by adding it to the function SYSTEM.
    %
    % X  : Input to the system of interest. Here , it can be
      specified as a sine
    % wave or a white Gaussian random signal.
    % Y  : Displacement response from the system.
    % DY: Velocity response from the system.
    %
    % User specified parameters for the simulation.
    nsave = 5000;              % Number of points returned.
    ntrans = 0;                % Number of points to capture transient.
    npts = nsave + ntrans;

    deltat = 0.0001;           % Sampling interval/time step.
    y0 = [0.0; 0.01];          % Initial conditions.

    time = 0.0 : deltat : deltat*(npts -1); % Array of sample times.

    % User specified input force.
    % x = randn([npts 1]) ';          % White Gaussian excitation.
    % omega = 2*pi*16;                % Frequency of harmonic excitation.
    % x = 20.0*sin(omega*time);       % Harmonic excitation.
    x = zeros([npts  1]) ';

    xt = x;
    xa = [xt xt(npts)];        % Augmented force vector for simulation.

    % Main engine.
    [t, yt] = rk4_FS(npts, time , y0, deltat, xa);
```

```
    y = yt(1, ntrans+1: end);          % Displacement.
    dy = yt(2, ntrans+1: end);         % Velocity.
end

function [tout , yout] = rk4_FS(npts , time , y0, deltat , xa)
    % RK4_FS Integrates a system of ordinary differential equations using
    % the fourth -order Runge -Kutta method with a fixed step size.
    % Initialize the output.
    tout = zeros(npts , 1);
    yout = zeros(length(y0), npts);
    tout(1) = time(1);
    yout(:,1) = y0;
    h = deltat;
    i = 1;

    % The main loop.
    while (i < npts)
      t = time(i);
      y = yout(:, i);

      % Compute the slopes.
      s1 = sim_system(t, y, deltat, xa);
      s2 = sim_system(t+h/2, y+h*s1/2, deltat, xa);
      s3 = sim_system(t+h/2, y+h*s2/2, deltat, xa);
      s4 = sim_system(t+h, y+h*s3, deltat, xa);
      y = y + h*(s1 + 2*s2 + 2*s3 +s4)/6;

      % Update.
      i = i + 1;
      yout(:, i) = y;
         tout(i) = tout(i-1) + h;
    end
end

function dydt = sim_system(t, y, deltat, xa)
    % Function SIM_SYSTEM. Just holds the specification of the system of
    % interest in first order form. Calls FORCE because RK4 needs to know
    % forces in between sample points.

    f = force(t, deltat, xa);

    % Standard SDOF linear system.
    m = 1.0;
    c = 20.0;
    k = 10000.0;

    dydt = [y(2); (f - c*y(2) - k*y(1))/m];
```

```
    % Duffing oscillator system.
    % m = 1.0;
    % c = 20.0;
    % k = 10000.0;
    % k3 = 5.0e9;
    %
    % dydt = [y(2); (f - c*y(2) - k*y(1) - k3*y(1)*y(1)*y(1))/m];
    % Van der Pol oscillator.
    % m = 1.0;
    % c = 0.2;
    % k = 1.0;
    %
    % dydt = [y(2); - c*y(2)*(y(1)*y(1) - 1) - k*y(1)];
end

function f = force(t,deltat ,xa)
    % Function FORCE. Just interpolates linearly between input sample points.
    % This is because RK4 needs to know values at mid-points.

    % Find where we are in the X array.
    ct = t/deltat;                  % Number of 'deltat 's into time array.
    it = floor(ct);                 % Index of last stored X value (-1).
    index = it + 1;
    rt = ct - it;                   % Distance into current deltat interval.

    % Interpolate to current value of X.
    f = xa(index) + rt*(xa(index+1) - xa(index));
end
```

Index

Q